This upper-division/graduate-level textbook provides an extensive survey of the seaweed literature. Guest essays by noted ecologists give a personal perspective on field studies. Tropical seaweeds and their habitats are included, as well as the better-known temperate communities. The mariculture chapter includes a case study of the carrageenan industry, and there is a taxonomic appendix. The book is a thoroughly rewritten version of the authors' and Mary Jo Duncan's successful earlier book, *The Physiological Ecology of Seaweeds* (1985).

Seaweed ecology and physiology

Seaweed ecology and physiology

Christopher S. Lobban
University of Guam

Paul J. Harrison
University of British Columbia

CAMBRIDGE
UNIVERSITY PRESS

PUBLISHED BY THE PRESS SYNDICATE OF THE UNIVERSITY OF CAMBRIDGE
The Pitt Building, Trumpington Street, Cambridge CB2 1RP

CAMBRIDGE UNIVERSITY PRESS
The Edinburgh Building, Cambridge CB2 2RU, United Kingdom
40 West 20th Street, New York, NY 10011-4211, USA
10 Stamford Road, Oakleigh, Melbourne 3166, Australia

First published 1994
Reprinted 1997
First paperback edition 1997

Printed in the United States of America

Library of Congress Cataloging-in-Publication Data is available.

A catalog record for this book is available from the British Library.

ISBN 0-521-40334-0 hardback
ISBN 0-521-40897-00 paperback

To our wives, María and Victoria,
for their patience, encouragement, and inspiration

and our parents, Olwyn and James, Beatrice and William,
for their forbearance and loving support

Contents

Contributors

Dr. Paul K. Dayton
Institute of Marine Biology
Scripps Institution of Oceanography
La Jolla, CA 92093

Dr. Marilyn M. Harlin
Department of Botany
University of Rhode Island
Kingston, RI 02881

Dr. Diane S. Littler
Department of Botany
Smithsonian Institution
Washington, DC 20560

Dr. Mark M. Littler
Department of Botany
Smithsonian Institution
Washington, DC 20560

Dr. Richard L. Moe
Herbarium
University of California
Berkeley, CA 94720

Dr. Piet H. Nienhuis
Netherlands Institute of Ecology
Yerseke
Netherlands

Dr. Trevor A. Norton
The Port Erin Marine Laboratory
University of Liverpool
Port Erin, Isle of Man
United Kingdom

Dr. Paul C. Silva
Herbarium
University of California
Berkeley, CA 94720

Dr. Robert T. Wilce
Department of Botany
University of Massachusetts
Amherst, MA 01003

Preface

The field of experimental phycology continues to grow, feeding on advances in other fields and sometimes, as in the past, contributing to them. The wealth of new literature alone would have warranted a revision of our original book, *The Physiological Ecology of Seaweeds*. However, the reasons for this revision – and its changes – go even deeper. In fact, the original book has been so thoroughly reworked and rewritten that we have given it a new title.

Seaweed Ecology and Physiology, like its predecessor, is intended primarily as a textbook. The rapid growth of knowledge in this field is at once exciting and daunting. Even more than in the first book, our method has been to select papers that help put together a coherent (if reticulate!) story. This book provides an entry to the literature, not a systematic literature review.

Our recent experiences in the tropics and an increasing literature on tropical algae have allowed us to redress the temperate bias of our earlier writing. Austral countries such as Australia, Chile, and South Africa have also been active in seaweed physiological ecology and have provided additional perspectives on seaweed biology.

Our teaching experiences suggested that the sequence of chapters could be improved. Chapters on communities and morphogenesis, which formerly served to review and tie together earlier themes, are now introductions to the organisms and their interactions. We have included an encapsulation of algal structure and life histories, but still expect that students using this book will have learned these subjects in more detail or will be learning about them concurrently. The chapter on mariculture has been greatly expanded because of the increased interest in aquaculture and algal biotechnology. Finally, we have invited several other phycologists to give their personal perspectives on some favorite habitats. We hope that these considerable changes have not destroyed the original merits of the book as a textbook.

This book has been greatly enhanced by the contributions of the essayists and of Paul Silva and Dick Moe, who volunteered the taxonomic appendix. To these colleagues we owe our especial thanks. Several students at the University of Guam helped compile the references and added the section numbers, especially Annie Dierking and Norman Wong. Sections of the text have been critically read by John Berges, Rob DeWreede, Louis Druehl, Ron Foreman, Tony Glass, Mike Hawkes, Catriona Hurd, Paul LeBlond, Sandra Lindstrom, and Valerie Paul.

C. S. L.　P. J. H.

1

Morphology, life histories, and morphogenesis

1.1 Introduction: the plants and their environments

1.1.1 Seaweeds

The term "seaweeds" traditionally includes only macroscopic, multicellular marine red, green, and brown algae. However, each of these groups has microscopic, if not unicellular, representatives. All seaweeds at some stage in their life cycles are unicellular, as spores or zygotes, and may be temporarily planktonic (Amsler & Searles 1980). Some remain small, forming sparse but productive turfs on coral reefs (Hackney et al. 1989). The blue-green algae are widespread on temperate rocky and sandy shores (Whitton & Potts 1982) and have occasionally been acknowledged in "seaweed" floras (e.g., Setchell & Gardner 1919; Newton 1931). They are particularly important in the tropics, where large macroscopic tufts of Oscillatoriaceae and smaller but abundant nitrogen-fixing Nostocaceae are major components of the reef flora (Hackney et al. 1989). Again, there are many unicellular blue-green algae. On the other hand, some benthic diatoms – normally not considered seaweeds – form large and sometimes-abundant tube-dwelling colonies that resemble seaweeds and presumably respond to the environment in much the same way (Lobban 1989). A deep-water green, *Palmoclathrus*, forms a morphologically complex thallus built from an apparently amorphous matrix with a nearly uniform distribution of cells (Womersley 1971; O'Kelly 1988), and a tropical chrysophyte, *Chrysonephos lewisii*, forms large, *Ectocarpus*-like thalli (Taylor 1960). On a smaller scale are the colonial filaments of some simple red algae, such as *Goniotrichum*. In this book we shall consider macroscopic and microscopic benthic environments and how algae respond to those environments.

Seaweeds are evolutionarily quite diverse. (In contrast, all vascular plants can be assigned to a single division, Tracheophyta.) The four traditional divisions (or phyla) – Cyanophyta, Rhodophyta, Phaeophyta, and Chlorophyta – are assigned to two or more kingdoms, depending on the systematist. Cyanophyta are clearly placed in the Kingdom Eubacteria, but the others are either in Plantae (because they are basically multicellular) or in Protista (because they are closely related to unicellular algae). A new kingdom, Chromista, has recently been proposed to encompass the "brown-algal line," namely, Phaeophyta, Chrysophyta, and Pyrrhophyta (Cavalier-Smith 1986). Other authors would recognize this group at the level of a division (Chromophyta). Taxonomic opinion is also divided over the classes, especially within Chlorophyta. Green seaweeds have been split into Chlorophyceae (uninucleate; also including freshwater genera) and Bryopsidophyceae (multinucleate), but recent studies, using new criteria, suggest that virtually all marine green seaweeds belong together (with some freshwater genera as well) in the Class Ulvophyceae (Mattox & Stewart 1984; Floyd & O'Kelly 1984; also see van den Hoek et al. 1988; Sluiman 1989).

Ocean vegetation is dominated by evolutionarily primitive plants: the algae. No mosses, ferns, or gymnosperms are found in the oceans, and only a few diverse angiosperms (the seagrasses) occur in marine habitats (though the latter are scarcely known). The water column is chiefly the domain of the phytoplankton – unicellular or colonial plants, including classes not represented in the benthos – but populations of floating seaweeds are common (Norton & Mathieson 1983). Rocky shores are abundantly covered with a macrovegetation that is almost exclusively seaweeds; in western North America, surf grass (*Phyllospadix* spp.) is an exception. On and around the larger plants are many benthic microalgae, including early stages of seaweeds. Muddy and sandy areas have fewer seaweeds, because most species cannot anchor there, though some siphonous greens (e.g., some species of *Halimeda* and

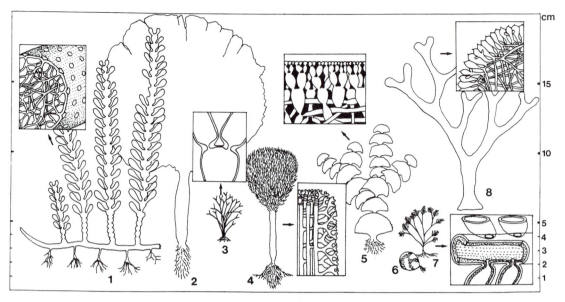

Figure 1.1. Thallus morphology and construction in siphonous green algae. Thalli drawn to scale; insets (not to scale) show principles of construction: (1) *Caulerpa cactoides* with network of trabeculae. (2) *Avrainvillea gardineri* (tightly woven felt of filaments). (3) *Chlorodesmis* sp.: bush of dichotomously branched siphons, constricted at the bases of the branches (inset). (4) *Penicillus capitus:* calcified siphons form a multiaxial pseudotissue in the stem (inset), but separate to form bushy head. (5) *Halimeda tuna:* segmented, calcified thallus of woven medulla and cortical utricles (inset). (6) *Halicystis* stage of *Derbesia,* a single ovoid cell (shown at gametogenesis). (7) *Bryopsis plumosa* gametophyte: pinnately branched free siphons. (8) *Codium fragile:* interwoven uncalcified siphons form multiaxial branches. (From Menzel 1988, with permission of Springer-Verlag, Berlin.)

Udotea) produce penetrating, rootlike holdfasts that may also serve in nutrient uptake (Littler et al. 1988). In such areas, seagrasses become the dominant vegetation, particularly in tropical and subtropical areas (Helfferich & McRoy 1980; Ferguson et al. 1980; Dawes 1981). There is also a paucity of freshwater macroalgae. Freshwater Rhodophyceae and Phaeophyceae are represented by relatively very few genera and species, and Ulvophyceae are also scarce, only a few genera (e.g., *Cladophora*) having penetrated fresh waters. That there are relatively few marine angiosperms may reflect the recent origin of the phylum and the problems of readaption to the sea, including the physiological problems imposed by the osmotic strength of seawater and its quite different ion composition as compared with soil (King 1981). But why so few freshwater Rhodophyceae, Phaeophyceae, and Ulvophyceae? Or, to put it another way (Dring 1982), what features of these groups have led to their being largely restricted to the sea? Perhaps the answer lies not so much in the characteristics of marine or freshwater habitats but in the characteristics of the brackish waters that lie between.

Most seaweeds, in contrast to phytoplankters, are multicellular most of the time. What does this imply for physiological ecology? Multicellularity confers the advantage of allowing extensive development in the third dimension of the water column. Such development can be achieved in other ways, however. Siphonous green al-

gae form large multinucleate thalli that are at least technically single cells (acellular rather than unicellular), supported by turgor pressure (*Valonia*), ingrowths of the rhizome wall (trabeculae) in *Caulerpa,* or interweaving of numerous narrow siphons (*Codium, Avrainvillea*) (Fig. 1.1). Colonial diatoms, both tube-dwelling and chain-forming, also build three-dimensional structures, as do zooxanthellae in association with corals. Multicellular algae often grow vertically away from the substratum; this habit brings them closer to the light, enables them to grow large without extreme competition for space, and allows them to harvest nutrients from a greater volume of water. Of course, there are creeping filamentous algae, even endophytic and endolithic filaments (e.g., *Entocladia*), as well as crustose plants such as *Ralfsia* and *Porolithon,* that do not grow up into the water column. Support tissue usually is not necessary for this upward growth, because most small seaweeds are slightly buoyant, and the water provides support. Support tissue is metabolically expensive, because it is nonphotosynthetic. However, strength and resilience are required to withstand water motion. Some of the larger seaweeds (e.g., *Pterygophora*) have stiff, massive stipes, but others (e.g., *Hormosira*) employ flotation to keep them upright. Many of the kelps and fucoids have special gas-filled structures, pneumatocysts; but in other seaweeds (e.g., erect species of *Codium;* Dromgoole 1982), gas trapped among the filaments achieves the same effect. (*Codium fragile* subsp. *tomentosoides* has become a nuisance in New England because it becomes buoyant enough to carry off the cultivated shellfish to which they have attached; Wassman & Ramus 1973.)

A second important feature of multicellularity is that it allows division of labor between tissues; such division is developed to various degrees in seaweeds. Nutrient (and water) uptake and photosynthesis take place over virtually the entire surface of the seaweed thallus, in contrast to the case for vascular land plants. Differentiation and specialization among the vegetative cells of algal thalli range from virtually nil (as in *Ulothrix*, where all cells except the rhizoids serve both vegetative and reproductive functions), through the simple but somewhat differentiated thalli seen in *Porphyra* blades (e.g., Kaska et al. 1988), to the highly differentiated photosynthetic, storage, and translocation tissues in a variety of organs, including stipe, blades, and pneumatocysts, that occur in fucoids and kelps (Fagerberg et al. 1979; Kilar et al. 1989) (see Fig. 1.5). Of course, no seaweed shows the degrees of differentiation seen in vascular plants. Even in vascular plants, the cells are biochemically more general than are animal cells: The organs of vascular plants (stems, leaves, roots, flowers) all contain much the same mix of cells, whereas animal organs each contain only a few specialized cell types. The low diversity of cells in an algal thallus means that each cell is physiologically and biochemically even more general than vascular-plant cells.

1.1.2 Environmental-factor interactions

Benthic algae interact with other marine organisms, and all interact with their physicochemical environment. As a rule, they live attached to the seabed between the top of the intertidal zone and the maximum depth to which adequate light for growth can penetrate. Among the major environmental factors affecting seaweeds are light, temperature, salinity, water motion, and nutrient availability. Among the biological interactions are relations between seaweeds and their epiphytic bacteria, fungi, algae, and sessile animals; interactions between herbivores and plants (both macroalgae and epiflora); and the impact of predators, including humans. Individual patterns of growth, morphology, and reproduction are overall effects of all these factors combined (Fig. 1.2).

An organism's physicochemical environment, consisting of all the external abiotic factors that influence the organism, is very complex and constantly varying. In order for us to discuss or study it, we need to reduce it to smaller parts, to think about one variable at a time. And yet, each of the environmental "factors" that we might consider – temperature, salinity, light, and so forth – is really a composite of many variables, and they tend to interact. The following paragraphs are intended to paint the whole picture, before we go on to study it pixel by pixel.

Factor interactions can be grouped into four categories: (1) multifaceted factors, (2) interactions between environmental variables, (3) interactions between environmental variables and biological factors, and (4) sequential effects.

Many environmental factors have several components that do not necessarily change together. Light quality and quantity, which are important in photosynthetic responses and metabolic patterns, both change with depth, but the changes depend on turbidity and the nature of the particles. In submarine caves, light quantity diminishes with little change in quality. Natural light has the further important component of day length, which influences reproductive states. Salinity is another complex factor, of which the two chief components are the osmotic potential of the water and the ionic composition. Osmotic potential affects water flow in and out of the cell, turgor pressure, and growth, while the concentrations of Ca^{2+} and HCO_3^- affect membrane integrity and photosynthesis, respectively. The hydrodynamic aspects of water motion are critical to thallus survival on wave-swept shores and to spore settling, and water motion also has important effects on the boundary layers over plant surfaces and thus on nutrient uptake and gas exchange. Nutrients must be considered not simply in their absolute concentrations but also in the amounts present in biologically available forms; concentrations of trace metals may create toxicity problems, particularly in polluted areas. Pollution, as a factor, may include not only the toxic effects of component chemicals but also an increase in turbidity, hence a reduction in irradiance. Emersion often involves desiccation, heating or chilling, removal of most nutrients (carbon can be an exception), and, frequently, changes in the salinity of the water in the surface film on the plants and in the free space between cells.

Interactions among environmental variables are the rule rather than the exception. Bright light is often associated with increased heating, particularly of plants exposed at low tide. Light, especially blue light, regulates the activities of many enzymes, including some involved in carbon fixation and nitrogen metabolism. Temperature and salinity affect the density of seawater, hence the mixing of nutrient-rich bottom water with nutrient-depleted surface water. Thermoclines can affect plankton movements, including migration of the larvae of epiphytic animals. Temperature also affects cellular pH and hence some enzyme activities. The carbonate equilibrium and especially the concentration of free CO_2 are greatly affected by pH, salinity, and temperature, while the availability of NH_4^+ is pH-dependent, because at high pH the ion escapes as free ammonia. Water motion can affect turbidity and siltation as well as nutrient availability. These are examples of one environmental variable affecting another. There are also examples of two environmental variables acting synergistically on plants; for instance, the combination of low

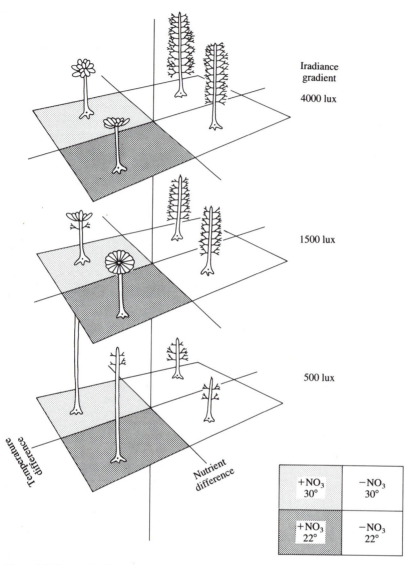

Figure 1.2. Interacting factors in *Acetabularia* growth and reproduction. Each matrix shows four conditions of nitrate availability and temperature; the matrices are arranged vertically along the irradiance gradient. (Original data for *A. calyculus* in illuminance units.) (Redrawn from Shihira-Ishikawas, in Bonotto 1988.)

salinity and high temperature can be harmful at levels where each alone would be tolerable. In several seaweeds, the combined effects of temperature and photoperiod regulate development and reproduction. Interactions between physicochemical and biological factors are also the rule rather than the exception. The environment of a given plant includes other organisms, as we have seen, with which the plant interacts through intraspecific and interspecific competition, predator–prey relationships, and basiphyte–epiphyte relationships. These other organisms are also affected by the environment, as are their effects on other organisms. Moreover, other organisms may greatly modify the physicochemical environment of a given individual. Protection from

strong irradiance and desiccation by canopy seaweeds is important to the survival of understory algae, including germlings of the larger species. Organisms shade each other (and sometimes themselves) and have large effects on nutrient concentrations and water flow. Other interactions stem from the way the biological parameters, such as age, phenotype, and genotype, affect a plant's response to the abiotic environment, as well as the effects that organisms have on the environment. The chief biological parameters that condition a given plant's response to its environment are age, reproductive condition, nutrient status (including stores of N, P, and C), and past history. By "past history" is meant the effects of past environmental conditions on plant development.

Genetic differentiation within populations leads to different responses in plants from different parts of a population. The seasons can also affect certain physiological responses, aside from those involved in life-history changes; these responses include acclimation of temperature optima and tolerance limits.

Finally, there are factor interactions through sequential effects. Nitrogen limitation may cause red algae to catabolize some of their phycobiliproteins, which will in turn reduce their light-harvesting ability. In general, any factor that alters the growth, form, or reproductive or physiological condition is apt to change the responses of the plant to other factors both currently and in the future. A good example of a sequential effect, and also biotic–abiotic interaction, was seen by Littler and Littler (1987) following an unusual flash flood in southern California. Intertidal urchins (*Strongylocentrotus purpuratus*) were almost completely wiped out, but the persistent macroalgae suffered little damage from the fresh water. Subsequently, however, there was a great increase in ephemeral algae (*Ulva, Enteromorpha*, Ectocarpaceae) because of the reduction in grazing pressure.

The complexity of the interactions of variables in nature often confounds interpretation of the effects even of "major" events, such as the recent El Niño warm-water period (Paine 1986). In laboratory experiments, usually one variable is tested at a time, and all other factors are held constant, or at least equal in all treatments. Variations in additional factors can confound the results. For example, Underwood (1980) criticized some field experiments designed to determine the effects of grazer exclusion because the fences and cages used to keep out grazers also affected the water motion over the rock surface and provided some shade. Reed et al. (1991) pointed out the potential for density effects to confound studies of abiotic factors. Schiel and Foster (1986) criticized many studies of subtidal ecology for methodological problems, including inadequate experimental design, use of pseudoreplicates, and lack of any measure of variance. They commented that "correlations between algal abundances and various physical and biological factors have been cited in dozens of studies, often with poor quantitative assessments. The existence of patterns and abundance of species constitutes evidence that these physical factors and biological interactions may affect the structure of these communities. They do not at the same time, however, demonstrate the importance or unimportance of these factors in producing observed patterns" (Schiel & Foster 1986, p. 273; see also Norton et al. 1982). The statistical designs of experiments are considerably more complicated when more than one variable is being assessed at a time; practical reviews in this area have been provided by Box et al. (1978), Green (1979), and Underwood (1981b).

Moreover, because plants are so different from humans and from the animals with which we are most familiar, any assumption made about plants needs to be checked by observation and experiment (Evans 1972). Drawing conclusions by analogy with other plants, or even with other algal groups, is no less fraught with potentially invalid assumptions. For instance, the planktonic stages of seaweeds are scarcely known, and one might be tempted to fill in missing information by making comparisons with other unicellular algae: phytoplankton. Yet, from the little we do know, there evidently are limits to the analogies, and not only because of differences between divisions: Seaweed propagules do not behave like phytoplankton, inasmuch as they have incomplete or inefficient photosynthetic systems and do not live long unless they settle (Santelices 1990b).

1.1.3 Culture versus nature

Several considerations confound the interpretation of field reality via laboratory studies. First, while laboratory studies can provide much more controlled conditions than are found in nature, they are limited in some important ways and contain some implicit assumptions, such as the following: (1) High nutrient levels do not alter the plants' responses to the factor under study. (2) The reactions of plants to uniform conditions (including the factor under study) are not different from their responses to the factor(s) under fluctuating conditions. To a certain extent these assumptions are valid. Culture media are very rich in nutrients, to compensate for lack of water movement and exchange, but that such substitution can give precisely the same results with all parameters is doubtful. Other culture conditions are also generally optimal, except for the variable under study, and the results may not elucidate the behavior of plants in the field, which are subject to competition and often suboptimal conditions (Neushul 1981). Another important difference between culture and nature is that in culture, species usually are tested in isolation, away from interspecific competition and grazing. Furthermore, culture conditions are uniform (at least on a large scale), whereas in nature there often are large and unpredictable fluctuations in the environment (Fréchette & Legendre 1978; Turpin et al. 1981). Microscale heterogeneity in culture conditions should not be overlooked (Allen 1977; Norton & Fetter 1981). In the culture flask, one cell may shade another, and cells form nutrient-depleted zones around them, creating a mosaic of nutrient concentrations through which cells pass. On the other hand, scale also needs to be considered at the large end – for instance, the amount of space needed for a patch of a given alga to establish itself (Schiel & Foster 1986).

Second, the use of taxonomic species to define ecological entities is a handicap:

The criteria used by the taxonomist for the delineation of taxa are chosen deliberately from the conservative and stable features of morphology that are not subject to marked genetic variation, polymorphism or phenotypic

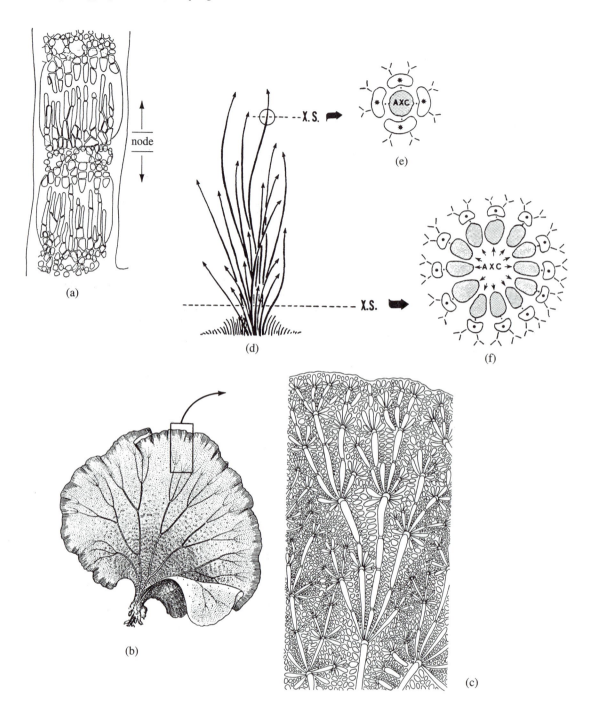

Figure 1.3. Filamentous thallus construction. (a) Small portion of a *Ceramium* axis with cortication growing upward and downward from a node between axial cells. (b, c) Formation of bladelike thallus from filaments in *Anadyomene stellata* (b, ×1.82; c, ×13.65). (d–f) Growth of *Dumontia incrassata* showing schematically the axial filaments and apical cells (arrows); cross-section in the uniaxial part of the thallus near the tip (e) shows a single axial cell (AXC) surrounded by four pericentral cells (*) that have in turn produced cortical cells; (f) cross section through base shows multiaxial construction with a core of axial cells, each with one pericentral cell. (g–n) Apical growth of *Gracilaria verrucosa*. (g) A primary apical cell (I) occurs at the tip of the main axis, and secondary apical cells (II, III, etc.) occur at the tips of lateral filaments. (h–m) Division of the apical cell (A.I), shown by dotted line in (h), gives rise to a subapical cell (SA.I:1) and a new apical cell (A.I:1)(i). In (i–j), the subapical cell is shown dividing to form an axial

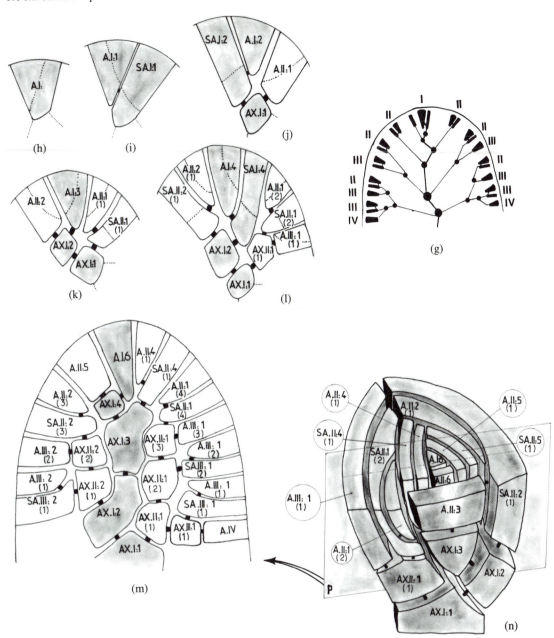

cell (AX.I:1) and a secondary apical cell (A.II:1), while the new apical cell (A.I:1) cuts off another subapical cell (SA.I:2) and becomes A.I:2. The lineages can be traced further with the help of the pit connections (represented as dark bars between cells). (n) The three-dimensional arrangement is complex because the apical cell divides on three faces. P is the plane of the vertical section in (m). (Part a from Taylor 1957; b and c from Taylor 1960, with permission of University of Michigan Press; d–f from Wilce & Davis 1984, with permission of *Journal of Phycology;* g–n from Kling & Bodard 1986, with permission of *Cryptogamie: Algologie.)*

change. These same criteria . . . may be quite inappropriate for describing the ecologically relevant differences between individuals, populations and communities. . . . The failure of taxonomic categories to fit as ecological categories is not surprising . . . yet it may be just the taxonomically useless characters that are mainly responsible for determining the precise ecologies of organisms [Harper 1982, p. 12]

See also Russell (1988). The taxonomy of widespread organisms must be approached with particular caution. When what appears to be a single species occurs in widely different latitudes or longitudes, its physiological and ecological parameters may be quite different (and the taxonomy may change as information accumulates). Incisive studies on the species concept have been published by Mann (1984) for diatoms and by Blackburn and Tyler (1987) for desmids.

For many topics, only one study or a few studies have been done, and a phenomenon demonstrated in a particular alga under certain conditions will not necessarily turn out to be the same in other algae or under other conditions. Lewin (1974, p. 2) commented about laboratory studies that "there is still a tendency . . . to over-generalize on the basis of investigations on no more than one or two examples." Equally, very few natural populations or communities have been studied often enough to assess how much variability is present from place to place (ecotypic variation). The kelp beds of southern California are exceptional in that they have been repeatedly analyzed by different people along the coast for over 25 years. The impression now emerging is that there is no typical kelp bed; environmental parameters differ from one kelp bed to another, and parameters such as specific growth rate versus nitrogen supply vary among *Macrocystis* populations, which have limited dispersal and genetic mixing (Kopczak et al. 1991).

Eventually, the isolated pixels have to be reassembled into models of nature. This can be done in part by experimentally assessing factor combinations, and in part through mathematical modeling (Newell 1979; McQuaid & Branch 1984; Kooistra et al. 1989). Alderdice (1972) and Newell (1979) suggested that an organism has a multidimensional "zone of tolerance," the boundaries of which are defined by its tolerance to all environmental variables. These boundaries depend not only on the species and genotype of the organism but also on its size, age, stage of life history, and previous environmental experience; the boundaries change as these change. Within the overall zone of tolerance there are smaller multidimensional zones that are defined by the local conditions under which the organism is operating; acclimation to other conditions, such as during seasonal changes, involves changes in the boundaries of these smaller zones. These zones can be visualized on paper as far as three axes (see Fig. 6.21), but computers can manipulate data along many axes.

In this first chapter we shall review the foundation of structures and life histories on which any understanding of seaweed physiological ecology must rest, and then trace events involved in the development of seaweed thalli from gametes or spores to reproductive individuals.

1.2 Seaweed morphology and anatomy
1.2.1 Thallus construction

Diversity of thallus construction in algae contrasts strongly with uniformity in vascular plants. In the latter, parenchymatous meristems (e.g., at the shoot

and root apices) produce tissue that differentiates in a wide variety of shapes. Among the algae, parenchymatous development is found in kelps, fucoids, Ulvales, Dictyotales, and others. However, the great majority of seaweeds either are filamentous or are built up of united or corticated filaments. Large and complex structures can be built up this way (e.g., *Codium magnum;* see photo by Dawson 1966). Cell division may take place throughout the plant, or the meristematic region may be localized. If localized, it is most commonly at the apex, but may be at the base or somewhere in between (intercalary).

A simple filament consists of an unbranched chain of cells attached by their end walls and results from cell division only in the plane perpendicular to the axis of the filament. Unbranched filaments are uncommon among seaweeds, except the blue-green algae (Oscillatoriaceae); two eukaryotic genera are *Ulothrix* and *Chaetomorpha*. Usually, some cell division takes place parallel to the filament axis to produce branches (*Cladophora, Antithamnion*) (see Fig. 1.15). Filaments consisting of a single row of cells (branched or not) are called uniseriate. Pluriseriate filaments, in genera such as *Percursaria, Bangia,* and *Sphacelaria* (Fig. 1.4a), are formed by vertical cell divisions in which the daughter cells do not grow out into branches but remain as compact parenchyma. Branches need not grow out free, but may creep down the main filament, forming cortication, as seen in *Ceramium* (Fig. 1.3a) and *Desmarestia*. In some of the more massive Rhodomelaceae, such as *Laurencia* and *Acanthophora*, the cortication becomes so extensive that the origin of the structure is obscured. A detailed study by Kling and Bodard (1986) of axis development in *Gracilaria verrucosa* (uniaxial) showed how complex – and difficult to interpret – pseudoparenchymatous growth can be (Fig. 1.3g–n; compare with Fig. 1.4d–m).

Many of the more massive seaweed thalli are multiaxial, produced by the adhesion of several filaments. This is common among the red algae (Fig. 1.3d–f; also see Fig. 1.44a,b) (Coomans & Hommersand 1990). Multiaxial construction is most readily seen in the less tightly compacted thalli of *Nemalion* or *Liagora*. The contrast between multiaxial and uniaxial growth can be seen within thalli of *Dumontia incrassata* (Fig. 1.3d–f), in which bases are multiaxial, but upper branches are uniaxial (Wilce & Davis 1984). Conversely, *Weeksia fryeana* is uniaxial at first and later becomes multiaxial (Norris 1971). The adhesion of filaments can also produce a pseudoparenchymatous crust (*Peyssonnelia, Ralfsia*) or blade (*Anadyomene;* Fig. 1.3b,c). Many siphonous green algae, including *Halimeda* and *Codium*, are formed by the interweaving of numerous filaments (Fig. 1.1). In the Corallinaceae, multiaxial apical growth forms the hypothallus (in crusts) or central medulla (in erect forms), while intercalary meristems on the lateral branches form the epithallus and perithallus (cortex in

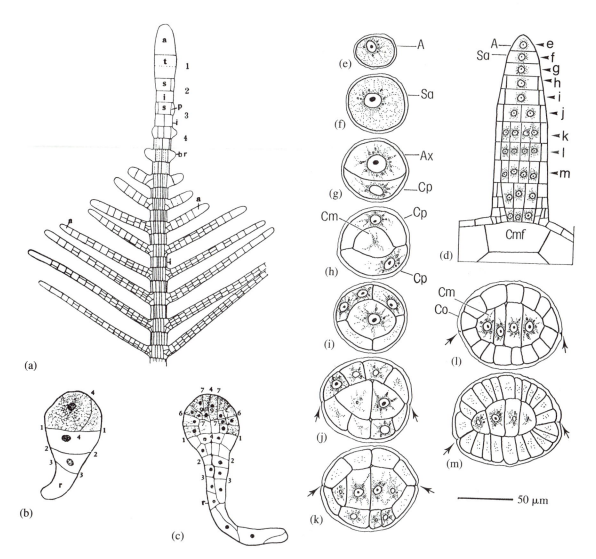

Figure 1.4. Parenchymatous development in seaweeds. (a) *Sphacelaria plumula* apex showing first transverse division (t), followed by pairs of cells (i, s), of which s forms branches, but i does not. (b, c) *Fucus vesiculosus* germination showing successive cell divisions (numbered) (divisions 5 and 8 in the plane of the page). (Parts a–c from Fritsch 1945, based on classical literature.) (d–m) *Dictyota:* development of parenchyma; (d) long section through adventive branch, showing locations of cross sections at each level (diagrammatic); (e–m) serial cross sections to show sequence of periclinal divisions. Arrows indicate junction between original two pericentral cells (first shown in h). For the sake of clarity, the proportions of the cells were changed; the adventive branch is actually half as long and twice as wide as shown. A, apical cell; Sa, subapical cell; Ax, axial cell; Cp, pericentral cell, Cm, medullary cell; Co, cortical cell. (Parts d–m from Gaillard & L'Hardy-Halos 1990, with permission of Blackwell Scientific Publications.)

erect axes) (Cabioch 1988). [Woelkerling (1988) provides new terminology for (crustose) corallines, emphasizing their filamentous development.] A potential disadvantage of pseudoparenchymatous growth is lack of cytoplasmic contact between adjacent cells in different filaments, a problem that red algae overcome (perhaps) through secondary pit connections (Raven 1986).

Cell division in two planes can alternatively result in a monostromatic sheet of cells, as in *Monostroma*

and some species of *Porphyra*. In the Delesseriaceae, marginal meristems produce the wings, while apical cells produce the axial filaments. Such solid tissues are called parenchyma and may become thicker through cell division in a third plane, as in *Ulva* and distromatic *Porphyra,* and in the kelps and fucoids (Fig. 1.4; also see Figs. 1.43 and 1.45). The ontogeny of the parenchyma in Dictyotales (Fig. 1.4d–m) has been followed in detail by Gaillard and L'Hardy-Halos (1990), who cite many sources, and by Katsaros and Galatis (1988).

Table 1.1. *Functional-form groups of macroalgae*

Functional-form group	External morphology	Internal anatomy	Texture	Sample genera
Sheet group	Thin, tubular, and sheetlike (foliose)	Uncorticated, one to several cells thick	Soft	*Ulva, Enteromorpha, Dictyota*
Filamentous group	Delicately branched (filamentous)	Uniseriate, multiseriate, or lightly corticated	Soft	*Centroceras, Polysiphonia, Chaetomorpha, Microcoleus*
Coarsely branched group	Coarsely branched, upright	Corticated	Fleshy-wiry	*Laurencia, Chordaria, Caulerpa, Penicillus, Gracilaria*
Thick, leathery group	Thick blades and branches	Differentiated, heavily corticated, thick-walled	Leather, rubbery	*Laminaria, Fucus, Udotea, Chondrus*
Jointed calcareous group	Articulated, calcareous, upright	Calcified genicula, flexible intergenicula with parallel cell rows	Stony	*Corallina, Halimeda, Galaxaura*
Crustose group	Prostrate, encrusting	Calcified or uncalcified parallel rows of cells	Stony or tough	*Lithothamnion, Ralfsia, Hildenbrandia*

Source: Littler et al. (1983b), with permission of *Journal of Phycology.*

Incomplete cytokinesis during tetraspore formation occurs in *Gracilaria tikvahiae,* leading to two-, three-, and four-nucleate spores. These give rise to chimeric germlings, detectable in crosses of color mutants because of different color segments (van der Meer 1977). The existence of chimeric plants – having several genotypes within one thallus – is a recently recognized phenomenon that has implications for understanding morphogenesis. Several cogerminating zygotes can become interwoven in *Codium fragile* (Friedmann & Roth 1977), *Dumontia incrassata* (Rietema 1984), and potentially in any multiaxial thallus, as well as in the parenchymatous *Smithora naiadum* (McBride & Cole 1972).

1.2.2 The Littler functional-form model

The construction of the thallus has importance for developmental physiology. Similar morphologies can be constructed in different ways; the overall morphology is important to ecological physiology. Among different algal classes, certain morphologies are repeated, which, as noted by Littler et al. (1983a), indicates convergent adaptations to critical environmental factors. On the other hand, species face divergent selection pressures: those favoring more productive, reproductive, and competitive thalli, versus those favoring longevity and environmental resistance (Littler & Kauker 1984; Russell 1986; Norton 1991). Many seaweeds show a variety of morphologies within one life history (see sec. 1.5). Heterotrichous plants with crustose bases and erect fronds within one generation (e.g., *Corallina*) and heteromorphic plants with crustose/filamentous and frondose generations (e.g., *Scytosiphon*) (Fig. 1.24) are both common. How can we assess the significance of morphology when we are faced with convergence between classes on the one hand and diversification within species on the other hand?

The functional-form model advanced by Littler and Littler in 1980, and subsequently tested extensively by them and by others, holds that the functional characteristics of plants, such as photosynthesis, nutrient uptake, and grazer susceptibility, are related to form characteristics, such as morphology and surface-area : volume (SA : V) ratios (Table 1.1). One can thus set up predictions of function from an examination of form. For example, a negative side of multicellularity is a reduced SA : V ratio for the organism. The effect of multicellularity is small in uniseriate filaments (where only the end walls adjoin other cells), and larger in massive parenchymatous forms. Rosenberg and Ramus (1984) demonstrated the predicted correlation with nutrient uptake: *Ulva curvata* (SA : V = 165 cm^2 cm^{-3}) had the highest uptake, *Fucus evanescens* and *Gracilaria tikvahiae* had about equal SA : V ratios and uptake rates, and *Codium decorticatum* (SA : V = 8.9 cm^2 cm^{-3}) had the lowest uptake rate. The decrease in area per unit volume is *relatively* small (300-fold) over a large (10^4-fold) increase in maximum dimension from a unicell to a large macrophyte, because the overall shape changes from isodiametric to laminar (Raven 1986).

Categorizing specific morphologies is not always simple, because there are no sharp boundaries between some groups. Littler and Littler (1983, p. 430) concluded that "functional-group ranking realistically should be regarded as recognizable units along a continuum, each containing considerable variations of form and concomitant functional responses." Algal turfs are adaptive forms on tropical reefs, but may comprise tight aggregates of small plants from several functional-form groups or (in so-called sparse turfs; see Fig. 15b) may have unicellular and filamentous components (Hackney et al. 1989). Alternative classifications of form can offer insight into specific functions. For example, Hay (1986) outlined some morphological types vis-à-vis light capture (see Fig. 4.11). Raven (1986) distinguished three basic life forms of plants in general, of which two are benthic. Haptophytes are plants attached to substrate particles that are large relative to the plant (e.g., most benthic microalgae and seaweeds on rocks). Rhizophytes are plants that penetrate substrata of relatively small particles (e.g., some siphonous green algae and many vascular plants). In rhizophytes, the shoots can be specialized for photon capture, and the rhizoids for nutrient uptake, whereas in haptophytes the shoots must also take up nutrients. Production of hairs by haptophyte shoots, especially when facing a low level of nutrients, is perhaps a way of improving a compromise situation (Raven 1986). Such classifications focus on functional morphology, rather than its developmental origin.

1.3 Seaweed cells

Although there is interaction between the morphology of the whole seaweed and the environment, the physiological responses to the environment, as well as the mechanisms by which the overall morphology is generated, occur within the individual cells. Cells are protected by walls and membranes and compartmentalized with membrane-bound organelles, and it is through these membranes and walls that contact with the environment must take place. The structures and compositions of cell components thus provide a necessary background to the study of physiological ecology.

Certain components and functions of algal cells are similar to (though not necessarily identical with) the systems worked out in other organisms (e.g., rats or bacteria). Mitochondrial structure and function, genetic material and its translation into proteins, and membrane structure (so far as it is understood) are fundamental features of eukaryotic cells. Algae are neither sufficiently different nor suitable as model systems for these aspects to have received particular study in the algae, although a unicellular freshwater green alga has been used to study the changes that take place in mitochondrial number and shape during the cell cycle (Chida & Ueda 1986). Nevertheless, there are differences, such as between animal and plant (and algal?) mitochondria (Douce & Neuburger 1989), between electron-transport components in brown and green algae (Popovic et al. 1983), and between chloroplast DNA arrangements in red algae and higher plants, as discussed later. Other cell components are distinctive in the algae; these include cell-wall composition and structure, flagellar apparatus, the cytoskeleton, and the thylakoid/photosystem structure. See Evans (1974), Bisalputra (1974), Brawley and Wetherbee (1981), and Pueschel (1990) for reviews of algal cytology; see Goodwin and Mercer (1983) and Hall et al. (1982) for reviews of higher-plant cell biology.

Algal cells may also contain unique structures. Brown-algal cells characteristically contain numerous refractive bodies called physodes (Fig. 1.5) that appear to be a kind of vacuole, but may not be membrane-bound (Ragan 1976; Pellegrini 1980; Clayton & Beakes 1983). These bodies originate in the plastids and contain phenolic and polyphenolic compounds that are active as antifouling agents (Craigie & McLachlan 1964) and herbivore deterrents (sec. 3.1.2 and 3.2.3). The *corps-en-cerise* that occur in some species of *Laurencia* have been shown to be storage vesicles for brominated compounds, which are abundant in this genus (Young et al. 1980). Other inclusions, such as the iridescent bodies shown in Figure 1.5, have largely unknown compositions and functions (nor is the physics of iridescence understood in these algae). Similarly, the functions of "gland cells," common in the red algae, are still largely uncertain (Pueschel 1990).

1.3.1 Cell walls

Cell walls do not merely provide rigidity. They are essential to cell growth and developmental processes, such as axis formation in zygotes and branching in growing plants; when walls are too weak, development may be impossible, as in a mutant form of *Ulva* (sec. 1.3.4). Walls are crucial in mating and in the release and adhesion of reproductive cells. The abundance of matrix material relative to fibrillar components, the extensive sulfation, and the extensive intercellular matrix are characteristics of seaweeds that suggest environmental adaptations (e.g., to wave force and desiccation) (Kloareg & Quatrano 1988).

Since the early days of electron microscopy, plant cell walls have been viewed as a meshwork of fibrils (usually cellulose) in an amorphous matrix (Mackie & Preston 1974). Little progress has been made in determining the structures of the bewildering array of matrix polysaccharides, except for some of commercial value. Some wall molecules, such as cellulose, are fibrillar and are sufficiently simple to be seen in the electron microscope (EM) and by X-ray crystallography. Indeed, some

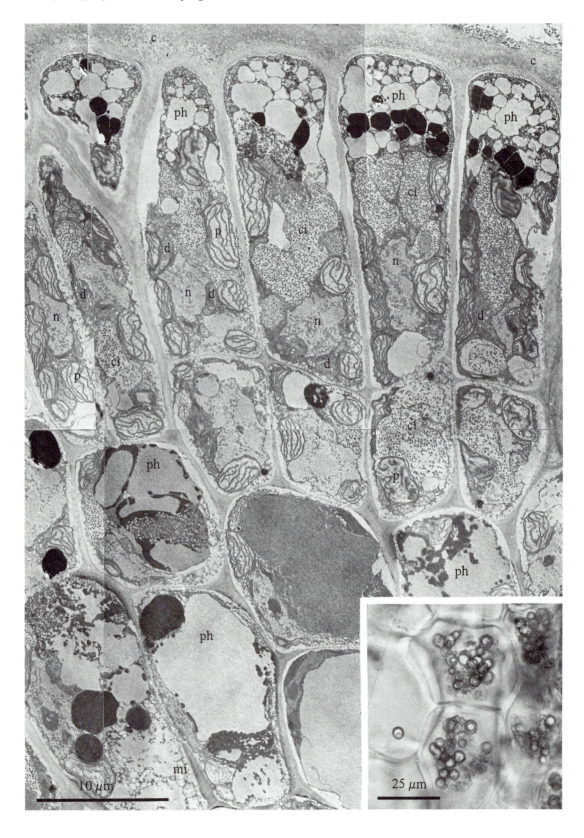

Table 1.2. *Taxonomic patterns in the skeletal polysaccharides in some green and red seaweeds*

Class Order	Genus	Cellulose	Xylan	Mannan
Ulvophyceae				
Siphonocladales		++	−	−
Cladophorales		++	−	−
Ulvales		++	−	−
Caulerpales		−	++	−
Dichotomosiphonales		−	++	−
Dasycladales		−	−	++
Codiales		−	−	++
Derbesiales	*Derbesia*	S−	−	++
		G+	+	−
	Bryopsis	S−	−	++
		G+	+	−
	Bryopsidella	+	+	−
Rhodophyceae				
Bangiales		G−	+	−
		S+	−	−

Note: −, absent; +, present; ++, abundant; S, sporophyte; G, gametophyte. Unless otherwise noted, all species tested in an Order were the same, but the literature is limited, and the exception in Derbesiales should caution against sweeping generalizations, even to genera.
Source: After Kloareg and Quatrano (1988), with permission of Aberdeen University Press, Farmers Hall, Aberdeen AB9 2XT, U.K.

of the first EM views of plant cell walls were of *Ventricaria*. However, algal walls can have other fibrillar molecules (Table 1.2), or several arrangements of cellulose fibrils, and there is also the array of complex matrix molecules (to be discussed in sec. 4.5.2). The most recent concept of wall structure (Fig. 1.6a) is highly speculative, as Kloareg and Quatrano (1988) acknowledge, and yet it still has little detail. Nevertheless, there is reason to suppose that wall structures and functions are as precise and interrelated as are the structures and functions of membranes and proteins (Craigie 1990a).

Cellulosic walls are made of layers of parallel celluose microfibrils. In some genera, such as *Chaetomorpha, Siphonocladus,* and perhaps *Ventricaria,* the fibrils in successive layers are oriented at steep angles to each other (90° = orthogonal). In such algae, hydrolysis of the microfibrils yields only glucose, as is to be expected if only cellulose is present. In other algae, or in certain walls, including aplanospores of *Boergesenia*

Figure 1.5. Cross section of the fucoid *Cystoseira stricta* showing differentiation of tissues. The cells at the top of the view are the outer, meristodermal cells; those at the bottom are promeristematic. Inset shows fresh section stained with caffeine to reveal physodes; c, cuticle; ci, iridescent body; d, Golgi body; mi, mitochondrion; n, nucleus; p, chloroplast; ph, physode. (From Pellegrini 1980, with permission of The Company of Biologists.)

forbesii, eggs of *Pelvetia fastigiata,* zygotes of *Fucus serratus,* and vegetative walls of *Spongomorpha arcta* and *Boodlea coacta,* the angle changes much more slowly, giving a helicoidal arrangement (Fig. 1.6b), and hydrolysis yields also some xylose and mannose. Neville (1988) argues that a hemicelluose (cellulose with flexible, bulky side chains) has the capacity of self-assembly into a helicoidal matrix, which then positions the cellulose fibrils that are visible in the electron microscope. In many algal walls, however, the microfibrils in each layer have no preferential orientation (Kloareg & Quatrano 1988).

Some algal walls also have an outer cuticle of protein (Hanic & Craigie 1969). The iridescence of some species of *Iridaea* has been explained by Gerwick and Lang (1977) as the result of a multilaminated cuticle, in which many thin layers of alternating higher and lower refractive indices produce interference, as in a soap bubble. Other algae are well known for impregnating their walls with carbonate.

Cellulose is the fibrillar material throughout the brown algae and most of the reds, but it is not the only fibrillar structural polysaccharide in algal walls (Table 1.2). Xylans also form microfibrils, and mannans form short rods. Moreover, some seaweeds feature a biochemical alternation of generations in which different ploidy levels have different fibrillar or matrix polysaccharides. For instance, the diploid thallus of

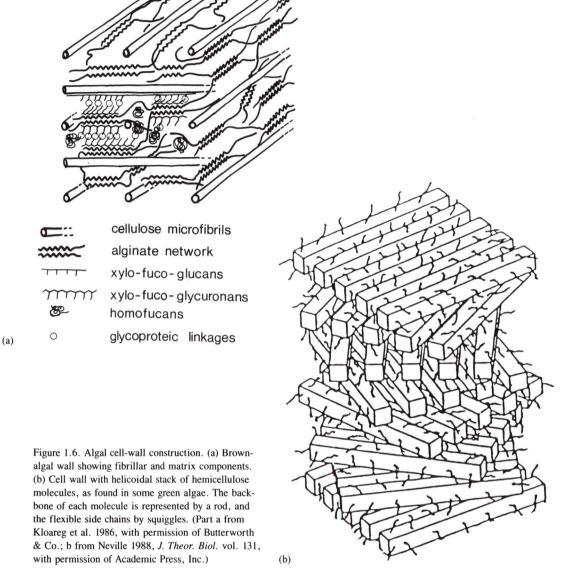

cellulose microfibrils
alginate network
xylo-fuco-glucans
xylo-fuco-glycuronans
homofucans
glycoproteic linkages

(a)

Figure 1.6. Algal cell-wall construction. (a) Brown-algal wall showing fibrillar and matrix components. (b) Cell wall with helicoidal stack of hemicellulose molecules, as found in some green algae. The backbone of each molecule is represented by a rod, and the flexible side chains by squiggles. (Part a from Kloareg et al. 1986, with permission of Butterworth & Co.; b from Neville 1988, *J. Theor. Biol.* vol. 131, with permission of Academic Press, Inc.)

(b)

Acetabularia has mannans, and yet the walls of the cysts (supposedly equivalent to gametophytes) have cellulose. In the *Derbesia-Halicystis-Bryopsis* group, gametophytes have celluose/xylan walls, and gametophytes have mannans, but parthenogenetic male and female sporophytes (presumably haploid) have the same wall composition and plant morphology as normal, diploid sporophytes (Huizing et al. 1979; Kloareg & Quatrano 1988). The fibrillar component of brown algae is consistently cellulose, and there is little variety in the red algae. Alternation in matrix carrageenans is found in some carrageenophytes (Gigartinaceae and Phyllopho-

raceae, e.g., *Chondrus,* but not Solieriaceae, e.g., *Eucheuma*). No reason for this biochemical difference between generations has been deduced.

The walls of blue-green algae are similar to the walls of Gram-negative bacteria and are strikingly different from eukaryote walls; more accurately, they are cell envelopes (Drews & Weckesser 1982). Outside the cell membrane is a "wall" comprising a peptidoglycan layer and an outer membrane, then a sheath or slime layer. The fibrillar component is the peptidoglycan, in which parallel sugar chains are cross-linked by short peptide chains. Work by De Vecchi and Grilli Caiola

(1986) indicates that the envelopes of *Anabaina* may contain various small molecules that are lost during routine fixation procedures.

The complexity and molecular specialization of wall surfaces are being revealed by the use of monoclonal antibodies and related techniques (Vreeland et al. 1987; Jones et al. 1988; Hempel et al. 1989). Different parts of a thallus are likely to have different wall structures. The high proportion of polyguluronic acid in adhesive alginate is well known (Craigie et al. 1984; Vreeland & Laetsch 1989) (sec. 4.5.2). The difference between rhizoidal and thallus poles has been detected even in germinating zygotes and regenerating protoplasts, again using antibodies to different carbohydrate fractions (Boyen et al. 1988). In a detailed study of *Fucus serratus* sperm, Jones et al. (1988) were able to distinguish several regions, including the tip of the anterior flagellum (crucial in egg recognition; sec. 1.5.4), the mastigonemes on the anterior flagellum, and the sperm body. Localization of certain wall components in certain regions of a cell has implications for the assembly process, but little is known of how carbohydrates are directed from their Golgi body to the appropriate piece of wall. The role of actin in zygote polarization in *Fucus* suggests, however, that this contractile protein may be involved (Kropf et al. 1989).

Assembly of cell-wall microfibrils is a complex process that must take place on both sides of the plasmalemma. The process has been dissected with inhibitors into three components: polymerization, orientation, and crystallization. Colchicine is well known for disrupting microtubules (much used in studies of mitosis); cycloheximide, which inhibits protein synthesis by 80S ribosomes, inhibits polyglucan synthesis; and two carbohydrate-binding dyes, Congo Red and Calcofluor White ST/M2R, inhibit crystallization of the glucan chains into microfibrils (Herth 1980; Quader 1981). *Ventricaria* and *Boergesenia* have been used as models to study wall synthesis because they readily form protoplasts on wounding (sec. 1.7.4). The protoplasts assemble an initial wall of randomly oriented fibrils 2–3 h after wounding and then begin oriented deposition (Itoh et al. 1986; Itoh & Brown 1988). Before wall regeneration begins, arrays of particles appear within the plasmalemma; these "terminal complexes" apparently polymerize glucose into cellulose chains and provide the orientation. The development of linear terminal complexes from circular clusters of particles (presumably itself oriented by microtubules) is more or less coincident with the change to oriented microfibril deposition.

1.3.2 Chloroplasts

Photosynthetic algal cells contain one or more chloroplasts (some *Acetabularia* species may have 10^7–10^8 per giant cell). In thick thalli, medullary cells, shaded from light and blocked from rapid gas exchange by overlying cortical cells, usually lack chloroplasts or have vestigial chloroplasts. Chloroplasts have characteristic shapes that are useful for taxonomy; they may be discoidal, stellate, band-shaped, or cup-shaped (Brawley & Wetherbee 1981; van den Hoek 1981; Larkum & Barrett 1983). All have photosynthetic pigments in thylakoids (and in red and blue-green algae, also on thylakoids), and the arrangements of thylakoids are taxonomically significant. Brown-algal thylakoids typically are three per lamella (Fig. 1.7b); in green algae they range from two to many; in red algae (Fig. 1.7a) and blue-green algae they are single. Some chloroplasts in the more advanced Florideophycidae have a peripheral thylakoid just inside the chloroplast envelope (Fig. 1.7a), and brown-algal chloroplasts have endoplasmic reticulum tightly associated on the outside (Fig. 1.7b). Some chloroplasts have a pyrenoid (again there are characteristic shapes) comprising chiefly RuBisCO (ribulose-1,5-bisphosphate carboxylase/oxygenase); in others, this key Calvin-cycle enzyme is dispersed in the matrix. Some pyrenoids (e.g., those of *Bryopsis maxima*) are also the sites of nitrate reductase (Okabe & Okada 1990).

Unfortunately, the significance to physiology of the shapes and arrangements of chloroplasts is not clear. To some extent the differences may simply reflect phylogeny – various ancestral types "fossilized" in the algal lines that developed from each. All must, however, fit the constraints of light harvesting, diffusion of inorganic carbon, concentration of carbon-fixing enzymes, and supply of adenosine triphosphate (ATP) and the reduced form of nicotinamide adenine dinucleotide phosphate (NADPH) (Larkum & Barrett 1983). Chloroplast shapes and sizes may reflect different evolutionary responses to the reduction in light absorbance by "packaged" pigments, as compared with pigments in uniform solution (the "package effect") (Osborne & Raven 1986; Dring 1990). A possible explanation for differences in thylakoid arrangement between red and blue-green algae (not stacked) and green and brown algae (stacked) was proposed by Larkum and Barrett (1983, pp. 48): ". . . algae harvest green/yellow light with reasonable efficiency if they have multiple appressed thylakoids. Thus in the [green and brown] algae . . . accessory light-harvesting pigments fill in only the blue-green region." Red and blue-green algae have pigments that absorb in the green/yellow region, but these are arranged on the thylakoid surfaces as phycobilisomes that prevent tight packing of the thylakoids (sec. 4.3.4).

Photophosphorylation takes place across the thylakoid membranes in the same way that respiratory phosphorylation is driven across the mitochondrial membrane. Electron transport in the membrane pumps protons across the membrane (to the inside of the thylakoids), and the electrochemical gradient drives protons back via ATPase particles. Algal thylakoids differ not so much in their electron-transport components

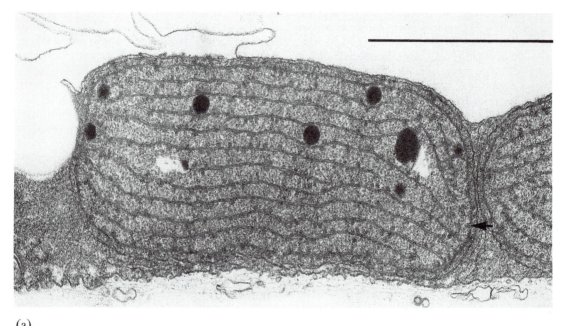

(a)

(b)

Figure 1.7. Algal chloroplasts. (a) Chloroplast of the red alga *Laurencia spectabilis* showing parallel single thylakoids and one thylakoid (arrow) surrounding the others, just inside the chloroplast membrane. (b) Chloroplast of a brown alga (*Fucus* sp.) showing characteristic triple thylakoids, the genome (G), and endoplasmic reticulum (ER) surrounding the organelle. Scale: 1 μm. (Courtesy of Dr. T. Bisalputra.)

as in their stacking arrangements and pigment systems (sec. 4.3.1).

Isolated chloroplasts have been very useful in higher-plant studies, but when chloroplasts are isolated from siphonous algae, they have a special integument around the chloroplasts that probably forms during disruption (i.e., when cells are damaged by herbivores or microscopists). This extra membrane encloses a small but perhaps significant amount of cytoplasm. Thus these cell fragments might more accurately be called cytoplasts (Grant & Borowitzka 1984). Isolated chloroplasts from *Codium* and *Caulerpa* do not swell or burst in distilled water. The integument may prevent the chloroplasts of these species from being digested when they are eaten by saccoglossan molluscs, thus allowing the chloroplasts to continue photosynthesis in a symbiotic relationship with the animal (Grant & Borowitzka 1984). Some siphonous green algae have two kinds of plastids: chloroplasts and colorless amyloplasts. This group, which does not separate taxonomically, includes *Caulerpa, Halimeda, Udotea,* and *Avrainvillea;* but *Codium, Derbesia,* and *Bryopsis* have only chloroplasts (van den Hoek 1981). Both plastid types in heteroplastidic genera have a distinctive set of concentric nonpigmented membranes at one end of the plastid, called a concentric lamellar system or (from its supposed function) thylakoid organizing center. Evidence for that function is inferential, and there is no explanation of how other chloroplasts organize without them.

Chloroplasts may migrate within a cell or siphon. Dramatic diel migration of chloroplasts takes place in *Halimeda* (Drew & Abel 1990): More than 100 chloroplasts from each surface utricle pass along cytoplasmic strands through narrow constrictions into medullary filaments. They end up below the carbonate exoskeleton (Fig. 1.8), leaving the plant looking bleached. Inward migration is triggered by the onset of darkness (at any time of the day). Outward migration begins before dawn, apparently on an endogenous rhythm. In *Caulerpa,* amyloplasts are even more mobile than chloroplasts and are transported on microtubules, whereas chloroplasts are moved by the actomyosin system (Menzel & Elsner-Menzel 1989b) (sec. 1.3.3).

Chloroplasts contain deoxyribonucleic acid (DNA), which is attached to an internal membrane (Tripodi et al. 1986). Two configurations of chloroplast DNA occur in seaweeds: scattered (but connected) small nucleoids in red and green algae, and a peripheral ring in brown algae (Coleman 1985). Chloroplasts can make some of their own components, but depend on the nucleus for others. They divide in the cells, but (in spite of a report to the contrary) are unable to replicate in cell-free suspensions (Grant & Borowitzka 1984). They are inherited, usually only from the maternal side, in sexual reproduction. In giant or coenocytic cells with numerous chloroplasts, not all chloroplasts are equal. Those at the apex in *Acetabularia* and *Caulerpa* are morphologically and physiologically different and more active (Dawes & Lohr 1978; Bonotto 1988). Indeed, in some *Acetabularia,* and *Batophora* species, many of the chloroplasts behind the apex lack DNA, although all chloroplasts in cysts have DNA. (This is seen in cultured plants and may not be so for wild plants; Bonotto 1988.) An absence of DNA in any chloroplast is surprising, because of its essential role in the organelle, and yet not surprising, because there is no mechanism comparable to the mitotic spindle for ensuring distribution of DNA to daughter chloroplasts (Lüttke 1988) (although prokaryotes divide with no problem). Presumably, in multiplastidic algae, such as *Acetabularia,* the presence of some chloroplasts without DNA does no harm, and many chloroplast functions can continue without new translation. In other plants, the arrangement of the nucleoid within the chloroplast effectively prevents accidental formation of organelles without DNA.

Among the proteins encoded by the chloroplast genome in higher plants is the large subunit of RuBisCO. However, in two red algae, the chrysophyte *Heterosigma carterae* and perhaps *Fucus,* both subunits are coded in the chloroplast (Keen et al. 1988; Cattolico & Loiseaux-de Goër 1989; Shivji 1991). The N-terminal end of the small subunit in *Fucus* uniquely lacks seven amino acids, in comparison with other organisms; Keen et al. (1988) have speculated that if, indeed, this subunit is coded in the chloroplast, the missing amino acids may represent those that in other plants constitute a signal peptide that guides the subunit through the chloroplast membrane and is then cut off. Red-algal chloroplast genomes also code for phycobiliproteins and some other photosynthetic proteins (Shivji 1991).

Chloroplasts (and mitochondria) are inherited by daughter cells during vegetative division and by offspring during reproduction. In sexual reproduction, the zygote's organelles come almost exclusively from the maternal side. Thus color mutations coded by chloroplast DNA show non-Mendelian inheritance (sec. 1.4.1). In some cases, such as with red algae, there is no paternal chloroplast to inherit, but even in isogamy and anisogamy the chloroplast of one strain is consistently destroyed soon after plasmogamy (e.g., Bråten 1983 on *Ulva mutabilis*). In one study it was found that sperm chloroplasts in *Laminaria angustata* zygotes remained small and did not divide, although they did survive, whereas mitochondria were enclosed in endoplasmic reticulum and digested (Fig. 1.9) (Motomura 1990). Surprisingly, the surviving centrioles came from the sperm, not the egg, in this species. Similar phenomena of recognition and elimination of one organelle in the presence of a homologue are known in diverse organisms (Koslowsky & Waaland 1987). Cell fusion also occurs when protoplasts are joined and during the repair process in *Griffithsia* (sec. 1.7.4). If *Griffithsia* filaments from two geographic isolates are joined, one will show a cytoplasmic incompatibility reaction, in which

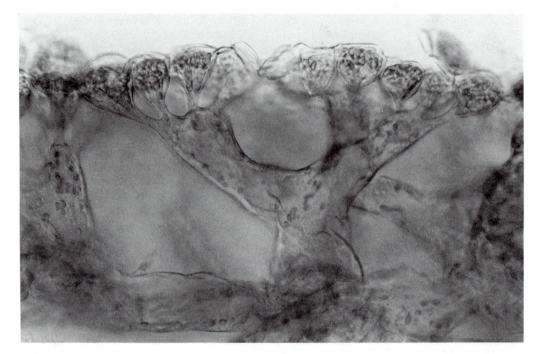

(a)

(b)

Figure 1.8. Migration of chloroplasts in *Halimeda*. (a) Daytime cross section shows surface (primary) utricles packed with chloroplasts. (b) Nighttime section shows that the chloroplasts have migrated below the calcified layer into the secondary utricles and medullary filaments. (From Drew & Abel 1990, with permission of Walter de Gruyter & Co.)

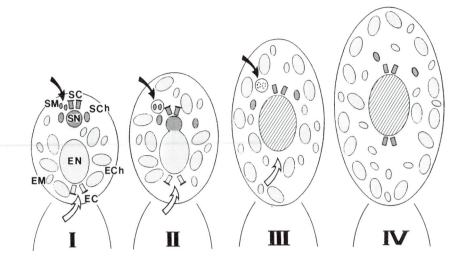

Figure 1.9. Cytoplasmic inheritance of mitochondria, chloroplasts, and centrioles in *Laminaria angustata* (schematic). After plasmogamy (I), sperm nucleus (SN), mitochondria (SM, curved black arrow), chloroplasts (SCh), and centrioles (SC) are incorporated into the egg cytoplasm. Before and after karyogamy (II), SM are enclosed by a membrane. After that (III), sperm mitochondria are completely digested in the membrane. Sperm chloroplasts remain (I–IV), but do not increase in size or develop. Sperm centrioles remain (I–III), whereas egg centrioles (EC, curved white arrow) disappear after karyogamy (III). Sperm centrioles duplicate and migrate to each mitotic pole in the early mitotic stage (IV). EM, egg mitochondrion; ECh, egg chloroplast. (From Motomura 1990, with permission of *Journal of Phycology.*)

its chloroplasts will be destroyed (Koslowsky & Waaland 1987). In all these cases, one set of organelles seems to act competitively against the other. In many cases, competition involves mitochondria as well, though no mitochondrial destruction was seen in the *Griffithsia* study.

1.3.3 Cytoskeleton and flagellar apparatus

The cytoskeleton in algal cells plays fundamental roles in germination, cell morphogenesis, cell motility, chloroplast movements, cytoplasmic streaming, and wound healing. The cytoskeleton consists primarily of networks and bundles of microfibrils. By analogy with animal cells, and on the basis of very few algal studies, these microfibrils are assumed to be composed of contractile actin microfilaments plus force-generating myosin filaments (La Claire 1989b). Microtubules, consisting of tubulin and dynein, are important in nuclear and chromosome movements, in large-scale movements of amyloplasts in *Caulerpa,* and in the structure of the flagellar apparatus. The cytoskeleton is associated with "a host of . . . proteins which serve to bind, crosslink, cap, sever, buffer, organize, and move the elements of the cytoskeletal framework" (Salisbury 1989a, p. 20). Actin filaments are regulated by the availability of Ca^{2+}, which itself probably is controlled by the protein calmodulin (La Claire 1989b).

An example of the way the cytoskeleton shapes cells is seen in the development of cysts in *Acetabularia* (Bonotto 1988; Menzel & Elsner-Menzel 1989a). During vegetative growth (Fig. 1.10), bundles of actin mi-

crofibrils are arranged along the axis of the cell (Fig. 1.11b,d). After the cap has formed, the diploid primary nucleus divides into several thousand haploid nuclei, each with a cluster of microtubules; Bonotto has likened their appearance to that of a comet tail (Fig. 1.11a). The nuclei migrate to the rays of the cap, where cyst formation ensues. There is a short "mixing phase" (Fig. 1.11e,f), during which nuclei swap positions and microfibrils swirl about. Then a circular domain forms around each nucleus, ultimately ringed by interwoven microfibrils and microtubules (Fig. 1.11k–n). Finally the rings contract and cyst walls form.

Flagella* of motile cells serve several purposes besides the important function of propelling cells (which for seaweed biology means getting gametes together and helping cells swim to the seabed). Of course, some algae reproduce well without flagellated cells. Flagella act as specialized recognition and adhesion organelles during mating (sec. 1.5.4), increase the cell surface area, and serve as a "walled . . . cell's window on the world" (Salisbury 1989a, p. 22). The mechanisms to respond to light, chemicals, and other cells presumably are built into the structure of the flagellar apparatus. Flagella themselves consist, as do all eukaryote flagella, of a 9+2 arrangement of microtubules (Fig. 1.12b). In some algae, including some seaweeds,

---

* The nonrotatory, 9+2 microtubular organelles in eukaryotes ought to be called cilia, according to Cavalier-Smith (1986). While accepting his arguments, we have for now retained the term more familiar in phycology.

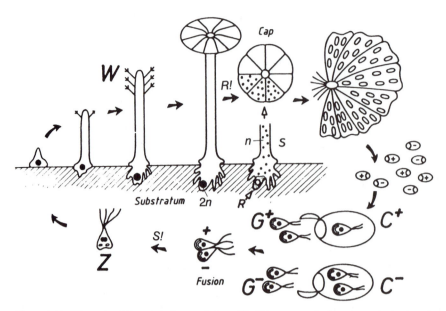

Figure 1.10. Life cycle of *Acetabularia acetabulum*. C^+, C^-, cysts producing, respectively, + and − gametes (G^+, G^-); R, rhizoid with a residual body (after meiosis); R!, meiosis; S, secondary nuclei; W, whorls of branches; Z, zygote. (From Bonotto 1988, with permission of BioPress.)

one or more of the flagella bear hairs (mastigonemes), and in some unicellular algae the flagella have scales. The central axoneme is connected to a basal body. The basal bodies are joined by striated fibers (and, in *Batophora*, by a capping plate, Fig. 1.12c) and are anchored into the cytoskeleton by four microtubular rootlets (one pair with two tubles, and one pair with three to five tubules). One of these rootlets is associated with the eyespot (in those cells that have them). The eyespot (= stigma) is a patch of lipid droplets, orange or red because of carotenoid pigments, on the side of the chloroplast, where the chloroplast is also closely associated with part of the cell membrane. The flagella also have striated "system II" roots reaching back around the nucleus (Fig. 1.12b). The two sets of striated fibers are contractile and are made of centrin, an acidic phosphoprotein with a molecular weight of about 20,000 (Salisbury 1989a,b).

The association of the eyespot and microtubular rootlets presumably is important in phototactic swimming. A swelling is present at the base of one flagellum in many phototactic motile cells; among the brown algae, each of those zoospores and gametes that has an eyespot also has a swelling on the posterior flagellum, and the swelling fits into a concavity on the eyespot (Kawai et al. 1990). In these cells also, the posterior flagellum is autofluorescent, apparently because of a flavin that is presumed to be a photoreceptor pigment (Kawai 1988). The action spectrum of phototaxis has peaks in the blue region (Fig. 1.13a) (Kawai et al. 1990). The function of the eyespot is to focus light onto the photoreceptor, either directly (like a lens) in the chromophytes

(Kreimer et al. 1991) or by constructive interference by stacked lipid layers (something like iridescence) in the green algae (Melkonian & Robenek 1984). Swimming *Ectocarpus* gametes and other chromophyte motile cells roll as they swim, and when they are moving at a sufficient angle to the light, the photoreceptor receives flashes of light as the cell rolls. This stimulation is thought to cause the posterior flagellum to beat, acting as a rudder (Fig. 1.13b). When the cell is swimming parallel to the light, the photoreceptor is continually shaded by the cell (Kawai et al. 1990).

Considerable taxonomic significance has been attributed to the absolute configuration of the microtubular flagellar root systems in the green algae (Fig. 1.14) (van den Hoek et al. 1988; Sluiman 1989). The Ulvophyceae, which include essentially all the green seaweeds, are characterized by an "11 o'clock/5 o'clock" arrangement of basal bodies (and see *Batophora*, Fig. 1.12b). (*Chlamydomonas* has a 1 o'clock/7 o'clock arrangement.) Microtubule arrangements during mitosis/cytokinesis are also taxonomically significant in separating most of the green algae (which have microtubules perpendicular to the division plane) from mosses and higher green plants that have a "phragmoplast" parallel to the division plane.

1.3.4 Cell growth

Cell growth is driven by water influx and is restricted by the cell wall. Plant cells are normally turgid, because water tends to flow into them by osmosis (Chapter 6). The layers of fibrils in the wall (sec. 1.2.3) resist swelling and stop net water influx. Cell growth is

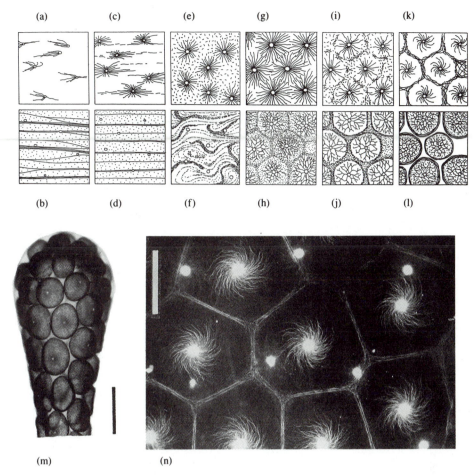

Figure 1.11. Cytoskeletal changes during cyst morphogenesis in *Acetabularia*. (a–l) Diagrams of the distribution of microtubules (upper row) and actin microfilaments (lower row); (a, b) nuclear migration in the stalk; (c, d) migration within the cap ray; (e, f) mixing phase; (g, h) disc stage; (i, j) ring stage; (k, l) dome stage. (m) Photomicrograph of a ray in the late dome stage, the cysts nearly formed. Scale bar = 500 μm. (n) Detail of ray at beginning of contraction (just after dome stage). Microtubules are associated with the nucleus and are in a ring delimiting the edge of the future cyst; they are interwoven with microtubules, not visible in this preparation. Scale bar = 50 μm. (Reprinted from Menzel & Elsner-Menzel 1989a, with permission of John Wiley/Alan R. Liss Inc.)

achieved by locally controlled loosening of the covalent and/or hydrogen bonds holding the fibrillar network in place. The exact mechanism in algae is not known (Nishioka et al. 1990). The established dogma for higher plants, that auxin stimulates a proton flux, which in turn breaks the bonds between microfibrils, has been critiqued by Hanson and Trewavas (1982).

The importance of structural control by the cell wall is illustrated by a mutant, *lumpy*, of *Ulva mutabilis* that grew as an aggregate of undifferentiated cells, rather than first forming a filament and later a holdfast plus blade. The large, round cells of this mutant contained 1.3 times as much polysaccharide as wild-type cells, but 80% of it was water-soluble, compared with only 50% for the wild type (Bryhni 1978). There may thus have been less cross-linking between the insoluble cellulose fibrils. This would have increased the plastic-ity of the cell walls, causing the cells to swell and preventing local differences in wall expansion that are required for normal, oriented morphogenesis.

Cell growth is commonly, but not necessarily, uniform throughout the wall (Garbary & Belliveau 1990). Species that feature localized growth are useful as experimental material (e.g., Garbary et al. 1988). The location of cell growth can be followed by labeling existing cell-wall polysaccharides with a fluorescent stain such as Calcofluor White M2R (a brightener at one time used in laundry detergents) (Waaland 1980; Belliveau et al. 1990). If cell growth occurs by extension of existing wall material, the dye will be uniformly diluted. If, on the other hand, cell growth occurs by localized synthesis of new wall, dark bands will appear on the cells when seen under ultraviolet (UV) light, because the new wall will not be stained.

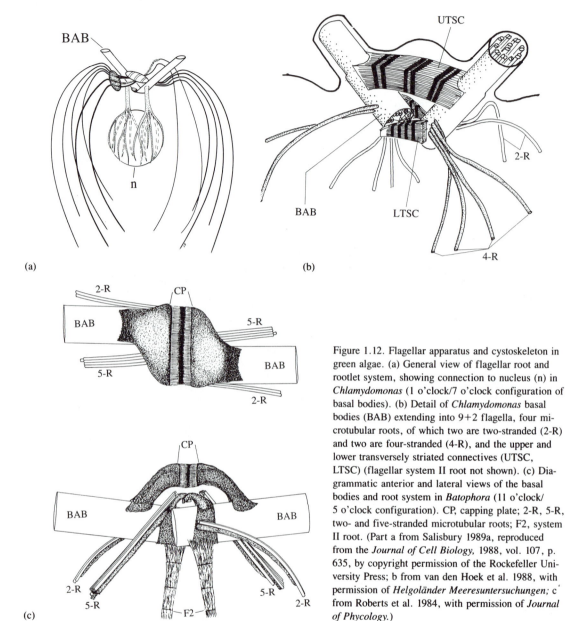

Figure 1.12. Flagellar apparatus and cystoskeleton in green algae. (a) General view of flagellar root and rootlet system, showing connection to nucleus (n) in *Chlamydomonas* (1 o'clock/7 o'clock configuration of basal bodies). (b) Detail of *Chlamydomonas* basal bodies (BAB) extending into 9+2 flagella, four microtubular roots, of which two are two-stranded (2-R) and two are four-stranded (4-R), and the upper and lower transversely striated connectives (UTSC, LTSC) (flagellar system II root not shown). (c) Diagrammatic anterior and lateral views of the basal bodies and root system in *Batophora* (11 o'clock/ 5 o'clock configuration). CP, capping plate; 2-R, 5-R, two- and five-stranded microtubular roots; F2, system II root. (Part a from Salisbury 1989a, reproduced from the *Journal of Cell Biology*, 1988, vol. 107, p. 635, by copyright permission of the Rockefeller University Press; b from van den Hoek et al. 1988, with permission of *Helgoländer Meeresuntersuchungen;* c′ from Roberts et al. 1984, with permission of *Journal of Phycology*.)

Intercalary cell extension in some Ceramiales, studied by Waaland and Waaland (1975), Garbary et al. (1988), and others, takes place through localized additions of wall material at each end of the cell (Fig. 1.15). The number and locations of the bands are characteristic of a species. In the *Antithamnion* illustrated, there is a strong basal growth band and a small apical band in axial cells, and only a basal band in determinate laterals. The location of band growth in this species is under apical control: If, for instance, the apex of a main axis is removed, the main growth band in those axial cells will switch to the other end of the cell, remaining basal relative to the nearest apex on an indeterminate lateral. This is an example of apical domi-

nance (sec. 1.7.3); there is no indication yet of how this control is effected.

Cell growth may follow or be followed by cell division (sec. 1.3.5). Meristematic cells divide and grow repeatedly; other cells may stop growth and enter a stage of differentiation. All of this development is ultimately under nuclear control, and the basis for this control is to be sought in nucleocytoplasmic studies, especially in *Acetabularia*.

1.3.5 Cell division

The brown algae, some green algae, and the Bangiophycidae have uninucleate cells, but coenocytic algae have many nuclei in cells or siphons, and thus karyo-

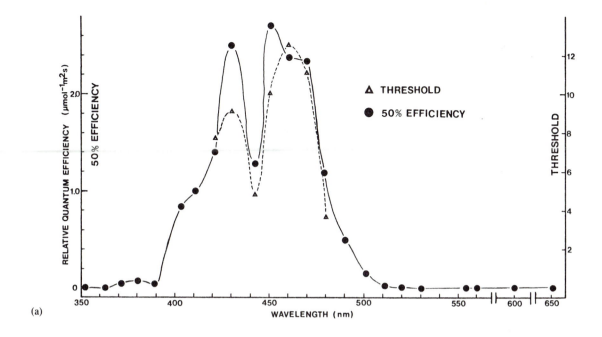

(a)

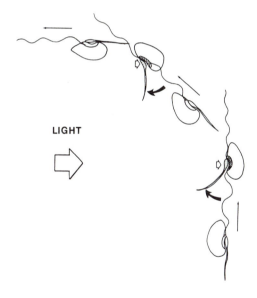

LIGHT

(b)

Figure 1.13. Phototaxis in *Ectocarpus* gametes. (a) Action spectrum based on the quantum irradiance required for threshold or 50% efficiency of the response (μmol of quanta = μE; see Chapter 4.) (b) Hypothetical photoorientation mechanism for positive phototaxis of gamete under unilateral illumination. The swelling (stippled) at the base of the posterior flagellum intermittently shades the photoreceptor (crosshatched). (From Kawai et al. 1990, with permission of Springer-Verlag, Berlin.)

kinesis and cytokinesis may be separated. Among the multinucleate taxa are the siphonous and hemisiphonous greens and many of the Florideophycidae.

The cytological details of cell division have been studied particularly in the green algae as a taxonomic tool. Altogether, eight types are recognized, but in the Ulvophyceae there are only two (Fig. 1.16) (van den Hoek et al. 1988). Both are characterized by having a persistent nuclear membrane (''closed'' mitosis) and persistent telophase spindle microtubules. In coenocytic taxa (Dasycladales, Bryopsidales, Cladophorales, as defined by van den Hoek et al. 1988), mitosis is not immediately followed by cytokinesis (type VI). In uninucleate taxa (Ulvales, Codiolales), a cleavage furrow

forms across the cell, and Golgi-derived vesicles are added to create the new cell wall (type V). In the division of the apical cell of *Acrosiphonia,* more of the nuclei are partitioned to the apical cell than to the subapical cell; the apical cell remains meristematic, whereas the other cell rarely divides again (Kornmann 1970). (Nuclei are equally distributed during intercalary divisions in this genus; Hudson & Waaland 1974.)

In contrast, little attention has been paid to cell division in the brown and red algae (Brawley & Wetherbee 1981; Scott & Broadwater 1990). Red algae are notable for the extensive evagination of the nuclear envelope that occurs during mitosis and for their polar rings, which perhaps substitute for centrioles (which the

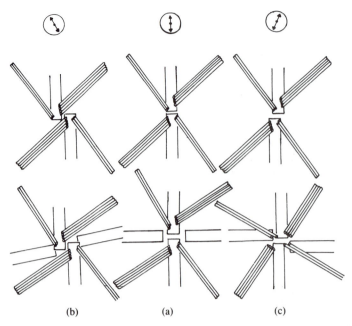

Figure 1.14. Diagrams illustrating the configurations of basal bodies in biflagellate (top row) and quadriflagellate (bottom row) motile cells of green algae. (a) 12 o'clock/6 o'clock configuration in the hypothetical ancestor; (b) 11 o'clock/5 o'clock configuration in Ulvophyceae; (c) 1 o'clock/7 o'clock configuration in *Chlamydomonas*. (From van den Hoek et al. 1988, with permission of *Helogoländer Meeresuntersuchungen.*)

red algae lack) as microtubule organizing centers. The Florideophycidae and sporophytes of the Bangiophycidae also characteristically produce pit plugs between cells (which have been used to trace cell lineages) and secondary pit plugs between neighboring cells (Brawley & Wetherbee 1981; Bold & Wynne 1985; Pueschel 1989, 1990) (see Fig. 1.22). Pit plugs are at least permeable to ions, though they are not permeable to certain dyes (Bauman & Jones 1986). In some green algae and brown algae, plasmodesmata connect neighboring cells; plasmodesmata become highly developed in the trumpet hyphae and sieve tubes of kelps.

Mitosis frequently occurs on a diurnal rhythm, with most cell division taking place at night (Austin & Pringle 1969; Kapraun & Boone 1987; Cannon 1989).

Cell division in the cyanobacteria involves invagination of the cell envelope, either all components together or first the cell membrane plus the peptidoglycan layer (Drews & Weckesser 1982). No microtubules are involved, and procaryotes also lack an actomyosin cytoskeleton (Prescott et al. 1990).

1.3.6 Heterocysts

Heterocysts are the usual nitrogen-fixing sites in blue-green algae (Wolk 1982). ("Heterocyst" has quite a different meaning as applied to coralline algae.) Typically, heterocysts are formed in low-nitrogen water (sec. 1.4.1, 5.5.1). Dinitrogen fixation is sensitive to oxygen, and the heterocysts provide an O_2-free environment. Recent evidence has shown that nonheterocystous

blue-green algae can fix N_2 (e.g., planktonic *Oscillatoria* spp. [*Trichodesmium*]) (Paerl & Bebout 1988), but again this depends on a low-O_2 environment. Oxygen is a by-product of normal (noncyclic) phosphorylation, but ATP is made in heterocysts via cyclic phosphorylation in which no oxygen (and also no $NADPH_2$) is formed. Additional energy is obtained from a disaccharide imported from the adjoining vegetative cell(s) (Fig. 1.17). Inward diffusion of external oxygen is sufficiently restricted by the thick heterocyst wall that cytoplasmic oxygenases can reduce it. Outside the ordinary wall are another two or three layers: an inner, laminated glycoprotein layer, which is probably the most effective in oxygen resistance, surrounded by polysaccharide, which may be divided into a homogeneous layer and a fibrillar layer (Wolk 1982). Among the common filamentous blue-green algae, the Oscillatoriaceae (*Schizothrix, Spirulina*) do not have heterocysts, but the Nostocaceae do; this family includes *Calothrix* and *Scytonema*. An estimate of nitrogen fixation on a coral atoll was comparable to that for a terrestrial legume pasture (Magne & Holm-Hansen 1975). Temperate blue-green algae also fix significant amounts of nitrogen (Whitton & Potts 1982).

1.4 Seaweed genetics and molecular biology
1.4.1 Classical and molecular-genetics studies of seaweeds

Two thin chapters on seaweed genetics in a 1976 book on algal genetics noted that such study is hindered

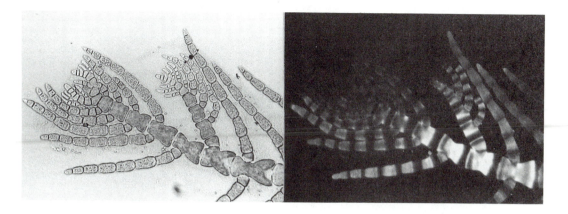

Figure 1.15a. Cell growth in *Antithamnion defectum,* visualized with Calcofluor White, as seen in bright field (left) and under UV light (right). A main axis with apical cell bears one indeterminate lateral and several determinate laterals. Under UV, dark bands of new, unstained wall are visible. The main axial and indeterminate lateral cells have two growth bands, the determinate laterals only one. Notice also the pit connections in the main axis. (From Garbary et al. 1988, reproduced by permission of the National Research Council of Canada from the *Canadian Journal of Botany,* vol. 66.)

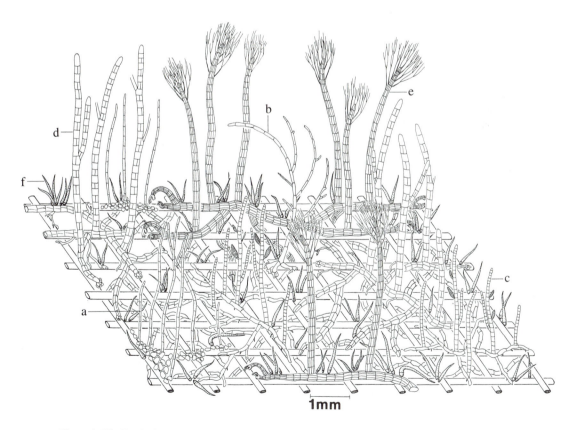

Figure 1.15b. Tropical algal sparse turf, shown growing on a 1-mm mesh plastic screen in the Smithsonian Institution's enclosed ecosystem. Genera shown include (a) *Pilinia,* (b) *Cladophora,* (c) *Giffordia,* (d) *Sphacelaria,* (e) *Herposiphonia,* and (f) *Calothrix.* (From W. H. Adey, 1991, *Dynamic Aquaria,* with permission of Academic Press.)

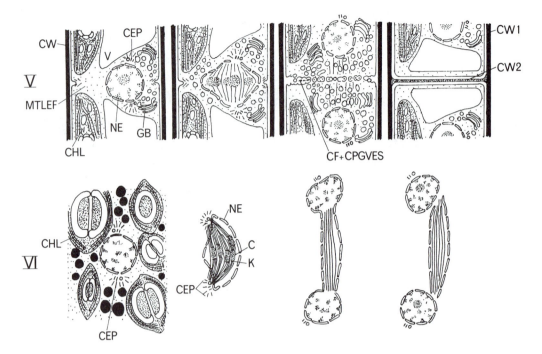

Figure 1.16. Mitosis-cytokinesis types in Ulvophyceae. Top row: Type V in *Ulothrix:* closed mitosis with persistent telophase spindle; cytokinesis by a cleavage furrow to which Golgi-derived vesicles are added. Bottom row: Type VI (*Valonia*): closed mitosis with a prominent peristent telophase spindle, causing a typical dumbbell shape; cytokinesis does not immediately follow mitosis. C, chromosome; CEP, pair of centrioles; CF, cleavage furrow; CHL, chloroplast; CPGVES, cell plate of Golgi-derived vesicles; CW, cell wall; CW1, CW2, wall layers; GB, Golgi body; K, kineto-chore; MTLEF, microtubules along leading edge of cleavage furrow of plasma membrane; NE, nuclear envelope; V, vacuole. (From van den Hoek et al. 1988, with permission of *Helogoländer Meeresuntersuchungen*.)

by the long generation times, the obligate photoautotro-phy of most seaweeds (in contrast to *Chlamydomonas* and *Euglena*), and the predominance of diploids, in con-trast to the convenient zygotic meiosis of *Chlamydomo-nas* or the ascomycete *Neurospora* (Fjeld & Løvlie 1976; Green 1976). The recent discovery that *Derbesia tenuissima* is dikaryotic, with karyogamy and meiosis both occurring in the sporangia, might be genetically exploitable. In the subsequent years, overall progress has continued to be extremely slow (van der Meer 1986a, 1990; Bonotto 1988), but the discovery of color mutants in the red algae and the enormous advances in molecular biology open up many new possibilities.

J. P. van der Meer's studies, largely on *Gracilaria tikvahiae*, began with the discovery of two spontaneous green mutants in gametophyte populations raised from spores (van der Meer & Bird 1977), which allowed a study of Mendelian inheritance. The green mutants, with reduced phycoerythrin, were stable, different from each other, and recessive (Table 1.3). Earlier, such color mutants probably had been dismissed as bleached or dis-eased curiosities, or had been noticed only after collec-tions has been fixed (van der Meer et al. 1984). As noted by van der Meer (1979), *G. tikvahiae* has several advantages in genetic studies besides its proclivity to mutate both spontaneously and under inducement by

chemical mutagens such as ethyl methanesulfonate. It grows readily at room temperature, is dioecious, pro-duces no asexual spores, but can be propagated vegeta-tively, and is of commercial importance as a source of agar. Besides a rainbow of color mutants, van der Meer has also accumulated morphological and repro-ductive mutants (van der Meer 1986a, 1990). Some of the mutants are in the chloroplast deoxyribonucleic acid (DNA) and show non-Mendelian inheritance: Tetrasporophytes have the phenology of the maternal gametophyte (Fig. 1.18) (van der Meer 1978). Color mutants have also been studied in *Champia parvula*, *Chondrus crispus*, and *Porphyra;* see Steele et al. (1986) for a review.

Color mutants have been used to show the exis-tence of mitotic recombination. Crossing-over of chro-mosomes normally occurs in meiosis but can also take place during mitosis (Fig. 1.19), with the result that one daughter cell in a heterozygous diploid gets both copies of one gene (wild type, +, in the example illus-trated), while the other cell gets both copies of the mu-tant gene (*grn*) (van der Meer & Todd 1977). The sex-determining gene (mt^m/mt^f) is also involved in the recombination, so that the color patches become diploid male and female gametophyte tissue and produce dip-loid gametes.

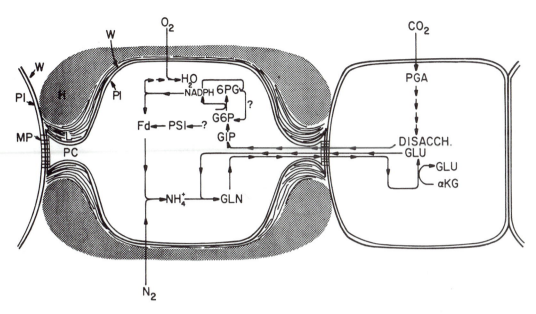

Figure 1.17. Blue-green-algal heterocyst (left) versus vegetative cell (right). Structural differences in the heterocyst include a laminated, glycolipid layer (L) outside the wall (W), plus a homogeneous polysaccharide layer (H). Micro-plasmodesmata (MP) join the plasma membranes (Pl) of the two types of cells at the end of the pore channel (PC) of the heterocyst. The biochemical interactions include movement of a disaccharide (DISACCH) from the vegetative cell to the heterocyst, where it may be metabolized to glucose-6-phosphate (G6P) and oxidized by the oxidative pentose phosphate pathway. Pyridine nucleotide (NADPH) reduced by this pathway can donate electrons to oxygen to maintain reducing conditions within the heterocyst and can reduce ferredoxin (Fd). Fd can also be reduced by photosystem I (PSI). Reduced Fd can donate electrons to nitrogenase, which reduces N_2 to NH_4 (nitrogen fixation). The ammonium combines with glutamate (GLU) to form glutamine (GLN), in which form it is transported to the vegetative cell, and the GLU is recycled to the heterocyst. (From Wolk 1982, with permission of Blackwell Scientific Publications.)

Table 1.3. *Results of crosses between wild-type (wt) plants and two spontaneous green mutants in* Gracilaria tikvahiae

Cross	Phenotype of tetrasporophyte	Phenotypes of F_1 gametophytes
Wt × wt	Wt	All wt
Green (B) × wt	Wt[a]	672 wt, 684 green
Green (M) × wt	Wt[a]	338 wt, 328 green
Wt × green (B)	Wt[a]	195 wt, 160 green
Green (B) × green (B)	Green	All green
Green (M) × green (B)	Wt[a]	360 wt, 367 light green, 689 green

Note: Female parent given first in the list of crosses. Green mutants from two locations are designated M and B.

[a] These plants had a greenish tinge on older fronds.

Source: Van der Meer and Bird (1977), with permission of Blackwell Scientific Publications.

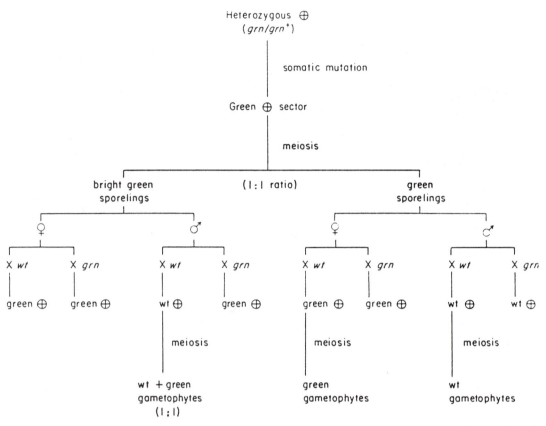

Figure 1.18. Non-Mendelian inheritance of a green somatic mutation in *Gracilaria tikvahiae:* wt, wild-type color (phenotype); *grn, grn*⁺, the mutation for green color and its normal allele; ⊕, tetrasporophyte. (From van der Meer 1978, with permission of *Phycologia.*)

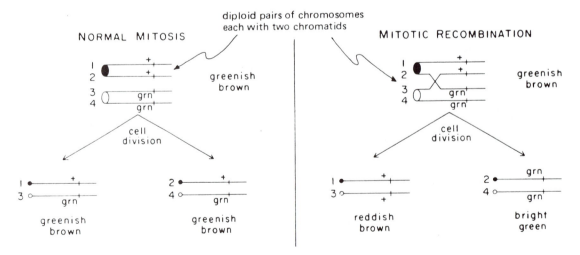

Figure 1.19. Mitotic recombination (right) compared to normal mitosis in heterozygous, diploid tetrasporophytes of *Gracilaria tikvahiae.* Diploid pairs of chromosomes are shown, each with two bivalents; the chromatids are numbered 1–4; +, wild-type color gene; grn, green mutant gene. (From van der Meer & Todd 1977, reproduced by permission of the National Research Council of Canada from the *Canadian Journal of Botany,* vol. 55.)

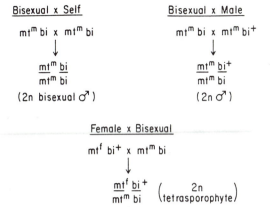

Figure 1.20. Inheritance of bisexual mutation in *Gracilaria*. Bisexuality resulted from a single recessive mutation in a gene (*bi*) that is different from the primary gene, *mt*, that controls male versus female differentiation. *bi*, *bi*+, the mutation for bisexuality and its normal allele; *mt*^{m}, *mt*^{f}, alleles of the primary sex-determining gene. (From van der Meer 1986, reproduced by permission of the National Research Council of Canada from the *Canadian Journal of Botany*, vol. 64.)

In addition to the primary sex-determining locus, there is a gene regulating the dioecious condition. A spontaneous bisexual mutant (*bi*) produced strange results in crosses, except with normal haploid females, and even then the F_1 had females, males, and bisexual plants in a 2 : 1 : 1 ratio (Fig. 1.20), suggesting that the mutation is expressed only in male plants (van der Meer et al. 1984). The bisexual allele cannot substitute for the female allele *mt*^{f}, and subsequent analysis (van der Meer 1986b) has suggested that the *bi*+ allele actually represses expression of female-specific genes. The significance of the *bi* mutation is that it allows the production, by selfing, of homozygous diploids.

Unstable mutants also occur (van der Meer & Zhang 1988). These are the result of transposition of genetic elements (transposons) during genome rearrangement. Insertion of a transposon disrupts the gene function, and removal restores it. The temporary change may be visible as an unstable mutation. Some transposons are autonomous; that is, they control their own insertion and excision. Transposition can also be a normal part of cell differentiation, however, as shown in the case of heterocyst development and nitrogen fixation in *Anabaena* (Golden & Wiest 1988): Two pieces of DNA are excised. One is a 55-kilobase-pair element called

fdxN; when this is excised, the heterocyst can form. The other is an 11-kb piece within the nitrogen-fixing gene *nifD*. Within the 11-kb piece, in turn, is a gene, *xisA*, that codes for a recombinase that clips out the 11-kb piece. The nitrogen-fixing gene is cut and spliced and contains the means for its own rearrangement. Because heterocysts are known to germinate into vegetative cells under some conditions, this transposon presumably can be reinserted.

Life-history studies of *Palmaria palmata* (dulse) (van der Meer & Todd 1980) and *Ahnfeltia plicata* (Maggs & Pueschel 1989) have benefited from color mutations. In the case of dulse, a color mutant made it possible to distinguish the tiny female gametophyte, which is quickly overgrown by the direct development of the tetrasporophyte. In *Ahnfeltia*, Maggs and Pueschel used green-mutant females crossed to wild-type males to demonstrate that carposporophytes (with wild-type coloration) arose from fertilized carpogonia. [Given the variations in life histories in the seaweeds and the pitfalls of assuming ploidy levels and sites of karyogamy and meiosis (sec. 1.5.2), this demonstration is not trivial.]

A different kind of genetic approach has been taken by Kapraun and co-workers, who have compared the DNA contents of several species of *Codium* and *Cladophora*; see Kapraun et al. (1988) for a review. Most species of *Codium* have $2n = 20$ (there is an aneuploid with $2n = 18$), but the sizes and DNA contents vary widely. Some correlation with ecological specialization is suggested: At one extreme is the restricted, stenohaline, stenothermal *C. intertextum*, and at the other the weedy *C. fragile* ssp. *tomentosoides*. The latter species is particularly interesting because it is haploid and parthenogenetic. Among vascular plants, parthenogenetic species (often also weedy) usually are polyploid. The amount of DNA in *C. fragile* ssp. *tomentosoides*, in fact, can be as much as five times that in the diploid species, so that it is *functionally* polyploid, a condition that is called cryptopolyploidy (Kapraun et al. 1988).

Genetic analysis has also been used in some population and taxonomic studies, as, for instance, in assessing intraspecific variation and character heritability (sec. 4.2.2) and interfertility between different or supposedly different species. However, few such studies have been thorough, especially lacking analysis of F_2 offspring, and unwarranted conclusions have sometimes been drawn (van der Meer 1986a). The importance of carrying experiments through to further generations is illustrated by Müller's work on *Ectocarpus siliculosus*. Most gametophyte populations worldwide can mate successfully (Müller 1979, 1988), but a few cannot (Müller 1976). However, some of the hybrid sporophytes can reproduce only asexually by mitospores because of nonfunctional pairing of chromosomes in meiosis.

The powerful tools of molecular biology are increasingly being used to tackle problems in seaweed biology (Olsen 1990). Many of the questions are not genetic, but the collected data may be useful later in answering genetic questions. Phylogenetic, biogeographic, and taxonomic questions are most commonly being asked. For instance, the code for the small subunit of cytoplasmic ribosomal RNA was worked out for *Costaria costata* by Bhattacharya and Druehl (1988) to evaluate its relatedness to other organisms. Chloroplast evolution has been studied by Cattolico and Loiseaux-de Goër (1989) and Kowallik (1989). Electrophoretic patterns of chloroplast DNA have been used to assess populations and species over geographic areas (Goff & Coleman 1988b) and to address kelp phylogeny (Fain et al. 1988). The relatedness of widely separated populations of a given morphological species or "closely" related species can also be assessed with single-copy DNA–DNA hybridization (e.g., Olsen et al. 1987; Bot et al. 1989b), but we are still a long way from correlating DNA divergence with ecotypic divergence (Bot et al. 1989b). Biogeographies and phylogenies of *Laminaria* and *Cladophora* species have been studied using DNA–DNA hybridization in van den Hoek's laboratory (Stam et al. 1988; Bot et al. 1989a).

The studies most directly relevant to genetics are those that have attempted to identify breeding populations (e.g., Bhattacharya et al. 1990), to link heteromorphic phases in apomictic populations (Parsons et al. 1990), or to map genomes, usually the chloroplast genome. Among the seaweed chloroplast DNAs (cpDNAs) at least partially mapped are those of *Pilayella littoralis* (Cattolico & Loiseaux-de Goër 1989), *Dictyota dichotoma* (Kowallik 1989), *Griffithsia pacifica* (Li & Cattolico 1987), and *Porphyra yezoensis* (Shivji 1991) (Fig. 1.21). Not surprisingly, the genes are for chloroplast components such as proteins in pigment complexes and, as mentioned earlier, photosynthetic enzymes such as RuBisCO and ATPase. There is evidence from *Acetabularia* that chloroplasts code for the kinase that phosphorylates thymidine for both chloroplast and nuclear DNA, showing that there is two-way dependence between these organelles (Bonotto 1988).

Chloroplast genomes in land plants show few differences in size, gene content, and gene order, whereas extreme divergence has already been seen among the few algae studied – even among species within the presumably ancient genus *Chlamydomonas* (van den Hoek et al. 1988). Divergence among algae is also evident in nuclear DNA homology. Again, different morphological (phenotypic) species of *Chlorella* have little or no homology, and different strains have comparatively low homology, as shown by single-copy DNA–DNA hybridization. Algae with more complex morphologies are likely to show greater DNA homologies, simply because we have more visual criteria on which to separate species. In widespread genera with simple morphology (e.g., *Cladophora*), DNA hybridization can be used to

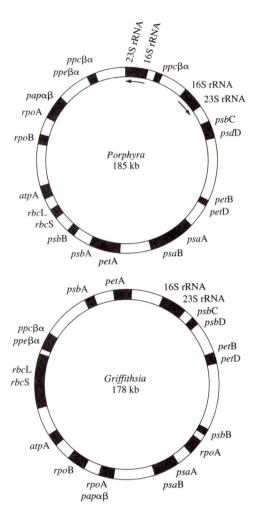

Figure 1.21. Organization of genes on the plastid genomes of *Porphyra yezoensis* and *Griffithsia pacifica*. Genome size is shown in the center of the maps. Arrows indicate approximate positions of inverted repeat structures in *Porphyra*. Approximate gene positions are shown as filled-in blocks. Gene designations: *psa*A, *psa*B, photosystem I apoprotein genes; *psb*A, *psb*B, *psb*C, *psb*D, photosystem II protein genes; 16S rRNA, 23S rRNA, ribosomal RNA genes; *rbc*L, *rbc*S, large and small subunit RuBisCO genes; *pet*A, *pet*B, *pet*D, cytochrome b_6/f complex genes; *atp*A, α subunit gene for CF_1 ATP synthase; *pap*$\alpha\beta$, *ppe*$\beta\alpha$, *ppc*$\beta\alpha$, phycobiliprotein α and β subunit genes. (Data from Li & Cattolico 1987, Shivji 1991; figures courtesy of M. Shivji.)

delimit subspecific groups (Bot et al. 1989b). The broadly divergent molecular biology of taxa within the concept of algae implies considerable biochemical differences in other molecules, but there must be much less divergence in the physiological machinery with which phenotypes interact with the environment. Thus, for instance, although the 5S ribosomal RNAs (rRNAs) from *Chlamydomonas* and *Ulva* have little similarity in nucle-

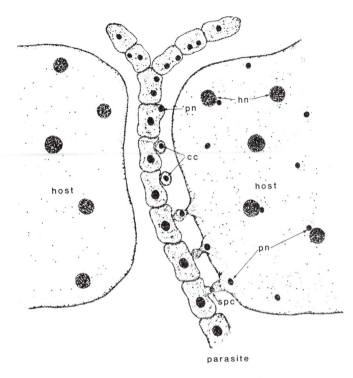

host

host

parasite

Figure 1.22. Parasitic attack by *Choreocolax* on *Polysiphonia* involves transfer of condensed parasite nuclei (pn) into the host cell. The parasite nucleus is first enclosed in a conjunctor cell, then transferred through a secondary pit connection (spc); hn, host nucleus. (After Goff & Coleman 1984.)

otide sequence, they carry out the same function in the framework of the larger ribosomal subunit.

In addition to nuclear, chloroplast, and mitochondrial genomes, algal cells may contain plasmids – small loops of DNA. The plasmids from two species of *Gracilaria* were only 1.8–6.5 kb, compared with 110–190 kb for chloroplast genomes in a range of red algae (Goff & Coleman 1988b). Their cellular locations and functions, if any, are still unclear. The plasmid complement appears to be a stable species character, not the result of infection (e.g., by viruses or parasites). However, it was found that two different plasmids from *Gracilariopsis lemaneiformis* did not hybridize to each other, nor to the cell's nuclear and organellar genomes, but did hybridize with nuclear DNA from other red algae that did not contain plasmids (Goff & Coleman 1988a).

Goff and Coleman (1984) have also examined the nuclear transfer that takes place when parasitic red algae establish on their hosts. Cells at the tip of penetrating *Choreocolax* filaments form secondary pit connections with their host *Polysiphonia* cells and then inject their own nuclei (Fig. 1.22). The nuclei are persistent, and infected cells enlarge and change, increasing their numbers of chloroplasts and mitochondria, and accumulating photosynthetic products. The carbon compounds pass into and sustain the parasite.

1.4.2 Nucleocytoplasmic interactions

Eukaryotes have three main compartments for gene expression: the nucleus/cytosol, with the perfo-

rated nuclear membrane partially separating them; the chloroplasts; and the mitochondria. In a large cell like *Acetabularia*, with thousands of chloroplasts and mitochondria, the organellar DNAs are significant components (Bonotto 1988). Nucleocytoplasmic interactions include the full range of interactions between these compartments.

Giant-celled, uninucleate algae, especially *Acetabularia* species, have provided cell-nucleus systems that can be experimentally manipulated, dating back to Hämmerling's classic studies in the 1930s, as reviewed by Bonotto (1988). The advantage of *Acetabularia*, besides the sizes of several common species, is that interspecific grafts can be made: Both nuclei and cytoplasm can be transferred between species (Bannwarth 1988).

Hämmerling concluded, long before messenger RNA (mRNA) was known, that "morphogenetic substances" were released from the nucleus into the cytoplasm, where they could be stored for some time, but were gradually used up. *Acetabularia* cells can still form a cap if, after reaching about one-third of their final length, the nucleus is removed. There are apico-basal and baso-apical gradients of morphogenetic substances (Fig. 1.23a). That these substances come from the nucleus has been shown by transplanting nuclei into opposite ends of enucleated cells (Fig. 1.23b,c). The type of cap formed by an enucleated stalk is characteristic of the species, but if a nucleus from another species is inserted, either as an isolated nucleus or by grafting on a basal fragment (Fig. 1.23d), the caps formed are first intermediate and then have the characteristics of

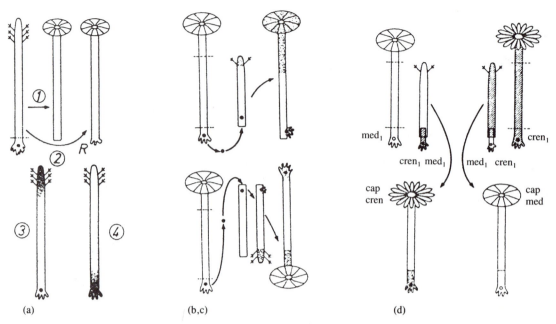

Figure 1.23. Polarity and morphogenesis in *Acetabularia*. (a) Nucleated (1) and enucleated (2) cells show typical polar growth, forming a cap at the apex and a reduced rhizoid at the base. "Morphogenetic substances" diffuse from the apex (3) and the base (4). (b, c) Induction of whorl and cap formation and of rhizoid development by a primary nucleus transplanted into the basal (b) or subapical (c) part of the stalk. (d) Grafting between two species (med$_1$ = *A. acetabulum;* cren$_1$ = *A. crenulata*). A nucleated base is grafted to an enucleate stalk; the resulting cap is typical for the species that supplied the nucleus. (From Bonotto 1988, with permission of BioPress.)

the nucleus donor species. This, and the fact that enucleated cells can form a cap only once, provides evidence that the morphogenetic substances are used up in cap formation.

Although Hämmerling's morphogenetic substances are now thought to be long-lived mRNA, there is still no proof of this (Bonotto 1988). Various experiments have tended to implicate mRNAs as the morphogenetic substances, but there are problems in interpreting the results (Green 1976; Bonotto 1988). One problem is that much RNA synthesis takes place in the chloroplasts. The extent to which chloroplast DNA may be involved in cell morphogenesis is unknown, but an interesting hint is that there is more DNA in apical chloroplasts than in basal or middle chloroplasts (Mazzo et al. 1977). Whatever the nature of the messages sent out by the nucleus, the cytoplasm has some control over when they are read.

1.5 Seaweed life histories
1.5.1 Introduction

The basic patterns of alternation of sporophyte and gametophyte (Fig. 1.24) must be regarded as a theme on which many variations are played. Each generation may reproduce itself asexually, and sexual reproduction should be taken to include meiosporogenesis as well as gametogenesis and mating (Clayton 1988). Asexual reproduction allows an economical population

increase, but no variation, whereas sexual reproduction allows variation but is more costly because of the waste of gametes that fail to mate (Clayton 1981; Russell 1986). Most seaweeds use both means of reproduction, and, as Russell (1986) has noted, where there are isogametes, these can function equally as asexual swarmers. [Indeed, parthenogenesis really can occur only (1) in oogamy or anisogamy, when female gametes, presumably specialized as such, develop without fertilization, and (2) in isogamy if the life history is heteromorphic and unfertilized gametes give rise to sporophyte morphology, as in *Derbesia* (sec. 1.3.1). Both of these possibilities seem to be rare events that have been seen only in culture, and their rates of occurrence and success in nature are unknown (Clayton 1988).] Russell (1986) also has reminded us that vegetative reproduction often is overlooked unless special propagules, such as in *Sphacelaria*, are involved. Plants may spread by stolons or rhizomes, giving a significant competitive edge in the space race (sec. 3.1.1). Some floating algal populations depend entirely on vegetative reproduction by fragmentation.

The life histories of seaweeds are known from culture studies on relatively few species, but this is one of the rapidly advancing fronts in phycology today. Sufficient variations have been discovered in the basic pattern – between species and within species – that today's generalizations must be viewed only as working

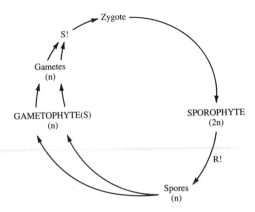

Figure 1.24. Basic pattern of diplobiontic alternation of generations. R!, meiosis (reduction division); S!, syngamy. Compare with Figure 1.28.

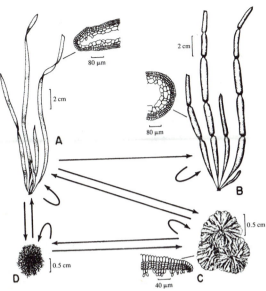

Figure 1.25. Life history and anatomical features of the *Scytosiphon simplicissimus* complex: (A) complanate form; (B) cylindrical form; (C) crustose form; (D) filamentous plethysmothalli. [From Littler & Littler 1983 (partly after Clayton), with permission of *Journal of Phycology.*]

hypotheses. Kraft and Woelkerling (1981), for instance, noted that less than 5% of red algae have been carried through their life histories in culture. Although a basic alternation of a sporophyte (typically diploid) and a gametophyte (typically haploid) is common among seaweeds (Fig. 1.24), various extras and shortcuts are known.* Indeed, a better generalization may be that almost any alternation is possible, and even no alternation at all. Moreover, the term "alternation" is a misnomer, in that it implies only two phases and a regular progression from one to the other; clearly that is not always the case (e.g., *Scytosiphon,* Fig. 1.25). Maggs (1988, p. 488) concluded that "life history patterns seem to be more labile than morphological features, and the role of life history variability in speciation, and in ecological success, should not be underestimated." This point has been explored at length by Russell (1986) and Clayton (1988).

1.5.2 Theme and variations

Three basic types of algal life histories are recognized (e.g., Bold & Wynne 1985). An alternation of two phases is called diplobiontic (Fig. 1.24). Genera such as *Enteromorpha, Chondrus,* and *Ectocarpus* have sporophytes and gametophytes that are vegetatively indistinguishable (not counting the carposporophytes of red algae, which are not free-living). Sometimes chemical differences occur between isomorphic phases, as in *Chondrus* (different forms of carrageenan in the walls). At reproduction, the two phases may become distinguishable by the reproductive structures. Heteromorphic generations usually fall into two different functional-form groups, such as erect fronds versus creeping fila-

* Life-history diagrams herein are as follows: *Acetabularia* (Fig. 1.10), *Scytosiphon* (Fig. 1.25), *Halimeda* (Fig. 1.26), *Derbesia* (Fig. 1.27), *Laminaria* (Fig. 1.29), *Porphyra* and *Nereocystis* (Fig. 1.32), and *Desmotrichum* (Fig. 6.12). See also Fig. 1.31.

ments or crusts (Figs. 1.25 and 1.32). A classic example is the well-known story of Drew's (1949) linking of *Conchocelis* (filamentous) and *Porphyra* (a blade) and its impact on the Japanese nori industry (sec. 9.2). Similar stories continue to unfold; for example, the crustose red *Erythrodermis allenii* was shown to be part of the *Phyllophora traillii* life history (Maggs 1989). Yet some seaweeds exist in only one phase; they have a haplobiontic life cycle (Figs. 1.10 and 1.26). Well-known examples are the Fucales and *Codium.* Diploid plants give rise to gametes by meiosis, and the zygotes grow directly back into diploid plants. (In *Halimeda* and other Udoteaceae, the site of meiosis is not known; if the large nucleus in the germling undergoes meiosis as it breaks up, as in Dasycladales, the plants will be haploid, but if meiosis occurs during gametogenesis, as in *Codium,* the plants will be diploid.) In fresh waters, vegetative plants are often haploid, with only the zygote being diploid; haploid, haplobiontic seaweeds appear to be quite rare. Some species of *Liagora,* such as *L. tetrasporifera,* that produce tetraspores in the carposporophyte may lack a free-living tetrasporophyte. Green algae with a "*Codiolum*" sporophyte have a unicellular, but still free-living and physiologically active, diploid generation. The cysts of *Acetabularia,* and a stage of the oogonia and antheridia of Fucales and Durvillaeales, can be regarded as highly reduced gametophytes retained on a diploid sporophyte (e.g., Maier & Clayton 1989), but they are not free-living, and these life histories are effectively haplobiontic (cf. carposporophytes, discussed later).

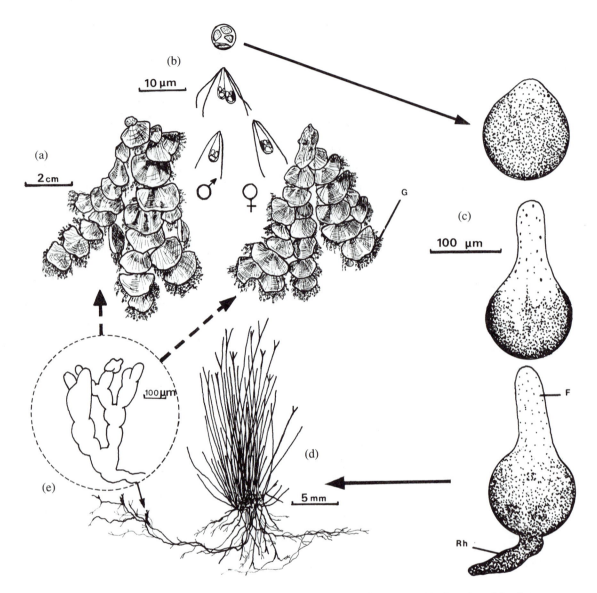

Figure 1.26. Haplobiontic life history and siphonous development in *Halimeda tuna*. Fertile male and female gameto-phytes (a), shown hanging downward, release biflagellate gametes from external gametangia (G) to form a zygote (b). (c,d) Bipolar germination of the zygote leads to rhizoids (Rh) and erect free filaments (F). Subsequently (e), buds form on horizontal filaments and grow into the calcified, segmented fronds (compare *Penicillus* in Fig. 1.45c). (From Meinesz 1980, *Phycologia,* with permission of Blackwell Scientific Publications.)

Ploidy levels are often assumed, but studies have sometimes demonstrated the unexpected. Most *Codium* species studied in the western Atlantic are diploid and reproduce via haploid gametes. However, *C. fragile* ssp. *tomentosoides* is haploid, reproducing parthenogeneti-cally (Kapraun & Martin 1987). The location of meiosis is particularly difficult to establish. In *Acetabularia*

(Fig. 1.10), the small secondary nuclei proved to be haploid (Bonotto 1988). In some *Porphyra* species, mei-osis has been shown, through color mutants, to occur during germination, not during formation of the con-chospore (Miura 1985; see also Guiry 1990). The site of karyogamy is considered more predictable, because enough chromosome and nuclear counts have been ac-

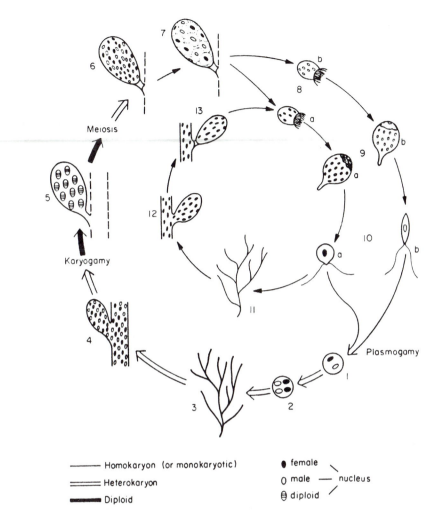

Figure 1.27. Life cycle of *Derbesia tenuissima*. Dikaryotic fusion product of male and female gametes (1) undergoes nuclear divisions (2) and forms siphonous thalli (3, 4). In the young sporangium, karyogamy takes place (5) and is immediately followed by meiosis (6). Spores are initially uninucleate (7), but by the time they are released they are multinucleate (but homokaryotic) (8a,b). The spores enlarge into gametophytes (*Halicystis* stage) (9) and form gametes (10). Female gametes may germinate parthenogenetically to form siphonous homokaryotic filaments (11, 12). (From Eckhardt et al. 1986, with permission of the British Phycological Society.)

cumulated that we can reasonably assume that it follows syngamy. However, with coenocytic species, caution is needed: *Derbesia marina* has recently been shown to be dikaryotic, like basidiomycete hyphae, with two different haploid nuclei, rather than one kind of diploid nucleus (Fig. 1.27) (Eckhardt et al. 1986).

Florideophycidae were included in the foregoing life-history generalizations in spite of the interpolation of a "carposporophyte." This structure can be regarded either (1) as an additional, diminutive diploid phase, epiphytic and parasitic on the female gametophyte, and producing spores by mitosis, or (2) simply as a mass of diploid spores produced from the original zygote. Because it is never free-living, this structure is not a generation comparable to the gametophyte and (tetra)-sporophyte. Some Bangiophycidae, including *Porphyra,* apparently also multiply the zygote, but the diploid cells form an indistinguishable part of the thallus, and Guiry (1990) refers to the products as zygotosporangia. The zygote in *Palmaria* develops into a large diploid phase, morphologically like the male gametophyte, that overgrows the tiny female gametophyte and produces spores by meiosis. Replication of the zygote is one of several ways to amplify the results of sexual reproduction,

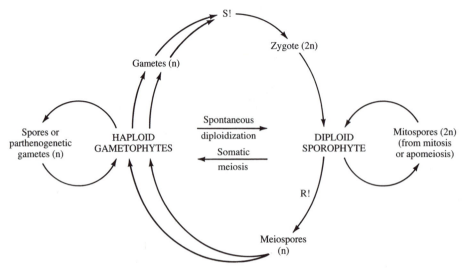

Fiugre 1.28. Some possible seaweed life-history progressions. Most species use only a small part of this range.

which potentially is restricted because spermatia are nonmotile (sec. 1.5.4) (Hawkes 1990).

Various morphological forms may be taken at a given ploidy level, and the same morphology may be formed at different ploidy levels. There is no necessary connection between ploidy and form, even though, in general, gametophytes are haploid and sporophytes are diploid. The DNA content of *Codium* spp. was discussed earlier (sec. 1.4.1). In several red algae, the amounts of DNA are the same in haploids and diploids (Goff & Coleman 1987, 1990). Several seaweeds are known in which an apparent alternation of generations occurs with no ploidy change (e.g., *Petalonia* and *Scytosiphon*) (Kapraun & Boone 1987). *Desmotrichum* (Fig. 3.17) can form filamentous microthalli or parenchymatous macrothalli, depending on temperature and day length. Also in response to temperature and day length, *Elachista stellaris* alternates between a diploid macrothallus, which produces meiotic spores, and a microthallus that can reproduce the macrothallus by spontaneous diploidization. Diploidization has also been reported in *Boergesenia forbesii*, which has an isomorphic life history (Beutlich et al. 1990); in this case it results in a preponderance of diploids in the population. Polyploidy can be developed in *Gracilaria tikvahiae* mutants; polyploid tetrasporophytes (e.g. 3*n*, 4*n*) were found to be robust, but polyploid gametophytes (again, 3*n*, 4*n*) were stunted (Zhang & van der Meer 1988).

Variations in these basic patterns include several reproductive shortcuts (Fig. 1.28). In sporangia that are expected to be meiotic, such as red-algal tetrasporangia and brown-algal unilocular sporangia, spores may be formed by mitosis instead, giving rise to more plants of the same ploidy level; this is called apomeiosis. For instance, in *Bonnemaisonia*, large gametophytes alternate

with a filamentous tetrasporophyte (*"Trailliella"*), except in the northern part of the range, where the tetrasporophyte is self-perpetuating through apomeiosis. Five types of complications can arise in red-algal life histories (Maggs 1988; see also Hawkes 1990): (1) formation of monosporangia, bisporangia, polysporangia, parasporangia, or vegetative propagules in a species that also forms tetraspores; (2) simultaneous occurrence of gametangia and tetrasporangia (mixed-phase reproduction); (3) bisexuality in a normally unisexual species; (4) direct development of tetrasporophytes from tetraspores (exclusively or mixed with gametophytes); (5) direct development of gametophytes from carposporophytes (exclusively or mixed with tetrasporophytes). In unilocular sporangia of the brown alga *Pilayella littoralis*, the first stages of meiosis were seen by Müller and Stache (1989), but no reduction in chromosome number took place, and apparently there were no haploid gametophytes.

Spontaneous diploidization and somatic meiosis can occur, giving ploidy changes within a thallus (usually without any morphological change). The presence of male and female tissue on tetrasporophytes can be due to mitotic recombination (sec. 1.4.1), with *no* change in ploidy level. In the filamentous browns, cells from plurilocular (mitotic) sporangia may be sexual (given the opportunity) or asexual. The presence of unilocular (meiotic) and plurilocular sporangia on a given plant may merely indicate simultaneous sexual and asexual reproduction.

Apospory is a process whereby diploid gametophytes are produced directly by sporophyte cells (i.e., without spores). Thus apospory differs cytologically from somatic meiosis in that no ploidy change occurs, but the morphological effect is the same. Apogamy is

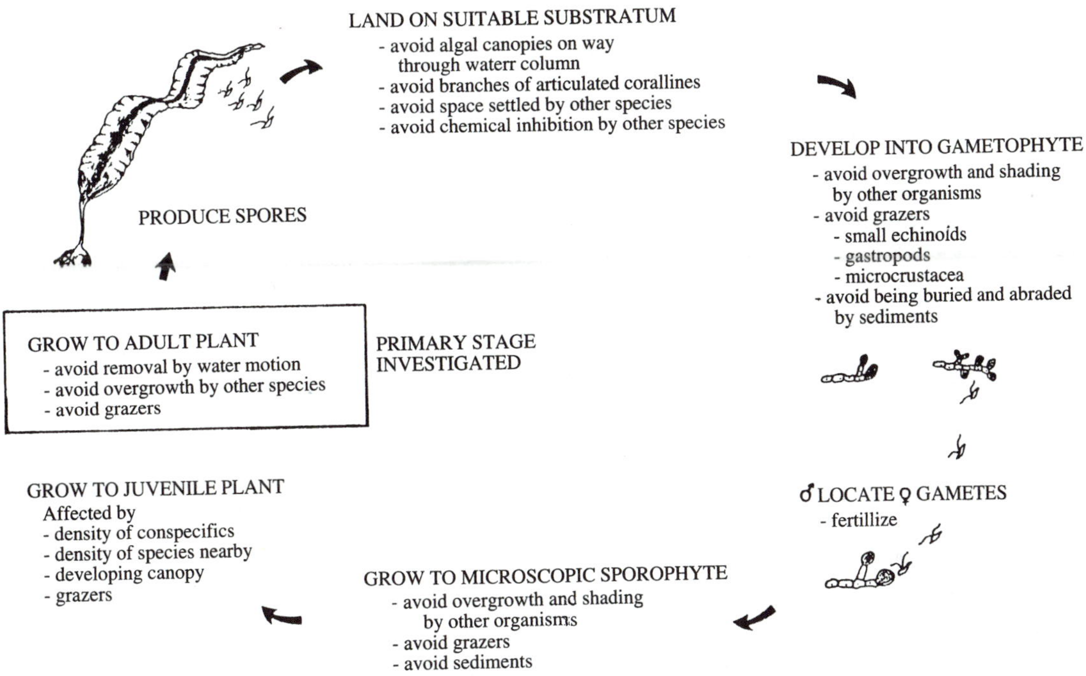

LAND ON SUITABLE SUBSTRATUM
- avoid algal canopies on way
  through waterr column
- avoid branches of articulated corallines
- avoid space settled by other species
- avoid chemical inhibition by other species

DEVELOP INTO GAMETOPHYTE
- avoid overgrowth and shading
  by other organisms
- avoid grazers
  - small echinoíds
  - gastropods
  - microcrustacea
- avoid being buried and abraded
  by sediments

PRODUCE SPORES

GROW TO ADULT PLANT
- avoid removal by water motion
- avoid overgrowth by other species
- avoid grazers

PRIMARY STAGE
INVESTIGATED

GROW TO JUVENILE PLANT
Affected by
- density of conspecifics
- density of species nearby
- developing canopy
- grazers

♂ LOCATE ♀ GAMETES
- fertilize

GROW TO MICROSCOPIC SPOROPHYTE
- avoid overgrowth and shading
  by other organisms
- avoid grazers
- avoid sediments

Figure 1.29. Life history of a laminarian alga, showing some of the major (chiefly biotic) environmenal hazards that must be overcome at each stage. In addition, success will be affected by abiotic factors such as light, temperature, water motion. (From Schiel & Foster 1986, *Oceanogr. Mar. Biol. Ann. Rev.*, with permission of Aberdeen University Press, Farmers Hall, Aberdeen AB9 2XT, U.K.)

the production of haploid sporophytes directly from gametophyte cells, and it differs from spontaneous diploidization in having no ploidy change. Apospory and apogamy are detectable only in heteromorphic life histories, such as those of *Alaria crassifolia* (Nakahara & Nakamura 1973) and *Desmarestia* species (Ramirez et al. 1986) (sec. 1.7.1). Perhaps a model alga can be found in which elucidation of events like these can lead to a genetic understanding of the variable relation between ploidy level and morphology, a line of inquiry that will lead ultimately to the roles and origins of alternating generations. Another approach to solving the morphology/ploidy riddle may be provided by fusion of protoplasts from heteromorphic generations (Butler et al. 1989).

1.5.3 Environmental factors in life histories

The life history of a species is a continuous interaction between the plants and their biotic and abiotic environments (Fig. 1.29). A seaweed begins life as a single undifferentiated cell, with the potential to produce the whole organism through the expression of its genetic information. The genotype interacts with the environment to produce the phenotype. The environment of a cell consists of the physical and chemical influences of the other cells in the plant, plus the environment of the plant itself. The environmental history of a plant, because it affects growth and form, in a sense becomes

recorded in the plant body (Fig. 1.2) (e.g., Waaland & Cleland 1972; Murray & Dixon 1975; Garbary 1979). Thus individual plants of the same genotype, planted in exactly the same place on the same day, but in different years, will grow into phenotypically distinct individuals (Evans 1972).

Successful growth and reproduction of plants are possible over quite wide ranges of form, size, and relative proportions of the parts (Evans 1972). Under exceptional circumstances, such as in moderate tidal rapids, seaweeds may become unusually large, but whether or not they have an intrinsic size limit is not known. Size is normally constrained by the environment. However, among the Desmarestiales there are examples, such as *Himantothallus grandifolius*, of closed growth. In these, the numbers and positions of blades are determined when the plant is only a few millimeters long, even though this species eventually reaches 10 m in length (Moe & Silva 1981). The successful competitors in a population will not necessarily be the largest, nor even the fastest growing.

The switch from vegetative growth to reproduction (which in most seaweeds involves very little growth) often depends on environmental factors such as temperature and light (Lüning & tom Dieck 1989; Lüning 1990). Kelp gametophytes, for example, may reproduce when they are only a few cells in size, or they may grow vegetatively almost indefinitely, depending on light quality and quantity. They have been used for studies of minimum irradiance requirements for growth and reproduction because of their extreme shade environment and ease of culture. A prerequisite for growth, of course, is that the energy trapped and carbon fixed must exceed the totals used in respiration. Chapman and Burrows (1970) showed that development of *Desmarestia aculeata* gametophytes depends on the mean daily irradiance [i.e., (irradiance × photoperiod)/24]. At the lowest irradiances tested, gametophytes did not mature, though they survived and were able to develop later when irradiance was increased. More detailed studies by Lüning and Neushul (1978) showed that various kelp gametophytes were saturated for vegetative growth at 4 W m^{-2} (about $20 \text{ μE m}^{-2} \text{ s}^{-1}$; see sec. 4.2.1 for explanation of units) but required two to three times that irradiance for reproduction. Blue light, alone or as part of white light, is required for kelp gametogenesis; in red light, gametophytes grow only vegetatively. The ability of these plants to grow vegetatively in extremely dim light and reproduce only when irradiance increases provides a mechanism for populations to retain space after the canopy of parent sporophytes is lost.

One of the most important ways in which algae (and all organisms) respond to their environment is in the timing of reproduction, because in reproduction lies the key to the survival of the species. Reproductive responses to the environment are particularly evident in algae with strongly heteromorphic generations, such as kelps and *Porphyra,* where different growth forms are adapted to different environments. When conditions are suitable for the growth of one form, vegetative growth or asexual propagation is likely to occur, whereas conditions poor for that form are likely to prompt a reproductive switch to the alternative morphology. However, because of the lead time sometimes needed for reproduction, some seaweeds may need to anticipate the changes in seasons, using some appropriate cue. (We can compare temperate deciduous trees, in which preparations for the winter cold period involve much growth and other activity, such as recycling nutrients from leaves and shedding leaves, and growing protective scales over buds.) Two kelps have recently been shown to have circannual rhythms as a means for timing growth (Lüning 1991; tom Dieck 1991); this mechanism may prove to be more widespread, but seaweeds can also (or instead) respond to environmental cues. Kain (1989) and other authors she cites distinguish season responders (plants that grow and reproduce when environmental conditions are suitable) from season anticipators (which grow and reproduce on an annual rhythm and in response to environmental triggers).

Temperature itself is an obvious seasonal cue in middle to high latitudes, and indeed some differences in the kinds of reproduction at different temperatures have been noted in seaweeds [e.g., *Ectocarpus siliculosus* (Müller 1963), *Sphacelaria furcigera* (Colijn & van den

Table 1.4. *Influences of temperature and day length on formation of upright fronds by* Scytosiphon lomentaria *var.* complanatus *in Nova Scotia*

Temperature (°C)	Day length (h)	Crusts with uprights (%)
0	14	100
5	14	100
10	8	100
10	12	100
10	16	100
15	12	3.8
15	16	0.3
20	12	0
20	16	0

Source: Correa et al. (1986), with permission of Blackwell Scientific Publications.

Hoek 1971), and some species of *Ulothrix, Urospora,* and *Monostroma* (Lüning 1980a; Tanner 1981)]. The formation of erect thalli in some isolates of *Scytosiphon lomentaria* var. *complanatus* studied by Correa et al. (1986) was dependent on temperature and independent of photoperiod (Table 1.4), in contrast to the better-known photoperiodism of the typical variety, as discussed later (Figs. 1.31 and 1.34). (In other isolates of *S. lomentaria,* this morphogenetic switch apparently is not responsive to either temperature or photoperiod.) In some species, different steps in reproduction have different temperature optima. In the conchocelis stage of *Porphyra tenera* in Japan, the temperature optimum for monosporangium formation is 21–27°C, whereas for monospore release it is 18–21°C (Kurogi & Hirano 1956; see also Dring 1974). Chen et al. (1970) found that conchosporangia of *P. miniata* from Nova Scotia were formed at higher temperatures (13–15°C in this case), but conchospores were released only with low temperatures (3–7°C) and short days. Dring (1974) commented that studies of "spore production" at different temperatures are liable to reveal only a compromise among the maxima for several processes.

Although temperatures are involved in reproductive cues, temperatures may undergo seasonal cycles, and such cues are erratic. A more dependable seasonal cue is day length (photoperiod) (Fig. 1.30a). As one progresses to farther northern and southern latitudes, day-length changes becomes increasingly pronounced, and it has been known since the 1930s that flowering plants respond to photoperiod, flowering when days are long (LD plants) or short (SD plants) (Noggle & Fritz 1983; Goodwin & Mercer 1983). (Some are day-neutral, and some require a sequence of SD+LD or LD+SD.) Photoperiodic responses were expected in temperate seaweeds, but not until 1967 was a true photoperiodic

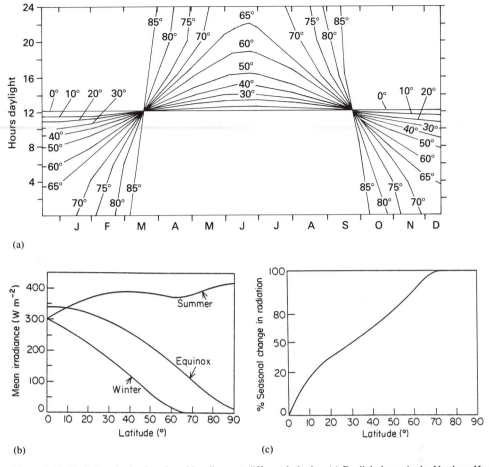

Figure 1.30. Variations in day length and irradiance at different latitudes. (a) Daylight hours in the Northern Hemisphere (Southern Hemisphere values may be obtained by six-month transposition of the abscissa scale). (b) Mean energy flux with a cloudless sky for the months containing the equinoxes and solstices (mean values for both hemispheres). (c) Percentage seasonal change in energy flux with a cloudless sky (recalculated from b). [Part a from Drew 1983, with permission of Clarendon (Oxford University) Press; b and c from Kain 1989, with permission of the British Phycological Society.]

response demonstrated, for the conchocelis phase of *Porphyra* (Dring 1967). By Dring's count in 1988, 55 seaweeds had been shown to respond to photoperiod. Many of these seaweeds have heteromorphic life histories, which is expected because the algae use the cue to switch to a different phase (which is assumed a priori to be better adapted to the conditions in the next season). A reproductive response in an isomorphic alga does not involve a switch in morphology (though perhaps there are subtle differences that are adaptive?); nevertheless, photoperiodic responses have now been found in isomorphic and even haplobiontic species (Dring 1988).

In many seaweeds, the response is "short day" (i.e., occurring when days are short and nights are long) (Fig. 1.31); however, Dring (1984a) noted that the apparent bias toward SD plants was partly due to inadequate controls in experiments claiming to show LD effects and that there was no reason to suppose that al-

gae would respond more often to SD than to LD. Higher plants are well known for measuring the length of the dark period, so a short-day plant is functionally a long-night plant. Higher plants are also well known for having a red/far-red-absorbing pigment, phytochrome, involved in their systems for measuring and responding to light/dark cycles. (The details of the mechanism remain uncertain; research on higher plants slowed sharply in the 1970s because of technical problems and has remained slow to date; Dring 1984a.) The presence of phytochrome has been demonstrated in algae, but as Dring (1988, p. 158) says, "in algae we seem to be looking at the early stages in the evolution of the phytochrome system." The responses by algae are not exactly the same as in flowering plants, and there are other pigment systems present in some species.

Photoperiodic effects have to be distinguished from the effects of the total irradiance received, which

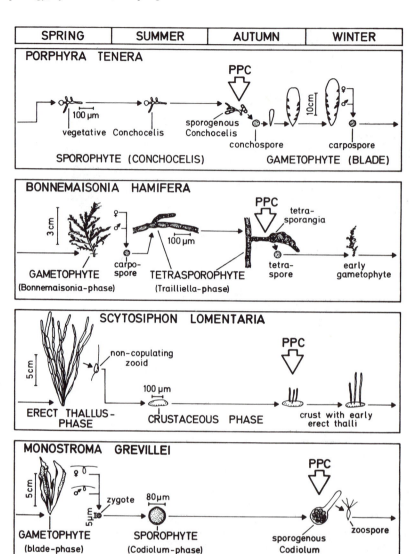

Figure 1.31. Annual cycles of four short-day algae. PPC, short-day signal. The responses are as follows: *Porphyra tenera* forms conchospores; *Bonnemaisonia hamifera* forms tetrasporangia; *Scytosiphon lomentaria* (in Europe) forms new erect thalli from the crust; *Monostroma grevillei* (codiolum stage) forms zoospores. (From Dring & Lüning 1983, *Encyclopaedia of Plant Physiology,* new series, vol. 16B, with permission of Springer-Verlag, Berlin.)

also changes seasonally (Fig. 1.30b,c) and can have strong effects on plant growth and development. The classic means of demonstrating a true photoperiodic effect is the night-break experiment, in which a long night (e.g., 16 h) is broken by a short period of weak light (in flowering plants, white light and red light are effective, but blue light and far-red light are not). The effect on SD plants is to spoil the inductive effect of the long night; the plant measures two short nights. In LD plants, a night break usually is inductive. Converse experiments, either (1) with a long day (16 h), broken by a short dark period, and a regular night, or (2) with the

cycle extended (e.g., to 32 h) to give a long night and a long day, are inductive in SD plants, just as is a regular light : dark 16 : 8 day. Phytochrome has two alternating forms, one that absorbs red light and one that absorbs far-red light, and when these wave bands are given in sequence in a night break, the last one determines the effect: Red light spoils the long night, but far-red light does not (and even counters the effect of red light). This night-break test has been used as evidence for the presence of phytochrome in seaweeds (e.g., Dring 1974; Rietema & Breeman 1982), and red/far-red spectral shifts have been used to detect phytochrome more directly (e.g., Dring 1967; van der Velde & Hemrika-Wagner 1978; also see Nultsch 1974). Yet phytochrome has been demonstrated conclusively in only a few green algae, in Dring's (1988) opinion.

Unfortunately, some LD flowering plants, as well as some seaweeds of both LD and SD species, are now known to be insensitive to night breaks, and the classic test has been abandoned as a criterion for a "true" photoperiodic response. Plants that respond to a night break clearly are measuring the length of the night, whereas those for which a night break makes no difference must be measuring the length of the day; the latter plants have been called "light-dominant" (Dring 1988; Breeman & ten Hoopen 1987).

A case study of *Acrosymphyton purpuriferum* is interesting for two reasons. First, the tetrasporophyte is an SD plant that measures day length, not night length; second, it is a warm-water species, and the genus in general is tropical, whereas the vast majority of seaweeds with photoperiodic responses are temperate. *A. purpuriferum* has a heteromorphic life history, with a crustose tetrasporophyte and an erect, fleshy gametophyte. *A. purpuriferum* from the Mediterranean Sea was shown by Cortel-Breeman & ten Hoopen (1978) to be present as gametophytes only in the spring and summer, and to have an SD tetrasporophyte. Night breaks with white, blue, or red light did not inhibit tetrasporogenesis, in contrast to what was expected at the time. Recently, Breeman & ten Hoopen (1987) concluded a more thorough study, with the following results: Tetrasporogenesis was inhibited when short days were extended for 8 h by weak light (below the photosynthetic compensation point). Threshold irradiance was far lower when the supplementary light periods preceded the main photoperiod than when they followed it; also, when the light used during the main photoperiod was "low," the threshold irradiance was lower than in "bright" light ($65 \ \mu\text{E m}^{-2} \text{ s}^{-1}$, but maximum surface irradiance is about $2,000 \ \mu\text{E m}^{-2} \text{ s}^{-1}$; sec. 4.2.1). Inhibition due to extended photoperiods was strongest in blue light, but red and yellow also caused some inhibition; far-red light caused no inhibition.

Circadian rhythms may complicate some photoperiodic responses, as was shown in *Nemalion helminthoides* (Cunningham & Guiry 1989). Formation of erect axes on plants from tetraspores depended on long days, but was completely inhibited by continuous light (in contrast to reproduction in *Sphacelaria rigidula;* ten Hoopen et al. 1983). When extended cycles were tried (by increasing the dark period), the best results were obtained with diurnal (24-h) or bi-diurnal (48-h) cycles; very few erect axes formed in a 32-h cycle (16 : 16 light : dark), and yet the plants were not measuring the night length, because breaks in long nights (8 : 16 light : dark) did not promote erect axes (nor did dark breaks in a long day inhibit them). As these authors note, virtually every photoperiodic-response permutation has been found in algae.

The photoperiodic effects of light on reproduction in seaweeds are not necessarily red/far-red effects; evidence has been accumulated for the involvement of a variety of pigments in algal photomorphogenesis (Dring 1988). Many are blue-light effects, such as the formation of uprights in *Scytosiphon* (Fig. 1.31) (Dring & Lüning 1975). Tetrasporangium formation in *Rhodochorton purpureum* takes place during short days and is inhibited by a night break of red light but not far-red light, and yet the red-light inhibition is not reversed by subsequent exposure to far-red light (in contrast to the case with flowering plants). Moreover, a night break by blue light is also inhibitory (Dring & West 1983).

Red-light effects may be mediated by phytochrome, although, as noted earlier, the presence of this pigment in nongreen algae is unproven; or red-light effects could come about through red light absorbed by phycobiliproteins (in red and blue-green algae), which have structures very similar to that of phytochrome. Blue-light effects (see Table 1.8) are attributed to "cryptochrome" in the brown algae (but cryptochrome may not be a single – or novel – pigment), but probably not in red algae. Among the more unusual effects of blue light is a rapid (10–20 min) release of eggs from oogonia when blue light is switched on (*Dictyota*) or off (*Laminaria*) (Dring 1984a). Dring (1988, p. 169) concluded that "photoperiodic responses in algae may be controlled by a variety of pigment systems analogous to the variety of pigments involved in algal photosynthesis." Let us leave this physiological mire and look at how photoperiodic effects apply in situ.

The more critical the timing of reproduction, the more complex the environmental cues need to be. Short days, for instance, occur in both autumn and spring, as well as through the winter. *Porphyra nereocystis* grows exclusively on the stipes of the annual kelp *Nereocystis luetkeana*. The host grows in early spring, and the epiphyte must get its spores onto the young stipe before the stipe becomes covered in other algae. Moreover, the stipes are high in the water column, whereas the sporophyte of *Porphyra* is on the bottom, in old shells. To time its spore release for spring, *P. nereocystis* responds to a dual photoperiod: prolonged short days followed by prolonged long days (Fig. 1.32) (Dickson & Waaland 1985). (Tests were run at 8 : 16 and 16 : 8 light : dark photoperiods, and critical photoperiods were not determined.) The response was also better in cooler water, typical of spring, than in warmer water, typical of autumn. The conchospores are released in slime strands that may produce a "bola" effect for increasing the chances of snagging and sticking to the slippery young kelp stipes. Other species of *Porphyra* have less critical photoperiodic control: For example, in *P. torta* from the same region (Puget Sound), conchospores can form in

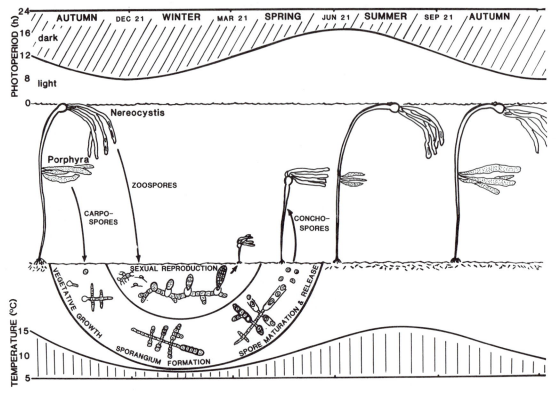

Figure 1.32. Life history and seasonal occurrence of the annual, epiphytic alga *Porphyra nereocystis* and its annual host plant, *Nereocystis luetkeana*. The top part of the diagram shows seasonal photoperiod variation at the Puget Sound study site; the lowest part of the diagram traces water temperatures. Carpospores from *Porphyra* blades form the shell-boring conchocelis stage, which releases conchospores in response to long days after short days, as the new annual crop of *Nereocystis* sporophytes elongates. Zoospores from *Nereocystis* form microscopic male and female gameto-phytes, and sexual reproduction (not photoperiodic) results in sporophytes. (From Dickson & Waaland 1985, with permission of Springer-Verlag, Berlin.)

any photoperiod, but they mature and are released only when there are short days (Waaland et al. 1987); this species is a winter annual on rocky intertidal substrates.

According to the results of culture experiments on the temperature and photoperiod requirements for repro-duction, certain predictions of reproductive timing in nature can be made. But how do conditions in the "real world," especially of the intertidal zone, affect temper-ature and photoperiodic responses? Few studies have at-tempted to find out, but Breeman and Guiry (1989) described how tides alter reproductive timing in *Bonne-maisonia hamifera* sporophytes. These are SD plants, requiring a narrow range of warm temperatures (Table 1.5). Lüning (1980a) had predicted reproduction only during a short time in early autumn, when the days be-come short enough but the sea is still warm (Fig. 1.33; Table 1.5). Whereas, in general, phenology in situ bore out the predictions (Breeman et al. 1988), two factors confounded the predictions: (1) High spring tides at the beginning or end of the day shortened the effective day length, allowing reproduction to start earlier than pre-dicted. (2) Low water of spring tides in the middle of

the day exposed plants to warm air temperatures, when water temperatures were below the threshold, and al-lowed reproduction to resume. Brief exposures to a suit-able combination of conditions sufficed for induction, and the reproductive period stretched from September into December. [At their study sites in Ireland, the times of high and low spring tides are always the same; that is not the case everywhere (sec. 2.1)]. In another example (Breeman et al. 1984), light was so reduced at high tide, because of turbidity and the fucoid canopy, that SD con-ditions prevailed all year for mid-intertidal populations of *Rhodochorton purpureum*.

"Short day" and "long day" obviously are rel-ative terms, and for a seaweed with a wide latitudinal range, what is a short day in higher latitudes may be a long day in lower latitudes. Compare, in Figure 1.34, for instance, the effects of 11-h days on *Scytosiphon* from Tjörnes (66° N) and from Punta Banda (32° N) (Lüning 1980a). Unfortunately, intraspecific differ-ences in critical day length do not always correlate with latitude, as Rietema and Breeman (1982) found in *Dumontia contorta*. Moreover, photoperiod responses

Table 1.5. *Effects of photoperiod and temperature on tetrasporangium formation in the trailliella phase of Bonnemaisonia hamifera*[a]

Parameter	Response to day length (at 15°C)										
Hours light per day	8	9	10	10.5	11	12	12.5	13	14	15	16
Percentage fertile	93	92	48	16	6	0	0	0	0	0	0
Parameter	Response to water temperature (at 8 h light per day)					
Temperature (°C)	10	12	15	17	20	23
Percent fertile	0	0	97	73	0	0

[a] 150 plants in each experiment were grown in enriched seawater (containing less than 20 μM NO_3^-).
Source: Lüning (1981b), with permission of Gustav Fischer Verlag.

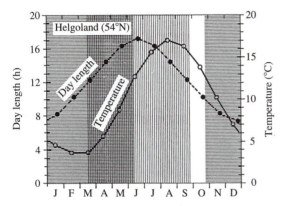

Figure 1.33. Predicted tetrasporogenesis in the trailliella phase of *Bonnemaisonia hamifera* in Helgoland, approximately the same latitude as the site in Ireland where Breeman and Guiry tested the prediction. The "window" for reproduction in September–October occurs between too-warm seas (vertical hatching) and too-short photoperiod (horizontal hatching). (From Lüning 1981b, *Ber. Deutsch. Bot. Ges.*, vol. 94, with permission.)

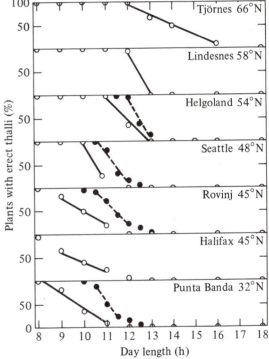

Figure 1.34. Effects of day length on erect-thallus formation by different geographic isolates of *Scytosiphon lomentaria* at 10°C (open circles) and 15°C (filled circles). Each value is based on a count of 250 plants. (From Lüning 1980a, with permission of The Systematics Association.)

sometimes are altered by temperature (Fig. 1.34) and may not be exhibited in the presence of high nitrogen levels (such as are created in standard culture media) (Table 1.5).

More and more studies are finding seaweeds in which reproduction, in at least one stage, is a function of both temperature and light; recent examples include studies by Maggs and Guiry (1987) and Anderson and Bolton (1989). In some cases the responses are quantitative (e.g., higher fertility at lower temperatures), in others qualitative (i.e., fertile vs. nonfertile). For instance, the initiation of growth of the macrothalli of *Dumontia contorta* is strictly controlled by day length, but the initials do not grow out unless the temperature is less than 16°C (Reitema 1982).

Seasons in the subtidal are mainly determined by nutrients and light, hardly at all by temperature, although temperature can still act as a trigger for morphogenetic or reproductive events (Kain 1989). Seaweeds that do not experience strong seasons of temperature and photoperiod may still require environmental cues for reproduction. The deep-water brown alga *Syringoderma floridana* has macroscopic sporophytes and microscopic gametophytes. Most of the two-celled gametophytes develop right on the sporophyte, because the zoospores have very limited motility (Henry 1988). In culture, sporogenesis was induced by low-temperature shock or by transfer to a nutrient-rich medium. Gametophytes matured and released gametes predictably 2 days after settlement of zoospores, at 20°C. Henry suggested that the arrival of a water mass high in nutrients (probably also relatively cool) or a low-temperature water mass followed by a warm one would induce simultaneous sporogenesis throughout a local population. Synchrony evidently is vital to plants with such small and short-lived gametophytes, and here temperature acts as a nonseasonal cue.

The photoperiod is not the only aspect of light that cues reproduction, and indeed some seaweeds are day-neutral. So far, gametophytes of two species of *Laminaria*, *L. saccharina* and *L. digitata*, have been shown to be insensitive to photoperiod. On the other hand, sporophyte sorus formation in *L. saccharina* and new frond formation in *L. hyperborea* require SD photoperiods (Lüning 1986, 1988). Some seaweeds, including some kelp and *Desmarestia* gametophytes, have minimum requirements for accumulated total daily irradiance or a certain irradiance intensity in order to reproduce (Chapman & Burrows 1970; Cosson 1977; Lüning & Neushul 1978), although *D. firma* gametophytes from South Africa are SD plants (Anderson & Bolton 1989). Other factors can also trigger reproduction in various cases. Gamete production in *Derbesia* and *Bryopsis* and egg production in *Dictyota dichotoma* have been shown to be controlled by endogenous rhythms of 4–5 days and 16–17 days, respectively (the latter a semilunar cycle) (Round 1981; Tanner 1981). Gametogenesis in *Dictyota diemensis* begins the day after a full moon and is completed with gamete release 10 days later (Phillips et al. 1990).

Sudden changes in the surrounding medium can induce reproduction in some simple seaweeds, a method exploited in culture work (Chapman 1973a). In one case, *Ulva mutabilis*, this may be because healthy vegetative thalli release substances that inhibit sporulation. Nilsen and Nordby (1975) showed that one of the substances is heat-labile, but they were not able to isolate and identify the compounds. Gametogenesis inhibitors in an *Enteromorpha* are complex glycoproteins (Jónsson et al. 1985). *Dictyosiphon foeniculaceus* requires a macrothallus-inducing substance, perhaps inositol, for development of the macroscopic stage, as well as an unknown substance to induce plurilocular sporangia on the microthallus (Saga 1986). (Both stages have the same ploidy level. *Dictyosiphon* is an obligate epiphyte, and the macrothallus-inducing substance is produced by bacteria present on host *Scytosiphon*.) Salinity shocks can also induce reproduction. The mechanisms are unknown, but might involve nutrient depletion or osmotic effects. Nitrogen availability has been shown to influence reproduction in a few instances, notably *Ulva* species: High nitrogen levels favor vegetative growth and asexual reproduction; low nitrogen levels stimulate gametogenesis (DeBoer 1981).

We know virtually nothing about reproductive phenology in the tropics. There are strong seasonal variations in growth and reproduction of the flora, but the cues are unknown (Price 1989). There are seasonal changes in the environment, albeit more subtle than those in midlatitudes. The more equable conditions are perhaps reflected in the apparently low numbers of tropical algae with heteromorphic life histories. Although a photoperiodic effect has been shown in temperate isomorphic species, and although there are day-length changes except very close to the equator (e.g., the range in Guam, at 13° N, is 11–13 h) (Fig. 1.30a), there is little reason to expect photoperiodic effects. It would be interesting to look for latitudinal effects in widely distributed heteromorphic species such as *Asparagopsis taxiformis* or *Tricleocarpa oblongata*. If tropical algae near their northern or southern limits (e.g., in Bermuda, Hawaii, or southern Queensland) show photoperiodic responses, what do they do near the equator? And if they are day-neutral, to what environmental cues do they respond?

1.5.4 Sexual reproduction

After reproduction has been initiated in response to environmental cues (i.e., the genes turned on and the type of reproduction determined), sporogenesis or gametogenesis takes place, and finally spores or gametes are released. Three types of sexual reproduction are traditionally recognized: isogamy, anisogamy, and oogamy (e.g., Bold & Wynne 1985; also see discussion by Rosowski & Hoshaw 1988). Oogamy involves a nonmotile female gamete or egg, but in the red algae the female reproductive system is much more complex than a mere egg (carpogonium), as discussed later. Among brown algae, many so-called isogamous and anisogamous species actually behave oogamously, with the female gamete settling before fertilization (Fig. 1.36), and Motomura and Sakai (1988) have now shown that *Laminaria angustata* eggs have vestigial flagella that are shed when the egg is released form the oogonium. Evidently our categories of reproduction are more convenient than they are accurate.

The timing of spore release may depend on conditions being suitable for settlement; the timing of gamete release may also enhance the chance for sexual fusion. Gamete release in intertidal *Ulva* occurs when the thalli are rewetted by the tide. In some populations there is a periodicity of release (which perhaps implies periodicity in gamete formation). On the Pacific coast of the United States, *Ulva* species release gametes at the beginning of the spring tide series, and spores 2–5 days later (Smith 1947), whereas *U. pertusa* in Japan releases gametes on the neap tides (Sawada 1972). Some *Monostroma* and *Enteromorpha* species also show periodicity of gamete or spore release. However, *Ulva* on the At-

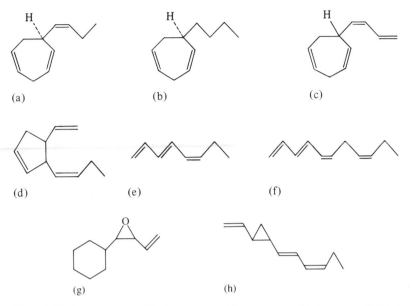

Figure 1.35. Sex attractants of the brown algae: (a) ectocarpene; (b) dictyotene C′; (c) desmarestene; (d) multifidene; (e) fucoserratene; (f) finavarrene; (g) lamoxirene; (h) hormosirene. (After Müller.)

Table 1.6. *Brown-algal pheromones of diverse species*

Species	Type of reproduction	Attractant	Structure in Fig. 1.35
Ectocarpus siliculosus	isogamous	ectocarpene	(a)
Sphacelaria rigidula	anisogamous	ectocarpene	(a)
Adenocystis utricularis	isogamous	ectocarpene	(a)
Cutleria multifida	anisogamous	multifidene	(d)
Dictyota dichotoma and *D. diemensis*	oogamous	dictyotene	(b)
Desmarestia viridis	oogamous	desmarestene	(c)
Laminariales (except *Chorda*)	oogamous	lamoxirene	(g)
Chorda tomentosa	oogamous	multifidene	(d)
Zonaria angustata	oogamous	multifidene	(d)
Fucus vesiculosus and *F. serratus*	oogamous	fucoserratene	(e)
Ascophyllum nodosum	oogamous	finavarrene	(f)
Durvillaea spp.	oogamous	hormosirene	(h)
Scytosiphon lomentaria	± isogamous	hormosirene	(h)
Colpomenia peregrina	± isogamous	hormosirene	(h)

lantic coast of the United States has no periodicity. Gamete release by *Derbesia tenuissima* in culture takes place at the beginning of the photoperiod. It is triggered by an instantaneous light-induced increase in turgor pressure that ruptures a weak area of the wall, forming a pore (Wheeler & Page 1974).

In species with unisexual gametophytes, coordination of gamete release and attraction of one gamete to the other increase the chances of successful syngamy. In Laminariales and Desmarestiales, with regular alternation of generations and with gamete production limited by the small size of the gametophytes, antheridia do not release their sperm until they detect the pheromone from mature female gametophytes; the same compound (lamoxirene, Fig. 1.35g) acts as antheridium releaser and sperm attractant (Müller et al. 1985; Müller 1989). In many brown algae, one gamete (female) releases a volatile attractant for the other (Table 1.6) (Müller 1981, 1989). The diverse taxa and reproductive types are reasons to expect further examples to be found. However, not all brown algae use pheromones. *Sargassum muticum* is monoecious, and fertilization takes place under a blanket of mucilage, while the oogonia are still strapped to the conceptacle by their mesochiton; presumably, self-fertilization can occur, which may contribute to the weediness of this species. *Himanthalia* eggs also apparently do not chemically attract sperm. Moreover, *Dictyopteris* and *Hincksia* [*Giffordia*] *mitchelliae* secrete compounds similar to those in Figure 1.35, from both gametophytic and sporophytic tissues, but these substances do not act as pheromones in these genera (Kajiwara et al. 1989; Müller 1989).

The various brown-algal pheromones (Fig. 1.35) are not distributed along taxonomic lines (Müller 1989) (Table 1.6). For instance, ectocarpene, characteristic of *Ectocarpus*, is also the attractant in *Sphacelaria rigidula* (Sphacelariales), *Adenocystis utricularis* (Dictyosiphonales), and *Analipus japonicus*. In *Analipus*, Müller et al. (1990) found evidence for a two-pheromone system, with hormosirene as the more active compound. All the compounds are simple, volatile hydrocarbons, either open-chain or cyclic olefinic hydrocarbons. Their insolubility and volatility prevent their concentrations building up in the water and enable the female gametes to maintain steep concentration gradients. The range of attraction probably is no more than 0.5 mm (Müller 1981). The quantities of attractant are minute: Five million *Macrocystis* eggs yielded 2.9 µg of lamoxirene (Müller et al. 1985). The compounds can be trapped onto adsorbent particles, which can then be used in bioassays.

The behavior of male gametes in the presence of an attractant varies from one species to another (Fig. 1.36). *Laminaria* sperm head straight for the egg. *Ectocarpus* males have a more complex pattern (Fig. 1.36a). In the water column they swim in straight lines, periodically changing direction abruptly. When they encounter a surface, they change to a wide, looping path along the surface. In the presence of attractant from the female gamete, the male changes to a circular path, the diameter of which decreases as the hormone concentration increases. Müller (1982) made an excellent film on algal pheromones.

Gamete recognition is a critical stage in sexual reproduction. Whereas attractants are general, recognition has to be species-specific. In *Fucus*, the same attractant works for at least three species, but syngamy is prevented because surface phenomena do not permit egg and sperm to unite. (Hybridization was thought to be frequent among *Fucus* species, but critical review of the evidence suggests that hybrids are not common; Evans et al. 1982). The processes of recognition and fusion have been studied in some Fucales, taking advantage of the fact that surface receptors are not obscured by cell walls (Evans et al. 1982). The egg membrane initially appears lumpy because of protrusion of cytoplasmic vesicles (Fig. 1.37a) (Callow et al. 1978). The spermatozoid probes the surface of the egg with the tip of its anterior flagellum, apparently seeking specific bind-

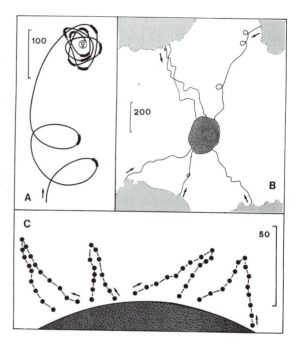

Figure 1.36. Different types of gamete approach in brown algae. (A) Chemo-thigmo-klinokinesis in *Ectocarpus siliculosus*; emphasized parts of male track indicate periods of hind-flagellum beat. (B) *Laminaria digitata*: impregnated silica particle as pheromone source in center, with tracks of individual sperm. (C) *Fucus spiralis*: return responses of individual sperm near a fluorocarbon droplet containing fucoserratene. Scales in micrometers. (From Müller 1989, with permission of John Wiley/Alan R. Liss Inc.)

ing sites (Friedmann 1961; Callow et al. 1978). Attachment takes place first by the flagellum tip, and later also by the body of the cell (Fig. 1.37b). Egg membrane surfaces carry special glycoproteins with fucose and mannose units in particular patterns that fit into carbohydrate-binding sites (ligands) on sperm membrane proteins, analogous to a lock-and-key mechanism (Bolwell et al. 1979, 1980). Some of the *Fucus* sperm surface domains that have been distinguished by monoclonal antibodies (Jones et al. 1988; sec. 1.3.1) probably are specific for egg recognition.

When one sperm has entered the egg, no more are needed. Indeed, polyspermy is lethal; fucoid germlings develop abnormally and die after a few days (Brawley 1987). In nature, only a small percentage of eggs are fertilized by more than one sperm, even though in monoecious species fertilization often takes place when oogonia and antheridia are newly released and the sperm concentration is likely to be high. A fast block to polyspermy has been shown in *Pelvetia* and *Fucus* (Brawley 1987); this block is Na^+-mediated and is replaced within about 5 min by a permanent change

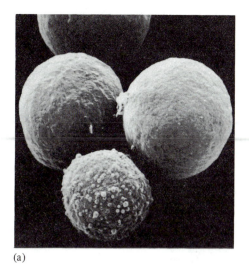

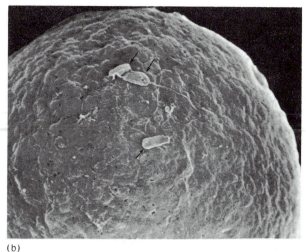

(a)   (b)

Figure 1.37. Scanning-electron-microscope views of eggs, sperm, and zygotes of *Fucus serratus*. (a) Group of cells 10 min after mixing eggs and sperm. Smooth cells have been fertilized and have formed a fertilization membrane; the rough cell in the foreground is an unfertilized egg ($\times$450). (b) Detail of fertilized egg with three sperm (arrows); the tip of the anterior flagellum of the middle sperm is embedded in secreted cell-wall material ($\times$1,600). (From Callow et al. 1978, with permission of The Company of Biologists.)

in wall structure that appears as a smooth membrane (Fig. 1.37a). Pheromones increase sperm concentrations around eggs in many brown algae, and a fast block to polyspermy is expected to be important in many species.

Sexual reproduction in red algae has been extensively studied by light microscopy because of its importance to systematics (Hommersand & Fredericq 1990). For physiology and ecology, the interesting features are (1) that spermatia are nonmotile and (2) that female gametes are specialized cells that are retained on the female gametophyte and develop in various complex ways, often involving many cell fusions and nuclear transfers. There is an evolutionary trend in red algae toward zygote amplification, usually in the form of carpospore production, which probably compensates for the lack of spermatium motility (Guiry 1987). At the extreme, many cystocarps may be initiated from a single fertilization (Fig. 1.38).

The traditional view of sexual reproduction in red algae holds that individual spermatia with sticky coats encounter trichogynes, but Dixon (1973) has pointed out the low statistical probability of such an encounter. Yet, clearly, fertilization does take place, and the red algae are a successful group (Fetter & Neushul 1981). Transfer of spermatia needs to be seen in terms of water movement, especially around the female plant. Hydrodynamic studies are difficult (Chapter 7), and no study has yet been done with an alga along the lines of Niklas and Paw U's (1982) elucidation of how different species of pollen grains are deposited in appropriate ovulate pine cones. The ability of spermatia to reach a trichogyne is improved when they are released in slime strands, as in

Tiffaniella snyderae (Fetter & Neushul 1981). The fibrillar mucilage is elastic; it stretches out in water flow, and when it attaches to the female plant it tends to sweep the surface and deposit spermatia on extended trichogynes. (An analogous means is used for pollen dispersal in the seagrass *Thalassia testudinum;* Cox & Tomlinson 1988.) There are cone-shaped appendages on spermatia from *Aglaothamnion neglectum* that are not sticky and bind only with trichogynes and hairs, though the binding is not species-specific (Magruder 1984).

The formation of carpogonal branches, support cells, and auxiliary cells (when these are on special branches) implies considerable cell morphogenesis. Even in such a simple plant as *Porphyra*, in which carpogonia are individual thallus cells, apparently little differentiated, the formation of carpogonia proves to be a remarkably dynamic process: A previously unsuspected large "procarpogonial mother cell" is formed (Fig. 1.39a) (Cannon 1989); this active, "pulsating" cell splits off carpogonia all around it, which then migrate to restore the monostromatic thallus.

As an example of more complex development, we can look at Kugrens and Delivopoulos's (1986) account of events in the parasite *Plocamiocolax pulvinata*. This species has, nevertheless, a relatively simple carposporophyte and no fusion cells. After fertilization, the diploid nucleus is transferred to an auxiliary cell. This splits to form multinucleate storage gonimoblast cells packed with starch and uninucleate generative gonimoblast cells that each contain a copy of the diploid nucleus and that divide to form carpospores. Carpospores grow at the expense of the storage gonimoblast cells, which gradually degenerate.

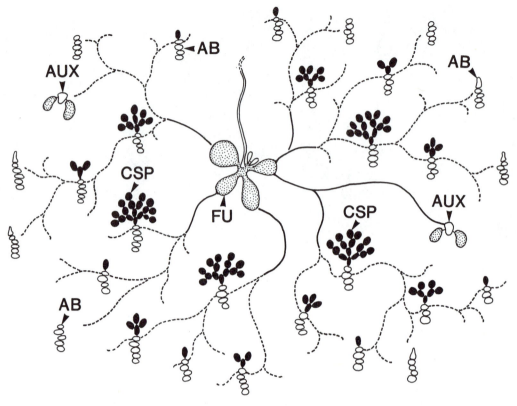

Figure 1.38. Multiple cystocarp production from a single fertilization in *Hommersandia maximicarpa*. Connecting filaments produced by the carpogonial fusion cell (FU) either fuse directly with an auxiliary cell (AUX) or branch and become septate (dotted lines) before contacting numerous accessory branches (AB). The gonimoblast filaments (drawn in reduced, diagrammatic form) are initiated by the contacted accessory branches and produce chains of carposporangia (CSP). TR, trichogyne. (From Hansen & Lindstrom 1984, with permission of *Journal of Phycology*.)

These complex postfertilization events pose many intractable mechanistic questions: What determines which cells will form carpogonial branches or which will become auxiliary cells? How do connecting filaments find the (often remote) auxiliary cells? What part do sterile branches play? The events usually are not observable in living, whole tissue and do not lend themselves to experimental manipulation. However, there are some genera, such as *Callithamnion*, in which the female reproductive system is exposed because there are no sterile branches or pericarp. O'Kelly and Baca (1984) were able to observe the timing of reproductive stages in *Aglaothamnion cordatum*, which in culture produced one new axial cell per day. Like all Ceramiales, this species has a four-celled carpogonial branch (Fig. 1.39b), and auxiliary cells are produced only after fertilization, in this case from the support cell and an additional auxiliary mother cell. Gamete fusion (including spermatium attachment, plasmogamy, transfer of the male nucleus down the trichogyne, and karyogamy) took 5–10 h. Carpogonia divided to form two daughter cells. Auxiliary cells formed after about 40 h, and at around 72 h were diploidized; that is, the original hap-

loid nucleus was partitioned off into a foot cell, and a diploid nucleus from a carpogonium daughter cell was transferred via a connecting cell.

In *Polysiphonia harveyi*, the mature carpogonial branch (but not the sterile branches) contains a channel of closely meshed, tubular, smooth endoplasmic reticulum that extends uninterrupted, except for pit connections, from the carpogonium to the support cell (Broadwater & Scott 1982). The function suggested for the channel is to transfer a message from the fertilized carpogonium to the support cell to initiate the auxiliary cell. In *Callithamnion*, the carpogonial branch cells seem to play no role, and instead there must be contact between some part of the enlarged carpogonium and an auxiliary mother cell for auxiliary cell formation (again implying hormonal communication) (O'Kelly & Baca 1984).

1.6 Settlement and germination

1.6.1 Settlement

Once the reproductive cells have been released from the parent generation, they must get to a surface and stick to it. Some cells, such as zooids of green and

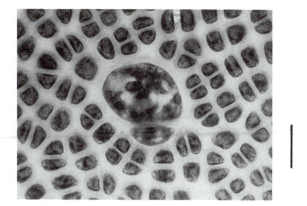

(a)

(b)

Figure 1.39. Carpogonium formation in red algae. (a) Procarpogonial mother cell in *Porphyra abbottiae* cutting off a smaller cell that will differentiate into a carpogonium. Scale bar = 10 μm. (b) Carpogonial branch of *Polysiphonia harveyi*. Electron-micrographic section and diagram of prefertilization appearance. AX, auxiliary cell; $CB_{1...3}$, carpogonial branch cells; CP, carpogonium; PR, pericarp; $ST_{1,2}$, sterile cells; SU, support cell; TG, trichogyne. (Part a from Cannon 1989; b from Broadwater & Scott 1982; both with permission of *Journal of Phycology*.)

brown algae, have a limited ability to swim. Others, such as red-algal spores, green- and brown-algal aplanospores, and multicellular propagules, as in *Sphacelaria* and *Sargassum muticum*, are nonmotile. All these structures are small enough to occupy the slow-moving and nonmoving layers of water that form against submerged objects (Chapter 7). Estimates of the thickness of such a nonmoving layer (only part of the total boundary layer) are 5–150 μm for various surfaces, whereas red-algal spore sizes, for instance, are 15–120 μm (Coon et al. 1972; Neushul 1972). In order to get into the "safe zone," where they have time to attach, cells must travel through moving water.

Nonmotile cells get to the seabed by strictly physical forces (Coon et al. 1972). Gravity tends to pull cells downward at ever-increasing speeds, but drag also increases with speed, so that a maximum (terminal) velocity is reached. This terminal velocity, V_t, depends partly on the density and radius of the spore. Coon et al. (1972) measured V_t for several species of red-algal spores using time-exposure photomicrographs. *Sarcodiotheca gaudichaudii* carpospores were fastest, sinking at 116 μm s^{-1}, but that is much less than typical water-current velocities. Neushul (1972) estimated that it would take a *Cryptopleura* carpospore 10 min to fall through perfectly still water from the cystocarp on the adult plant to the seabed. However, turbulence may tend to keep cells in suspension unless the surface roughness is suitable for depositional eddies to form as discussed later.

Motile cells may be better able to reach the seabed, but their swimming speeds are slow compared with the velocities of water currents. North (1972) recorded *Macrocystis* zoospore velocities of approximately 5 mm s^{-1}, and Suto (1950) reported speeds for various zooids of only 125–300 μm s^{-1}, but swimming clearly is an advantage over merely sinking, in terms of attachment efficiency. Many zoospores, including those of *Macrocystis*, swim randomly, changing direction frequently. Some zoospores can orient with respect to light; some of these are negatively phototactic and swim toward the seabed, but others (e.g., *Enteromorpha*) are positively phototactic, sink very slowly, and spend a long time in the plankton (Amsler & Searles 1980; Hoffman & Camus 1989). In *Monostroma*, gametes from the leafy intertidal phase are negatively phototactic, according to Suto (1950), and so they settle in the subtidal, to form shell-boring "gomontia" sporophytes. In fact, in *M. grevillei*, and also in *Ulva lactuca*, gametes are initially positively phototactic; they become negative upon pairing (Kornmann & Sahling 1977). Zoospores, which must move back to the intertidal, are positively phototactic. Conceivably, motile cells may be able to make limited choices during site selection (e.g., by chemotaxis or chemoperception), though evidence for this is weak yet (Amsler & Neushul 1990). As Amsler and Neushul point out, motility is energetically ex-

pensive and must be oriented to be useful. Because the zoospores of most kelps are not phototactic, they presumably must use some other cue for orientation. Experiments on algae parallel to those of Lawrence et al. (1987) and Lawrence and Caldwell (1987) on bacterial settlement behavior would help us understand the biological components of successful settlement. Motile bacteria are able to move upcurrent against flow velocities greater than their maximum swimming speed. They manage this by remaining attached while moving across the surface, a paradox that Lawrence et al. (1987) called "motile attachment."

Several seaweeds have evolved interesting means of improving the chances of spore settlement. *Nereocystis* blades float far above the seabed, but the entire sorus, which sinks readily, is shed before the spores are released. Sorus shedding takes place for a few hours around dawn, giving the spores the best chance for photosynthesis and survival. Spore release begins before sorus abscission, continues as the sorus sinks, and is completed within about 4 h (Amsler & Neushul 1989a,b). *Postelsia*, which grows in very high energy intertidal habitats, releases its spores when the first wavers of the incoming tide splash over the plants; water and spores flow down channels in the drooping blades and drip onto the rock and parent-plant holdfast. Spores settle in about 30 min, before the tide completely covers them (L. D. Druehl & J. M. Green, unpublished data). *Fucus* releases its eggs still held together in the oogonium; the mass of eight eggs sinks faster than would a single egg. *Sargassum muticum*, which has become a weed in several parts of the world, has a very effective settling mechanism: Eggs released from the conceptacles remain attached to the outside of the receptacle, where they are fertilized and develop into small germlings, usually without rhizoids, before they drop to the seabed. As a result of their relatively large size (mean 156 μm), these propagules sink at an average rate of 530 μm s^{-1} in still water (Deysher & Norton 1982), some 5–10 times faster than unicellular spores. Once rhizoids start to grow out, they increase the drag and slow the sinking rate (Norton & Fetter 1981).

Whereas these seaweeds have special means of improving spore settlement, other have special problems. Parasitic algae (*Harveyella, Plocamiocolax*) and epiphytes with particular hosts (*Microcladia californica, Polysiphonia lanosa*) must encounter the appropriate alga in order to grow. Most such specialists are red algae, which is surprising, given their nonmotility. The problems will be different for plants on long-lived or abundant hosts (*P. lanosa* spores are very likely to fall on *Ascophyllum*) than for plants on scattered or ephemeral hosts. In the latter, timing and other strategies must be important, as shown in *Porphyra nereocystis* (sec. 1.5.3).

Surface properties greatly affect settlement success, whether cells are motile or not. These properties

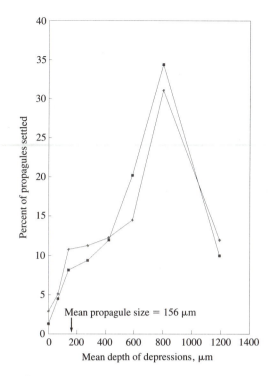

Figure 1.40. Effect of substratum roughness on settlement of *Sargassum muticum* propagules. The substratum consisted of sand-coated microscope slides on a surface with a jet of water flowing over it (hydrodynamic characteristics shown in Fig. 7.4). Two independent experiments were run; within each experiment, several water velocities were used, and the results were pooled. (Drawn from data in Norton & Fetter 1981, with permission of Cambridge University Press.)

include roughness and surface energy. Clean glass slides (a favorite experimental surface in the past) are unnatural surfaces, to which macroalgal cells do not adhere well. Natural surfaces, in contrast, usually are rough. Evidence from a number of experiments shows that surface roughness, even though it increases turbulence, is an important factor in settlement. Essentially, cells are deposited by eddies, in the same way that sand grains are deposited on the lee side of a sand dune. Norton and Fetter (1981) built a ''waterbroom'' to study the effect of surface roughness on settling of *Sargassum muticum* propagules in moving water. A jet of water from a fixed nozzle flowed onto and along a plate into which microscope slides could be recessed. Sand grains sorted to particular sizes were used to create rough surfaces. Norton and Fetter found that settlement of *Sargassum* propagules was best on a surface with a mean depression depth of 800 μm (Fig. 1.40), no matter what the water speed (range of 0.22–0.55 m s^{-1}). The propagules attached not because of sinking but because of turbulent deposition. The reason suggested for the low rate of attachment in the depression of largest size was that that

size would be big enough to be swept clean by water flow, rather than creating depositional eddies. Small algae already growing on the seabed create an algal turf that provides a place for reproductive cells to lodge. Of course, where cells can settle, so can sediment. The reduced water velocity in eddies also provides a more favorable settling environment, while at the same time providing turbulence and good nutrient availability to growing plants.

Within the nonmoving boundary layer, viscous forces are most important. The ability of cells (i.e., their mucilage) to stick to a surface depends on the surface energy, sometimes called surface tension or wettability. High-energy surfaces are hydrophilic (wettable), and low-energy surfaces are hydrophobic. The surface energy depends on the nature of the substratum, including any coatings. Any material submerged in the ocean will quickly be coated by a biofilm of bacteria and their associated mucilage, which will increase the surface energy and make the surface much more suitable for macroalgal settlement (Fletcher et al. 1985; Dillon et al. 1989). In contrast, treatment of surfaces with hydrophobic coatings, such as silicone elastomers or silanes, reduces algal settlement and is an effective antifouling technique (Fletcher et al. 1985; Callow 1989).

The foregoing discussion assumes propagules that are free in the water column. However, significant numbers of propagules may reach the seabed in the fecal pellets of grazers (Santelices & Paya 1989). Herbivores ingest vegetative and reproductive tissues, sometimes preferring the latter (Santelices et al. 1983), and spores and tissue fragments often survive passage through their guts. Such fragments can form swarmers or protoplasts that will give rise to new plants, especially in opportunistic algae like *Enteromorpha*. Cells in fecal pellets have several advantages: The pellets are heavy, sinking 8–22 times faster than *Sargassum* propagules and 40–100 times faster than algal spores; the stickiness of pellets greatly improves attachment; the pellets provide protection against desiccation in the intertidal zone, giving sensitive germlings a chance to establish; and nutrient availability may be higher in the pellets.

Zoospore or gamete ultrastructure and settling have been studied in a few species, including *Enteromorpha intestinalis* (Evans & Christie 1970; Callow & Evans 1974), *Scytosiphon* (Clayton 1984), and a variety of Laminariales (Henry & Cole 1982). These cells initially lack a wall and have among their organelles numerous cytoplasmic vesicles that contain adhesive material (Fig. 1.41a). Attachment first takes place by the tip of the anterior flagellum in kelps, and presumably by all four flagella in *Enteromorpha* swarmers. Within a short time the flagella are withdrawn into the cell, and the cytoplasmic vesicles are released, adhering the cell to the substratum (Fig. 1.41b). Cell-wall secretion begins once the spore has attached.

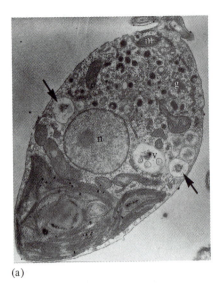
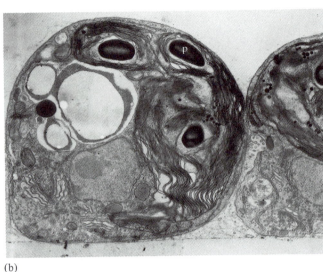

(a)   (b)

Figure 1.41. Ultrastructure of swimming (a) and newly settled (b) zoospores of *Enteromorpha intestinalis*. In the anterior of the swimming cell can be seen numerous vesicles filled with adhesive (arrows). Also visible are part of the nucleus (n) and flagella bases, Golgi body (g), vacuole (v), and mitochondrion (m). (b) A mass of secreted adhesive lies in the triangle between the two cells, and there are virtually no vesicles remaining in the cell. (The attachment surface is parallel to the bottom of the photograph); c, chloroplast; p, pyrenoid. Scales: (a) ×9,000; (b) ×10,000. (From Evans & Christie 1970, with permission of the Annals of Botany Company.)

After spores have contacted and stuck to a surface, they begin to improve their adhesion by hardening the adhesive and by developing rhizoids. The best-studied cells are fucoid zygotes. Fucoid eggs are initially covered by a mucilaginous layer of alginates and fucoidan that attaches them to the oogonium wall. The eggs are expelled from the conceptacle still enclosed in this layer, which is called the mesochiton. In *Fucus* and *Himanthalia* the mesochiton soon breaks down, and the zygotes attach by the zygote wall. In *Pelvetia canaliculata* the mesochiton persists, probably to protect zygotes from drying out in the very high shore habitat of this species (Moss 1974b; Hardy & Moss 1979). The mesochiton, rather than the zygote wall, attaches the pairs of *Pelvetia* zygotes to the substratum. Within 24 h of settling, the *Pelvetia* zygote develops a firm alginate wall inside the mesochiton. Each zygote divides once or twice and then pushes out a group of up to four rhizoids, each from a single cell. These rhizoids grow down into the substratum, entering minute crevices, if these are available, and the mesochiton splits open. The time between fertilization and the formation of rhizoids in this species is about 1 week. For various seaweeds, the surface energy of the substratum affects the morphology of the germlings, especially their rhizoids. Many species (but not all), when on the preferred high-energy surfaces, form compact, well-attached basal filaments or rhizoids, whereas on low-energy surfaces the filaments or rhizoids spread widely and are poorly attached (Fletcher et al. 1985).

Hardening of the attachment mucilage in various seaweeds apparently involves the formation of cross-links between polymer molecules, especially Ca^{2+} bridges between alginate chains or between sulfate ester groups of fucoidan (see Fig. 4.25). The rhizoid wall and the rest of the zygote wall have different alginates, as shown by antibody labeling (Boyen et al. 1988). *Fucus* embryos grown in sulfate-free seawater form normal rhizoids, but cross-linking cannot occur, and the rhizoids cannot adhere to the substratum (Crayton et al. 1974). Moreover, sulfation is necessary for intracellular transport of fucan (sec. 1.6.2). Attachment of single cells mechanically released from *Prasiola stipitata* thalli also requires sulfation of a cell-wall polysaccharide; inhibitors of sulfation (such as molybdate) and of protein synthesis prevent attachment (Bingham & Schiff 1979). The protein may be complexed with the polysaccharide or may be an enzyme involved in the sulfation process.

Attachment and hardening take time, and thus experiments designed to dislodge settled cells have shown that with a given water pressure, the numbers of cells that are washed away decrease the longer the cells have been allowed to settle (Christie & Shaw 1968). Other experiments have shown that the hydrodynamic force that a settled cell can withstand increases with time. *Ascophyllum* zygotes apparently cannot settle unless there is an adequate period of very calm water – something that rarely happens (Vadas et al. 1990). Thus, when patches of *Ascophyllum* are cleared, those areas usually are not recolonized by that fucoid.

The normal course of events is for reproductive cells to settle, attach, and then germinate. However, Kain (1964) noted that *Laminaria hyperborea* zoospores sometimes lost their motility and started to germinate while still in the water column; that did not preclude their subsequent attachment. North (1976) took advantage of the fact that very young sporophytes of *Macrocystis pyrifera* are sticky and was able to transfer germlings from cloth-culture substrata to the seabed.

1.6.2 Germination

Germination is an oriented process. Cells and plants have polarities, especially apico-basal polarities that distinguish holdfast from frond. Eggs and young zygotes of fucoids are symmetrical, as are red-algal spores and other nonmotile spores (probably), in contrast to the polar eggs of most animals and higher plants and flagellated reproductive cells in algae. They then become polarized by environmental gradients, and after 10–12 h (in fucoids) the axis is fixed (Evans et al. 1982; Quatrano et al. 1985; Quatrano & Kropf 1989). Fucoidan secretion and later rhizoid outgrowth normally occur on the side in contact with the substratum, with light being the primary gradient, but in the laboratory they can be induced to grow out on other parts of the cell, by altering the polarity of the cell. Various natural and artificial gradients, including pH, Ca^{2+}, K^+, and the proximity of other zygotes, can serve as orientation cues in *Fucus.*

When a *Fucus* zygote is placed in an orienting gradient, a polar axis begins to form parallel to the gradient. Division of the cell into rhizoid- and frond-forming cells will later take place perpendicular to this axis. For the first 8–14 h, the polar axis can be reoriented if the gradient is moved, but after that it is fixed. During the labile period, changes take place in membrane patches on the side where the rhizoid will form (e.g., the shady side in a light gradient). These membrane patches become fixed to the underlying cytoplasm by microfilaments; cytochalasin B, which disrupts microfilaments, inhibits axis fixation. The membrane patches then generate an influx of calcium ions, which remain near the rhizoidal pole and are essential for rhizoidal growth. There is continuing debate between Jaffe's group (e.g., Speksnijder et al. 1989) and Quatrano's group (e.g., Kropf 1989) whether or not the calcium gradient is essential to axis fixation. The first known macromolecular asymmetry is of filamentous actin (F-actin). F-actin microfilaments accumulate at the rhizoidal pole during axis fixation (Fig. 1.42a). A cell wall is also required for axis fixation. The most recent working hypothesis for axis fixation (Kropf 1989; Quatrano & Kropf 1989) is that there are bridges across the membrane at the presumptive rhizoidal pole, from the cell wall to the cytoskeleton (Fig. 1.42a). Brownlee (1990) postulated a cascade of molecular changes

(Fig. 1.42b) that would amplify the initial photoreceptive stimulus.

After axis fixation, several cellular changes can be seen. Numerous extensions of the nuclear membrane project toward the rhizoidal pole, and there is an accumulation of vesicles at the pole, apparent in the light microscope as a "cortical clearing." The vesicles, derived from the Golgi apparatus, are filled with sulfated fucoidan that is deposited in the cell wall at the pole and serves to anchor the cell to the substratum. Negative charges on the sulfate ester groups or perhaps on the vesicle surfaces may be needed to draw the material toward the positively charged pole. Sulfation of the fucan requires new enzyme synthesis; if synthesis is prevented by cycloheximide, or if SO_4^{2-} is lacking, there is no movement of fucan to the rhizoidal pole.

Fucoid eggs and zygotes have some important advantages for studies of embryogenesis (Quatrano 1980). Their patterns of polar development and embryogenesis are morphologically similar to those of many algae (some exceptions will be discussed later) and angiosperms; the zygotes are fairly large (75 μm, still small enough to require a fine hand to insert an electrode) and can be collected in quantity aseptically, and they develop in synchrony in defined media. Unfortunately, they have provided a model system to such an extent that embryogenesis in other algae has not been studied except morphologically. Certainly at the morphological level there is great variety. Even within the Fucales, *Himanthalia* shows a different pattern, which Ramon (1973) suggested was not oriented by light. Not all germinating spores or zygotes first divide parallel to the substratum (or perpendicular to the light gradient). Horizontal germination is common, in which a single filament (germ tube) or basal crust is formed. In some species, such as *Coelocladia arctica* (Dictyosiphonales), the protoplast migrates into the germ tube, leaving the spore wall empty (Pedersen 1981). Other brownalgal zoospores push out several lobes that are then cut off as cells; this stellate kind of germination leads to a monostromatic crust. Dixon (1973) described five types of germination among the Rhodophyta. Spores of Corallinaceae divide in patterns characteristic for particular species (tetraspores and carpospores show the same pattern) (Chamberlain 1984).

Germination patterns are not exclusively genotypic; environmental conditions may cause changes. Germination of carpospores of *Bangia fuscopurpurea* may be unipolar, leading to the conchocelis stage, or bipolar, leading directly back to the erect stage. The type of germination in this case is apparently regulated by the photoperiod: Unipolar germination occurred when the day length was greater than 12 h, and bipolar when it was less than 12 h (Dixon & Richardson 1970). In *Gigartina exasperata* sporelings, a density-dependent dimorphism was seen by Sylvester and Waaland (1984).

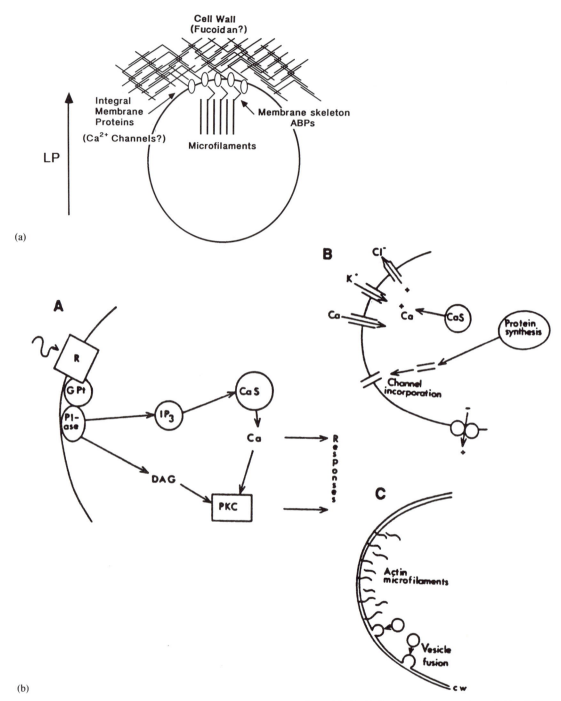

(a)

(b)

Figure 1.42. Polar-axis fixation in *Fucus* embryos. (a) Kropf's model of axis fixation. A transmembrane bridge is localized at the presumptive rhizoid site on the shaded hemisphere in unilateral light (LP, light-pulse direction). Microfilaments are envisioned to be linked to integral membrane proteins via the membrane skeleton, which probably contains actin-binding proteins (ABPs). The integral membrane proteins associate on the outside with cell-wall constituents. The membrane-spanning molecules may be Ca^{2+} channels. (Drawing not to scale.) (b) Brownlee's summary of molecular events from perception of light stimulus to fixation of polar axis. (A) Activated photoreceptor (R) interacts with G-protein (GPt) on the inner face of the plasma membrane. This stimulates phosphoinositidase (PIase) activity and the hydrolysis of phospholipids to form inositol triphosphate (IP_3) and diacylglycerol (DAG). Localized IP_3 formation causes localized calcium release from internal stores (CaS). DAG and calcium stimulate protein kinase C (PKC)

The role of surface energy in germling morphology has thus far barely been pointed out in the literature.

1.7 **Thallus morphogenesis**
1.7.1 Cell differentiation

During normal development of a germling of an erect thallus, the basal cell forms the rhizoids, while the apical cell forms the frond. The cells resulting from further divisions are locked into their developmental patterns if only because the thallus is part of their regulatory environment. Before the first division of a new individual, the components of the cytoplasm become unequally distributed, so that after division the nuclei of the two cells are in different environments. As the plant develops, the changing environment of each cell (and its nucleus) is critical in determining when particular genes will be expressed. Although the traditional view is that differentiation involves selective reading of a DNA "blueprint" that remains intact, recent molecular evidence shows that there can be structural changes in genomes during differentiation (see the example of heterocysts in sec. 1.4.1).

Cell differentiation in simple thalli is most obvious if there is an apical cell. *Sphacelaria* apices have large cells with clearly defined functions (Ducreux 1984). Organelles in the apical cell are concentrated near the tip, and the nucleus is also in the distal half of the cell. In many species the apical cell undergoes regular mitosis to form a symmetrical subapical cell. This cell, in turn, divides to produce two cells with different morphogenetic potentials: The upper one of the pair (the nodal cell) will branch, but the other (internodal) cell will not (Figs. 1.4a and 1.43). The asymmetry of the apical cell apparently is essential to its role as an apical cell and is also dependent on contact with older cells, as shown by regeneration experiments (Ducreux 1984). If the apical cell is cut off, the subapical cell will become polarized and take over as a new apical cell (before or after dividing) (Fig. 1.43b,c). An isolated subapical cell will form a new axis (Fig. 1.43d), whereas an isolated apical cell will retain its polarity and continue to divide as before (Fig. 1.43e).

When thalli or cells are injured, cells may dedifferentiate or redifferentiate. Cells isolated from the parent individual may behave as zygotes or spores and may

Figure 1.42 (*cont.*)
activity. PKC and calcium activate a cascade of responses. (B) These responses include stimulation of channel activity and locally increased positive-current entry. This, together with increased calcium entry, serves to amplify the signal. (C) Axis fixation involves calcium-activated microfilament localization and polysaccharide-vesicle fusion (fucoidan release), as shown in figure a. (Part a from Kropf 1989, *Biological Bulletin*, with permission of Woods Hole Marine Biological Laboratories; b from Brownlee 1990, copyright Cambridge University Press.)

form a whole new organism; the ability of a cell to do this is called totipotency. Single cells from simple thalli such as *Prasiola* (Bingham & Schiff 1979) readily regenerate the whole thallus, but cells from complex algae, such as cortical cells from *Laminaria* (Saga et al. 1978), can also regenerate the thallus under appropriate culture conditions. There probably are few cells in algal thalli that are irreversibly differentiated; anucleate sieve elements of *Macrocystis* obviously would provide one example. Yet cells released from contraints imposed by neighboring cells do not always grow into a plant of the same generation. Examples are seen in the phenomenon of apospory: The diploid sporophytes of three kelp species raised in stagnant culture for 3–4 months became bleached, leaving only isolated epidermal cells alive (Nakahara & Nakamura 1973). A few of these epidermal cells germinated on the thallus, giving rise to gametophytes, which were shown to be diploid. Similar results have been obtained by isolating cells of moss sporophytes (Bold et al. 1980).

Development of isolated protoplasts depends on cell totipotency. Currently there is much interest in plant protoplasts, both as means for propagating desirable crop phenotypes (e.g., Kloareg et al. 1989) and as experimental systems for somatic hybridization and (potentially) genetic engineering. (Cell fusion sometimes occurs in nature between genetically different individuals, particularly in crustose plants or germlings; Maggs & Cheney 1990.) Protoplast isolation is achieved by digesting the existing walls with a mixture of enzymes, often including natural extracts from marine herbivores. Until recently, the viability and regenerative capacity of protoplasts were low (van der Meer 1986a), but greater success is now being achieved as, for example, by Polne-Fuller and Gibor (1984) working with *Porphyra*, Ducreux and Kloareg (1988) with *Sphacelaria*, Fujimura et al. (1989) with *Ulva*, and Butler et al. (1989) with *Laminaria* (sec. 9.2.4 and 9.8). Regeneration of cells from different parts of the thallus has illustrated the point that even such "simple" thalli as those of *Porphyra* and *Sphacelaria* are differentiated, and under the conditions used so far, not all protoplasts are equal.

Some cells from seaweeds, if isolated, can regenerate into a whole plant, but others cannot. Blade cells of *Ulva mutabilis* are unable to form rhizoidal cells, but form vesicular thalli one cell thick (Fjeld & Løvlie 1976). However, isolated rhizoidal cells of this species can form the whole plant. A repressor is present in the thallus cells, as shown by Fjeld's study of a mutant, *bubble* (*bu*), that behaves like isolated blade cells (Fjeld 1972; also see Fjeld & Løvlie 1976). The mutant gene is recessive and chromosomal. Curiously, *bu* spores from meiotic sporangia on heterozygous plants (bu^+/bu) develop a partly or completely wild-type phenotype in the first generation. When these are propagated asexually, subsequent generations are completely the mutant type. The explanation appears to be that there are repressor or

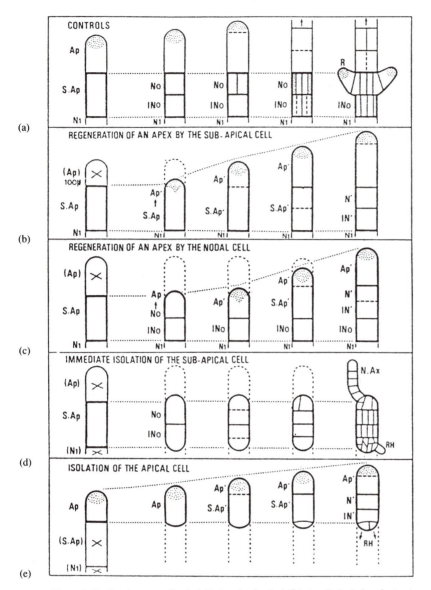

Figure 1.43. Development of apical (Ap) and subapical (SAp) cells in *Sphacelaria cirrosa*. (a) Normal ontogenesis of the subapical cell on control axes. (b, c) Regeneration of an apex after removal of the apical cell. If the subapical cell has formed recently (b), it transforms itself into an apical cell. If it is older (c), it undergoes a first division, and the nodal cell regenerates the apical cell. (d) Development of the subapical cell when isolated immediately after formation: Modified development sequence leads to formation of a complete new axis. (e) An isolated apical cell continues normal sequence of cell divisions, except for the development of rhizoid initials (RH). N.Ax., newly formed axis; No, nodal cell resulting from transverse partitioning of the subapical cell; INo, corresponding internodal cell; N1, IN1, . . . , successive nodal and internodal segments; R, branch. Abbreviations in parentheses indicate cells removed. (From Ducreux 1984, with permission of *Journal of Phycology*.)

rhizoid-forming genes in normal blade cells and *bubble* mutant cells and that this repressor is removed during sporogenesis, so that spores can form rhizoids when they settle. The bu^+ wild-type gene is thus responsible for removal of the repressor, and its transcription takes place before meiosis, so that the de-repressor is present in the cytoplasm of *bu* spores, as well as in wild-type spores. (This substance is not diluted through many cell generations, because the number of rhizoidal cells is small.) The rhizoid-forming genes of both types are re-repressed early in development, but the mutant gene cannot de-repress them when it forms spores.

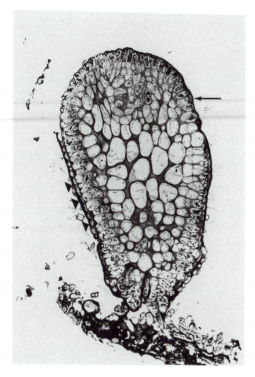

Figure 1.44. Apical meristem (arrow) in a fucoid germling (*Hormosira banksii*): long section through an 8-week-old plant as seen in the electron microscope (×124). The outer wall, already colonized by microscopic epiphytes, is starting to slough off (arrowheads). At this stage, cortex and medulla are already differentiated, but the apical pit (primary cryptostoma) is just forming at the meristem. Compare with drawings of earlier stages in Figure 1.4b,c. (From Clayton et al. 1985, *Phycologia*, with permission of Blackwell Scientific Publications.)

1.7.2 Development of adult form

During the initial stage of growth of the fucoid embryo, cell divisions occur throughout the thallus, but then an apical cell, or a group of cells, becomes organized (Fig. 1.44) and is responsible for all subsequent growth (Bold & Wynne 1985; Clayton et al. 1985). In *Fucus* the apical cell has been shown to divide only rarely, with the surrounding tissue active as the meristem (Moss 1967); nevertheless, the apical cell may still play a controlling role (Moss 1965; cf. Clayton & Shankly 1987 on *Splachnidium*). In *Hormosira banksii* (Clayton et al. 1985), the apical group of four cells does divide, including vertically (longitudinally) to initiate branches; the apical cell is also very active in *Cystophora*, as discussed later (Fig. 1.46). There is a meristematic region around the apical cells in these spe-

cies. The growing regions of fucoids probably are the most complex among seaweeds. In most seaweeds, subsequent growth depends on more diffuse growth regions (e.g., in kelps) or on specific, often apical, cells (sec. 1.2.1).

In thallose red algae, similar shapes are produced by a variety of different means, and external morphology is independent of anatomical construction and internal cell shape. Miller (1988) suggested that the cuticle and extracellular matrix may play roles in coordinating overall form during ontogeny.

Thallus morphogenesis in multicellular seaweeds requires not only cell division but also cell adhesion. Though this may seem a trivial point, a few green algae, including *Ulva lactuca*, *Enteromorpha* species, and *Monostroma*, require external morphogenetic factors for cell adhesion (Provasoli & Pintner 1977; Provasoli et al. 1977; Tatewaki & Provasoli 1977). In the normal course of development of *Monostroma oxyspermum*, the biflagellate swarmer produces a filament that divides in three planes to give a little sac, which subsequently ruptures, yielding a flat, monostromatic sheet (Tatewaki 1970). However, if placed in axenic culture, the germinating swarmer will form only a 2-cell thallus consisting of an apical cell (which will slough off cells during subsequent divisions) and a basal, rhizoidal cell. Normal morphology can be restored by addition of exudates of axenically cultured brown and red seaweeds, by growing *Monostroma* in bialgal axenic culture with a red or brown seaweed, and by extracts of seven marine bacteria (out of over 200 isolates tested) in the genera *Caulobacter*, *Cytophaga*, *Flavobacterium*, and *Pseudomonas* (Tatewaki et al. 1983). In nature, *Monostroma* does not bear a diverse microflora, but is colonized only by a *Pseudomonas* species. If the morphogenetic factor, which apparently is a small polypeptide residing in the cement between the cells, is added to rhizoidal cells, they will coalesce and form a complete plant, whereas apical cells will coalesce to form sheets without rhizoids (Provasoli et al. 1977). A continuing supply of the factor is needed to maintain thallus integrity. *Ulva* apparently induces its epiphytic bacteria to produce the morphogenetic substance, because the bacteria cultured in isolation stop producing it (Provasoli & Pintner 1980).

The formation of erect uniaxial or multiaxial thalli from crustose germlings or microthalli requries that one or a number of erect filaments have their tips converted into meristems (Fig. 1.45) (Dixon 1973; Rietema & Klein 1981). Formation of the meristems (macrothallus initials) and the outgrowth of the erect thalli are separate events, and in *Dumontia contorta* (*D. incrassata*) these events are controlled by different environmental cues (Rietema 1982, 1984). Production of the initials in this species depends solely on photoperiod: Short days (long nights) are required. Outgrowth of the initials also requires short days, but in addition the

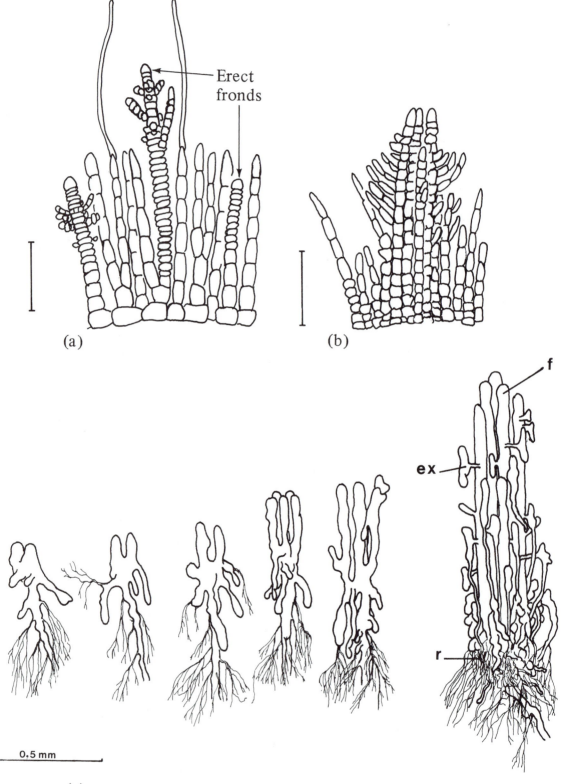

(a)

(b)

Erect
fronds

f

ex

r

0.5 mm

(c)

temperature must be 16°C or lower. The formation of erect multiaxial thalli in *Codium,* which has a feltlike filamentous juvenile stage, involves the coalescence of the filaments, a process that Ramus (1972) found to require water motion. *Codium fragile* filaments twist together into knots that develop polarity and thus become primordia (of macrothallus initials). If primordia are kept in shaken cultures, they will develop the characteristic adult thallus structure, but if placed into calm water, they will revert to nonoriented filamentous growth.

The larger seaweeds, especially Laminariales and Fucales, have several different tissue and cell types, including photosynthetic epidermis, cortex, medulla, sieve tubes, and mucilage ducts. There is also evidence for a certain amount of specialization over a thallus, at least in the larger browns (Fagerberg et al. 1979; Arnold & Manley 1985; Kilar et al. 1989). For instance, studies by Fagerberg et al. (1979) of the morphological and physiological differences between stipes and blades of *Sargassum* species showed that blades have relatively more epidermis and cortex, whereas stipes have more thick-walled medulla. There are also greater areas of thylakoid and mitochondrial cristae per unit volume of blades than of stipes, and correspondingly greater rates of photosynthesis and respiration. Many differences can be seen among cell types in cytological studies of these and simpler brown algae (e.g., Gaillard et al. 1986; Clayton & Shankly 1987; Katsaros & Galatis 1988).

Although coenocytes such as *Caulerpa* and *Bryopsis* are technically single cells, there are differences between regions of their cytoplasm, allowing the same kinds of differentiation that occur in multicellular thalli. Not all nuclei look alike (e.g., in *Cymopolia barbata;* Liddle et al. 1982), and undoubtedly there are many further differences that are not ultrastructurally obvious. Further, cytoplasm and organelle movements are under cellular control. In *Caulerpa prolifera* there is a concentrated "meristemplasm" at the growing tips (Dawes & Lohr 1978), and there are diurnal movements of cytoplasm between shoot and rhizome (Dawes & Barilotti 1969).

Filamentous thalli can be prostrate, erect or heterotrichous (i.e., both). The ratio between prostrate and erect filaments has been used as a taxonomic criterion in groups such as Ectocarpales, but such ratios are envi-

Figure 1.45. Development of erect thalli from crustose bases or microthalli. (a) Formation of uniaxial frond in *Gloiosiphonia capillaris.* (b) Formation of multiaxial macrothallus in *Platoma bairdii* (Scales a,b: 25 μm.) (c) Stages in the development of *Penicillus capitatus* from the free siphonous *Espera* stage (cf. *Halimeda* in Fig. 1.26d). Several erect filaments (f) form lateral expansions (ex) that grow upward and downward, resulting in a multiaxial stipe. (Parts a and b from Dixon 1973, with permission of the author; c from Meinesz 1980, *Phycologia,* with permission of Blackwell Scientific Publications.)

ronmentally variable (Russell 1978). Plants with reproductive structures have been assumed to be full-grown, but plants can reproduce over very wide ranges of form, size, and relative proportions of parts. Because species of Ectocarpales are opportunistic, the timing of their reproduction, and therefore their size at maturity, is liable to be quite flexible; indeed, their life cycles overall are very flexible (Wynne & Loiseaux 1976). Young plants or microscopic stages of larger genera can easily be mistaken for full-grown specimens of the smaller genera (as the genera are currently conceived).

Branching is a characteristic developmental step in many algae and is important in establishing the final morphology of a plant (Waaland 1990; Coomans & Hommersand 1990). Branching patterns are consistent enough in many species (e.g., among Ceramiales) to be used as taxonomic criteria. In some species (e.g., of *Sargassum;* Chamberlain et al. 1979), branching is under apical control. Two patterns of branching in main axes are monopodial and sympodial. In monopodial branching, the primary axis is maintained, whereas in sympodial branching the apex of the main axis is continually replaced by lateral axes that become temporarily dominant. Such patterns are used taxonomically, as, for instance, to distinguish genera in Chordariaceae and to separate Dasyaceae from other families of Ceramiales (Bold & Wynne 1985). Two types of sympodial branching have been distinguished by Norris et al. (1984). However, the resulting appearance of the axes may not be a reliable guide to the ontogeny, as shown by Klemm and Hallam (1987) in a detailed study of *Cystophora* species. Apical cells in these fucoids are lenticular in long view and divide longitudinally to form a short series of cells to one side and then a series to the other side (Fig. 1.46a–c). As these daughter cells divide further, they form branches that initiate new apical cells, but the original apical cell continues to form the main axis (Fig. 1.46d,e). This "swinging" of the main apex gives a shape that looks sympodial but is actually monopodial.

In many other plants, branching is irregular and is controlled as much by environment as by genotype. *Enteromorpha* species are particularly variable in response to salinity (Norton et al. 1981; Pringle 1982). In *Ascophyllum nodosum,* branch initials are formed, some of which grow out into vegetative or reproductive lateral shoots, while others remain dormant. This pattern, reminiscent of buds in flowering plants, suggests internal control. In other large seaweeds, such as *Egregia* and *Macrocystis,* branches (new fronds) arise when lateral blades resume indeterminate growth; again there is the suggestion of internal control in the timing of these outgrowths, but no direct evidence.

Pneumatocysts (gas vesicles) provide buoyancy for a number of the large brown algae (Dromgoole 1990). They develop in various positions on the frond, but all are essentially hollows in the medulla of the stipe

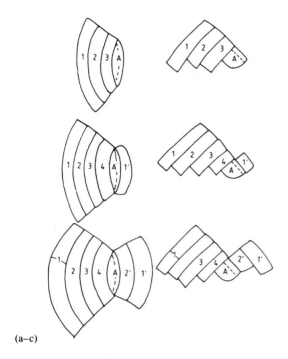

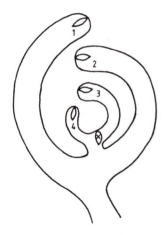

(a–c)

Figure 1.46. Apical-cell divisions and axis development in *Cystophora*. (a–c) Diagrams of apical-cell divisions as seen in long view (left) and cross section (right), showing sequence of divisions first to one side (1, 2, 3, 4) and then to the other (1′, 2′). (d, e) Diagrams of frond tips showing "swing" of main apex (×) as a result of sequential division series. New apical cells in the branches (1–6) arise de novo. (From Klemm & Hallam 1987, *Phycologia*, with permission of Blackwell Scientific Publications.)

(d)   (e)

or stipe/lamina. Some plants produce one large pneumatocyst (e.g., *Nereocystis*); others produce numerous small vesicles, sometimes on special lateral branches (e.g., *Sargassum*). In the austral fucoid *Durvillaea,* air chambers form a honeycomb within the thallus. Oxygen and nitrogen, in roughly the same proportions as in air, form the bulk of the gas, but there are also small, variable amounts of CO_2 and, in those kelps with a single large pneumatocyst, carbon monoxide (CO) (Foreman 1976; Dromgoole 1990). The O_2 and CO_2 derive partly from the metabolic activities of the cells in the pneumatocyst wall, and diurnal changes in the composition and pressure of pneumatocyst gases have been shown in *Carpophyllum* (Dromgoole 1981). However, equilibration takes place between the gases in the pneumatocyst and

in the surrounding water (or air); this is the source of the nitrogen in the vesicles and the major source of O_2 and CO_2 (Hurka 1971). The CO probably is a by-product of degradative metabolism involved in the formation of the pneumatocyst, and its concentration quickly diminishes when pneumatocyst growth ceases. Although pneumatocyst morphology may change with environment (e.g., pressure), critical experiments on causes and effects are still lacking, and the relationship between form and function is largely speculative (Dromgoole 1990).

Morphogenesis can be affected by several environmental factors, including light, nutrients, gravity, and herbivory. Light quantity can have marked effects on branching and elongation patterns (e.g., Fig. 1.2), and it causes differences in the assimilator morphology

Table 1.7. *Nonphotosynthetic effects of blue light on marine macroalgae*

Description of response	Genus
1. Photoorientation responses:	
Induction of polarity in germinating zygotes	*Fucus*
Negative phototropism of haptera	*Alaria*
Negative phototropism of rhizoids	*Griffithsia*
Chloroplast displacement	*Dictyota, Alaria*
2. Effects on carbon metabolism and growth:	
Stimulation of protein synthesis and mobilization of reserves	*Acetabularia, Dictyota*
Stimulation of dark respiration	*Codium*
Stimulation of uridine diphosphate glucose phosphorylase	*Acetabularia*
3. Effects on vegetative morphology:	
Induction of two-dimensional growth	*Scytosiphon*
Induction of hair formation	*Scytosiphon, Dictyota, Acetabularia*
4. Effects on reproductive development:	
Stimulation of cap formation	*Acetabularia*
Induction of egg formation	*Laminaria, Macrocystis*
Stimulation of egg release	*Dictyota*
Inhibition of egg release	*Laminaria*
5. Photoperiodic effects:	
Blue light alone effective as night break	*Scytosiphon*
Blue and red light effective as night break	*Ascophyllum, Rhodochorton*
Blue light effective as day extension	*Acrosymphyton*

Source: Dring (1984b), with permission of Springer-Verlag, Berlin.

of *Caulerpa racemosa* that have been given varietal status (Peterson 1972). Specific wave bands of light also can affect morphogenesis. Preliminary evidence has been adduced that red or far-red light has effects on the growth of kelp stipes (*Nereocystis leutkeana* and *Laminaria saccharina;* Duncan & Foreman 1980; Lüning 1981b). Red light causes specimens of the red alga *Calosiphonia vermicularis* to grow shorter and bushier than they do under white or blue light (Mayhoub et al. 1976).

Blue light seems to be generally required for normal development of seaweeds, so far as is known. In work on *Fucus* species, McLachlan and Bidwell (1983) even found that light in the range 575–625 nm (provided by orange and yellow-green filters) was deleterious to excised apices and to embryos after about 2 weeks of growth. This was not a photosynthetic problem, because tissues survived darkness without harm. But longer and shorter wavelengths allowed continued growth and prevented necrosis. The basis for the importance of blue light is not understood. In phytoplankton and other plants, blue light generally promotes synthesis of protein, RNA, and DNA, whereas red light promotes carbohydrate synthesis (Voskresenskaya 1972; Raven 1974). Normal growth obviously depends on a balance between these two extremes.

Several photomorphogenetic effects are due to blue light (Table 1.7) (Lüning 1981b; Dring 1984b;

Schmid 1984). Some blue-light effects may be indirect; indeed, the fact that there seems to general improvement of growth suggests that the effects are complex. For example, hair production in these cases may be a response to the increased nutrient demands of vigorous growth. Nutrient shortage is known to cause hair formation in several species, including *Acetabularia acetabulum, Ceramium rubrum, Fucus spiralis,* and *Codium fragile* (DeBoer 1981; Norton et al. 1981; Benson et al. 1983). The hairs are a means of increasing the absorptive surface area of the thallus. However, blue light and red light directly affect hair formation in *Acetabularia acetabulum* (Schmid et al. 1990). If this species is grown in red light, no hairs form, and growth gradually slows. If a pulse of blue light is given and then growth in red light is continued, hair whorls are produced. Blue light induces the response; the red light is used solely in photosynthesis, and there is no evidence for a red/far-red receptor.

A role of nutrition in morphogenesis has been shown in *Petalonia fascia* (Hsiao 1969) and *Scytosiphon lomentaria* (Roberts & Ring 1972). In *Petalonia* from Newfoundland, Hsiao found that zoospores from plurilocular sporangia on the blade could form protonemata (sparsely branched uniseriate filaments), plethysmothalli (profusely branched filaments), or *Ralfsia*-like thalli, any one of which could reproduce itself via zoo-

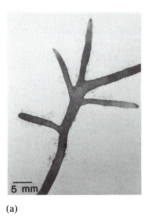
(a)

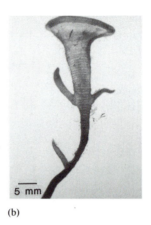
(b)

(c)

Figure 1.47. Morphological plasticity in *Padina jamaicensis.* (a) In heavily grazed areas, a prostrate, branching thallus with single apical cells forms a dense turf. (b) After grazing is reduced, the typical fan-shaped, calcified thallus begins to form; a row of apical cells develops along the tip of the thallus. (c) Foliose form on older turf after 8 week of reduced herbivory. The fan-shaped blades have produced concentric rings of tetrasporangia (t) on their upper surfaces. (From Lewis et al. 1987, with permission of the Ecological Society of America.)

spores or give rise directly to the blade. Protonemata and plethysmothalli survived in iodine-free medium, but formation of *Ralfsia*-like thalli or blades requied iodine (5.1 mg L^{-1} and 508 μm L^{-1}, respectively). Plethysmothalli formed blades in progressively shorter times as the iodine concentration increased. Roberts and Ring (1972) found that changes in the proportions of filamentous and crustose microthalli correlated with nitrogen and phosphorus levels.

Growth of kelp haptera is oriented by negative phototropism, not geotropism, with blue light being the most strongly orienting part of the spectrum (Buggeln 1974). Thigmotropism takes over when the elongating hapteron touches the substratum (Lobban 1978). Several other examples of phototropism were reviewed by Buggeln (1981). Orientation of unicellular rhizoids is more rapid and easier to interpret than orientation of multicellular haptera. Unilateral irradiance is detected by some pigment as yet unidentified. The information can be stored for several hours, with the response exhibited in subsequent darkness. However, gravity may be the stimulus for rhizoid orientation in *Caulerpa prolifera*. When rhizomes are inverted, rhizoid initiation is preceded by a movement (sinking) of amyloplasts toward the lower side, and rhizoid initials contain numerous amyloplasts (Matilsky & Jacobs 1983). Amyloplasts play a role in root-tip orientation in angiosperms and may interact with the movement of growth substances (Salisbury & Ross 1985).

A morphogenetic switch induced by fish grazing has been found in *Padina jamaicensis* (Lewis et al. 1987). When grazing is intense, plants grow as uncalcified, straplike, creeping branches formed from single apical cells. In the absence of herbivory, a marginal row of apical cells forms, and the typical erect, calcified, fan-shaped thallus develops (Fig. 1.47).

Morphogenesis depends partly on attachment, if only because orientation depends on consistency in the direction of environmental cues. Unattached seaweeds may remain in place as loose-lying plants, with little change in morphology, if there is little water movement. If there is extreme water motion, plants will be tossed ashore. Moderate water motion, if it tumbles the thalli, can lead to growth in all directions to form balls, a habit technically called aegagropilous (Norton & Mathieson 1983). Such plants are often distinct from their attached counterparts, as may be see in the characters of *Ascophyllum nodosum* (Table 1.8). Aegagropilous forms of coralline algae (''rhodoliths'') are harvested as maerl in Europe (Johansen 1981). Freshwater *Cladophora* balls (*marimo*) are living national monuments in Japan. Species such as *Chondrus crispus* and *Gracilaria tikvahiae* in cultivation tanks also form balls. Many filamentous algae form hemispherical tufts that are restricted by the substratum from growing downward; when free, these will easily form balls. As balls develop from fragments, abrasion and grazing damage will tend to increase their compactness by promoting regeneration and proliferation (Norton & Mathieson 1983).

Free-floating populations of all kinds are remarkable for reproducing almost solely by fragmentation (vegetative propagation). This is equally true for plants whose reproductive structures are complex (fucoids) or simple (*Cladophora*). The reproductive structures typical of attached plants are rare and also have not formed in culture, even with conditions in which detached specimens have become fertile. The reasons for sterility are not clear; Norton and Mathieson (1983) discussed various possibilities, of which the ''most intriguing'' was that unattached algae become locked into a juvenile state, in which they are unresponsive to environmental reproductive triggers. Fragmentation may occur simply

Table 1.8. *Differences between attached* Ascophyllum nodosum *and the aegagropilous ecad* mackaii

Attached *Ascophyllum nodosum*	Ecad *mackaii*
Attached to rock by a large basal holdfast	Unattached; no sign of holdfast
Thallus flattened, branching in one plane	Thallus ± terete; plant globose; branching in various planes, but radially arranged
± Regular seasonal periodicty of growth, dichotomy, and bladder production	Apical branching frequent but not seasonal; gas bladders infrequent or absent
Lateral nodes produced seasonally give rise to branches for receptacles	No lateral nodes produced
Receptacles rounded and undivided, borne on lateral branches; gametes mostly viable	Receptacles absent, or, if present, pointed, often divided, borne apically; few gametes viable
Apical cell four-sided, surrounded by promeristem	Apical meristem three-sided; promeristem absent
Peripheral meristematic epidermis produces cortical tissue and increases thallus girth	Peripheral epidermis not meristematic
Epiphytic *Polysiphonia lanosa* often common	*Polysiphonia lanosa* rare or absent

Source: Norton and Mathieson (1983), with permission of Elsevier Science Publications.

by breakage of large balls or following senescence near the center. In the case of a *Pilayella* population studied by Wilce et al. (1982), fungal infection by the chyrid *Eurychasma dicksonii* played a key positive role.

1.7.3 Algal growth substances

Growth is an oriented process: Polarities in cells and thalli are established from the start and are maintained throughout development. Apical dominance, the influence of the apical meristem in controlling shoot development, is well known in vascular plants, where it is due to growth substances, especially auxins, produced by the shoot apex. Presumably, apico-basal polarity and apical dominance in seaweeds are also caused by growth regulators.

The term or terms to be used for such growth regulators have generated much debate, as may be seen from the dialogue between Bradley (1991) and Evans and Trewavas (1991). This is not merely a semantic argument, because much inference is based on analogy: By assigning a growth regulator to a class of compounds, we endow that regulator with the characteristics of the class. The word "*hormone*," orginally coined for animal morphogenetic substances, implies substances produced in specific sites (glands) and having specific targets. In plants, this term is best restricted to compounds like rhodomorphin (sec. 1.7.4) and the sex attractants (sec. 1.5.4). Higher-plant *growth substances* are produced in unspecialized tissues and work together on many aspects of morphogenesis. External chemicals required for growth, such as vitamins and the compounds required for cell adhesion in *Ulva* and *Monostroma*, are not considered to be growth regulators, although like nutrients and light they certainly af-

fect growth. On the other hand, growth substances may be produced by epiphytic bacteria or fungi, rather than by the seaweed – virtually all studies have used wild plants or bacterized cultures, rather than axenic cultures. In principle, as Buggeln (1981) suggested, the origin of the substance is unimportant provided it has a specific action on seaweed growth.

Many studies have looked at the effect of applying higher-plant growth regulators, such as auxins and cytokinins, to seaweeds. Many other studies have attempted to extract and characterize seaweed compounds with growth-regulatory effects. Augier published a comprehensive review of such studies in a series of papers (e.g., Augier 1978). Several reviewers have pointed out the problems with such studies, most recently Bradley (1991) and Evans and Trewavas (1991).

Various procedures can be used to identify compounds; those that allow identification of chemical structures are considered best [e.g., gas chromatography with mass spectrometry (GC-MS) or nuclear magnetic resonance (NMR)], whereas GC alone or high-performance liquid chromatography (HPLC) will give "presumptive evidence" (Bradley 1991). With such chemical tests, auxin, abscissic acid (ABA), and cytokinins have been identified in seaweeds (Table 1.9). For instance, Jacobs et al. (1985) identified auxin in *Caulerpa paspaloides,* where it apparently mediates the growth of new rhizoids when rhizomes are reoriented (sec. 1.7.4). However, none of these techniques can show that the compound is active as a growth substance, nor could they allow recognition of a growth substance that did not fall into one of the classic substance groups. Critical questions in demonstrating a growth-regulatory role include the following (Bradley 1991):

Table 1.9. *Identification of plant growth substances in marine algae*

Type of growth substance	Seaweeds(s)	Method of identification
ABA	*Ascophyllum nodosum*	GC-MS
	Ulva lactuca	GC-MS
Auxin (IAA, PAA[a])	*Undaria pinnatifida*	MS
Auxin (PAA and a phenolic compound)	*Enteromorpha compressa*	GC-MS
Auxin (PAA)	*Pelagophycus porra, Sargassum muticum*	GC-MS
Auxin (IAA); ABA; cytokinins	*Ascophyllum nodosum*	GC
Auxin (IAA)	*Caulerpa paspaloides, C. prolifera*	GC-MS, HPLC
Auxin	*Prionitis lanceolata*	NMR
Cytokinins	*Durvillaea potatorum*	GC-MS, HPLC
Cytokinins	*Porphyra lanceolata, Sargassum muticum*	GC-MS
Cytokinins	*Valoniopsis pachynema, Caulerpa taxifolia, Udotea indica*	MS, NMR

Source: Bradley (1991), with permission of *Journal of Phycology.*
[a] PPA, phenoxyacetic acid.

—Is the compound present in the responsive part of the plant, rather than throughout?

—Is the reaction produced by a specific compound or a group of compounds?

—Does removing the compound lead to abnormal growth, and adding it back restore normal growth?

One potential means for demonstrating specificity would be the use of compounds having structures similar to that of the putative growth regulator, but having, in higher plants, no growth-regulatory activity or producing opposite effects. The specificity for indole-3-acetic acid (IAA) can be tested with a number of anti-auxins, including (2,4-dichlorophenoxy)acetic acid (2,4-D), 3,5-D, and naphthoxyacetic acids (NOAA). It was found that growth inhibition of *Alaria esculenta* blades by these compounds could not be distinguished from the response to IAA, and so the response in this plant probably is not auxin-specific (Buggeln 1976).

Several questions confound the interpretation of experiments in which seaweeds are exposed to higher-plant growth regulators in nature or in culture. Are the substances ever taken up by the seaweeds, or are they removed by microorganisms? If they are taken up by microbes, do these organisms release other substances, such as vitamins, that affect seaweed growth? If the growth regulator is taken up by the seaweed, a negative response to it may mean that there was already an optimal concentration of it in the alga; on the other hand, a positive response does not prove that the specific compound is a native regulator in the seaweed. In many of the earlier experiments, the doses applied were very high, raising the question of toxicity.

Nevertheless, some experiments on adding growth regulators to seaweed are relevant to our earlier discussions. The auxin IAA at 1 ppm inhibited growth of excised lateral fronds of *Sargassum muticum,* suggesting that this compound might be the agent of apical dominance in this alga, as it is in higher plants (Chamberlain et al. 1979). IAA has also been shown to promote rhizoid formation in several seaweeds in culture, among them *Bryopsis, Ulva,* and *Caulerpa* (Moss 1974a). However, a substratum of sand rich in microorganisms was just as effective as added IAA in promoting rhizoids. A cytokinin, kinetin (6-furfurylaminopurine), induced *Ectocarpus fasciculatus* in axenic culture to produce erect axes; without the cytokinin, only prostrate axes were produced (Pedersén 1973). Specificity for particular chemicals was not established in any of these experiments.

1.7.4 Wound healing and regeneration

Thallus damage is a fact of life for plants. For seaweeds, the major sources of injury are herbivores, sand abrasion, and wave forces. Plants must at least be able to heal the injury; in many cases, regrowth occurs in seaweeds, as in land plants. Different events are necessary for wound healing in multicellular seaweeds and in coenocytic seaweeds. In cellular seaweeds there is no need for the cut cells to recover, and sealing of the wound involves changes in the underlying cells. In siphonous algae, rapid retraction of the cytoplasm often is accompanied by new wall formation.

Defense against wounding in siphonous algae involves several organizational levels, as reviewed by Menzel (1988). First, at the structural level, thallus architecture (Fig. 1.1) provides reinforcement and flexibilty to giant cells. Second, these algae are chemically defended (sec. 3.2.3), which reduces herbivore attacks and microbial invasion. Third, the cytoplasmic organi-

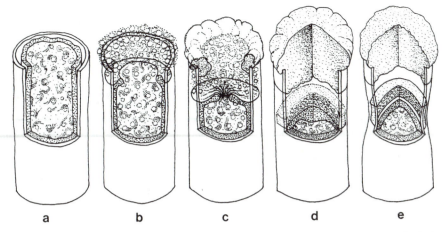

a   b   c   d   e

Figure 1.48. Wound healing in *Bryopsis*. Cutaway diagrams show changes in cell contents in the hour following a wound. (a) Undamaged siphon has a peripheral layer of cytoplasm and a large central vacuole filled with plug precursor material. (b) At 15–30 s after the siphon is cut, the cytoplasm begins to retract and form a concentric closure. Plug precursor is expelled; it swells and adheres to the edge of the cut wall. (c) After about 1 min, the cytoplasmic contraction is almost complete; the plug precursor coagulates, and the wound plug begins to form. (d) At 5–10 min after wounding, the wound plug begins to develop internal and external layers. (e) Within an hour, new cell wall has formed under the internal plug and begins expanding. (From Menzel 1988, with permission of Springer-Verlag, Berlin.)

zation includes an active, motile cytoskeleton for almost instantaneous repair of damage to the cytoplasm. Finally, at the biochemical level, the production and distribution of polymeric materials allow rapid plugging of the wound and repair of the wall. The basic sequences of events are much the same in the various families, but there are differences in the details of timing and plug composition.

In most siphonous green algae (i.e., Dasycladales and Caulerpales), the sequence of events is as follows (Fig. 1.48) (Menzel 1988): The membranes (tonoplast and plasmalemma) are repaired extremely quickly (within 1–2 s), and the cell begins to restore the ionic balance (especially Ca^{2+}) upset by the rupture. Restoration of turgor pressure usually takes only 10–30 s. The cytoplasm at the wound contracts and is pulled away from the wound by actin microfibrils. Plug materials that were stored in the vacuole (cell lumen) are extruded; they stick to the cut surface and polymerize to form a solid plug. The material differs from one family to another; in some it is mostly protein (e.g., in *Bryopsis*); in others, polysaccharide (e.g., *Caulerpa*); in still others, perhaps a mixture. Phenolic compounds, such as coumarins, with peroxidase and perhaps other enzymes, are also involved. The basis for solidification is cross-linking, regardless of the type of material, but the mechanism is not known. Release of the plug precursors seems to mix two or more components together, like making epoxy glue or polyurethane foam.

A different process is seen in Siphonocladales, an Order that is characterized by unique segregative cell divison (e.g., *Ventricaria*, *Ernodesmis*, *Boergesenia*, *Valonia*) (La Claire 1982a,b). In most cases, no plug is

formed; rather, the cytoplasm retracts from the wound – again the work of actin microfibrils (La Claire 1989a) – and then closes around the central vacuole, in one or a few pieces, or breaks up into many protoplasts. The latter process looks much like segregative cell division and the production of gametangia. Cells of one species, *Valonia aegagropila*, surprisingly do not recover from wounding.

Wound healing in multicellular algae has been most thoroughly studied in *Fucus vesiculosus* (Moss 1964; Fulcher & McCully 1969, 1971), *Sargassum filipendula* (Fagerberg & Dawes 1977), and *Kappaphycus alvarezii* (Azanza-Corrales & Dawes 1989). In the fucoids, the thin, perforated cross-walls of the medullary filaments are plugged after about 6 h with newly synthesized sulfated polysaccharide (presumably fucoidan). Later there is general accumulation of polysaccharide at the wound surface. Medullary cells adjacent to the damaged cells round off and become pigmented. After about a week they give rise to lateral filaments, which elongate and push through to the wound surface, where they branch repeatedly to form a protective layer. According to Fulcher and McCully (1969), these filaments are short-lived and are full of antibiotic polyphenolics. Cortical cells undergo longitudinal division (parallel to the wound surface), and the outer cells assume the cytological and functional characteristics of epidermal cells (e.g., they become pigmented). Cells of the medulla may also contribute to the formation of new epidermis. There is no formation of undifferentiated callus tissue.

Eucheuma is in a different Division and has multiaxial pseudoparenchymatous growth, and yet the stages of its recovery from wounding are remarkably

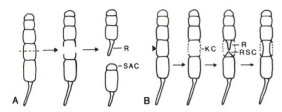

Figure 1.49. Cell regeneration (A) versus cell repair by cell fusion (B) in *Griffithsia*. When the filament is severed, a rhizoidal cell (R) and a new-shoot apical cell (SAC) form, and two separate filaments develop. If an axial cell is killed (KC), the rhizoidal cell fuses with a repair-shoot cell (RSC) and makes a new living link in the filament. (From Waaland 1989, with permission of John Wiley/Alan R. Liss Inc.)

similar to those of the parenchymatous fucoids (Azanza-Corrales & Dawes 1989): Cut cells lose their contents, while proteinaceous and phenolic substances accumulate at the pits of cortical and medullary cells just below the cut. After a few days, cellular extensions begin to grow from underlying cells, proliferate, and form a layer of new, pigmented cortical cells below the wound.

Wound healing is commonly followed by either regeneration or proliferation. The simplest kinds of regeneration involve uniseriate (branched or unbranched) filaments having apical growth. After the wound has healed, growth continues, as in the example of *Sphacelaria* (Fig. 1.43). The physiologically best known cases of wound healing and regeneration are in *Anotrichium tenue* (*Griffithsia tenuis*) and *Griffithsia pacifica* (Waaland & Cleland 1974; Waaland 1975, 1989, 1990): If filaments are severed, a rhizoid is produced from the base of the apical portion, and a new apical cell is regenerated on the basal portion (Fig. 1.49A). If, instead, an axial cell is killed, but the wall remains intact, the filament repairs itself (Fig. 1.49B). A regenerating rhizoid is produced by the apical fragment, and a special repair-shoot cell, not an apical cell, is produced by the basal fragment. This repair-shoot cell is induced by species-specific hormones called rhodomorphins, which diffuse out of the regenerating rhizoid. The repair-shoot cell grows toward and fuses with the regenerating rhizoid. Intraspecific fusions have worked between haploids and diploids, males and females, but no repair can be induced between fragments of different species. The structures of rhodomorphins have yet to be elucidated, but the rhodomorphin from *G. pacifica* has been shown to be a glycoprotein of molecular weight about 17,000, probably with a terminal mannose residue (Watson & Waaland 1983, 1986). Glycoprotein hormones are common in animals, but rare in plants (the sexuality inducer of *Volvox* is also a glycoprotein) (Waaland 1989). Evans and Trewavas (1991) have argued that recognition factors such as rhodomorphin and the glycoprotein incompatibility factor in *Brassica* species (mustards) are not

hormones or plant growth substances, because they are species-specific.

Some algae produce proliferations from cut surfaces; these are lateral outgrowths of cortical filaments, as in red algae *Schottera* and *Gigartina* (Perrone & Felicini 1972, 1976) and the brown *Dictyota* (Gaillard & L'Hardy-Halos 1990). The type of tissue produced in the reds – rhizoidal or bladelike – depends on the position of the wound with respect to the apex or base of the thallus. In other words, there is a correlation with an internal thallus polarity. Proliferations in *Schottera nicaeensis* clearly show these correlation effects (Fig. 1.50a) (Perrone & Felicini 1972). In general, leafy outgrowths arise from the apical sides of cut surfaces, and rhizoidal outgrowths arise from basal sides. When the apex of the thallus in intact, horizontal cuts near the base give leafy proliferations; near the apex they give rhizoids. Regeneration of thallus segments results in leafy outgrowths from the apical end, and rhizoids from the basal end. In *Dictyota* the apical cell and the base of the thallus control the positions, numbers, and sizes of adventive fronds (Gaillard & L'Hardy-Halos 1990). ("Adventive" signifies that the fronds arise from an existing frond rather than from a zygote.)

In *Caulerpa*, excised "leaves" regenerate rhizomes and rhizoids from the basal end and new leafy shoots from the apical end (Fig. 1.50b). Rhizomes regenerate first rhizoids from the apical end, and later rhizoids from the basal end plus rhizome and leafy shoots from the apical end (Jacobs 1970). In Jacobs's experiments, "leaf" segments 30 mm long formed only rhizoids; if 40 mm long, half the specimens also formed a rhizome and a new leaf; if 50 mm long, all regenerated completely. However, leafy-shoot production and rhizoid production from the rhizome of *Caulerpa* also respond to gravity, as shown by Jacobs and Olson's (1980) experiments, in which uninjured thalli were turned upside down. Rhizoids were produced from the new lower side, and leafy shoots from the new upper side (the rhizome did not twist, so polarity had been reoriented).

The process of regeneration in *Fucus* is unusual in that distinct embryos, rather than lateral branches, are formed. During the process of wound healing in this genus, epidermal cells in certain regions of the wound begin to divide perpendicular to the wound surface, forming groups of branch initials (visible macroscopically after 4–6 weeks in culture), which develop directly into adventive embryos (Fulcher & McCully 1969, 1971). The midrib region of the thallus regenerates much more rapidly than the wings (Moss 1964), correlating with the abundance in the midrib region of medullary filaments, which are primarily responsible for formation of new epidermis. Regeneration from vegetative branches always gives rise to vegetative shoots. Regeneration of strips cut from the discolored frond beneath spent receptacles of the dioicous species *F. vesiculosus*, although extremely slow, results in branches

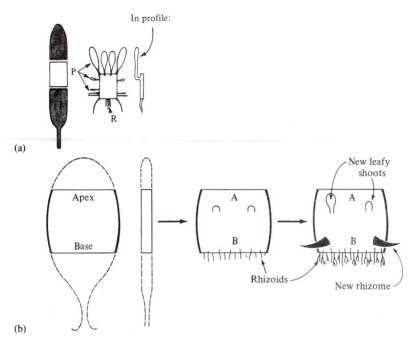

Figure 1.50. Polarity and regeneration in larger seaweeds. (a) Bladelike proliferations (P) and rhizoidal branches (R) in the red alga *Schottera nicaeensis*. (b) Regeneration from a portion of a "leaf" of *Caulerpa prolifera*. In both cases, leafy shoots form at the original apical end, and rhizoids form at the basal end. (Part a from Perrone & Felicini 1972, with permission of Blackwell Scientific Publications; b from Jacobs 1970, with permission of the New York Academy of Sciences.)

with small receptacles at their tips. Branches regenerated from strips cut from male thalli bear male receptacles, and those from female thalli bear female receptacles (Moss 1964).

This concludes our survey of the structure, development, and life history of seaweeds as individuals. In the following chapters we shall examine the communities and habitats in which they live, the biotic factors they face, and the ways in which they are affected by abiotic factors.

1.8 Synopsis

Benthic ocean vegetation is dominated by multicellular Chlorophyta, Rhodophyta, and Phaeophyta, together with some Cyanophyta and colonial algae of similar functional form. The term "seaweeds" represents an ecological grouping of disparate taxa. Moreover, the microbenthos includes reproductive cells and early stages of seaweeds, as well as microalgae. Seaweeds of all Divisions show a range of morphologies, including filamentous, pseudoparenchymatous, and parenchymatous. Their anatomy ranges from virtually no differentiation between cells to the complex tissues of kelps and fucoids. Seaweed cells differ from the cells of "higher" plants in general by their broader range of metabolic functions. Some special features of algal cells have to do with their wall chemistries, the variety of chloroplast structures and pigmentation, the different

arrangement of flagella in motile cells, and the details of cell division. Seaweed genetics, especially using color mutatnts in some red algae, have shown Mendelian and non-Mendelian inheritance. The techniques of molecular biology are being applied to seaweeds and are showing a diversity that reflects the range of Divisions and Kingdoms represented.

Seaweed life histories can follow several patterns, depending on the species and the environment. An alternation between two free-living stages – one a haploid gametophyte, the other a diploid sporophyte – is common, but many variations exist. Some seaweeds have dissimilar sporophytes and gametophytes; others have only one free-living stage. There is no direct relation between ploidy level and morphology, and so many variations of life cycles are possible, including changes between microthalli and macrothalli of the same chromosome number.

The life of a seaweed is a complex sequence of interactions between its genetic information and its external stimuli and constraints. Development, from the initial polarization of the spore or zygote to the production and release of reproductive cells, is a highly coordinated process. Light (quality and quantity), photoperiod (usually the length of uninterrupted darkness), and temperature are the principal environmental cues. In contrast to the phytochrome system in higher plants, seaweeds respond to several wave bands of light, using several receptor pigments. Minor roles in morphogenesis may be played by gravity and water motion. Cells in a thallus also receive influences, probably both physical and chemical, from other cells. Cells released from these constraints may exhibit totipotency and regenerate an entire thallus, or some genes may remain repressed so that certain parts (e.g., rhizoids) cannot be regenerated.

Sexual reproduction may be isogamous, anisogamous, or oogamous, although some brown algae with morphologically similar gametes are functionally oogamous. Syngamy is regulated by cell recognition mechanisms on cell/flagella surfaces. Motile gametes in brown algae may be attracted to each other or to a stationary egg by volatile pheromones. In red algae, sexual reproduction often involves complex postfertilization development of a carposporophyte for zygote amplification.

Settlement of spores or other reproductive structures depends a great deal on water motion (turbulence and eddies), notwithstanding the limited capacity of some cells for oriented swimming. Following settlement and attachment, spores must produce a firm holdfast. At first they are susceptible to being resuspended. Germination is an oriented process, allowing for differentiation into rhizoidal cells and thallus-forming cells.

Erect thalli characteristically have an apicobasal polarity, which is expressed in the position and kind of regenerative outgrowths on wounded thalli and sometimes in apical dominance. Growth-regulating substances similar to or identical with higher-plant growth regulators are almost certainly present, although most of the evidence is still circumstantial, and some of these compounds may in fact be produced by epiphytic microorganisms. Vitamins and cell binding factors are also produced by the microbiota. Other substances, such as rhodomorphin, probably are unique to the algae.

Wound healing is an important function in seaweeds, which are continually subjected to damage by grazers and abrasion. In siphonous algae, rapid plugging of the wound takes place to prevent cytoplasm loss. In multicellular algae, cut cells usually die, and wound healing is accomplished by the underlying cells. Regeneration commonly takes place from cut surfaces, with either frondlike or rhizoidlike tissue produced as a function of the distance from the dominant apex.

2

Seaweed communities

2.1 Seaweed communities

Seaweeds exist as individuals, but they also live together in communities with other seaweeds and animals – communities that affect and are affected by the environment. In Chapter 1 we reviewed the morphologies, life histories, and developmental processes of seaweeds as species. In this chapter we consider the patterns and processes in marine benthic communities as a starting point for later factor-by-factor dissection of the environment. We open with overviews of three major habitats and the seaweeds in them: rocky intertidal zone, tropical reefs, and kelp forests. We hope that these personal essays by some noted algal ecologists will also give the reader a glimpse of the phycologist at work and a sense of the excitement of physiological ecology. Near the end of the chapter, three more ecologists tell about some less well known habitats: salt marshes, seagrasses, and the Arctic.

2.1.1 Essay: The rocky intertidal zone
*Trevor A. Norton**

Few habitats are so frequently visited by ecologists as the rocky intertidal zone, for it offers intermittent access to a fascinating variety of organisms. It must be unique, however, in that it is invariably examined when most of its inhabitants are out of their element. The number of ecologists who study the shore at high tide when its residents are active and operational could, I suspect, be counted on the arms of a starfish. This is

* Trevor Alan Norton is Professor of Marine Biology, University of Liverpool, England, and director of the Port Erin Marine Laboratory, Isle of Man. He is also Chairman for Aquatic Life Sciences for the Natural Environment Research Council and President of the International Phycological Society.

a pity, for it is the shore when underwater that is the shore in action (Fig. 2.1).

The term "rocky intertidal zone" may slightly mislead the reader, for shores are rarely composed exclusively of bedrock. Many have pebble-littered gullies or sand-carpeted pools, and below the low water mark the rock often gives way to sand or mud. The proximity of such mobile substrata greatly enhances the abrasiveness and therefore the ecological importance of waves.

Even where stable bedrock predominates, the effective substratum may not be rock at all. The mid-shore region is usually covered with closely packed barnacles, with little rock visible between them. Tide pools are often lined with encrusting pink and purple Corallinaceae, which may also carpet the lowermost levels of the shore. By occupying the rock so comprehensively, these organisms replace it. They become the substratum to which other organisms must attach, and yet little is known about their ecological significance as substrata. Do the propagules of other organisms settle preferentially on some crusts and shun others? Can the crusts shed the settled propagules of some fouling species, but not all? The interactions between these little-studied substrata and other shore dwellers may be major influences on the patterns of intertidal vegetation.

Water motion has long been recognized as a major determinant of intertidal communities. A stroll along the coast will reveal striking differences in the vegetation of exposed promontories and that of sheltered bays. An awakening of interest in biomechanics has demonstrated that seaweeds do not confront the waves, but rather yield to them. Immense mechanical strength is less useful than pliability, elasticity, and an ability to conform to the flow (Norton et al. 1982; Koehl 1986; Denny 1988).

Ecologists talk glibly of exposed or sheltered shores, but even the most wave-battered shores may have some relatively protected places. The drama of the

Figure 2.1. The intertidal seaweed community is not the two-dimensional world it appears to be at low tide. When under water, the plants form an erect and dynamic canopy. *Ascophyllum nodosum* (center and left) and *Fucus vesiculosus* (right) are buoyed up by pneumatocysts. (Photo taken on Isle of Man, U.K., by Tim Hill, © Tim Hill, and reproduced by permission of the photographer.)

surf and the striking gradients caused by waves and tides must not blind us to the myriad of microenvironments that occur on rocky substrata. Variations in topography are particularly influential: The vegetation of a shaded gully is obviously different from that of adjacent sun-baked rock. Crevices that etch the surface offer damp havens in which spores and juvenile plants may find some protection from desiccation and grazing.

The roughness of the rock surface even on a microscopic scale may also be important, for some seaweed propagules are "caught" in tiny depressions, especially those whose depth just exceeds the diameter of the spores (Norton & Fetter 1981). Attachment within such depressions also enhances the germlings' ability to survive water motion, but the optimum depression size for avoiding dislodgement may be different from that which most favors settlement (Norton 1983). Seaweeds almost invariably inhabit the nooks and crannies of the environment, but the luxuriance of their subsequent growth masks the fact that initially they anchored in a crevice or pocket.

Although many seaweeds have quite restricted distributions in the intertidal region, they initially colonize a wider range and are then progressively pruned back to their eventual equilibrium zone by environmental factors (Schonbeck & Norton 1978). The propagules of most seaweeds have little or no control over their destiny. When released into the chaos of the sea, they are at the mercy of waves and currents. Clearly, to disperse far

beyond the zone that they will be able to inhabit is very wasteful of propagules, and some intertidal plants seem able to limit dispersal to a meter or so (Dayton 1973; Deysher & Norton 1982). Propagules of *Fucus* species can colonize up to at least 23 m from the parent plants on a shore devoid of seaweeds (Burrows & Lodge 1950), but the situation in a dense stand of *Fucus* may be different, for propagule extrusion and release occur only following desiccation and rewetting of the receptacles. The surrounding tangle of seaweeds may baffle the first waves of the returning tide, allowing many propagules to settle close by. As yet, we know far too little of the sizes and movements of the clouds of propagules that drift across the shore and give rise to the attached communities (Hruby & Norton 1979; Norton 1992).

The life of intertidal algae is a succession of setbacks. They are regularly abandoned by the sea and left to be dried by the wind. The return of the tide rehydrates them and replenishes their nutrients, but it also brings fouling organisms and mobilizes grazers once more.

Most intertidal seaweeds are truly aquatic plants and do not need to be exposed to air. Mild desiccation may slightly stimulate their photosynthesis, but significant drying usually causes a substantial decline. More important, the growth rate decreases as a result of even mild desiccation, and repeated exposures are even more deleterious than a single, more severe episode (Hodgson 1984). Illuminated intertidal *Fucus* plants grow significantly only when submerged; irradiating them while they are emersed (but unstressed) is ineffective (Schonbeck & Norton 1979c). Growth, not photosynthesis, is the net integration of the plant's physiological activities and a major contributor to its ecological performance.

I know of only one seaweed that requires exposure to air: *Pelvetia canaliculata,* an extreme-high-shore dweller. It decays if transplanted to the lower shore or kept permanently submerged in culture (Rugg & Norton 1987). Is it unique in this respect? Does *Pelvetiopsis limitata* (the Pacific coast equivalent of *Pelvetia canaliculata*) "drown" if kept submerged? When I attempted to test this question experimentally, I found that permanently submerged plants were overwhelmed by epiphytes before a possible physiological decline became apparent. The stresses of life on the upper shore are often emphasized, but the benefits of "desiccation cleaning" are never mentioned.

All other upper-shore species that have been tested benefit from long submergence (e.g., Edwards 1977). The frequently repeated claim that *Fucus spiralis* is an obligate high-shore plant is based on a misinterpretation of the results of a very early experiment. In fact, when the shore is denuded, *F. spiralis* can colonize the lower intertidal zone and thrive there (Burrows & Lodge 1951).

Competition on the shore may be fierce and often sets the lower limits of algal zones. Certainly, removing adjacent species allows plants to invade the space provided, even in a zone from which they are normally excluded (Schonbeck & Norton 1980a). In many ways seaweeds are ideal for studies of competition, for they inhabit a two-dimensional surface and often occur in abutting monospecific stands. However, we rarely know with certainty the resource for which the plants are competing (Carpenter 1990) or the exact mechanism that determines which plants will win. Often the winner grows faster than its rivals, suggesting that overgrowth and shading are the means of conflict (Schonbeck & Norton 1980a), but this may not always be the case. Our knowledge of possible chemical "warfare" between seaweeds is very rudimentary.

The result of competition often is the elimination of the inferior competitor, rather than coexistence. It is not clear why the loser loses so completely. As seaweeds do not etiolate, once overshadowed they have no way of catching up. But is the loser kept below its compensation point until it enters an irreversible physiological decline, or, deprived of growth, does it fall to grazers?

Dense beds of large seaweeds, such as the fucoids of temperate shores and the *Sargassum* species of warmer waters, greatly modify the character of the intertidal zone. They provide a sheltering breakwater when the tide is in and a moist protective blanket when it is out. They also furnish an immense surface area for colonization by epiphytes. Seaweeds do not merely influence the habitat – to a large extent they *are* the habitat.

Seaweed stands provide a home for a variety of herbivorous snails, limpets, sea urchins, and chitons, whose grazing activities can greatly reduce seaweed populations (Hawkins & Hartnoll 1983b). Micrograzers such as amphipods and isopods may also be influential, but have been far less well studied.

Several workers have shown that many herbivores do not feed indiscriminately, but consume some algae in preference to others, and such selective removal may determine which species remain to dominate the habitat (Norton et al. 1990). Most researchers have concentrated on the consumption of macroscopic seaweeds, but many grazers may exert their greatest influence on the vegetation by browsing on the microscopic germlings of macroalgae, eliminating some seaweeds before they become apparent to an observer. To be too small is often to be ignored by ecologists. Intertidal rock sometimes bears a dense film of microalgae and the tiny early stages of seaweeds. Although they cast no more than a bloom upon the rock, the rapid turnover of such small algae means that they may constitute an enormous unseen larder that has been little studied (Hawkins et al. 1989; Voltolina & Sacchi 1990). There is also evidence that some grazers capable of eating both microscopic and macroscopic food may prefer microalgae to larger plants (Jernakoff 1985; Watson & Norton 1985).

Our horizons are set by our experience. We know the shores that we visit regularly, and we tacitly assume that they are "typical." I grew up beside a rocky coast infested with a large variety of grazers: several species of periwinkles, top shells, and limpets. The first time I crossed the Atlantic to explore shores that I thought would be just like those at home, I was surprised to find hoardes of *Littorina littorea* as virtually the only grazer. What to the locals was typical, to me seemed abnormal.

The paradox is that although shores in different parts of the world are washed by quite different tides and baked by a hotter or cooler sun, nonetheless beneath the obvious differences in their dominant inhabitants there run threads of similarity. The Stephensons have conducted the most detailed comparisons of rocky shores worldwide (Stephenson & Stephenson 1972), and 40 years ago they drew attention to the underlying similarities. They emphasized the "universal" occurrence of a barnacle zone and a littorinid snail zone, but they assumed that the seaweeds overlaid this fundamental structure and obscured it. But for seaweeds, too, there are similar niches to be exploited worldwide, and often the same genera and even the same species fill the role in widely separated geographic areas. In many warmer seas the lower intertidal zone is characterized by a dense low turf of red algae, and closely related species dominate this turf in every ocean (Kain & Norton 1990).

The rocky intertidal zone has always evoked a fascination. No matter how many of its mysteries we solve, I suspect that young biologists will continue to be drawn to the magical tidal margins of the sea.   ◆

2.1.2 Essay: Tropical reefs as complex habitats for diverse macroalgae
*Mark M. Littler and Diane S. Littler**

Beneath the vast expanse of warm azure waters, tropical biotic reefs comprise spectacularly complex ecosystems on limestone bases, derived mainly from the fossilized remains of calcareous algae and coelenterate corals. Such reefs occur around the globe within the 22°C isotherms (north and south). Reef systems have evolved an extremely high level of biological diversity, including many uniquely specialized macroalgae. The calcite ($CaCO_3$) cement produced by coralline algae consolidates calcareous (aragonitic) skeletons of coral animals and other calcifiers, along with terrigenous debris, and leads to reef formation. The nonarticulated coralline algae may also form a seaward intertidal ridge that buffers wave shock, thereby reducing erosion and destruction of the more delicate corals and softer organisms typical of reef-flat habitats. A diverse group of calcified green algae deposit the aragonite form of calcium carbonate, which is responsible for much of the sand and lagoonal sediments within the reef-flat and deeper fore-reef areas. For example, skeletal sand-sized components from some tropical Atlantic reef sediments are composed of up to 77% *Halimeda* fragments. Tropical reefs are remarkable for their development of massive structure in conjunction with high primary productivity; algae are responsible for much of the former and all of the latter.

In 1966, while completing degrees at the University of Hawaii, we became intrigued by the challenges of understanding the complex interactions structuring tropical reefs. Our studies of biotic reefs have taken us on adventures to Micronesia (Guam, Palau, Enewetak), the Australian Great Barrier Reef, the Galapagos Islands, Tahiti, Republic of the Seychelles, Kenya, Panama, Brazil, Belize, Mexico, Greater Antilles, Lesser Antilles, French Guyana, Bahamas, Florida, and Bermuda. Most of our ongoing research is centered in the Florida Keys and in Belize where we are investigating algal–animal interactions and the long-term interactions of nutrients and herbivory in reference to the Relative Dominance Model we developed (Fig. 2.3).

---

* Mark and Diane Littler have spent much time studying tropical reefs worldwide. They are cited in the *Guiness World Book of Records* for their discovery of the deepest plant life on earth. In addition to their major contributions to ecological and systematics research on reef algae, they have an interest in underwater photography and have published a color guidebook to Caribbean seaweeds (D. Littler et al. 1989).

There are three major reef types, based on their location – fringing, barrier, and atoll – and all have basically the same ecological zones (Fig. 2.2a). The most seaward portion of a typical reef is the fore-reef slope that grades upward to the reef crest. Where wave action is consistently high, the reef crest develops into an intertidal algal ridge generally dominated by *Porolithon* and *Lithophyllum* (Fig. 2.2a,b). The most massive algal ridges are found on Pacific atolls, although they are present intertidally on any reef system consistently exposed to high wave energy. Shoreward of the algal ridge is the shallow reef flat where limestone-boring organisms rework the calcareous matrix. In this habitat, slower-growing corals, various coralline algae, and frondose algae dominate. The reef flat usually grades upward toward the shoreline to form an intertidal reef platform dominated by Cyanophyta, where storms may cast calcareous sediment, rubble, and boulders. This material accumulates, particularly on windward barrier and atoll reefs, to form low islands (known as cays/keys in the Caribbean, motus in the South Pacific).

Various calcareous and noncalcareous groups of algae tend to predominate within different reef habitats. The relative dominance of frondose algae, calcareous algae, and corals appears to be related directly to biological factors such as competition and grazing, in addition to being influenced indirectly by abiotic factors, including nutrient levels, wave action, irradiance, desiccation, and temperature. Where herbivory is reduced or nutrient levels are elevated, biotic reefs shift from coral to algal domination. Such shifts from coral dominance to fleshy algal dominance have been related to excess nutrient increases and other stresses for reefs off Venezuela (Weiss and Goddard 1977), on the Abaco reef system, Bahamas (Lighty et al. 1980), and in Kanoehe Bay, Hawaii (Banner 1974; Smith et al. 1981). Unfortunately, the effect of modern mankind on tropical reefs has been to decrease herbivorous fishes through netting and trapping while simultaneously adding nutrients via sewage and agricultural pollution. Unless curbed, this anthropogenically induced shift from coral to algal domination on reefs will continue at an accelerating pace.

Noncalcareous algae. Frondose macroalgae normally are rare on reefs because of grazing by herbivorous fishes and sea urchins. Filamentous algae on the shallow fore-reef slope are also kept inconspicuous by intensive grazing in these spatially heterogeneous habitats. Where there is much turbulence or little topographic shelter from higher-order carnivores on tropical reefs, herbivore activity is reduced, and larger standing stocks of macrophytes develop (e.g., *Sargassum, Turbinaria, Acanthophora, Eucheuma*). Deeper sand plains often contain isolated rubble fragments that provide suitable substrata for strikingly attractive frondose gen-

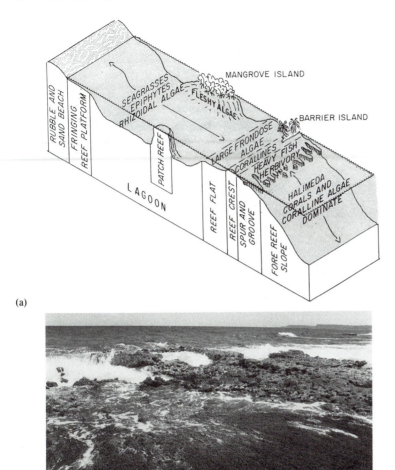

(a)

(b)

Figure 2.2. Coral reefs. (a) Sectional view through a tropical continental shelf containing characteristic barrier reef and mangrove systems. Dominant macrophyte groups are indicated for the various habitats. (b) Photograph of *Porolithon-Lithophyllum* ridge at Pago Bay, Guam. (Part a from D. S. Littler et al., 1989, with permission of Cambridge University Press; b by María Schefter, © 1992, María Schefter.)

era such as *Halymenia, Kallymenia, Dasya,* and *Gracilaria,* which can reach considerable size in these refuges. Chemical defenses among macroalgae reach their greatest diversity and frequency in tropical-reef habitats (Hay & Fenical 1988), and some genera (e.g., *Halimeda, Stypopodium, Laurencia, Dictyota*) often are abundant even where grazing is high. Such algal populations may contribute a major portion of the total primary productivity to some tropical reefs. However, it is the sparse mats of fast-growing, opportunistic filamentous algae (see Fig. 1.15b) that usually are responsible for the very high primary productivity per unit area in most biotic reefs. Proportionately, sparse filamentous mats are considerably more productive per unit of algal biomass than are dense stands of the larger macroalgae,

because of their high surface-to-volume ratios. Herbivorous fishes, by their scraping mode of feeding, continuously provide new substrata and thereby select for opportunistic microalgal forms, as well as long-lived scrape-tolerant coralline algae.

Fixation of atmospheric nitrogen by blue-green algae such as *Calothrix crustacea* (e.g., Wiebe et al. 1975) within filamentous microalgal assemblages also is an important feature that enhances reef productivity and nutrition. The greater productivity of benthic reef communities versus planktonic oceanic systems is in large part due to this nitrogen fixation, as well as to unusually efficient nitrogen and phosphorus recycling within the symbiotic populations (Johannes et al. 1972). Macroalgae and corals may also be closely associated with blue-

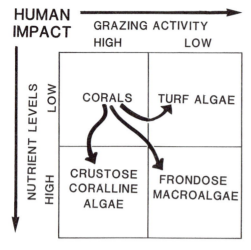

HUMAN IMPACT

GRAZING ACTIVITY

HIGH   LOW

NUTRIENT LEVELS

LOW

HIGH

CORALS   TURF ALGAE

CRUSTOSE CORALLINE ALGAE   FRONDOSE MACROALGAE

Figure 2.3. Diagram of the relative dominance paradigm. Potentially predominant space-occupying groups of primary producers are emphasized as a function of long-term nutrient levels and disturbance. Human activities tend to reduce grazing animals and increase nutrient levels, thereby shifting reefs from coral to algal domination (arrows). (Modified from Littler & Littler 1984).

green algae. These blue-green-algal associations fix nitrogen at rates equal to those recorded for the richest nitrogen-fixing terrestrial systems (e.g., alfalfa fields).

Other important algae in reef ecosystems are erosive agenic species that contribute to the breakdown of reef structure. Such penetrating or boring algae play an important role in bioerosion. The commonest rock-boring algae are Cyanophyta that attack skeletal materials differentially; the aragonitic coral skeletons are more susceptible, and the denser calcitic deposits of coralline algae are more resistant. One systematic study of penetrating algae in the Indo-Pacific recorded 20 species distributed among Cyanophyta, Chlorophyta, Phaeophyta, and Rhodophyta (Weber-van Bosse 1932), and 33 species of carbonate-boring algae have been reported from tropical China (Chu & Wu 1983). Much research remains to be done on the biology of this interesting group of endolithic marine plants.

Calcareous algae. Calcareous algae have long been recognized as predominant contributors to both the bulk and frame structures of the majority of reef limestone deposits. Such deposits often have been associated with petroleum reserves, and this relationship has brought the calcifying seaweeds to the attention of geologists, paleobiologists, and ecologists.

The order of prominence for the reef-forming organisms that provided bulk during the development of the reef at Funafuti Atoll, Tuvalu (formerly Ellice Islands) (8° 30′ S, 179° 10′ E), was as follows: (1) nonarticulated coralline algae, (2) *Halimeda,* (3) foramini-

fera, and (4) corals (Finckh 1904). Subsequent ecological work (e.g., Littler 1971) and paleontological studies (e.g., Easton & Olson 1976) have substantiated the predominant role of coralline algae in cementing coarse and fine-grained sediments produced by calcareous green algae, molluscs, and foraminifera, along with the bulkier deposits provided by hermatypic (reef-building) corals.

Some of the adaptive advantages of calcification in reef algae include mechanical support and minimization of damage from sand scour, wave shock, and herbivory, as well as reduction of fouling epiphytes (by means of carbonate sloughing). Also, by providing their own substrata, calcareous algae may increase the stability and quality of their attachment sites.

The calcifying Rhodophyta grow on solid substrata intertidally and subtidally down to at least 268 m, but reach maximum abundances in shallow, physically disturbed areas. There is evidence that some corallines require physical disturbances such as wave shock or herbivory to prevent their overgrowth by fleshy algae. Coralline algae, in contrast to most fleshy algae, have relatively low primary productivity because of their high structural commitment. Interestingly, calcification rates appear to differ little among reef-flat communities consisting of diverse kinds of calcifiers (Wanders 1976), whether they be corals, nonarticulated coralline algae, or turfs of articulated corallines. Reef-building Corallinales are able to grow at greater depths in weaker light than other primary producers (Littler et al. 1986). *Porolithon* and certain *Lithophyllum* species that dominate algal ridges (e.g., *L. moluccense*) are somewhat exceptional in that they can withstand considerable desiccation and exposure to the highest sunlight irradiances.

The calcareous Chlorophyta predominate mainly in protected shallow areas on soft bottoms (which are unsuited for most other macroalgae), and they occur only in subtropical and tropical regions, often in association with seagrasses. *Halimeda* is also common on the deeper fore-reef slopes. Psammophytic (sand-dwelling) algae such as *Udotea, Penicillus,* and some *Halimeda* species can translocate nutrients from rich sediment-pore waters by means of their unique bulbous rhizoidal systems (Williams & Fisher 1985). Few quantitative studies have been done on any aspect of the ecology of the calcareous Chlorophyta, with the exception of the widely studied genus *Halimeda.* Numerous *Halimeda* species are abundant on protected reef-flat and fore-reef habitats, occurring over a broad depth range on both hard and soft substrata. Other psammophytic forms are associated with shallow seagrass beds and mangroves. Recently, impressive banklike mounds composed of living *Halimeda* and its sediments (dating back to 5,000 years B.P.) have been discovered in back-reef regions of the Great Barrier Reef, Australia (Davies & Marshall 1985).

Deep-water reef algae. Submersible vessels have greatly expanded our knowledge of the distributional limits for marine organisms, but macroalgae have received only incidental attention until recently. The record depth (268 m) for an attached living marine macrophyte was discovered during our own ecological surveys of a seamount off San Salvador Island, Bahamas (Littler et al. 1985). These studies from a submersible, in conjunction with shore-based productivity measurements, revealed unsuspected abundances and potential importances of other deep-water tropical macroalgae. Four zonal assemblages were present on the seamount over the depth range from 81 to 268 m: a *Lobophora*-dominated group (81–90 m), a *Halimeda* assemblage (90–130 m), a *Peyssonnelia* group (130–189 m), and a crustose coralline zone (189–268 m). The zonation pattern observed (i.e., reds > greens > browns with increasing depth) is quite similar to that recorded in Malta by Larkum et al. (1967).

Dominant members of the diverse multilayered macrophyte community on top of the San Salvador seamount (at 81 m) showed net productivity levels comparable to those for shallow-water seaweeds, although receiving only 1–2% of the light energy available at the surface. Deep-water macroalgal communities produce at rates comparable to those for some shallow reef systems, but lower than those for most seagrasses or typical carbonate reef-flat habitats. Calcification rates in deep-water *Halimeda* species are, significantly, similar to those reported for shallow forms of the genus.

We still know very little about the physiological ecology, population biology, and community dynamics of algae that affect the ecology and biogenesis of biotic reefs. Until recently, few workers had directed their efforts toward determining the functional and ecological roles of algae on living reefs. We are at a stage where descriptive (correlative) and mechanistic (experimental/causative) approaches must be combined to produce more conceptual theoretical perspectives, which will accelerate our predictive understanding of algal roles in reef biology. General ecological theories are already being modified as a result of experimental studies of tropical algal biology.

The lure of tropical reefs lies in their unsurpassed natural beauty. There is no terrestrial counterpart to the underwater scenery of a rich biotic reef; the vibrant colors and intricate structures of the plants and animals are unique to the marine environment. Reefs are among the few places where one can observe a complex community of plants and animals interacting naturally, seemingly little disturbed by human presence. A burgeoning awareness of the attractiveness of tropical marine plants as experimental organisms for the elucidation of ecological and reef-building processes offers exciting prospects for the future of reef research.   ◆

2.1.3 Essay: Kelp forests
Paul K. Dayton*

My interest in kelp communities grew from my thesis work in the intertidal habitat where I studied algal ecology. I considered several types of biological relationships in the intertidal communities, including canopy effects and the roles of herbivores and their predators. Competitive dominance was important, but the expression of the dominance was much affected by wave exposure. I became intrigued by kelp forests, which seemed to offer many important parallels with terrestrial forests, while still allowing some of the manipulative opportunities of intertidal systems.

The sublittoral zone on most temperate rocky shores is dominated by kelps – large brown algae of the Order Laminariales – or the morphologically similar fucoids (Figs. 2.1 and 2.4). Both canopy types include fronds suspended in the water column by some form of flotation, fronds supported above the substratum by semirigid stipes, and fronds lying on or immediately above the substratum. Their high productivity and the often-extensive vertical structure formed by their fronds provide food and habitat for many of the species that occur in these regions. Because of this, they are also quite important to human fishermen and divers (Dayton 1985; Chapman 1986; Schiel & Foster 1986). Large kelp forests can considerably reduce alongshore currents and cross-shore water motion (Jackson & Winant 1983).

Like all other populations, kelps are affected by many biotic and abiotic factors. Biotic influences usually include several types of grazers and competition. Kelp systems have many grazers, including polychaetes, arthropods, molluscs, and vertebrates, but they seem to differ from most other plant associations in that the predominant plants can be severely overgrazed by a single type of herbivore (strongylocentrotid sea urchins). The potentially devastating impact of urchins must have a strong influence on evolutionary trends in kelps.

Important abiotic factors include nutrients, appropriate substrata for settlement, and, most important, light. Kelps compete among themselves and with other kinds of algae for limiting resources, especially when they are small. The effect of this competition is highly variable, but it is always potentially there and must be considered. In most situations, one species (usually with floating canopies) can conspicuously dominate the kelp community. One of the most interesting research

* Paul Dayton is a professor of marine ecology at the Scripps Institution of Oceanography of the University of California, San Diego. He has studied kelp forests all over the northeastern Pacific and in Chile, Argentina, Australia, and New Zealand. He also conducts a long-term benthic ecology program at McMurdo Sound, Antarctica.

Figure 2.4. *Macrocystis* fronds form a complex under-water forest that provides a habitat for many other organisms and greatly affects the physicochemical conditions within. Here, the heterogeneity of under-water light is evident (see Fig. 4.2c). (Photo courtesy of John Pearse, © 1983, John Pearse.)

topics regarding kelp communities is an evaluation of the environmental pressures that determine the degree of domination that one particular species can exert. As in the intertidal zone, this varies across physical and grazing gradients.

Southern California has some of the largest and most diverse kelp habitats in the world, with rich flora and fauna in the canopy, understory, and turf. To some extent this richness is a result of the presence of a rela-tively large, shallow shelf offering extensive habitats at appropriate depths. The Channel Islands afford much of the coastline some shelter from northern storms. There is enough coastal structure to cause small-scale up-welling areas that, with the Southern California Eddy, result in a fairly predictable nutrient regime. Large kelp forests occur on most hard and cobble substrata at depths of 20–25 m. The floating canopies in these for-ests include *Macrocystis pyrifera, Egregia menziesii,* and two species of *Pelagophycus,* as well as two fucoids

(*Cystoseira osmundacea* and *Sargassum muticum*). There are two stipitate species, *Pterygophora califor-nica* and *Eisenia arborea,* and two prostrate species, *Laminaria farlowii* and *Agarum fimbriatum.*

Associated with the kelp forests are many species of fish. Although none is absolutely dependent on the kelps, the abundances of several species of fish are much enhanced by the kelp forest. The kelp habitat serves fish both as protection from predators and as a foraging area for the many invertebrates associated with the kelp. There are herbivorous fishes, but these have little direct impact on the kelp. A more common but little-studied role of canopy fishes is to remove inverte-brates such as scallops and pericarid crustacea that po-tentially can sink or heavily graze the plants (Tegner & Dayton 1987).

Sea-urchin species include *Strongylocentrotus franciscanus, S. purpuratus, Lytechinus anamesus,* and *Centrostephanus coronatus;* the first three all have the potential for overgrazing and locally eliminating the kelps. At one time the southern California coast was in-habited by sea otters (*Enhydra lutris*), which elsewhere have been shown capable of controlling the densities of invertebrate herbivores. Observations of sea otters and abundant abalone and sea-urchin remains in very old Native Indian middens suggest that predation on otters by early humans may have reduced their community im-pact long before Europeans rendered them almost ex-tinct in the early nineteenth century. The spiny lobster, *Panularis interruptus,* and the sheephead wrasse, *Semi-cossyphus pulcher,* are capable predators on sea urchins and may have reduced urchin densities in southern Cal-ifornia before they, too, fell victim to man's greed.

My initial research asked questions similar to those in the intertidal zone, emphasizing plant–plant and plant–animal interactions. Many such relationships were observed, and interesting patterns of population dynamics and stability were defined. As the work con-tinued, we soon came to appreciate the importance of interdecadal climatological differences, as the relatively benign 1970s were followed by a decade characterized by a massive El Niño/southern oscillation event, as well as some extremely destructive storms. These types of massive disturbances, as well as periods of intense over-grazing by sea urchins, appear to be characteristic of kelp habitats worldwide.

Among the interesting and yet little-studied ques-tions about kelps are the evolutionary aspects of their life cycle, in contrast, say, with the fucoids that are also present in the subtidal forests. The microscopic game-tophytes of Laminariales are difficult to study, espe-cially in the field, so most of the research has focused on the sporophyte. In Fucales, in contrast to kelps, mei-osis takes place in the conceptacles and gamete fertili-zation on the conceptacle surface; the propagules that disperse are diploid. Perhaps the major ecological significance of this is in the dispersal biology. Because the male and female gametophytes of Laminariales must be in close proximity to ensure fertilization, they suffer a much stronger dilution factor than do the Fucales, whose $2n$ propagules disperse and can develop independent of any dilution factor. Thus, in theory, fucoids should be more effective colonizers. The tiny gametophyte stage of kelps may have an advantage in avoiding environmental stresses, such as sand abrasion and grazing, but this evolutionary question has been little studied.

These issues contribute to much larger biogeographic-scale questions that also are only recently receiving much attention. The biogeography of kelps is very interesting, as it represents a blend of history, dispersal, and regional evolution. Perhaps the most interesting and fundamental biogeographic observation is that the Northern Hemisphere is dominated by Laminariales, whereas the Southern Hemisphere has more prominent Fucales. The North Pacific has 26 genera and at least 64 species of Laminariales; the North Atlantic has 5 genera and 11–18 species, whereas all of the Southern Ocean has only 4 genera and 10–12 species; see the review by Estes and Steinberg (1989). In addition to there being many more species, many of the North Pacific species have considerable morphological plasticity and can assume different ecological roles in different habitats. In addition to having much less morphological diversity, some of the North Atlantic *Laminaria* appear to be interfertile. Rather than dominating their communities as they do in the north, the Southern Ocean Laminariales often seem to occupy disturbed areas and appear more as fugitive species. The Southern Ocean habitats are dominated by many species of Fucales, many of which have flotation devices and might be expected to disperse well in the Westwind Drift and to be able to colonize relatively efficiently with their diploid propagules. Nevertheless, the Southern Hemisphere is characterized by a surprising algal provincialism. The sub-Antarctic islands are characterized by *Macrocystis*, *Lessonia*, and ephemeral populations of *Desmarestia*. The Antarctic Convergence forms an interesting biogeographic barrier below which the strongly endemic Antarctic algal flora lacks true kelps – the brown algae being represented by species of Desmarestiales. It is interesting to speculate where these Antarctic endemics occurred during much of the Pleistocene, when the Antarctic habitat was covered with ice. The few ice-free sub-Antarctic islands could not have offered much area to serve as refuges for so many species. These life-history and biogeographic patterns are fascinating, and there is a real need for innovative research in this area.

I now realize that the structural parallels between terrestrial and kelp communities are misleading. There are many differences between kelps and the higher plants in terrestrial habitats. Perhaps the most fundamental is that terrestrial habitats have many intricate co-evolved relationships. In contrast, whereas kelps do have structural and chemical adaptations to herbivores, both the kelps and their herbivores tend to be more generalized. Although there has been little research, there is almost no evidence of important co-evolved relationships for kelps, such as pollination or dispersal vectors and root symbionts. Other important research topics in kelp habitats involve the classic issues of diversity, succession, and stability. Although they lack terrestrial complexity, kelp habitats have several advantages in the study of such topics because they respond more quickly to experimental manipulation. Also, because the relationships are simpler, it is possible to identify questions that may have communitywide implications.   ♦

2.2 **Intertidal zonation patterns**
2.2.1 Tides

Almost all marine shores experience tides, although tidal amplitudes vary greatly from place to place. The pattern of high and low waters also varies from place to place, depending on the interaction between the tide waves and the standing waves caused by water slopping back and forth in the ocean basins; for details, see texts such as those by Gross (1982), Thurman (1988), and Denny (1988). More important to seaweed ecology and physiology are the changes at one place. Besides the progression from neap to spring tides twice a month, the times of high and low water change during the lunar month, and often from season to season (e.g., Mizuno 1984; Price 1989; Matson 1991); this is important because desiccation stress in the intertidal zone and water-temperature stress in tide pools and reef flats are increased when summer low tides occur during the day. Sea levels in the tropical Pacific can also change because of El Niño/southern oscillation (ENSO) events (sec. 6.2.6).

Tidal cycles are of three types (Fig. 2.5). Diurnal tides have one high and one low per day; this is an unusual type, occurring in parts of the Gulf of Mexico. Semidiurnal tides rise and fall twice a day, with successive highs and lows more or less equal in height; this type is common along open coasts of the Atlantic Ocean. Mixed tides occur twice a day but have clearly unequal highs and lows (Fig. 2.5b). Mixed tides are characteristic of Pacific and Indian Ocean coasts, as well as in smaller basins such as the Caribbean Sea and the Gulf of St. Lawrence (Gross 1982). In addition, there are storm tides, with irregular periods, usually of several days, caused by barometric-pressure changes and winds. Where the tide range is less than 1 m, such as on the Swedish west coast or the Caribbean coast of Panama, atmospheric-pressure changes and onshore and offshore winds may combine to produce very irregular and unpredictable changes in water level (Fig. 2.6). Such shores are called atidal (Johannesson 1989).

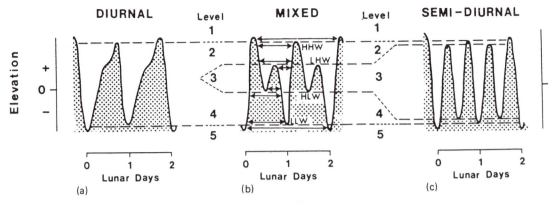

Figure 2.5. Physical division of the seashore. Daily (first-order) critical tide levels (CTLs) in diurnal, mixed, and semi-diurnal tide regimes during a neap tidal cycle. Stippling indicates submergence; arrows indicate duration of continuous exposure or submergence immediately above and below CTLs. CTLs divide the intertidal region into four or five levels. (From Swinbanks 1982, reprinted with permission of Elsevier Biomedical Press, Amsterdam.)

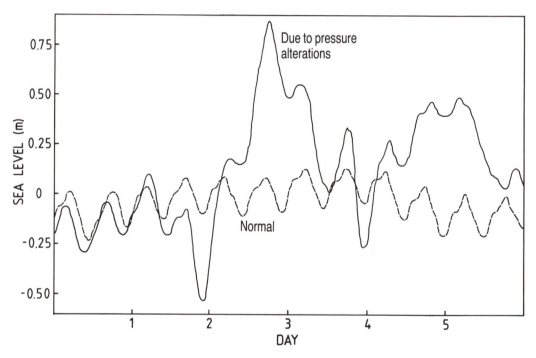

Figure 2.6. Sea-level changes on an atidal shore (Tjärnö, Sweden). One period features more regular, tide-dominated variations; the other is characterized by dramatic changes due mainly to atmospheric-pressure alterations. For example, during day 2, the water level changed nearly 1.5 m (0 m = mean sea level). (From Johannesson 1989, with permission of Munksgaard International Publishers.)

2.2.2 Zonation

Two features of intertidal seaweed vegetation are readily apparent: Distinct bands of particular species or associations run parallel to the shoreline, and there are variations in the flora over short horizontal distances. Despite the extent of global shorelines, there are general patterns upon which local variations are superimposed. A vertical strip of the shore may logically be divided into zones determined by the organisms present: a *Fucus*

zone, a barnacle zone, and so forth. Alternatively, it may be divided according to tide levels, either in general terms (high, middle, and low intertidal), or on the basis of critical tide levels.

The Stephensons' now-familiar zonation scheme, which is based on biological features, is shown in Figure 2.7a, together with the effect of wave action as first shown by Lewis (1964); see also Brattström (1980) and Pérès (1982a,b). Wave action decreases the duration of

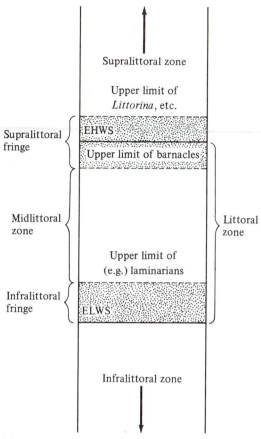

(a)

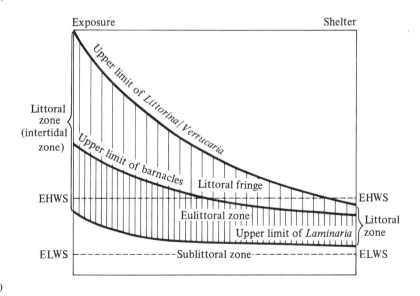

(b)

Figure 2.7. Biological division of the seashore. (a) The Stephensons' universal scheme for intertidal zonation. (b) Lewis's scheme for intertidal zonation, illustrating the effects of wave exposure in broadening and raising the zones (toward the left of the diagram). EHWS, extreme high water of spring tides; ELWS, extreme low water of spring tides. (Part a from Stephenson & Stephenson 1949, with permission of Blackwell Scientific Publications; b from Lewis 1964, with permission of the author.)

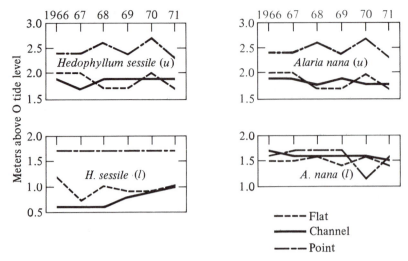

Figure 2.8. Year-to-year changes in upper and lower limits for two intertidal kelp species on three transects at an exposed site on the west cost of Vancouver Island, British Columbia. A gently shelving platform, a rocky point, and a narrow channel are compared (see also Fig. 2.10). (From Druehl & Green 1982, with permission of Inter-Research.)

atmospheric exposure of a spot on the shore (Fig. 2.7b), as first shown by Lewis (1964). Because zone boundaries are determined partly by the duration of exposure to the atmosphere (sec. 6.4), the important criterion is not the theoretical exposure time determined from tide tables but the actual exposure time (see Fig. 2.10 and sec. 2.2.3). Many examples of intertidal zonation can be found in the books by Lewis (1964) and Stephenson and Stephenson (1972) and in the review by Pérès (1982b). Tropical examples are given by Brattström (1980, 1985) and Thomas (1985); Round (1981) reviewed tropical reef zonation. We expect that readers will have opportunities to study the zonations in their areas.

There are difficulties inherent in defining zones on the basis of the organisms found in them, however. First, there is variation in space. On the small scale, most shorelines consist of irregular rocks and boulders and are likely to present very confusing patterns of organisms, with zones breaking down into patches. On the large scale, flora and fauna change geographically. From their broad experience with intertidal zonation, the Stephensons expressed the problem this way: "The zonation on a thoroughly broken shore has to be described in terms of the prevailing arrangement of species on all the rocky facets of the same type at the same level. [Probably] the best instrument for appreciating these interrupted zonations is the suitably trained human eye" (Stephenson & Stephenson 1972, pp. 15–16). In the two decades since then, some ecologists have turned to computer analysis of communities. (This also requires a suitably trained eye!) For instance, Murray and Littler (1974) determined the percentage cover of each species in each quadrat and applied statistical clustering to determine similarities between quadrats. Clusters did not correspond to "eyeball" descriptions.

Variability in time is also important. Aside from seasonal and successional changes in the vegetation, the timing of harsh conditions with respect to settling and the timing of the clearing of space with respect to reproduction of potential settlers (and competitors) add to the heterogeneity in the communities; see, for example, Archambault and Bourget (1983) and Dethier (1984). The patchiness in the pattern is not static. Vertical limits of species vary from year to year (Fig. 2.8), perhaps dependent on variations in emersion/submersion histories. Relative abundances and distributions of species that are nearly equal as competitors will change from time to time, and longer-term changes are also known to occur (Lewis 1980). Understanding such long-term changes will require long-term observations, while an understanding of the short-term changes still will require much work on how various factors affect the critical stages of the life cycles of the plants. Observations of the presence of a plant at a given site can be interpreted to mean that conditions there have been suitable for its growth since it settled. Absence, on the other hand, says little: Perhaps conditions were unsuitable at some time, or perhaps the reproductive bodies of the species were unable to reach the site. Unsuitable conditions may have been present always, or only at some brief time in the past, if the plant was ever there.

Finally, the concept of "community" is a problem (Chapman 1986). At one extreme is the view that plant communities are closely integrated units, almost superorganisms. At the other extreme is the view that communities consist merely of coincidental associations of plants. Marine botanists tend to view seaweeds on rocky shores as forming natural associations of characteristic species with clear zonation (Russell & Fielding 1981). Pérès (1982b) and others working in the Mediterranean, and Taniguti in the Far East (e.g., Taniguti 1987), view seaweed communities in terms of characteristic organisms and use formal Zurich-Montpellier classification into orders, alliances, and associations (biocoenoses if animals are included) to describe the vegetation (Chapman 1986). For instance, a community dominated by *Udotea* and *Peyssonnelia* might be classified as a Udoteo-Peysonnelietum. As Chapman points out, such phytosociological classification is justifiable at the level of biocoenoses (although the sizes of the units seem to vary widely among phytosociologists), but ranking into alliances and orders, in the style of Linnean taxonomy, implies a genetic-evolutionary history of vegetation that is not present.

2.2.3 Critical tide levels

Shoreline communities can be studied in reference to physical parameters rather than the biological parameters described earlier. The concept of critical tide levels (CTLs) was introduced in one form by Colman in 1933, but Doty's (1946) modified version is most widely recognized. Doty's attempt to equate changes in vegetation with CTLs has been criticized by those who stress the importance of biological factors in determining zonation (e.g., Chapman 1973b). CTLs can be used simply as a frame of reference for describing zonation patterns, as has been proposed by Swinbanks (1982) in his revision of Doty's scheme. In some situations they may be useful for explaining zonation.

CTLs are levels in the intertidal zone at which there are marked increases in the duration of exposure or submergence. They occur at crests and troughs in daily, monthly, annual, or even longer-term tidal cycles. In a mixed tidal cycle (Fig. 2.5b), successive high and low waters are of different heights, so that there are several approximate doublings of duration of emersion – for example, from just below to just above lower (or higher) high water (LHW, HHW). Thus, on any given day the shore can be divided into five tidal zones. Shores with diurnal or semidiurnal tides have fewer zones (Fig. 2.5a,c). These diurnal zones are produced by so-called first-order CTLs. The weekly progression from spring to neap tides gives second-order CTLs. Swinbanks's scheme is summarized in Figure 2.9.

Correlations between zone boundaries and CTLs may be expected, because the stress of exposure to the atmosphere increases with the duration of exposure, and seaweeds have differing abilities to recover from the stress. Colman's (1933) CTLs were based on average percentage exposures, which did not take into account duration and time of exposure. Moreover, Underwood (1978) subsequently showed that Colman's exposure curve was inaccurate. Doty (1946) defined CTLs differently, using maximum durations of exposure or submergence; this is the scheme that Swinbanks (1982) amplified (Fig. 2.9). When all the four possible orders of CTLs are considered, there is such a profusion of

them that the probability of one of them coinciding with a zone boundary is high. On the other hand, several factors complicate possible correlations, most notably wave action (Fig. 2.10), seasonal changes in the time of low water, and the ability of organisms to become acclimated during periods of subcritical conditions. Further, a critical period may apply to reproduction, settlement, or germling survival, rather than to adults, so that one cannot simply take a correlation as evidence of a relationship. Some element of chance is eliminated because organisms are cued to seasons: Late spring, for example, is a time of reproduction for some species and also a time of some annual extreme CTLs and the first warm, sunny weather of the year.

The major shortcoming of this physical scheme is that it depends on predicted water levels; it does not allow for wave action or storm tides. Sheltered habitats and even gently shelving shores in wave-exposed areas may be exposed close to predicted times (Fig. 2.10a). However, exposure times on steep-sided, wave-beaten promontories must be measured, and indeed in such places the concept of CTLs becomes meaningless owing to the unpredictability of wave height from day to day.

Druehl and Green (1982) attempted to correlate vertical distribution with actual submergence/emergence durations. Using a ''surf-sensor'' devised to record submergence events, they examined the submergence histories of three contiguous areas (within 50 m of each other) with differing slopes in a wave-exposed location and the relations between seaweed vertical distributions and corrected exposure times. Actual submersion/emersion curves differed from predicted tidal curves in two ways: (1) The intertidal range was greater, and the lowest emerged level was higher, than predicted; a gently shelving transect was closest to the predicted, and a rocky point the most different (Fig. 2.10a). This compares to the general observation depicted in Figure 2.7b. (2) The harmonic pattern of the tidal curves was disrupted, again the most on the rocky point (Fig. 2.10b,c). Thus waves tend to blur the positions of CTLs. Druehl and Green (1982) found that the upper limits of seaweeds correlated most closely with accumulated time submersed, whereas the lower limits correlated best with the duration of the longest single exposure. They did not directly test CTLs as causative factors, but nevertheless argued in favor of physical factors controlling species limits in the intertidal.

An appropriate direct test, suggested by Swinbanks (1982), is a series of experiments testing growth, survival, and reproduction just above and just below a CTL that is likely to be important. So far there are no such data for seaweeds. A correlation that Swinbanks pointed out as a good candidate for testing is the upper limit of *Pelvetia canaliculata*. This coincides with the lowest spring higher high water, where maximum exposure in May–June jumps from approximately 12 to 24

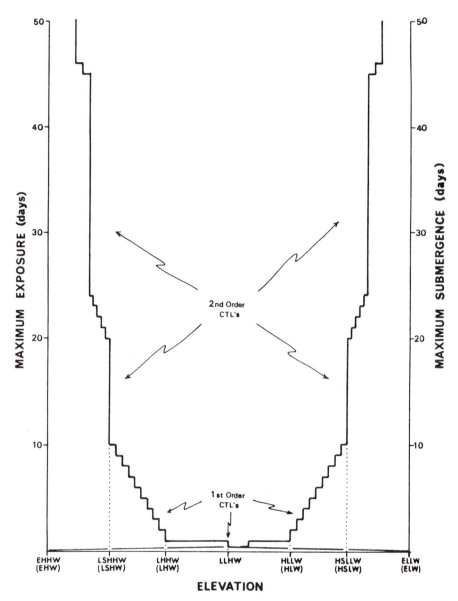

Figure 2.9. Curves showing maximum continuous exposure and submergence with respect to tidal elevation, illustrating the idealized sequence of first- and second-order CTLs for all types of tides. The CTLs used in this scheme are labeled for a mixed tide (as in Fig. 2.5*b*); the stepped curve rising to the left indicates increasing exposure; that rising to the right indicates increasing submergence. Abbreviations for tide levels: EHHW, extreme higher high water; ELLW, extreme lower low water; HLLW, highest lower low water; HSLLW, highest spring lower low water; LHHW, lowest higher high water; LLHW, lowest lower high water; LSHHW, lowest spring higher high water. (From Swinbanks 1982, with permission of Elsevier Biomedical Press, Amsterdam.)

days (Fig. 2.9) [data of Schonbeck & Norton (1978) re-analyzed by Swinbanks (1982)]. In early spring, *P. canaliculata* germlings develop higher on the shore than the normal population limit, and they are pruned back in summer.

On atidal shores, the effect of the irregular water-level fluctuations is to increase the variation in sub-

mersion/emersion times compared with tidal shores (Johannesson 1989). A 10% submergence time on a tidal shore will mean a few hours every day underwater, but on an atidal shore the time may be very differently distributed. The risk of long exposure times may result in algae on Swedish shores living at a greater percentage submergence than on tidal shores (see *Verrucaria* zones

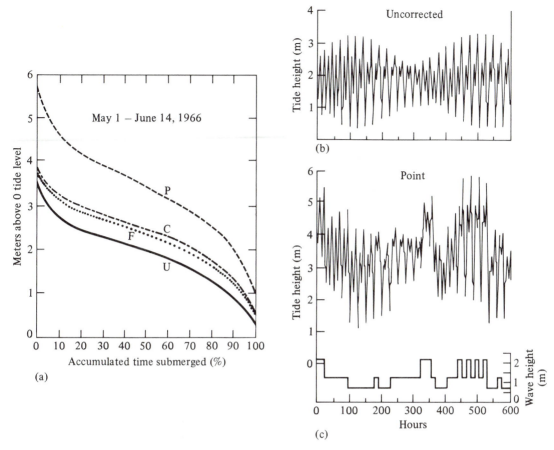

Figure 2.10. Submergence–emergence data from a site on Vancouver Island, British Columbia. (a) Measured accumulated time submerged as a function of elevation above zero tide for a rocky point (P), a channel (C), and a gently shelving rock face (F), all within 50 m of one another, compared with data from 6-min tidal predictions (U). (b) Predicted tide heights over a lunar cycle. (c) Actual tide heights and wave heights at the rocky point. The wave-height data are derived from twice-daily observations. (From Druehl & Green 1982, with permission of Inter-Research.)

in Fig. 2.11). In effect, this brings us back to the idea of critical exposure levels, though Johannesson does not espouse that theory.

The physical factors and their physiological effects are only part of the explanation of zonation patterns. There are also various biological controls that in some cases determine zone limits: grazing, competition, and other interspecies and intraspecies interactions. These contrasting groups of factors and the experiments needed to distinguish between them have been thoughtfully reviewed in essays by Norton (1985) and Underwood (1985), in a book dedicated to J. R. Lewis.

2.3 **Submerged zonation patterns**
2.3.1 Tide pools

Tide-pool communities are difficult to characterize because of the intrinsic variety in pool conditions. Physical conditions within a pool will change from seawater conditions, the extent depending on the size of the

pool (especially surface : volume ratio) and its height on the shore (thus duration of exposure), as well as on the extrinsic factors of atmospheric conditions.

High pools tend to be very stressful because of changes in temperature and salinity (sec. 6.1.3), pH, and oxygen concentration. They tend to be populated by single tolerant species such as blue-green algae or certain greens (e.g., *Chaetomorpha aerea*, *Enteromorpha* spp., *Rhizoclonium* spp.) (Wolfe & Harlin 1988a,b; Kooistra et al. 1989). Communities in mid- and low-shore pools, where physical conditions are progressively less rigorous, are more diverse, resembling the communities of the low littoral and sublittoral fringe. As on the exposed rocks, there are seasonal changes in the flora, which are nonetheless characteristic for each pool (Dethier 1982; Wolfe & Harlin 1988a). In deep pools there may be zonation within the pool (Kooistra et al. 1989); community composition may depend on numbers of littorine grazers, or the presence of allelopathic in-

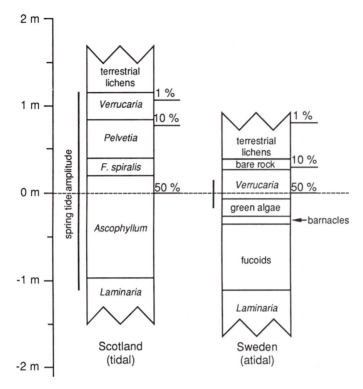

Figure 2.11. Zonation patterns, relative to mean sea level (0 m), for two sheltered rocky shores, one in Scotland (tidal) and one in Sweden (atidal). The zone of ephemeral greens on the Swedish shore is maintained by ice scour. Notice that the fucoid zone is subtidal on the atidal shore. Mean ranges of spring tides are indicated by vertical bars. Percentage submergence over a year is given at three levels. (From Johannesson 1989, with permission of Munksgaard International Publishers.)

teractions (see Chapter 3). The upper zones in pools are affected by high temperatures, a phenomenon related to exposure. Lower zones are more strongly regulated by irradiance, as in the sublittoral. Competitive interactions are found in all zones.

2.3.2 Sublittoral zonation

Descriptions of subtidal zonation require scuba equipment or even submersibles and therefore have been fewer than intertidal descriptions (e.g., Shepherd & Womersley 1970; Littler et al. 1985). Most experimental studies have been of plant–herbivore interactions; little is known of the relationships between subtidal plants and abiotic factors, or of plant–plant interactions (Schiel & Foster 1986). Much of the autecological work has been on large laminarians – and much of that only on the sporophytes (Fig. 1.29), which dominate the vegetation in kelp beds (sec. 2.1.3). Moreover, studies have often been limited to sites of particular interest (e.g., urchin barrens), so that the scale of effects seen is often poorly known (Schiel & Foster 1986). As in the intertidal region, cluster analyses have been used to classify sublittoral quadrats (e.g., van den Hoek et al. 1975; Lindstrom & Foreman 1978; Field et al. 1980). Péres (1982a,b) and Golikov (1985) both considered zona-

tion of entire ocean basins, including much that is "aphytal" (Péres's term). They used light and temperature/salinity, respectively, to define zones, but their schemes are too coarse to be of much use in understanding algal zonation.

A more useful general division of the sublittoral into zones dominated by different functional-form groups was proposed by Vadas & Steneck (1988): an uppermost zone of thick, leathery macrophytes, a middle zone of coarsely branched, sheetlike, or filamentous algae, and a deep zone of crustose algae. The smaller functional groups occur at all depths; the dominance and zonation come about because of the successive elimination of larger forms with depth. Vadas & Steneck compared literature on tropical sites (e.g., Littler et al. 1986a) with their study in the Gulf of Maine to show that this pattern is widespread. They also noted that within the genus *Halimeda*, the depth distribution of plant habits parallels the functional-form groups, with a gradation from erect to low-lying species (Hillis-Colinveaux 1985). Those studies included very deep algae, down to the limit of algal growth. The full development of this zonation depends on the availability of substrate, while the depths of the zones depend primarily on water clarity.

2.4 Some other seaweed habitats and communities

Virtually any substratum in salt or brackish water is habitat for seaweeds. Rocks, wood, sand, glass, shell, and other nonliving substrata are most typical, along with the surfaces of other algae and other living submerged plants and the shells of living molluscs. Endolithic algae are important in reef processes (sec. 2.1.2). Some remarkable seaweed habitats have been reported: the tissues of sea pens, where some *Desmarestia* gametophytes grow (Dube & Ball 1971); the beaks of parrotfish, which provide a "moving reef" for pioneer species such as *Polysiphonia scopulorum* and *Sphacelaria tribuloides* (Tsuda et al. 1972); a symbiotic association with a sponge (Price et al. 1984); the face and belly of the Hawaiian monk seal (Kenyon & Rice 1959); and the necks of green turtles (Tsuda 1965).

Physiological ecology must account for the success of seaweeds wherever they grow, whether in well-studied habitats such as the rocky intertidal zone or (at the other extreme) the unusual habitats listed earlier. The essays in sections 2.4.1–2.4.3 explore further the range of seaweed habitats.

2.4.1 Essay: Seaweeds in estuaries and salt marshes
Piet H. Nienhuis*

Estuaries. Estuaries occur all over the world where seawater mixes with river water. Writer Paul Gallico captured the ambience of one such habitat in his story "The Snow Goose" (quoted by Long & Mason 1983): ". . . one of the last wild places in England, a low, far-reaching expanse of grass and half-submerged meadowlands ending in the great saltings and mudflats near the restless sea. . . . It is desolate, utterly lonely, and made lonlier by the calls and cries of the wild birds that make their homes on these marshes." Tidal estuaries in temperate climates are dominated by fringing salt marshes; on tropical shores, mangrove swamps are the ecological equivalent.

Tidal estuaries show a marked longitudinal gradient in environmental factors from the sea to the river. Generally, salinity and wave action decrease, whereas silt, turbidity, light extinction, and nutrient concentrations all increase. These environmental factors show enormous variations over a tidal cycle and even more over the year.

* Piet H. Nienhuis developed a love of ecology as a child and has made a lifetime study of salt marshes. He is now a research manager at the Netherlands Institute of Ecology and professor of ecology at the University of Nijmegen, The Netherlands.

The environmental gradients are reflected in the numbers of macroalgae to be found in the estuary, these numbers decreasing upstream. Estuaries harbor an impoverished seaweed flora, shifting from a marine, epilithic assemblage rich in species to a restricted number of euryhaline species attached to silt and sand farther inland. Near the sea the perennial zonation pattern of large fucoid algae and subtidal laminarians is evident. This pattern, as on rocky shores, is shaped by tolerance for desiccation at the upper borders and by biological competition and light extinction at the lower borders. Further inland, brown and red algae disappear, leaving the green and blue-green algae in a single-layered, indistinct zonation pattern, restricted to the intertidal zone and consisting mainly of annual species (Wilkinson 1980). In nontidal, brackish, and sheltered lagoons, where turbidity is low and light penetrates deeper, the sublittoral macroalgal vegetation may be abundant, but the species diversity is always low compared with open seacoasts.

Species poverty is a characteristic of many estuaries and brackish waters. The lowest numbers of species are generally found in salinities of 4–8‰ (Nienhuis 1980; Arndt 1989). Why are the numbers so low in brackish waters, compared with either fresh waters or the sea? A plausible explanation for estuaries that came into existence after the last ice age is that the inconstancy of the habitats over geological time is combined with a strong gradient in physiological stress for marine organisms (Sanders 1968; Wolff 1972). In the Baltic, one of the best-studied estuarine areas, many of the so-called genuine brackish-water algae appear to be physiologically adapted phenotypes or ecotypes of euryhaline seaweeds (Wallentinus 1991). Areas such as the Black Sea and Caspian Sea that are consistently brackish and have remained undisturbed by ice ages over hundreds of thousands of years have many genuine brackish-water species, many of which are endemic (Remane & Schlieper 1971).

The variable seaweed abundance and the absence of ecological competitors in temperate estuaries have made these places open niches for colonists. Well-documented examples from European coasts are *Codium fragile* ssp. *tomentosoides,* and *Colpomenia peregrina,* both invaders from the Pacific. *Colpomenia* is known as the "oyster thief" because its globose thalli, which often attach to oysters, fill with air and become buoyant, floating away with the animal (Farnham 1980). Invading *C. f. tomentosoides* creates a similar problem in the northeastern United States. *Sargassum muticum* was recently introduced to European estuaries and is still spreading vigorously (Critchley et al. 1987).

Salt marshes and their algae. To the superficial observer, salt marshes may look dull and unattractive, places where one sinks ankle-deep in soft, black mud. Connoisseurs know better; for them the salt marsh offers

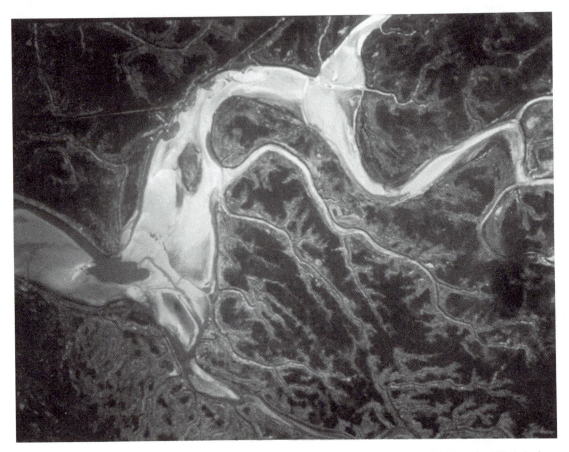

Figure 2.12. Aerial view of a salt marsh in The Netherlands showing the typical pattern of tidal creeks. This is in the Westerschelde estuary, in Antwerp harbor. (Photograph courtesy of P. H. Nienhuis, © 1991, P. H. Nienhuis.)

a complex pattern of meandering tidal creeks (Fig. 2.12), elevated creek banks, and back marshes, with a delicate zonal vegetation pattern from the lower to the upper marsh. Typically, salt marshes have three zones of flowering plants and their associated algal communities. At the lowest level, within full reach of the tides, a sparse growth of a few species of higher plants forms the pioneer zone or low marsh. This intergrades at higher levels with the richer flora of the mature zone or middle marsh. At the highest levels, the species of the mature zone are partially replaced by species from nonsaline habitats, which can, however, withstand brief and infrequent submergence in salt water. In many estuaries, salt marshes have been embanked and "reclaimed," or otherwise exploited by humans. Yet in some of the most populous and most heavily industrialized areas of the developed world, stretches of salt marsh are among the few remaining habitats not altered by humans.

Salt marshes are to be found in temperate regions around the globe. The east coast of the United States has extensive stretches of diversified salt marsh alternating with sandy beaches, protected shallow waters, and barrier islands. Among the studies along this coastline are the 20 years of intensive work on the structure and functioning of Great Sippewisset Salt Marsh, Massachussets (Teal 1986), Teal's earlier work on Sapelo Island, Georgia (Teal 1962), and studies of the Great Bay Estuary System, Maine/New Hampshire (e.g., Mathieson et al. 1981). My own experience encompasses the northwest coast of Africa, bordering the Sahara Desert, where extensive stretches of intertidal mudflats and salt marshes harbor tremendous prairies of seagrasses, and along the coasts of Europe, where many large and diverse salt marshes occur. In The Netherlands, salt marshes once formed a major part of the coastline, but human expansion has drained many of them to make farmland (polders) or industrial areas, or dredged them for harbors.

There are no trees in temperate marshes comparable to the mangroves of tropical salt swamps, but there are, in Europe, micromangroves, shrubs of *Halimione portulacoides,* with perennial wooden stems. The red-algal genus *Bostrychia* grows attached to these stems just as it does to the woody prop roots of tropical mangroves.

The macroalgal vegetation of temperate salt marshes is characterized by ephemeral and often morphologically plastic species, which were major taxonomic problems until revisions between 1960 and 1980 cleared the way for ecological work. About 100 benthic macroalgal taxa have now been recognized in salt marshes along the eastern Atlantic coasts, but only 11 species (or species groups) are dominant and frequent (Nienhuis 1987).

The salt marsh is a harsh environment for aquatic plants. The loose-lying algae in the lower marsh are at the mercy of tides, wind, and waves. Algal vegetation is less diverse where siltation is rapid than in places with stable substrata. In the upper marsh, rain, sun, and frost are predominant abiotic factors. Upper marshes may become cattle ranges, with consequent trampling and manuring of the soil. Sensitive species such as *Bostrychia scorpioides* and *Fucus vesiculosus* disappear, as do the woody vascular plants and their epiphytes, and open spots are formed that may easily fill with filamentous green algae such as *Rhizoclonium riparium, Percursaria percursa,* and *Enteromorpha torta.*

Several dominant salt-marsh algae show adaptations to extreme environmental conditions, such as desiccation and extreme salinities, that make them good species for ecophysiological work. For instance, along a gradient from sea to river, *Fucus vesiculosus* loses its air bladders, holdfasts, and sexual reproduction; it lives in the upper marsh as a loose-lying dwarf form, strongly reduced from the open-coast form. Comparable reduction can be seen in *Enteromorpha prolifera,* a highly polymorphic species, revised by Bliding (1963). In the lower marsh and on intertidal flats it forms highly branched thalli over a meter long. Higher up, it loses its branches, and the cells, otherwise square or rectangular and arranged in rows, grow out irregularly, with the cell walls becoming thicker and the chloroplast shifting to one side of the cell. In the upper marsh there remain only short, inflated pieces of thallus. The change from wet to dry conditions from low to high marsh is paralleled in the middle marsh from spring to summer: Typical *E. prolifera,* vigorously growing in spring, gradually changes into *E. "intestinalis"* on the same spot.

Salt-marsh algae have several reproductive strategies: (1) The perennial red alga *Bostrychia scorpioides* lives in the relatively constant micromangrove habitat among woody halophytes. (2) Species persistent on stable sediments survive with a below-ground network of thallus parts (e.g., *Vaucheria* spp.). (3) Pioneers show extremely rapid development under favorable conditions (e.g., *Ulothrix* spp.). (4) Opportunistic algae, such as *Enteromorpha,* show a seasonal periodicity in the lower marsh and a periodicity governed by local factors (frost, desiccation) in the higher marsh. (5) Several competitive, fast-growing blue-green algae flourish in temporally unstable habitats, while their mucous sheaths enable them to survive under extremely dry conditions.

The population dynamics and autecology of salt-marsh algae have been neglected items in phycology. Yet the few macroalgae that can thrive in the fluctuating, unpredictable, and always stressful habitat are very interesting subjects for ecological and physiological studies: Physiological adaptation mechanisms, polymorphism across the land–water ecotone, metabolism under extreme conditions, ubiquitous geographic distribution – almost all these areas are terra incognita. ♦

2.4.2 Essay: Seagrass beds as habitats for algae
*Marilyn M. Harlin**

For over two decades I have been intrigued with seagrass leaves as they affect or are affected by algae, especially those that grow on their surfaces. My interest was initiated by finding little published information on what makes this surface a good place to live. Because seagrasses are not algae, they were out of the domain of phycologists. Yet because they are marine, they lay outside the research interest of the terrestrial botanists. Still, I was fascinated by these surfaces that were heavily laden with both plant and animal life, and I espoused ideas about their relationships – nutritional and physical. Some of these ideas I tested, and some have been tested by other investigators. Today much more is known than when I began my studies. Epiphytes now have a bona fide position in models of seagrass beds (Wetzel & Neckles 1987). I have extended my own investigations to include interactions between unattached algae and seagrass, and I am currently looking at mechanisms of algal attachment. Numerous questions remain, however. Some of these are posed at the end of this essay, and the alert reader will think of many more. My intent is to challenge the reader to test them.

Seagrass leaves as substrata for algae. Seagrasses are angiosperms, though not in the grass family. Their leaves provide substrata for colonization by other organisms, up to 18 m^2 of surface per square meter of bed area (*Thalassia;* Taylor et al. 1973). Except for species of *Syringodium,* seagrass leaves are flat, thus maximizing their photosynthetic surface, diffusion of gases, and nutrient uptake (Phillips & Meñez 1988). Leaves of *Posidonia oceanica* have been reported to stay attached up to 30 weeks, but the leaves of other species turn over much more rapidly. A sequence of epiphytes can be followed on these ephemeral substrata. Novak (1984) reported coccoid bacteria and diatoms as the first organisms to appear on new leaves. This community is followed by animals and macroalgae in a "hierarchy of competitive efficiency," with macroalgae at the top.

---

* Marilyn M. Harlin is a professor in the Department of Botany at the University of Rhode Island, Kingston, RI 02881.

Casola et al. (1987) examined the changes in leaf structure with age (inner to outer leaves, base to apex), and Cinelli et al. (1984) described the relationship to water depth. The composition of the epiphytic community will be determined by the time at which a leaf becomes available, the presence of propagules, and the habitat in which the plant appears.

Seagrass leaves appear to be particularly suitable surfaces for algal settlement, as evidenced by the fact that they support a great many species of algae and invertebrates. In a review chapter (Harlin 1980), I listed 356 species of macroalgae, 153 species of microalgae, and 177 species of invertebrates with their respective host genera. These lists were not comprehensive, even at the time they were compiled. For example, Main and McIntire (1974) had identified 221 diatom taxa, of which only the 36 most abundant were included in my list. Algae appear less on artificial seagrass (plastic strips) and seaweeds (Harlin 1973; Sieburth & Tootle 1981). No one knows all the reasons that make seagrasses especially suitable. Possibilities include surface shape, texture, and chemistry.

There is no solid evidence that seagrasses resist colonization by active means, although they contain water-soluble phenolic acids (Zapata & McMillan 1977). Harrison and Durance (1985) showed that water extracts from green *Zostera* leaves reduce carbon uptake in epiphytic diatoms, and Cariello and Zanetti (1979a) showed that extracts from *Posidonia oceanica* were antibiotic.

Epiphytes frequently are distributed more heavily on the tips and margins (edges) of the flat blades (Harlin 1980). The tips are older and thus have had longer periods for colonization. In addition, at least in *Posidonia*, the phenolic compound chicoric acid is present in younger leaves (Cariello & Zanetti 1979b). I suggest four hypotheses to explain the "edge effect": (1) An inhibitor of algal growth is released more from the laminar (flat) surface. (2) A metabolite is released at the edge to stimulate growth. (3) The edge is an easier place for algae to adhere initially. (4) It is more difficult for grazers to hold on to the edge, and therefore algae escape being eaten. All of these ideas need to be tested. A combination could be responsible, or a mechanism entirely different from those suggested.

Except for the expansion of habitat, the relation of seagrass leaves to epiphytes is speculative. Uptake of nutrients from the water column by leaves and epiphytes probably is enhanced as the moving substratum breaks up laminar flow and thus the barrier to diffusion. There is some evidence that seagrasses act as nutrient conduits from sediment to leaf and on to epiphytes. *Zostera* releases dissolved organic carbon, which is potentially available to epiphytic algae. Epiphytic nitrogen-fixing blue-green algae provide a source of nitrogen for other algal epiphytes, an important function where nitrogen is limiting.

Seagrass leaves support sessile invertebrates such as bryozoa, hydrozoa, sponges, and tunicates, which compete with algae for space. The leaves also shelter populations of animals such as molluscs and crustacea that crawl along the leaf surface, grazing attached algae (Montfrans et al. 1984; Lewis 1987). Grazers remineralize surficial algae, converting organic material into nutrients that can be taken up by other algae. In doing so, they leave a surface cleansed of epiphytes and available for recolonization at a time of different spore set. The renewed substrata can thus support a different community structure.

Impact of epiphytic algae on seagrasses. One consequence of epiphyte biomass is to shade the host plant (Sand-Jensen & Borum 1984). Light attenuation up to 90% has been measured (Borum et al. 1984), although the average is 10%. Thus, the compensation point of seagrass is increased, the compensation depth is decreased. The presence of an epiphytic load on the leaves increases drag and reduces movement through the water column. In doing so, epiphytes reduce the leaves' ability to remove inorganic nutrients, causing the plant to rely more on absorption of nutrients by roots. Howard and Short (1986) demonstrated that seagrass survivorship is enhanced by grazing activity.

Competition between seaweeds and seagrasses. Do unattached seaweeds compete with seagrasses, and if so, for what are they competing? Only one genus of seagrass, *Phyllospadix* (five species), grows obligately in the rocky intertidal zone, where it can potentially compete for space with intertidal algae (Turner 1985). All other species grow in sandy and silty sediments, usually subtidally. Tropical siphonous green algae, such as *Caulerpa*, *Penicillus*, and *Halimeda*, are held in sediments by rhizoids. Williams (1984) used ^{15}N to show that *Caulerpa cupressoides* takes up ammonium from interstitial waters and translocates this nutrient to photosynthetic portions of the alga, in the same manner as seagrasses. Thus, in nitrogen-poor waters, algae and seagrasses could compete for both space and nutrients. Williams (1988) also showed that during storms, *Halimeda incrassata* retained more leaf area than did seagrass within the same community. Usually, however, seagrass roots and rhizomes are buried in silt, where algae cannot attach.

Seagrasses have a greater requirement for light than do algae, partly because of the respiratory demands of their underground tissues, and are less efficient in removing nitrogen from the water column. In shallow estuaries rich in nitrogen, masses of unattached *Ulva lactuca* or *Enteromorpha* species reduce light below that required by *Zostera* during its growing period. Here we see competition for light, in which the green alga wins. That was the outcome in a coastal lagoon in Rhode Island, where we (Harlin &

Thorne-Miller 1981) experimentally enriched the water with nitrate, ammonium, and phosphate. When nutrients from a water-treatment plant in Connecticut were diverted, *Ulva* no longer predominated, and *Zostera* expanded its population (French et al. 1989). Red algae such as *Gracilaria tikvahiae* frequently lie in loose mats at the base of seagrass leaves. In northern temperate estuaries, they do not appear to compete with seagrasses. But in Tampa Bay, Florida, seagrass beds are disappearing where red algae have built up dense biomasses (Dawes 1985).

Questions to be explored. Unanswered are numerous questions concerning the relationships between seagrasses and algae. Examples include the following: What makes the surface of a seagrass a suitable habitat? Does the host provide a nutritional advantage or merely an extension of habitat? Why do algae appear more often on the margins than on the flat surfaces of leaves? Do variations in metabolites or phenolics alter the composition of the algal community? To what extent do grazers preferentially select algal species and thereby control community structure? How do seagrasses interact with unattached algae, such as the benthic seaweeds and phytoplankton? Are they competing, and if so, for what resource(s)? Are they playing complementary roles in maintaining an ecosystem? In the past two decades, investigators have become increasingly curious about seagrass leaves as a habitat for algae. ◆

2.4.3 Essay: The Arctic subtidal as a habitat for macrophytes
Robert T. Wilce*

The attached algal flora at high latitudes is impoverished in terms of the number of species, macrophyte stature and size, and occurrence of populations. Observations of floristic impoverishment and environmental severity are common generalizations one can make about most coastal areas of the north polar sea and its peripheral, more southerly extensions. Indeed, the greatest portion of the Arctic sublittoral is unsuitable for

---

* Robert T. Wilce is a professor emeritus in the Department of Biology, University of Massachusetts, Amherst, MA 01003. He has worked in the Arctic and sub-Arctic studying marine macrophytes during all seasons of the year. Beginning in 1954 he worked in northern Labrador and Ungava Bay and repeated prior studies in northwest Greenland, Ellesmere Island, and most recently in northern Baffin Island and the north coast of Alaska. No other phycologist has worked as extensively in the north. His research goals have been and remain to understand the composition of the macrophyte flora, to describe the modes of macrophyte adaptation to the Arctic marine environment, and to learn more of the origins of that flora.

macrobenthos development. This is especially true of immense coastal areas of Eurasia, much of central and northwestern Canada, and northern Alaska. In these areas the shallow sublittoral (<10 m) is characteristically brackish for much of the light year, and the bottom is mostly soft sediments – both environmental features that discourage attached algal colonization. Where these conditions persist, a few small attached macrophytes occur, along with loose-lying perennating algae often present in large populations. On the other hand, an appreciable Arctic algal flora occurs where rocky substrata are free of extensive sediment deposition and freshwater runoff does not reduce salinities appreciably.

Throughout the Arctic the marine flora contains more than one hundred and fifty species, most with a circumpolar distribution, and most also found throughout the north temperate North Atlantic. The following paragraphs briefly characterize the sublittoral portion of this northern algal flora, including geographic limits and a brief description of environmental features most responsible for the character and composition of this cold-water flora. These generalizations are based on many years of personal study of Arctic macrophytes in northern Alaska, central and northeastern Canada, northwestern Greenland, and northernmost Norway.

Prior to the 1960s, knowledge of Arctic sublittoral attached algae was based entirely on information and specimens obtained from dredging. Dredged samples provided material for floristic descriptions and "broad-brush" concepts of the sublittoral environment (Wilce 1959). But dredging provided little information, beyond conjecture, on most aspects of community structure and the general ecology of northern sublittoral macrophytes. Scuba dramatically improved our understanding of macrophyte ecology in the top 30 m, and it enhanced collections, although most of the species had previously been collected from dredging efforts. Diving as a tool of research began for me in the early 1960s in a small cove of Disko Bay, western Greenland. Repetitive in situ observations and collections from 0–35 m depth throughout many areas of the north are the basis for the following generalizations. My knowledge of deep-growing vegetation, down to some 90–100 m, continues to be based on my own dredging and the work of earlier phycologists. As a diver, I am repeatedly made aware of the accuracy of detail based on dredging. My predecessors working in northern areas have proved largely correct concerning depth records and ecological generalizations. Performed properly, dredging and the use of a Petersen Grab provide specimens and a rather accurate picture of some aspects of the bottom habitat, including depth.

What is it like to dive in the Arctic? It is uncomfortably cold and generally poorly lighted even during the nonwinter months. Water temperatures below 10 m rarely rise above 0°C and for the most of the year remain at −1.85°C. In the 1960s, wet suits were the standard uniform, making long dives almost unbearable. With the advent of dry suits, life in ice-covered water improved dramatically. Still, one remains uncomfortable and constrained beneath firm ice or floe ice. The real problems of working as a phycologist in the Arctic are those of logistics: transportation to a dive site by dog sled or small Eskimo boats, or the occasional luxury of a helicopter, and the constant problem of filling air tanks under difficult conditions. The problems of securing a place to live, a base camp, and those associated with the necessities of food and heating oil are always paramount. So much preparation required to collect and study the lowly macrobenthos!

Arctic sublittoral macrophyte vegetation is banded where best developed. More commonly, it lacks uniformity and may be widely spotty and discontinuous. Four distinct bands of sublittoral vegetation are recognizable, at least in a composite description. Each band is depth-related and is characterized by distinct species, species assemblages, and growth forms. Vegetation cover is largely discontinuous, with much barren rock substratum uncolonized for no apparent reason. One or more algal bands may be absent or sparsely represented by characteristic species at any one site. Invariably, species richness and biomass in the Arctic sublittoral are highest between 4 and 7 m. At this depth there is sufficient light penetration to support extensive kelp and understory forms, and, equally important, in most coastal non-fjord or embayment habitats this vegetated area lies below disastrous ice scour and low salinities of inshore waters. In addition, most Arctic sublittoral macrophytes are perennial. Several, such as *Agarum clathratum*, *Laminaria solidungula*, *Alaria esculenta*, and especially a variety of crustose forms, appear to be long-lived, perhaps considerably more than one or two decades. Several perennial species, such as *Leptophytum laeve*, *L. foecundum*, *Cruoria arctica*, the tetrasporophyte stage of *Turnerella pennyi*, *Sorapion simulans*, and a *Pseudolithoderma* species, perennate in small populations at depths in excess of 90 m. While relatively few perennial or annual species occur below 50–60 m, large populations of kelp and associated understory species perennate between 30 and 50 m. Lastly, two adaptive features of the Arctic sublittoral macrophyte are requisite: adaptation to low light in a narrow spectral band for a short season, and adaptation to an essentially constant low water temperature ($-1.85°C$). Characterization of four variably distinct vegetated bands of the sublittoral follows.

Immediately below the level of the ELWS (extreme low water of spring tides), a barren zone extends downward about 0.75 m. This uppermost sublittoral zone is devoid of all macroscopic biota and contrasts sharply in appearance with the lowest littoral and uppermost vegetated bands of the sublittoral. The extreme severity of ice scour and the persistence of very low salinities during much of the short ice-free season appear responsible for this sublittoral barren zone of northern rocky coastal areas.

From the barren band downward to about 3 m is zone I, where conditions for macrophyte development are also severe. Here long periods of low salinity, much ice scour, and invertebrate grazing dictate a restricted species composition and vegetation density. Ice scour is less severe than in the barren zone, but still the stress in this environment regulates population numbers and algal form. Ice scour rarely eliminates species from the zone. Zone I contains three conspicuous, erect, bushy or blade-forming species: *Fucus evanescens*, *Devaleraea ramentaceum*, and *Neodilsea integra*. *Fucus* occurs in two spatially distinct populations along high Arctic rocky coasts. In sheltered areas, *Fucus* may form dense populations in zone I, where it is responsible for the highest biomass values. A second *Fucus* population occurs in the most protected and lowest littoral rock niches, similar to the occurrence of this fucoid on temperate and sub-Arctic North Atlantic shores. All three species occur throughout zone I, but *Fucus* becomes less abundant with increased depth. Owing to severe ice scour in zone I, these and other perennial macrophytes occur only on sides of rock or in sheltered rock crevices. During most of the year they dominate an otherwise sparsely vegetated band. *Haemescharia polygyna*, a perennial, noncalcareous red crust, grows on the sides of rocks and lends a distinctive touch of color to the generally drab lower zone I vegetation. Even more of the lateral rock surfaces are colored by whitish-to-light-pink calcareous crusts including *Lithothamnion glaciale*, *Clathromorphum circumscriptum*, and *C. compactum*. More delicate thalli, such as *Pilayella littoralis*, *Dictyosiphon foeniculaceus*, *Stictyosiphon tortilis*, *Desmarestia viridis*, and gelatinous masses of the tube-forming diatom *Berkeleya rutilans*, cover much of zone I during late July and August with a light brown, filmy, filamentous, stringy cover. *Haplospora globosa* appears in August as large (10–15 cm), dark brown "clouds" of delicate branching axes mixed with several annual species of smaller size. Except for patches of the algae just described, the general appearance of zone I after the midsummer ice breakup is one of bleakness. The sparse vegetation at this time is entirely dark, the thalli twisted and distorted, with no evidence of annual growth increments. By late summer, in habitats of low wave energy, zone I becomes covered with dun-colored masses of filmy annual species. In areas of increased wave energy, annuals are weakly developed, and this portion of the Arctic sublittoral remains drab and depauperate except for the few perennials mentioned.

In zone II, from about 3 to 6 m below ELWS, there is less scour and a more stable salinity, and this area of the sublittoral commonly supports more species and a greater biomass than zone I. Three kelps, *Lami-*

naria saccharina, L. longicruris, and *Agarum clath-ratum,* and annual populations of *Chorda tomentosa* form a conspicuous canopy in this zone. With increasing depth, two more kelps, *Alaria esculenta* and *Agarum,* become increasingly abundant. Unlike boreal and sub-Arctic populations of *A. esculenta,* Arctic representatives of this species in zone II and deeper areas perennate only as infrequent clumps of several, often large, thalli, with long, rigid, robust stipes and deeply pigmented, tattered blades. The habitat, longevity, and, to some extent, the form of Arctic *A. esculenta* differ from these aspects of the species in southerly regions.

In the lower part of zone II, one meets the greatest diversity and biomass in the Arctic sublittoral. Even so, typically, the bottom is covered with a spotty distribution of macrophytes over what appears to be heavily grazed, barren rocky substratum. Only in rare instances is the substratum in zone II abundantly and continuously colonized by attached macrophytes and animals. One may repeatedly ask, What is responsible for the spotty character of the vegetation in zone II? Why is there so much "free" substratum in areas devoid of ice scour and with only small populations of grazers? Macrophyte species beneath kelp (i.e., the understory) are more numerous but form relatively little biomass. Leafy and blade-forming reds such as *Phycodrys rubens, Phyllophora truncata, Turnerella pennyi,* and a number of wiry and small thalloid forms (e.g., *Sphacelaria arctica, S. plumosa, Stictyosiphon tortilis, Desmarestia viridis,* and *Chaetomorpha melagonium*) all predictably occur beneath the kelp canopy. To these conspicuous species one must add another 25–30 species of small stature, largely filamentous or crustose. Zone II, in all of the Arctic, supports a recognizable western North Atlantic algal flora, dominated by several species of kelp common from Cape Cod, Massachusetts, northward. Differences in Arctic and boreal kelp vegetation are few. In the Arctic are fewer species, less biomass, the absence of digitate kelp species, and the frequent dominance of the northern kelp, *Laminaria solidungula.* Only a few Arctic endemics are present in zone II, and the entirety of the Arctic. In general, phycologists with a knowledge of the temperate North Atlantic marine flora would easily recognize the character and composition of zone II vegetation.

In zone III, from about 7 to 20 m, *Laminaria solidungula* and *Agarum clathratum* dominate the vegetation. Both species may occur as dense mixed populations or as virtually monospecific canopies, with only several of the seemingly most tenacious understory species remaining. Populations of *L. solidungula* may cover the bottom with totally prostrate, large, leathery masses of blades and stipes. Or the seascape may consist of low dense forests of *Agarum* with stipes stiff and erect, supporting long, deeply pigmented, drooping, perforate blades. Occasionally, a deep-water form of *Alaria escu-*

lenta and small plants of *L. saccharina* may occur, but they are minor components of the canopy of zone III. Conspicuous understory species in this zone are fewer than in zone II, with the most common being *Turnerella pennyi, Callophyllis cristata, Polysiphonia arctica, Scagelia pylaisaei, Ptilota serrata, Sphacelaria arctica,* and *Chaetomorpha melagonium.* Two generalities are evident concerning the vegetation in zone III. First, one meets with relatively few species, but these occur in predictable communities. Second, one notes the striking appearance of relatively pure populations of the dominant canopy species.

In zone IV, between about 25 and 30 m, canopy vegetation is sparse and discontinuous; *Laminaria solidungula* and *Agarum clathratum* remain dominant, with occasional large populations of *Desmarestia aculeata* in upper portions of this zone. *Agarum* is the last of the large erect macrophytes to disappear, with increased depth, in the sublittoral. The depth limit of this species is unknown, but I have seen large forests of this kelp between 40 and 50 m and have dredged representatives of the species from 75–90 m. Its unknown mode of adaptation to perennation in Arctic deep-water environments and the enigma of its common occurrence in the American and western Greenland Arctic, but not in northeastern Greenland or the remainder of the Arctic, help to identify *Agarum* as future tool of research.

In zone V, from about 30 to 90 m, a number of small forms, as well as noncalcareous (fleshy) and calcareous crusts, occur as the foundation layer of the macrophyte vegetation at all depths. Most of these species have disappeared below about 20 m. Only brown crusts, *Sorapion simulans* and *Pseudolithoderma* species, and several calcareous crusts persist below 20–25 m. Of the dominant calcareous crusts, *Lithothamnion glaciale* is least tolerant of depth. *Leptophyllum laeve* is the deepest-growing and, with *L. foecundum,* perhaps the most abundant calcareous crust in the Arctic sublittoral. *Clathromorphum compactum,* a fourth calcareous Arctic crust, is the least common of these crusts, but is ecologically important in local areas as it forms thick, smooth calcareous pavements over considerable areas of rocky substratum in zones II and III. Generalizations regarding zone IV vegetation include the spotty character of *Agarum* clumps, with *Ptilota, Callophyllis,* and *Scagelia* mingled among *Agarum* haptera. Most exhilarating and interesting to the diver at these depths is the occurrence of large (to 80+ cm in diameter), deep red blades of *Turnerella pennyi* that appear entirely "stuck" to vertical rock surfaces and beneath overhanging rock projections. Blade attachment is a single holdfast, and yet, in situ, the blades remain fully extended and closely pressed to the rock substratum. Uniformly present in nonsilted, low-light, deep-water habitats are populations of three commonly associated species that form a thin reddish turf on small cobble surfaces. Two of these species are rarely recorded, the cyanobacterium *Pleurocapsa violacea* and the minute filamentous green *Uronema curvatum*, while the third is an undescribed cyanobacterium in the genus *Radaisia*.

Phycologists with a knowledge of the western North Atlantic marine flora easily recognize several floristic features and numerous species this flora has in common with the polar sea sublittoral flora. One is comfortable with these similarities between the northern marine flora and that of the western North Atlantic. Also, one is soon conscious of the absence of almost all northern Pacific taxa throughout the entirety of the north polar ocean, and as well as the absence of many European species that fail to extend into the polar seas. The North Atlantic character of the Arctic marine vegetation is further enhanced by a small degree of endemism. Only a few endemic genera and half a dozen species have distributions limited to the Arctic (e.g., *Platysiphon verticillatus*, *Punctaria glacialis*, *Acrocystis groenlandica*, *Ralfsia ovata*, *Jonssonia pulvinata*, *Haemescharia polygyna*, *Cruoria arctica*, and few others). Another distinctive feature of the flora is the unusual depth penetration of a number of species. The occurrence and perennation of macrophytes in deep sublittoral habitats in Arctic coastal waters remain the more enigmatic considering the minuscule amounts of light present in these environments during short periods of each year.

The excitement of discovery awaits future generations of phycologists who will work with Arctic macrophytes – discovery of unknown forms of adaptation that permit polar macrophytes to thrive in an unusually harsh environment. Only one species among the entire flora of more than 150 species, *Laminaria solidungula*, has been studied in situ and in the laboratory. The results of those studies are surprising, in view of kelp biology. *L. solidungula* adapts to polar seasonal oscillations of light and nutrients by storing assimilates, made during a short low-light period, to use for growth during a long dark season, when essential nutrients peak! *L. solidungula* adapts to its environment using a strategy of growth dramatically different from any previously recognized in other *Laminaria* species (Chapman & Lindley 1980; Dunton & Schell 1986). What other forms of physiological and reproductive adaptions remain to be discovered in Arctic macrophytes? Most basic concepts and principles of physiological ecology remain unknown in the context of Arctic marine algae. The prospects for exciting and significant research are clear, considering the salient features of the flora stated at the outset of this essay. Namely, what are the factors responsible for the paucity of species in the polar sea and for their form and strategies of growth and reproduction? But perhaps most enigmatic, considering its development during the past 60 million years, is why the Arctic Ocean flora remains so completely dominated by North Atlantic macrophytes (Wilce 1990).   ◆

2.5 Community analysis
2.5.1 Vegetation analysis

Zonation descriptions of the type just given provide only a superficial view of the whereabouts of the most conspicuous and dominant organisms. Transect diagrams can hardly give even the crudest impression of abundances, neither the relative abundances of species nor changes in the abundance of one species as a function of tide level or of time. To properly understand population and community structure and the controlling forces, quantitative data and methods of vegetation analysis are needed. As Schiel and Foster (1986, p. 268) noted, "the structure of a community is essentially a numbers game." The abundance of an organism may be measured in several ways. The percentage of the substrate covered by the species or association is an appropriate measure where the vegetation is dense and uniform, as, for example, with algal turf or encrusting species. The number of individuals may be counted if they are clearly separate. The biomass (usually dry weight) or energy content of the population may be measured. Each method can be difficult or impossible with some species. For example, one cannot count the numbers of individuals in a turf (or even in an *Ascophyllum* bed), because each of the overlapping plants sends up numerous erect shoots; on the other hand, it is difficult to determine the biomass of encrusting species or to ascertain the percentage cover of widely separated individuals. Biomass estimates are useful for energy studies, but are of no use for population biology or demography (Schiel & Foster 1986).

The basis for any quantitative study of an area is a sound sampling procedure (Gonor & Kemp 1978; Russell & Fielding 1981; DeWreede 1985). The samples must be representative of the population as a whole and must, among other things, reflect the heterogeneity of the population. Samples are most commonly chosen by placing quadrat frames in the study area, employing some means for ensuring the randomness required for statistical analysis. The vegetation in the quadrat can be counted or photographed, which will allow assessment of the same area on subsequent dates, or it can be collected for one-time biomass and species determination. Achieving randomness in placing the quadrat is difficult because the terrain usually is irregular and the vegetation patchy. "Putting [quadrats] in 'representative' or 'typical' places is not random sampling" (Green 1979). As an alternative to mathematically random sampling, many people manage *unbiased* sampling at fixed intervals along lines (transects) stretched horizontally or vertically across the shore (Russell & Fielding 1981). Another practical consideration is the size of the quadrat, which, through preliminary work, must be chosen appropriate to the population or community to be studied and to the type of substrate (Russell 1972; Pringle 1984; Littler & Littler 1985; Chapman 1985, 1986; Wil-

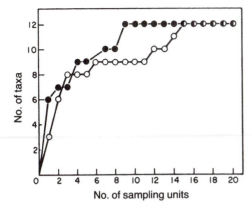

Figure 2.13. Effect of area sampled on number of taxa counted microscopically. Numbers of epiphytic blue-green algal taxa at two localities are related to the numbers of sampling units. The sampling unit was the first 1 cm of an average shoot of *Cladophora prolifera;* area about 2.5 mm². (From Wilmotte et al. 1988, with permission of *British Phycological Journal.*)

Table 2.1. *An approach to the investigation of successional patterns and mechanisms of species replacement*

Step 1: Observe the natural regime of disturbance to which an algal assemblage is subjected; make quantitative measurements if possible

Step 2: On the basis of these observations, design and conduct a multifactorial experiment to reveal the patterns of succession that occur under a variety of realistic regimes of disturbance that differ in
  1. Intensity
  2. Areal extent
  3. Frequency of occurrence
  4. Season of occurrence
  5. Various combinations thereof

Step 3: Formulate and test specific hypotheses concerning mechanisms of successional-species replacements; this may involve studies of
  1. Interspecific competition
  2. The impact of grazing
  3. The tolerance of species to physical stress

Source: Foster and Sousa (1985), with permission of Cambridge University Press.

motte et al. 1988). In general, the number of taxa found in a quadrat will increase to a plateau as quadrat area increases (Fig. 2.13). Too-small quadrats will underestimate species diversity, and too-large quadrats will generate unnecessary work and reduce the number of replicates possible.

Seaweeds are by no means uniformly distributed over the areas they occupy; they are patchy at all scales (e.g., Sousa 1984). Great variation between transects in a given area is also evident. Community analysis can nevertheless show that the apparent discontinuities between zones are real – for example, the breaks between the eulittoral and the littoral and sublittoral fringes (Bolton 1981).

For analysis of community structure, a combination of species incidence (presence or absence) and abundance may be used. Several methods of statistical analysis are available and have been used with varying success by different authors; see McIntyre and Moore (1977), Lindstrom and Foreman (1978), Russell (1980), John et al. (1980), and Russell and Fielding (1981) for details, and see Underwood (1981a) for a review. For example, a method recommended by Russell (1980) produced no discrete groupings when used by Lindstrom and Foreman (1978) on their quadrats. The latter authors, after trying six methods of interpreting their data, concluded that "there is no single best method . . . and the use of several techniques is recommended since the results of one may aid in interpreting the results of another." Each method requires certain kinds of data, makes certain assumptions, and provides certain kinds of information. If different statistical tests are valid, they should give the same results. The null hypothesis

should be tested in assessing statistical outputs, and rigorous testing of hypotheses has been advocated by some ecologists (e.g., Dayton & Oliver 1980; Vadas 1985). On the other hand, a test of theories of causality in ecology by formal hypothesis testing may not be possible or appropriate. As Quinn and Dunham (1983, p. 613) noted, "sensibly stated hypotheses in the methodology of most field investigations . . . are not intended to be mutually exclusive, in any sense exhaustive, or global in their application. It is not possible in [principle] to perform a 'critical test' or experiment to distinguish between the truth of 'alternative hypotheses' if the proposed causal processes . . . occur simultaneously."

Communities are not static: Succession is a normal process in communities that must be taken into account in an assessment of community structure (Foster & Sousa 1985). Successional stages probably rarely lead to a climax community; disturbances of many kinds clear space on small and large scales. A three-step research program for studying successional patterns (Table 2.1) allows one to evaluate the mechanisms of temporal change. Three classical models of succession (Connell & Slatyer 1977) are (1) the facilitation model, in which earlier colonizers make the habitat more suitable for subsequent settlers and less suitable for themselves, (2) the tolerance model, in which later stages simply grow up more slowly through the earlier settlers, without any positive effects of the earlier stage, and (3) the inhibitional model, in which early settlers prevent

Table 2.2. *Life stages of plants: a generalized angiosperm, a generalized alga, and* Durvillaea potatorum

Stage	Generalized angiosperm	Generalized alga	*Durvillaea potatorum*
Unicelluar	Unfertilized spore	Planktonic spore	Planktonic spore
Embryonic	Seed	Benthonic spore/germling	Germling
Growth	Seedling/sapling	Postembryonic/juvenile	Postembryonic/juvenile
Generative	Flowering	Spore development	Plants with conceptacles
Senescence	Senescence	Breakdown of form	Loss of vitality

Source: Cheshire and Hallam (1989), with permission of Walter de Gruyter & Co.

succession. Sometimes one of these models will apply to marine communities, as in the succession on boulders in southern California, where earlier-settling *Ulva* inhibited later-successional algae (Sousa 1979), and the inhibition of *Ecklonia* recruitment by dictyotalean turf (even after its removal) in New South Wales (Kennelly 1987b). In other communities, the complexities in the seasonality of recruitment, growth, and mortality – as well as indirect interactions – have led to interactions that have changed with time (Van Tamelen 1987).

2.5.2 Population dynamics

In seaweed vegetation analysis, little attention has been paid to population dynamics, or demography, as distinct from biomass or productivity studies. Demography is the study of changes in the *numbers of organisms* in populations and of the factors influencing them (Harper 1977; Russell & Fielding 1981; Chapman 1986). Methods for demographic study have been reviewed by Chapman (1985). Plants, including multicellular algae, pass through several stages in their lives (Table 2.2). For a population-dynamics description, we need to know about propagule pool size, germling and adult mortality, age at first reproduction, reproductive life span, proportion of individuals reproducing at a given time, fecundity and fecundity–age regression, and reproductive effort versus growth and predator defense (Solbrig 1980; Chapman 1986). No study of seaweeds has included all these parameters (Chapman 1986), but several studies have produced data on certain aspects. Particularly lacking are data on age-specific fecundity. Part of the problem lies in determining the ages of seaweeds, which has been possible in only a few cases, such as *Laminaria* species (Chapman 1985; Klinger and DeWreede 1988); the methods involved have been reviewed by Cheshire and Hallam (1989). However, even calendar ages may have little bearing on the role of the individual in the population: "periods of dormancy or pauses in the development may result in stands in which plants with widely disparate calendar ages are morphologically similar and fulfil equivalent roles" (Cheshire & Hallam 1989, p. 200).

The number of spores produced, as estimated from sporangial density or in situ spore counts, for instance (Joska & Bolton 1987), does not necessarily represent *fertility*, because the viability of the spores may be low (e.g., Clayton 1981), or in the case of gametes, the success of syngamy may be low.

Reproductive cells are generally shed into the water column. Some of the cells immigrate back into an existing population; others may find a new population. Several studies have shown that dispersal of algal spores is very limited, and yet recruitment at considerable distances can take place; contrast the results of Anderson and North (1966) with those of Ebeling et al. (1985). Some authors have suggested that rafting of reproductive thalli helps disperse spores; others (e.g., Reed et al. 1991) argue for dispersal of individual spores. Successful recruitment depends, in the first place, on suitable physicochemical factors for reproduction, but then propagules/germlings face various hazards before successfully becoming established (as visible recruits). Reed et al. (1991) studied variations in spore settlement by collecting spores on glass slides at different distances from parent-plant stands and then culturing the collection for several days to allow settled spores to grow into recognizable gametophytes (of kelps) or germlings (of filamentous brown algae). They found that whereas kelp dispersal indeed occurred over a very short range most of the time, storms greatly increased dispersal. In contrast, *Ectocarpus* regularly showed long-range dispersal; its spores, unlike those of the kelps studied (*Macrocystis, Pterygophora*), are phototactic and tend to remain in the water column longer, rather than settling.

The survival rate for spores or microscopic stages probably is exceedingly low, but the size of the microscopic population is largely unknown (Chapman 1985). Plants may produce millions or even billions of spores, and yet there is very little recruitment. Kelps produce about 10^{10} spores per plant per year; yet this represents only a tiny fraction of their primary production (ca. 0.17%; Joska & Bolton 1987). On the other hand, *Ascophyllum nodosum* puts about 50% of its biomass into receptacles each year (Josselyn & Mathieson 1978); yet in a 3-year study, Cousens (1981) found not one new recruit in his study plots. Direct observations of microscopic stages have confirmed the enormous loss rate (Chapman 1984; Vadas et al. 1990).

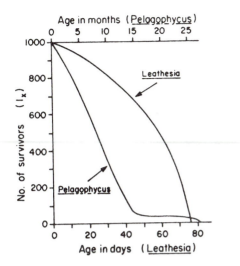

Figure 2.14. Survivorship curves for a summer annual (*Leathesia difformis*) and a perennial brown alga (*Pelagophycus porra*). (From Chapman 1985, with permission of Cambridge University Press.)

Mortality rates for some perennial species (*Laminaria, Pelagophycus, Pelvetia*) have been shown to be high in the younger age classes and low in the older age classes (Fig. 2.14). The shape of the curve reflects the fact that juveniles are being recruited into established populations and are competing with adults. For kelps, there must be densities of spores sufficiently high that male and female gametophytes will be close enough for reproduction. Yet these densities are much greater than those at which sporophytes, even at the "recruitment" stage (i.e., when first visible), can grow. Reduction in numbers by intraspecific competition and other agents is therefore inevitable in the prerecruitment period of such plants (D. C. Reed 1990). Frond densities of *Macrocystis pyrifera* at various stages of sporophyte growth have been modeled as functions of irradiance and temperature (Burgman & Gerard 1990).

For demographic study, *Pelvetia fastigiata* has several advantages (Gunnill 1980a). Its life cycle is direct (haplobiontic), and it is the only fucoid in the very high intertidal, so that all juvenile fucoids in the zone can reasonably be assumed to be *P. fastigiata*. In Gunnill's study, as is usual in such studies, the number of zygotes that settled could not be determined; "recruitment" referred to plants about 10 mm long and already at least several weeks old. Recruitment took place throughout the year, but most recruits appeared 3–6 months after peak release of eggs and sperm from conceptacles. Most losses were of recruits rather than established plants. Plants that survived to 1.5 years, the time of first reproduction, had long life expectancies; 9% of the recruits survived that long.

In the annual *Leathesia difformis*, Chapman and Goudey (1983) found higher mortality among adults, due to crowding as the plants grew larger (Fig. 2.14). Juveniles entered an environment essentially free of intraspecific competition. Mortality due to "self-thinning" (intraspecific competition) and interspecific competition is examined further in section 3.1.3.

Mortality occurs also from grazing (sec. 3.2) and from abiotic conditions. Algal spores and zygotes tend to settle indiscriminately all over the shore, except insofar as some zooids are phototactic and can migrate upward or downward. Some cells or germlings in the intertidal region are subsequently killed by adverse physical conditions (Lüning 1980b; Santelices et al. 1981; Underwood 1981b; Seapy & Littler 1982). For example, *Costaria costata* and other kelps in British Columbia develop in the intertidal during spring, when low tides occur during the early hours. In early summer, warmer weather and daytime low tides combine to produce temperature/desiccation stress lethal to individuals above the sublittoral fringe (L. D. Druehl & M. J. Duncan unpublished data). The situation for the red alga *Gastroclonium coulteri* in central California is similar, and Hodgson (1981) showed that desiccation, rather than light or temperature stress, was the damaging factor. However, seasonal changes in grazer distributions may also be factors (Lubchenco & Cubit 1980; Santelices et al. 1981). Unfortunately, as Chapman (1986) noted, the many studies of the impact of grazing on seaweed populations have not related herbivore densities to seaweed densities, but only to percentage cover or biomass.

In an environment populated to its carrying capacity, the level of competition will be high, and the demand for resources will approximate the supply. Natural selection there will tend to favor competitive ability (including grazer resistance), with a slow growth rate and delayed reproduction as the price.* In unstable areas, where there is a low level of competition because resources exceed demand, selection will favor rapid growth, early reproduction, and short life spans. In an unpredictable environment, where a population or community may be suddenly removed, species with high growth rates (opportunistic species) will have the advantage and thus become primary settlers. However, populations with very high growth rates tend to overshoot the carrying capacity (there is a delay in the feedback that controls growth), and this typically results in a population "crash." This crash, in turn, allows slower-growing species to take over. Such late-successional species fare best in stable environments. There is, however, a continuum of characteristics between these extremes.

The attributes that hypothetically should increase the fitness of opportunistic and late-successional macroalgae have been more completely formulated by Littler

---

* In view of the confusion over the meanings of the terms "*r*-" and "*K*-selection" (Boyce 1984), we have avoided the use of these terms here.

Table 2.3. *Attributes that would seem, a priori, to improve the fitness of opportunistic macroalgae (representative of young or temporally fluctuating communities) versus late-successional macroalgae (characteristic of mature or temporally constant communities)*

Opportunistic forms	Late-successional forms
1. Rapid colonizers on newly cleared surfaces	1. Not rapid colonizers (present mostly in late seral stages); invade pioneer communities on a predictable seasonal basis
2. Ephemerals, annuals, or perennials with vegetative shortcuts to life history	2. More complex and longer life histories; reproduction optimally timed seasonally
3. Thallus form relatively simple (undifferentiated), small, with little biomass per thallus; high thallus SA : V	3. Thallus form differentiated structurally and functionally, with much structural tissue (large thalli high in biomass); low thallus SA : V ratio
4. Rapid growth potential and high net primary productivity per entire thallus; nearly all tissue photosynthetic	4. Slow growth and low net productivity per entire thallus unit because of respiration of nonphotosynthetic tissue and reduced protoplasm per algal unit
5. High total reproductive capacity, with nearly all cells potentially reproductive; many reproductive bodies, with little energy invested in each propagule; released throughout the year	5. Low total reproductive capacity; specialized reproductive tissue, with relatively high energy contained in individual propagules
6. Calorific value high and uniform throughout the thallus	6. Calorific value low in some structural components and distributed differentially in thallus parts; may store high-energy compounds for predictably harsh seasons
7. Different parts of life history have similar opportunistic strategies; isomorphic alternation; young thalli just smaller versions of old	7. Different parts of life history may have evolved markedly different strategies; heteromorphic alternation; young thalli may possess strategies paralleling those of opportunistic forms
8. Can escape predation because of their temporal and spatial unpredictability or by means of rapid growth (satiating herbivores)	8. Can reduce palatability to predators by complex structural and chemical defenses

Source: Reprinted with permission from Littler and Littler (1980), *American Naturalist*, vol. 116, pp. 25–44, © 1980, The University of Chicago Press.

and Littler (1980) (Table 2.3), and the costs and benefits of each attribute compared (Table 2.4). By stating the hypotheses in this form, Littler and Littler were able to test some of the predictions; see also Littler et al. (1983a). As expected, they found that thin, rapidly growing, short-lived algae were characteristic of unstable (temporally fluctuating) environments, whereas coarse, slower-growing, long-lived algae were characteristic of stable environments. Some species, however, through morphologically or ecologically dissimilar alternate phases, have attributes of both extremes (Santelices 1990b). For example, *Mastocarpus papillatus, Scytosiphon* (Fig. 1.23), and *Petalonia* all have a crustose stage (late-successional) and an erect phase (opportunistic). In terrestrial environments, stress-resistant plants tend to have late-successional characteristics, but among marine algae the more stress-resistant species, such as blue-greens, *Ulva,* and *Enteromorpha,* are opportunistic species (Littler & Littler 1980).

Two classes of factors constrain plant establishment and growth in Grime's (1979) theory of plant strategy. *Stresses* are more or less continuous suboptimal conditions that restrict plant productivity, such as shortages of water or nutrients, or suboptimal temperature or salinity. *Disturbances* are discontinuous events, such as mechanical or chemical destruction of tissue. Plants can be categorized as (1) competitors, which occupy habitats of low stress and low disturbance, (2) stress tolerators, or (3) ruderals (opportunists), which occupy highly disturbed areas. There is no successful strategy for areas where stress and disturbance are both high; so in such areas no plants can grow.

2.6 Synopsis

Seaweeds live in complex communities in which they respond to a wide variety of ever-changing biotic and abiotic factors. Many different habitats are occupied by seaweeds, including rocky intertidal and subtidal

Table 2.4. *Hypothetical costs and benefits of the attributes listed in Table 2.3 for opportunistic and late-successional species of macroalgae*

Opportunistic forms	Late-successional forms
Costs	
1. Reproductive bodies have high mortality	1. Slow growth and low net productivity per entire thallus unit result in long establishment times
2. Small and simple thalli are easily outcompeted for light by tall canopy-formers	2. Low and infrequent output of reproductive bodies
3. Delicate thalli are more easily crowded out and damaged by less delicate forms	3. Low SA : V ratios relatively ineffective for uptake of low nutrient concentrations
4. Thallus relatively accessible and susceptible to grazing	4. Overall mortality effects more disastrous because of slow replacement times and overall lower densities
5. Delicate thalli are easily torn away by shearing forces of waves and abraded by sedimentary particles	5. Must commit relatively large amounts of energy and materials to protect long-lived structures (energy that is thereby unavailable for growth and reproduction)
6. High SA : V ratio results in greater desiccation when exposed to air	6. Specialized physiologically, and thus tend to have a narrow range of morphology
7. Limited survival options because of less heterogeneity of life-history phases	7. Respiration costs high because of maintenance of structural tissues (especially during unfavorable growth conditions)
Benefits	
1. High productivity and rapid growth permit rapid invasion of primary substrates	1. High quality of reproductive bodies (more energy per propagule) reduces mortality
2. High and continuous output of reproductive bodies	2. Differentiated structure and large size increase competitive ability for light
3. High SA : V ratio favors rapid uptake of nutrients	3. Structural specialization increases toughness and competitive ability for space
4. Rapid replacement of tissues can minimize predation and overcome mortality effects	4. Photosynthetic and reproductive structures relatively inaccessible and resistant to grazing by epilithic herbivores
5. Can escape from predation because of their temporal and spatial unpredictability	5. Resistant to physical stresses such as shearing and abrasion
6. Not physiologically specialized, and tend to have a broader range of morphology	6. Low SA : V ratio decreases water loss during exposure to air
	7. More available survival options because of complex (heteromorphic) life-history strategies
	8. Mechanisms for storing nutritive compounds, dropping costly parts, or shifting physiological patterns permit survival during unfavorable but predictable seasons

Source: Reprinted with permission from Littler and Littler (1980), *American Naturalist,* vol. 116, pp. 25–44, © 1980, The University of Chicago Press.

zones, coral reefs, and salt marshes. The seaweeds themselves may be major components of the habitats, as in kelp beds. Environmental factors impose patterns on the communities; these patterns are particularly evident in intertidal zonation. Intertidal organisms have differing degrees of tolerance to atmospheric exposure. Tides produce a gradient in atmospheric exposure with abrupt changes, called critical tide levels. Water levels are also affected by storms and atmospheric pressure, and in "tideless" areas the water level becomes the main determining factor. The longer an organism is exposed, the more severely it may suffer stresses such as heat, salinity shock, or desiccation. The time a seaweed spends out of water depends not only on its height relative to the tides but also on the amount of wave action; in wave-exposed (high-energy) environments, zones are higher and wider than in sheltered locations. Plants in tide pools are not exposed to atmospheric stress, but may experience more or less severe changes in water temperature and salinity, depending on the height of the pool in the intertidal zone and its surface-to-volume ratio. The major gradient in the subtidal zone is irradiance. Zonation is much less obvious subtidally, but far fewer studies have been done there.

Descriptions of zonation patterns give no information on population structures and how they change with time or in response to the environment. Vegetation analysis provides a quantitative basis essential to environmental monitoring. Communities are not static: Succession takes place on any newly exposed surface, and plants and animals interact competitively, often facilitating or inhibiting later stages of succession.

Vast numbers of propagules are produced by seaweeds, but only small numbers survive to become mature plants. Competition takes place within and between species, and grazing and physical factors account for much of the mortality among juveniles. Each environment has a carrying capacity, and populations are maintained generally at or below that capacity. Some species experience rapid growth and have short life spans; these tend to be pioneer or opportunistic species. Other species grow more slowly, reproduce later, and put more energy into maintenance and defense; these tend to dominate later in the succession and to form persistent canopies. Highly competitive plants can occupy habitats in which stresses and disturbances are both low. Opportunistic plants are found where disturbances are frequent but stress is low. Some slow-growing plants can tolerate stress, but cannot survive if there are also frequent disturbances.

3

Biotic interactions

The environment of an organism includes both biotic and abiotic (physiochemical) factors. Communities of marine organisms encompass not only the seaweed communities but also the animal communities, of which the benthic grazers and their predators are most important to seaweed ecology. Thus, the biotic interactions of seaweeds include not only competition with other seaweeds (both within and between species) and with sessile animals but also predator–prey relations at several trophic levels; the mix of such interactions will change as the individual changes with age and environmental history.

Biotic interactions are complex, and their study often requires large-scale and long-term observations and manipulations in the laboratory, as well as in the field; see this series of minireviews: Olson and Lubchenco (1990), Carpenter (1990), Paine (1990), Maggs and Cheney (1990). Two quotations from reviews that are both methodological and philosophical will serve to introduce two major topics of this chapter – competition and herbivory: "The objectives of studies of competition include, first, a demonstration that competition occurs, second, identification of the mechanism by which it occurs, and, third, determination of the importance of competition to the ecology of species or communities" (Denley & Dayton 1985). "Approaches and solutions to the problems and measurement of herbivory are strongly influenced by the nature of the questions being asked. Many algal–herbivore interactions have significance in an ecological context only when examined by a whole-community approach, often involving experimental manipulation of seemingly-unrelated parameters" (Vadas 1985).

3.1 Competition

The term "competition" implies that a common resource is potentially limiting (Denley & Dayton 1985; Carpenter 1990). Competition will be most important in a community structure where abiotic stresses and distur-

bances are reduced (Carpenter 1990), but the competitive abilities of organisms will, in general, be affected by abiotic conditions. Two kinds of competition have been recognized, though they are not mutually exclusive (Fig. 3.1): *Exploitative competition* involves a scramble for a limiting resource (e.g., space, light, nutrients) without direct antagonism between organisms. *Interference competition* results from interactions between organisms that may not relate directly (if at all) to any limiting resource; examples include whiplash and allelopathy. If interference competition is taking place, however, exploitative competition must also be potentially possible (Denley & Dayton 1985).

In spite of the fact that the presence of competition has long been recognized (it is a cornerstone in Darwin's theory of natural selection), competitive interactions, particularly marine plant–plant and plant–animal interactions, have not been well documented and are poorly understood. However, "the study of competition in seaweeds is at a crossroads" (Olson & Lubchenco 1990, p. 4). New ecological theories that incorporate spatial heterogeneity and clonal growth have emerged from studies of land-plant populations. "A focus on the functional significance of seaweed traits can help integrate laboratory and field studies by directing our attention to the ecological consequences of traits and to the underlying mechanisms of ecological patterns" (Olson & Lubchenco 1990, p. 4).

3.1.1 Interference competition

Competition for space is most obvious among crustose species and between turf-forming species and erect algae. Crustose coralline algae of a given species may fuse and lose their individual identities when their thalli meet, or they may form "minimal borders," whereas between species, distinct borders are formed, with one species sometimes overgrowing the other (Paine 1990).

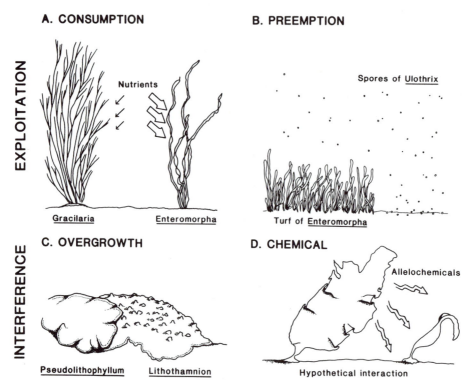

A. CONSUMPTION

B. PREEMPTION

EXPLOITATION

Nutrients

Spores of <u>Ulothrix</u>

<u>Gracilaria</u>   <u>Enteromorpha</u>

Turf of <u>Enteromorpha</u>

C. OVERGROWTH

D. CHEMICAL

INTERFERENCE

Allelochemicals

<u>Pseudolithophyllum</u>   <u>Lithothamnion</u>

Hypothetical interaction

Figure 3.1. Schematic representations of some mechanisms of competition between seaweeds. (A) Consumption of nutrients mediates the interaction between *Gracilaria tikvahiae* and *Enteromorpha* species in the field. (B) A dense turf of *Entermorpha intestinalis* preempts space, preventing colonization by spores of *Ulothrix pseudoflacca* in culture. (C) *Pseudolithophyllum muricatum* overgrows *Lithothamnion phymatodeum* in the field on a smooth, artificial surface in the absence of grazers. (D) An alga uses allelochemicals to exclude a competitor (hypothetical interaction). (Drawing by Charles Halpern, from Olson & Lubchenco 1990, with permission of *Journal of Phycology*.)

Turf-forming algae often are able to preempt space in the intertidal zone, as the following examples illustrate: In central Chile (Santelices et al. 1981), *Codium dimorphum* forms a thick, spongy crust and is able to overgrow and exclude most other lower-intertidal algae during fall and winter. During spring and summer, irradiance and temperatures increase, and the occurrences of low tides shift to the daytime. The *Codium* crust is bleached in the lower intertidal and killed in the mid-intertidal (i.e., becomes noncompetitive), and new borders are created, along which grazers can attack. During fall, winter, and early spring, *Codium* reinvades the mid-intertidal.

In New England and eastern Canada, the low shore is dominated by a belt of *Chondrus crispus* (lower) and *Mastocarpus stellatus*. At the upper edge of its belt, where it mixes with *Mastocarpus*, *Chondrus* is killed or damaged by freezing temperatures during winter low tides, whereas *Mastocarpus* is not harmed (Dudgeon et al. 1989).

In southern California, a low-intertidal turf comprising *Gigartina canaliculata*, *Laurencia pacifica*, and *Gastroclonium subarticulatum* outcompetes the brown alga *Egregia laevigata*. The kelp recruits only from

spores and only at certain times of the year, whereas the red algae expand vegetatively at all seasons via prostrate axes, encroaching on any space that becomes available. During one study, a 100-cm² clearing in the middle of a *G. canaliculata* bed was completely filled in 2 years. The turf traps sediment, which fills the spaces between the axes and prevents the settlement of other algal spores on the rock (Sousa 1979; Sousa et al. 1981). *E. laevigata* may settle if clearings become available at the right time, but by the time it becomes adult and reproductive (it lives only 8–15 months), the turf will have encroached all around and will prevent the kelp from replacing itself in that area. In the mid-intertidal zone, a turf of *Corallina* is persistent and dominant; Stewart (1989) showed how its life history, physiology, and morphology adapt it to weather, tides, and sand (sec. 7.2.3).

Larger algae do not form dense turfs, but some, such as the fucoid *Halidrys dioica*, are nevertheless able to secure space from other species by spreading vegetatively. Clear space is maintained around *Lessonia* plants on wave-swept shores in Chile by the whiplash effect of the laminae (Santelices & Ojeda 1984). In dense beds, scouring of the shore by these blades completely inhibits recruitment of juveniles. In pools, from which adults are absent, urchin grazing eliminates recruitment. However, in patchy *Lessonia* beds, juveniles can be recruited into openings that are large enough to ensure reduced whiplash and yet small enough to feature reduced grazing pressure.

Competition for space also takes place between algae and sessile animals. According to Foster (1975), the outcome of this competition will depend on three factors: irradiance, the presence or absence of grazers, and the presence or absence of predators seeking sessile animals. Under very low irradiance, sessile animals will dominate, whether or not they face predators, but under moderate irradiance the algae may predominate over the sessile animals if predators reduce the numbers of animals. Foster speculated that the lower limits of sublittoral algal growth may be partly regulated by the inability of the algae to compete against sessile animals, rather than by the inadequacy of the light flux to the algae per se. An interesting aside to this is that in California, the ocean goldfish, *Hypsypops rubicunda*, establishes nests of filamentous red algae on poorly lit vertical walls dominated by sessile animals. It does this by systematically removing the animals from the area, with the result that the algae can successfully compete for space (Foster 1972) (sec. 3.2.2).

Competition for space takes place among *Postelsia*, mussels, and a red-algal turf (chiefly *Corallina officinalis*) on the northeast Pacific coast (Paine 1988). Mussels will outcompete the turf that becomes established in clearings in the mussel bed. If clearings occur when nearby *Postelsia* are fertile, new sporophytes will grow up on the rock, but they are annuals and will be replaced by the turf. Although *Postelsia* can attach to turf, it cannot persist on that substrate, and apparently it relies on waves to remove mussels and provide new bare rock on which it can reestablish. In other mussel beds, such as those in Chile, various algae coexist with the shellfish. The mussels ingest algal spores (some of which may survive), but they protect germlings from desiccation and probably fertilize them. Algal populations are also regulated by small grazers among the mussels (Santelices & Martínez 1988). In all these examples, the influences of external factors, especially abiotic factors, are apparent.

3.1.2 Epiphytism and allelopathy

In the "space race," one solution is to grow as an epiphyte. Epiphytism is a common way of life, though relatively few algae are obligate epiphytes. In some locations, most of the algae are epiphytes. Such a place is the intertidal zone in southern California, where two species of *Corallina* occupy over 60% of the total rock substrate, and at any given time and site some 15–30 species of seaweeds are epiphytic on the coralline turf (Stewart 1982). However, even as this way of life solves the space problem for the epiphyte, it creates problems for the anchor species (or basiphyte).

Epiphytes can include filamentous seaweeds, small coralline crusts such as *Melobesia*, encrusting bryozoans such as *Membranipora*, and other sessile animals such as *Spirorbis* and *Obelia*. Whether plant or animal, the effects of epiphytes generally are to shade the anchor species, impede gas and nutrient exchange and thereby decrease its growth rate, and increase drag on the fronds (Sand-Jensen 1977; Harlin et al. 1985; Silberstein et al. 1986). Moreover, heavy encrustation of kelp blades by bryozoans can lead to loss of the blades to predatory fishes, which cannot get the animal prey without also taking the alga; small *Macrocystis* beds have been destroyed in this way (Bernstein & Jung 1979). On the other hand, dense epiphytes on the *Corallina* turf described earlier may help alleviate desiccation stress (Stewart 1982), and epiphytism may allow shade-loving plants to grow in sunnier locations, by creating shade (e.g., *Cladophora rupestris;* Wiencke & Davenport 1987).

Various factors can mitigate these effects: For instance, small grazers often take off the epiphytes while leaving the macroalga intact (sec. 3.2.1). Some algae, including *Ulva, Enteromorpha,* and *Cladophora,* may avoid epiphytism simply because of their very rapid growth and their changes in pH at the thallus surface caused by a rapid metabolic rate (den Hartog 1972a). When the growth of such species slows, epiphytes soon cover them. More rapid turnover of *Macrocystis* blades results in relatively little epiphytism, as compared with a long-lived blade (*Rhodymenia californica*) in the understory, on which several layers of epiphytic animals can develop (Bernstein & Jung 1979). Differences in carrageenan composition make gametophytes of *Chondrus crispus* and *Iridaea cordata* less susceptible than sporophytes to infection by endophytic *Acrochaete operculata* (Correa & McLachlan 1991).

Many macrophytes deter epiphytes either through periodic sloughing of their surfaces or by production of antibiotic chemicals (Gonzalez & Goff 1989); such phenomena are discussed later in this section. The process of biofouling is of interest not only to botanists studying host–epiphyte relations but also to marine engineers, who must try to control fouling of ship hulls and other submerged structures.

The process of production and release of compounds that inhibit other algae is termed allelopathy (Harlin 1987). Sometimes the affected alga is a competitor, sometimes an epiphyte. A mixed culture of three crustose algae studied by Fletcher (1975) showed that the growth of two reds (*Porphyrodiscus simulans* and *Rhodophysema elegans*) was inhibited along margins adjacent to *Stragularia spongiocarpa*. Crustose germlings of *Chondrus crispus* (but not the closely related and morphologically similar *Mastocarpus stellatus*) can inhibit diatom growth (Khfaji & Boney 1979).

Brown-algal phenolics, stored in physodes (sec. 1.3), appear to serve as antifouling compounds. The

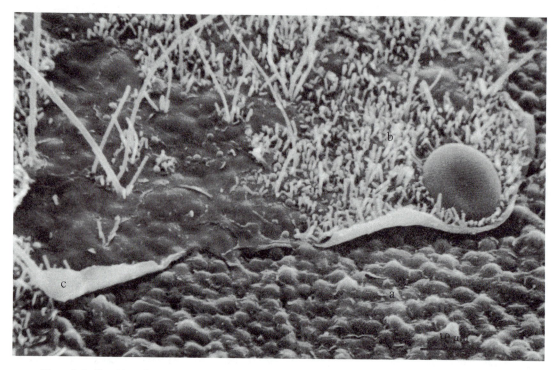

Figure 3.2. Sloughing of outer cell-wall layers, with removal of epiphytes. Scanning EM view of the surface of *Chondrus crispus* showing the cuticle (c), with bacteria (b) sloughing away, leaving a clean algal surface (a). Scale = 10 μm. (From Sieburth & Tootle 1981, with permission of *Journal of Phycology.*)

phenolic content of *Sargassum* species in the Sargasso Sea is highest in the growing tips of plants near the center of the sea, and only those tips are free of epiphytes (Conover & Sieburth 1964; Sieburth & Conover 1965). Phenolics conceivably could be the cause of inhibition of epiphytic animals on distal tissue of *Laminaria* (Al-ogily & Knight-Jones 1977), and there may also be a correlation in Rawlence's (1972) observation that physodes were absent from tissues of *Ascophyllum* near invading rhizoids of *Polysiphonia lanosa*, but these casual observations need experimental support.

Antibacterial activity by seaweeds has been shown in several studies (e.g., Hornsey & Hide 1976; Caccamese et al. 1985; Calvo et al. 1986; Moreau et al. 1988). Various distribution patterns for such activities occur in the thalli, when any activity is present, and the highest level of activity is not correlated with growing tissue. The compounds implicated include brominated phenolics, coumarins, and halogenated lipids (Harlin 1987; Jones 1988). Unfortunately, in this kind of study, the tissues often are dying under experimental conditions, causing substances to leach from the cells; hence some of the experimental results may not reflect the natural situation (Fenical 1975). Harlin (1987, p. 245) ended her review by pointing out that "given the metabolic cost of synthesizing an allelochemical and also the dilution factor in the aquatic milieu, . . . evo-

lutionary significance of allelopathy to the releasor seems unlikely."

Several algae have been found to slough their outer layers, which has the effect of ridding the thallus of epiphytes (Fig. 3.2). *Enteromorpha intestinalis* continuously produces new wall layers and sloughs off the outer layers of glycoprotein (McArthur & Moss 1977). This species has an unusual wall structure, and its ability to remain free of epiphytes probably is essential in minimizing drag on the thallus and allowing the alga to colonize ship hulls, where hydrodynamic forces are great. In *Halidrys siliquosa* (Moss 1982) and *Himanthalia* (Russell & Veltkamp 1984), the old walls of the epidermal cells are shed as "skins" following production of a new wall underneath. The whole outer epidermal layer may be shed by *Ascophyllum nodosum* (Filion-Myklebust & Norton 1981), although Moss (1982) has questioned this. The epiphyte *Polysiphonia lanosa* may retain its hold on *Ascophyllum* by having rhizoids that penetrate below the sloughing layer, a mechanism that Gonzalez and Goff (1989) recently showed to be the means whereby *Microcladia coulteri* defeats its hosts' attempts to slough it off. However, most *P. lanosa* attaches in wounds and lateral pits, where there is no sloughing (Lobban & Baxter 1983; Pearson & Evans 1990). Two epiphytes specific to *Himanthalia* receptacles retain their hold by penetrating into the cryptostomata. *Laminaria* species continuously slough the distal part of the blade. As a result, epiphytes are restricted to a few that can grow and reproduce before they reach the end of the production belt (Russell 1983). *Laminaria* stipes are long-lived, in contrast, and bear a much richer epiphytic flora (e.g., Tokida 1960).

Corals also must defend themselves against overgrowth by seaweeds, although the balance on the reef is predominantly a function of nutrients and grazers (Fig. 2.3). The prostrate, foliose brown alga *Lobophora variegata* can overgrow adult scleractinian corals, but the corals are able to reduce the growth of adjacent fronds, presumably chemically. However, heavy grazing on and rapid turnover of *Lobophora* blades are more important means by which the corals remain clear (de Ruyter van Steveninck et al. 1988).

3.1.3 Exploitative competition

Intraspecific exploitative competition is manifest in the effects of density on plant size and survival. Populations of newly recruited juveniles may be very dense, whereas by the time the plants mature there will be far fewer of them (Fig. 2.14); part of the decline is due to intraspecific competition (i.e., self-thinning). Plants in dense stands tend to be small, whereas those in more disperse populations may be larger. On logarithmic plots of mean plant weight versus plant density, data points lie on or below a line of slope $-\frac{1}{2}$ (Fig. 3.3), and this line seems to represent a boundary for weight (w) × density (d). The equation for the line is $w = Kd^{-3/2}$, where K is a constant equal to 4.3 for perennial herbs. The equation can be rewritten $\log w = 4.3 - 1.5 \log d$. Below the boundary line, any combination of mean plant weight and density can occur. Although data for seaweeds are scarce, Cousens and Hutchings (1983) have shown that these data also generally fit the $-\frac{1}{2}$-power law (Fig. 3.3). Self-thinning often is attributed to light attenuation by the plant canopy. However, there is no evidence as yet that thinning via competition drives the $-\frac{1}{2}$-power law (R. Cousens, personal communication). Density-dependent competition occurs between *Macrocystis pyrifera* and *Pterygophora californica*, as well as within each species (D. C. Reed 1990; Reed et al. 1991). In mixed-species experiments, *Pterygophora* consistently inhibits *M. pyrifera* recruitment.

Interspecific exploitative competition between seaweeds has been studied in the laboratory only twice (Russell & Fielding 1974; Enright 1979), although there have been several studies on phytoplankton (e.g., Kayser 1979, Maestrini & Bonin 1981, Tilman et al. 1982). Russell and Fielding devised a "triangular" method, in which three species were grown in pairs (AB, BC, AC). In this way any differences in yield of a species when in culture with each of the other two, respectively, are likely to be due to competition from the other species. Competitive ability was judged from the sizes of the differences. The limitation of this method is that, if no dif-

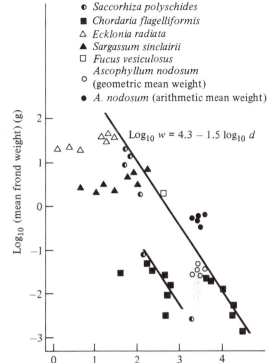

Figure 3.3. Relationships between mean frond weight and frond density for stands of various marine macrophytic algae growing in unialgal stands. Data for *Saccorhiza polyschides* and *Chordaria flagelliformis* are for time courses of growth of stands. Both lines have a slope of $-\frac{1}{2}$. The upper line has the equation $\log w = 4.3 - 1.5 \log d$. (From Cousens & Hutchings 1983, reprinted by permission from *Nature*, vol. 301, pp. 240–1, © 1983, Macmillan Journals Limited.)

ference is seen, it may mean that there is no competition with either of the other two species or that the species is equally affected by the other two. The three species used were *Ectocarpus siliculosus*, *Erythrotrichia carnea* and *Ulothrix flacca*. Russell and Fielding found that *Ectocarpus* was involved in many competitive interactions. For example, as shown in Table 3.1, the yield of *Ulothrix* at 10° C was lower in the presence of *Ectocarpus* than in the presence of *Erythrotrichia*, except in three of the nine combinations (high irradiance with U:Co inoculum ratio 3:1; and low irradiance with U:Co::1:3 and 2:2). Under some conditions every species became noncompetitive: *Ulothrix* tended to be noncompetitive in dim light whatever the temperature, whereas *Ectocarpus* and *Erythrotrichia* became noncompetitive in dim light only at high temperatures. All three species lost competitive ability in reduced salinity, and the loss was greater than anticipated from results with monocultures. *Ulothrix* at 15° C and approximately 38 μE m^{-2} s^{-1} grew well against *Erythrotrichia* but was nearly extin-

Table 3.1. *Yields*[a] *of* Ulothrix *in co-culture with its competitors* Ectocarpus *(Ec) and* Erythrotrichia *(Er)*[b].

(Temperature) °C	Approximate irradiance ($\mu E\ m^{-2}\ s^{-1}$)	Competitor	Ratio *Ulothrix* : Competitor					
			1 : 3		2 : 2		3 : 1	
10	38	Ec	12	13	26	24	37	36
		Er	25	26	31	34	35	36
	19	Ec	4	4	15	14	25	23
		Er	15	15	31	27	32	34
	6	Ec	4	6	9	8	10	10
		Er	5	5	9	9	14	15
15	38	Ec	0	0	0	0	8	8
		Er	17	20	25	23	27	23
	29	Ec	2	1	5	6	8	9
		Er	8	8	10	12	16	17
	6	Ec	2	2	7	8	10	10
		Er	4	5	8	9	8	10

[a] Yield is given in relative units related to packed cell volume
[b] Salinity in all experiments was 34‰. Boxed-in parts of the table show significant differences in yields of *Ulothrix* in culture against each of the two competitors.
Source: Russell and Fielding (1974), with permission of Blackwell Scientific Publications.

guished by *Ectocarpus* (Table 3.1). Russell and Fielding concluded that, "the presence of a competitor . . . seems to sharpen the sensitivity of a species to its environmental conditions and to reduce the amplitude of those conditions in which the species is vigorous."

Viewed against this study, the results of experiments in which the competitive abilities of different species have been assessed from growth in monoculture may be misleading. For example, although Kain (1969) showed that *Saccorhiza polschides* grew faster at 17°C than at 10°C, whereas *Laminaria hyperborea* and *L. digitata* grew equally at both temperatures (see Fig. 6.10), this does not mean that, growing together, all would survive in accordance with the predictions based on monoculture.

Enright (1979) studied competition between *Chondrus crispus* and one of its epiphytes, *Ulva lactuca*, in a flow-through seawater system. The two species were planted in several ratios, as well as alone. The possible outcomes of the experiment, called a de Wit replacement series, are shown in Figure 3.4, where the two species are called *A* and *B:* If there is no interaction (Fig. 3.4a), the yield of each species is directly proportional to the amount sown. Interaction between species may increase the proportion of *A* (Fig. 3.4b) or *B* (Fig. 3.4c), or it may increase or decrease the

yields of both species (Fig. 3.4d,e). Enright tested the growth of the two species under several light and temperature combinations (Fig. 3.5). Under all conditions, *Ulva* grew faster than *Chondrus* in monoculture, and it grew better still (average 1.6×) in mixed culture. *Chondrus* in mixed culture showed only 60% of its monoculture growth rate. Dominance by *Ulva* was greater at higher irradiance and temperature. Enright concluded that in all of these experiments the eventual outcome would have been monocultures of *Ulva*, because her results did not show any evidence of competitive equilibrium. This conclusion seems to conflict with field observations on these two species and on other epiphyte–basiphyte associations, where basiphytes do in fact survive. The answer to the paradox may lie in the role of grazers in nature, as indicated in an example of the effect of snails on *Chondrus* and *Enteromorpha* in tide pools (sec. 3.2.1). However, the outcome of experiments using de Wit replacement series can depend on the initial density of plants, and furthermore, such ratio diagrams give no information about competitive interaction in the field, where initial densities are not experimentally constrained (Inouye & Schaffer 1981).

Continued experimentation with mixed cultures should help to determine the nature of competitive in-

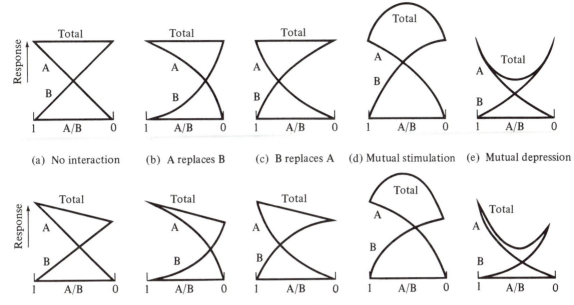

Figure 3.4. Possible outcomes of de Wit replacement-series experiments. The initial ratio of species A to species B goes from 1 to 0 in each experiment. In the top row, the yields of the two species in monoculture are identical; in the bottom row, species B is lower-yielding than species A. The vertical axis is any measure of species response (e.g., fresh weight). Interactions (a) to (e) apply to both rows. (From Bannister 1976, with permission of Blackwell Scientific Publications.)

teractions, including the extents to which different resources are important, and perhaps whether or not extracellular products are important in any positive or negative ways.

3.2 **Grazing**
3.2.1 Impact of grazing on community structure and zonation

Every community has herbivores; their impact on seaweed populations can be devastating, as in the replacement of kelp beds by urchin barrens in California and Nova Scotia; yet in many communities algae and herbivores coexist (e.g., Vásquez et al. 1984). In most situations, less than 20% (often less than 10%) of the available algal biomass passes through herbivores; most goes to the detrital food web (Vadas 1985). Herbivory can affect the competitive balance among seaweeds, and several studies have shown that algal species diversity is highest with intermediate grazing pressure (Paine 1977; Lubchenco 1978), but this effect may be by no means universal; the evidence was reviewed by Chapman (1986). Herbivores may be categorized into three groups: fish (large foraging ranges, low densities), urchins (intermediate ranges and densities), and mesograzers (small foraging ranges, but high densities) (Carpenter 1986). Herbivorous fish occur chiefly in warmer waters (40° N to 40° S) and are especially significant in the tropics (Horn 1989).

Seaweed–urchin–predator interactions often are complex and controversial. In the Strait of Georgia, British Columbia, the green urchin *Strongylocentrotus*

droebachiensis is at its southern limit and frequently fails to recruit because of water temperatures that are too high (Foreman 1977). Its successful recruitment led to temporary reductions in both biomass and species diversity (Fig. 3.6). The urchins removed the foliose macrophytes first, then filamentous forms, and finally, to a small extent, crustose algae and corallines. The period of time required for the community parameters to be restored to their pre-urchin levels (not necessarily the exact pre-urchin flora) was 4–6 years in the mid-subtidal regions dominated by *Agarum* or *Laminaria*, but a little less, 3–5 years, for shallower subtidal communities (Foreman 1977).

The urchin–kelp relationship in Nova Scotia, Canada, has been interpreted and reinterpreted, with opinion shifting between cyclic and noncyclic change (Fig. 3.7). Elner and Vadas (1990), in tracing these shifting paradigms, were critical of too much weak inference and reliance on "common sense" (in lieu of experimental testing) in the research papers. Initially, the system was predicted to have a 3–4 year cycle driven by lobster abundance (lobsters supposedly are predators on urchins) (Fig. 3.7a). When barrens persisted longer than 4 years, their formation was reinterpreted as an irreversible consequence of overfishing for lobsters (Fig. 3.7b). As the role of lobsters came into question, the explanation shifted to complex behavioral interactions with several predators (Fig. 3.7c). Most recently, on the basis of an urchin die-off associated with warm water, the pattern was projected to be a 15–20 year cycle regulated by temperature (Fig. 3.7d).

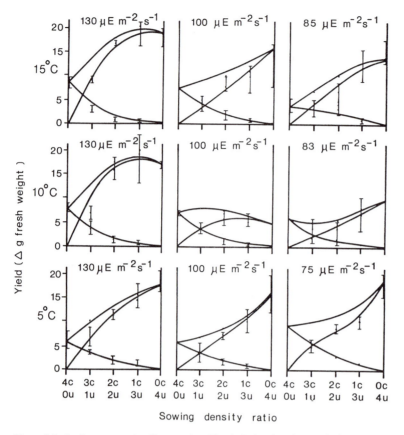

Figure 3.5. Replacement-series diagrams (see Fig. 3.4) showing changes in fresh weights of *Chondrus crispus* (c) and *Ulva lactuca* (u) when grown at a range of initial densities and at several temperatures and irradiances. Means and standarad deviations of measurements are shown on the lines for each species; summed yields are also given. The biomass of *Ulva* in the experiment at 10°C, 100 μE m^{-2} s^{-1}, declined because the thalli sporulated and then disintegrated.) (From Enright 1979, with permission of Science Press.)

Severe attacks by urchins on kelp beds have received considerable attention because of the commercial value of the plants and some associated animals, such as abalone and lobster. Urchins eat mostly plants, but are considered omnivores (Lawrence 1975), and they have various (and variable) feeding "strategies." Some are sedentary, others nomadic; a species may behave differently depending on its density (Vadas 1990). Highly mobile species (or individuals) can be selective in what they eat, whereas sedentary species must be generalists. In the tropics, where grazing pressure from fish is high, search costs are high, and urchins there tend to be sedentary, in contrast to the situation in midlatitudes, where the algal biomass is high and herbivorous fish few (Vadas 1990).

The purple urchin, *Strongylocentrotus purpuratus*, and the large red urchin, *S. franciscanus*, have devastated large areas of kelp forest in southern California (Leighton et al. 1966). The increase in urchin numbers was triggered, according to Tegner (1980), by the harvesting of the various species of abalone, which began on a large scale in the 1930s. Abalones do not eat ur-

chins, but are competitive herbivores. In stable ecosystems, both abalones and urchins collect drift seaweed and rarely forage on attached plants. Removal of the abalones resulted in a vastly increased food supply for urchins; as a result, urchin numbers exceeded the level at which natural predators (fish, starfish, spiny lobster) could control them. When the drift kelp supply was outstripped, the urchins moved onto attached plants, and the populations did not crash. Sportfishing for the sheephead, an urchin predator, has also contributed to the problem, as has the ability of urchins to use dissolved amino acids from sewage. The role of otters in the kelp–urchin interaction along the west coast of North America is still controversial. Schiel and Foster (1986) concluded from a review of the literature that whereas otters certainly reduce urchin numbers, other factors such as winter storms and turbulence along the coastline are more important, at least in central California.

Fish and urchins are competitive herbivores, both important in structuring tropical-reef algal communities. Intense grazing pressure leads to a sparse turf of algae only a few millimeters high, comprising unicells

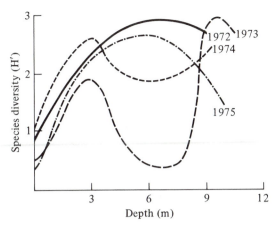

Figure 3.6. Recovery of an urchin-grazed community in British Columbia after the urchins had moved on. Grazing took place in 1973 and markedly reduced the species diversity (H') in the 3–9-m depth range. By 1974, H' was nearly restored to its 1972 (pregrazing) value, except for a patch in mid-depth dominated by *Nereocystis*. By 1975, recovery was virtually complete (but the species present in 1975 were not exactly the same as before grazing). (From Foreman 1977, with permission of *Helgoländer Meeresuntersuchungen*.)

and filaments, or crustose species (Hackney et al. 1989) (see Fig. 1.15b). Productivity is high, and turnover very rapid (4–12 days), but the biomass is extremely low (Fig. 3.8) (Adey & Goertemiller 1987; Klumpp & McKinnon 1989; Williams & Carpenter 1990).

The effects of urchins in the tropics were dramatically demonstrated by a mass mortality in the Caribbean in 1983, when 93–100% of the *Diadema antillarum* died throughout the region (de Ruyter van Steveninck & Bak 1986; Hughes et al. 1987; Carpenter 1988; Lessios 1988). In areas of high fishing pressure, such as Jamaica and Curaçao, *Diadema* controlled shallow fore-reef algal communities, and crustose corallines dominated (Littler et al. 1987a; Morrison 1988). Experiments before the die-off (e.g., Hay et al. 1983) had shown that when the urchins were excluded, the corallines became overgrown by filamentous and fleshy algae. As expected, there was a great increase in biomass, but a reduction in productivity, following the mass mortality. The reduced productivity came about because of the lower efficiencies of thicker algae; see Littler and Littler's functional-form groups (Table 1.1, Fig. 4.27) (sec. 4.3.2, 4.7.2). Yet the magnitude of the biomass increase was unexpected. Apparently it was due to a difference between natural and artificial removals. When a small area was experimentally cleared, fish from surrounding areas (where they were competing with urchins) moved in and grazed the test plots, whereas after the die-off the fish had an overabundance of algal food (Lessios 1988).

The conclusions from both tropical and temperate studies of grazing have been questioned because of methodological problems such as pseudoreplication and lack of error estimates. Schiel and Foster (1986, p. 289) pointed out that ''where variability is obscured because of sampling design used or measures of variability do not exist, it is often considered to be insignificant. . . . Better conceived and designed experiments and sampling would result in better evidence, so that equally plausible alternative hypotheses can be rigorously tested.'' Also at issue in studies of grazing impact are the effects of caging. Whereas, hypothetically, cages change only the grazing pressure, in fact they also affect water motion, irradiance, and temperature, particularly as the cages become fouled with algae (Underwood 1980; Underwood et al. 1983; Kennelly 1983; Vadas 1985).

Mesograzers compose an important group of herbivores, but they have rarely been studied because their size and activity make density manipulation in the field difficult. This group includes amphipods, copepods, and polychaetes (Brostoff 1988). (In some publications they have inappropriately been called micrograzers.) Brawley and Adey (1981) were able to study the effects of gammarid amphipods in a coral-reef microcosm (a model reef in a big aquarium), and they found that filamentous pioneer algae such as *Hincksia rallsiae, Bryopsis hypnoides*, and *Centroceras clavulatum*, among others, were quickly grazed down, allowing a chemically defended species, *Hypnea spinella*, to dominate. More recently, the importance of caprellid and gammarid amphipod grazing of epiphytic algae was shown by Brawley and Fei (1987). Grazing prevented epiphytic overgrowth of *Gracilaria*, even in the presence of fish predation on the amphipods. Similar results by D'Antonio (1985b) and Brostoff (1988), including field studies have shown the importance of mesograzers in epiphyte removal from macrophytes. Another role of mesograzers is illustrated in the facilitation of release and dispersal of spores from cystocarps of *Iridaea*. The mature cystocarps are the preferred food of the amphipod *Hyale media*, but it eats only a fraction of the spores it breaks out. Many spores stick to the animal's legs and are dispersed; some of the spores that are eaten also survive and are later deposited (Buschmann & Bravo 1990).

The role of mesograzers in nitrogen availability also needs to be assessed; there is growing appreciation of the significance of nitrogen pulses from animals, particularly in oligotrophic tropical waters with tight nutrient cycling (e.g., Nelson 1985; Williams & Carpenter 1988) (see Chapter 5).

Another overlooked grazer/predator is the human. In many regions of the world, seaweeds and herbivorous invertebrates (e.g., abalones) are collected for food. In Chile, human predation in the intertidal zone is extensive, both on animals (e.g., the limpet *Fissurella picta*) and on algae (*Iridaea boryana* and *Durvillaea*) (Castilla & Bustamente 1989). By harvesting limpets for food, fishermen inadvertently increase the stock of *Iridaea*, which is harvested as a cash crop for export to

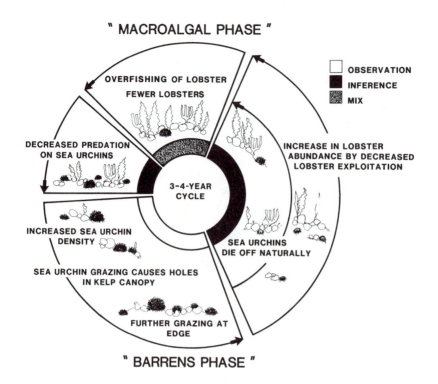

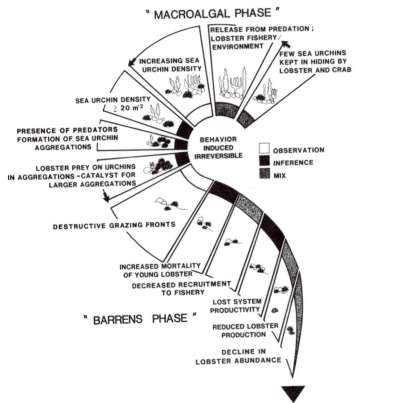

(a)

(c)

Figure 3.7. Cyclic or noncyclic? Changing views of the urchin–kelp interactions in Nova Scotia, Canada. (a) A 3–4 year cycle driven by lobster abundances was the model from about 1970 to 1975. (b) The second model (ca. 1975–8) held that the urchin barrens were irreversible consequences of overharvesting of lobsters. (c) For a few years (1978–83), more complex interactions were viewed as the causes of irreversible decline. (d) Finally (since 1984), the model has again become cyclic, now driven by urchin recruitment variations. (Reprinted from Elner & Vadas 1990, *American Naturalist* vol. 136, pp. 108–25, © The University of Chicago Press.)

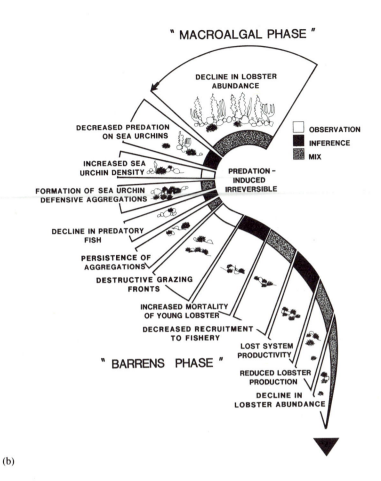

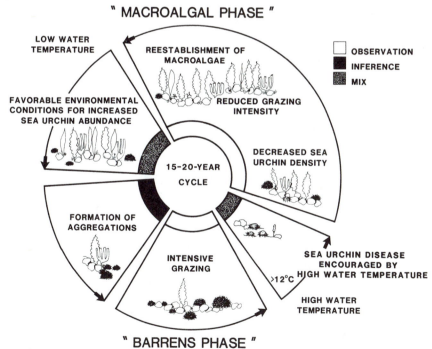

(b)

(d)

INTENSITY OF MACROGRAZING PRESSURE

	Low To Absent	Moderate To High	High	Most Extreme (Bare Sub-stratum)
Predominant Benthic Algal Component	Macroalgae	Algal Turfs	Crusts	
Standing Crop	Very High	Low	Very Low	
Primary Productivity (Per Unit Area)	Very High	Moderate	Low	
Primary Productivity (Per Unit Biomass)	Low	Very High	Very Low	
Predominant Types Of Adaptations	Maintain Upright Morphology	Growth	Resistance	

Figure 3.8. Diagrammatic representation of grazing continuum model for tropical reefs (from absence of herbivory to the highest grazing pressure encountered in the shallow reef environment). Various characteristics of the three benthic algal communities are listed, including the predominant adaptations that allow persistence in these environments. (From Hackney et al. 1989, with permission of Blackwell Scientific Publications.)

Japan. In a reserve where fishing was prevented, the limpet eradicated the red alga (Moreno et al. 1984).

Moderate grazing pressure may increase the diversity of algal species by preventing dominance of large, canopy species. For instance, in tide pools and in the subtidal zone on the coast of Washington state, the presence of sea urchins maintains a population of small, ephemeral algae. When grazers were removed experimentally, a succession of algae ended in dominance of the pools by the kelp *Hedophyllum sessile* and dominance of the subtidal rocks by *Laminaria* species (Paine & Vadas 1969; Duggins 1980). Diverse algae are maintained in the territories of tropical damselfish by selective grazing (Hixon & Brostoff 1983; Sammarco 1983); beyond those territories, where fish grazing is severe, and in fish-exclusion cages after natural succession had taken place, algal diversity is lower (sec. 3.2.2). Limpets and snails have major roles in controlling the seasonal abundances of *Collinsiella, Rhodomela,* and diatoms in Washington tide pools (Dethier 1982).

Similarly, interspecies competition may be affected by grazing pressure. In New England, intertidal rock pools tend to be dominated by either *Chondrus crispus* or *Enteromorpha. Chondrus* is successful when high densities of *Littorina* keep the rapidly growing *Enteromorpha* in check. Removal of the snails from a *Chondrus*-dominated pool resulted in a takeover by *Enteromorpha.* The snails were controlled, in turn, by primary and secondary carnivores: shore crabs and seagulls (Lubchenco 1978).

Crustose coralline algae compete for space by overgrowing each other (Fig. 3.1), but limpets influence the outcome in tide pools in Washington. Thicker crusts overgrow thinner crusts. In high pools, frequent but low-intensity grazing rasps away the surface cells of *Lithophyllum impressum* (see Fig. 3.9), but deeply wounds *Pseudolithophyllum whidbeyense,* which lacks a multilayered epithallus. *Lithophyllum* wins. In lower pools, meristems of both crusts are injured, but *Pseudolithophyllum* rapidly overgrows its wounds and the neighboring *Lithophyllum* (Steneck et al. 1991).

With all these effects of grazers on community structure, one expects to find some zonation patterns attributable to grazing. In the intertidal zone the major grazers are limpets and littorinids. Among the examples of zone control attributed to these grazers is the upper limit of a red *Laurencia-Gigartina* belt in parts of Britain by mid-shore *Patella* populations (Lewis 1964). Remarkably sparse grazers are capable of preventing the establishment of foliose algae (Underwood 1980). Removal of the grazers from plots in the animal-dominated middle-to-upper-intertidal zone in New South Wales allowed foliose algae to establish, but many did not survive as adults owing to the physical conditions of the intertidal (Underwood 1980).

An interesting difference in plant–animal interactions between North American and southeastern Australian rocky intertidal zones has been pointed out by Chapman (1986). In North America, low-shore algal-dominated zones occur only where carnivory by whelks and seastars impacts the competitively superior barnacles and mussels. In Australia, in contrast (Underwood & Jernakoff 1981), seaweeds competitively displace limpets, which feed only on microscopic algae/stages, and there is a dense lower-intertidal belt of algae. The algal zone ends abruptly in the mid-intertidal, where it is superseded by barnacles or grazing molluscs (depending on wave action).

Zonation may also be influenced by a combination of grazing and competition. For instance, in southern Chile, a low-intertidal zone of crustose corallines and *Ulva rigida* normally separates a lower zone of *Ahnfeltiopsis furcellatus* from an upper zone of *Iridaea boryana*. But this band is maintained by mollusc grazing; when the herbivores were removed, *Gymnogongrus* and *Iridaea* outcompeted the slow-growing corallines and ephemeral *Ulva*, so their zone limits changed (Moreno & Jaramillo 1983). Competitive ability depends on growth conditions (sec. 3.1), and grazing pressure can change seasonally. Thus intertidal zonation is maintained in a state of dynamic flux, resulting in fluctuating zone limits, as shown in Figure 2.8.

A striking pattern in algal distribution is the preponderance of smaller and more calcified plants in the tropics, as compared with temperate waters. Suggested reasons for this pattern include more severe desiccation, warmer water temperatures, lower nutrient availability, and changes in herbivory. Gaines and Lubchenco (1982) argued that, of these, only the changes in herbivory can account for the latitudinal change in seaweed form. The pertinent characteristics of the tropical herbivore fauna include the great diversity and high biomass of herbivorous fish and urchins, year-round foraging, and the well-developed visual capabilities and mobility of the fish.

Interactions among grazing, competition, and physical factors are illustrated in the following examples. In the New England sheltered intertidal zone, *Chondrus crispus* dominates the sublittoral fringe, outcompeting fucoids. Farther north, in the Gulf of St. Lawrence, ice scour alters the interaction in favor of fucoids, and in Scotland the heavy limpet grazing has the same result (Lubchenco 1980). Wave action and predation affect the competition between *Chondrus* and competitively dominant mussels (*Mytilus edulis*) in New England; persistence of *Chondrus* in sheltered areas depends on starfish predation on mussels. In wave-exposed areas, starfish are absent, and mussels outcompete *Chondrus* (Lubchenco & Menge 1978).

3.2.2 Seaweed–herbivore interactions

The periodic community distrubances that often are so dramatic are merely extreme examples of a phenomenon that every seaweed (except, perhaps, rock-boring species) may experience: predation. At the individual level we can explore the factors that influence grazing pressure and the responses that seaweeds have evolved to resist that pressure.

The expected herbivore damage has three components: (1) the probability that an individual plant will be encountered by a herbivore, (2) the probability of the plant being eaten, at least partially, if it is encountered, and (3) the cost, in terms of fitness, of the grazing damage (Lubchenco & Gaines 1981). Minimum damage will occur to plants that are out of contact with grazers, are unpalatable, or are able to sustain damage without significant cost. The effect on the alga of grazing will depend on the amount and kind of tissue lost and on the timing, particularly with respect to reproduction. Grazing may sometimes increase productivity by reducing self-shading in algal turfs; Williams and Carpenter (1990) suggested this as the reason that tropical turfs grazed by urchins are 2–10 times more productive than turfs grazed by fish or mesograzers. Loss may also be counterbalanced by the survival and dispersal of algal spores during digestion by herbivores. Santelices et al. (1983) showed that the spores of opportunistic species survive much better than those of late-successional species. At one extreme, great damage to *Macrocystis* can be done by urchins, which eat little of the plant but cause the loss of entire fronds by chewing through the bases of the stipes. At the other extreme, some crustose coralline algae are susceptible to grazing by limpets but do not suffer much harm from it. The weakly calcified epithallial layer is removed, but it is replaced by the meristem beneath, which survives. Indeed, in the absence of limpet grazing, the epithallial layer becomes too thick and the plant is overgrown by epiphytes and dies (Fig. 3.9). [Larvae of the limpet *Acmaea testudinalis* selectively settle on the coralline *Clathromorphum circumscriptum*, the preferred food (Adey 1973; Steneck 1982; Steneck & Paine 1986). Perhaps there has even been co-evolution between crustose corallines and their molluscan grazers (Steneck 1985).]

The means available (at least hypothetically) to seaweeds for minimizing grazing damage have been enumerated by Littler and Littler (1984, 1988; see also Duffy & Hay 1990) (Table 3.2) as follows:

1. occupation of refuge habitats that are physically unfavorable or unavailable to herbivores
2. unpredictable spatial and temporal distributions
3. close association with unpalatable organisms

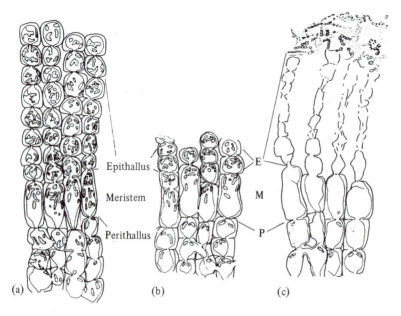

Figure 3.9. Structure of *Clathromorphum circumscriptum* in cross section, showing (a) field specimen that has had moderate limpet grazing, (b) maintenance of the meristem even under heavy limpet grazing, and (c) death of cells due to overgrowth by diatoms in the absence of grazing. (From "A limpet–coralline alga association: adaptations and defenses between a selective herbivore and its prey," by R. S. Steneck, *Ecology,* vol. 63, 1982, pp. 507–22, © The Ecological Society of America. Reprinted with permission.)

4. proximity to the territories of carnivorous predators
5. location close to highly territorial animals
6. cryptic appearances or mimicry
7. rapid growth involving the replacement of vegetative and reproductive tissues, while simultaneously satiating the appetites of grazers
8. allocation of materials and energy toward herbivore defenses

The first five conditions are avoidance mechanisms, or "noncoexistence escapes." The last three are tolerance mechanisms or "coexistence defenses" (Table 3.2). We shall examine some of these mechanisms in detail. (Such mechanisms are often called "strategies," but this term may have a connotation that seaweeds can adapt to carry out some behavior. There is no suggestion, of course, that any seaweed spore can seek out or be attracted to carnivorous predators or territorial animals, for instance. Proximity is the result of chance. Nevertheless, seaweeds lucky enough to settle in such places have greatly increased chances of survival).

Refuge habitats. The first strategy takes advantage of the fact that not all places at all times feature high grazing pressure. Even on tropical reefs, where herbivory is generally high, the grazing pressure varies markedly in time and space (Hay 1981a,b). Vertical migrations of herbivores occur (Vadas 1985), and there are places that herbivores cannot reach. For example, in

temperate waters, germlings of *Fucus vesiculosus,* which are more susceptible to periwinkle grazing than are older plants, escape predation when they are in crevices that the snails cannot reach (Lubchenco 1983).

Crustose or boring stages (or crustose parts of heterotrichous plants) are more resistant to attack by grazers, being either out of reach of the grazers or not readily removed by them. (These stages are also more resistant to removal by abiotic agents, such as wave action and ice. For instance, the crustose base of *Corallina* is important in its survival in physically disturbed environments; Littler & Kauker 1984.) Shell-boring algae, such as gomontia and conchocelis stages, are especially well protected against grazing. Tropical-reef carbonate is a major habitat for boring, filamentous blue-green, green, and red algae. Numerous species are present (certainly not all are alternate stages of macrophytes), making up over 90% of the algal biomass within the coral (zooxanthellae contribute relatively very little) (Littler & Littler 1984). Grazing by parrotfish does, nevertheless, reach these algae. Upright stages, in contrast, may be totally removed through damage to their relatively narrow stipes. The disadvantage to the crustose or boring stage is the increased susceptibility to being overgrown, but this problem may be mitigated by grazers removing the epiphytes.

A refuge may occur by virtue of being too far away from grazers' shelters. Visually striking halos occur around the edges of clearings in mussel beds on the

Table 3.2. *Herbivore-resistance mechanisms shown by seaweeds*

Resistance mechanism	Examples
Noncoexistence escapes	
1. Temporal escapes	
Short or alternating life histories	Annuals (e.g., *Liagora*, *Trichogloea*) and seasonally alternating heteromorphic forms (*Porphyra*, *Gigartina*)
Opportunistic colonization of unpredictable new substrata	Thin sheetlike, and filamentous forms (e.g., *Polysiphonia*, *Centroceras*, *Enteromorpha*)
2. Spatial escapes	
Refuge habitats	Intertidal habitats (*Ahnfeltia*, *Chnoospora*), high-energy environments (*Sargassum*), sand plains (*Gracilaria*), and crevices (*Amansia*)
Association with unpalatable organisms	Next to toxic or stinging organisms such as *Gorgonia* and *Millepora* (*Liagora*, *Laurencia*)
Association with carnivorous predators	Next to grouper or snapper lairs (*Acanthophora*)
Association with territorial animals	Within damselfish territories (e.g., *Polysiphonia*)
Coexistence escapes	
1. Crypsis	Inconspicuous forms (some *Gracilaria*) and small forms (*Plectonema*)
2. Mimicry	Coral mimics (*Eucheuma arnoldii*)
3. Structural defenses	
Morphologies that minimize accessibility	Upright large algae (e.g., *Turbinaria*), branched crustose corallines (*Neogoniolithon*), and turf-forming colonies (*Halimeda*)
Textures that inhibit manipulation and feeding	Encrusting species (*Lobophora*) and spiny forms (*Turbinaria*)
Materials that lower food quality	$CaCO_3$ in corallines (*Porolithon*) and calcareous green algae (*Penicillus*, *Neomeris*)
4. Chemical defenses	
Toxins, digestion inhibitors, unpalatable substances	*Halimeda*, *Caulerpa*, *Stypopodium*, *Dictyota*
Lowered energetic content	Coralline algae
5. Satiation of herbivores	Rapid replacement of lost tissues (*Ulva*)

Source: Littler and Littler (1988), with permission of Cambridge University Press.

west coast of North America (Fig. 3.10) because grazing limpets are limited in how far they can roam from their hiding places during low tide (Paine & Levin 1981). A similar phenomenon occurs around tropical patch reefs, which provide shelter for urchins during the day (Ogden et al. 1973). In the first case, desiccation avoidance drives the limpets back; in the case of the urchins, the driving force is predatory fish.

Refuge habitats also include areas with too much wave action for grazers. An example of this comes from Pace's (1975) work on seasonal migrations of urchins in British Columbia. During winter, urchins move deeper into the subtidal, leaving a belt with temporarily reduced grazing pressure. Strong wave action on tropical-reef crests reduces grazing pressure from fish and urchins and allows a "climax" coralline algal community (*Lithophyllum congestum* and *Porolithon pachydermum* in the Caribbean) to build intertidal algal ridges (Adey & Vassar 1975) (see Fig. 2.2b).

Unpredictable spatial or temporal distributions: Lubchenco and Cubit (1980) and Slocum (1980) have advanced the hypothesis that heteromorphic algal life histories may allow a response to grazing. Lubchenco and Cubit (1980) worked with several winter ephemerals (*Ulothrix*, *Urospora*, *Petalonia*, *Scytosiphon*, *Bangia*, and *Porphyra*) in areas where the grazing intensity varied seasonally. When grazers were removed experimentally, upright forms of the algae were found at times of the year when normally only the small phase

Figure 3.10. Halos around the edges of algal patches within mussel beds on the exposed Washington coast are caused by grazers (limpets and chitons) that shelter among the mussels during low tide and graze only 10–20 cm from shelter. They are unable to forage in the middle of the large patch, but can graze the whole of the small patch (the two patches are the same age). The fleshy algae in the middle of the larger patch are *Porphyra* (center), surrounded by *Halosaccion*. The halo is occupied by barnacles and *Corallina vancouveriensis*. See also Figure 7.12. Photo courtesy of Robert T. Paine, © 1991 by Robert T. Paine.)

would occur. Slocum (1980) found that the relative amounts of crusts and blades of *Mastocarpus papillatus* depended on grazing intensity, and Dethier (1982) concluded that seasonal changes in abundance may be largely or entirely due to herbivore abundance and feeding rates. If variations are unpredictable, the population survival will be highest with continuous production of both morphs, although only one will survive at a time (Lubchenco & Cubit 1980). "Bet-hedging" (as Slocum calls it) maximizes the total progeny of each such generation. However, Littler and Littler (1983) have suggested that grazing pressure has received disproportionate emphasis (from studies such as those by Lubchenco and Cubit and Dethier) and that although crustose stages are more grazer-resistant, they also are less susceptible to sand scour and burial, wave shock, desiccation, and temperature, thus representing a phase appropriate for various "harsh" conditions. Clearly, some conflicts are yet to be resolved, because the reproduction of some species or populations of *Porphyra* and *Scytosiphon* is known to be cued by light or temperature, and one would not expect to find erect forms developing out of season. In some species the upright form is produced directly, not as a re-

sult of reproduction, whereas in other cases the upright form has a perennating base or holdfast that is capable of regeneration if older shoots are lost to grazing (or other factors, such as ice scouring).

Association with unpalatable organisms. Protection may be gained from proximity to unpalatable algae or even toxic animals (Hay 1988). The dictyotalean brown, *Stypopodium zonale*, is one of the most toxic seaweeds, and delectable algae such as *Acanthophora spicifera* have been found to suffer much less grazing damage when close by *S. zonale* (Littler et al. 1986b, 1987b). Even plastic models of *S. zonale* have had significant salutary effects on neighboring algae. A parallel example was found in North Carolina, where understory *Gracilaria tikvahiae* was protected by *Sargassum* (Pfister & Hay 1988). The costs of competition with the neighbor were outweighed by the benefit of grazer protection – survival was at the expense of a slower growth rate. The reason that toxic animals (e.g., a gorgonian coral and a fire coral) repel herbivores apparently relates to the generalist feeding of the fish. In fact, parrotfish have been observed to "graze" corals as well as algae (Littler et al. 1987b).

(a)

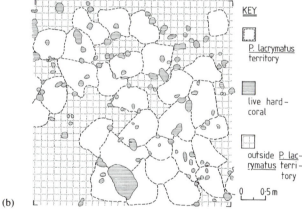

KEY

☐ P. lacrymatus territory

▦ live hard-coral

▢ outside P. lac-rymatus terri-tory

0   0·5 m

(b)

Figure 3.11. Farmer-fish territories. (a) A common modification in the growth form of massive *Porites* (coral) colonies in territories of *Stegastes nigricans* on the Great Barrier Reef, Australia. The damselfish maintains dense algal turf on the sides of the lobes (1.15 cm across in this photograph). (b) Diagrammatic map showing the distribution and sizes of adult territories of *Plectroglyphidodon lacrymatus* on a section of reef on Matupore Island, Papua, New Guinea. (Part a from Done et al. 1991, with permission of Springer-Verlag, Berlin; b from Polunin 1988, with permission of Elsevier Science Publishers.)

An even closer association that perhaps may confer a protective benefit is seen in the symbiosis between a red macroalga, *Ceratodictyon spongiosum,* and a possibly toxic sponge, *Sigmadocia symbiotica* (Price et al. 1984).

Association with highly territorial animals. The case of the garibaldi nests discussed in section 3.1.1 as an example of competition is also an example of the benefits of falling into association with a territorial animal. This temperate-water damselfish is a carnivore,

but it drives away all other fish, including herbivores (which otherwise would try to eat its nest).

More complex behavior is shown by a group of tropical damselfish known as farmer fish. These fish maintain algal assemblages inside their territories that differ markedly in species abundance and species composition from those of surrounding areas (Fig. 3.11a) (Lobel 1980; Lassuy 1980; Sammarco 1983; Hinds & Ballentine 1987; Klumpp & Polunin 1989). Contiguous territories can cover large areas of reef flats (Fig. 3.11b) and contribute substantial prouctivity, most of which is consumed by the fish (Brawley & Adey 1977; Klumpp et al. 1987; Polunin 1988). Farmer fish are selective regarding which algae they will allow to grow, weeding out certain species. Nitrogen-fixing blue-greens form about 50% cover in the territories on Australian reefs (Sammarco 1983), but in fish-exclusion cages the cover drops to less than 10%. Given the low concentrations and high turnover rates for nitrogen in tropical waters (Adey & Goertemiller 1987), nitrogen fixing by blue-greens is undoubtedly important to the farm community. Like ocean goldfish, farmer fish chase away trespassers and thereby reduce grazing pressure. (However, schooling herbivorous fish can drive off the farmer fish and clean out the algae in a matter of minutes.) Larger territories allow better growth of the species preferred by the fish; in smaller territories the fish must eat more of the less-preferred species (Jones & Norman 1986).

Two contrasting examples of farmer fish were studied in the Gulf of California, Mexico, by Montgomery (1980a,b). The giant blue damselfish, *Microspathodon dorsalis*, maintains a virtual monoculture of a *Polysiphonia* species, which has a low standing crop, but high productivity. The Cortez damselfish, *Eupomacentrus rectifraenum*, in contrast, maintains a multispecies algal mat of high biomass and low overall productivity, but feeds selectively on *Ulva*, which features high productivity and rapid colonization. Within these territories, the crustose coralline and coral covers typically are low; filamentous algae overgrow the existing corals and corallines and prevent resettlement (Potts 1977). It would be interesting to know something of the population dynamics of food and nonfood algal species inside and outside such territories, especially in relation to reproductive potential.

An example of "gardening" by an invertebrate was reported by Branch (1975), who showed that the limpet *Patella longicosta* in South Africa establishes *Ralfsia expansa* "lawns" by grazing away the established crustose corallines. It then grazes off competing algae and grazes the *Ralfsia* in strips, so that productivity is actually enhanced because of edge growth.

Allocation of materials and energy to defenses. Whereas plants that experience rapid growth tend to rely on that rapid growth to satiate herbivores, slower-growing plants tend to allocate resources to defenses (Coley et al. 1985). Such defense responses include growth in shapes and sizes that will minimize accessibility, growth in textures that will inhibit herbivore manipulation and feeding, deposition of structural materials such as $CaCO_3$ that will decrease palatability, calorific content, or nutritional value, and, finally, production of toxins, digestion inhibitors, or unpalatable compounds within the cells (Littler & Littler 1984), also called allelochemistry (Harlin 1987). Two topics, in particular, have received much attention in the published

studies: chemical deterrence of herbivore feeding (sec. 3.2.3) and grazer feeding preferences.

Many herbivores that have been tested have shown certain preferences among food species and, in some cases, the ability to detect the preferred species. Sea urchins, for example, will move upstream in a current of water that has passed over a kelp plant, but their feeding behavior is influenced by both preference and availability (Vadas 1977). Tests of the abilities of algal compounds to attract herbivores usually have involved offering crude algal extract, or simply passing water over a food alga, but Carefoot (1982) tested the responses of sea hares to specific algal compounds. Starch, glutamic acid, and aspartic acid, which might leach from algae or be released through damage, attracted sea hares and induced them to feed on agar wafers.

Feeding-preference studies involve giving an animal a choice between two (sometimes more) algal species and measuring the amount of each that is eaten. There are several potential pitfalls in this kind of study (Vadas 1985). First, the age of the alga may affect its palatability. *Littorina* snails readily eat *Chondrus* sporelings less than a few weeks old, but they have little taste for older plants. The vulnerability of the youngest stages obviously is very important in setting the distribution pattern of a species (Sousa et al. 1981; Cheney 1982). Second, there may be intraspecific resource partitioning such that preferences differ among individuals, such as individuals of different ages. Finally, the amount of alga eaten may not give the best indication of the importance of the species to the grazer. *Littorina littorea* in New England preferred *Enteromorpha* in terms of the biomass they consumed, but they grew best on a mixed diet of *Enteromorpha*, *Fucus*, and *Chondrus* (Cheney 1982).

The probability of the grazer feeding on the seaweed once the encounter has taken place will depend on the form and palatability of the seaweed, but these characteristics are difficult to predict or measure. Steneck and Watling (1982) predicted the susceptibilities of seaweeds to molluscan grazers (not including opisthobranchs) by arranging both into functional groups. Algal groups were based on morphology and toughness (Fig. 3.12), similar to those of Littler and Littler (Table 1.1). Some algae, of course, pass through two or more such groups as they grow, and species with heteromorphic life histories straddle two groups. Molluscan grazers were grouped according to the type of radula and grazing action they employed: "brooms" (e.g., keyhole limpets), "rakes" (e.g., *Littorina littorea*), "shovels" (e.g., true limpets), and "multipurpose tools" (e.g., chitons). Algal groups 1 and 2 are readily scraped off the substrata by grazers with broomlike or rakelike radulae, whereas the largest or most expansive algae (groups 5 and 7) are consumed primarily by the other two groups of herbivores, which can occupy the alga and gouge out tissue (Fig. 3.12). Intermediate-size al-

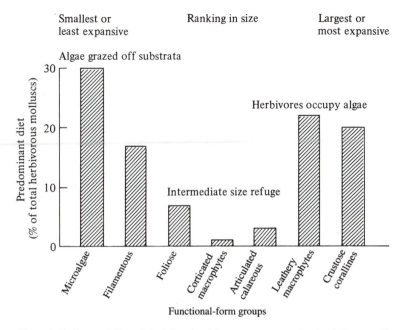

Figure 3.12. Susceptibility of algal functional-form groups to grazing by different molluscs, mesured as the relative importances of the algal groups in the diets of 106 herbivorous species. (Modified from Steneck & Watling 1982, with permission of Springer-Verlag, Berlin.)

gae, of moderate toughness (groups 3, 4, and 6), appear to have a size refuge from these molluscan grazers, being too large to be grazed off the substrate and too small to be occupied. On the other hand, Padilla (1985, 1989) has argued that thallus form (as used in the functional-form models) alone is a poor predictor of structural resistance to limpet grazing. Her results were obtained by scraping algae with limpet radulae in a way intended to mimic the limpet feeding mechanism. In her 1985 study she found that the most easily grazed alga was the calcified crust, *Pseudolithophyllum whidbeyense,* whereas grazing of *Iridaea splendens* and *Hedophyllum sessile* (both leathery macrophytes and thick blades) required more force. Steneck and Watling (1982) classified these algae as fifth and seventh in grazing difficulty, but they also showed that these two groups rank high in the predominant diets of limpets (Fig. 3.12). However, R. S. Steneck (personal communication) pointed out that "grazing difficulty" should have been defined as including both morphological and material characteristics. He also expressed concern with Padilla's method. Given that a limpet will immobilize algal thalli with its foot and lips while its radula gouges the plant, Steneck questioned the validity of laterally scratching a large, dissected radula over an unsupported piece of an elastic alga such as *Iridaea.*

Specialization for single food sources by small, slow-moving grazers such as these molluscs is usually found in groups (such as sea hares) whose individuals are small enough to occupy and gouge the larger algae, which are, from the small grazer's vantage, abundant and long-lived food supplies. Small nudibranchs have been seen feeding *inside* the large coenocytic cells of *Boergesenia forbesii* on reef flats in Guam (P. J. Hoff, C. Lobban, et al., unpublished observations). *Elysia hedgpethi* is small and slow-moving, and yet selective in choosing only siphonous green algae such as *Codium* and *Bryopsis* (Greene 1970). *Aplysia californica* is able to become less specialized as it grows and its mouthparts toughen; new recruits grow only on *Plocamium* (Pennings 1990b).

3.2.3 Chemical ecology of alga–herbivore interactions

Boreal waters tend to feature herbivores that are slow-moving (such as sea urchins and limpets) or those that if quick are also small (such as amphipods). Morphological grazer resistance (toughness or calcification) seems to be more common than chemical defense by toxic or distasteful compounds. However, a few examples have been found (the best-known being some *Desmarestia* spp.) that have a high sulfuric acid content (vacuolar pH $\approx$ 1) (McClintock et al. 1982). That acidity seems to deter sea urchins, unless there is no other food source available.

Some temperate brown algae produce polyphenolics or phlorotannins, which can be present in high concentrations (up to 20% of dry weight). These compounds are sometimes called tannins, but the algal compounds are chemically distinct from the tannins of

Figure 3.13. Chemical defenses. (a) A simple fucol, a dimer of phloroglucinol. (b) Udoteal, a fish deterrent from *Udotea*. (Part a after Ragan & Glombitza 1986; b after Paul et al. 1982.)

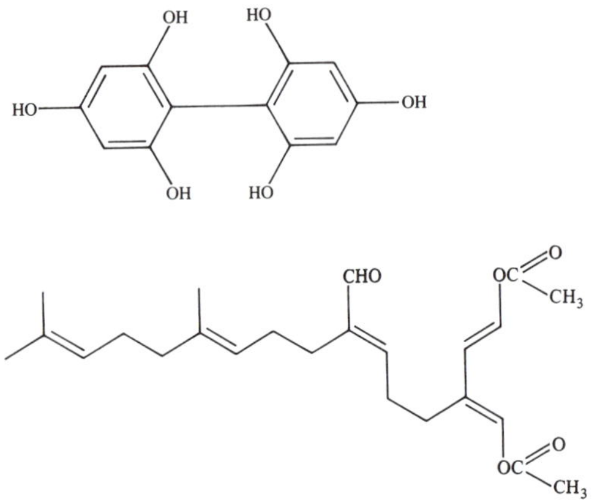

higher plants (Hay & Fenical 1988). There are many polymeric compounds based on phloroglucinol (Fig. 3.13a) (Ragan & Glombitza 1986). One of the ways in which phenolics produce a toxic effect is by precipitating proteins (indeed, they cause problems in extracting enzymes from kelps and fucoids; Keen & Evans 1988). They are readily extruded and are abundant in the *Fucus* ''juice'' that is released when *Fucus* is kept in plastic bags (Ragan & Glombitza 1986). Both laboratory studies and field observations have shown that snails and urchins avoid phenolics or phenolic-laden brown algae. For instance, Geiselman and McConnell (1981) showed that an *Ulva*-agar blend can be made unpalatable by mixing in some fucoid extract. Phenolic exudates also have allelopathic effects in tide pools, where exuders such as *Petalonia* and *Laminaria* inhibit ephemerals such as *Chaetomorpha* (Wolfe 1988). However, polyphenolics do not constitute the whole defense strategy in brown algae, nor are they produced exclusively for defense. Temperate *Sargassum* species have more phenolics than tropical species, even though they face much less grazing pressure than do those in the tropics (Steinberg 1986). Ilvessalo and Tuomi (1989) found evidence that the polyphenolic content in some temperate seaweeds correlates with internal nitrogen content.

Phenolics appear to deter invertebrate herbivores, but not fish. Austral brown algae, subjected to fish as well as invertebrate grazing, produce both polyphenolic and nonpolyphenolic secondary metabolites (Steinberg & van Altena 1992). In the tropics, many fish are herbivores. Surgeonfish, parrotfish, and some damselfish are highly mobile, visual predators. The abundant tropical algal turf, so striking a contrast to the dense temperate intertidal algal beds, is perhaps due to heavy grazing pressure, as noted in the preceding section. Large algae in the tropics generally are well de-

fended chemically and morphologically. Fucoids such as *Turbinaria* and *Sargassum* are tough and spiny, but do not have polyphenolics, even though those compounds are deterrents to tropical fish. Van Alstyne and Paul (1990) have postulated that their lack of polyphenolics may be the result of ineffective production or storage in warm, nitrogen-poor tropical waters. Tropical Dictyotaceae produce complex mixtures of toxic chemicals, such as the diterpenoid pachydictyol-A (Hay & Fenical 1988), but these are not herbivore deterrents (Steinberg & Paul 1990). Caulerpalean greens produce toxic compounds, and many are also calcified (Paul & Fenical 1986). *Caulerpa* species are not calcified, some *Avrainvillea* and *Udotea* are lightly calcified, and *Halimeda* species are highly calcified. Among the red algae, notable toxin producers include *Halymenia, Laurencia,* and many others (Hay & Fenical 1988).

The toxic substances produced by seaweeds compose part of the spectrum of natural compounds that deter or attract other organisms. These chemicals are referred to as secondary metabolites because they are not involved in the major metabolic pathways. Several hypotheses have been advanced to account for the presence of these compounds, whose functions often are still obscure. In the view of Williams et al. (1989), secondary metabolites are not useless or waste compounds, but rather increase the producer's fitness by acting on receptors in competing organisms. Over the past decade, study of the ecology of secondary metabolites from algae has begun (Norris & Fenical 1985; Bakus et al. 1986; Van Alstyne & Paul 1988; Hay & Fenical 1988). Whatever their basic structures (terpenoid, phenolic, fatty acid, etc.), these compounds often carry reactive groups such as aldehyde, acetate, alcohol, or halogens. The exact placement of these groups can greatly affect the biological activity of a compound. Such abundance and diversity of halogenated compounds are characteristics of marine organisms not shared by terrestrial plants (Howard & Fenical 1981).

Terpenoids are among the most prevalent feeding deterrents in tropical seaweeds. They are lipids, structurally related to compounds such as carotenoid pigments (compare the structures of udoteal, Fig. 3.13b, and fucoxanthin, Fig. 4.5b) (Howard & Fenical 1981). Caulerpales and Dictyotales are particularly rich in terpenoids, and some of these compounds have been shown to discourage fish feeding. A correlation has been shown between the levels of deterrent compounds in reef areas heavily grazed by herbivores and the levels in low-grazing seagrass beds where grazing pressure is low (Paul & Fenical 1986). However, as Van Alstyne and Paul (1988) have pointed out, this may reflect predator-induced defenses, localized selection for defended plants, or some factor unrelated to predation intensity. In addition to local differences between individuals, there are variations within individual plants. Young parts, such as newly formed buds of *Halimeda*, and reproductive parts tend to be more heavily defended.

Several interesting strategies are used by *Halimeda* in defense against herbivores. Chloroplast migration (sec. 1.3.2) may protect against surface grazers such as saccoglossan molluscs, which try to capture chloroplasts for their own use (Drew & Abel 1990). Many species of *Halimeda* (but not all) put out new buds at night, when herbivorous fish are inactive. The new growth is unpigmented at first and has high concentrations of halimedatrial. Around dawn, the tissue turns green, perhaps because of chloroplast migration from the parent segment (Drew & Abel 1990). Calcification begins when photosynthesis begins. After 48 h, the toxin level will have decreased, and calcification and morphological toughness will provide adequate defense. However, not all *Halimeda* species behave this way. New segments of *H. macroloba* remain white throughout the next day and occasionally are produced during the day (Hay et al. 1988a). What does this tell us about resource allocation? Paul and Hay (1986) have suggested that tropical-reef algae will commonly have multiple defenses, because of the diversity and abundance of herbivores; no single defense is likely to be effective against all herbivores.

Some herbivores, far from being deterred by secondary metabolites, actually seek them out and use them for their own defense. An example of this is *Elysia halimedae*, which feeds preferentially on *Halimeda macroloba*. It takes halimedatetraacetate, which is effective against fish feeding, and simply converts one aldehyde group to an alcohol, to produce a deterrent for its own protection, as well as for its eggs (Paul & Van Alstyne 1992). *Aplysia californica* strongly prefers *Laurencia pacifica* and *Plocamium cartilagineum* (Pennings 1990a); it has been found that juveniles kept on a diet of *Ulva* have low terpene contents and are more palatable to predators. Other examples of specialist herbivores include several other saccoglossans on several other siphonous greens (Norris & Fenical 1985; Hay et al. 1990) and one that feeds exclusively on the blue-green alga *Blennothrix lyngbyacea* (Paul & Pennings 1991).

Defense against fish herbivory is clearly important in a tropical setting, but fish are by no means the only predators. In fact, mesograzers, which use the algae for shelter as well as for food, survive better among chemically defended algae, because there they are not likely to be accidentally ingested by herbivorous fish. Tests by Hay et al. (1988b) showed that a polychaete living characteristically on *Dictyota* was not deterred by dictyol-E or pachydictyol-A, which did deter a herbivorous fish. On the algal side of the equation, grazing by such mesoherbivores may not be so significant as to provoke a deterrence response, or, as discussed in section 3.2.1, it may actually be beneficial by removing epiphytic microalgae.

Ragan and Glombitza (1986, pp. 225–6) have suggested that brown-algal polyphenolics can be considered in the context of "optimal defense theory" (Rhoades 1979):

(a) organisms evolve defenses in direct proportion to their risk from predators and in inverse proportion to the cost of defense, other things being equal;

(b) within an organism, defenses are allocated in direct proportion to the risk of the particular tissue and the value of that tissue in terms of fitness . . . and in inverse proportion to the cost of defending the particular tissue;

(c) commitment to defense is decreased when enemies are absent and increased when organisms are subject to attack;

(d) commitment to defense is a positive function of the total energy and nutrient budget . . . and is negatively related to energy and nutrients allocated by the organism to other contingencies.

The costs of defenses in seaweeds have never been measured. However, at the time of Ragan and Glombitza's review, evidence from brown algae supported all these correlative hypotheses except item (c). For instance, perennial fucoids, which are at higher predation risk than are ephemerals such as ectocarpoids, have higher phenolic levels. Reproductive sporophylls, which are of more importance to fitness than are blades in Alariaceae, are much tougher and have more phenolics (Steinberg 1984). Both defenses are also present in haptera and stipes, on which the survival of the plant depends. More recently, evidence regarding item (c) has come from wounding studies on *Fucus gardneri* (Van Alstyne 1988, 1990, as *F. distichus* subsp. *edentatus*) and *Ascophyllum nodosum* (Lowell et al. 1991): Wounded *Fucus* produced more polyphenolics over a 2-week period and reduced their reproductive output (fitness). Herbivorous snails (*Littorina sitkana*) shifted their preference from newly clipped plants to unclipped plants during the same period. As a result, clipped plants lost about 50% less tissue to grazers. One would expect chemically defended tropical green algae to show similar responses, but attempts to demonstrate this have thus far been unsuccessful (V. J. Paul and K. L. Van Alstyne unpublished data).

This brings us to the question of costs versus benefits. As to costs, Ragan and Glombitza (1986) have suggested that the production of broad-spectrum, weak (but stable) polyphenolics, which might last 100 days, is more economical than production of high-activity molecules that are short-lived (e.g., 1 day). In terrestrial plants, tannins are characteristic of slow-growing plants (often resource-limited), whereas high-turnover compounds are typical of more rapidly growing species when those species are chemically defended (Coley et al. 1985). However, turnover rates, rates of release (if any), and levels of energy required for maintenance of the deterrent(s) vary markedly among species (Hay &

Fenical 1988). Hay and Fenical have questioned whether energy is "the currency in which costs should be measured or [whether] there is any significant difference in the total resources expended on each type of defense once the costs of synthesis, turnover, storage, and possible functions other than herbivore deterrence are considered" (1988, p. 135).

As to benefits, seaweeds in areas of predictably low herbivory are not chemically defended. This and other bits of circumstantial evidence suggest that the production of deterrents affects the fitness of the individual or species (Hay & Fenical 1988). Unfortunately, adducing clear evidence for this is very difficult, as Van Alstyne and Paul (1988) noted, because it is experimentally difficult to determine the changes in growth or reproduction attributable to deterrent production. One key question might be answerable for some species: If grazing pressure causes an increase in deterrent compounds (as in *Fucus;* Van Alstyne 1988), how much energy does the plant put into its enzymatic machinery and deterrent molecules, and how much benefit does it derive in terms of tissue saved from grazing?

3.3 Symbiosis

At one end of the spectrum of biotic relationships among seaweeds is epiphytism. Some instances of epiphytism are quite specific; for other seaweeds, epiphytism is simply one solution to the space problem. Endophytism also represents a range of space solutions. At the other extreme are parasitic relationships. The distinctions among epiphytism, endophytism, and parasitism are nutritional: Parasites derive their carbon largely or exclusively from their hosts. Symbiosis (literally, "living together") implies a closer relationship than simple epiphytism, but includes a range of mutualistic partnerships and parasitism. However, the nutritional relationships often are assumed rather than demonstrated. Few of the red algae that are considered parasitic have received critical study (Goff 1982).

3.3.1 Mutualistic relationships

The range of mutualistic associations between benthic marine algae and other organisms is quite broad if the microalgae are included. Microalgae, including blue-green algae, green algae, dinoflagellates, and a few diatoms, form several groups of symbioses, such as lichens, hermatypic corals, and associations with diverse other invertebrates and other algae. Putative symbioses involving multicellular seaweeds are less common (perhaps even unusual) and include some internal fungal partners and some external sponge partners.

Several seaweeds have been found characteristically to contain endophytic fungi. The associations include some little-known pairs such as *Blidingia minima* var. *vexata* + *Turgidosculum ulvae*, *Prasiola borealis* + *Guignardia alaskana*, and *Apophlaea* spp. + *Mycosphaerella apophlaeae* (Kohlmeyer & Kohlmeyer 1972;

Kohlmeyer & Hawkes 1983). Better known are the associations between the ascomycete *Mycosphaerella ascophylli* and two fucoids, *Ascophyllum nodosum* and *Pelvetia canaliculata*. Specimens of these fucoids more than 5 mm long are invariably infected with *Mycosphaerella,* but specimens less than 3 mm high lack the fungus (Kohlmeyer & Kohlmeyer 1972; Kingham & Evans 1986; Garbary & Gautam 1989). The fungus grows throughout the host thallus, without penetrating the cells or forming haustoria. Fungal fruiting bodies (perithecia) are formed chiefly, but not exclusively, in receptacles of the hosts, where they appear to the naked eye as small black dots. The fungus also occurs in vegetatively reproducing salt-marsh *Ascophyllum* (J. Higgins, personal communication). Perithecia form on vegetative apices at about the same time as receptacles are produced, but persist about a month longer (Garbary & Gautam 1989). Infection takes place after *Ascophyllum* has germinated; the eggs apparently are not infected, even though perithecia are present in receptacles. *Pelvetia canaliculata* germlings are free of hyphae for their first year. Kohlmeyer and Kohlmeyer (1972) have pointed out that these associations are not lichens, because their morphologies are dominated by the sexually reproducing algae, and no secondary lichen compounds are formed.

Mycosphaerella undoubtedly obtains carbon from the host; in axenic culture it can grow on laminaran and mannitol, but not alginic acid (Fries 1979). *M. ascophylli* is similar to saprophytic marine fungi in requiring pH 7–8 and a high NaCl concentration, but differs in having a lower temperature optimum and a requirement for the vitamins thiamine and biotin. The presence of the fungus in *Pelvetia canaliculata* changes the ratio of mannitol : volemitol from 1.7 : 1.0 (in plants less than 5 mm long and in plants treated with fungicides) to 0.8 : 1.0 (Kingham & Evans 1986). (Volemitol is restricted to *P. canaliculata* from Europe; it is not found in other species of *Pelvetia*, nor in *Ascophyllum;* Kremer 1979.) The compounds, if any, that the fucoids get from the ascomycete are unknown, and no benefit to the algae has yet been found. Indeed, the fungus is potentially a pathogen; Rugg and Norton (1987) have suggested that uncontrolled fungal growth is the cause of the lethal rotting of *P. canaliculata* that is kept submerged. Lichen fungi will also overrun their association if growth conditions are good for them.

Thamnoclonium dichotomum and two species of *Codiophyllum*, southern Australian red algae, are consistently covered by particular sponges (Scott et al. 1984; also see Bold & Wynne 1985). Scott et al. (1984) have suggested that the sponges are merely epiphytic, because the algal growing points show no sign that the sponge affects the morphology of the alga, or vice versa. The sponges bond themselves to the algae with collagenlike fibrils and do not modify the structure or ultrastructure of underlying tissues. Nevertheless, the constancy of the associations implies more than mere epiphytism. Moreover, most algae guard against extensive epiphytism (sec. 3.1.2) and would be expected not to tolerate such overgrowth unless it were beneficial. Scott el al. (1984) have noted an absence of grazing marks, and sponges typically are chemically defended (Bakus 1981). These associations, therefore, could represent extreme cases of grazer avoidance by association with unpalatable organisms (as discussed in sec. 3.2.2). The sponge may benefit from algal exudates, as has been suggested for a similar symbiosis between *Phyllophora sicula* and a sponge in the Mediterranean (Sciscioli 1966). Price et al. (1984) were able to culture the red alga *Ceratodictyon spongiosum* from a tropical Indo-Pacific symbiosis; it grew well in the laboratory without the sponge, but when outplanted, it disappeared, for unknown reasons.

3.3.2 Algal parasites

The nutrition of parasitic algae is, by definition, at least partly heterotrophic. Various colorless or weakly pigmented algae, particularly in the Rhodophyta, grow on other seaweeds and are assumed to receive organic carbon from their hosts. Obligate epiphytes, particularly *Polysiphonia lanosa*, which has penetrating rhizoids, sometimes have also been thought to be partially parasitic. However, the presence of penetrating rhizoids may be simply a means for an epiphyte to anchor to a basiphyte (Gonzalez & Goff 1989). *Polysiphonia lanosa* is at most auxotrophic (Harlin & Craigie 1975; Turner & Evans 1977); it may get minor but significant compounds (e.g., growth factors) from *Ascophyllum* (or *Mycosphaerella?*), but is fully capable of satisfying its own carbon requirements. Garbary et al. (1991) and Pearson and Evans (1990) have concluded that *P. lanosa*'s virtual restriction to *Ascophyllum nodosum* is an ecological problem of recruitment, not a physiological relationship; it can grow on *Fucus vesiculosus* if spores settle in wounds (Pearson & Evans 1990). Some colorless pustules that had been classified as parasitic algae have now been shown to be galls of host tissue, presumably induced by bacteria (e.g., ''Lobocolax''; McBride et al. 1974; also see Evans et al. 1978).

There seem to be no parasitic brown or green seaweeds. Though there are endophytes such as the brown *Streblonema* and the green *Acrochaete* (Andrews 1976; Correa & McLachlan 1991), these species are photosynthetic and have not been shown to derive nutrients from host tissues. Among the red algae, in contrast, are many parasitic species, of two different types. Of some 65 parasitic species, about 80% are closely related to their hosts (Bold & Wynne 1985); these are called adelphoparasites; examples include *Gracilariophila* and *Janczewskia*. The remainder, alloparasites, are not closely related (usually in different Orders); examples include *Harveyella*, *Choreocolax*, and other Choreocolacaceae. It seems possible that alloparasites evolved from epiphytes by gradually developing dependency, whereas adelphoparasites may have been derived through in situ germination of mutant tetraspores or carpospores (Bold & Wynne 1985). This would seem a natural extension of the apparent support of carposporophytes by female gametophytes. The ability of red algae to form secondary pit connections may have been an important factor in the evolution of parasitic species. Certainly, the transfer of parasite nuclei into host cells (Fig. 1.22) is a key part of the parasitism.

Although adelphoparasites are much more numerous, alloparasites have received the most ecological and physiological attention. In fact, the adelphoparasite *Janczewskia gardneri* is moderately pigmented as an adult, and Court (1980) found no translocation to it from its closely related host *Laurencia spectabilis*.

Harveyella mirabilis is an alloparasite of *Odonthalia* and *Rhodomela* that has been thoroughly studied by Goff (reviewed in her 1982 paper) and Kremer (1983). *Harveyella* cells penetrate among the host cells via grazing wounds and cause proliferation of host cells to form pustules, over which the cortex of *Harveyella* forms its reproductive structures. Contiguous cells of the host thallus and isolated host cells in the pustule of *Harveyella* export assimilates to *Harveyella* cells, but the secondary pit connections formed between parasite and host are plugged, probably precluding translocation through them. Goff (1979) proposed that translocation could take place across the cell-wall matrix, but the exact pathway and the composition of the translocated material have not yet been determined in this association. *Harveyella* relies completely on its host for carbon, which it receives as digeneaside, perhaps with some amino acids (Kremer 1983).

Three possible mechanisms have been suggested by Goff (1979) to explain how *Harveyella* induces its host to release organic carbon: (1) by changing the permeability of the host plasmalemma; (2) by modifying the host's cell-wall polysaccharide synthesis such that compounds normally deposited in the wall will remain soluble; (3) by mechanically or enzymatically damaging the host's cell wall, leading to lysis and release of metabolites.

Other host–parasite pairs have been less extensively studied, but enough is known to indicate that each relationship is unique and must be assessed separately. *Gracilaria verrucosa* transfers carbon as floridoside to its nonpigmented alloparasite *Holmsella pachyderma* (Evans et al. 1973), and *Polysiphonia lanosa* exports mannoglycerate to its weakly pigmented alloparasite *Choreocolax polysiphoniae* (Callow et al. 1979).

3.4 **Synopsis**

Biotic interactions include competition within and between species for space, light, nutrients, or any limiting resource. Predator–prey relations at several levels affect seaweeds directly or indirectly. Interference competition for space takes place among algae and between algae and sessile animals; the outcome can be influenced by both herbivores and carnivores. Exploitation competition for light and nutrients takes place among algae. Epiphytism solves the space problem for the epiphyte, but creates a competition problem for the anchor species. Some algae produce chemicals or slough their outer layers to inhibit the growth of epiphytes.

Three groups of herbivores are important in seaweed communities: fish (only in warmer waters), urchins, and mesograzers (invertebrates with small ranges but high densities). Damage to seaweeds by grazers will depend on the occurrence of an encounter between the two, on how much is eaten or broken off, and on what parts are lost and when, especially as the losses affect reproduction and hence the fitness of the individual. Opportunistic species and species with heteromorphic life histories may avoid grazers because of their unpredictable distribution in space and time. Other species avoid grazing by living close to toxic or territorial animals. Some seaweeds tolerate grazing, but reduce its impact by producing toxic or acidic feeding deterrents or by having tough (including calcareous) cell walls. In a few instances, grazing is beneficial to seaweeds.

Many symbiotic and parasitic associations occur in the sea, and some involve seaweeds. A few seaweeds harbor internal fungal symbionts, and there are even smaller numbers of seaweeds that are constantly coated by symbiotic sponges. Almost nothing is known of the nutritional relationships. Parasitic algae are confined to the Rhodophyta, and they parasitize other, usually closely related, red algae. Transfers of various carbon compounds have demonstrated the dependence of the parasites, though some associations that once appeared parasitic have been proved otherwise.

4

Light and photosynthesis

Seaweeds grow in circumstances that feature exceptionally diverse and dynamic lighting climates. The water clarity and the continual ebb and flood of tides have profound effects on the quantity and quality of the light that reaches seaweeds, adding greatly to the variation already present in the irradiance at the earth's surface. The primary importance of light to seaweeds is in providing the energy for photosynthesis, energy that ultimately is passed on to other organisms. Light also has many photoperiodic and photomorphogenetic effects, as we saw in Chapter 1. Thus light is the most important abiotic factor affecting plants, and also one of the most complex.

The principles of photosynthesis are similar in algae and higher plants, and indeed some principles (e.g., the Calvin cycle) were worked out using algae. Most of the catalytic proteins involved in the thylakoid reactions of red algae, for instance, are homologous with those in all other photosynthetic plants, but some are analogous (Raven et al. 1990). There are, moreover, several important features of seaweeds and their habitats that stand in sharp contrast to those in higher plants, the land plants, and it is on these that we shall focus. Such features include the diversity of pigmentation among marine algae and the diversity of the light climate in the oceans, the nature of the carbon supply in the sea, and the diversity of photosynthetic products in different algal classes. We assume that the common details of photosynthetic mechanisms and pathways have been covered in introductory courses; they will be reviewed only briefly in the following section. Textbooks on plant physiology and biochemistry offer extensive treatments of all aspects of angiosperm photosynthesis (e.g., Hall et al. 1982; Goodwin & Mercer 1983; Salisbury & Ross 1985). The accounts of light harvesting and carbon metabolism presented here owe much to the detailed reviews by Larkum and Barrett (1983) and Kerby and Raven (1985), respectively, which students should consult for more information and references. A review of photosynthesis, respiration, and nitrogen-fixing interrelationships in cyanophytes has been published by Scherer et al. (1988).

4.1 An overview of photosynthesis

Photosynthesis encompasses two major groups of reactions. Those in the first groups, the "light reactions," involve the capture of light energy and its conversion to chemical potential as ATP and NADPH. The light reactions, in turn, consist of three processes: energy absorption, energy trapping, and generation of chemical potential. Those in the second group, the "dark reactions," include the sequence of reactions by which this chemical potential is used to fix and reduce inorganic carbon.

Light travels, paradoxically, as packets (photons or quanta) and as waves. The energy of a photon is inversely related to its wavelength; in other words, blue light (wavelength around 400 nm) has more energy than red light (wavelength around 700 nm). Pigment molecules absorb the quantum of energy when hit by a photon. They are in a chemically excited state for a small fraction of a second before releasing that energy. The close packing of pigment molecules into light-harvesting pigment complexes allows most of the energy to be passed from molecule to molecule until it gets to the reaction center. (A small amount of energy is dissipated in each transfer, and so the wavelength of the photon transferred becomes longer each time.) An electron is boosted out of the chlorophyll in the reaction center, giving the electron a large redox potential. It is captured by an electron acceptor and passed along an electron-transport chain consisting of several compounds capable of undergoing redox reactions (e.g., $Fe^{2+} \leftrightarrow Fe^{3+}$). Some compounds also pick up or pass on protons (H^+) with the electron. The reaction centers and electron-transport chain are arranged in the thylakoid membrane in such a way that a proton gradient develops across the membrane, the membrane being impermeable to H^+. This gradient is relieved through complex ATPase particles, in which the energy of the gradient is used to phosphorylate adenosine diphosphate (ADP).

There are two photosystems in most cells, and they are connected by electron-transport compounds, as is usually drawn on a redox scale in the "Z scheme." Photosystem II (PS-II) is able to split water, producing electrons, protons, and waste oxygen. The electrons are finally used to reduce $NADP^+$ in noncyclic phosphorylation: Electrons boosted out of chlorophyll a in PS-II (P680) are given another boost in PS-I (P700). In cyclic phosphorylation, which can take place, for example, in blue-green algal heterocysts all the time and other algal cells sometimes, only PS-I is active, and electrons cycle back to PS-I, generating ATP but not NADPH. One of the proteins in red algae and other "chromophyte" algae that is analogous to green-plant proteins (rather than homologous) is the electron-transfer compound between the cytochrome b_6–f complex and the oxidizing end of PS-I: In these algae, cytochrome c takes the place of plastocyanin (Raven et al. 1990). The two catalysts are similar in many ways, but cytochrome c uses iron (Fe), whereas plastocyanin uses copper (Cu) as the prosthetic group.

Although the two photosystems are held in the membrane, they are not locked together in a 1 : 1 ratio as the Z scheme implies. Rather, there is a differential distribution of PS-I and PS-II between stacked and unstacked parts of the thylakoids, and there also are variable numbers of pigment complexes that may pass energy to both reaction centers, rather than to only one. The Z scheme as an electron-flow diagram is thus too simple, in that it shows how the redox potential drives the process, but it does not have the dynamic aspects of the physical mechanism. In blue-green algae, photosynthetic electron transport and respiratory electron transport take place in the same compartment and are intimately linked, but the arrangement of the components on the available membranes is not known (Scherer et al. 1988). Blue-green algae can exhibit bacterial flexibility and switch to anoxygenic photosynthesis, using sulfide rather than water as the electron donor (Paden & Cohen 1982). This could be significant not only in hot springs but also in salt marshes, mangrove swamps, and other places where there are anoxic sediments and cyanophyte mats, but its presence and magnitude in such habitats are scarcely known (cf. Whitton & Potts 1982).

ATP and NADPH are the end products of the light reactions; they contain the energy saved from light and are used to fix inorganic carbon in the Calvin cycle in the stroma of the chloroplast. The Calvin cycle does not directly require light (although some of the enzymes involved are stimulated by light). The steps in the Calvin cycle are called the "dark" reactions of photosynthesis, but in fact the pathway functions only during the light. The Calvin cycle is the only means of net carbon fixation in most plants; it is called C_3 photosynthesis because the first product of CO_2 incorporation is a three-carbon acid, 3-phosphoglycerate (3-PGA). Some plants in dry areas and/or areas that receive high levels of irradiance have an additional series of reactions in the cytoplasm that fix CO_2 into 4-carbon acids, but in these C_4 plants no net CO_2 fixation results from the extra steps, because the C_4 compounds are broken down to give CO_2 again, which is then refixed in the Calvin cycle. In ordinary C_4 plants, the extra steps, called the Hatch-Slack pathway, serve to concentrate CO_2 at the chloroplast, and both pathways function in the daytime. In plants with crassulacean acid metabolism (CAM plants), C_4 fixation occurs at night (when stomata can be open), and release of CO_2 to the Calvin cycle in the same cells occurs during the day, when stomata are closed. However, in some seaweeds, especially the large browns, there is a different process called light-independent carbon fixation, which results in net carbon fixation in darkness. We shall further explore the question whether algae are C_3 or C_4 plants in section 4.4.4.

In the Calvin cycle, carbon dioxide (1 carbon) and ribulose-1,5-bisphosphate (RuBP: 5 carbons) react to give two molecules of PGA, the enzyme being, of course, RuBisCO (ribulose-1,5-bisphosphate carboxylase/oxygenase). Subsequent steps, using ATP and NADPH in a pathway almost the reverse of glycolysis, convert PGA to glyceraldehyde-P and dihydroxyacetone-P and then put these together to give fructose-1,6-bis-P. Some fructose-bis-P is siphoned off to produce various low-molecular-weight free sugars or sugar alcohols (in higher plants, sucrose is formed). Other fructose-bis-P molecules, together with glyceraldehyde-P and dihydroxyacetone-P, enter a complex series of carbon transfers that regenerate RuBP. For net production of one hexose molecule, six turns of the cycle and six CO_2 molecules are needed.

Different algal groups are characterized by different light-harvesting pigments and photosynthetic products; this variety, together with the different light climate that exists underwater, contrasts markedly with the story of photosynthesis as known from green land plants. As Larkum and Barrett (1983) and Miller and Spear-Bernstein (1989) have noted, study of algae should help to provide a more nearly complete and more balanced understanding of the mechanisms of photosynthesis.

4.2 Irradiance

4.2.1 Measuring irradiance

"Light" refers to the narrow region of the electromagnetic spectrum whose wavelengths are visible to the human eye, plus the ultraviolet and infrared wavelengths. However, there are important differences between the spectral sensitivities of the human eye and plant photosynthetic pigments. Our visual pigment,

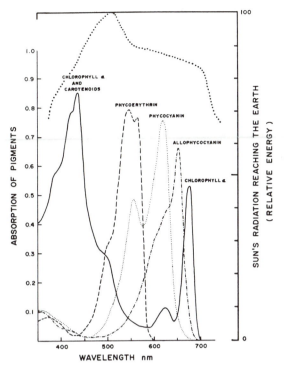

Figure 4.1. Spectrum of solar energy at the earth's
surface (upper dotted curve), and absorption spectra of
algal pigments. (From Gantt 1975, *BioScience*
25:781–8. Copyright © 1975 by the American Institute of Biological Sciences.)

rhodopsin, has one major peak, in the green region (556
nm), with absorption decreasing sharply on either side.
The chlorophylls and other light-harvesting pigments
have different absorption peaks, and together they absorb across a broad region of what is called photosynthetically active radiation (PAR) (Fig. 4.1). PAR is
defined as wavelengths of 350 nm (or, more commonly,
400 nm) to 700 nm, but there is some evidence that photosynthetic absorbance extends down to 300 nm in *Ulva
lactuca* and the tetrasporophyte of *Bonnemaisonia hamifera* (Halldal 1964).

The light meters used by lighting engineers and
photographers are designed with the same spectral sensitivity (bias) as our eyes, and their measure of light is
called illuminance. They do not give an accurate measure of the light available to plants, even on land. Underwater, where the spectral quality of light changes,
such instruments are particularly inappropriate. Unfortunately, they are still used (they are inexpensive); they
give data in the units of either foot-candles or lux. The
meters that are appropriate for such use are uniformly
sensitive to all wavelengths. Instrumentation for measuring light must be designed and used with care, as
Drew (1983) has explained quite well.

Photosynthesis is a quantum process, and so the
most useful measurement is the number of photons of

PAR received by a unit of algal surface. This measure is
called the photon-flux density (PFD). *Irradiance* is,
strictly speaking, a measure of the amount of *energy*
falling on a flat surface. Underwater, scattering contributes upwelling light, which becomes a significant component when downwelling light is dim (i.e., at depth),
or over reflective coral sand (Kirk 1989). Instruments
with spherical collectors can measure light from all directions, called ''scalar irradiance'' if energy is measured, or ''photon-flux fluence rate'' (sometimes
shortened to ''fluence'') if photons are measured. In
any given study in which light is measured, only one of
these terms can be accurate. As we overview many studies that have measured light all sorts of different ways
(including illuminance!), we face a quandary over which
term to use. Ultimately, the significant value for algae is
the total number of photons they can *trap*, which no
measure of incident light can convey, because of reflection and transmission of some photons. On the other
hand, simple terms like ''light intensity'' have been criticized as referring to the light emitted by the source,
rather than the light received. We shall use the term ''irradiance'' to mean the amount of PAR falling on a surface, regardless of whether we are citing data for quanta
or energy, and regardless of whether or not upwelling
light is included (cf. Drew 1983).

The units most commonly used to report PFD
are microeinsteins per square meter per second
($\mu E\ m^{-2}\ s^{-1}$), where an einstein is a mole of photons.
Some authors use simply mole (or micromole), because
the einstein is not a Système International unit. However, as Lüning (1981a) has pointed out, that removes
any clue that the quantity measured is light (as opposed
to, say, nitrate). The energy of light is measured in
joules or watts, joules being units of total quantity,
and watts units of flux (Smith & Tyler 1974). Thus,
$1\ W\ m^{-2} = 1\ J\ m^{-2}\ s^{-1}$. Illuminance units cannot be
precisely converted into irradiance units, except for
specified light sources, but Lüning (1981a) has given
some approximate values. ''Full sun'' (i.e., irradiance
in the middle of the day) shines approximately 2,000
$\mu E\ m^{-2}\ s^{-1}$ on the earth's surface, depending on latitude and season.

4.2.2 Light in the oceans

Light reaching the earth's surface already shows
variations as a result of scattering and absorption by the
atmosphere, but those are relatively small compared
with the changes it undergoes in water. The total irradiance is also affected by the sun's angle, being less as
the sun approaches the horizon. Four factors having to
do with light as it interacts with the oceans require special attention: (1) its heterogeneity over time, (2) the effect of the water surface on light penetration, (3) its
spectral changes with depth, and (4) its attenuation with
depth and the question of the depths to which seaweeds
can grow (Larkum & Barrett 1983; Drew 1983).

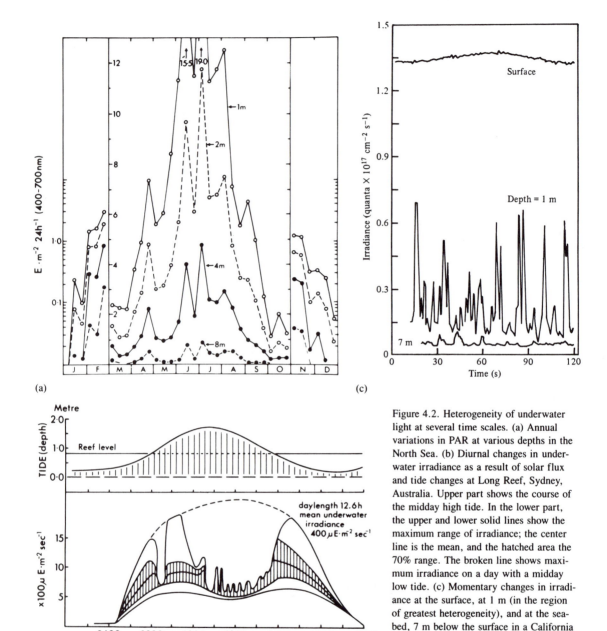

(a)

(b)

(c)

Figure 4.2. Heterogeneity of underwater light at several time scales. (a) Annual variations in PAR at various depths in the North Sea. (b) Diurnal changes in underwater irradiance as a result of solar flux and tide changes at Long Reef, Sydney, Australia. Upper part shows the course of the midday high tide. In the lower part, the upper and lower solid lines show the maximum range of irradiance; the center line is the mean, and the hatched area the 70% range. The broken line shows maximum irradiance on a day with a midday low tide. (c) Momentary changes in irradiance at the surface, at 1 m (in the region of greatest heterogeneity), and at the seabed, 7 m below the surface in a California giant kelp bed. (Part a after Lüning & Dring 1979; b after Larkum & Barrett 1983; c courtesy of and © 1983 Valrie Gerard.)

Heterogeneity. The underwater light climate varies seasonally, both predictably because of changes in day length and solar angle and unpredictably because of cloudiness and turbidity from storm waves, runoff, and seasonal plankton blooms. This variability is much greater than that seen on land. There are few data to illustrate the magnitude of the changes, but Lüning and Dring's (1979) study at Helgoland in the North Sea (54° N) was the most thorough. Instantaneous readings were taken every 20 min and extrapolated to give daily totals (Fig. 4.2a). Approximately 90% of the annual total light in the subtidal zone was received from April through September; during winter, the photic zone was very shallow. That study, carried out in temperate latitudes and turbid waters, needs to be replicated in other areas.

Tropical waters generally are much clearer than temperate waters, but the silt load produced by poor land-management practices is increasingly a problem, particularly during the rainy season. There are also

Table 4.1 *Effects of solar altitude and sea state on reflection from the surface*

Condition	Percentage reflectance at solar altitude						
	90–50°	40°	30°	20°	10°	5°	0°
Calm sea	3	4	6	12	27	42	100
Rough sea (Beaufort scale 4)	3	4	6	10	11		

Source: Drew (1983), with permission of Clarendon Press.

some natural tropical situations in which turbidity is a major factor, although this is unusual. On the Carribean coast of Panamá, predictable winds during the dry season create extreme turbidity on sand plains on the reef, as described by Hay (1986, p. 649): ". . . dives were generally made only on days when conditions were relatively calm. During the few dives conducted under the normally rough conditions of dry season, light levels were so low that I could not see my light meter even when holding it against my mask." He estimated that under rough conditions the irradiance was less than $1 \ \mu E \ m^{-2} \ s^{-1}$; compensation irradiances for seaweeds on the sand plain were $2–6 \ \mu E \ m^{-2} \ s^{-1}$ (*Gracilaria* spp., *Spyridia hypnoides, Solieria tenera*). Sand plains in other Carribean areas (e.g., U.S. Virgin Islands) do not undergo seasonally predictable periods of turbidity (Hay 1986).

Large diurnal fluctuations in irradiance also occur, because of changes in clouds, tides, turbidity, and the angle of the sun (Fig. 4.2b). Finally, there are momentary changes in irradiance due to waves and to canopy movement (Fig. 4.2c). The light-harvesting systems of seaweeds must be able to cope with these ranges of irradiance; there must be mechanisms to allow adaptations on each of these time scales (Henley & Ramus 1989).

The light climate in the intertidal zone is even more complex that in the subtidal, but nevertheless it has been modeled by Dring (1987). The complexity stems from three factors: (1) the water type (as discussed later), which may or may not change significantly from time to time at a given site; (2) the tide range, which has a monthly progression (sec. 2.2.1); (3) the timing of high and low tides in relation to the diurnal changes in irradiance, which are functions of both the progression in the times of the tides and the changes in day length. (Dring's model assumes that extreme low tides at any site always occur at approximately the same time of day, but that is not true in all regions, as noted in section 2.2.1. Where the time of ELWS changes, another variable must be added to the model.) He found that (for British shores) the lower intertidal receives less light in summer than in spring because the high tide is more likely to occur during the day as day length increases. Dring demonstrated that in estuaries with tur-

bid water and large tidal ranges, critical low light levels may occur in the intertidal rather than subtidally. For instance, kelp growth is considered to be limited at 1% of surface irradiance, as discussed later; in the Avon estuary (U.K.) this level becomes intertidal, and kelps disappear, because they are also limited in upward extent by exposure to air. *Fucus serratus,* which normally occurs into the subtidal zone, has an intertidal lower limit in the upper Avon estuary, again because of inadequate mean irradiance.

Effect of the surface. The sea surface plays a large role in the underwater light climate (Campbell & Aarup 1989; Mobley 1989). Some of the light hitting the sea surface is reflected; the percentage reflected will depend on the angle of the sun to the water, and hence also on the state, or roughness, of the water. Reflection from a smooth sea with the sun near its zenith is only about 4% of the total light (sun plus sky), whereas with a sun altitude of 10°, reflection is 28% (Table 4.1). With an overcast sky, reflection is about 10% regardless of the altitude of the sun. Waves will increase light penetration when the sun is low by increasing the angle between the water surface and the sun (Table 4.1), but whitecaps and bubbles in rough seas will increase reflection and can cut the light entering the water by as much as 50%. Under sunny skies, waves can cause considerable heterogeneity (glitter) in the subsurface light field by temporarily focusing the sun's rays to certain spots and away from others, especially in the top few meters of water (see Fig. 2.4) (Larkum & Barrett 1983; Drew 1983), with significant (though diverse) effects on growth and photosynthesis in phytoplankton (Fréchette & Legendre 1978; Queguiner & Legendre 1986) and a seaweed (*Chondrus crispus;* Greene & Gerard 1990).

Spectral changes with depth. As solar energy penetrates the oceans, it is altered in both quality and quantity. The attenuation results from absorption and scattering, as in the atmosphere. Water itself absorbs maximally in the infrared and far red, above 700 nm; far red (ca. 750 nm) penetrates only to depths of 5–6 m (Smith & Tyler 1976). Absorption of radiant energy by particles in the water, such as phytoplankton, will depend on the pigments they contain. Scattering by

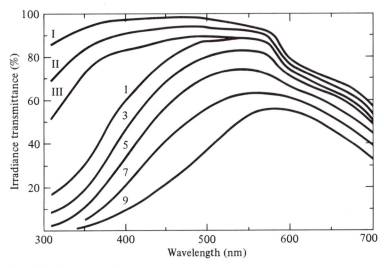

Figure 4.3. Percentages of transmittance downward of irradiance of various wavelengths in Jerlov's different optical types of water. The types range from clear oceanic (type I) to turbid coastal (type 9). (From Jerlov 1976, reprinted with permission from *Marine Optics* © 1976, Elsevier Science Publishers.)

particles larger than 2 μm contributes to attenuation by increasing the length of the optical path of quanta once they are in the water, and thereby increasing the opportunities for absorption. Smaller particles and sea salts do not contribute appreciably to attenuation in the visible region (Jerlov 1970, 1976; Larkum & Barrett 1983).

Attenuation of light by various processes and particles results in seawaters of different optical properties, and these have been classified by Jerlov (1976): The oceans have been divided into two broad categories: green coastal waters and blue oceanic waters, with subdivisions in each of these categories (Fig. 4.3). Jerlov (1976) has distinguished five ocean-water types and nine coastal-water types. Figure 4.3 shows that in the clearest water (oceanic type I), the maximum transmittance (about 98.2% of surface irradiance) is at approximately 475 nm. The Jerlov water type that transmits the least solar energy is coastal type 9. In such water the maximum transmittance occurs at about 575 nm (green) and is only 56% of the total irradiance at the surface. The characteristic green color of coastal waters is due to absorption at shorter wavelengths by plant pigments and by yellowish dissolved organic substances (*Gelbstoff*) that absorb strongly in the blue wave bands. *Gelbstoff* comes from terrestrial humic material brought to the seas by rivers, and it is also produced in the sea by algae. More recent measurements with sensitive spectroradiometers have shown that Jerlov overestimated light penetration, especially in the blue region and in coastal waters (Pelevin & Rutkovskaya 1977). As a result of the wavelength dependence of transmittance (Fig. 4.3), both the quality and quantity of light will change with depth in any given water mass (Fig. 4.4). Light pene-

tration in very turbid estuaries is even less than that through Jerlov's worst water type, and Dring (1987) has extrapolated two theoretical new coastal-water types, 11 and 13.

Most physiological experiments and culture studies do not depend on the sun for irradiance, but employ artificial light, usually from fluorescent tubes. There are many different types of fluorescent tubes. "Gro-lux" tubes are favored by horticulturists, but for work on algae, "daylight" tubes are best because (despite their name) they have a spectral distribution close to that of natural light at a few meters of depth in clear coastal waters (Drew 1983). Fluorescent tubes give low irradiance; Drew has discussed ways of producing higher irradiance and of using filters to more closely imitate particular light fields.

The electromagnetic spectrum continues from PAR into shorter wavelengths as ultraviolet (UV) light (Lüning 1981a). The near UV is not damaging (except at high irradiances, when it may be part of a general photoinhibitory effect), but shorter wavelengths (up to about 310 nm) are absorbed by proteins and nucleic acids and can be lethal. UV-B extends from about 280 to 315 nm; it is a significant component of sunlight and is transmitted well, especially through oceanic-water types (Fig. 4.3). Biologically effective dosage rates of UV(-B) are delivered on reef flats and other shallow subtidal zones and in the intertidal zone (Smith & Baker 1979; Fleischmann 1989). Seaweeds and marine animals (e.g., corals) protect themselves either through self-shading (e.g., in *Ecklonia*, where the meristems are below the blades) or by production of massive amounts of UV-absorbing substances (e.g., in growing tips of *Eucheuma*) (Wood 1987, 1989).

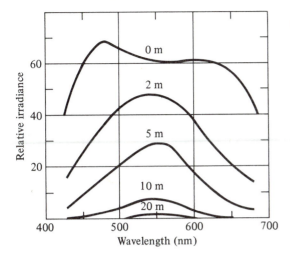

Figure 4.4. Energy spectra of natural light at various depths in the northern Baltic Sea. (From Jerlov 1976, reprinted with permission from *Marine Optics* © 1976 Elsevier Science Publishers.)

Limits to growth. Irradiance in the sea is reduced with increasing depth, ultimately to zero. Most of the ocean floor (>90%; Russell-Hunter 1970) is permanently dark and has no algal growth, but what is the depth limit to which seaweeds can grow? How little light is enough? The compensation depth, at which irradiance allows enough photosynthesis to just outweigh respiration (all integrated over time), is a significant but elusive number. A value of 1% of surface irradiance (i.e., of "full sun") has commonly been used in oceanography to define the bottom of the euphotic zone (Steemann-Nielsen 1974), though Lüning and Dring (1979) have reported 0.5–1% as the lower limit for Laminariales and 0.05–0.1% as the lower limit for multicellular algae. Lüning (1981a) has suggested that the reason seaweeds can adjust their metabolism to extremely low light is that the flux is more nearly constant and predictable than the light climate for plankton.

Ideally, given the extreme variations in surface irradiance, and especially for studies of long-lived seaweeds, we need to know annual total irradiance. Surface irradiance in the Mediterranean is about 3,000 MJ m^{-2} yr^{-1} or about 12.6 × 10^3 E m^{-2} yr^{-1}. Lüning and Dring (1979) measured the annual irradiance as 1.3 MJ m^{-2} yr^{-1} at the lower limit for seaweeds off Helgoland. However, measurements usually are made on a per-second basis, with surface irradiance, as mentioned, 1,500–2,000 μE m^{-2} s^{-1} (e.g., Littler et al. 1985). These are midday values, and means for the whole day obviously will be lower. For instance, Osborne and Raven (1986) estimated the mean daily surface irradiance in Britain to range from 1,000 μE m^{-2} s^{-1} in June to 75 μE m^{-2} s^{-1} in winter. The deepest known seaweeds, at 268 m on an seamount off

the Bahamas, received a maximum of 0.015–0.025 μE m^{-2} s^{-1}, or a mere 0.0005% of surface irradiance (Littler et al. 1985). This, as Raven et al. (1990) have pointed out, is scarcely more than the irradiance at the sea surface on a night with a full moon. Under Artic ice, a habitat in which Wilce (1967) had suggested that there might be so little light that algae there would have to exist heterotrophically, Chapman and Lindley (1980) and Dunton and Schell (1986) showed that there was adequate light for growth of *Laminaria solidungula* at its depth limit (20 m).

There are not yet enough data to determine if there are "annual quantum requirements" for major algal zones, as proposed by Lüning (1981a), but the depth range for seaweeds, where the substratum allows growth, evidently extends deeper than has commonly been thought. Searching for depth records in clear waters requires the use of submersibles, because algae grow beyond scuba limits, and dredging always involves a measure of uncertainty about the depth and whether or not the plant was attached and growing there.

4.3 **Light harvesting**
4.3.1 Pigments and pigment-protein complexes

Three kinds of pigments are directly involved in algal photosynthesis: chlorophylls, phycobiliproteins, and carotenoids (Meeks 1974; Goodwin 1974; Ragan 1981; Rowan 1989; Dring 1990). Chlorophyll *a* is essential in the reaction center and is found in all algae. The other pigments, along with the bulk of the chlorophyll *a*, funnel energy to the reaction centers. Three additional chlorophylls occur in seaweeds. Chlorophyll *b* is found in Ulvophyceae (and in other Chlorophyta and higher plants); chlorophylls c_1 and c_2 (Fig. 4.5a) occur in Phaeophyceae (and other lines of "brown" algae). All are tetrapyrrole rings with Mg^{2+} chelated in the middle; chlorophylls *a* and *b* each have a long fatty-acid tail ($C_{20}H_{39}COO—$) that chlorophylls c_1 and c_2 lack. Chlorophylls *c* absorb blue light more strongly and red light less strongly than do chlorophylls *a* and *b*. In addition to the chemically different chlorophylls, these pigments, especially chlorophyll *a*, bind to proteins in various ways to create even more variety, as seen in their absorption spectra (especially the red peak). Chlorophyll *d* has been reported sporadically in some red algae, but it probably has no function in photosynthesis (Rowan 1989).

Seaweeds have a wide variety of carotenoids, of which at least three are involved in light harvesting, in contrast to the case for higher plants. Carotenoids are C_{40} tetraterpenes; carotenes are hydrocarbons, and xanthophylls contain one or more oxygen molecules (Fig. 4.5b). Larkum and Barrett (1983) distinguished three groups of algae on the basis of the abundance and roles of their carotenoids: (1) Red and blue-green algae have several minor carotenoids. Those associated with thylakoids are confined to the reaction centers, where they

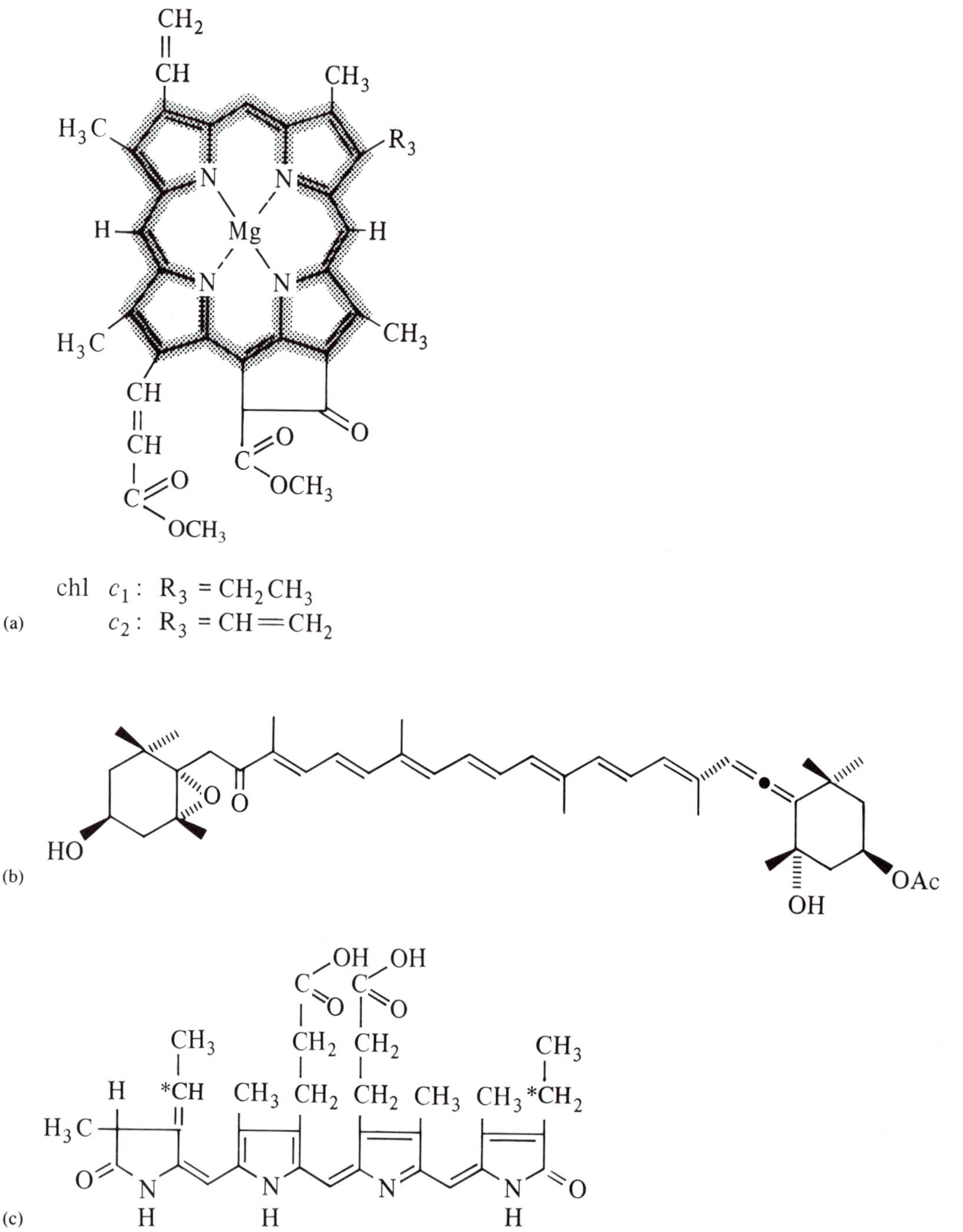

chl c_1: $R_3 = CH_2CH_3$
c_2: $R_3 = CH=CH_2$

(a)

(b)

(c)

Figure 4.5. Algal pigment types. (a) Chlorophylls (chl c), showing the conjugated double-bond system (stippled). (b) Carotenoids (fucoxanthin). (c) Phycobilins (chromophore of phycocyanobilin). Sulfide bridges to protein cys residues form at carbons marked*. (Redrawn after various sources.)

have a protective role. The light-harvesting complexes in these algae consist of phycobiliproteins. (2) In most green algae, carotenoids (β-carotene and perhaps lutein) play a role in light harvesting but are not the predominant antenna pigments. (3) There is a mixed group in which carotenoids play a major role in light harvesting. This group includes the brown algae and other "Chromophyta" with fucoxanthin (Fig. 4.5b), plus some siphonous green algae and some deep-water greens from other orders (e.g., *Ulva japonica, U. olivascens*) with

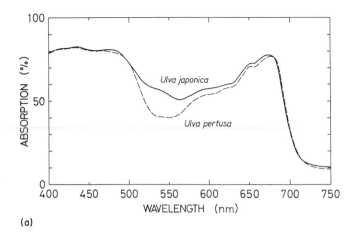

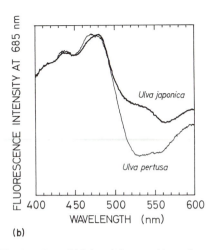

(a)   (b)

Figure 4.6. Comparisons of the absorption and fluorescence spectra of *Ulva japonica*, which has siphonaxanthin, and *U. pertusa*, which lacks it. (From Kageyama et al. 1977, with permission of Y. Yokohama.)

siphonaxanthin. Violaxanthin is a major xanthophyll in brown algae, but it has not yet been shown to contribute to light harvesting (except in Eustigmatophyceae) (Dring 1990). *Ostreobium*, a small siphonous green alga that lives in coral skeletons beneath the zooxanthellae layer, uses xanthophylls to harvest the very dim, spectrally unusual light available there (Fork & Larkum 1989). Fucoxanthin, β-carotene, and siphonaxanthin are unusual carotenoids because in vivo they absorb into the green region. The roles of accessory pigments in photosynthesis can by shown by comparing absorption spectra with action spectra (photosynthetic rate per unit of irradiance as a function of wavelength) (Figs. 4.6 and 4.14). In all groups, some β-carotene is bound in a close and oriented association with PS-I, indicating efficient energy transfer to this reaction center, but it probably is not involved in transferring energy to PS-II. The second major role of carotenoids is in protecting the reaction-center chlorophyll from photooxidation. For instance, in the xanthophyll cycle (known in green algae), zeaxanthin can capture oxygen to form violaxanthin, which can be reduced again by ascorbic acid (Goodwin & Mercer 1983; Rowan 1989).

Phycobiliproteins are characteristic pigments of red and blue-green algae; they also occur in Cryptophyceae. Unlike chlorophylls and carotenoids, they are water-soluble, not lipid-soluble, and form particles (phycobilisomes) on the surfaces of thylakoids, rather than being embedded in the membranes. (In Cryptophyceae they are in the lumen of the thylakoid, a unique location for photosynthetic pigments; Spear-Bernstein & Miller 1989.) Phycobiliproteins consist of pigmented phycobilins, which are linear tetrapyrroles (Fig. 4.5c), covalently bound in various combinations to protein complexes. There are two principal phycobilins, phycoerythrobilin (red) and phycocyanobilin (blue). The pro-

tein complexes always have two different polypeptide chains, α and β, usually in a 1 : 1 ratio, but with as many as six of each chain in the phycobiliprotein. Three main classes of phycobiliprotein occur: Phycoerythrins (PE) absorb in the green region (495–570 nm) and generally have two phycoerythrobilin molecules on each α chain and four on each β chain. (In two PEs there is a third polypeptide bearing a third phycobilin: phycourobilin.) Phycocyanins (PC) and allophycocyanins (APC) generally have one phycocyanobilin on each protein chain. There are many specific molecules (Rowan 1989). Phycocyanins absorb in the green-yellow region (550–630 nm), and allophycocyanins in the orange-red region (650–670 nm). Representative absorption spectra of these pigments are shown in Figure 4.1. Each of the three types occurs in red and blue-green algae, although the proportions vary, as is evident from the predominant algal colors.

Pigments are arranged in or on the thylakoid in very particular ways. The reaction centers are special chlorophyll-protein complexes, surrounded by a core antenna of chlorophyll *a*. Most of the chlorophyll *a* and other accessory pigments are organized into light-harvesting complexes in the membrane or, in the red and blue-green algae, as phycobilisomes on the membrane. These arrangements act as funnels to pass energy to the reaction centers. At each level the nonphycobilin pigments are noncovalently bound to proteins, giving many different combinations with different absorption spectra. We shall examine each of these three complexes in turn, again based on the detailed discussions by Larkum and Barrett (1983) and Rowan (1989).

Each of the two photosystems (PS-I, PS-II) has a unique chlorophyll composition. The reaction-center chlorophylls, which are the molecules directly involved in electron transfer (as opposed to energy transfer), are

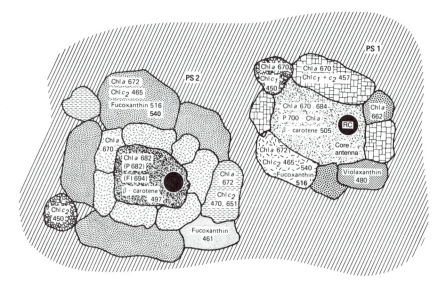

Figure 4.7. Schema of spatial arrangement of pigment–protein complexes in RC-I and RC-II supracomplexes of the brown alga *Acrocarpia paniculata*. (From Larkum & Barrett 1983, with permission of Academic Press.)

designated RC-I (or P700) and RC-II (or P680). PS-I has been most successfully studied, whereas the water-splitting PS-II is less well understood. Some of the cell's chlorophyll *a* is closely associated with the reaction center in PS-I to form core complex I (CC-I), formerly called P700-chlorophyll-*a*-protein complexes. The ratio of chlorophyll *a* to P700 in CC-I is about 40 : 1. CC-I also contains small amounts of β-carotene (α-carotene in siphonous green algae). The composition of CC-II is uncertain, especially in algae. Also tightly associated with the reaction-center pigments, and often extracted in the complexes, are the electron-acceptor molecules and, in the case of PS-II, the water-splitting enzymic machinery. Core-complex pigments transfer energy from light-harvesting pigment complexes (LHC-I, LHC-II) to the reaction centers.

The light-harvesting complexes of green and brown algae contain most of the chlorophyll *a* and all of the chlorophyll *b*, fucoxanthin, and other light-harvesting pigments (except a small amount of the β-carotene, as noted earlier. LHC-II is the larger of the light-harvesting complexes and can contain as much as half the total pigment in Chlorophyta. Several supermolecular complexes have been isolated from algae, and their relationships to LHC-I or LHC-II are often uncertain (Rowan 1989). In green algae (and higher plants) there is a chlorophyll *a*/*b*-protein (often referred to in the literature as LHCP, light-harvesting chlorophyll-protein). A chlorophyll-*a*/*b*-siphonaxanthin-protein complex occurs in those greens that have siphonaxanthin (Larkum & Barrett 1983; Levavasseur 1989). Two different sizes of LHC-II, not correlated with the presence or absence of siphonaxanthin, have been found in Ulvophyceae (Fawley et al. 1990). Chlorophyll-*a*/c_2-

fucoxanthin-protein and chlorophyll-*a*/c_1/c_2-violaxanthin-protein complexes occur in brown algae. Katoh and Ehara (1990) have deduced that this complex in *Petalonia fascia* contains 128 molecules of chlorophyll *a*, 27 chlorophyll *c*, 69 fucoxanthin, and 8 violaxanthin.

Within the light-harvesting complexes the pigments apparently are arranged in clusters (Fig. 4.7). Most of the available evidence implies that each photosystem has its own antenna(e), but as Larkum & Barrett (1983) have noted, rapid changes in underwater light quality and quantity would make it advantageous for algae to control the distribution of energy between photosystems. This might take place through passage of excess energy from PS-II to PS-I (spillover) or, more probably, through control of distribution from the antennae. Both these hypotheses ignore the fact that PS-I and PS-II units often are physically separated in the thylakoids, as discussed later. Recent models have incorporated phosphorylation of pigment-protein complexes as central to the delivery of energy to alternate photosystems (Williams & Allen 1987).

Phycobiliproteins form clusters on the surfaces of thylakoids (Fig. 4.8), where they are more amenable to observation (Cohen-Bazire & Bryant 1982; Larkum & Barrett 1983; Glazer 1985; MacColl & Guard-Friar 1987). Hemidiscoidal phycobilisomes occur in most blue-green algae and a few red algae, whereas the usual shape in red algae (and a few blue-greens) apparently is globular (e.g., Talarico 1990). However, most of the studies have been done on only a few species in culture. In field material of *Porphyra umbilicalis*, and perhaps other red seaweeds, both shapes of phycobilisomes occur, which may provide a means of response to fluctuating irradiance (Algarra et al. 1990).

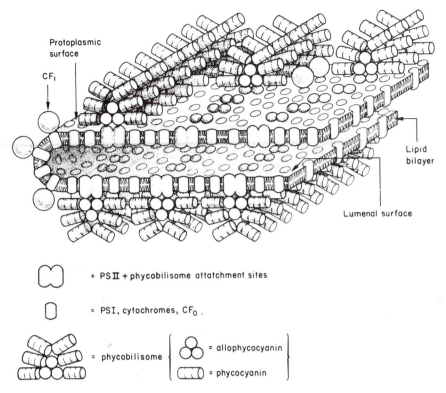

Figure 4.8. A model of the thylakoid membrane from cyanelles of *Cyanophora paradoxa*, showing the phycobilisomes attached to PS-II and the CF_1 coupling factor at ATPase attached to a unit comprising PS-I, cytochromes, and the CF_0 subunit of the coupling factor. If phycoerythrin were present, it would be on the outer ends of the phycocyanin rods. (From Giddings et al. 1983, with permission of the American Association of Plant Physiologists.)

The arrangement of the pigments parallels the pathway of energy transfer, which can also be traced in the pigment spectra (Fig. 4.1), because energy is inversely proportional to wavelength: PE → PC → APC → chlorophyll. Phycobilisomes also contain a number of colorless proteins, at least some of which are involved in assembly of the phycobilisomes. Most phycobilisomes are associated with PS-II; reaction-center I receives some light directly from the remaining phycobilisomes, but most via spillover from PS-II, not via pigment-protein phosphorylation (Biggins & Bruce 1989; Gantt 1989, 1990).

Pigment-protein complexes have been isolated by gently fractionating chloroplasts and thylakoids. Freeze-fracture techniques and electron microscopy have been used to try to determine the physical arrangements of photosystems and light-harvesting complexes. In higher plants, which have stacked regions of the thylakoids called grana, marked differences in abundances of photosystem particles have been found between stacked and unstacked regions: PS-II is predominantly in stacked regions, and PS-I in unstacked regions. Thylakoid appression occurs in all algae except red and blue-green algae (which have phycobilisomes) (Fig. 1.7). In green algae the pattern of stacking can resemble that in higher

plants, whereas in brown algae the thylakoids are charcteristically in threes, so that the only "unstacked" surfaces will be those on the outside of the stack plus the single girdle lamella that lies just under the chloroplast envelope. The distributions of photosystems and antennae are not known in algae, but it seems probable that there is differential distribution.

Several theoretical advantages of differential distribution have been advanced (Larkum & Barrett 1983, pp. 140–1); on the other hand, such a distribution raises questions about how the two photosytems manage to interact. As Larkum and Barrett (1983, p. 142) noted, "any discussion of the role of thylakoid appression must be set against the fact that [it] does not occur in Cyanobacteria and Rhodophyta. . . . Thus appressed thylakoids cannot confer any unique property. Rather, the two approaches, appressed thylakoids with membrane-located light harvesting complexes and non-appressed thylakoids with large phycobilisomes located in the stroma space, seem to serve a similar end although employing different mechanisms." Although pigments in phycobilisomes harvest green light very efficiently, phycobilisomes are nutritionally expensive (because of the protein), and their bulkiness relative to photosystem particles seems to preclude efficient packing on the

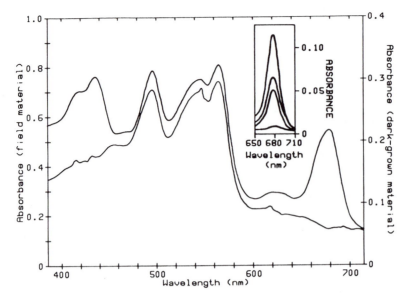

Figure 4.9. In vivo absorbance spectra of field and dark-grown *Delesseria sanguinea,* showing the absence of chlorophyll peaks (lower line) in the dark-grown specimen. Inset shows development of chlorophyll *a* peak at 678 nm after 1, 3, 5, and 7 days at an irradiance of 5 μE m^{-2} s^{-1}. (From Lüning & Schmitz 1988, with permission of Blackwell Scientific Publications.)

thylakoid. In each approach, space in the chloroplast is apportioned to photosystems, pigment packages, ATP/NADPH generation, and Calvin-cycle components.

One of the ways in which algal light-harvesting systems have been thought to differ from those in land plants is in not needing light to produce pigments. However, that is proving to be partly a misconception. Whereas angiosperms grown in darkness from seed are nonpigmented (and show other peculiarities in a syndrome called etiolation) (Salisbury & Ross 1985), those algae that can be grown in the dark (not many!) often are pigmented. A specific step in chlorophyll synthesis in higher plants has an absolute requirement for light, whereas in these algae it does not. The examples from algae are in fact few, mostly heterotrophic microalgae (Kasemir 1983), which is not surprising, because algal propagules lack the food reserves of seeds and simply do not grow in the dark. The notion that seaweeds do not need light for pigment synthesis goes back to the old erroneous notion that deep-subtidal seaweeds grow where there is no light; witness the following excerpt from Harvey (1841, p. x): "At enormous depths, to which the luminous rays, it is known, do not penetrate, species exist as fully coloured as those along the shore. They therefore, in this respect, either differ from all other plants, or perhaps, what are called the *chemical* rays, in which seem to reside the most active principles of solar light, may be those which cause colour among [seaweeds], and these *may* penetrate to depths which luminous rays do not reach." In one well-known study of dark growth of a seaweed (Lüning 1969), new blades of *Laminaria hyperborea* formed in darkness from stored

reserves apparently were fully pigmented. However, two red algae, *Maripelta rotata* and *Delesseria sanguinea,* are unable to form chlorophyll in darkness (Fig. 4.9) (Lüning & Schmitz 1988), In another study, *Delesseria* did form phycobiliproteins, though phycobilisomes were not organized, in contrast to the case for dark-grown *Tolypothrix* (Cyanophyta) (Ohki & Gantt 1983). Regreening in *Delesseria* took place in 5 μE m^{-2} s^{-1} (Fig. 4.9, inset). No green algae have been studied this way, and much more work is needed on red and brown algae using species with perennating parts.

4.3.2 Functional form in light trapping

The light-harvesting ability of seaweeds depends not only on the types and amounts of pigments and their disposition within the thylakoid but also on higher levels of organization: the arrangement of chloroplasts in the thallus and the morphology of the thallus. The arrangements of the chloroplast(s) in algal cells are quite varied, but in cells with large vacuoles or many small chloroplasts the chloroplasts tend to be distributed around the periphery of the cell. Similarly, pigmented cells tend to be distributed around the periphery of thicker thalli, with medullary cells having few or no chloroplasts. These arrangements obviously facilitate the passage of light to the pigments, but they may also be essential to maintain adequate inorganic carbon fluxes (Larkum & Barrett 1983).

Many seaweeds are constructed of thick, optically complex tissues (Osborne & Raven 1986; Ramus 1990). The packaging of pigments into light-harvesting complexes, chloroplasts, and cells reduces light absorption, as compared with in vitro absorption by a pigment solution (the "package effect"). On the other hand, refraction and reflection increase the light path within a thallus, enhancing the chance of photon capture. Some seaweeds have internal air spaces (e.g., *Enteromorpha*) or internal layers of calcium carbonate (e.g., *Halimeda*) that tend to increase the backscatter of light, in some cases perhaps even acting as "light guides" (Ramus 1978). The theories on light absorption tend to assume a relationship between pigment concentration (especially chlorophyll *a*) and absorption; this relationship holds for absorption of monochromatic light by pigment solutions (Beer's law), but not for the ordered, hierarchical structure of seaweeds in polychromatic light. The net effect of pigment and thallus properties can be seen in thallus *absorptance*, the fraction of incident light absorbed. Figure 4.10 shows absorptance spectra for *Ulva* (70 μm thick) and *Codium* (3 mm thick), with different amounts of pigment. Absorptance of *Codium* changes little over a large change in pigment concentration, whereas that of *Ulva* is significantly affected by pigment concentration. Algae such as *Codium*, which absorb virtually all incident light, have been called optically black.

The functional form of a seaweed must affect its light-harvesting ability, but the groupings suggested in Chapter 1 may not be the most incisive for studies of light harvesting, even if they are useful indicators of photosynthetic rates and productivity (sec. 4.7.2). Hay (1986) suggested a different set of growth forms: somewhat restated, they are as follows (Fig. 4.11):

1. Monolayers with flat, opaque thalli, including fleshy umbrella-shaped algae such as *Constantinea*, and also (though Hay did not discuss these) crustose calcareous forms and noncalcareous crusts
2. Multilayered, translucent thalli, such as *Ulva*, *Halymenia*, and *Padina*
3. Multilayered thalli with flat but narrowly dissected blades or terete branches; here we might also include filamentous thalli (which, again, Hay did not consider)
4. Multilayered thalli with midribs supporting thin blades (e.g., *Sargassum*); here we might also include *Macrocystis*, in which the blades are supported by floats along a stipe rather than by a midrib (Lobban 1978a).

As with Littlers' functional-form groups, a species may fall into more than one category (e.g., by genotypic variation, as shown in *Gracilaria tikvahiae* by Hanisak et al. 1988). Superimposed on the morphological characters are biochemical/physiological differences in light-harvesting ability (e.g., "sun" vs. "shade" plants, ecotypic variation; Algarra & Neill 1987; Gerard 1988; Coutinho & Yoneshigue 1990).

Hay (1986) hypothesized that in well-lighted habitats, multilayered seaweeds (with a high thallus-area : substratum-area ratio, in higher plants called

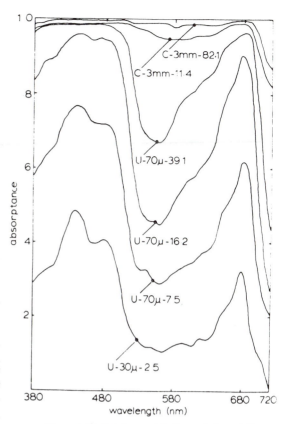

Figure 4.10. Relative contributions of pigmentation and thallus construction to light absorption. Absorptance spectra for *Codium* (C), an opaque seaweed, and *Ulva* (U), a translucent seaweed. Increasing pigment concentration (given in nanomoles per square centimeter) in 70-μm-thick *Ulva* thalli greatly increases absorptance (fraction of incident light absorbed). *Codium* thalli absorb essentially all the incident light, even with pigment concentrations much lower (on a surface-area basis) than those for *Ulva*. (From Ramus 1978, with permission of *Journal of Phycology*.)

leaf-area index) would have a growth-rate advantage over monolayered seaweeds with thallus-area ratios near unity, whereas in dimly lighted habitats the reverse should be true. If broad, flat thalli are translucent, they may be multilayered; translucency depends on thin thalli and adequate light, and in very dim light even thin thalli may absorb all the light entering them. Thicker thalli can be multilayered only if their branches are narrow enough to cast little shade on the branches below. Because theory on seaweed functional form is recent and incomplete, both Hay (1986) and Lobban (1978a) drew on terrestrial literature, especially Horn's (1971) theory of tree-canopy structure. However, while some of Horn's ideas are useful starting points, his theory is difficult to transfer to seaweeds because it assumes that leaves and branches remain in fixed relation

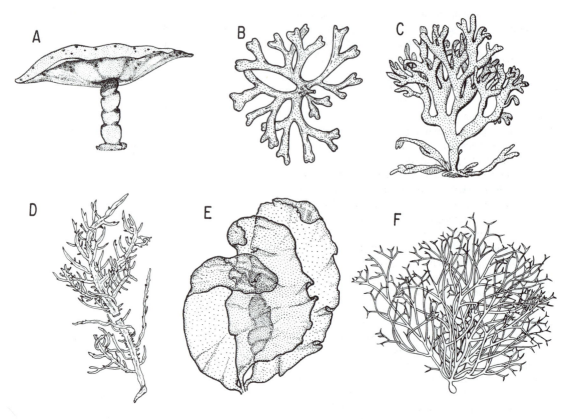

Figure 4.11. Hay's classification of nonfilamentous, upright seaweeds into functional-form groups for light-harvesting capacity. (A, B) Monolayered species with flat, opaque thalli (e.g., *Maripelta, Codium*). (C) Multilayered, with mid-ribs supporting thin margins (*Fucus*). (D) Multilayered, with flat but narrowly dissected thalli (*Desmarestia*). (E) Multilayered, with translucent blades (*Porphyra, Ulva*), (F) Multilayered, with thin, terete branches (*Cladophora, Hypnea*). (From Hay 1986, with permission of Cambridge University Press.)

to each other. Water motion, by constantly rearranging branches and blades, makes even a thick canopy of large blades such as *Macrocystis* effectively multilayered (Lobban 1978a), at least in well-lighted habitats. The extent and effect of self-shading in *Macrocystis* canopies have been examined by Gerard (1984b) and Jackson (1987). The morphology of narrowly branched seaweeds with erect thalli (Fig. 4.11f) is more analogous to a *leafless* tree or to vascular herbs such as *Equisetum* and *Asparagus* than to any arrangement of flat leaves. This point can be emphasized with the surface-area index of *Corallina elongata*. Algarra and Neill (1987) determined it to be 1,413 m^2 m^{-2} for sun plants and 224 m^2 m^{-2} for shade plants, in extreme contrast to the values for kelps and fucoids (4–7 m^2 m^{-2}), temperate deciduous forests (3–6 m^2 m^{-2}), and tropical forests (7–17 m^2 m^{-2}).

Peckol and Ramus (1988) criticized Hay's hypothesis when they found that multilayered – not monolayered – species were most abundant in a deep, low-light community. Competition for space and nitrogen were also important factors there, and the community presumably reflected the combined results of these op-posing selective pressures. The success of plants in the field must always be the result of many factors, and that does not invalidate Hay's hypothesis about the relationship between growth rate and irradiance.

Another complication not addressed in Hay's or the Littler's theories is the growth habit of turfs. Dense turfs or mats (common on well-lighted flats) consist of filamentous algae with a high "leaf-area index," but so crowded that self-shading becomes significant (Williams & Carpenter 1990). Sparse turfs may be closer to the theoretical functional-form group of filamentous algae.

4.3.3 Photoacclimation and growth rate

The rate of photosynthesis depends on the irradiance available, or ultimately the irradiance absorbed. The relationship between photosynthesis and irradiance, the *P–I* curve, is useful for comparing the physiology of light harvesting in different plants. A generalized *P–I* curve is shown in Figure 4.12a. In extremely dim light, respiration is greater than photosynthesis. When photosynthesis balances respiration, the irradiance level is at the compensation point, I_c. The rate of photosynthesis

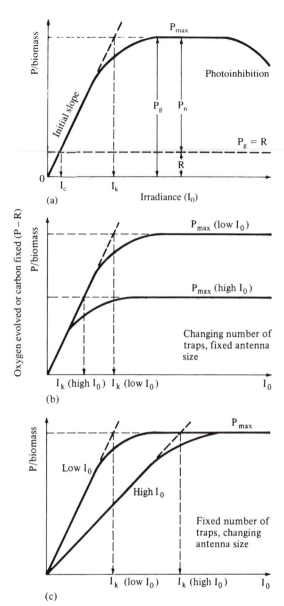

Figure 4.12. Model light-saturation curves for net photosynthesis (P) versus incident irradiance (I_o). (a) General model defining P_{max}, maximum photosynthesis; P_g, gross photosynthesis; P_n, net phtosynthesis; R, respiration; I_c, compensation irradiance; I_k, saturating irradiance level. Horizontal dashed line represents zero net photosynthesis ($P_g = R$). (b) Model of adjustment of the photosynthetic unit to extreme low and high I_0 by changing the number of PSUs but not their size. (c) Model of adjustment to extremes of irradiance by changing the size of a fixed number of PSUs. (From Ramus 1981, with permission of Blackwell Scientific Publications.)

increases linearly at first, and the initial slope, α, is a useful indicator of quantum yield. At higher irradiances, photosynthesis becomes saturated (P_{max}), limited by the "dark" reactions. Saturating irradiance, I_k, is defined as the point at which the extrapolated initial slope crosses P_{max}. I_c and I_k are not correlated with functional form (Arnold & Murray 1980).

The levels of irradiance needed to reach compensation are some 2–11 μE m^{-2} s^{-1} for shallow-water seaweeds (Arnold & Murray 1980; Hay 1986), but are much lower in dim habitats (sec. 4.2.2). Saturating irradiances show some correlation with habitat, but generally are low compared with full sun, suggesting that seaweeds are all more or less "shade" plants (Reiskind et al. 1989). Intertidal species require 400–600 μE m^{-2} s^{-1} (ca. 20% of full sun), upper and midsublittoral species 150–250 μE m^{-2} s^{-1}, and deep-sublittoral species less than 100 μE m^{-2} s^{-1} (Lüning 1981a). Diatoms under ice, in very dim light, saturated at 5 μE m^{-2}s^{-1} and were already photoinhibited at 25 μE m^{-2} s^{-1} (Palmisano et al. 1985). I_c and I_k change with assay temperature, but these short-term changes may be compensated by changes in pigmentation during growth at different temperatures (Davison et al. 1991) (sec. 6.2.2).

Very high irradiance may cause photoinhibition, especially in phytoplankton that are being vertically mixed (Neale 1987) or in seaweeds under desiccation stress (Herbert & Waaland 1988; Herbert 1990), but also in submerged plants (Huppertz et al. 1990). Photoinhibition involves damage to some components of the photosystems (especially PS-II), such as the membranes or electron-transport proteins, but the exact locations of damage and the mechanisms of resistance and repair in diverse plants remain uncertain.

Blue light affects photosynthetic rates and may stimulate photosynthetic capacity, in both the short term and long term. Photosynthesis in red light by *Laminaria digitata* is stimulated by 2 min of blue light (Fig. 4.13), but only for about 1 h (Dring 1989). Although blue light is known to affect Calvin-cycle enzymes, and this might be the cause of longer-term stimulation, the short-term effect in this kelp may be on the bicarbonate-uptake mechanism. When algae are shifted from high to low irradiance, they suffer an "energy crisis" (Falkowski & LaRoche 1991), to which they may respond by diverting macromolecule biosynthesis from lipids and carbohydrates to proteins (for light-harvesting complexes) and then back to lipids (for photosynthetic membranes). When algae are shifted to high light, their pigment decreases via dilution (through growth) and degradation. These changes are called photoacclimation (Falkowski & LaRoche 1991).

The new pigment may be apportioned in at least two ways, and Ramus (1981) has used P–I curves to assess these. P–I curves are mathematical models of photosynthetic rates (Jassby & Platt 1976;

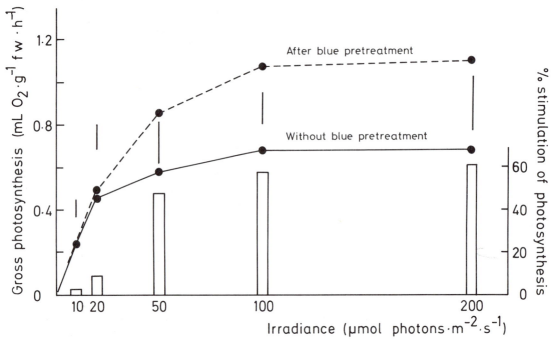

Figure 4.13. Gross photosynthesis versus irradiance in red light for *Laminaria digitata* with and without pretreatment with blue light. Histograms show percentage increase in photosynthesis at each irradiance resulting from blue pretreatment. Plants were cultivated in red light at 50 μE m^{-2} s^{-1} (= micromoles of photons per square meter per second). (From Dring 1989, with permission of *Journal of Phycology.*)

Nelson & Siegrist 1987; Madsen & Maberly 1990). The rate can be standardized to units of biomass or chlorophyll. As models, they can be used to predict the change in rate if one of the parameters, such as pigment concentration, changes. For instance, an alga may acclimate to low irradiance by increasing the number of reaction centers or the size of the antennae. Ramus (1981) used the concept of a photosynthetic unit (PSU), consisting of a reaction center and its associated light-harvesting pigment complex. We now recognize that there is not a 1 : 1 ratio between PS-I and PS-II or between them and their antennae, but the concept of the PSU is still useful. Ramus used the model to predict that if the *number* of PSUs is increased (Fig. 4.12b), P_{max} will increase, and more light will be needed to saturate photosynthesis (higher I_k). On the other hand, if the *size* of PSUs increases, without changing their number (Fig. 4.12c), P_{max} will not change, but less light will be required to saturate photosynthesis; the PSUs in this case will be more efficient. Few seaweeds have been studied to determine whether or not they adjust the number or size of the PSUs (or both). Among those that have been studied, *Ulva lactuca* changes its PSU number, whereas *Porphyra umbilicalis* changes both (Ramus 1981). In the unicellular marine red alga *Porphyridium purpureum*, the size of the PSU (PS-II associated phycobilisome) changes and the number remains constant, but high-light plants have *higher* P_{max}

values in spite of having the smallest PSUs (Levy & Gantt 1988).

Such *P–I* models, as Ramus (1990) pointed out, assume homogeneous suspensions of unicells in random motion, in which light scattering is quite different from what takes place within seaweed thalli. Photon gradients in seaweeds are steep and are greatly affected by the strong backscattering, except in very thin, homogeneous plants like *Ulva* and *Porphyra*. Thus, Ramus argued that for ecological studies of seaweeds we need to relate photosynthesis to *absorbed* irradiance, I_a, not to incident irradiation, I_0, and he defined the "optical-absorption cross section" as a parameter of light-harvesting ability.

Ramus (1990) developed a model of plant production relating growth yield to irradiance in order to give an autecological perspective. He defined photon growth yield (PGY) as a quantitative parameter to assess the efficiency of light utilization, where

$$PGY = \mu / I_a$$

where μ is specific growth rate, as discussed later (sec. 5.7.1). In comparing *Ulva* and *Codium*, he found that the growth of *Codium* saturated quickly and then became photoinhibited, plotted against both I_0 and I_a (Fig. 4.14a,b), but that at intermediate irradiances *Codium* used incident light more efficiently than did *Ulva* (Fig. 4.14c). [Specific growth rate, μ, is a measure of growth

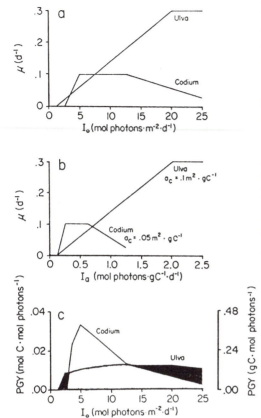

Figure 4.14. Growth-rate–light relations for *Ulva* and *Codium*. (a) Specific growth rate (μ) versus incident light (I_0). (b) Specific growth rate versus light absorbed (I_a), where $I_a = I_0 a_c$ (a_c = carbon-specific absorption cross section). (c) Photon growth yield (PGY) versus I_0. (From Ramus 1990, with permission of Kluwer Academic Publishers.)

relative to nutrient or energy availability. It can be related to *P–I* curves, as shown earlier, or to nutrient-uptake curves (sec. 5.7.1), or it can be dissected further, as was done by Raven (1986). If carbon is used as the nutrient in modeling μ versus irradiance, μ is in moles of carbon assimilated per mole of plant carbon per second.]

4.3.4 Action spectra and "chromatic adaptation"

Each pigment-protein complex has a characteristic absorption spectrum (Fig. 4.1); the sum of all the pigments gives thallus absorption spectra (e.g., Fig. 4.6), which, however, are complicated by morphological effects. The spectra shown are based on measurements using narrow-wave-band light, as in Engelmann's (1883, 1884) and Haxo and Blinks's (1950) classic experiments. However, natural light, even in deep water, is broad-band, and this significantly changes the wavelength dependence of photosynthesis (Fig. 4.15). When

a small amount of green light (546 nm) supplemented the monochromatic light in Fork's (1963) experiments on red algae, significant increases in photosynthesis occurred below 500 nm and above 650 nm, out of proportion to the amount of extra light, and the action spectrum more closely corresponded to the absorption spectrum. The supplementary wavelength chosen by Fork is close to the peak of the deep-water spectrum (Fig. 4.3 and 4.4).

Engelmann's experiments led to a theory relating seaweed zonation to their pigment complements. Although this theory of complementary chromatic adaptation has been adequately disproved over the past 15 years (Ramus 1978, 1982, 1983; Dring 1981, 1990), its "compelling logic" (Ramus 1982) and century-long acceptance have kept it popular as a story in biology textbooks (Saffo 1987). Engelmann's theory proposed that green seaweeds, carrying only chlorophylls, would be restricted to shallow water; the brown algae, with fucoxanthin, could extend farther; and the red algae, with phycobilins filling in a supposed "green window," would have the greatest vertical distribution and would be the only ones in deep (green) water. The theory implied that red algae had evolved phycobilins for survival in deep water. The theory would really apply only where light is limiting, not in the intertidal zone (Dring 1981, 1990).

The theory rested on three main assumptions, all false (Saffo 1987): (1) the vertical distribution of red, green, and brown seaweeds would be as just described, (2) light would be the only factor that would affect their zonation, and (3) the pigment complement of red algae would make them more efficient in the submarine light quality. There were many opponents of the theory from the beginning. Oltmanns (1892) particularly insisted that the zonation pattern had been overstated and that total irradiance had a greater role in establishing the vertical distribution of attached seaweeds than did the spectral quality. Lubimenko and Tichovskaya (1928) concluded that the total quantity of pigment, rather than specific pigments, determined an alga's capability to photosynthesize in the low light at a depth of 50 m.

The generalized zonation pattern that Engelmann sought to explain had been promulgated in an 1844 treatise on marine biology. Engelmann recognized that red algae could be abundant in shady intertidal habitats, where light would be dim, but not green, but he dismissed that as anomalous (Saffo 1987). Deep-water dredging brings up crustose corallines, but dredging has many limitations, and only by using scuba and submersibles can we be sure about the depths at which algae are attached and growing. In many places, red algae are the deepest-growing seaweeds, but we now know that they do not necessarily or exclusively occupy the greatest depths at which seaweeds can be found. Green algae grow at or near the limit of seaweed depths off Malta, Hawaii, Eniwetak Atoll, and in the Bahamas, for exam-

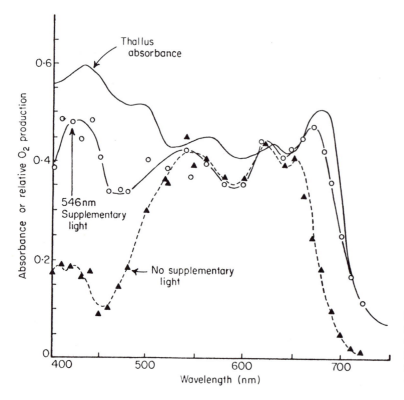

Figure 4.15. Absorptance and action spectra for *Porphyra perforata*, with and without supplementary background light of 546 nm. (From Ramus 1981, after Fork 1963, used with permission of D. C. Fork.)

ple (Gilmartin 1960; Larkum et al. 1967; Doty et al. 1974; Lang 1974; Littler et al. 1985). Some green algae (e.g., *Palmoclathrus* and *Rhiliopsis*) are found *only* in deep water. The numbers of species and biomasses of red relative to green algae do not increase with depth (Schneider 1976; Titlyanov 1976).

Light is by no means the only environmental variable, even in the subtidal zone (Chapter 2). Closely related species, with similar pigments and functional forms, often have different depth limits. The functional form of a seaweed affects not only its light-harvesting ability but also its nutrient and carbon fluxes (Pekol & Ramus 1988). All these abilities and more contribute to growth and competitive abilities. Sand-Jensen's comment (1987, p. 100) about carbon uptake applies equally to light harvesting: "Many characteristics have been suggested as adaptations to enhance the external supply of inorganic carbon. . . . The plant, however, is an integrated functional unit that can exploit an array of environmental variables, so rather than looking at those characteristics purely as adaptations to enhance the carbon gain, they can be viewed as general characteristics of the physiology and ecology of the plant."

As far as light-harvesting ability, the particular pigment complement is unimportant if a seaweed is optically black (Ramus 1978, 1981). Thick thalli, or those within which there is much scattering and reflection of light, are the most thorough absorbers, because the op-

tical path is longer and the chances of a photon striking a plastid are greater. Moreover, although green algae absorb relatively poorly in the green region of the spectrum, they do absorb there, as can be seen in Figure 4.6, and absorptance in the yellow-green region increases disproportionately when the chlorophyll concentration is increased (Larkum & Barrett 1983).

We must reject Engelmann's hypothesis because green algae grow as deep as red algae, light is not the sole factor in zonation, and the pigmentation of red algae does not better adapt them to subtidal light regimes. Rather, the quantity of light is crucial, and seaweeds harvest dim light by increasing all accessory pigments and making adaptations in thallus morphology and orientation.

Nevertheless, changes in pigment complements or ratios under different light qualities, such as phenotypic adjustment (acclimation) to colored light, have been reported, especially in blue-green algae (Bogorad 1975; also see Ramus 1982; Grossman et al. 1989). The effects of light quality must be separated from the effects of low quantity by comparing equal energies of colored lights. In *Calothrix*, which has two types of phycocyanin, two different phycocyanin operons have been identified. One, *cpc1*, encodes the two PC1 subunits under all colors of light, the other, *cpc2*, encodes the two PC2 subunits and their linker proteins and is triggered by red wavelengths (Tandeau de Marsac et al. 1988).

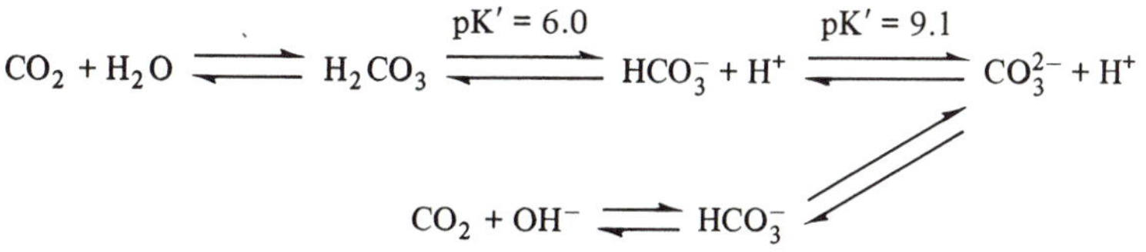

$$CO_2 + H_2O \ \rightleftharpoons \ H_2CO_3 \xrightarrow[\ \ ]{pK' = 6.0} HCO_3^- + H^+ \xrightarrow[\ \ ]{pK' = 9.1} CO_3^{2-} + H^+$$

$$CO_2 + OH^- \ \rightleftharpoons \ HCO_3^-$$

Figure 4.16. The carbonate equilibrium.

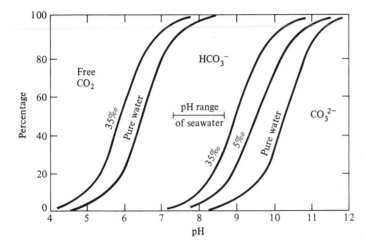

Figure 4.17. Percentage distribution of different forms of inorganic carbon in seawater as a function of pH in three different salinities. (Modified from Kalle 1945, *Der Stoffhaushalt des Meeres*, with permission of Akademische Verlagsgesellschaft, Geest und Portig KG.)

4.4 Carbon fixation: the "dark reactions" of photosynthesis

4.4.1 Inorganic-carbon sources and uptake

Seaweeds use inorganic carbon as virtually their sole carbon source. However, some blue-green algae can be grown heterotrophically (organic carbon only, in darkness) or mixotrophically (organic and inorganic carbon, in light), though the significance of that for the situation in nature remains uncertain (Smith 1982). Some opportunistic seaweeds (*Ulva, Hincksia*) have limited ability to take up and use organic-carbon sources, such as glucose, acetate, and leucine, if they are available (Schmitz & Riffarth 1980; Markager & Sand-Jensen 1990). Parasitic seaweeds are at least partially heterotrophic (Court 1980; Kremer 1983) (sec. 3.3.2).

Seawater has inorganic-carbon properties different from those in air or fresh water. Compared with fresh water, its alkalinity is high, its pH is high and stable, and its salinity is generally high. Plants in air can get inorganic carbon only as CO_2, but it diffuses to them rapidly as they take it up. Plants in water can also get CO_2, and the concentration is similar to that in air, but diffusion is 10^4 times slower. However, CO_2 in water is part of the carbonate buffer system (Fig. 4.16), and inorganic carbon also is potentially available as bicarbonate (HCO_3^-). The relative proportions of the forms of inorganic carbon depend on pH and salinity, as shown in Figure 4.17, and also temperature (Kalle 1972; Kerby & Raven 1985). In seawater of pH 8 and salinity 35‰, about 90% of the inorganic carbon occurs as HCO_3^-. Absolute values are approximately 10 µM CO_2 and

more than 2 mM HCO_3^-; this compares with CO_2 at 300 ppm (ca. 13 µM) in air (Kremer 1981a). Carbon limitation of photosynthesis in the sea had long been considered rare because the concentration of inorganic carbon seems high, and replenishment supplies from carbonates in shells and rocks seem limitless. Recent evidence, however, shows that carbon limitation may be relatively common, especially among subtidal species (Holbrook et al. 1988; Surif & Raven 1989) and in tide pools (Maberly 1990).

There has been much debate over whether or not seaweeds can use bicarbonate. RuBisCO requires CO_2 as a substrate; if seaweeds take up only CO_2, they are dependent on diffusion from relatively low concentrations, as compared with the carbon available in bicarbonate. CO_2 diffuses readily across cell and chloroplast membranes (or diffusion may be facilitated; sec. 5.3.3), but HCO_3^-, like other ions, does not. Marine algae have been assumed to use the abundant bicarbonate source of inorganic carbon, but proving that they do and demonstrating how they do it have been difficult tasks, in terms of both technique and interpretation, because HCO_3^- and CO_2 exist together in equilibrium (Kerby & Raven 1985). Bicarbonate in the unstirred layer next to a thallus will passively supply CO_2 as the equilibrium is pulled by CO_2 uptake.

There are two ways in which seaweeds could actively use bicarbonate. In both ways, the enzyme carbonic anhydrase would greatly speed equilibrium between HCO_3^- and CO_2; this enzyme has a turnover number of 36 million molecules of CO_2 per minute in human blood (Lehninger 1975). Its location in algae can be extracellular, cytoplasmic, and/or possibly in the chloroplast, but in most species this is not known. (1) Seaweeds could actively take HCO_3^- into the cells by a specific porter or by a general anion-exchange protein, there (perhaps at the chloroplast envelope) to convert it to CO_2^-, or (2) they could use extracellular carbonic anhydrase to load the top end of the CO_2 diffusion gradient (Kerby & Raven 1985; Smith & Bidwell 1989). Any mechanism that concentrates CO_2 inside the cell has to work against the leakiness of the plasmalemma, which allows a backflux. If HCO_3^- is taken into the cell, it affects the electrochemical potential across the membrane, so it must be either cotransported with a cation (e.g., H^+), or exchanged for another anion (e.g., the OH^- released when HCO_3^- is dissociated) (Raven & Lucas 1985). Bicarbonate users typically raise the pH around their thalli – some to over 10.5 (Maberly 1990). The expulsion of OH^- relates bicarbonate uptake to $CaCO_3$ precipitation in some calcified algae (Borowitzka 1982; Pentecost 1985) (sec. 5.5.3).

Uptake of bicarbonate has been shown in a variety of intertidal and subtidal seaweeds, including fucoids, kelps, *Palmaria palmata, Chondrus crispus*, an *Ulva* species, and several tropical species (Beer & Eshel 1983; Bidwell & McLachlan 1985; Brechignac et al. 1986; Cook & Colman 1987; Holbrook et al. 1988; Surif & Raven 1989). However, several subtidal red algae cannot use HCO_3^- (Maberly 1990). The mechanisms by which these different seaweeds use bicarbonate are unknown, except in *Chondrus crispus*, in which external carbonic anhydrase dehydrates bicarbonate, and CO_2 diffuses passively across the plasmalemma (Smith & Bidwell 1989). A trend toward more avid use of bicarbonate by intertidal fucoids, as compared with subtidal browns, was tentatively identified by Surif and Raven (1989) and confirmed by Maberly (1990). Photosynthesis of Fucaceae at light saturation is not carbon-limited (Surif & Raven 1989), but the subtidal taxa of browns they studied, and all but one species studied by Holbrook et al. (1988), were carbon-limited at light saturation (i.e., adding more HCO_3^- to the medium increased P_{max}). The *Ulva* species studied by Beer and Eshel (1983) was saturated with both HCO_3^- and CO_2, but in *Enteromorpha* (Beer & Shragge 1987), adding CO_2 to HCO_3^--saturated plants increased their photosynthetic rate, indicating that bicarbonate does not completely offset CO_2 limitation in this intertidal species.

Although seaweeds are essentially marine plants, intertidal species can photosynthesize when emersed, some at rates equal to submerged rates (Kremer & Schmitz 1973; Bidwell & McLachlan 1985; Johnston & Raven 1990; Surif & Raven 1990; Madsen & Maberly 1990). They must use CO_2 because the bicarbonate in the capillary layer of water remaining on the plant surface will be quickly exhausted (Kerby & Raven 1985). High concentrations of CO_2 can substitute for an ab-

sence of HCO_3^- (Bidwell & McLachlan 1985), and intertidal species can acclimate physiologically to high levels of CO_2 (Johnston & Raven 1990). Surif and Raven (1990) concluded from study of brown algae that intertidal species are nearly carbon-saturated in air because of their CO_2-uptake ability, in contrast to subtidal species, which show severe carbon limitation in air. However, Bidwell and McLachlan's (1985) data showed that photosynthesis of *Fucus vesiculosus* was not nearly saturated by air levels of CO_2, a discrepancy that Surif and Raven could not explain (Fig. 4.18). In *Fucus spiralis*, photosynthetic characteristics such as I_k, I_c, and optimum temperature will change when the plant is emersed, and provided that its water content remains high, this high-intertidal plant photosynthesizes in air as well as it does in water, if not even better (Madsen & Maberly 1990).

Shell-boring algae, such as the conchocelis stage of *Porphyra*, the gomontia stage of *Monostroma*, and the numerous species endolithic in corals, live in particularly favorable carbon environments. *P. tenera* sporophytes can satisfy most of their carbon and calcium requirements from the shells, presumably by acidifying $CaCO_3$ to give Ca^{2+} and HCO_3^- (Ogata 1971).

4.4.2 Photosynthetic pathways in seaweeds

The primary photosynthetic pathway is variously called the photosynthetic carbon-reduction cycle (PCRC), the reductive pentose phosphate pathway, C_3 photosynthesis, and the Calvin cycle. The steps as far as fructose-6-phosphate appear to be common to all algae and higher plants, but from there, characteristic low-molecular-weight and polymeric carbohydrates are produced in different algae (Craigie 1974; Raven 1974; Kremer 1981a). Some of the low-molecular-weight compounds play important roles in osmoregulation (sec. 6.3.2).

Few enzymes of the Calvin cycle have been studied in seaweeds because of the difficulty of extracting active enzymes from the mucilages and phenolics that abound in seaweeds (Kerby & Raven 1985). RuBisCO, whose three-dimensional structure was established from an angiosperm by Chapman et al. (1988), has 16 large and small subunits, L_8S_8. It is activated by Mg^{2+}, CO_2, light, and a specific activase (Salvucci 1989). In a species of the diatom *Cylindrotheca*, which has significant C_4-like incorporation of CO_2 via phosphoenolpyruvate (PEP) carboxylase, RuBisCO is stimulated by malate and aspartate (C_4 end products) and inhibited by PEP (C_4 substrate) (Estep et al. 1978). Oxygen is a competitive inhibitor of the carboxylase activity of RuBisCO and acts as a substrate for the oxygenase activity, leading to photorespiration. RuBisCO has been purified from very few marine algae, but early indications are that its variants are structurally and functionally similar in chromophytes and rhodophytes, but different in chlorophytes (Newman et al. 1989). One possible reason is

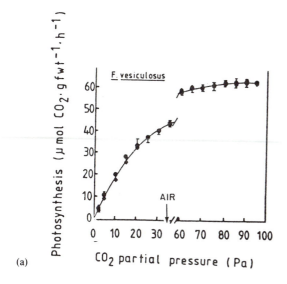

(a)

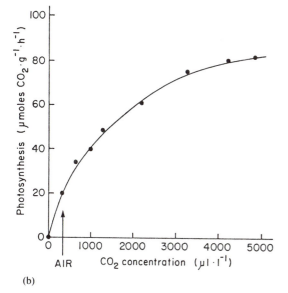

(b)

Figure 4.18. Is photosynthesis of intertidal seaweeds CO_2-limited or nearly saturated at ambient CO_2 concentration? Data from Surif and Raven (1990) (a) and Bidwell and McLachlan (1985) (b), both using *Fucus vesiculosus* measured by infrared gas analysis, gave contradictory results, as yet unexplained. [Other conditions: (a) $t = 10°C$, $I_0 = 500 \mu E$ m^{-2} s^{-1} (data at 2 kPa and 20 kPA O_2 not significantly different); (b) $t = 15°C$, $I_0 = 350 \mu E$ m^{-2} s^{-1}.] (Reprinted with permission of Springer-Verlag, Berlin, and Elsevier Science Publishers BV, respectively.)

that in chlorophytes the small subunit is encoded in the nucleus and is more variable that the chloroplast-encoded large subunit (in chlorophytes), and more variable than both subunits in other algae. RuBisCO is a major component of pyrenoids, but many species lack these structures, and in any case active RuBisCO may be in the chloroplast stroma.

The principal low-molecular-weight photosynthetic product in the majority of the green algae examined so far, as in higher plants, is sucrose, a disaccharide of glucose and fructose. There have been scattered (and sometimes conflicting) reports of green algae producing glucose and fructose, rather than sucrose, or producing considerable quantities of fructose alone (Kremer 1981a). *Caulerpa simpliciuscula* deposits its hexoses largely in β-linked glucans and sugar monophosphates, rather than in sucrose and starch (Howard et al. 1975). The Phaeophyceae are noted for their production of the sugar alcohol mannitol (Fig. 4.19). *Himanthalia* also forms altritol (Wright et al. 1985), and *Pelvetia canaliculata* produces a C_7 alcohol, volemitol (Kremer 1981a). Mannitol is formed by reduction of fructose-6-phosphate (Ikawa et al. 1972), apparently in the chloroplasts (Kremer 1985). Volemitol probably arises in a similar manner from sedoheptulose-7-phosphate, a Calvin-cycle intermediate (Kremer 1977). Cyanophycean low-molecular-weight carbohydrates include glucose, fructose, and trehalose (1,1-α-glucosyl-α-glucose). Amino acids are also significant, especially in nitrogen-fixing species, and may be stored in polymeric form as cyanophycin granules (Smith 1982).

There are three groups of Rhodophyceae, categorized on the basis of their low-molecular-weight photoassimilates (Fig. 4.19) (Kremer 1981a). In the Ceramiales, digeneaside is formed, except in *Bostrychia* and the related genus *Stictosiphonia*, in which the sugar alcohols D-sorbitol and (in most species in these general) D-dulcitol are formed instead (Kremer 1981a; Karsten et al. 1990). In all other Orders the principal product is floridoside. Floridoside consists of glycerol plus galactose, whereas digeneaside consists of glyceric acid plus mannose. The enzymatic steps by which these characteristic carbohydrates are made have yet to be worked out, but one may expect that galactose and mannose come from fructose-6-phosphate by epimerization and that they are primed by being esterified to a nucleotide phosphate (UTP, GTP) before being coupled to glycerol or glyceric acid. The glycerol/glyceric acid presumably is derived from one of the C_3 compounds early in the Calvin cycles (or glycolysis), such as glyceraldehyde-P. The pathway of dulcitol and sorbitol formation may parallel that for mannitol formation.

Photorespiration (or photosynthetic carbon-oxidation cycle, PCOC) is a consequence of the oxygenase activity of RuBisCO, in which RuBP is split, with the introduction of O_2 rather than CO_2. One of the early products is the C_2 acid glycolate. In angiosperms this is oxidized, through a complex pathway involving peroxisomes and mitochondria, in such a way that three-fourths of its carbon is salvaged and used in the Calvin cycle (Kerby & Raven 1985). Glycolate oxidase is

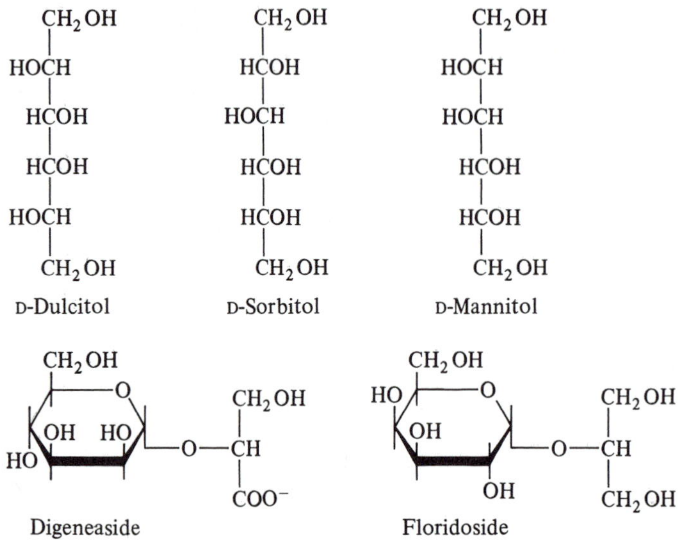

CH$_2$OH
|
HOCH
|
HCOH
|
HCOH
|
HOCH
|
CH$_2$OH

D-Dulcitol

CH$_2$OH
|
HCOH
|
HOCH
|
HCOH
|
HCOH
|
CH$_2$OH

D-Sorbitol

CH$_2$OH
|
HOCH
|
HOCH
|
HCOH
|
HCOH
|
CH$_2$OH

D-Mannitol

Digeneaside

Floridoside

Figure 4.19. Some low-molecular-weight carbohydrates from red and brown seaweeds.

located in the peroxisomes; as glycolate is oxidized, oxygen is reduced to H_2O_2 (hydrogen peroxide), which in turn is immediately reduced to water by catalase in the peroxisomes. Alternatively, glycolate can be oxidized via glycolate dehydrogenase in mitochondria. A survey by Suzuki et al. (1991) found different glycolate dehydrogenases in diatoms and Chlorophyceae, and oxidase+catalase in brown algae, but was inconclusive on Ulvophyceae and Rhodophyceae.

The extent of photorespiration in seaweeds is uncertain. Evidence from gas-exchange experiments is difficult to interpret (Bidwell & McLachlan 1985; Kerby & Raven 1985) and may not represent the extent of the biochemical cycle if refixation of CO_2 is possible (sec. 4.4.4). Glycolate excretion occurs in some marine algae, but, again, the natural extent has been controversial (Tolbert & Osmond 1976; Burris 1977; Harris 1980); in any case, it does not measure the glycolate that goes through the oxidation pathway. Kerby and Raven (1985), after thoroughly reviewing the gas-exchange data, concluded that RuBisCO oxygenase activity and photorespiration are to varying degrees suppressed under normal conditions by carbon-concentrating mechanisms (sec. 4.4.1). Photorespiration does not seem to be significant in the metabolism of thin algal turfs on coral reefs (Hackney & Sze 1988).

4.4.3 Light-independent carbon fixation

Seaweeds, like all plants, contain enzymes that can interconvert C_3 and C_4 compounds by carboxylation and decarboxylation. Several produce oxaloacetic acid (OAA), the Krebs-cycle intermediate that accepts carbon from glycolysis. Because Krebs-cycle intermediates are used for biosynthetic pathways, anaplerotic ("filling-up") reactions are needed to keep the cycle turning. Several carboxylating enzymes add CO_2 (HCO_3^- may be the substrate of some) to the β-carbon of

PEP or pyruvate and hence are called β-carboxylases, and two particularly important ones in seaweeds are PEP carboxykinase (PEPCK; Fig. 4.20) and PEP carboxylase (PEPC):

$$PEP + CO_2 + H_2O \xrightarrow[\text{PEPC}]{Mg^{2+}} OAA + P_i$$

$$PEP + CO_2 + \begin{Bmatrix} GDP \\ IDP \\ ADP \end{Bmatrix} \xrightleftharpoons[\text{PEPCK}]{Mn^{2+}} OAA + \begin{Bmatrix} GTP \\ ITP \\ ATP \end{Bmatrix}$$

OAA is rapidly converted to malate or (by transamination) to amino acids such as aspartate. While such reactions are general as anaplerotic components of respiration and biosynthesis, their ability to fix CO_2 has been exploited by some plants. C_4 and CAM vascular plants use β-carboxylases to fix CO_2 onto pyruvate, but, as noted earlier, this does not yield net carbon fixation in the long term.

If seaweeds are supplied with labeled carbon in the dark, they fix it into various characteristic products different from those derived from the Calvin cycle. Red and green macroalgae in darkness form only small amounts of amino acids. Brown algae, with considerably higher rates of light-independent fixation, form malate, aspartate, citrate, and alanine, via PEPCK (Kerby & Evans 1983). Although present throughout the brown algae, light-independent carbon fixation (which takes place in the light as well as in darkness) occurs to a significant extent only in young tissues of kelps and fucoids, where it can account for over 20% of the total carbon fixed (Kremer 1981b; Johnston & Raven 1986). The significance of β-carboxylation in kelps and fucoids is seen in two situations. In kelps, photosynthesis is highest in old tissue, and β-carboxylation is most active in meristems. Mannitol is remobilized from mature or old tissue to provide energy and carbon skeletons

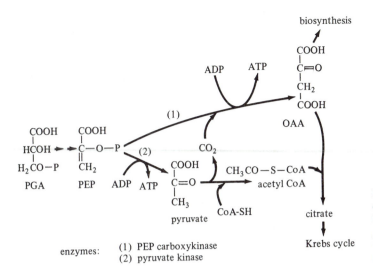

enzymes: (1) PEP carboxykinase
           (2) pyruvate kinase

Figure 4.20. Utilization of PEP via PEP carboxykinase (1) in light-independent carbon fixation, or via pyruvate kinase (2) in glycolysis. (Modified from Kremer 1981b, with permission of Springer-Verlag, Berlin.)

for growth in the meristems. PEPCK permits salvage of the CO_2 lost when pyruvate is converted to acetyl coenzyme A (acetyl–CoA) in glycolysis (Fig. 4.20). There is no difference in energy output, because ATP is generated from PEP whether PEP is used to fix CO_2 or is converted to pyruvate in glycolysis (Kremer 1981b). One mole of mannitol yields 2 moles of PEP and thus can refix 2 moles of CO_2, and the CO_2 fixed into C_4 compounds does not have to be released again to regenerate C_3 substrate, so there can be net carbon fixation – in contrast to CAM and C_4 metabolism. This does not mean that dark fixation is greater than CO_2 loss by respiration, and in fact Johnston and Raven (1986) showed that in *Ascophyllum nodosum* it is not, but it can conserve some carbon lost from respiration. In fucoids, photosynthesis and β-carboxylation are both maximum in growing tips, and β-carboxylation could serve as a carbon-concentrating mechanism, as it does in C_4 and CAM vascular plants. The evidence for this possibility is addressed in the following section.

4.4.4 C_3 versus C_4 characteristics of seaweeds

Against the background of inorganic-carbon uptake and the various metabolic pathways, we can now examine the question of the type of photosynthesis in seaweeds, as compared with the categories used for angiosperms. At the outset we should note (Bowes 1985, p. 202) that "because they inhabit such a different environment, especially regarding access to HCO_3^-, and are so biosystematically diverse, it may be specious to expect marine macrophytes to conform to any terrestrial categories."

A graph of photosynthesis versus inorganic-carbon concentration does not pass through the origin: A finite amount of CO_2/HCO_3^- is needed to give zero net carbon fixation. This is because the oxygenase activity of RuBisCO is a function of the concentration ratio $CO_2 : O_2$ (sec. 4.4.2). C_4 pathways in angiosperms

serve as CO_2-concentrating mechanisms, increasing the CO_2 concentration by decarboxylating a C_4 acid. C_4 plants show the following characteristics: (1) formation of a C_4 acid as a primary product of photosynthesis, using PEPC, which is insensitive to O_2; (2) anatomical separation of β-carboxylation and the Calvin cycle (Kranz anatomy); (3) very low CO_2 compensation (Γ), because of recapture of any CO_2 lost in photorespiration, as well as reduced photorespiration; (4) no O_2 inhibition of photosynthesis (no Warburg effect); (5) no consistent postillumination burst of CO_2 or gulp of O_2 (Johnston & Raven 1987; Holbrook et al. 1988; Reiskind et al. 1988). The last three are external symptoms of the internal mechanism. However, any mechanism that concentrates CO_2 at RuBisCO will cause at least effects (3) and (4).

Several algae have recently been shown to have the C_4-like symptoms of low Γ and no Warburg effect (Bidwell & McLachlan 1985; Bowes 1985; Johnston & Raven 1987; Holbrook et al. 1988; Reiskind et al. 1988), whereas others have C_3 symptoms (Kremer 1981a; Bowes 1985; Kerby & Raven 1985). High PEPCK activity, apparently not merely anaplerotic, and little O_2 inhibition of photosynthesis occur in some brown algae (as discussed earlier) and at least one green (*Udotea flabellum*), whereas a lack of PEPCK and the absence of O_2 sensitivity were found concomitantly in other greens (e.g., *Codium*) and at least one brown (*Sargassum* sp.) (Bowes 1985; Holbrook et al. 1988). In a more detailed study of *Udotea* and *Codium*, Reiskind et al. (1988, 1989) showed that in *Udotea*, malate and aspartate were early photosynthetic products from PEPCK (in light) and that these turned over quickly, with the label apparently passing into C_3 compounds. Those authors inferred that a PEPCK C_4-like pathway is the mechanism reducing photorespiration in this plant. *Codium* had typical C_3 products and photorespiration, but some HCO_3^- use partly mitigated photorespiration. A small change in acidity in *Ascophyllum nodosum* between day and night due to light-independent carboxylation by PEPCK suggests a CAM-like pattern of C_4 acid use, but the activity is low relative to RuBisCO activity (Kerby & Evans 1983; Kerby & Raven 1985; Johnston & Raven 1986). Significant fluctuations in malate or other C_4 acidity have not been found in other seaweeds surveyed (Reiskind et al. 1989). Another potential means for reducing photorespiration is to reduce the O_2 concentration – and thus oxygenase activity. Because O_2 is produced by PS-II, photosynthesis by PS-I could generate ATP without releasing oxygen. This mechanism is well known in blue-green-algal heterocysts, where oxygen-sensitive nitrogen-fixation enzymes work. It may also operate in red-algal pyrenoids. Red algae with pyrenoids have RuBisCO concentrated there. Thylakoids cross the pyrenoid, but phycobilisomes and PS-II are absent from these parts of the thylakoids, perhaps giving a locally low oxygen concentration (McKay & Gibbs 1990).

We must conclude that while the question of a C_4-like pathway in some algae remains open, most seaweeds, if not all, have C_3 pathways, and many species more or less suppress photorespiration by CO_2 concentration via HCO_3^- (Reiskind et al. 1989; Raven et al. 1990).

4.5 Seaweed polysaccharides

The low-molecular-weight products of photosynthesis serve many ends, including immediate use in respiration and use as carbon skeletons for functional or structural components of the cells. Some go into storage compounds. The algae are noted for their diversity of storage and structural polysaccharides, which, like pigments, are characteristic of higher taxonomic groups. In this final section on cell metabolism, we shall examine these polysaccharides.

4.5.1 Storage polymers

Carbon may be stored in monomeric compounds such as mannitol, but much is stored in polymers. One advantage of polymers for storage is that they have smaller effects on the osmotic potential than would the same amount of carbon in monomeric form. Several characteristic storage polysaccharides are found in red, brown, and green seaweeds and cyanophytes, but most are branched or unbranched chains of glucose units (glucans) (Craigie 1974; McCandless 1981; Smith 1982). Most green algae, like higher plants, store starch, a mixture of branched molecules (amylopectin) and unbranched molecules (amylose) (Fig. 4.21). Amylose consists of α-D-glucose units linked $1 \rightarrow 4$; amylopectin also has $\alpha(1 \rightarrow 6)$ branch points. Amylose is insoluble in water, forming micelles in which the molecules are helically coiled, whereas amylopectin is soluble. Dasycladales, such as *Acetabularia*, do not always store

starch, but often store inulin, a fructose polymer (Percival 1979). Blue-green algae store a highly branched compound more like glycogen than amylopectin. Red algae store primarily floridean starch, a branched glucan similar to amylopectin except for having a few $\alpha(1 \rightarrow 3)$ branch points. McCracken and Cain (1981) showed that primitive red algae, including five marine species, also have amylose in their starch. Brown algae store laminaran, which, like starch, comprises a branched, soluble molecule and an unbranched, insoluble molecule (these compounds are called simply soluble and insoluble laminaran). The glucose in laminaran is in the β form, however, and the links are $\beta(1 \rightarrow 3)$ and $\beta(1 \rightarrow 6)$ (Fig. 4.21c). Furthermore, some laminaran molecules, called M-chains, have a mannitol molecule attached to the reducing (C-1) end.

Storage compounds show quantitative changes correlated with season (really with growth), plant part, and reproductive condition. These changes have been particularly well documented in commercially valuable kelps, fucoids, and Gigartinales. Studies such as those by Black (1949, 1950) and Jensen and Haug (1956) on *Laminaria* species and fucoids were primarily concerned with fluctuations in the valuable wall-matrix polysaccharides and iodine, but they also documented changes in mannitol and laminaran contents. More recently, the relationships among growth, storage, and sometimes nitrogen availability have been worked out for *Hypnea musciformis* (Durako & Dawes 1980), *Eucheuma* species (Dawes et al. 1974), and *Laminaria longicruris* (sec. 5.8.3) (Chapman & Craigie 1977; Gagné et al. 1982). Buildup of carbohydrate in matrix polysaccharides during periods of low growth or nitrogen starvation has also been shown, as, for instance, in *Chondrus crispus* (Neish et al. 1977) and *Eucheuma* (Dawes et al. 1977). Although most of the few known cases of such storage seem to have resulted from conditions that permitted photosynthesis but not growth, there is no evidence that seaweeds can retrieve this extracellular carbon. However, the colonial prymnesiophyte *Phaeocystis pouchetii* has been shown to deposit and reassimilate matrix polysaccharides on a light/dark cycle (Veldhuis & Admiraal 1985).

4.5.2 Wall-matrix polysaccharides

Marine algae are notable for the large amounts of matrix polysaccharides they produce, and marine plants (including seagrasses) are notable for forming sulfated polysaccharides, in contrast to freshwater plants (including algae) and land plants. These features suggest that such compounds are produced in response to the marine environment, but their roles are not as yet clear (Kloareg & Quatrano 1988). Each Class produces a range of characteristic compounds. Several of these compounds are of commercial value as phycocolloids (Chapter 9).

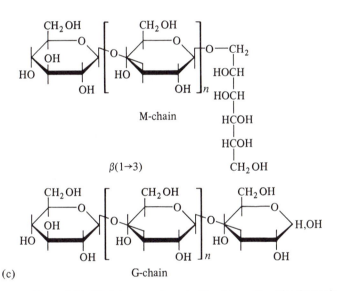

(a)

α(1→4) chain

(b)

(c)

M-chain

β(1→3)

G-chain

Figure 4.21. Algal storage polysaccharides: (a) amylopectin; (b) amylose; (c) two types of laminaran chains, M with mannitol attached to the reducing end, G with glucose at the reducing end. [Parts a and b from Lehninger, *Biochemistry*, 2nd ed., © 1975, Worth Publishers, New York; c reprinted from Percival & McDowell 1967, *Chemistry and Enzymology of Algal Polysaccharides*, © 1967, Academic Press Inc. (London), Ltd.]

The matrix polymers present a much more complex picture than either fibrillar polysaccharides (sec. 1.3.1) or storage compounds. Many are quite variable, so that terms such as "carrageenan" cover a range of similar but not identical molecules. In general, these polymers, like amylose, are thought to form helices that are aggregated in various ways in the gel state (Fig. 4.22) (Rees 1975; Kloareg & Quatrano 1988). Knowledge of polysaccharides, particularly their native conformations is still fragmentary. Polysaccharides are still described in terms of their solubility characteristics and component units. Kloareg and Quatrano (1988) outlined the principles that determine the shapes of polysaccharides and summarized what is known about the functional significance of the shapes.

Chlorophyceae produce highly complex sulfated heteropolysaccharides, each molecule of which is made up of several different residues. The major sugars are glucuronic acid, xylose, rhamnose, arabinose, and galactose, made up in several combinations. For instance, xylogalactoarabinans in Cladophorales and Codiales have a backbone of (1 → 4)-linked L-arabinose blocks

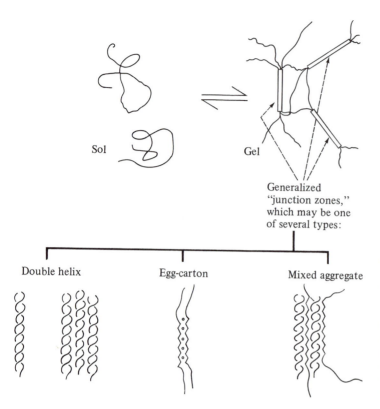

Sol

Gel

Generalized
"junction zones,"
which may be one
of several types:

Double helix   Egg-carton   Mixed aggregate

Figure 4.22. Some mechanisms by which a carbohydrate chain may bind to another of similar or different structure with the formation of a gel network. Lower series of drawings, from left to right, correspond to isolated double helix (e.g., ι-carrageenan), aggregated double helices (e.g., agarose), "egg carton" (e.g., calcium polyguluronate), mixed aggregate (e.g., agarose-galactomannan). (From Rees 1975, with permission of the author.)

separated by D-galactose, with D-xylose and D-galactose on the ends of chains (Kloareg & Quatrano 1988). Considerably less is known about these green-seaweed polysaccharides than about those from reds and browns, perhaps because of their complexity and because none has yet found commercial application.

The matrix polysaccharides of the brown algae were once thought to be relatively simple: alginic acid consisting of mannuronic and guluronic acids, and fucoidan consisting of fucose (Percival & McDowell 1967). However, "fucoidan" is now known to cover a wide range of compounds, the simplest being nearly pure fucan (e.g., in *Fucus distichus*). The most complex (e.g., in *Ascophyllum nodosum*) are heteropolymers containing fucose, xylose, galactose, and glucuronic acid, in which the uronic acid may form a backbone and the neutral sugars form extensive branches (Larsen et al. 1970; McCandless & Craigie 1979). Even fucan is complex, having $\alpha(1 \rightarrow 2)$ and $\alpha(1 \rightarrow 4)$ links between fucose units, as well as $\alpha(1 \rightarrow 3)$ branches and varying degrees of sulfation on the remaining hydroxyl groups (Fig. 4.23a).

Alginic acid consists of two uronic acids (sugars, with a carboxyl group on C-6: $\beta(1 \rightarrow 4)$-D-mannuronic acid (M) and its C-5 epimer α-$(1 \rightarrow 4)$-L-guluronic acid (G) (Fig. 4.23b). In most chains the residues occur in blocks as $(-M-)_n$, $(-G-)_n$, and $(-MG-)_n$ (Kloareg & Quatrano 1988; Gacesa 1988). The strengths of alginates depend on Ca^{2+} binding, with guluronic acid having a much greater affinity for Ca^{2+} than does mannuronic acid. This effect depends on the more zigzag conformation of polyguluronic acid, which allows Ca^{2+} to fit into the spaces like eggs in an egg carton (Fig. 4.25b). Stiff, guluronate-rich alginates are typical of holdfasts; elastic, mannuronate-rich alginates predominate in the blades of kelps (Cheshire & Hallam 1985).

Most red-algal matrix polsaccharides are galactans in which $\alpha(1 \rightarrow 3)$ and $\beta(1 \rightarrow 4)$ links alternate (Craigie 1990a). Variety in the polymers comes from sulfation, pyruvation, and methylation of some of the hydroxyl groups and from the formation of an anhydride bridge between C-3 and C-6 (Fig. 4.24). The important commercial groups of red-algal polysaccharides are the agars and the carrageenans.

Agars consist of alternating β-D-galactose and α-L-galactose with relatively little sulfation. The best commercial agar, neutral agarose, is virtually free of sulfate. Some more highly sulfated polymers with the agar structure do not gel and indeed are not referred to as agars (e.g., funoran from *Gloiopeltis* spp.).

In carrageenans, β-D-galactose alternates with α-D-galactose not α-L-galactose), and there is much more sulfation. The sulfate groups project from the outside of the polymer helix; this polyelectrolyte surface makes the molecule more soluble (Rees 1975). Conversion of a 6-sulfate group to an anhydride bridge, as in ι- and κ-carrageenan, yields a stronger gel. Classically, two carrageenan fractions, λ and κ, were distinguished on

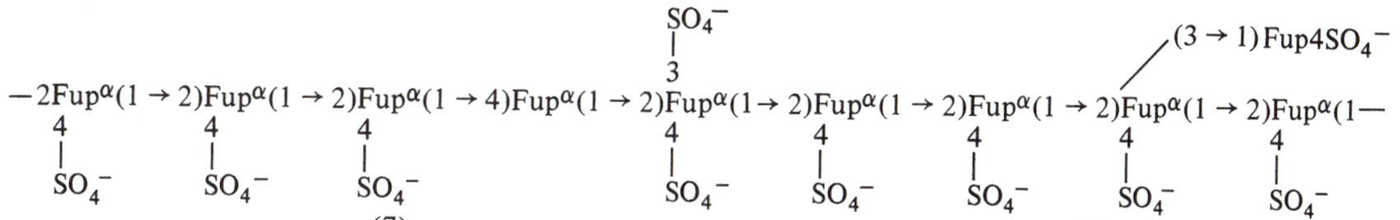

$$\overset{\displaystyle SO_4^-}{\underset{3}{|}}$$

$$-2Fup^\alpha(1 \to 2)Fup^\alpha(1 \to 2)Fup^\alpha(1 \to 4)Fup^\alpha(1 \to 2)Fup^\alpha(1\to 2)Fup^\alpha(1 \to 2)Fup^\alpha(1 \to 2)Fup^\alpha(1 \to 2)Fup^\alpha(1-$$

with SO_4^- at position 4 on residues, and $(3 \to 1)Fup4SO_4^-$ branch

(7) (8)

Fup = L-fucose

$\alpha(1 \to 2)$ $\alpha(1 \to 4)$ $\alpha(1 \to 3)$

(a)

—103 nm—   —87 nm—

(-M-)$_n$ (-G-)$_n$

(b)

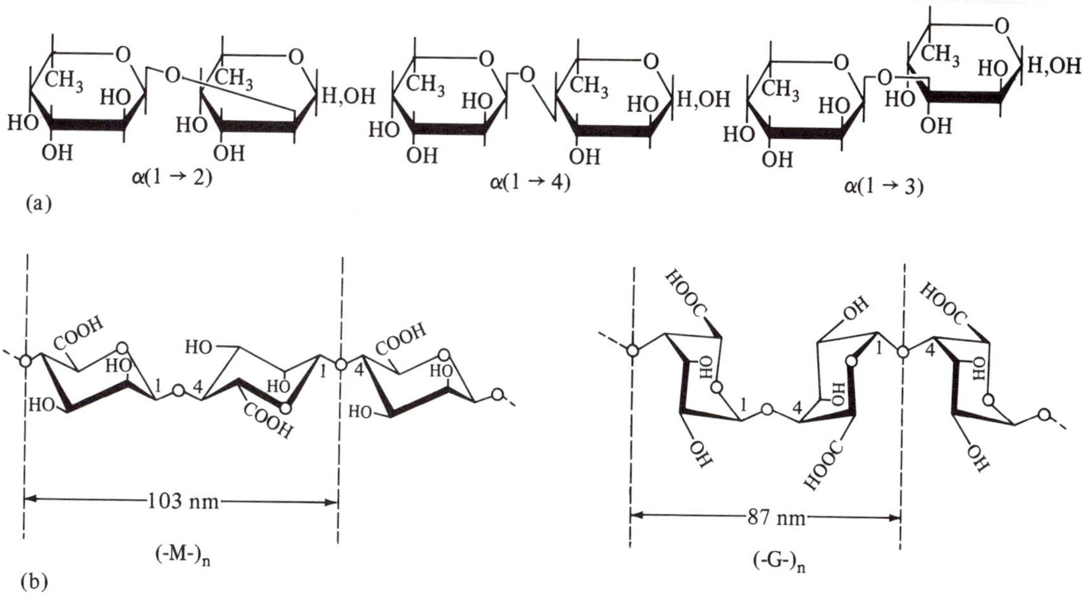

Figure 4.23. Brown-algal cell-wall matrix polysaccharides: (a) fucan, showing overall structure of part of a chain, and details of three kinds of linkage; (b) portions of alginic acid (left, polymannuronic acid; right, polyguluronic acid). [Part a reprinted from Percival & McDowell 1967, *Chemistry and Enzymology of Algal Polysaccharides*, © 1967, Academic Press Inc. (London), Ltd.; b from Mackie & Preston 1974, with permission of Blackwell Scientific Publications.]

the basis of their solubility in KCl. Now several varieties are recognized, grouped into κ and λ families based on the presence or absence, respectively, of sulfate on C-4 of the β-galactose residues (Fig. 4.24) (McCandless & Craigie 1979; Kloareg & Quatrano 1988). The more sulfated molecules form stronger gels. The conversion of galactose-6-sulfate to the 3,6-anhydride results in a stiffer gel because it takes a "kink" out of the chain, allowing more extensive double-helix formation and thus a more compact gel (Percival 1979). κ-carrageenan is predominantly found in the gametophytes, and λ-carrageenan in the tetrasporophytes, of *Chondrus crispus* and some other Gigartinaceae and Phyllophoraceae. The significance of this biochemical alternation for the physiology of the plants is not yet clear, especially given the spectrum of carrageenans that are known to exist. Other Gigartinales, such as *Eucheuma*, produce only one kind of carrageenan in both phases (Dawes 1979).

Gel-forming polysaccharides no doubt add rigidity to the cell wall, while perhaps providing a certain amount of elasticity necessary in the aquatic environ-ment. The strength of some gels depends on binding Ca^{2+} or other divalent cations that can cross-link polymer chains. However, the properties of the polymers that tend to increase divalent-ion binding are diverse. Gel strength and Ca^{2+} binding of fucoidan depend on sulfation (e.g., in sulfate-free medium, *Fucus* embryos are unable to adhere to the substrate). However, less highly sulfated agars gel better, and sulfate esters interfere with calcium binding in polysaccharides of *Ulva lactuca* (Haug 1976). In *Ulva*, borate becomes complexed with rhamnose residues in the polymers, and Ca^{2+} stabilizes that complex (Fig. 4.25a). Sulfate prevents complexing with borate, and hence results in a weaker gel. Although a change in sulfation, or isomerization between uronic acids, could alter wall strength in seaweeds (as might, for example, be useful for spore release), thus far there is little evidence that seaweeds can make such changes once the polymers are outside the cells (sec. 4.5.3).

The roles of the nongelling matrix polysaccharides are less clear. There have been suggestions that

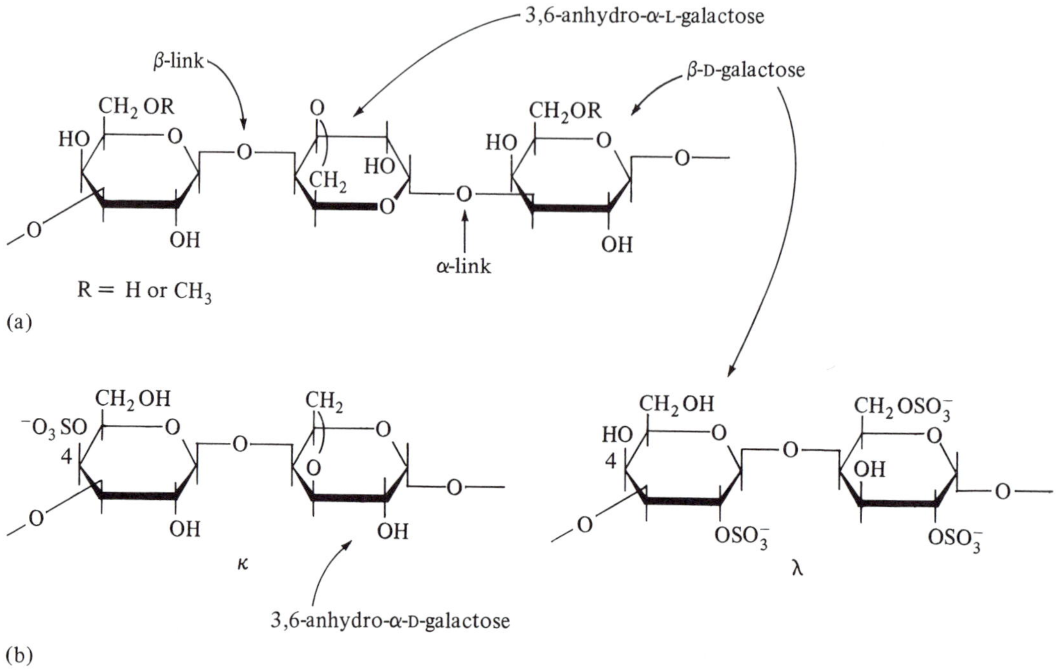

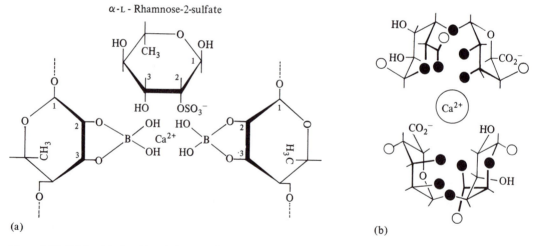

Figure 4.24. Red-algal cell-wall matrix polysaccharides: (a) agar; (b) κ- and λ-carrageenans showing differences in sulfation, and the C-4 of β-D-galactose used in classifying carrageenans.

Figure 4.25. Calcium binding (a) to a rhamnose-borate complex from *Ulva* (also showing free rhamnose-2-sulfate) and (b) between two chains of polyguluronic acid (the egg-carton model); oxygen atoms shaded black are involved in binding to Ca^{2+}. (Part a from Percival 1979, with permission of the British Phycological Society; b from Rees 1975, with permission of the author.)

they provide protection for the cells against desiccation by binding water and that they may serve as a kind of ion-exchange material. Evidence for these possibilities is scant and controversial (Kloareg & Quatrano 1988). Many inorganic ions are required to balance the fixed charges on intercellular-matrix polysaccharides; logically, these must lower the water potential (sec. 6.3.1) and help keep thalli hydrated (Smidsrød & Grasdalen 1984; Shephard 1987). In view of the very limited knowledge of their structures and locations, it is not surprising that the roles of the matrix polysaccharides are not well understood.

4.5.3 Polysaccharide synthesis

The various cell-wall and storage polysaccharides are synthesized at several sites in the cell. Cell-wall materials probably are made in Golgi vesicles, which pass to the outside of the cell by reverse pinocytosis. The enzyme UDP-galactosyltransferase, which can transfer galactose to fucoidan, has been located in Golgi bodies of *Fucus serratus* (Coughlan & Evans 1978). Starch granules in the cytoplasm of the red alga *Serraticardia maxima* contain the starch-synthesizing enzyme ADP-glucose : α-1,4-glucan α-4-glucosyltransferase (Nagashima et al. 1971). In other algae, starch is stored, and presumably made, in the chloroplasts.

Polysaccharide synthesis seems generally to involve the addition of a nucleotide diphosphate-linked monomer to a primer or an existing chain (Turvey 1978). The route by which mannose is incorporated into alginic acid was postulated by Lin and Hassid (1966) to be

D-mannose $\rightarrow$ D-mannose-6-P $\rightarrow$ D-mannose-1-P $\rightarrow$
GDP-D-mannose $\rightarrow$ GDP-D-mannuronic acid $\rightarrow$
polymannuronic acid $\rightarrow$ alginate

The basic pathway has been confirmed, but details remain uncertain (Gacesa 1988). Each different bond type in a polymer is made by a different enzyme. Thus amylopectin synthesis requires one enzyme to make the $(1 \rightarrow 4)$ links and another to make the $(1 \rightarrow 6)$ branch points (Haug & Larsen 1974). Complexity seems to be added after polymer synthesis. Alginic acid is made initially as polymannuronic acid; then residues are converted to guluronic acid by epimerization at C-5. Addition of sulfate and methyl groups and the formation of anhydride bridges in various polysaccharides also take place after polymerization (Percival 1979). Wong and Craigie (1978) partially characterized an enzyme from *Chondrus crispus* (also known from other red algae) that forms the 3,6-anhydride bridge on galactose residues sulfated at C-6, with the release of the sulfate. The structure of the resulting κ-carrageenan is shown in Figure 4.24. There is evidence from another red alga, *Catenella caespitosa*, for extracellular turnover of sulfate ester on λ-carrageenan after deposition. The sulfate appears to be carried by a methylated cytidine monophosphate (de Lestang-Brémond & Quillet 1981; Quillet & de Lestang-Brémond 1981).

4.6 **Carbon translocation**

In the simplest seaweeds, each cell is virtually independent of the others for its nutrition. However, many seaweeds contain nonpigmented cells in the medulla that evidently are supplied with photoassimilates by the pigmented cortical or epidermal cells. Parasitic algae receive organic carbon via short-distance translocation from their hosts. In those algae with an apico-basal gradient there is clearly movement of growth-regulating substances within the thallus. Such short-distance transport might take place through plasmodesmata, which in some green and perhaps most brown algae traverse the cross-walls and join the cytoplasm into a continuous symplast. In the red algae, pit plugs potentially provide a route between cells, although they usually are bounded on both sides by a membrane (Brawley & Wetherbee 1981; Pueschel 1990).

Long-distance translocation evolved in higher plants as an adaptation to habitats where light and CO_2 were available in air, whereas water and minerals were available chiefly in the soil. In seaweeds, the whole outer surface is photosynthetic and is involved in nutrient absorption, and so there is no need of translocation for exchange of materials between different regions of the plants. However, translocation can also serve to redistribute photoassimilates from mature (i.e., nongrowing), strongly photosynthetic areas to rapidly growing regions. This role is useful only where there is a localized growing region and a relatively large or distant mature region. Kelp sporophytes and fucoids have such structure, and translocation is well established especially in the Laminariales (Schmitz 1981; Buggeln 1983; Moss 1983; Diouris & Floc'h 1984). Transport of mineral ions is well established for Fucales and Laminariales (sec. 5.6).

Translocation in virtually all Laminariales takes place through the sieve elements, as has been shown by autoradiography (Steinbiss & Schmitz 1973). The sieve elements lie in a ring between the cortex and medulla throughout the stipe and blades. Longitudinal files of sieve elements branch and interconnect and are connected to cortical cells (Buggeln et al. 1985). The structure of sieve elements (in particular, the characteristically perforated sieve plates on their end walls) shows a trend from the smaller, simpler kelps (e.g., *Laminaria*) to the largest, most complex (*Macrocystis*). The pores become less numerous but larger in larger kelps, and thus more effective in transport. The sieve elements of *Laminaria* are filled with cytoplasm, organelles, and numerous small vacuoles, whereas those in *Macrocystis* more closely resemble vascular-plant sieve elements in having a peripheral layer of enucleate cytoplasm and a very large central vacuole or lumen (Fig. 4.26). The pores in the sieve plates are lined with callose (a β-1,3-linked glucan), which, as in vascular plants, can be deposited to block the pores. This mechanism prevents great loss of sieve-tube sap in case of injury and may also regulate routes of translocation. The only kelp known to differ from the description just given is *Saccorhiza dermatodea* (Emerson et al. 1982). Here translocation takes place in the medulla, through highly elongated cells (solenocysts) that are cross-connected by smaller cells (allelocysts). The ultrastructure of solenocysts is similar to that of the sieve elements of other genera.

Kelps translocate the same organic materials that they make in photosynthesis; apparently there is no selectivity in sieve-tube loading. Mannitol and amino acids (chiefly alanine, glutamic acid, and aspartic acid) each account for about half of the exported carbon. In addition, there are numerous other compounds (Manley 1983). These materials move at velocities ranging from less than 0.10 m h^{-1} in *Laminaria* to about 0.70 m h^{-1} in *Macrocystis*, the speed generally correlating with the size of the sieve-plate pores. The *rate* of translocation

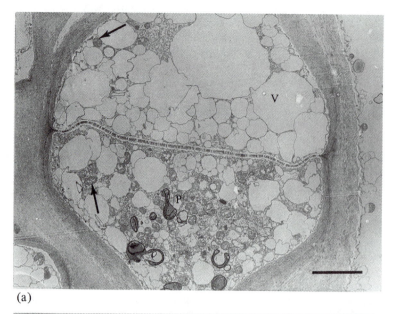

(a)

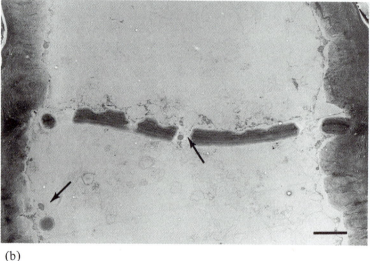

(b)

Figure 4.26. Longitudinal sections through sieve tubes with sieve plates of (a) *Laminaria groenlandica* and (b) *Macrocystis integrifolia*. Mitochondria can be seen in both (arrows); plastids (P) and vacuoles (V) are indicated in *Laminaria*. Note the many narrow pores through the sieve plate of *Laminaria* compared with the few, large pores in *Macrocystis*. Scale = 5 μm. (Part a from Schmitz & Srivastava 1974, with permission of Wissenschaftliche Verlagsgesellschaft mbH., Stuttgart; b from Schmitz 1981, with permission of Blackwell Scientific Publications.)

depends not only on the velocity but also on the amount of material that is moving and hence on the concentrations of solutes and the cross-sectional area of transporting sieve elements. Rates range from* 1 g_{dw} h^{-1} cm^{-2} in *Alaria* to 5–10 g_{dw} h^{-1} cm^{-2} in *Macrocystis*. Rates in most vascular plants are 0.2–6.0 g_{dw} h^{-1} cm^{-2}, although some translocate at much higher rates (Schmitz 1981).

* g_{dw}, gram dry weight.

Photoassimilates follow a source-to-sink pattern of translocation. Sources include mature tissue; sinks include intercalary meristems and, to a lesser extent, sporophylls and haptera. In species with only one blade, the pattern is simple: Mature distal tissue exports, and meristematic tissue at the blade–stipe junction imports. The pattern becomes complicated in *Macrocystis*, which has numerous blades in various stages of growth and maturity, as well as young fronds developing from old fronds. In this genus, the changing import–export pattern of a blade as it matures is very similar to the pattern found in dicotyledonous angiosperms. Young blades only import, but as they near full size, export begins. Initially export is upward, to the meristem from which the blade developed; later it is also downward, to fronds developing from the base of the parent frond. In *M. integrifolia*, which lives in a seasonally variable habitat, downward export in autumn also appears to carry photosynthates to the base of the plant for storage (Lobban 1978b,c). When growth stops in kelps, translocation also stops (Lüning et al. 1973; Lobban 1978c).

The mechanism of translocation in kelps has not yet been determined; this is hardly surprising in view of the controversy over mechanisms in vascular plants, which have been far more intensively studied. The structure of *Macrocystis* sieve elements (Fig. 4.26b) would be consistent with a mass-flow mechanism, whereby sieve-tube loading in the source would cause an osmotic influx of water, which in turn would push the solutes to the sink, where assimilates would be unloaded. Even the vesiculate structure of *Laminaria* sieve tubes (Fig. 4.26a) does not preclude mass flow (Schmitz 1981; Buggeln 1983).

The ecological significance of translocation is that it allows more rapid growth of localized meristems. This is especially important in plants like *Macrocystis*, which may be attached in deep water, where the new frond initials are shaded by both the water column and the surface canopy of blades. In populations of *M. pyrifera* growing in stratified water where the surface layer is poor in nutrients, translocation may also serve to carry nitrogen (as amino acids) to the surface canopy (Wheeler & North 1981). In perennial species of *Laminaria*, translocation serves to move stored carbon from the mature blade to the meristem when new growth begins; this start of new growth is triggered by photoperiod in *L. hyperborea* (Lüning 1986) and by nitrogen availability in several other species (e.g., *L. longicruris*; Gagné et al. 1982).

4.7 Photosynthesis rates and primary productivity

Photosynthesis is a major, easily measured metabolic process and is routinely used as a gauge of environmental effects on seaweeds. It is also the basis of the primary productivity, and its accurate measurement is important for ecological studies. Primary productivity is the rate of net incorporation of carbon into organic compounds. It includes carbon retained in the plants and organic carbon released as exudates or pieces of tissue (or of entire plants, if population productivity is being considered). It does not include carbon returned to the environment as CO_2. The development of ecologically sound estimates of primary production is enormously difficult, whether based on short-term photosynthesis or on long-term yield. On the one hand, intrinsic variability in photosynthesis and respiration, the complex

changes in I_0 and total daily irradiance, and the effects of numerous other environmental factors on metabolism are extremely difficult to measure or model (sec. 4.7.2). On the other hand, biomass data must account for grazing and other tissue losses (Murthy et al. 1986; Ferreira & Ramos 1989) (sec. 4.7.3). The many generalizations and assumptions required to generate a carbon budget for a plant or a community combine to produce estimates with wide confidence intervals (sec. 4.7.4). Such estimates are important, however, for resource management at several trophic levels, and much effort has gone into developing models, especially for kelp beds, which are major commercial resource systems.

4.7.1 Measurement of photosynthesis and respiration

The classical equation for photosynthesis and respiration,

$$6CO_2 + 6H_2O \underset{R}{\overset{PS}{\rightleftharpoons}} C_6H_{12}O_6 + 6O_2$$

implies some connection between CO_2 fixation and O_2 release (and vice versa) that is not there. Several processes take up or release C, and several take up and release O_2; the measured value is a composite of all these, dominated (one hopes) by photosynthetic carbon reduction and water-splitting electron transport. There is a fundamental question about what is being measured in gas-exchange or ^{14}C-fixation experiments.

Whereas photosynthesis consumes CO_2 and produces O_2, respiration uses O_2 and releases CO_2. Because both processes occur when light is available, the gas-exchange rate in light, which measures net or apparent photosynthesis, must be corrected for respiration *in the light*, to give gross photosynthesis. The correction for respiration usually has been taken from the dark rate. There has been considerable disagreement over the rates of respiratory gas exchange in the light and in darkness, but the weight of data now indicates that it is much lower in the light, although the metabolic pathways operate continuously (Raven & Beardall 1981; Noggle & Fritz 1983; Kelly 1989). In higher plants, most of the CO_2 released in the light comes from photorespiration. A small (but unknown) amount of O_2 is consumed in the light by pseudocyclic photophosphorylation, or the Mehler reaction, in which electrons (from water) pass to oxygen, generating ATP but resulting in use of half the O_2 released from water (Raven & Beardall 1981). These processes often are relatively small, but they can become significant, as when photosynthesis is low, or oxygen concentration high. Respiration is routinely measured (in dark bottles), but other processes are generally ignored.

The two standard methods for measuring photosynthesis and respiration in water are oxygen release and ^{14}C uptake in light and dark BOD (biological

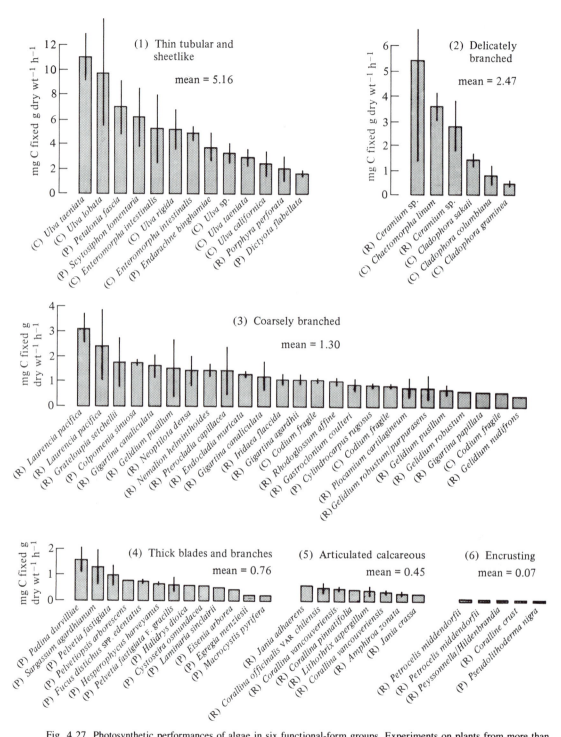

Fig. 4.27. Photosynthetic performances of algae in six functional-form groups. Experiments on plants from more than one of six collecting sites, in southern California and Baja California, are shown separately. Mean rate in milligrams of C fixed per gram dry weight per hour is shown for each group. C, Chlorophyta; P, Phaeophyta; R, Rhodophyta. (From Littler & Arnold 1982, with permission of *Journal of Phycology*.)

oxygen demand) bottles or other suitable containers. Each method has its particular advantages and limitations, which are discussed in technical manuals, including those by Strickland (1960), Vollenweider (1969), and Hellebust and Craigie (1978), and in certain research papers, including those by Littler (1979), Carpenter and Lively (1980), Peterson (1980), Sondergaard (1988), and Thomas (1988). Infrared gas analysis (IRGA) is being increasingly used to measure CO_2 exchange in air (Bidwell & McLachlan 1985; Surif & Raven 1990). A thorough review of all methods used in photosynthesis measurement is given by Geider and Osborne (1992).

Standard BOD bottles hold 300 mL; this imposes constraints on photosynthesis/respiration measurements of seaweeds. At one extreme, turf algae are too small and too entangled with other species (including epiphytes, which become significant at this scale) to be measured individually as plants or even as species; in this case one must measure community rates (e.g., Atkinson & Grigg 1984; Hackney & Sze 1988). At the other extreme, very large plants must be sampled using relatively small tissue pieces, although large enclosures have also been used (e.g., Hatcher 1977; Atkinson & Grigg 1984). Sampling, especially from a plant like *Macrocystis,* and the complication of wound respiration are major obstacles to calculation of accurate rates (Littler 1979; Arnold & Manley 1985; Knoop & Bate 1988a).

The simple equation given earlier also implies a 1 : 1 ratio between CO_2 fixed and O_2 released. In practice, this ratio is rarely 1.0. Thus, if estimates of productivity (in terms of C) are to be based on O_2 measurement, they must be corrected for that ratio, the photosynthetic quotient (PQ). Measurements of PQ have frequently been made for phytoplankton and have often shown its value to be approximately 1.2 (Strickland & Parsons 1972). However, there can be considerable variation, depending in part on whether the fixed C is going mainly into carbohydrates (as in the foregoing equation, fats, or proteins and whether or not photosynthetic energy is being directed to uses such as NO_3^- reduction. Few measurements of PQ have been made for seaweeds, but Hatcher et al. (1977) reported a range in *Laminaria longicruris* from 0.67 to 1.50.

Photosynthetic rates may be expressed on the basis of several denominators, with different results. Ramus (1981) recommended expressing photosynthesis as carbon flux per unit of chlorophyll (also called the assimilation number; Kelly 1989), even while recognizing that the amount of chlorophyll and the rate of photosynthesis do not bear a constant relationship even at saturating irradiances. Respiration probably is best expressed on the basis of total protein, because it is a process taking place in the cytoplasm portion of the cells, rather than on the basis of dry weight, which includes the cell walls. "The rate of a process yields maximum information about the process itself if it is

expressed on the basis of a plant characteristic which limits or at least strongly influences the process" (Sesták et al. 1971). Nevertheless, direct comparison of rates is possible only within carefully controlled groups of experiments, because the complex processes of photosynthesis and respiration are affected by a great many variables, including irradiance, age of the tissue, nutrient levels, temperature, and pH.

4.7.2 Intrinsic variation in photosynthesis

Photosynthetic rates are affected by many abiotic factors besides light, and these will be examined in the following chapters. There are also some biotic factors, intrinsic in the individuals, that affect photosynthetic rates: morphology, ontogeny, and circadian rhythms.

Seaweed photosynthetic rates are strongly influenced by thallus morphology (Fig. 4.27), particularly the surface-area : volume ratio and the proportion of nonphotosynthetic tissue (sec. 1.2.2) (Littler & Littler 1980; also see Littler & Arnold 1982; Littler et al. 1983a). In those studies, the sheet group had the highest mean net photosynthesis; this is not surprising, because, in general, all the cells are photosynthetic, have direct access to carbon supplies in the water, and cause little self-shading. On the basis of surface-area : volume ratios, filamentous algae should have had higher productivity, but clumping (which may have advantages in desiccation resistance, etc.) decreases the effective surface area (Littler & Arnold 1982). There was a twofold decrease in mean rate between each of the first five groups, while the encrusting species had extremely low rates. There was, of course, considerable overlap from group to group; for instance, the rate for *Laurencia pacifica* (group 3) was higher than that for *Porphyra perforata* (group 1).

Changes in photosynthetic characteristics take place as a tissue or plant ages. In a complex plant like *Macrocystis,* ontogenetic gradients occur both along each blade and between blades along each frond. As an example, Figure 4.28 shows net photosynthesis in three categories of blade ages. Steep gradients in the rate per gram of dry weight are seen. However, we can also see how the trend changes depending on the denominator. Data expressed on the basis of chlorophyll *a* and those based on area are different because the ratio of photosynthetic meristoderm to cortical/medullary tissues decreases toward the base of the blade. In the thin tips there is relatively more photosynthetic tissue; in the thicker blade bases (also meristematic) there is more respiratory tissue. Obviously a single tissue disc cannot be representative of a blade, nor a single blade representative of the plant.

Similar ontogenetic changes have been recorded in other kelps (e.g., Küppers & Kremer 1978), *Fucus* (McLachlan & Bidwell 1978; Khailov et al. 1978; Küppers & Kremer 1978), and *Sargassum* (Kilar et al. 1989). The proportion of nonphotosynthetic tis-

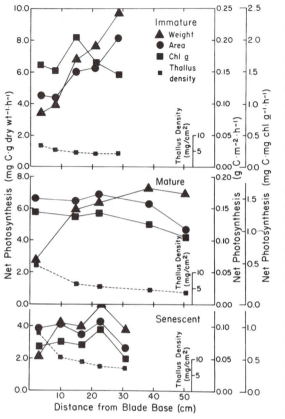

Figure 4.28. Variations in photosynthesis along *Macrocystis* blades of various ages, also showing differences due to normalizing to dry weight (triangles), area (circles), and chlorophyll *a* (large squares). The thallus density (milligrams dry weight per square centimeter) is presented (small squares) for each disc position. (From Arnold & Manley 1985, with permission of *Journal of Phycology.*)

sue can also vary between individuals (e.g., wiry versus fleshy plants of *Gigartina canaliculata*) (Littler & Arnold 1980).

Changes in photosynthetic rate over a day do not necessarily parallel the changes in I_0, partly because of photoinhibition, photoinduction, and diurnal rhythms. The prediction based on instantaneous measures would be for photosynthesis to parallel the increase in I_0 from dawn until I_0 exceeded I_k, to remain level until I_0 again fell below I_k in the afternoon or evening, and then to decrease toward dusk. However, morning and afternoon *P–I* curves do not always correspond (Ramus & Rosenberg 1980; Ramus 1981; Knoop & Bate 1988b; Gao & Umezaki 1989a,b). There is often a noontime depression of photosynthesis, with only partial recovery in the later afternoon, and as much as 70% of daily photosynthesis may take place before noon. The effect disappears in low light. The underlying cause is unknown, but it could be saturation of Calvin-cycle intermediates (Gao & Umezaki 1989b). Three subtidal red algae show slow

photo*induction,* with photosynthetic capacity in saturating light increasing twofold to sixfold over the first 2–4 h of the light period, partially reversing during the latter part of the photoperiod (Knoop & Bate 1988b). The duration of this photoinduction period and the size of the change depend partly on the prior light history (photodose), an observation with implications for the storage and preincubation conditions of samples for photosynthesis measurements.

Some plants show diurnal changes in photosynthetic rate even under uniform laboratory conditions (Kageyama et al. 1979; Mishkind et al. 1979; Ramus 1981). These changes in photosynthetic rates are due to endogenous clocks. Circadian (diel) rhythms affect several aspects of cellular activity, enzymes, or cell division in a number of species. They have periods of 21–27 h (hence circa-dian) and are entrained to 24 h by light stimuli such as the time of dawn or dusk (Hillman 1976; Sweeney & Prézelin 1978). The diurnal changes are due to an endogenous rhythm and are not the results of environmental variables. This can be shown by transferring plants to continuous (usually dim) light and taking samples for measurement of light-saturated photosynthetic rates at various times parallel to the original light–dark cycle. Characteristically, the rate is highest during a time corresponding to the middle of the light period and lowest in what would have been the middle of the dark period (Fig. 4.29). Over several days of continuous illumination the period of the rhythm gradually changes because there is no entraining stimulus (Mishkind et al. 1979; Oohusa 1980). If the light–dark cycle is reversed, the phase of the endogenous rhythm also shifts, as shown in Figure 4.29b.

There are several potential contributors to a cycle in photosynthetic rate. In *Ulva* there is a diel migration of chloroplasts between the sides and faces of the cells (Britz & Briggs 1976), but this does not regulate the diel rhythm of photosynthesis (Nultsch et al. 1981; also see Larkum & Barrett 1983). Rather, the probable site of circadian control is the rate-limiting step in electron transport, probably one of the steps between plastoquinone and PS-I (Mishkind et al. 1979). Further points of control may lie in certain enzymes. Yamada et al. (1979), using the brown alga *Spatoglossum pacificum,* showed rhythms in several enzymes of the Calvin cycle, including RuBP carboxylase, fructose-1,6-bisphosphate phosphohydrolase, mannitol-1-phosphate phosphohydrolase, and ribose-5-phosphate isomerase. However, the relationships of these enzyme activities to the circadian clock are likely to prove complex. For example, RuBP carboxylase (in higher plants) is activated by light-driven Mg^{2+} fluxes (Jensen & Bahr 1977). Yet these control points are not the mechanism of the rhythm, but rather its expression.

The current model of the mechanism of circadian rhythms is Sweeney's (1974) membrane model, based on studies of *Acetabularia,* which holds that the rhythms are generated by a feedback loop consisting of active

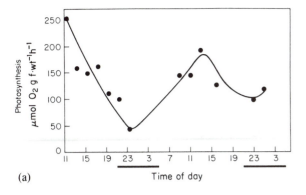

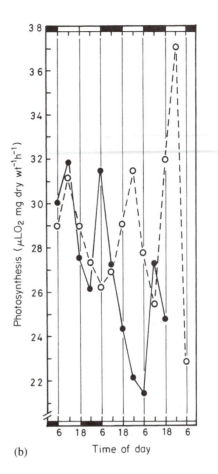

Figure 4.29. Photosynthetic rhythms. (a) Light-saturated photosynthetic rate for *Ulva lactuca* kept in continuous dim light (2 mW cm^{-2}). Along the abscissa, Eastern Standard Time and the hours of darkness in the collection environment. (b) Diurnal changes in photosynthetic capacity of *Porphyra yezoensis*. The solid line shows plants transferred to continuous light after the first day (light–dark cycles shown on lower abscissa). The broken line shows plants transferred to a reversed light–dark cycle (upper abscissa) and shows the reversal of the endogenous rhythm. (Part a from Mishkind et al. 1979, with permission of the American Society of Plant Physiologists; b from Oohusa 1980, with permission of Walter de Gruyter & Co.)

transport across organelle membranes and the distribution of molecule(s) between organelles and cytoplasm. The organelle membrane is predicted to be slowly permeable to some molecule or molecules, X. Thus diffusion tends to equalize the concentrations of X in the organelle and cytoplasm. When equilibrium is reached, active import of X by the organelle is initiated and continues until a certain critical internal concentration of X causes a change in membrane properties so that active transport stops. In the absence of active uptake, diffusion gradually reestablishes equilibrium. Light is known to affect membrane properties and could thus reset the rhythm (e.g., by initiating active transport of X). Each rhythm may depend on different key substances in the appropriate organelle; for example, Mg^{2+} might affect RuBP carboxylase activity. Alternatively, X might be potassium ion (Sweeney & Prézelin 1978). Among the features of *Acetabularia* behavior that led Sweeney to her model is that the rhythm of photosynthesis continues in enucleated fragments, yet if a nucleus and cytoplasm from two plants with diametrically opposed rhythms are combined, the rhythm gradually shifts from that of the cytoplasm to that of the nucleus. According to the model, this can be explained on the basis of both nucleus and chloroplasts transporting X: The nucleus of *Acetabularia* has a relatively large volume, and whether

it is leaking or importing X it will have a major effect on the concentration of X in the cytoplasm. Hence the flux of X between cytoplasm and chloroplast, and the timing of the photosynthetic rhythm, will gradually change.

In addition to the intrinsic changes in the hourly rate of photosynthesis, this rate in nature is not constant over a day, nor from day to day, nor season to season, because of changes in irradiance and seasonal changes in the plants (Sakanishi et al. 1989) (sec. 6.3.2). Because only a tiny fraction of the season or year or lifetime of the organism is experimentally measured, errors in extrapolation from typical 4- or 8-h measurements of photosynthesis can lead to large errors in productivity estimates.

4.7.3 Carbon losses

The rates of exudation by various seaweeds under most natural conditions remain a matter of debate. Considerable controversy was generated by various studies that attempted to assess exudation under experimental conditions. Some experiments showed that exudation was 30–40% of net assimilation in seaweeds (Khailov & Burlakova 1969; Sieburth 1969). Other experiments showed it to be much less, ranging from a small percentage down to less than 1%, except under stress (Moebus & Johnson 1974; Moebus et al. 1974; Harlin &

Craigie 1975; Brylinski 1977; Carlson & Carlson 1984). In all those experiments, exudation or its absence may have been an artifact of the method. The functions of any exuded materials (in the biology of the exuders) are not known.

Tissue loss is more easily documented and understood than exudation. Tissue loss can result from extrinsic and intrinsic causes. Extrinsic causes include direct or indirect grazing damage (sec. 3.2), physical abrasion (sec. 7.2.3), and microbial degradation (e.g., breakdown of old *Laminaria* tissue by the ascomycete *Phycomelaina laminariae;* Schatz 1980). Seaweeds rarely shed significant amounts of tissue, but examples of loss from intrinsic causes include shedding of old fruiting branches (e.g., by fucoids) and erosion of old tissue at the tips of *Laminaria*. Loss of tissue from populations also includes loss of entire plants.

Surprising quantities of tissue are abscised by *Fucus* and *Ascophyllum* following reproduction. In the former, not only the receptacles but also the internodes beneath are shed (Knight & Parke 1950). *Ascophyllum* receptacles can account for half the biomass of the plant (Josselyn & Mathieson 1978). On the other hand, kelp spore production, while amounting to 3×10^{10} spores per year, accounts for only 0.17% of *Ecklonia* annual production (Joska & Bolton 1987). *Laminaria* blades continuously lose old tissue from the tips, and the entire blade can be turned over one to five times per year (Mann 1972). *Laminaria hyperborea* differs in that a distinct new blade is produced in late winter at the base of the old blade. Carbon, and perhaps nitrogen, will be salvaged from the old blade for the new growth before the old blade is shed. Lüning (1969) demonstrated that the new blade could form in darkness with the old blade present, but if the old blade was cut off, there was virtually no new growth, even in light. Subsequently, Lüning et al. (1973) demonstrated translocation of newly fixed carbon from old to new blades, but no one has yet shown with tracers that *old* carbon or amino acids are salvaged. The analogy with senescence of angiosperm leaves suggests that salvage might occur in kelps, because there is translocation.

4.7.4 Autecological models of productivity and carbon budgets

The degree of difficulty in estimating productivity is contingent on the complexity of the plant or population. The productivities of *Laminaria* and other linear kelps have frequently been modeled from length increments, because the meristem produces a "moving belt" of tissue (Parke 1948; Mann 1972; Dieckmann 1980; Gagné & Mann 1987). The method relies on some equation relating length increase to biomass. Simple length–weight regressions (Mann 1972) are not adequate because the relationship changes over the growth cycle. Kain (1979) found a 10-fold difference when she calculated annual production from the data of Mann

(1972) and Hatcher et al. (1977) for the same population of *L. longicruris*. The data of Hatcher et al. came from a carbon budget based on photosynthesis measurements. Various formulas have been tried to improve the accuracy of biomass estimates. Gagné & Mann (1987) tested four models for *Laminaria longicruris* and concluded that the best estimates were obtained simply by multiplying linear growth by the weight per unit length of a section from the uniformly wide part of the blade. In this species and some other kelps the tapering meristematic region is a small portion of the overall strap-shaped blade. In more triangular thalli, more complex calculations must be used, and in genera such as *Ecklonia* (Mann et al. 1979), *Eisenia*, and *Macrocystis*, this method becomes cumbersome or impossible.

Ferreira and Ramos (1989) wanted to combine short-term estimates from photosynthetic rates with long-term estimates from biomass. For three estuarine species, they measured biomass monthly and ran their calculations for each month based on the starting biomass and adding productivity modeled from irradiance estimates (taking account of hourly tide height and water turbidity) and *P–I* relations (including an assumption about the photosynthesis of *Fucus* in air). They simplified their calculations by having a constant day length of 16 h (the study was done in Portugal) and including zero net productivity for the 8 h of darkness (not accounting for respiratory losses). Although the model did not account for losses during a month, it did not carry over any error into the following months, because they started each month's calculation with the actual biomass.

Modeling the productivity of a *Macrocystis* bed involves also modeling the large, three-dimensional environment it occupies, because the huge plants significantly affect the physical environment (light, currents) and the chemistry of the water moving through the bed, just as they are affected by these factors. A model is being developed by Jackson and co-workers in a series of steps, including modeling of the physical environment (Jackson 1984), the relations among fronds (Jackson et al. 1985), and the relation between irradiance and yield (Jackson 1987). The 1987 model gave results for biomass and production that were "in general agreement" with field measurements, but with differences in the predicted timing of growth maxima and minima. The simple seasonal cycle predicted on the basis of irradiance is complicated by other factors, with nutrients probably most important. Jackson (1987) listed five aspects that could be incorporated into future models, including the adaptations of blades to lower irradiances, the tissue losses due to grazing and decay, the effects of clouds on the light climate, and the effects of nutrients. A computer model of sporophyte densities as controlled by temperature/nutrients (correlated) and irradiance has been developed by Burgman and Gerard (1990).

The ultimate goal of these autecological models is to be able to predict the effects of harvesting, pollution, and other disturbances on the resource. During their development, they point out where more detailed data are needed. Beyond productivity estimates for individual species, ecosystem modelers relate production to energy flow through the community, including carbon budgets for bacteria, phytoplankton, and consumer animals at all trophic levels.

4.7.5 Ecological impact of seaweed productivity

Carbon or energy fixed by seaweeds is sooner or later passed to other trophic levels. Herbivores eat the plants directly; detritivores and filter feeders eat particles produced by grazing, wave action, or dissolution of organic materials; bacteria and fungi remineralize whatever is left (Fig. 4.30a). Kelp beds in particular can contribute to temperate near-shore secondary production well beyond the kelp-bed ecosystem. Duggins et al. (1990) found kelp-derived carbon throughout the nearshore food web in Alaska (Fig. 4.30b). The flow of energy from primary producers into other organisms is important in the dynamics of ecosystems, especially for ecosystem and resource management.

Ecosystem modelers develop predictive mathematical models. System variables (e.g., primary productivity), transfer functions (e.g., energy transferred to another trophic level), and forcing functions (inputs such as sunlight that are not affected by the system) are put into the model to derive constants or parameters for the ecosystem (Walters 1971). Compartmental models deal primarily with quantities of energy and materials in different compartments of an ecosystem (e.g., the "biomass budget-box model" of Polovina 1984) (Fig. 4.31). This type of model can be extended to show energy flow, using symbols and concepts from electrical-circuit diagrams (Odum 1972) (Fig. 4.32).

Simulation of the dynamics of an ecosystem requires the inputs of production, biomass, and energy flow in and between all trophic levels. This much information may not be available, especially in the tropics, but a biomass budget-box model requires less information and can still give results useful for management (Polovina 1984; Grigg et al. 1984). Polovina modeled a reef in northwestern Hawaii, using productivity : biomass ratios from the literature for phytoplankton and seaweeds (70 and 12.5, respectively), plus data (including diet) for other trophic groups. The budget diagram is shown in Figure 4.31. A net primary productivity of 4.3×10^6 kg km^{-2} yr^{-1} (wet weight) was predicted on the basis of the number of trophic levels, the productivity of the top level, and the efficiency of transfer between levels. The phytoplankton productivity was only one-tenth that of the benthos, and that is negligible given the confidence limits of the prediction. Atkinson and Grigg (1984) then estimated benthic primary productivity, enclosing whole coral knolls in plastic and measuring community gas exchange. (Zooxanthellae are significant contributors to benthic algal production.)

The estimate was 6.1×10^6 kg km^{-2} yr^{-1} ($\pm 50\%$), close enough to the model prediction to give confidence in the model as a whole (Grigg et al. 1984). Only about 5% of the net production passed up the food chain to nonbenthic reef consumers (top row of boxes in Fig. 4.31).

Electrical-analogue models have been developed for some kelp beds (e.g., Mann 1972; Field et al. 1977; Branch & Griffiths 1988). The model for South African *Laminaria pallida* beds (Fig. 4.32) extends the picture of the fate of kelp production shown in Figure 4.30a. Forcing functions include the effects of solar energy (the only one for which a value is known), waves, and wind. Work gates are processes that affect energy transfer between trophic levels. Waves move the kelp and increase its light harvesting (work gate 2); they also promote tissue breakage, which can provide material for debris feeders (work gate 4) and leads to export of plants and pieces from the kelp bed (work gate 3). Waves also affect the ability of climbing grazers to feed on the kelp (work gate 4) and contribute to erosion of POM and DOM from the tips of kelp blades (work gate 5). Southwest winds cause upwelling and lead to increased growth and light-harvesting ability (work gate 1).

These models portray the kelp bed or coral reef as a closed system. Import and export fluxes may be significant, however, as in *Macrocystis* beds, but open models require even more information than closed models.

4.8 **Synopsis**

Photosynthetically active radiation has wavelengths from 350 or 400 nm to 700 nm. The quantity of light arriving, or the flux, is called irradiance and is measured in microeinsteins per square meter per second or microwatts per square meter. The quality and quantity of light in the oceans are highly variable. Irradiance changes at several time scales, and waves strongly affect the penetration of light into the sea. Once in water, light is attenuated by scattering and absorption, with the red end of the spectrum being more strongly reduced. Several ocean-water types have been defined on the basis of the light quality in them.

Photosynthetic pigments in plants always include chlorophyll *a*. In addition, each group of algae has a characteristic array of accessory pigments, including other chlorophylls, carotenes, xanthophylls, and, in red and blue-green algae, phycobilins. Chlorophylls and phycobilins occur in pigment-protein complexes. Phycobilins are clustered in phycobilisomes on the thylakoid membrane surface; other accessory pigments and the rest of the photosynthetic apparatus are integral parts of the membrane.

The arrangements of thylakoids within the chloroplasts differ among Divisions, but thylakoid membranes and the processes of photosynthesis are essentially the same in all plants. The photosynthetic apparatus consists of two different reaction centers with slightly different

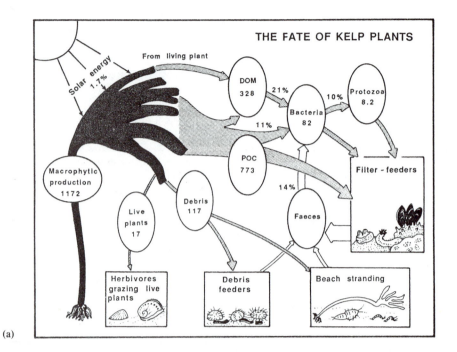

(a)

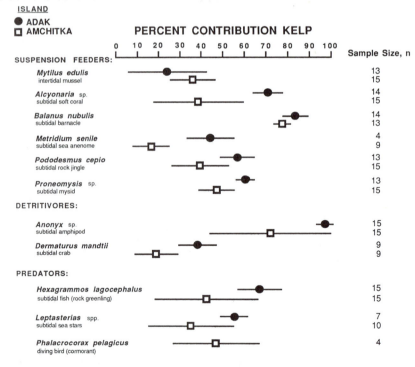

(b)

Figure 4.30. Fate of seaweed productivity in the near-shore ecosystem. (a) Kelp production and consumption (grams C per square meter per year). Percentages on the arrows indicate conversion efficiencies between different components. (b) Percentages of carbon photosynthesized by kelps found in tissues of consumers at Alaskan islands with extensive subtidal kelp beds. Kelp-derived carbon was identified by its $\delta^{13}C$ signature. (Part a from Branch & Griffiths 1988, *Oceanogr. Mar. Biol. Ann. Rev.*, with permission of Aberdeen University Press, Farmers Hall, Aberdeen AB9 2XT, U.K.; b from Duggins et al. 1990, *Science* 245:170, © 1989 by the AAAS.)

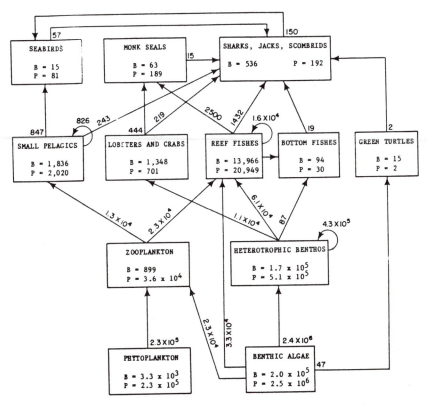

Figure 4.31. Biomass budget schematic for major prey–predator pathways on a Hawaiian coral reef. Annual production denoted as P, and mean annual biomass as B, with values in units of kilograms per square kilometer based on a habitat area of 1,200 km^2. Numbers on the arrows are biomasses of prey consumed (kilograms per square kilometer). (From Polovina 1984, with permission of Springer-Verlag, Berlin.)

red-absorption peaks. Reaction centers are linked to core complexes, and these in turn to light-harvesting complexes; at each level, pigments are bound to proteins. The accessory pigments and most of the chlorophyll *a* are arranged in antennae, with those absorbing at longer wavelengths thought to be closer to the reaction center.

The amount of light harvested by a thallus (its absorptance) depends on both pigment concentration and thallus morphology. Significant morphological aspects are the thickness and the arrangement of branches relative to self-shading. The rate of photosynthesis is strongly dependent on irradiance level. At the compensation point, photosynthesis equals respiration. At the saturation point, photosynthesis is maximum. At very high irradiance, the photosynthetic rate may decline again because of photoinhibition.

Seaweeds can acclimate to differences in light quality and quantity. The quantity of pigment or the density of photosynthetic units can be increased, and sometimes the ratio of accessory pigments to chlorophyll *a* changes. However, the old idea that seaweeds change color to complement the color of the light in their habitat has been discredited.

Carbon fixation constitutes the "dark reactions" of photosynthesis. Carbon is available in seawater chiefly as HCO_3^-. There are small amounts of CO_2, the proportion depending on pH and salinity. Parasitic algae rely on carbon from their hosts, but other seaweeds have little or no ability to use exogenous organic carbon, particularly at environmental concentrations.

CO_2 is fixed by RuBP carboxylase (RuBisCO) in the Calvin cycle. HCO_3^- is converted to CO_2 with the aid of carbonic anhydrase. In the brown algae, especially in young tissue of kelps and fucoids, light-independent carbon fixation occurs, in light and dark, via PEP carboxykinase, which also uses free CO_2. RuBisCO also acts as an oxygenase, leading to photorespiration, but most seaweeds seem able to suppress this by various means of concentrating CO_2 near RuBisCO.

Various principal products of photosynthetic carbon fixation are found among the seaweeds. Green algae form chiefly sucrose; brown algae form mannitol; red algae form chiefly floridoside and digeneaside. In addition, all groups accumulate some amino acids.

In the kelps, the low-molecular-weight compounds can be translocated from mature regions to meristematic regions. The driving force probably is a water potential gradient set up by the loading of organic molecules into the sieve elements in the source and their unloading in the sink.

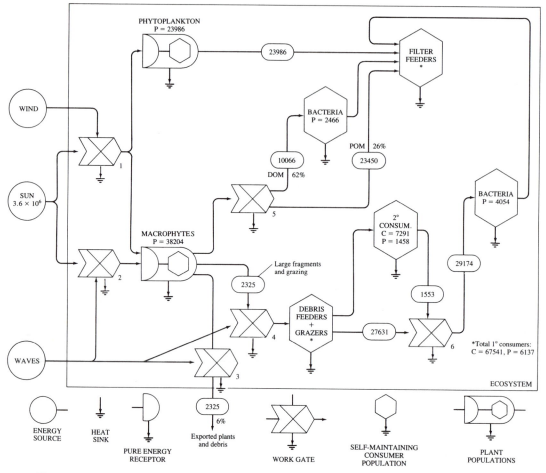

Figure 4.32. Energy flow through a kelp bed on the west coast of Cape Peninsula, South Africa. Units are kilojoules per square meter per year. Work gate 6 = feces loop. DOM; dissolved organic matter; POM particulate organic matter. (Modified from Field et al. 1977, and Newell et al. 1982.)

The initial products of photosynthesis may be polymerized into polysaccharides for storage or cell walls. As with low-molecular-weight products, each Division has a characteristic group of compounds. Storage compounds are starch (green algae) or floridean starch, and laminaran in the brown algae. Seaweed cell walls have a fibrillar layer consisting of cellulose, mannan, or xylan. Various characteristic mucilaginous polysaccharides are also found in seaweed walls. These include some of commercial value, such as agars, carrageenans, and alginates. Chemical changes in these polymers can make them weaker or stiffer gels. The native structures of these compounds are not well known. Different stages in the life history may have different wall polysaccharides.

Photosynthesis is the basis of primary productivity, and its measurement also is often used to indicate plant responses to environmental variables, because it is affected by temperature, pH, circadian rhythms, ages of the tissues, and irradiance. Photo-synthetic rates can be measured by following CO_2 uptake or O_2 release by illuminated tissue, but account must be taken of respiratory use of O_2 and release of CO_2 in the light. Respiration usually is measured in darkness, but it also occurs, to a much smaller extent, in the light.

The estimation of primary production requires measurements not only of photosynthesis, respiration, and photorespiration (if any) but also of tissue loss and organic-carbon exudation rates. Seaweeds can be divided into functional-form groups, such as crustose forms and thin sheets, which have characteristic levels of productivity.

Carbon budgets and energy-flow models show how seaweed productivity is passed to higher trophic levels. In environments with large seaweeds, such as kelps, carbon from primary production may be spread far and wide; in others, such as coral reefs with generally small and microscopic producers, production may be tightly cycled chiefly within the system.

5

Nutrients

Seaweeds require inorganic carbon, water, light, and various mineral ions for photosynthesis and growth. This chapter will examine the mechanisms of uptake, the nutrient requirements, and the metabolic roles of essential nutrients (excluding C, H, and O). The importance of nutrient uptake and growth kinetics will be discussed in terms of their effects on chemical composition, growth, development, and distribution of macroalgae. Particular emphasis will be placed on nitrogen, because it is the element most frequently limiting to seaweed growth. Even though seaweeds are larger than phytoplankton and usually are attached to a substratum, their nutritional requirements are very similar, and therefore some discussion of phytoplankton nutrition is also included when little or no information exists for seaweeds.

5.1 Nutrient requirements

5.1.1 Essential elements

The development of defined culture media for growing algae axenically has allowed the testing of a variety of elements to determine which are essential. The criteria to define an absolute requirement for an element were established by Arnon and Stout (1939):

1. A deficiency of the element makes it impossible for the alga to grow or complete its vegetative or reproductive cycle.
2. It cannot be replaced by another element.
3. The effect is direct and is not due to interaction with (e.g., detoxification of) other, nonessential elements, stimulation of epiflora, or the like (Levitt 1969).

C, H, O, N, P, Mg, Cu, Mn, Zn, and Mo are considered to be required by all algae (O'Kelley 1974; DeBoer 1981); S, K, and Ca are required by all algae, but can be partially replaced by other elements; Na, Co, V, Se, Si, Cl, B, and I are required only by some algae. All the major constituents of seawater, except for Sr and F are required by macroalgae (DeBoer 1981). There is a tendency to consider the requirements of all algae to be similar, but the heterogeneity of macroalgae makes it difficult to generalize about their nutritional requirements.

Up to 21 elements are required for the main metabolic processes in plants (Table 5.1), but more than double that number are present in seaweeds. The mere presence of an element in a seaweed is not proof that the element is essential, nor is the amount present indicative of the relative importance of the element. Generally, essential (and nonessential) elements are accumulated in the tissues of algae to concentrations above their concentrations in seawater, giving rise to concentration factors of up to 10^3 (Phillips 1991) (Table 5.2). Some elements are absorbed in excess of an alga's requirements, whereas others are taken up but not utilized. The elemental composition of the ash in macroalgae is similar to that of phytoplankton. The effects of nutrient supply on chemical composition are discussed in section 5.8.1, and the metabolic roles of elements in section 5.5.

5.1.2 Vitamins

Some seaweeds require trace amounts of one or two organic-carbon compounds for normal growth; these compounds do not act as carbon sources for algal growth. This type of nutrition is referred to as auxotrophy, and the organic compounds are vitamins. Phytoplankton also require the same vitamins as seaweeds, but most higher plants synthesize their own vitamins and do not depend on environmental sources.

The three vitamins that are routinely added to culture media are B_{12} (cyanocobalamin), thiamine, and biotin (Fig. 5.1). Of these, B_{12} is the most widely required by seaweeds and this may be related to the fact that B_{12} is present in seawater in lesser amounts (ca. 1 ng L^{-1}) than are thiamine (ca. 10 ng L^{-1}) and biotin (ca. 2 ng L^{-1}). Thiamine is known to be required only by *Acetabularia acetabulum* (the only macroscopic

Table 5.1. *Functions and compounds of the essential elements in seaweeds*

Element	Probable functions	Examples of compounds
Nitrogen	Major metabolic importance in compounds	Amino acids, purines, pyrimidines, amino sugars, amines
Phosphorus	Structural, energy transfer	ATP, GTP, etc., nucleic acids, phospholipids, coenzymes (including coenzyme A), phosphoenolpyruvate
Potassium	Osmotic regulation, pH control, protein conformation and stability	Probably occurs predominantly in the ionic form
Calcium	Structural, enzyme activation, cofactor in ion transport	Calcium alginate, calcium carbonate
Magnesium	Photosynthetic pigments, enzyme activation, cofactor in ion transport, ribosome stability	Chlorophyll
Sulfur	Active groups in enzymes and coenzymes, structural	Methionine, cystine, glutathione, agar, carrageenan, sulfolipids, coenzyme A
Iron	Active groups in porphyrin molecules and enzymes	Ferredoxin, cytochromes, nitrate reductase, nitrite reductase, catalase
Manganese	Electron transport in photosystem II, maintenance of chloroplast membrane structure	
Copper	Electron transport in photosynthesis, enzymes	Plastocyanin, amine oxidase
Zinc	Enzymes, ribosome structure(?)	Carbonic anhydrase
Molybdenum	Nitrate reduction, ion absorption	Nitrate reductase
Sodium	Enzyme activation, water balance	Nitrate reductase
Chlorine	Photosystem II, secondary metabolites	Violacene
Boron	Regulation of carbon utilization(?), ribosome structure(?)	
Cobalt	Component of vitamin B_{12}	B_{12}
Bromine[a] Iodine[a]	Toxicity of antibiotic compounds(?)	Wide range of halogenated compounds, especially in Rhodophyceae

[a]Possibly an essential element in some seaweeds.
Source: DeBoer (1981), with permission of Blackwell Scientific Publications.

marine green alga studied thus far); no requirement has yet been found for biotin among the seaweeds. One Chlorophyceae, 1 Phaeophyceae, and 10 Rhodophyceae require vitamin B_{12} (DeBoer 1981). Because only a few seaweeds have been studied, it is not yet safe to draw generalizations from the limited data available.

Thiamine and B_{12} are both complex molecules (Fig. 5.1), as compared with biotin, and some phytoplankton require only part of the molecule. In the case of thiamine, the requirement may be specific for the thiazole moiety or for the pyrimidine moiety, or the alga may require both moieties, the whole molecule, or homologues of thiazole or pyrimidine (Provasoli & Carlucci 1974). In the case of B_{12}, the requirements may be satisfied with the corrin group alone, by the corrin group plus any nucleotide side chain, or by the whole B_{12} molecule. Algae requiring only the corrin group

would have the greatest advantage, because many bacteria produce cobalamins of some kind, but only 15–30% produce the entire B_{12} molecule.

Because there is no rapid analytical technique to determine vitamin concentrations, relatively few measurements have been made. The recent development of HPLC techniques for vitamins may stimulate more vitamin research in the future. Currently, a bioassay generally is used to measure vitamins, but it is laborious, and suitable bioassay organisms are not known for some moieties (Provasoli & Carlucci 1974). Further difficulties arise in the analysis of vitamin B_{12} because some phytoplankton produce a substance that binds the vitamin and makes it unavailable for uptake (Swift 1980). The exogenous vitamins required by some seaweeds may be obtained from seawater, or they may come directly from microorganisms living on seaweed

Table 5.2 *Concentrations of some essential elements in seawater and in seaweeds*

Element	Mean concentration in seawater (mmol kg^{-1})	(μg g^{-1})	Concentration in dry matter Mean (μg g^{-1})	Range (μg g^{-1})	Ratio of concentration in seawater to concentration in tissue
Macronutrients					
H	105,000	10,500	49,500	22,000–72,000	2.1 × 10^0
Mg	53.2	1,293	7,300	1,900–66,000	1.8 × 10^{-1}
S	28.2	904	19,400	4,500–8,200	4.7 × 10^{-2}
K	10.2	399	41,100	30,000–82,000	1.0 × 10^{-2}
Ca	10.3	413	14,300	2,000–360,000	2.9 × 10^{-2}
C	2.3	27.6[a,b]	274,000	140,000–460,000	1.0 × 10^{-4}
B	0.42	4.50	184	15–910	2.4 × 10^{-2}
N	0.03	0.420[a,c]	23,000	500–65,000	2.1 × 10^{-5}
P	0.002	0.071	2,800	300–12,000	2.4 × 10^{-5}
Micronutrients					
Zn	6 × 10^{-6}	0.0004[a]	90	2–680	4.4 × 10^{-5}
Fe	1 × 10^{-6}	0.00006[a]	300	90–1,500	1.0 × 10^{-5}
Cu	4 × 10^{-6}	0.0002[a]	15	0.6–80	1.7 × 10^{-4}
Mn	0.5 × 10^{-6}	0.00003[a]	50	4–240	2.0 × 10^{-5}

[a]Considerable variation occurs in seawater (Bruland 1983).
[b]Dissolved inorganic carbon.
[c]Combined nitrogen (dissolved organic and inorganic).
Source: DeBoer (1981), including concentrations of elements in seawater from Bruland (1983), with permission of Blackwell Scientific Publications.

surfaces. In addition, a few phytoplankters are known to produce vitamins (the ones not required for their own growth) during part of the growth cycle (Provasoli & Carlucci 1974).

5.1.3 Limiting nutrients

More than a century ago an agriculturalist named Liebig stated that "growth of a plant is dependent on the minimum amount of foodstuff presented," and this statement has come to be known as "Liebig's law of the minimum." The nutrient available in the smallest quantity with respect to the requirements of the plant will limit its rate of growth, if all other factors are optimal. Evidence that N, P, and Fe, and also Cu, Zn, Mn, and C, might limit algal growth comes from the fact that the concentrations of these elements in seawater vary considerably because of biological activity, and the concentrations of these elements in tissues are 10^4 to 10^5 greater than their concentrations in seawater (Table 5.2). Recent experiments have shown that among these possible limiting elements, nitrogen most frequently limits the growth of seaweeds (Topinka & Robbins 1976; DeBoer & Ryther 1977) and phytoplankton (Ryther & Dunstan 1971; Howarth 1988); phosphorus is the second most limiting nutrient (Manley & North 1984; Lapointe 1986, 1987; Lapointe et al. 1992).

The concentration of a nutrient will give some indication whether or not the nutrient is limiting, but the nutrient's supply rate or turnover time is more important in determining the magnitude or degree of limitation. For example, if the concentration of a nutrient is low, but the supply rate is slightly less than the uptake rate by the algae, then the algae will be only slightly nutrient-limited.

The possibility of growth limitation by two nutrients simultaneously, or dual nutrient limitation, has been explored. The original concept proposed by Droop (1973) was for a multiplicative effect by the two nutrients, but experimentation has revealed that it is an "either/or" effect, with only one nutrient limiting the growth of a species at one time (Droop 1974; Rhee 1978). However, the ratio of two nutrients (e.g., N/P) required by one algal species may be quite different from the ratio required by another species. Consequently, one species may be nitrogen-limited, while another species may be phosphorus-limited, illustrating limitation by a single nutrient, with competition for different nutrient resources (Tilman et al. 1982).

5.2 **Nutrient availability in seawater**

The concentrations of various elements in seawater can differ by up to six orders of magnitude (Table

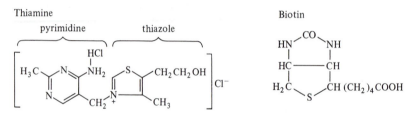

Vitamin B$_{12}$
(Cyanocobalamin)

Figure 5.1. Chemical structures of the three vitamins required for growth of macroalgae. (Reprinted with permission from Bauernfeind & DeRitter 1970, *Handbook of Biochemistry,* 2nd ed. © CRC Press, Inc.)

5.2). Those in the nanomolar (nM) range (<0.01 µg g^{-1} in Table 5.2) are considered micronutrients or trace elements (e.g., Fe, Cu, Mn, Zn) for nutritional purposes. Elements occurring at higher concentrations frequently are referred to as macronutrients (e.g., C, N, P). A variety of units can be used to express concentrations. Generally, marine scientists deal in terms of micromolar (µM) concentrations, although microgram-atoms per liter (µg-at. L^{-1}) is still used. Freshwater nutrient concentrations generally are expressed as micrograms per liter (µg L^{-1}) or parts per billion (ppb). To convert from micrograms per liter to micromoles, one must multiply micrograms per liter by the atomic weight of the element. For example, the atomic weight of nitrogen is 14, so 14 × 1 µg L^{-1} = 1 µM nitrogen. For combined forms, 1 µM nitrogen = 1 µg-at. L^{-1} NO$_3^-$ or NH$_4^+$, but 2 µg-at. L^{-1} urea, because urea has two nitrogen atoms per molecule. The methods for measuring nutrients that are particularly important for macroalgal growth (e.g., nitrate, ammonium, and phosphate) have been reviewed by Wheeler (1985). Nitrate, nitrite, ammonium, and phosphate concentrations vary from 0 to 30, 1, 3, and 2 µM, respectively, for most pristine temperate areas.

The aspects of nutrient cycles have been discussed elsewhere (Redfield et al. 1963; Riley & Skirrow 1965; Riley & Chester 1971; Parsons & Harrison 1983; Bruland 1983); only a few basic principles will be reviewed here. Nitrogen is the element that most frequently limits algal growth in the sea, and the important ions for seaweed growth are nitrate and ammonium. The important features of the nitrogen cycle are summarized in Figure 5.2.

Processes that bring nitrogen into the euphotic zone (Fig. 5.2) where it can be used directly in autotrophic production by the plants include the following: (1) physical advection, primarily in the form of nitrate from below the nutricline; (2) atmospheric input of ammonia and nitrate either via rain (Paerl 1985; Paerl et al. 1990) or through N$_2$ fixation by bacteria and cyanobacteria; (3) ammonium from bacterial decomposition of organic matter in sediments, the magnitude of which may be enhanced by the activity of burrowing animals or by a salinity intrusion into the interstitial waters of

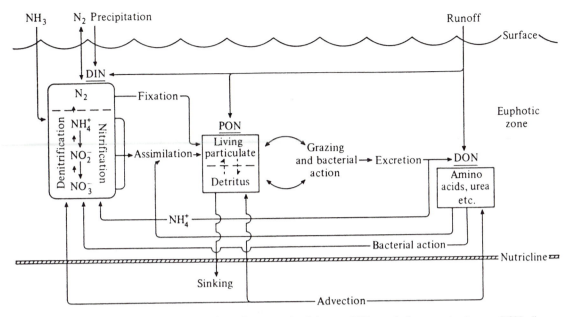

Figure 5.2. Schematic representation of the nitrogen cycle of the sea. PON, particulate organic nitrogen; DON, dissolved organic nitrogen; DIN, dissolved inorganic nitrogen. (From Turpin 1980, with permission of the author.)

the sediments; and (4) in coastal areas, inputs of nitrogen from land drainage, sewage, and agricultural fertilizers. Regeneration of nitrogen in the water column occurs as a result of two largely separate processes; one involves bacteria, and the other results from excretion by marine fauna, particularly ammonium by the zooplankton community (Fig. 5.2). In some small, shallow eutrophic estuaries, decomposition of extensive mats of *Enteromorpha* or *Chaetomorpha* may dominate the nitrogen cycle for short periods in summer (Owens & Stewart 1983; Sfriso et al. 1987; Lavery & McComb 1991a).

Red, green, and brown seaweeds are unable to make direct use of nitrogen gas (N_2), even though it is 20 times more abundant than nitrate. However, nitrogen fixation is associated with *Codium decorticatum* (Rosenberg & Paerl 1981) and *C. fragile* (Gerard et al. 1990). Nitrogenase activities in these situations have been attributed both to nitrogen-fixing bacteria (e.g., *Azotobacter*) and to heterocystous cyanobacteria (e.g., *Calothrix*) that occur as epibionts, and not to the seaweed. The measured rate of nitrogen fixation is equivalent to about 7% of the daily nitrogen requirement of a seaweed, but there is no evidence that the host plants obtain significant amounts of this fixed nitrogen (Gerard et al. 1990). *Codium isthmocladum* does not have this associated nitrogen-fixing flora; further studies will be required to determine if this example is an exception. Cyanobacteria belonging to the genus *Oscillatoria* occur as epiphytes on the blades of *Sargassum fluitans* and *S. natans* (Phlips & Zeman 1990). Certain species are nitrogen-fixers (sec. 1.3.6), and it is believed that they

contribute significantly to the *Sargassum* community (Phlips et al. 1986). Nitrogen-fixers are also important nitrogen contributors to coral reefs (Larkum et al. 1988).

Chemical and physical methods for determining the total dissolved inorganic phosphorus species (called "phosphate") have indicated that the total dissolved phosphorus in seawater consists of a rather heterogeneous group of inorganic-phosphate species and phosphorus-containing organic compounds. At the pH of seawater, phosphorus exists primarily in three free ionic species in equilibrium. At pH 8.2 and 20°C, HPO_4^{2-} accounts for 97% of the free ions, PO_4^{3-} for less than 1%, and $H_2PO_4^-$ for 2.5% (Turner et al. 1981). Orthophosphate ions (PO_4^{3-}) can form metallophosphate complexes (e.g., with Ca^{2+} and Mg^{2+}), or they can combine with organic compounds, and therefore free orthophosphate represents less than one-third of the total inorganic phosphate in seawater. The standard technique for determining inorganic phosphate with the molybdate reaction tends to overestimate the true concentration, because some forms of organic phosphorus are hydrolyzed by reagents in the phosphate analysis (Cembella et al. 1984). Several techniques have been used to distinguish among the forms of organic phosphorus in seawater, but there are problems with all the techniques, and the separation is not always clear (Strickland & Parsons 1972; Cembella et al. 1984).

Iron, the fourth most abundant element in the earth's crust, is one of the least soluble metals in oxygenated waters. For this reason, iron concentrations may limit the growth of some seaweeds, depending on their requirements. Recent reports (Martin & Fitzwater

1988) suggest that iron is limiting to phytoplankton growth in the central regions of the equatorial and northeastern Pacific Ocean, where the input of iron from atmospheric dust is low. The speciation of iron in seawater is complex, and that is a major impediment to our understanding of iron uptake by algae (Byrne & Kester 1976; Wells 1991). At the pH of seawater (ca. 8.2), ferric ion combines with hydroxyl ions to form ferric hydroxide, which is relatively insoluble ($K_{sp} \simeq 10^{-38}$ M). Therefore, much of the iron is maintained in solution only by the formation of complexes with natural chelators or ligands such as humic acid. Recent measurements in the western North Pacific indicate that dissolved iron concentrations are 2.2–3.8 µg L^{-1}, of which 80–90% is organically bound iron (Sugimura et al. 1978). The results from many past experiments are now known to be inconclusive because of the following problems:

1. Iron contamination during experiments is likely unless the investigators use special clean-room techniques.

2. Adsorption of iron onto the surfaces of the experimental containers (especially glass) can lead to the erroneous conclusion that the seawater being tested is iron-deficient (Lewin & Chen 1971; Paasche 1977).

3. Addition of a chelator, such as ethylenediaminetetraacetic acid (EDTA) may stimulate algal growth, again suggesting that iron limitation has been overcome, because the chelator has made the iron more readily available. However, the chelator may have served only to eliminate or reduce the adsorption of iron onto the container walls or to detoxify other metals such as copper and zinc by forming a metal complex (Huntsman & Sunda 1980; Sunda 1991).

4. An increase in particulate iron cannot be ascribed to uptake by the algae, because it may be due to abiotic processes in the medium or passive adsorption onto the cell surface.

5. Freshly added iron may form a precipitate in the medium and consequently co-precipitate and adsorb other metals, reducing their availability to algae (Huntsman & Sunda 1980; Sunda 1991).

6. The ratio between chelator and trace metal is important. It is now clear that the activity of the free metal ion generally determines availability or toxicity, particularly for Cu, Zn, and Mn. For iron, a few studies have shown that phytoplankton may be able to take up the free ion (probably Fe^{2+}), as well as chelated forms, providing the iron is not bound too tightly by the chelator. When a chelator forms a complex with a free metal ion, its activity is reduced. The chelator acts as a metal buffer, releasing more ions into the medium as other ions are taken up by the algae, but under optimal conditions never allowing toxic levels to be reached. However, if too much chelator is present in relation to the trace-metal concentration, the activity of the metal may be reduced below that required for optimum growth. The strength with which the chelator binds the metal ion is also important in determining trace-metal availability (Huntsman & Sunda 1980; Sunda 1991). Turner et al. (1981) have reviewed some of the general rules that control the speciation of trace elements in seawater.

Natural copper concentrations range from 1 to 4 µM in coastal waters (Schmidt 1978; Boyle 1979; Lewis & Cave 1982), whereas open-ocean values are about three orders of magnitude lower (0.5–4 nM) (Bruland 1983). However, it is the concentration of the free ion that is important, not the total copper concentration. There is no easy method for determining the concentration of the free ion in seawater (chloride ions interfere with the cupric-ion electrode), and therefore bioassays are used to determine "biologically available" copper, which is equated with the free-ion concentration (Sunda & Guillard 1976; Whitfield & Lewis 1976; Lewis & Cave 1982). Recently, differential pulse anodic stripping voltammetry has been used to measure "labile" Cu and to examine the degree to which copper is associated with organic chelators (Coale & Bruland 1988). It was found that more than 99.7% of the total dissolved copper in the surface waters of the central northeastern Pacific was associated with strong organic complexes, suggesting that copper bioavailability, rather than toxicity, may be of concern in these waters.

Early reports indicated that addition of the synthetic chelator EDTA could alleviate poor growth of phytoplankton in newly upwelled seawater (Barber & Ryther 1969). Subsequently, additions of iron or manganese were also shown to stimulate growth (Barber et al. 1971). Experiments under tightly controlled conditions have verified that the toxicity of copper to phytoplankton is determined by cupric-ion activity and that chelated copper is not directly toxic (Sunda & Guillard 1976). The cupric-ion concentrations in some waters that are low in natural chelators (e.g., recently upwelled water) have been shown to be toxic to some phytoplankton species (Cross & Sunda 1977; Brand et al. 1986; Sunda 1991). When these recent observations are applied to the problem of determining the causes of poor phytoplankton growth in recently upwelled water, we see that earlier hypotheses (Barber & Ryther 1969) must be revised. Unlike the concentrations of other trace metals such as Cd, Cu, Ni, and Zn, which increase with depth in the oceans, manganese concentrations usually decrease (Bender et al. 1977). Therefore, newly upwelled water may have relatively high cupric-ion and low manganous-ion concentrations, producing a Mn^{2+}/Cu^{2+} ratio that is unfavorable for phytoplankton growth. As the water "ages," the cupric-ion concentration decreases, and the manganous-ion concentration increases, so that the toxic effect of the water is overcome without the intervention of chelators (Sunda et al. 1981). In laboratory experiments it was found that copper detoxification occurred upon the addition of metal chelators or other metal micronutrients (Mn and Fe), suggesting that copper toxicity was due to copper's competitive inhibition of other metal-requiring enzyme systems (Brand et al. 1986).

Measurements of the copper-complexing ability of exudates of several macroalgae suggest that the exudates may aid in complexing and possibly detoxifying copper when it is present in high concentrations (Sueur et al. 1982). Similarly, Ragan et al. (1979) found some detoxification of zinc by brown-algal polyphenols.

5.3 Pathways and barriers to ion entry
5.3.1 Adsorption

Ions enter a cell by moving across the boundary layer of water surrounding the cell (sec. 7.1.3). The route of ion entry into cells is to the cell surface, and then they pass through the cell wall and plasmalemma into the cytoplasm. The thickness of the boundary layer may affect the uptake rate of an ion, because if turbulence around the thallus is low, the boundary layer is thick, and uptake may be limited by the rate of diffusion across this layer. The cell wall, unlike the plasmalemma, does not generally present a barrier to ion entry. When a macroalga is placed in nutrient medium, there may be an initial rapid uptake that does not require energy (i.e., is independent of metabolism) and usually lasts less than a minute. This observation is generally attributed to diffusion into the apparent free space that is exterior to the plasmalemma. Ions can readily be removed from the apparent free space by washing the alga in nutrient-depleted medium or keeping it in this medium for some time.

Some ions, especially cations, may not reach the plasmalemma because they become adsorbed to certain components of the cell wall (Kloareg et al. 1987). Polysaccharides and proteins have sulfate, carboxyl, and phosphate groups from which protons can dissociate, leaving a net negative charge on these compounds in the cell wall. In effect, these macromolecules act as cation exchangers, and consequently large numbers of cations can be adsorbed from the environment. Haug and Smidsrød (1967) suggested that the concentrations of Ca, Sr, and Mg in brown algae are largely the result of ion exchange between seawater and the acid polysaccharide alginate in the cell walls; the amounts of these ions in the cytoplasm and vacuole compose a relatively small portion of the total in the whole thallus. The relative affinities of alginic acid for different metals are as follows (Haug 1961):

$$Pb > Cu > Cd > Ba > Sr > Ca > Co$$
$$> Ni, Zn, Mn > Mg$$

This cation adsorption by seaweeds and all other multicellular plants is referred to as a Donnan exchange system, and the volume in which these cations are confined is the Donnan free space. When radioactively labeled

roots of higher plants are transferred to well-stirred distilled water, there is a brief but significant efflux of the label. When this is complete, transfer of roots to a well-stirred nonlabeled solution of the nutrient results in further efflux of the tracer. These observations indicate that the apparent free space has two components. The first release (efflux) in distilled water represents ions that are free to diffuse out of the extracellullar volume of the root. This volume is called the water free space. The second release into nonlabeled nutrient medium represents the efflux from the Donnan free space. The water free space plus the Donnan free space compose what is called the apparent free space, and it is the volume of the cell wall into which ions can diffuse or be adsorbed (Lüttge & Higinbotham 1979; Glass 1989). The apparent free space can be measured quantitatively in physiological experiments, and this term is generally used in a physiological context, whereas the term "apoplast" is used in discussing translocation and is based partially on morphology. In higher plants, the apparent free space extends to the endodermis of the roots; in seaweeds it includes the cell walls and all intercellular spaces. Only a few measurements of apparent free space have been made in seaweeds, but these measurements indicate that the space can be significant, up to 20–30% of the cell volume for the red alga *Porphyra perforata* (Eppley & Blinks 1957; Gessner & Hammer 1968; Hammer 1969).

5.3.2 Passive transport

Algal and higher-plant plasma membranes consist of lipid bilayers interspersed with two types of proteins (Fig. 5.3a). Intrinsic proteins completely traverse the membrane and most likely control transport across the bilayer. Extrinsic proteins only partially penetrate into the bilayer. The central, hydrophobic region of the bilayer, which is composed of the hydrocarbon tails of membrane phospholipids, is a barrier for charged ions.

Nonelectrolytes (uncharged particles) diffuse through membranes at a rate proportional to their solubilities in lipid and inversely proportional to their molecular sizes. Dissolved gases move more freely than most solutes. As a result, many important gases (e.g., CO_2, NH_3, O_2, and N_2) cross lipid bilayers by dissolving in the lipid portion of the membrane, diffusing to the other lipid–water interface, and then dissolving in the aqueous phase on the other side of the membrane. Uncharged molecules such as water and urea are also highly mobile. However, molecules such as NH_3 may be trapped inside cells when they are converted to ions (R. H. Reed 1990). The pK_a for NH_3/NH_4^+ is 9.4, and in seawater with a pH of 8.2, only 5–10% of the total ammonium is present as NH_3 (hence, it is preferable to use the term "ammonium" rather than "ammonia" when referring to concentrations in seawater). At a higher pH, which may be found in dense cultures or restricted tidal pools, the percentage of NH_3 can increase to 50% (at pH 9.4) or more, allowing rapid diffusion

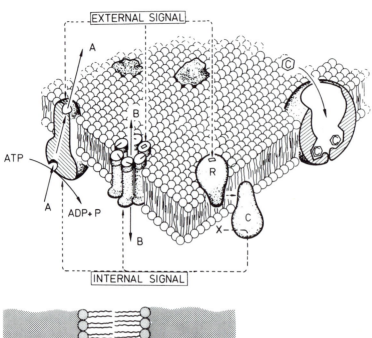

(a)

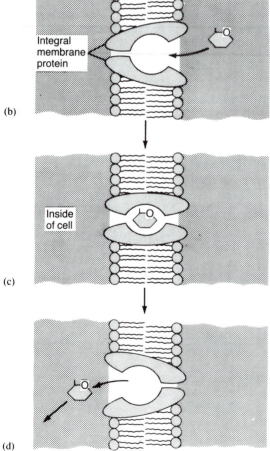

(b)

(c)

(d)

Figure 5.3.(a) Mechanisms of ion transport and signal transduction in biological membranes. Three-dimensional diagram of the fluid-mosaic model, showing the integral transport protein embedded in the lipid bilayer. From left to right: A, ion pump; B, ion channel; R, C, coupling proteins for signal perception and transduction; C, carrier. (b–d) Possible mechanism for facilitated transport. An integral membrane protein provides a stereospecific hydrophilic channel for a polar solute through an otherwise hydrophobic membrane. (b) Solute binds to the outer surface of the transport protein. (c) A change in protein conformation moves solute into the hydrophilic channel. (d) A further change in conformation releases solute on the other side of the membrane. Because the carrier functions equally well in both directions, the process is readily reversible. The direction of solute movement through the membrane will depend on the relative concentrations of solute on the two sides of the membrane. (Part a reprinted from Hedrich & Schroeder 1989, with permission from *Trends in Biochemical Sciences;* b–d after Becker & Deamer 1991, *The World of the Cell.*)

into the cell. Because the pH of the cytoplasm is only 7–7.5, most of the NH_3 that enters is protonated to NH_4^+ and cannot diffuse back across the membrane. This mechanism can account for approximately 10% of net uptake at high pH (Walker et al. 1979). Passive accumulation of ammonia by diffusion and acid trapping in the cytoplasm/vacuole could theoretically result in an accumulation increased by a factor of 10^3 or more in marine diatoms (Wheeler & Hellebust 1981). Presumably the same mechanism operates in macroalgae.

Uncharged molecules diffuse down a free-energy or chemical-potential gradient; hence the term "downhill transport." The rate of diffusion varies with the chemical-potential gradient or the difference in activities (approximately equivalent to the concentrations) across the plasmalemma [Table 5.3, equation (1)]. The permeability coefficient of the membrane (P) is proportional to the diffusion coefficient of the molecule and inversely proportional to the thickness of the membrane [Table 5.3, equation (2)].

Ions usually have a much lower permeability than uncharged molecules such as CO_2 or urea. The charge on the ion makes it difficult for the ion to penetrate a membrane that is electrically polarized and contains charged groups that either repel or attract (immobilize) the ions. In addition, ions usually are strongly hydrophilic, and their particle sizes frequently are increased by a substantial layer of the water of hydration. Both of these properties tend to decrease the rate of diffusion. Because ions are charged, the driving force for ion movement arises from electrical-potential differences (E) and/or chemical-potential differences ($\Delta\mu$), and the combined effect is referred to as an electrochemical-potential gradient ($\Delta\mu$). Plant membranes typically maintain electrical-potential differences (inside negative) of 100–200 mV across the plasma membrane. Such electrical gradients favor the entry of cations, but oppose anion penetration. The electrochemical driving force across a membrane can be calculated for any ion using the Nernst equation [Table 5.3, equation (3)]. The net transmembrane electrical-potential difference at equilibrium is the sum of the $\Delta\bar{\mu}$ values for all the ions involved, modified by the permeability coefficient (P) for each ion. Because the concentration gradient and the electrical gradient can independently influence the movement of an ion across the membrane, it is possible that the electrical-potential difference could be large enough to produce diffusion of an ion even against a gradient in concentration. There are two generalities that apply: First, the absorption of an anion is always active if internal intracellular concentrations exceed the external concentration, because membrane potential differences are negative inside; this is true because under these conditions absorption is against both electrical and chemical differences. Second, the type of absorption for cations can be evaluated only after careful evaluation of the concentration and electrical terms that apply.

Table 5.3. *Nutrient-transport equations*

(1) $J = P\Delta C$

(2) $J = \dfrac{D}{X}\Delta C$

(3) $\Delta\bar{\mu}_{i0} = (RT \ln a_o + ZF\Psi_o) - (RT \ln a_i + ZF\Psi_i)$, and at equilibrium,

$E = \dfrac{RT}{ZF} \ln(a_o/a_i)$

(4) $V = V_{max} \dfrac{S}{K_s + S}$

(5) $V = \dfrac{V_{max}}{(1 + K_s/S)(1 + i/K_i)}$

Symbols: a_o/a_i, chemical potential difference between outside and inside of membrane; ΔC, concentration gradient across membrane; D, diffusion coefficient of molecule; E, Nernst electrical-potential difference; F, Faraday constant; i, inhibitor concentration; J, flux of molecules; K_i, half-saturation constant for inhibitor; K_s, half-saturation constant for substrate; P, membrane permeability coefficient; R, gas constant; S, substrate (nutrient-ion) concentration; T, temperature (Kelvin); V, initial uptake rate; V_{max}, maximum uptake rate at saturating substrate concentration; X, membrane thickness, Z, charge per ion (valence); $\Delta\bar{\mu}_{io}$ electrochemical-potential gradient between inside (i) and outside (o) of the membrane; Ψ, electrical potential.

Passive diffusion occurs without the expenditure of cellular metabolic energy; however, the electrical gradient that may drive passive cation movement is the result of cellular metabolism. In addition, no carriers or binding sites are involved in diffusion, and therefore it is nonsaturable. Unlike the situation for active transport, an increase in temperature has little effect on diffusion. Hence, although such fluxes occur according to the strict definition of passive transport as movement down the free-energy gradient, they are nevertheless indirectly dependent upon the expenditure of metabolic energy. Thus, definitive conclusions concerning active transport can be made only after calculation of $\Delta\bar{\mu}$ for a particular ion, even though sensitivity to metabolic inhibitors may indicate active transport.

5.3.3 Facilitated diffusion

Facilitated diffusion resembles passive diffusion in that transport occurs down an electrochemical gradient, but frequently the rate of transport by this process is faster. In facilitated diffusion, carriers are thought to bind the ion at the outer membrane surface and to assist in providing a pathway across the membrane to the inner surface (Fig. 5.3b–d). Facilitated diffusion exhibits properties similar to those of active transport in that (1) it can be saturated, and transport data fit a Michaelis-

Menten-like equation (sec. 5.3.4), (2) only specific ions are transported, and (3) it is susceptible to competitive and noncompetitive inhibition. However, in contrast to active-transport mechanisms, any energy expenditure required for transport must be indirect. More recently, ion channels (Fig. 5.3a), which are macromolecular pores that traverse the cell membrane, have been invoked to explain the high transport rates that occur down an electrochemical-potential gradient (Hedrich & Schroeder 1989).

5.3.4 Active transport

Active transport is the transfer of ions or molecules across a membrane against an electrochemical-potential gradient. For this reason, active transport has also been termed ''uphill'' transport. Because external concentrations of inorganic nutrients (e.g., NO_3^- and PO_4^{3-}) typically are in the micromolar range, and intracellular concentrations of these same nutrients are in the millimolar range, passive diffusion along an electrochemical gradient is unlikely to be important, because of the large concentration difference between the inside and outside of the cell (R. H. Reed 1990). To conclusively demonstrate active transport, as opposed to free or facilitated diffusion, the following criterion must be satisfied: Active uptake is energy-dependent. A change in the uptake rate should occur after the addition of a metabolic inhibitor (e.g., dinitrophenol) or a change in temperature, because both of these factors influence energy production (but see sec. 5.3.2, which discusses indirect energy-dependent passive transport). Other properties of active transport, such as unidirectionality (in or out of the cell), selectivity regarding the ions transported, and saturation of the carrier system (exhibiting Michaelis-Menten kinetics), are not definitive criteria for active transport because they are also characteristic of facilitated diffusion. Active transport typically is much slower than channel-mediated (i.e., diffusive through channels) transport; channels may allow 10^6 ions per second, compared with 10^3–10^5 ions per second for active transport.

On the basis of the few electrochemical measurements that have been made on algae, primarily in the giant cells of the freshwater alga *Chara*, and also by analogy with vascular plants, transport in seaweeds is thought to be primarily active. Obtaining electrochemical measurements with the use of microelectrodes is difficult for most macroalgae because of the relatively small sizes of individual cells.

The isolation of membrane ATPases and the correlations between ion fluxes and ATPase activities suggest that ATP is the energy source for ion transport (Fig. 5.3a). The primary transport reaction in which ATP is consumed is the transport of H^+ across cell membranes by H^+-pumping ATPases. The resulting H^+ electrochemical-potential gradient is assumed to be consumed in driving secondary ion transport via

coupled transport. Coupled transport may arise from the transfer of different ions at separate sites on a given carrier in opposite directions (antiport or countertransport) or the same direction (symport or cotransport). In microalgae, proton-linked (H^+-linked) cotransport of sugars and thiourea has been demonstrated (Syrett 1981). In many animal cells, transport generally takes place by cotransport of sodium ions (Na^+) rather than hydrogen ions (H^+). Some preliminary evidence suggests that this is also true for transport by marine microalgae (Syrett 1981). This is not surprising, because seawater is high in Na^+ and low in H^+.

5.4 Nutrient-uptake kinetics

The kinetics of nutrient uptake will depend on which uptake mechanism is being used. If transport occurs solely by passive diffusion, then the transport rate will be directly proportional to the electrochemical-potential gradient (external concentration) (Fig. 5.4a). In contrast, facilitated diffusion and active transport will exhibit a saturation of the membrane carriers as the external concentration of the ion increases. The relationship between the uptake rate of the ion and its external concentration is generally described by a rectangular hyperbola, similar to the Michaelis-Menten equation for enzyme kinetics (Fig. 5.4b). The equation is given in Table 5.3: equation (4). K_s (equivalent to K_m) is called the half-saturation constant, and it is the substrate concentration at which the uptake rate is half its maximum. The lower the value of K_s, the higher is the affinity of the carrier site for the particular ion. The transport capabilities of a particular macrophyte are generally described by the parameters V_{max} and K_s. Healey (1980) and Harrison et al. (1989) have suggested that the slope of the initial part of the hyperbola (the linear portion) or the ratio $V_{max} : K_s$ may be a more useful parameter for comparing the competitive abilities of various species for a limiting nutrient; this is similar in concept to the use of the slope, a, in the P–I curve (Fig. 4.12). One of the problems in using K_s is that its value is not independent of V_{max} (i.e., when V_{max} decreases, the value of K_s will also decrease, even though the initial slope of the hyperbola remains the same) (Fig. 5.4b). The kinetic parameters may be estimated graphically from the rectangular hyperbola, but generally the Michaelis-Menten equation is rearranged to yield a straight line, from which K_s and V_{max} can be calculated more accurately by linear regression (Fig. 5.4c) (Atkins & Nimmo 1980). The advantages and disadvantages of the three possible linear plots have been examined by Dowd and Riggs (1965); the plot of S/V versus S or V/S versus V is clearly superior to the plot of $1/S$ versus $1/V$. Computer programs are now available for a direct fit of the data to a rectangular hyperbola. These obviate the necessity of making assumptions about the data and also circumvent the problem inherent in all transformations (Wilkinson 1961).

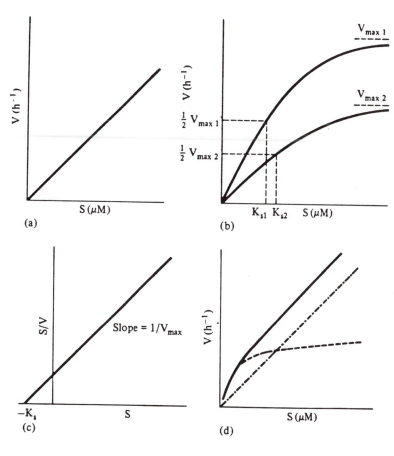

Figure 5.4. Hypothetical plots of nutrient-uptake rates (V) and concentrations of the limiting nutrients (S) for (a) passive diffusion only, where V is directly proportional to S, and (b) facilitated diffusion or active transport, in which V_{max} in example 2 is half of the V_{max} in example 1, resulting in a concomitant decrease in K_s. (c) Linearized plot of the data in (a) to illustrate how the kinetic parameters V_{max} and K_s are determined graphically. (d) Passive diffusion plus active transport (solid line), and active transport (dash line) with the passive-diffusion component (dot-and-dash line) subtracted.

Active uptake may not follow the simple saturation kinetics described here (Cornish-Bowden 1979). Studies with higher plants have revealed that the patterns of uptake may be biphasic or multiphasic in nature (Epstein 1972; Nissen 1974). In the case of biphasic kinetics, a plot of V versus S reveals two rectangular hyperbolas, which frequently are referred to as the high- and low-affinity systems. At low substrate concentrations, the high-affinity system operates, exhibiting a high degree of ion specificity and a low value for K_s, whereas at high concentrations the low-affinity system is operative, and it exhibits much less ion selectivity and a very high value for K_s. Others have denied the evidence for multiphasic uptake kinetics, claiming that it is not statistically significant (Borstlap 1981). To date there is no strong evidence for biphasic uptake kinetics in macroalgae, but Serra et al. (1978) clearly demonstrated biphasic kinetics of nitrate uptake in the marine diatom *Skeletonema costatum*. Deviation from saturation kinetics may also occur if active uptake and diffusion occur simultaneously. Diffusion is not likely to be important at low substrate concentrations, but it may be significant at concentrations well above environmental concentrations. In this latter case, the total uptake rate is composed of active uptake plus diffusion; the resultant uptake pattern is shown in Figure 5.4d, and it has been shown to occur for ammonium uptake in *Macrocystis*, *Gracilaria tikvahiae*, and *Agardhiella subulata* at ammonium concentrations greater than 25 μM (Haines & Wheeler 1978; D'Elia & DeBoer 1978; Friedlander & Dawes 1985). Under conditions where the nutrient is limiting, an uptake rate may not follow a Michaelis-Menten hyperbola, because uptake may be controlled by the internal nutrient level rather than by external nutrient concentrations (Fujita et al. 1989).

5.4.1 Measurement of nutrient-uptake rates

There are two main techniques for measuring nutrient-uptake rates: (1) radioactive- or stable-isotope uptake and (2) disappearance of nutrient from the medium, measured colorimetrically (Harrison & Druehl 1982). One of the problems of using either radioactive or stable isotopes is that different parts of the thallus will accumulate the isotope at different rates, and samples from different areas of the thallus should be taken and averaged in order to obtain a whole-thallus uptake rate (the most useful measurement for ecological purposes). When comparing the stable-isotope method with the nutrient-disappearance method, O'Brien and Wheeler (1987) found good agreement. That contrasts with the findings of Williams and Fisher (1985), who reported that more ammonium disappeared from the

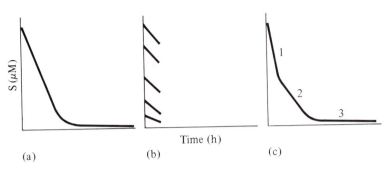

Figure 5.5. Hypothetical time series showing nutrient disappearance from the medium using the following methods: (a) the perturbation method, where saturated uptake is linear with time; (b) the multiple-container, constant-incubation-time method; (c) the perturbation method, where saturated uptake is nonlinear with time. Phase 1 is the enhanced or rapid uptake, possibly due to filling of intracellular pool(s); phase 2 may represent an assimilation rate (also referred to as V_i); phase 3 is the final depletion of the limiting nutrient from the medium. (From Harrison & Druehl 1982, with permission of Walter de Gruyter & Co.)

medium than could be accounted for by the incorporation of ^{15}N in *Caulerpa;* isotope dilution of the ammonium pool was ruled out. They suggested that either (1) a secondary ammonium sink, such as wall sorption or bacterial uptake, reduced ammonium concentrations or (2) ^{15}N was lost as labeled dissolved organic nitrogen or was volatilized during ^{15}N sample preparation. In all three of these techniques, very short incubation times of probably less than 10–15 min would yield an estimate of gross uptake rate (influx), whereas long incubation times (more than 6 h) would give rates that would approximate net uptake, taking account of efflux of the nutrient from the thallus back into the medium.

Nutrient-uptake rates for seaweeds are measured in the laboratory by incubating epiphyte-free tissue discs or, preferably, whole plants in filtered natural seawater to which nutrients (except the one under study), trace metals, and vitamins have been added at saturating levels. In order to eliminate the effect of rapid diffusion into the apparent free space, plants frequently are pre-incubated in saturating nutrients for a few minutes and then placed in appropriate experimental concentrations of the nutrient; see Reed and Collins (1980), Harrison and Druehl (1982), and Harlin and Wheeler (1985) for further discussion. There are two basic approaches to following the disappearance of the nutrient from the culture medium (Harrison et al. 1989). The first is to spike the culture with the nutrient of interest and follow the nutrient disappearance for several hours until nutrient exhaustion occurs; this method gives the time course of the uptake rate (Fig. 5.5a). However, if the uptake rate varies with time (Fig. 5.5c), as is frequently the case when ammonium (Probyn & Chapman 1982; Rosenberg et al. 1984; Thomas & Harrison 1987) or nitrate (Thomas & Harrison 1985; Thomas et al. 1987a) is limiting, then a second method involving a short incubation period must also be used to correctly estimate the maximal uptake rate (Harrison et al. 1989). The second

method involves the use of many containers with different concentrations of the nutrient, and each with a different specimen of the species being studied. The incubation period in this case usually is short (10–60 min) but constant for all concentrations (Fig. 5.5b). If the uptake rate is constant over time (Fig. 5.5a), then the choice of method is not important. Because uptake rates frequently vary with time, Harrison et al. (1989) suggested putting a time superscript on V_{max} values (e.g., $V_{max}^{0-10\ min}$ means maximal rates measured over the first 10 min of the incubation).

The reason that the uptake rate varies is that under nitrogen limitation, intracellular nitrogen pools may be low, and the initial enhancement in uptake rate over the first 10–60 min may represent a pool-filling phase (Fujita et al. 1988). As the pools fill, the decrease in uptake rate may be due to feedback inhibition (Harrison et al. 1989). Therefore, this latter uptake rate does not represent the true transmembrane transport that is free from feedback inhibition; this effect is discussed in more detail in section 5.5.1. The method where the nutrient is followed until it is depleted from the medium is not recommended for estimation of V_{max} and K_s because the nutritional past history of the thallus is changing with time. Nevertheless, this method is useful in determining the assimilation rate (Harrison et al. 1989), the rate at which intracellular nitrate or ammonium is incorporated into amino acids and proteins (Fig. 5.5c); this rate has been termed V_i, the internally controlled uptake rate (Conway et al. 1976). In summary, a time course of nutrient-uptake rates should always be run before deciding which method to use in measuring nutrient-uptake rates.

Whereas the limiting nutrient (especially ammonium and phosphate) may enhance uptake rates upon readdition to NH_4^+ or P-limited algae, C fixation is often shut down. For microalgae this depression in C fixation may last for minutes (Turpin 1983), and in macroalgae it may take 24 h for the photosynthetic rate to recover (Williams & Herbert 1989).

Although uptake determined from thalli in the incubation medium is generally attributed solely to the thalli, in some cases this may not be true, because most thalli have epiphytes such as microscopic algae and bacteria on their surfaces. These epiphytes are not easily removed. Antibiotics such as streptomycin and penicillin have been employed to inhibit bacterial uptake (Harlin & Craigie 1978), but it is unlikely that there is an antibiotic concentration that can completely inhibit the bacterial uptake of nutrients without also affecting the alga.

Nutrient-uptake rates may be expressed in four commonly used sets of units. Uptake may be normalized to surface area (μmol cm^{-2} h^{-1}), wet weight (μmol g$_{ww}^{-1}$ h^{-1}), dry weight (μmol g$_{dw}^{-1}$ h^{-1}), or, finally, nutrient content in the plant, which simplifies to a specific uptake rate (h^{-1}). If conversion factors are not given, nutrient-uptake data in the literature that are expressed in different units cannot be accurately compared.

5.4.2 Factors affecting nutrient-uptake rates

Physical factors. Light affects nutrient uptake indirectly through photosynthesis, which (1) provides energy (ATP) for active transport, (2) produces carbon skeletons that are necessary for incorporation of nutrient ions into larger molecules (e.g., amino acids and proteins), (3) provides energy for the production of charged ions that establish Donnan potentials, and (4) increases the growth rate and thus increases the need for nutrients. There is strong evidence for the effects of irradiance on uptake rates for nitrate (Fig. 5.6a) and other ions (Floc'h 1982). The nitrate-uptake data roughly fit a rectangular hyperbola for *Macrocystis* (Wheeler 1982) and also for many microalgal species (Syrett 1981). Note, however, that the curve intersects the y axis, indicating that nitrate uptake in the dark is substantial. In contrast, ammonium uptake in *Macrocystis* has been found to be independent of irradiance (Wheeler 1982) (Fig. 5.6a). Photoperiod affects nitrate uptake, possibly because of the diel periodicity in activity and synthesis shown by the nitrate reductase enzyme (Syrett 1981, 1988). Diel periodicity of nitrate uptake is more pronounced for nitrate-sufficient algae than for nitrogen-limited algae (i.e., dark uptake increases to equal a higher percentage of light uptake when algae are N-limited).

Temperature effects on active uptake and general cell metabolism approximate a Q_{10} value of 2 (i.e., a 10°C increase in temperature leads to a doubling of the rate). For a purely physical process, such as diffusion, temperature has less effect, and the Q_{10} value is 1.0–1.2. Several studies have indicated that the effect of temperature on ion uptake is ion-specific and is depen-

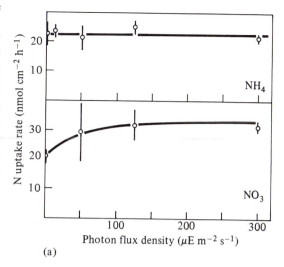

(a)

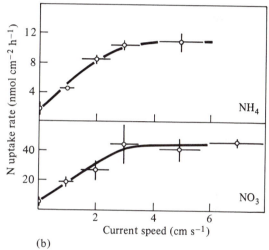

(b)

Figure 5.6. Nitrate- and ammonium-uptake rates for *Macrocystis pyrifera* mature blade discs as functions of (a) irradiance and (b) current speed. (From Wheeler 1982, with permission of Walter de Gruyter & Co.)

dent on the algal species. For example, a marked decrease in nitrate-uptake rate was observed for *Laminaria longicruris* (Harlin & Craigie 1978), but not for *Fucus spiralis* (Topinka 1978), as the temperature was lowered. This apparent discrepancy may be explained by the fact that these two species have different temperature optima for nutrient uptake. Because temperatures can fluctuate daily, it would be interesting to study how rapidly uptake rates can respond to changes in temperature.

Water motion is another factor that is important in the movement of ions to the surface of the thallus (Wheeler & Neushul 1981) (see Chapter 7). In areas of low turbulence, or in unstirred laboratory cultures,

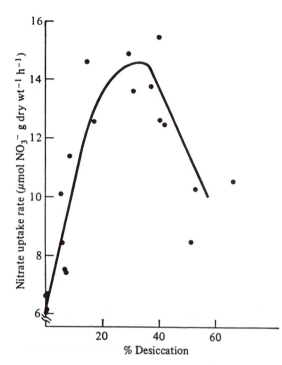

Figure 5.7. Nitrate-uptake rate as a function of the percentage of desiccation for *Fucus distichus*. Plants were desiccated to different degrees and then placed in medium containing 30 μM NO_3^-. The uptake was determined over a 30-min interval. (From Thomas & Turpin 1980, with permission of Walter de Gruyter & Co.)

transport across the boundary layer is limited by the rate of diffusion, not necessarily by the concentration of the nutrient in the medium (Pasciak & Gavis 1974; Wheeler & Neushul 1981). Such diffusion-limited transport rates have been demonstrated for single-cell phytoplankters (Pasciak & Gavis 1974), but the effect may be more pronounced for multicellular algae in quiescent waters, especially if the thallus is thick rather than filamentous (hence a lower surface-area : volume ratio). Wheeler (1980) and Gerard (1982a) demonstrated that the giant kelp *Macrocystis* encountered transport limitation for carbon and nitrogen when the current over the fronds was less than 3–6 cm s^{-1} (Fig. 5.6b).

Desiccation is another factor to be considered. Exposure to air during a low tide frequently results in loss of water from the thallus, depending on the season (sec. 6.4). Recent studies have shown that mild desiccation (10–30% water loss) enhances short-term (10–30 min) nutrient-uptake rates, as compared with hydrated plants, for several intertidal seaweeds when they are submerged in nutrient-saturated seawater (Fig. 5.7) (Thomas & Turpin 1980; Thomas et al. 1987b). This enhanced uptake response occurred when growth was limited by the nutrient in question and when the thallus had been exposed to repeated periodic desiccation for sev-

eral weeks. The relative degree of enhancement of the nitrogen-uptake rate, the percentage desiccation that produced maximal uptake rates, and the tolerance to higher degrees of desiccation were positively related to tidal height for five intertidal macroalgae (Thomas et al. 1987b). Low-intertidal species such as *Gracilaria pacifica* showed no enhancement of nitrogen uptake following desiccation. In contrast, two high-intertidal species, *Pelvetiopsis limitata* and *Fucus distichus*, showed a twofold enhancement of nitrate and ammonium uptake following desiccation of more than 30% and continued uptake even following severe desiccation (50–60% of original fresh weight); the period of enhanced nitrate uptake was much longer (20–60 min) than that for enhanced ammonium uptake (10–30 min). Using *Gracilaria pacifica*, Thomas et al. (1987a) showed that when plants from the low intertidal (1.0 m) were transplanted to high-intertidal sites (1.8 m), they developed enhanced nitrogen-uptake rates following desiccation, whereas high-intertidal plants transplanted to the low intertidal did not maintain their enhanced nitrogen-uptake abilities. That study also demonstrated both intraspecific and interspecific adaptations that were dependent on intertidal height.

In contrast to those studies demonstrating enhanced nitrogen uptake following desiccation for high-intertidal algae, Hurd and Dring (1991) found that phosphate uptake was not enhanced by desiccation, even in high-intertidal fucoid algae such as *Fucus spiralis* and *Pelvetia canaliculata*. Because the same species were used in both the nitrogen and phosphorus studies, these contrasting results may be explained by the fact that the nitrogen-uptake studies of Thomas et al. (1987b) were conducted at near-saturating nitrogen concentrations, and they measured short-term uptake rates (10–30 min), whereas phosphate uptake (Hurd & Dring 1991) was measured at subsaturating phosphate concentrations and over several hours. Hurd and Dring (1991) did find that the degree of tolerance to desiccation (measured by how quickly species recovered their maximal phosphate-uptake rates after losing 50% of their water through desiccation) increased with increasing shore height of the fucoid algae.

Chemical factors. Chemical factors such as the concentration of the nutrient being taken up and the ionic or molecular form of the element will affect uptake rates. For example, nitrogen in the form of ammonium often is taken up more rapidly than nitrate, urea, or amino acids in many seaweeds (DeBoer 1981). Uptake rates can also be influenced by the concentrations of other ions in the medium. Ammonium may inhibit nitrate uptake by as much as 50% in many seaweeds (DeBoer 1981) and microalgae (Dortch 1990). In contrast, *Gelidium, Macrocystis,* and *Laminaria* take up nitrate and ammonium at equal rates when they are supplied simultaneously (Bird 1976; Haines & Wheeler

1978; Harrison et al. 1986). Phosphate (>1 μM) has been shown to inhibit nitrate uptake in a marine diatom (Terry 1982). On the other hand, high concentrations of nitrate (>100-μM) were thought to inhibit PO_4^{3-} uptake in *Ulva* and to explain why polyphosphate concentrations in this seaweed decreased during growth on high levels of nitrate (Lundberg et al. 1989). Ca^{2+} and Na^+, as well as Mg^{2+} and K^+, are mutually antagonistic. Exogenous inhibitors (or pollutants) may affect membrane carriers by altering their activity or rate of synthesis. Intracellular ion concentrations in the cytoplasm and vacuoles will also influence uptake rates. Wheeler and Srivastava (1984) found that the nitrate-uptake rate was inversely proportional to the intracellular nitrate concentration in *Macrocystis integrifolia*.

Biological factors. Biological factors that influence uptake rates include the surface-area : volume (SA : V) ratio, hair formation, the type of tissue, the age of the plant, its nutritional history, and interplant variability (e.g., Gerard 1982b). Uptake determinations on whole thalli are preferable for ecological measurements, but if a thallus is too large, then a portion must be used. Cutting the thallus to produce tissue segments may alter the uptake rate, either through a wounding response and increased respiration (Hatcher 1977) or by elimination of the translocation system or active-sink region (Penot & Penot 1979). A comparison of excised sections of *Macrocystis* tissue with whole blades showed a marked decrease in uptake rate by the cut sections (Wheeler 1979).

Many seaweeds are perennials, and therefore their natural populations often consist of different age classes. The nitrate- and ammonium-uptake rates for three age classes of *Laminaria groenlandica* decreased with increased age; the uptake rate (per gram of dry weight) for first-year plants was three times greater than that for second- and third-year plants (Harrison et al. 1986). Distinct differences in nutrient-uptake abilities also occur between early life-history stages and mature thalli of the same species. Ammonium- and nitrate-uptake rates for *Fucus distichus* germlings were 8 and 30 times higher, respectively, than those for the mature thalli (Thomas et al. 1985). The germlings showed saturable uptake kinetics, but the mature thalli did not, indicating that the uptake kinetics of germlings are more like the kinetics of phytoplankton than are the kinetics of the mature plants. The presence of ammonium inhibited nitrate uptake in mature plants, but not in germlings. These characteristics of nutrient uptake indicate that the germlings are better adapted for procurement of the limiting nutrient than are mature seaweeds. This sizable difference in uptake abilities probably is due to the large proportion of storage and support tissues in the adult plants, tissues that do not actively require nitrogen. It is also interesting that older fronds and stipes retain some ability to take up ammo-

nium, but entirely lose their capacity to take up nitrate. Young tissue, which is metabolically active, appears to need both nitrate and ammonium to meet its greater nitrogen requirements.

Different parts of a plant may also take up nutrients at different rates. The stipe has demonstrated the lowest nitrogen-uptake rate in *Fucus spiralis*, in keeping with its low metabolic activity (Topinka 1978). Schmitz and Srivastava (1979) showed that mature regions of *Macrocystis* took up and translocated phosphate to meristematic sink regions. Davison and Stewart (1983, 1984) obtained similar results for nitrogen in *Laminaria digitata*. They demonstrated that 70% of the nitrogen demand by the intercalary meristem in this species was supplied by transport of nitrogen assimilated by the mature blade, probably in the form of amino acids. In *Macrocystis*, Jackson (1977) deduced from indirect evidence that mature blades deeper in the water column may serve an important role by taking up nitrogen from the relatively nutrient-rich deep water and translocating it (as amino acids) to the blades in the nutrient-poor surface water. Additional evidence from Gerard (1982c) supports the suggestion that at least some nitrogen can be moved from deeper, mature–senescent blades to shallower, growing blades.

Members of the Caulerpales inhabit the soft bottoms of oligotrophic tropical waters and have well-developed rhizoidal holdfasts that grow into the sediments. Williams and Fisher (1985) found that *Caulerpa cupressoides* not only is adapted to use ammonium in interstitial waters but also meets virtually all of its nitrogen requirements from this source. Significant translocation of ^{15}N label occurred from the rhizoid to the blade, indicating that ammonium taken up by the rhizoids in the sediments was available for light-driven organic-nitrogen production in the blades (Williams 1984).

The uptake rate of ammonium is a function of the nutritional past history of the plant. When *Gracilaria foliifera* and *Agardhiella subulata* were grown under conditions in which they were nitrogen-limited, the C/N ratio in the thalli was greater than 10 (by atoms), and the plants showed higher rates of ammonium uptake at a given ammonium concentration than did plants that were not nitrogen-limited (C/N ratio less than 10) (D'Elia & DeBoer 1978).

The SA : V ratio and the shape of the thallus are also important factors influencing nutrient-uptake rates. Rosenberg et al. (1984) found that for both nitrate and ammonium, the maximal uptake rate (V_{max}) and the initial slope of the uptake–substrate-concentration curve (*a*, which is an index of the affinity for the limiting nutrient) were positively related to the SA : V ratio for four intertidal seaweeds. Unfortunately, their choice of seaweeds, *Ulva* (a short-lived, opportunistic algae with a high SA : V ratio) and *Codium* (a long-lived, late-successional alga with a low SA : V ratio), probably overemphasized the importance of this ratio. In a more extensive study of 17 macroalgae, Wallentinus (1984) measured higher uptake rates for nitrate, ammonium, and phosphate in short-lived, opportunistic, filamentous, delicately branched or monostromatic forms (*Cladophora glomerata*, *Enteromorpha procera*, etc.) that had high SA : V ratios. The lowest uptake rates occurred among late-successional, long-lived, coarse species with low SA : V ratios (*Fucus vesiculosus*, *Phyllophora truncata*, etc.). An increase in the nutrient-uptake has also been found to parallel an increase in the SA : V ratio associated with increasing the number of hairs protruding from the thallus of *Ceramium rubrum* (DeBoer & Whoriskey 1983) (see section 5.8.2 for further discussion of hairs).

5.5 Uptake, assimilation, and metabolic roles of essential nutrients

There have been few studies on nutrient metabolism in macroalgae, as compared with the number of studies in microalgae. Consequently, for many aspects of nutrient assimilation and metabolic roles, we refer to the research on phytoplankton in order to avoid leaving large gaps in certain areas, acknowledging that the situations may be rather different in seaweeds. "Uptake" refers to transport across the plasmalemma, and "assimilation" refers to a sequence of reactions in which inorganic ions are incorporated into organic cellular components. The key metabolic roles of the elements are summarized in Table 5.1 and are discussed in the following sections.

5.5.1 Nitrogen

Nitrogen uptake and assimilation in macroalgae have been reviewed by Hanisak (1983). In the few seaweeds that have been studied, the uptake rate for ammonium generally has exceeded that for nitrate at ecologically relevant concentrations. At very high concentrations (> 30–50 μM), ammonium may be toxic to some seaweeds (Waite & Mitchell 1972). In some seaweeds (e.g., *Codium fragile*), ammonium shows saturation kinetics, suggesting active transport (Hanisak & Harlin 1978). For other species of macroalgae, the ammonium-uptake rate does not saturate as the ammonium concentration is increased, but instead increases linearly (Fig. 5.4d). This linear increase in uptake rate at high ammonium concentrations may represent a second transport mechanism, perhaps diffusion via ion channels. This suggestion is supported by the good hyperbolic fit (linear regression of transformed data in Fig. 5.4c) obtained when the linear component is subtracted from the total uptake rate (Fig. 5.4d). However, definitive experiments have not been conducted to confirm this suggestion. Linear ammonium-uptake rates over a range of ammonium concentrations up to 60 μM have also been reported for *Fucus distichus* (Thomas et al. 1985), for several kelp, such as *Macrocystis pyrifera*

(Haines & Wheeler 1978) and *Laminaria groenlandica* (Harrison et al. 1986), for the red algae *Gracilaria pacifica* (Thomas et al. 1987a), *G. tikvahiae* (Friedlander & Dawes 1985), and *Chondrus crispus* (Amat & Braud 1990), and for chlorophytes such as *Enteromorpha* (Fujita 1985), *Ulva*, and *Chaetomorpha* (Lavery & McComb 1991b).

Although long-lived radioisotopes of ammonium are available, a close chemical homologue, ^{14}C-labeled methylamine (CH_3NH_2), has been used to study ammonium-uptake kinetics in *Macrocystis pyrifera* (Wheeler 1979). However, if ammonium is present even at 1 μM, then separate experiments must be performed to quantitatively assess the degree to which ammonium inhibits methylamine uptake (Wheeler & McCarthy 1982).

Uptake of nitrate generally exhibits saturation kinetics (Harlin & Craigie 1978; DeBoer 1981). However, there have been several reports of a linear increase in nitrate uptake with increasing concentrations for *Laminaria groenlandica* (Harrison et al. 1986), *Chaetomorpha linum* (Lavery & McComb 1991b), *Gracilaria pacifica* (Thomas et al. 1987a), and *Chondrus crispus* (Amat & Braud 1990). Concentrations in intracellular (cytoplasmic and/or vacuolar) pools as much as 10^3 times greater than that in the surrounding seawater strongly suggest an unfavorable electrochemical gradient for transport, and therefore transport is likely to be primarily an active process. Even for microalgae, the actual mechanism of plasmalemma transport is not known (Syrett 1981, 1988), although Tischner et al. (1989) have proposed that nitrate reductase might act as a transport protein.

Nitrate uptake has been studied in only a few cases. In *Codium fragile*, the maximal uptake rate for nitrite was found to be similar to that for nitrate, but lower than that for ammonium (Hanisak & Harlin 1978). Recently, Brinkhuis et al. (1989) observed rapid nitrite uptake lasting only a few minutes, followed by a release of some of that nitrite back to the medium, and then a sustained uptake rate for several hours.

Whereas there have been several reports of growth on dissolved organic-nitrogen compounds, such as urea and amino acids, there has been only one definitive study on the uptake kinetics of amino acids. Schmitz and Riffarth (1980) examined the uptake of 17 different amino acids by the filamentous brown alga *Hincksia mitchelliae* and found the highest uptake rate for *L*-leucine. That uptake rate for leucine was observed to be light-independent, with a very low affinity (K_s = 30–120 μM) and a low maximal uptake rate (0.03 μmol g_{dw}^{-1} h^{1}). They suggested that exogenous amino acids contribute less than 5% of the nitrogen demanded by this alga. Urea is an excellent source of nitrogen for many seaweeds (Nasr et al. 1968; Probyn & Chapman 1982; Thomas et al. 1985), but poor growth has been reported for other species (DeBoer et al. 1978). Assessment of urea as a source of nitrogen in bacteria-contaminated culture studies is complicated because the bacteria may break down the urea to ammonium, which can be utilized by the seaweed. Urea uptake is generally thought to be active because it exhibits saturable uptake kinetics, although it should have a high rate of diffusion because it is a small, uncharged molecule.

Recent investigations have demonstrated that some phytoplankton are able to use various forms of dissolved organic nitrogen without initial transport into the cell (Palenik & Morel 1990; Palenik et al. 1991a). This novel mechanism involves the use of cell-surface enzymes to degrade these forms of nitrogen to NH_4^+ which is then taken up by the cell. Amino acids and other forms of dissolved organic nitrogen are now thought to be important for the nitrogenous nutrition of phytoplankton (Flynn & Butler 1986; Antia et al. 1991). The role of dissolved organic nitrogen in seaweed growth and production must be explored further.

Examples of values obtained for nitrogen-uptake kinetic parameters are summarized in Table 5.4. The K_s values are generally high (2-40 µM), up to 10 times higher than for phytoplankton, and the V_{max} values range from 3 to 188 µmol g_{dw}^{-1} h^{-1} for both nitrate and ammonium. Because specific uptake rates (where the rate is normalized to the nitrogen content of the thallus) generally are not reported for seaweeds, comparisons with phytoplankton are seldom possible. It is important to recall from section 5.4 that V_{max} frequently varies with the length of the incubation period that was used to measure it (Thomas & Harrison 1987). Therefore, V_{max} values for ammonium measured over 0–10 min may be much higher than values obtained over 0–60 min, because uptake rates decrease with time as the intracellular pools of N-limited algae fill. In contrast, nitrate uptake by N-limited algae may increase over 1–2 h, as added nitrate induces more rapid uptake (Thomas & Harrison 1987). Whereas V_{max} and K_s values are the traditional parameters to characterize nutrient-uptake kinetics for most N-limited phytoplankton, other parameters are required to fully characterize the uptake kinetics – for example, $V_{max}^{0-10 \text{ min}}$, $V_{max}^{0-60 \text{ min}}$, $V_{max} : K_s$ ratio [the initial slope (α) of the rectangular hyperbola], and the assimilation rate $V_{max}^{60-240 \text{ min}}$ (Harrison et al. 1989). The time superscript on the V_{max} term denotes the time period over which the measurement was made; it is important to use this time superscript, because V_{max} often changes with the length of the incubation period. The assimilation rate refers to the V_{max} value when the internal pools are full (i.e., no changes in V_{max} with time at saturating N concentrations). These time periods vary with species, nutritional history, and conditions, and they must be determined for each new situation.

Preliminary studies indicate that nitrogen assimilation (the incorporation of NO_3^- and NH_4^+ into amino acids and proteins) in seaweeds is similar to that in phytoplankton, although these processes in phytoplankton

usually are faster. Because no macroalgae are known to fix N_2 (except those containing symbionts; see sec. 5.2), the inorganic sources for these plants are nitrate, nitrite, and ammonium. Ammonium is already reduced and can be directly incorporated into amino acids, but other ions must first be reduced intracellularly to ammonium. Eight electrons are necessary to reduce nitrate (oxidation state +5) to ammonium (oxidation state −3), and the reduction occurs in two main steps.

The first step is the reduction of nitrate to nitrite, catalyzed by nitrate reductase (Davison & Stewart 1984; Thomas & Harrison 1988):

$$NO_3^- + NAD(P)H + H^+ \rightarrow$$
$$NO_2^- + NAD(P)^+ + H_2O$$

This enzyme has been isolated and purified from many species of microalgae (Syrett 1981; Tischner et al. 1989; Solomonson & Barber 1990), and it is also present in seaweeds (Weidner & Kiefer 1981; Wheeler & Weidner 1983; Thomas & Harrison 1988; Brinkhuis et al. 1989). Nitrate reductase (NR) is a relatively large molecule ($2-5 \times 10^5$ Da) that has three functional domains, such as molybdopterin, cytochrome b_5 and flavin adenine dinucleotide (FAD). The electron donor usually is NADH, but there are some microalgae in which the donor is NADPH instead (Lee 1980; Syrett 1981; Solomonson & Barber 1990). The enzyme is thought to occur in the cytoplasm, although there is some evidence that it may be associated with chloroplast membranes (Solomonson & Barber 1990). It may also be involved in membrane transport of nitrate (Tischner et al. 1989), although others disagree (Warner & Huffaker 1989).

Techniques for measuring NR activity in vitro have been worked out for microalgae (Eppley 1978) and adapted to seaweeds (Weidner & Kiefer 1981; Thomas & Harrison 1988; Chapman & Harrison 1988; Brinkhuis et al. 1989). There are two issues to consider in enzymatic assays in plants: (1) Is the enzyme extracted completely and in good condition? (2) Are the assay conditions optimal for maximal activity? Measurement of NR in seaweeds is often difficult, because tough, often rubbery, thalli are difficult to grind, and the presence of phenolics (Ilvessalo & Tuomi 1989; Gregenheimer 1990) may inactivate the enzyme and interfere with the protein determinations that are used to normalize NR activity. The problems are particularly pronounced in the Phaeophyta, but they have been overcome by adding polyvinylpyrrolidone or bovine serum albumin to bind the phenolics (Haxen & Lewis 1981; Thomas & Harrison 1988; Gregenheimer 1990), or by purification steps to remove alginate, which also interferes with enzyme activity (Kerby & Evans 1983). Proteases may also attack the enzyme, but this can be stopped by adding the protease inhibitor chyostatin (Long & Oaks 1990). The second issue involves optimum assay conditions, which are species-dependent.

Table 5.4. *Examples of nitrogen-uptake kinetic "constants" for seaweeds*

Alga	Temperature (°C)	NO$_3^-$ $K_s \pm SE$ (μM)	NO$_3^-$ V_{max} ($\mu mol\ g_{dw}^{-1}\ h^{-1}$)	NH$_4^+$ $K_s \pm SE$ (μM)	NH$_4^+$ V_{max} ($\mu mol\ g_{dw}^{-1}\ h^{-1}$)	References
Phaeophyceae:						
Fucus spiralis	5	6.6 ± 0.9		6.4 ± 2.0		Topinka (1978)
	10	6.7 ± 0.8		5.4 ± 2.0		
	15	7.8 ± 1.4		9.6 ± 2.6		
	15[a]	12.8 ± 3.5		5.8 ± 1.8		
Fucus distichus	15	1–5	25	3–5	60	Thomas et al. (1985)
Laminaria longicruris	15[b]	4.1	9.6			Harlin & Craigie (1978)
	10[c]	5.9	7.0			
Macrocystis pyrifera	16	13.1 ± 1.6	30.5	5.3[f] ± 1.0	23.8[f]	Haines & Wheeler (1978)
	6–9			50[g]	23.6[g]	Wheeler (1979)
Chordaria flagelliformis	11	5.9 ± 1.2	8.8			Probyn (1984)
Laminaria groenlandica	13		20[d]		20	Harrison et al. (1986)
			6[e]		6	
Rhodophyceae:						
Gracilaria foliifera	20	2.5 ± 0.5	9.7	1.6[f]	23.8[f]	D'Elia & DeBoer (1978)
Agardhiella subulata	20	2.4 ± 0.3	11.7	3.9[f]	15.9[f]	D'Elia & DeBoer (1978)
Hypnea musciformis	26	4.9 ± 3.9	28.5	16.6 ± 1.8		Haines & Wheeler (1978)
Chondrus crispus	17			35.5	62	Amat & Braud (1990)
Porphyra perforata	12		15		40–50	Thomas & Harrison (1985)
Chlorophyceae:						
Chaetomorpha linum	25		60		230	Lavery & McComb (1991b)
Enteromorpha prolifera	12–14	2.3–13.3	75–169	9.3–13.4	39–188	O'Brien & Wheeler (1987)
Enteromorpha spp.	15	16.6	129.4			Harlin (1978)
Codium fragile	6	1.9 ± 0.5	2.8	1.5 ± 0.2	13.0	Hanisak & Harlin (1978)
	24	7.6 ± 0.6	9.6	1.4 ± 0.2	28.0	
Ulva lactuca	20			40.7 ± 8.5	138 ± 78	Fujita (1985)
Ulva rigida	25	20–33	60–90		60	Lavery & McComb (1991b)

[a] Dark.
[b] Summer tissue.
[c] Winter tissue.
[d] First-year plants.
[e] Second- and third-year plants.
[f] Mechanism-1 uptake (high-affinity system).
[g] Uptake of methylamine, an analogue of NH_4^+.

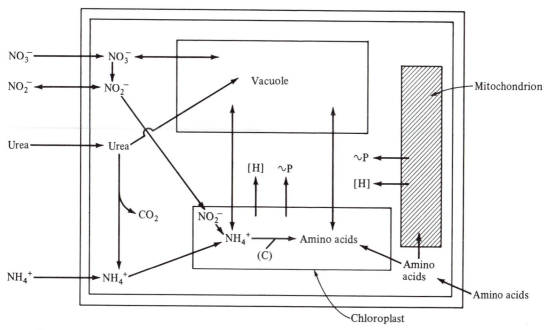

Figure 5.8. Main features of nitrogen uptake and assimilation in a eukaryotic algal cell. (From Syrett 1981, reproduced by permission of the Ministry of Supply and Services, Canada.)

For example, when the optimum pH for NR activity in *Porphyra* was used for *Enteromorpha*, activity was reduced by 50% (Thomas & Harrison 1988).

"In vivo" NR assays involve incubating the algae in medium with a very high nitrate concentration (20 mM) and *n*-propanol (to permeabilize the cell membrane) in the dark and measuring the appearance of nitrite in the medium. Some researchers (Dipierro et al. 1977; Davison et al. 1984; Corzo & Niell 1991) have recommended the "in vivo" assay because it preserves all systems in the natural state, and thus it should give a more realistic estimate of activity. However, because *n*-propanol causes disruption of membranes and lipids, this is not truly an "in vivo" assay at all. Thomas and Harrison (1988) did further studies and detected a major drawback: Activity varied with the length of time the plant had been incubated in the *n*-propanol medium, indicating that permeability was limiting the reaction rate, not NR activity. There also are potential problems of diffusion of nitrate into the tissue, especially if too much *n*-propanol is used (causes bleaching of tissues). Brinkhuis et al. (1989) have also pointed out that nitrite, which is released into the medium, might be taken back up by the thallus. Thus, the in vitro NR assay is the method of choice.

NR has been used to explore the metabolic activities of various tissues of the kelp *Laminaria digitata* (Davison & Stewart 1984). When longitudinal and transverse profiles of NR activities within the thallus were measured for plants growing at their maximum growth rate, the highest NR activities occurred in the mature blades, and activities declined toward the basal meristematic region. These observations are consistent with the suggestion of maintenance of meristematic growth by internal transport of organic nitrogen from the mature blade to the meristem (the sink). The transverse profile of NR activities in the stipe is similar to that reported for the enzymes of carbon assimilation (Kremer 1980). The higher NR activity in the outer, highly pigmented meristoderm tissue probably is associated with the high level of photosynthetic activity that occurs there. Activities in the stipe and holdfast are low.

An alternative to reducing nitrate to nitrite in the cytoplasm is to store nitrate in the vacuole (Fig. 5.8). Analysis of intracellular nitrate pools indicates that substantial amounts of nitrate may accumulate in the cytoplasm/vacuole in some intertidal macroalgae, especially when NR is relatively inactive (Thomas & Harrison 1985; Hwang et al. 1987). On the basis of earlier studies on phytoplankton, NR had been thought to provide a good index of nitrate uptake, but more recent work has shown a poor correlation between uptake and enzyme activity that may underestimate uptake by a factor of 10 (Syrett 1981; Dortch 1982). Storage of nitrate in the vacuole, or enzyme instability caused by proteases, may account for this poor correlation. If the same pattern holds true for seaweeds, then NR activity should be used only as an indication of the ability to use nitrate. It is important to realize that enzyme activity can serve only as a measure of the maximum capacity of the step, not as a measure of an instantaneous rate.

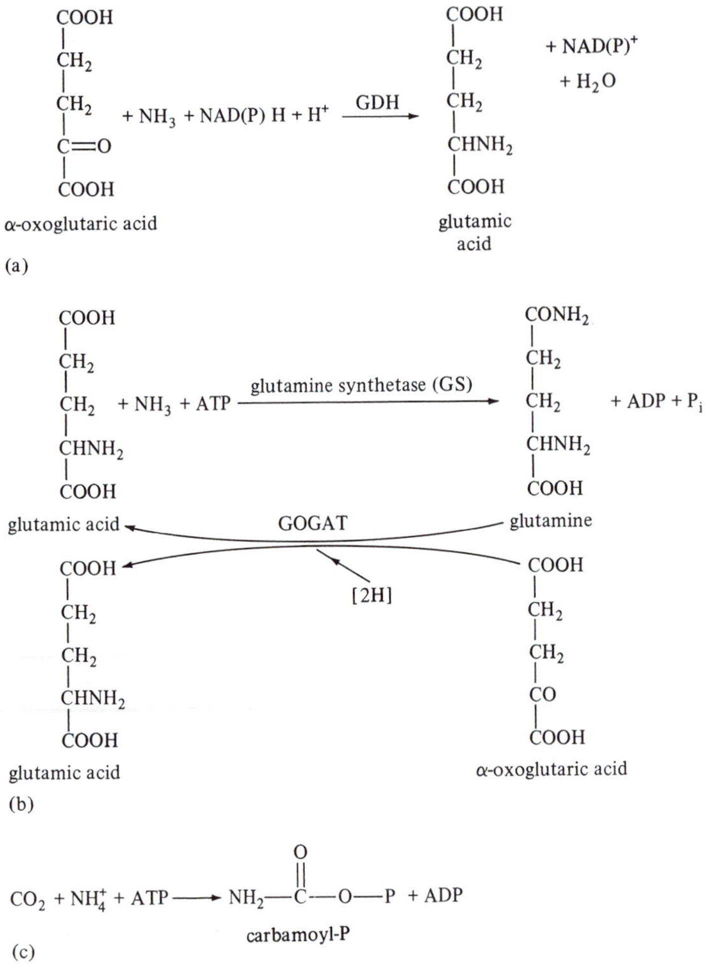

(a)

```
COOH                              COOH
|                                 |
CH2                               CH2        + NAD(P)+
|                                 |          + H2O
CH2    + NH3 + NAD(P)H + H+  GDH   CH2
|                          ─────►  |
C=O                               CHNH2
|                                 |
COOH                              COOH
α-oxoglutaric acid                glutamic
                                  acid
```

(b)

```
COOH                                              CONH2
|                                                 |
CH2                                               CH2
|          glutamine synthetase (GS)              |
CH2    + NH3 + ATP ─────────────────────►         CH2      + ADP + Pi
|                                                 |
CHNH2                                             CHNH2
|                                                 |
COOH                                              COOH
glutamic acid          GOGAT                      glutamine

COOH                    [2H]                      COOH
|                                                 |
CH2                                               CH2
|                                                 |
CH2                                               CH2
|                                                 |
CHNH2                                             CO
|                                                 |
COOH                                              COOH
glutamic acid                                     α-oxoglutaric acid
```

(c)

$$CO_2 + NH_4^+ + ATP \longrightarrow NH_2\text{—}\overset{\displaystyle O}{\overset{\|}{C}}\text{—}O\text{—}P + ADP$$

carbamoyl-P

Figure 5.9. Pathways of incorporation of ammonia. GDH, glutamate dehydrogenase; GOGAT, glutamine-oxoglutarate aminotransferase. (Parts a and b based on Syrett 1981, reproduced by permission of the Ministry of Supply and Services, Canada.)

After nitrate is reduced to nitrite, nitrite is transported to the chloroplasts for reduction to ammonium (Fig. 5.8). This reduction is catalyzed by nitrite reductase (NiR), according to the following reaction:

$$NO_2^- + 6Fe_{red} + 8H^+ \rightarrow NH_4^+ + 6Fe_{ox} + 2H_2O$$

(where Fe_{red} and Fe_{ox} are the reduced and oxidized forms, respectively, of ferredoxin). Although the six electrons probably are added sequentially in pairs, the two possible intermediates, nitric oxide and hydroxylamine, have not been detected. Both are toxic, and they are presumed to remain enzyme-bound (Bidwell 1979). NiR has an iron prosthetic group and uses ferredoxin as a cofactor to supply the electrons. There have been almost no studies on NiR activities in macroalgae, but it certainly deserves attention equal to that given to NR.

The control of ammonium assimilation is often held to be rate-limiting for nitrogen incorporation (Syrett 1981, 1988; Hellebust & Ahmad 1991). The regulation of these enzyme systems is enormously complex (e.g., Rigano et al. 1979). Ammonium incorporation

into amino acids was once thought to begin with the formation of glutamic acid via glutamate dehydrogenase (GDH) (Fig. 5.9a). However, in the early 1970s, work with bacteria revealed an alternate pathway (Tempest et al. 1970) that subsequently was shown to operate in the leaves of higher plants, in cyanobacteria, and in green algae (Miflin & Lea 1976). In this pathway (Fig. 5.9b), glutamine, rather than glutamate, is the first product of ammonium assimilation. Glutamic acid is formed later by a second reaction in which the amide group of glutamine is transferred to α-oxoglutaric acid (α-ketoglutarate) from the Krebs cycle, forming two molecules of glutamate (Syrett 1981; Hellebust & Ahmad 1991). The enzyme catalyzing the formation of glutamine is glutamine synthetase (GS), and the enzyme catalyzing the second reaction is glutamine-oxoglutarate aminotransferase (GOGAT) (also referred to by its older name, glutamate synthase).

Recent labeling and enzyme-activity studies have shown that the GS-GOGAT pathway is the primary route for ammonium incorporation into amino acids in the fronds of *Macrocystis angustifolia* (Haxen & Lewis

1981), *Hincksia mitchelliae* (Schmitz & Riffarth 1980), and *Laminaria digitata* (Davison & Stewart 1984). The activity of GDH in *Macrocystis* evidently is not significant in the assimilation of ammonium produced from nitrate reduction, even if a large concentration of ammonium is created by inhibiting GS with a specific inhibitor such as methionine sulfoxide (Haxen & Lewis 1981). An investigation of chloroplasts isolated from the siphonous green alga *Caulerpa simpliciuscula* revealed that both pathways of ammonium assimilation (i.e., GS-GOGAT and GDH) were present (McKenzie et al. 1979). In fact, GDH has an unusually low K_m for ammonium (0.4–0.7 mM), indicating that the enzyme could provide an alternative means for ammonium incorporation into amino acids in this species (Gayler & Morgan 1976). Studies on phytoplankton have suggested that under conditions of high external ammonium, the GDH pathway is operative; it has a low affinity for intracellular ammonium ($K_m \simeq 5$–28 mM) and does not require ATP, but it does use NAD(P)H and therefore energy (Syrett 1981). In the presence of low ammonium concentrations, or in nitrate-grown cells, the GS-GOGAT pathway operates; it has a very high affinity for intracellular ammonium ($K_m \simeq 29$ μM for a marine diatom; Falkowski & Rivkin 1976) and requires ATP. This view is attractive, but it lacks conclusive evidence.

Another, minor pathway for ammonium is the carbamoyl phosphate pathway (Fig. 5.9c), in which ammonium is incorporated into carbamoyl phosphate and then into citrulline and arginine and thence into pyrimidines (including thiamine) and biotin. However, the ammonia for carbamoyl phosphate synthesis is derived from the amide group of asparagine or glutamine, rather than from free ammonium (Beevers 1976).

The effects of light on the assimilation of nitrate and ammonium in seaweeds have not been well studied. More extensive studies on phytoplankton have revealed that the interactions are complex (Table 5.5) and are dependent on the metabolic state of the cells (Table 5.6). Both nitrate assimilation and ammonium assimilation can take place in darkness, provided that sufficient carbon reserves are available. Reduced ferredoxin is also required for NiR; therefore, there must be catabolic reactions that can lead to ferredoxin reduction.

Urea usually is assimilated by first being broken down into carbon dioxide and ammonium via the enzyme urease, and then the free ammonium is incorporated into amino acids via the GS-GOGAT pathway, as reviewed by Antia et al. (1991). There has been one study suggesting that urea may be directly assimilated into organic nitrogen (Kitoh & Hori 1977). Purified urease contains nickel, and the development of urease activity in some marine diatoms has a strong dependence on the nickel concentration in the medium (Syrett 1981). A urease enzyme assay exists for phytoplankton, but it has not been applied to macroalgae, except for a

Table 5.5. *Possible interactions of light with inorganic-nitrogen metabolism in algae*

Photosynthetic (chloroplast) effects

Generation of reduced ferredoxin which then:
  Reduces NO_2^- (and N_2 and NO_3^- in blue-greens)
  Reduces NAD(P)H and hence NO_3^- in eukaryotic algae
  Drives GOGAT reaction of NH_4^+ assimilation
  Activates/inactivates enzymes via thioredoxin
Generation of ATP through photophosphorylation, which then:
  Is used to drive transport mechanisms for NO_3^-, NO_2^-, NH_4^+
  Drives GS reaction of NH_4^+ assimilation
  Stops the reoxidation of mitochondrial NADH by O_2, thus making this NADH available for NO_3^- reduction
  Drives N_2 fixation in blue-greens
Photosynthetic fixation of CO_2 makes C acceptors available for NH_4^+ assimilation, thus removing feedback inhibition by organic N compounds of NO_3^- (NO_2^-) uptake

Other effects

Phytochrome (red-light) effects?
Direct enzyme activation/inactivation by blue light, possibly mediated through flavoproteins
Effects of light quality on protein synthesis

Source: Syrett (1981); reproduced with permission of the Ministry of Supply and Services, Canada.

brief account of urease in *Ulva lactuca* (Bekheet et al. 1984). Microalgae belonging to the Chlorophyceae (except Mesotaeniales, Ulvales, and Zygnematales) contain urea amidolyase instead of urease (Syrett 1981; Syrett & Al-Houty 1984).

Assimilation of nitrogen by nitrogen-deficient or nitrogen-limited cells is limited by the rate of protein synthesis, as suggested by the early work of Syrett (1956). This has been substantiated recently by evidence for the accumulation of internal pools of nitrate, ammonium, and free amino acids after the addition of nitrogen to nitrogen-limited cultures of phytoplankton (DeManche et al. 1979; Dortch 1982) and seaweeds (Haxen & Lewis 1981; Thomas & Harrison 1985). Such pools would not accumulate if the rates of protein synthesis were equal to or greater than the rates of membrane transport and subsequent metabolism to amino acids. Several species of *Laminaria* accumulate nitrate, and the tissue levels of nitrate account for a significant portion of the total nitrogen (Chapman & Craigie 1977). However, in many phaeophytes, internal nitrate levels are low and never account for more than 5% of the total tissue nitrogen (Buggeln 1978; Wheeler & North 1980;

Table 5.6. *Interactions of light and metabolic states in determining NH_4^+ or NO_3^- assimilation by green microalgae*

	Metabolic state of cells		
Process	Carbon-starved ($-CO_2$)	Normal growth ($+CO_2$)	Nitrogen-starved
Storage of C compounds	Nil	Very low	High
Rate of assimilation of			
NH_4^+ (light)	0	+	++++
NH_4^+ (dark)	0	0	+++
NO_3^- (light)	+[a]	+	++
NO_3^- (dark)	0	0	+[b]
NH_4^+ inhibition of NO_3^- uptake	0	+[c]	±[c]

[a]Ammonium accumulates.
[b]Nitrite accumulates.
[c]With dependence on NH_4^+ concentration.
Source: Syrett (1981); reproduced with permission of the Ministry of Supply and Services, Canada.

Asare & Harlin 1983). Hence, storage of nitrogen as nitrate is not widespread in the Phaeophyceae. Amino acids (especially alanine) and proteins appear to form the major nitrogen-storage pools in *Gracilaria tikvahiae* (Bird et al. 1982), *G. foliifera* (Rosenberg & Ramus 1982), and *Macrocystis pyrifera* (Wheeler & North 1980). In contrast, citrulline and arginine are abundant in *Gracilaria secundata* (Lignell & Pedersén 1987) and *G. verrucosa* (Bird et al. 1980). High amounts of citrulline and the dipeptide citrullinylarginine have been reported to play major roles as nitrogen-storage compounds in *Chondrus crispus* (Laycock et al. 1981), *Gracilaria*, and other red seaweeds (Mayazawa & Ito 1974; Laycock & Craigie 1977). Thomas and Harrison (1985) found that intracellular nitrate and ammonium pools in *Porphyra perforata* composed only 10% of the pools of free amino acids. When plants were starved of nitrogen, the nitrate decreased to undetectable levels in 5 days, while the ammonium and amino acid pools decreased by only 50%.

The amount of nitrogen storage can vary considerably in closely related seaweeds or in a given species growing in different localities. *Gracilaria tikvahiae* can store enough nitrogen to allow it to grow at maximal rates for several days without nitrogen (Lapointe & Ryther 1979; Fujita 1985), whereas *Gracilaria secundata* has very limited nitrogen-storage abilities (Lingell & Pedersén 1987). The difference between these two species may correlate with habitats; *G. secundata* grows in a eutrophic environment, whereas *G. tikvahiae* may be subjected to various periods of nitrogen limitation. Because many seaweeds can store nitrogen, this characteristic has been used to advantage in controlling epiphyte growth in aquaculture systems. Weekly pulses of 0.5-mM NH_4^+ controlled epiphyte growth in *Gracilaria conferta* growing under low irradiances and

yielded the highest growth rates for this macroalga (Friedlander et al. 1991).

Little is known about catabolism and turnover of cellular protein. Recycling of nitrogen may occur via the photorespiratory pathway (Singh et al. 1985) and by catabolism of specialized nitrogen-storage compounds such as guanine (Pettersen 1975) and arginine (Wheeler & Stephens 1977) in the Chlorophyceae. Pigments and associated proteins such as phycoerythrin may serve as nitrogen-storage compounds in the Rhodophyceae (Gantt 1980; Bird et al. 1982). Although enzymes such as NR are constantly being turned over (Velasco et al. 1988), this nitrogen is thought to be efficiently recycled and therefore of minor importance.

5.5.2 Phosphorus

Phosphorus is not generally considered to be a limiting nutrient in the marine environment, but there are significant exceptions, such as the western North Atlantic (Lapointe 1986; Lapointe et al. 1992), the southern coast of China (Harrison et al. 1990), and parts of the Mediterranean (Azov 1986). There have been few studies on phosphorus-uptake kinetics in seaweeds, despite the fact that phosphorus concentrations frequently are near the limit of detection when nitrogen is exhausted from the seawater. Preliminary studies indicate that phosphorus is taken up actively by *Porphyra* (Eppley 1958), and saturation kinetics have been examined in the red alga *Agardhiella subulata*, yielding a V_{max} of 0.47 µmol g_{dw}^{-1} h^{-1} and a K_s of 0.4 µM (DeBoer 1981). The kinetic parameters for two chlorophytes, *Ulva* and *Chaetomorpha*, that dominate a eutrophic estuary were considerably higher; K_s values for *Ulva* and *Chaetomorpha* were 3.5 and 10 µM, respectively, and V_{max} values were 8.5 and 20.8 µmol g_{dw}^{-1} h^{-1}, respectively (Lavery & McComb 1991b). In *Gracilaria tikvahiae*, the uptake rate showed three phases of uptake, two saturation phases at lower concentrations (0–0.2 and 0–2 μM) and a linear phase (0–11 μM) (Friedlander & Dawes 1985).

Hurd and Dring (1990) studied phosphate uptake in relation to zonation and season in five intertidal fucoid algae. In their time series of uptake experiments they found an initial (30 min) rapid uptake of phosphate, followed by almost no uptake (30 min), and then intermediate rates over several hours in *Pelvetia* and two species of *Fucus*. *Ascophyllum* had a constant slow uptake rate over 6 h. The fucoid algae were divided into two distinct groups: *Pelvetia* and *Ascophyllum* took up small amounts of phosphate over a tidal cycle, and the ratio of their uptake rates was 6 : 1 (*Pelvetia : Ascophyllum*). Their uptake abilities were related to their positions on the shore (*Pelvetia* was in the highest zone) and to their low requirements for nutrients (low growth rate). The three species of *Fucus* formed the second group. They were able to take up larger amounts of phosphate during a tidal cycle, the order of uptake being *F. spiralis* > *F. vesiculosus* > *F. serratus,* and their rates were directly related to their positions on the shore. Hurd et al. (1993) observed that all species of *Fucus* produced hyaline hairs on the apical region and upper mid-thallus in late winter and shed them in autumn; those hairs were shown to enhance nutrient uptake. A reduced uptake rate for *F. spiralis* in July was due to desiccation damage and the grazing of hairs by littorinid snails.

Growth-enrichment studies by Lapointe (1986) utilizing in situ cage cultures and a shipboard flowing-seawater culture system were conducted in the summer with whole-plant populations of pelagic *Sargassum natans* and *S. fluitans* in the western Sargasso Sea. Phosphorus enrichment doubled their growth and photosynthetic rates, as compared with nitrogen enrichment and no enrichment. Nitrogen-fixing cyanobacteria on the surface of *Sargassum* (Carpenter 1972) may contribute some nitrogen and thus leave this macroalga short of phosphorus. This suggestion will require experimental verification. Further enrichment studies and tissue analysis with *Gracilaria tikvahiae* in the Florida keys demonstrated that *Gracilaria* was phosphorus-limited in the summer and nitrogen- and phosphorus-limited in the winter (Lapointe 1985, 1987). This confirms earlier observations in which phosphorus limited phytoplankton productivity more frequently than did nitrogen in the near-shore northeastern Gulf of Mexico (Myers & Iverson 1981). O'Brien and Wheeler (1987) also suggested that *Enteromorpha prolifera* off the coast of Oregon may have been phosphorus-limited in November. Greatly elevated C : P and N : P ratios further confirm phosphorus limitation (Lapointe 1987). Ratios for C : P > 1,800 and N : P > 120 for unenriched *G. tikvahiae* were threefold higher than the median values of 550 and 30, respectively, reported in a survey of C : N : P ratios

in benthic marine plants (Atkinson & Smith 1983). Winter phosphorus limitation has also been suggested for *Macrocystis* sporophytes in the Pacific Ocean (Manley & North 1984).

Algae acquire phosphorus principally as orthophosphate ions (PO_4^{3-}). Other sources are inorganic polyphosphates and organic-phosphorus compounds. Sugar phosphates reportedly are taken up intact by bacteria (Rubin et al. 1977), but most eukaryotic algae require extracellular enzymatic hydrolysis to remove the sugar before phosphate uptake (Nalewajko & Lean 1980). Polyphosphates may also require extracellular cleavage; the freshwater macroalga *Cladophora glomerata* breaks down pyrophosphate and triphosphate (both common in detergents) and takes up the phosphorus as orthophosphate (Lin 1977). Some seaweeds can use some organic forms of phosphate, such as glycerophosphate, by producing extracellular alkaline phosphatase (Walther & Fries 1976). The ability of cells to enzymatically cleave the ester linkage joining the phosphate group to the organic moiety is conferred by the activity of phosphomonoesterases (commonly called phosphatases) at the cell surface. Two groups of these enzymes have been distinguished on the basis of their pH optima, their phosphate repressibility, and their cellular locations (Cembella et al. 1984). Alkaline phosphatases are phosphate-repressible, are inducible, have alkaline pH optima, and generally are located on the cell surface or are released into the surrounding seawater. Acid phosphatases are phosphate-irrepressible, are constitutive, have acidic pH optima, and generally are found intracellularly in the cytoplasm. Both types may be found simultaneously in algal cells, with alkaline phosphatases aiding in the uptake of organic-phosphorus compounds, and acid phosphatases playing crucial roles in cleavage and phosphate-transfer reactions in metabolic pathways within the cell. The essential feature of phosphatases that allows them to participate efficiently in cellular metabolism is their ability to be alternately induced and repressed, depending on metabolic requirements. When external inorganic-phosphate concentrations are high, synthesis of alkaline phosphatase is repressed, and cells exhibit little ability to utilize organic-phosphorus compounds. Generally, after the external inorganic phosphate has been exhausted, intracellular phosphorus from stored polyphosphates and orthophosphate is used up quickly (Lundberg et al. 1989), followed by increased alkaline phosphatase activity. Weich and Granéli (1989) found that alkaline phosphatase activity in *Ulva lactuca* was stimulated by phosphorus limitation and by light. The magnitude of the increase was species-specific and depended on the availability of organic phosphates and the degree of phosphate limitation experienced by the cells (Healey 1973; Cembella et al. 1984).

Inorganic phosphate transported across the plasmalemma enters a dynamic intracellular phosphate pool

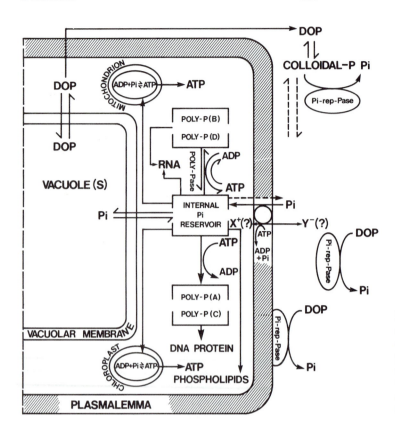

Figure 5.10. Main features of phosphorus uptake and assimilation in a microalgal cell. DOP, dissolved organic phosphate; Pi, inorganic phosphate. (From Cembella et al. 1984, reprinted with permission from *Critical Reviews of Microbiology* 10:317–91, © 1984 CRC Press, Boca Raton, Fla.)

from which it is incorporated into phosphorylated metabolites (Chopin et al. 1990) or stored as luxury phosphorus in vacuoles or in polyphosphate vesicles in algae (Lundberg et al. 1989) (Fig. 5.10). Some of the cytoplasmic phosphate pool may leak back out of the cell and reappear as external phosphate. Phosphorus-deficient algae possess the ability to incorporate phosphate extremely rapidly, and the amount taken up usually exceeds the actual requirements of the cell. The excess is built into polyphosphates by the action of polyphosphate kinase:

$$ATP + (polyphosphate)_n \rightarrow$$
$$ADP + (polyphosphate)_{n+1}$$

(Kuhl 1974; Cembella et al. 1984). An important difference between phosphorus metabolism in vascular plants and that in algae is the formation of these polyphosphates; seaweeds known to form polyphosphates include species of *Ulva, Enteromorpha, Ceramium,* and *Ulothrix* (Kuhl 1962; Lundberg et al. 1989). Lundberg et al. (1989) used high-resolution NMR with ^{31}P and found relatively short (6–20 PO_4^{3-} units) polyphosphates stored in the vacuole in *Ulva lactuca.* In contrast, they found that the brown alga *Pilayella* stores phosphorus mainly as phosphate (not polyphosphate) in its vacuole. This latter storage mechanism is considered to be an exception in higher plants and algae (Raven 1984). The storage compounds are classified into cyclic and linear

polyphosphates (Fig. 5.11). These two types cannot be easily separated by simple extraction procedures, but they can be divided into four categories (A–D, Fig. 5.10) on the basis of sequential extraction techniques (Cembella et al. 1984). Intracellular polyphosphate acts as a noncompetitive inhibitor of phosphate uptake in phosphate-limited cultures of the freshwater unicellular green *Scenedesmus* (Rhee 1974). The kinetics can be described by an equation similar to that for enzyme kinetics under noncompetitive inhibition [Table 5.3, equation (5)].

Phosphorus plays key roles in many biomolecules, such as nucleic acids, proteins, and phospholipids (the latter are important components of membranes). However, its most important role is in energy transfer through ATP and other high-energy compounds in photosynthesis and respiration (Fig. 5.10) and in "priming" molecules for metabolic pathways.

5.5.3 Calcium and magnesium

Calcium is important to all organisms for maintenance of cellular membranes. Calcium is required inside the cell by only a few enzymes (α-amylase, phospholipase, and some ATPases). In fact, many enzymes (e.g., PEP carboxylase) are inhibited by Ca^{2+}. Calcium, along with other divalent cations, has been shown to activate ATPase in three calcareous algae (Okazaki 1977). Activation by calcium was stronger

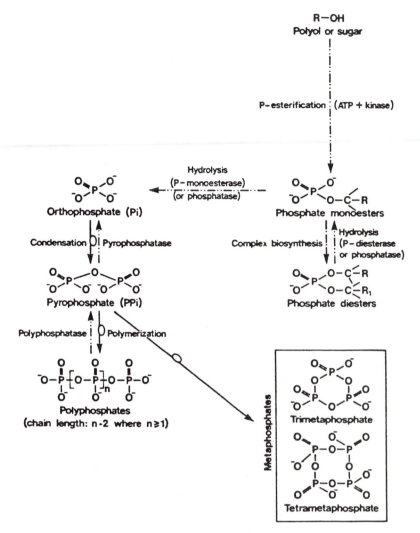

Figure 5.11. Interconversion of inorganic- and organic-phosphate storage compounds. (From Cembella et al. 1984, reprinted with permission from *Critical Reviews of Microbiology* 10:317–91, © 1984, CRC Press, Boca Raton, Fla.)

than that by other divalent cations, which in fact competed with Ca^{2+}. Calcium has roles in morphogenesis and phototaxis, and evidence for Ca^{2+} concentration as a messenger in light signaling is seen in the example of calcium in the development of cellular polarity in fucoid eggs. The hypothesis is that Ca^{2+} channels are localized at the presumptive rhizoid end of the light gradient, causing a net influx of Ca^{2+} into the dark end, with a net efflux on the light or thallus end (Quatrano & Kropf 1989) (see sec. 1.6.2 for further discussion). There is considerable evidence that Ca^{2+}, together with the regulatory protein calmodulin, is an exceedingly important second messenger in plant cells.

There are more than 100 genera of calcareous algae, and they are widely distributed taxonomically (e.g., Chlorophyceae, Rhodophyceae, and Phaeophyceae) (Table 5.7). The activities of calcareous algae are clearly seen in the formation of reefs and atolls (Barnes & Chalker 1990). Encrusting red algae of the Order Corallinales aid in cementing reefs together and stabilizing them against wave action. Calcified segments of *Halimeda* may contribute a major portion of the sediment on an atoll, and thus it plays a role in atoll formation (Barnes & Chalker 1990).

Calcium is deposited as calcium carbonate ($CaCO_3$), sometimes along with small amounts of magnesium and strontium carbonates. Calcium carbonate occurs in two crystalline forms, calcite (hexagonal-rhombohedral crystals) and aragonite (orthorhombic), which never occur together in a single alga under natural conditions. Aragonite is the form most commonly deposited, and it is the form that precipitates abiotically. The extensive literature on the morphological and physiological aspects of calcification has been reviewed by

Table 5.7. *Sites and forms of algal mineralization*

Sites	Forms	Examples of taxa
Extracellular		
Cell-wall surface	Concentric bands of fine aragonite needles on cell surface	*Padina* (Dictyotaceae)
	Surface encrustation of calcite crystals	*Chaetomorpha* (Cladophoraceae)
Intercellular	Fine aragonite needles in intercellular spaces in utricles	*Halimeda, Udotea* (Halimedaceae), *Neomeris* (Dasycladaceae)
	Intercellular crystals of aragonite and/or calcite that may form small bundles	*Liagora* (Liagoraceae), *Galaxaura* (Galaxauraceae)
Sheath	Bundles of aragonite needles in external sheath	*Penicillus, Udotea* (Udoteaceae)
	Irregular bundles of needle-like crystals, usually aragonite	*Plectonema* (Oscillatoriaceae)
Within cell walls	Calcite crystals, often clearly oriented	*Lithophyllum, Lithothamnion* (Corallinaceae)
Intracellular	Calcified plates (coccoliths) of various forms, usually calcite; formed within Golgi vesicles	*Emiliania, Cricosphaera* (Prymnesiophyceae)

Source: Simkiss and Wilbur (1989), with permission of Academic Press.

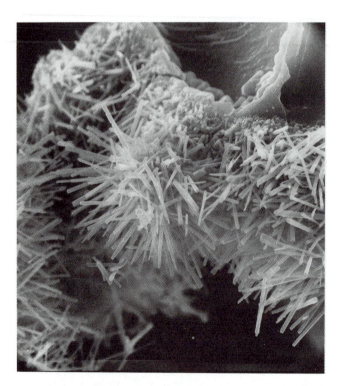

Figure 5.12. Scanning electron micrograph of the extracellularly mineralizing calcareous alga *Halimeda* (species unknown) showing the thin, unoriented crystals of aragonite ($\times 2{,}880$). A portion of the cell wall is visible in the upper right-hand corner. (From Weiner 1986, reprinted with permission from *Critical Reviews in Biochemistry* 20:365–408.)

Borowitzka (1987, 1989), Johansen (1981), Pentecost (1985), and Simkiss and Wilbur (1989). The morphology of algal mineralization takes a variety of forms, depending on the taxon evolved, but there are two basic types: extracellular and intracellular (Table 5.7).

The best-studied example of extracellular deposition is aragonite in the intercellular spaces in the green alga *Halimeda*. Its outer surface is made up of vertical, swollen filaments called utricles (Figs. 1.1 and 1.8). Their outermost walls become fused, forming a barrier that prevents flow of seawater into the intercellular spaces. Aragonite is precipitated by *Halimeda* and other Chlorophyta as crystals of varying shapes, and without a preferred orientation (Fig. 5.12). The crystals begin as small granules on the fibrous material of the intercellular space (nucleation sites), and then they grow until the

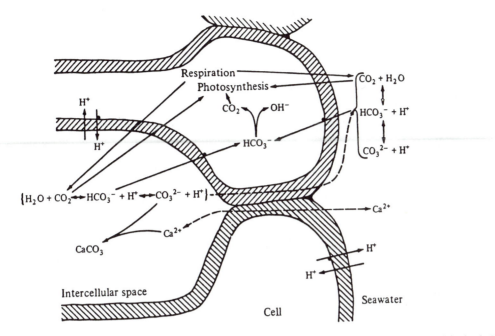

Figure 5.13. Schematic representation of the postulated ion fluxes affecting calcium carbonate precipitation in *Halimeda*. A black dot at the plasmalemma indicates that the flux is postulated to be active. (From Borowitzka 1977, with permission of *Oceanography and Marine Biology: An Annual Review*.)

intercellular space is almost filled. Ca^{2+} and HCO_3^- enter the intercellular space either by diffusion through the cell walls of the fused filaments or by active uptake into the filament and then an efflux into the intercellular space. During photosynthesis, CO_2 is taken up from the intercellular space, resulting in increases in intercellular pH (i.e., $HCO_3^- \rightarrow CO_2 + OH^-$) and CO_3^{2-} concentration, with subsequent deposition of aragonite (Fig. 5.13). This hypothesis for calcification in *Halimeda* is not applicable to all aragonite depositors, for in the brown alga *Padina* there are no intercellular spaces, and aragonite is precipitated in concentric bands on the outer surface of the thallus. Moreover, there are other seaweeds with apparently suitable morphology (intercellular spaces) that do not calcify (e.g., *Enteromorpha*; Borowitzka 1977).

The Rhodophyceae, in contrast to Chlorophyta, are primarily calcifiers of cell walls, although some do deposit $CaCO_3$ within intercellular spaces as well (Cabioch & Giraud 1986). Calcification of the wall is so extensive that the cells become encased, except for the primary pit connections at the end. In the Rhodophyceae, calcification occurs in Corallinaceae (calcite) and in some members of Squamariaceae (aragonite), Gigartinaceae (aragonite), and Bangiales (aragonite). Of these algae, members of the Corallinaceae are both the most abundant and the best-known.

The Corallinales deposit calcite containing high levels of magnesium (6% Mg) within the cell walls of their vegetative cells, except for the meristematic, genicular, and reproductive cells, which are noncalcified. Calcification commences in the middle lamella and rapidly spreads throughout the cell wall. Along the middle lamella the crystals are arranged along the growth axis, whereas near the plasmalemma the calcite crystals are oriented with the crystal axis at right angles to the plasmalemma (Fig. 5.14). The calcite crystals are in close association with the organic cell-wall material, and some cell-wall components probably act as a template for deposition and orientation of new crystals. Studies of the compartmentation of exchangeable Ca^{2+} in the thallus of *Amphiroa foliacea* have indicated that there are at least two major compartments for binding and exchanging organic calcium, presumed to be the COO^- and $O\text{-}SO_2\text{-}O^-$ groups of the acidic polysaccharide wall compounds (Borowitzka 1979). Because aragonite is the normal crystalline form of calcium carbonate precipitated from seawater, the organic wall material of the corallines is presumed to be responsible for deposition as calcite. These results further support the proposed role of some specific cell-wall components of coralline aglae in influencing the crystallography of the $CaCO_3$ precipitated, in addition to suggesting that they contain compounds (probably polysaccharides) that block crystal growth sites and stop calcification. This latter fact probably is the reason that most marine macrophytes are *not* calcified.

Little is know about the actual mechanism of calcification in the coralline algae except that it is directly proportional to photosynthesis, is stimulated by light,

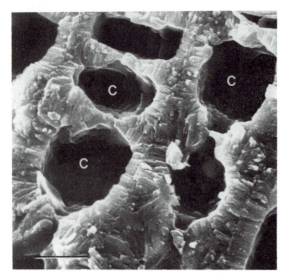

Figure 5.14. Scanning electron micrograph of a fractured thallus of the crustose coralline *Lithothamnion australe* showing the calcite-impregnated cell walls. This alga has especially large calcite crystals, which can clearly be seen to be oriented at right angles to the cells (c). Scale = 5 μm. (Reprinted from Borowitzka 1982, with permission of Elsevier Scientific Publications.)

and is highest in young tissue. There is considerable diversity in the proposed calcification models for corallines. The optimum pH for photosynthesis in *Amphiroa* species lies between 6.5 and 7.5. Above pH 8, photosynthesis is greater than would be expected based solely on the utilization of CO_2. This suggests that they take up HCO_3^- at a higher pH. Calcification is also affected by CO_3^{2-} concentration and by light. The dark calcification rate is approximately 50% of the rate in saturating light at normal pH, and this is considered to be the nonmetabolically influenced component of the coralline calcification process (Borowitzka 1989). Two models of calcification in the Corallinales have been proposed by Pentecost (1985) (Fig. 5.15a,b). One is slightly modified from the scheme proposed by Digby (1977), in which HCO_3^- is taken up and converted to CO_3^{2-}, which is released extracellularly and combines with Ca^{2+} to form $CaCO_3$. In another model, an efflux of H^+ reacts with HCO_3^- in seawater to produce CO_2, which then diffuses into the cells for photosynthesis (Fig. 5.15b). An efflux of OH^- maintains electrical neutrality in the cell, raises the pH and the CO_3^{2-} concentration in the intercellular spaces, and leads to localized calcification. This second model obviates a HCO_3^- uptake mechanism and the high intracellular pH necessary to produce CO_3^{2-}, as required by the first model. Calcium transport may be by a Ca-ATPase enzyme in some calcareous red algae (Okazaki 1977).

Orthophosphate inhibits calcification in corallines (Brown et al. 1977), but the mechanism remains unknown. Thus, when coralline algae are grown in phosphorus-enriched culture medium (30–150-μM PO_4^{3-}), they are only weakly calcified. These observations suggest that the growth of corallines could be inhibited in phosphate-polluted coastal waters.

In summary, data from Ca^{2+}-efflux studies and stable-isotope studies, and the $CaCO_3$ crystal isomorph deposited, all suggest that much of the carbon for the $CaCO_3$ is of seawater origin, although long-term pulse-chase studies have indicated that respiratory CO_2 is also incorporated into the $CaCO_3$ in *Halimeda*. The precipitation of $CaCO_3$ requires sufficient Ca^{2+} and a metabolically induced rise in pH. In seawater, this pH shift is easily achieved by photosynthetic CO_2 uptake; however, in fresh water, H^+ efflux or OH^- uptake by the algae may also be necessary. $CaCO_3$ nucleation appears to be also influenced by the organic matter in the plant cell wall, although to date no detailed information on the chemical nature of this material is available. Once nucleation is achieved, $CaCO_3$ deposition continues under the influence of localized pH and carbonate changes caused by photosynthetic CO_2 uptake.

Unlike many animals, algae do not require $CaCO_3$ or related salts for skeletal support; in fact, $CaCO_3$ deposits may more often be a liability rather than an asset. $CaCO_3$ deposits, especially if they are extracellular, inhibit nutrient uptake by creating diffusion barriers and also limit light penetration into the thallus, thus reducing photosynthesis and inhibiting growth. The high pH near the plasmalemma in calcareous algae could also hinder phosphate uptake, because at high pH, $H_2PO_4^-$ is converted to HPO_4^{2-} and then PO_4^{3-}. Only in the coralline algae is there any evidence that the $CaCO_3$ may confer a benefit by reducing the damage caused by grazers. Thus algal calcification is best seen as a by-product of photosynthesis, one that many algae have evolved mechanisms to avoid (Simkiss & Wilbur 1989).

Magnesium is an essential component of chlorophyll, forming a metalloporphyrin. It can play a role in binding charged polysaccharide chains to one another because it is a divalent cation. It is a cofactor or activator in many reactions, such as nitrate reduction, sulfate reduction, and phosphate transfers (except phosphorylases). It is also important in several carboxylation and decarboxylation reactions, including the first step of carbon fixation, where the enzyme ribulose-1,5-bisphosphate (RuBP) carboxylase attaches CO_2 to RuBP. Magnesium also activates enzymes involved in nucleic acid synthesis and binds together the subunits of ribosomes. There are several means by which magnesium may act (Bidwell 1979): (1) It may link enzyme and substrate together, as, for example, in reactions involving phosphate transfer from ATP. (2) It may alter the equilibrium

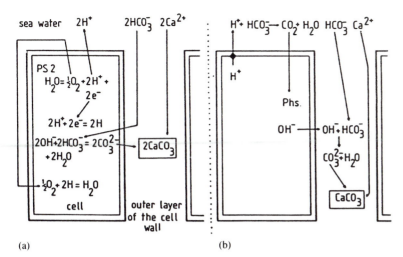

Figure 5.15. Models of calcification in the Corallinaceae. (a) Scheme proposed by Digby (1977), slightly simplified. (b) Alternative model with localized efflux of H^+ and OH^-. (From Pentecost 1985, with permission of the American Society of Plant Physiologists.)

constant of a reaction by binding with the product, as in certain kinase reactions. (3) It may act by complexing with an enzyme inhibitor.

5.5.4 Sodium and potassium

Sodium, potassium, and chloride ions are never likely to be limiting to macroalgal growth in the marine environment (in contrast to nitrogen and possibly phosphorus). However, interest in their uptake rates is associated with understanding osmoregulatory processes (R. H. Reed 1990) (sec. 6.3.2). Reed and Collins (1980) used radioisotope-equilibration techniques to study the influx and efflux of K^+, Na^+, and Cl^- in *Porphyra purpurea*. The initial rapid uptake of these ions was due to extracellular adsorption. Plasmalemma transport showed saturation kinetics, with cells discriminating in favor of K^+ and Cl^- and against Na^+. Hence, there was an active uptake and a passive loss of K^+ and Cl^- and an active efflux of Na^+. The kinetics of ^{86}Rb exchange have been used to study intracellular compartmentation of K^+ in *Porphyra* (Reed & Collins 1981). More recently, several detailed studies of electrophysiology and compartmental analyses have been conducted on *Enteromorpha intestinalis* and *Ulva lactuca*, as reviewed by Ritchie and Larkum (1987).

The role of potassium in ionic relations is nonspecific, as it is only one of several monovalent cations involved. Potassium has a more specific role as an enzyme activator (O'Kelley 1974); many protein-synthesis enzymes do not act efficiently in the absence of K^+, but the way in which K^+ binds to the enzymes and affects them is not well understood. It is known to bind ionically to pyruvate kinase, which is essential in respiration and carbohydrate metabolism (Bidwell 1979). Although rubidium may, in some cases, or to a certain extent, substitute for potassium, West and Pitman (1967) have shown that the rates of uptake of $^{86}Rb^+$ by *Ulva lactuca* and *Chaetomorpha coliformis* are very

much slower than uptake rates for $^{42}K^+$. Ritchie (1988) also found that ^{86}Rb was a poor analogue for K^+ in experiments with *Ulva lactuca*.

5.5.5 Sulfur

Coughlan (1977) found that sulfate-uptake rates in *Fucus serratus* showed Michaelis-Menten-type saturation kinetics ($K_m = 6.9 \times 10^{-5}$ M), but she also noted inhibition by selenate, molybdate, tungstate, and especially chromate. The kinetics of this inhibition were not worked out. In the economically important red alga *Chondrus crispus*, sulfate uptake was found to be multiphasic, similar to the pattern in higher plants (Jackson & McCandless 1982). The value of K_m was about 3 mM. Sulfate uptake by the unicellular marine red alga *Rhodella maculata* was biphasic, with a K_m of 22 mM for the low-affinity system and 63 mM for the high-affinity system (Millard & Evans 1982).

Most algae can meet all of their sulfur requirements by reducing sulfate, the most abundant form of sulfur (25 mM) in the aerobic marine environment (O'Kelley 1974). For this reason, few studies have been conducted on the utilization of other forms of sulfur, such as sulfite and organic-sulfur-containing compounds. Sulfur nutrition and utilization in algae have been reviewed by O'Kelley (1974), Schiff (1980, 1983), and Raven (1980). Before sulfate can be incorporated into various compounds, it must be activated, since it is a relatively unreactive compound (Schiff & Hodson 1970). The enzyme ATP sulfurylase catalyzes the substitution of SO_4^{2-} for two of the phosphate groups of ATP, to form adenosine-5'-phosphosulfate (APS) (Fig. 5.16). APS can have another phosphate added from another ATP to form adenosine-3'-phosphate-5'-phosphosulfate (PAPS), which is believed to be the starting point of sulfate ester formation in many systems and for sulfate reduction. (Like nitrogen, sulfur is incorporated into proteins in its most reduced form.) Although the

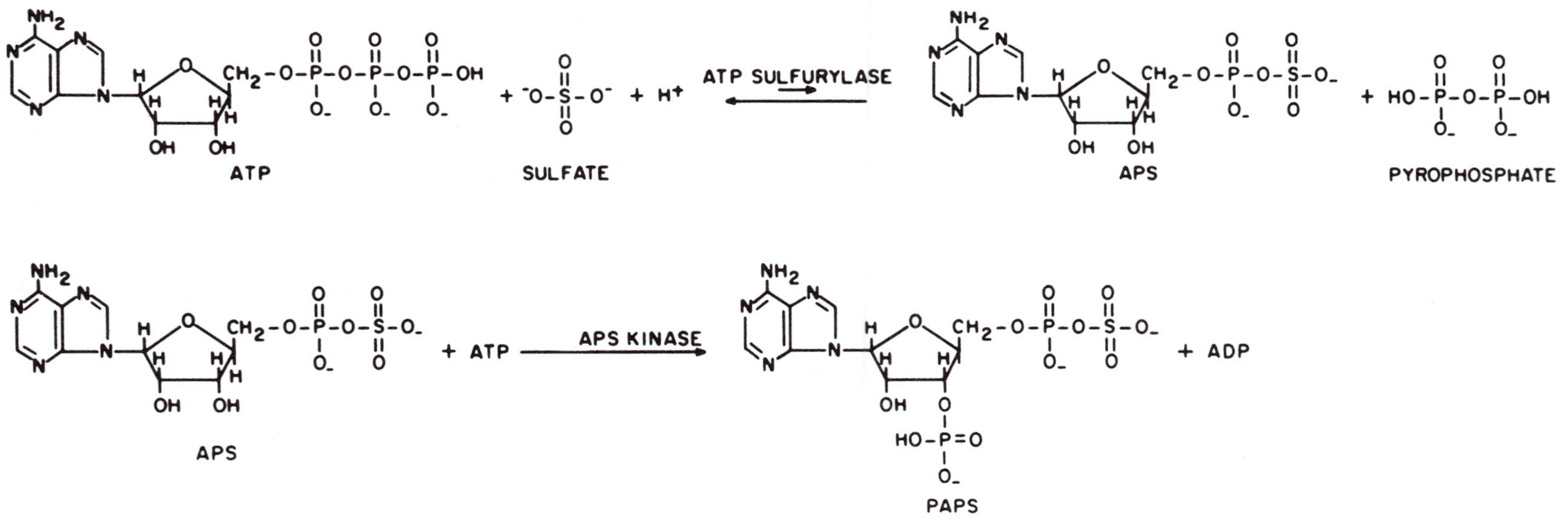

Figure 5.16. Sulfate activation. APS, adenosine-5'-phosphosulfate, PAPS, phosphoadenosine phosphosulfate. (From Schiff & Hodson 1970, with permission of the New York Academy of Sciences.)

majority of the studies on algal sulfate reduction have been done on freshwater unicells, a sulfite (SO_3^{2-}) reductase has been demonstrated in several marine red algae (e.g., *Porphyra* spp.) and green algae (O'Kelley 1974). The significance of an enzyme with the ability to reduce sulfite was questioned by Schiff and Hodson (1970), who pointed out that sulfite is extremely reactive and therefore may react with any nonspecific reductase; they believe that thiosulfate ($S_2O_3^{2-}$) is the intermediate, rather than sulfite. Coughlan (1977) recently showed that the $SO_4^{2-} \rightarrow APS \rightarrow PAPS$ activation system is also operative in *Fucus serratus*, but that the bulk of the sulfate is attached to fucoidan without further reduction.

Much of a cell's sulfur is incorporated into proteins. Two sulfur-containing amino acids, cysteine and methionine, are important in maintaining the three-dimensional configurations of proteins through sulfur bridges, as well as in linking chromophores to the protein in phycobiliproteins (Fig. 4.5). Parts of the biologically important molecules biotin, thiamine (Fig. 5.1), and coenzyme A also involve sulfur. However, many algae, especially seaweeds, produce commercially valuable sulfated polysaccharides that are important in thallus rigidity (e.g., carrageenan in red algae) and adhesion (e.g., fucoidan in brown algae) (sec. 4.5.2 and 1.6.1). The uptake of sulfate by *Fucus serratus* has been studied in this connection (Coughlan 1977). Other sulfur-containing compounds in seaweeds include, in red algae, taurine and its derivatives (sulfur at the sulfite level of reduction) (O'Kelley 1974; Ragan 1981). Some species of *Desmarestia* have so much sulfuric acid in their vacuoles that the pH is close to 1; this may serve as a grazer deterrent (sec. 3.2.3). Crystalline sulfur has been found in *Ceramium rubrum* and has been shown to be responsible for the toxicity of this alga to the bacterium *Bacillus subtilis; C. rubrum* is unusual in having a high free-sulfur content (Ikawa et al. 1973).

5.5.6 Iron

There is little information on the chemical forms of iron occurring in seawater, and little is known about their availability to algae. Iron availability appears to be a function of the chemical lability of the colloidal and particulate phases of iron; the "available" iron likely comprises those phases that dissolve (or dissociate) on time scales similar to algal generation times. Current evidence suggests that iron availability can be estimated using chelating agents (e.g., oxine) to measure the lability of iron in seawater. Although "dissolved" iron generally is more labile than "particulate" iron, much of the "dissolved" iron may be unavailable for uptake. Conversely, large proportions of the concentrations of labile iron often occur in the "particulate" fractions. Thus, processes affecting the transport and removal of particulates probably are important when considering iron nutrition among algae.

Recent evidence that phytoplankton in the sub-Arctic Pacific and in the Antarctic Ocean may be iron-limited has stimulated much interest in the role of iron in phytoplankton productivity (Martin & Fitzwater 1988; Martin et al. 1990). In these open-ocean environments, deposition from atmospheric dust appears to be the principal source of iron (Martin et al. 1990). There is evidence that photolysis of colloidal and particulate iron may be an important pathway for converting crystalline, unavailable iron into biologically utilizable forms (Wells 1991; Palenik et al. 1991b).

Despite its abundance, much of the earth's iron is in the Fe(III) form and is not readily available to organisms. In the presence of oxygen, aqueous Fe^{2+} is rapidly oxidized to Fe^{3+}, which readily forms its insoluble ferric hydroxide at the pH of seawater. In fact, most organisms possess special mechanisms to acquire iron for growth (Boyer et al. 1987).

Up to 90% of the iron in seawater is complexed with organics, such as humic or fulvic acids, or with special biologically produced chelators called siderophores (Boyer et al. 1987). Siderophores are defined as low-molecular-weight compounds produced by prokaryotic and eukaryotic microorganisms and higher plants under iron limitation. They serve as (virtually) Fe-specific extracellular chelators to aid in the solubilization and assimilation of Fe^{3+}. Their chemistry and physiology have been reviewed recently (Moody 1986; Boyer et al. 1987; Römheld 1987). Microbial siderophores can be divided into three major categories: catechols (restricted to bacteria), hydroxamates, and phytosiderophores (known from higher-plant roots; Römheld 1987). There are several well-known hydroxamate siderophores: a series of ferrichromes produced by bacteria and fungi, rhodotorulic acid produced by yeast, schizokinen from a cyanophyte (*Anabaena* sp.), and prorocentrin from a marine dinoflagellate (Trick et al. 1983). Synthetic chelators such as EDTA are used in culture media to keep iron in solution. Citrate, although usually not regarded as a siderophore, can solubilize iron, but with a much lower affinity than siderophores.

Although an actual mechanism for iron uptake is unknown, one possibility is that a siderophore forms a complex with Fe^{3+}. This complex may be taken across the membrane or reduced at the cell surface, liberating Fe^{2+}, with concomitant recycling or extracellular release of the siderophore. Because the complexing strength of the hydroxamic acids is much greater for Fe^{3+} than for Fe^{2+}, the reduction of the Fe^{3+} complex provides a means of releasing the complexed iron and freeing the siderophore to pick up more Fe^{3+}. Other possible mechanisms have been discussed by Neilands (1973) for microorganisms, Boyer et al. (1987) for cyanophytes, and Römheld (1987) for higher plants.

Phytoplankton apparently take up both Fe^{2+} and Fe^{3+} (Anderson & Morel 1980, 1982; Finden et al.

1984). The ferrous ion was previously thought to be present in very low amounts, but more recent evidence suggests that it can be formed by (1) oxidation rates ($Fe^{2+} \rightarrow Fe^{3+}$) that are 150 times slower than previous estimates (Anderson & Morel 1982), (2) chelators that can chemically or photochemically reduce Fe^{3+} to Fe^{2+} (Huntsman & Sunda 1980), and (3) direct photoreduction of Fe^{3+} (Wells 1991). Other suggested iron sources have been iron oxides, which, often in association with colloids, adsorb onto the cell surface and subsequently dissolve, releasing iron for transport into the cell (Wells et al. 1983). However, because the dissolution kinetics of iron oxide are not well understood, the possible importance of this process is unknown.

To date there have been only two studies of iron uptake by seaweeds. Saturation kinetics were reported for iron uptake by *Macrocystis pyrifera* (Manley 1981). Labeled iron exchanged more slowly from the free space than did other divalent cations. The slow exchangeability of iron may reflect the higher affinity of the cell wall and intercellular constituents for ferric ion. Manley (1981) also found that when bathophenanthroline disulfonate (BPDS) was added to the culture medium, inhibition of iron uptake was immediate and drastic. Because BPDS has a high affinity and specificity for chelating the ferrous ion, there may be a reduction of Fe^{3+} to Fe^{2+} before iron is taken up by *Macrocystis* (Anderson 1984). This is similar to the situation for phytoplankton (Anderson & Morel 1982) and higher plants (Brown 1978; Römheld 1987). After iron is taken up by *Macrocystis*, it enters the sieve tubes, where it reaches 11 times the external concentration. Manley (1981) postulated that the iron was chelated by some organic compound and was translocated to juvenile fronds. This is similar to the case for higher plants, in which ferrous ion is oxidized to ferric in the xylem and chelated with citrate (Brown 1978).

There have been no studies on seaweeds and only a few for phytoplankton that have determined the concentration of iron that limits algal growth. Brand et al. (1983) and Doucette and Harrison (1990) reported that the half-saturation constant for iron-limited growth for 10 species of neritic phytoplankton ranged from 10^{-23}-M to 10^{-21}-M Fe. Iron limitation in a marine dinoflagellate resulted in reduced numbers of chloroplasts and some degeneration of lamellar organization (Doucette & Harrison 1990). The ratio of the maximum to the minimum Fe per cell ranged from 10 to 100. Cyanophytes that fix N_2 have a much higher iron requirement because of the iron in the nitrogenase enzyme. Iron is stored in higher plants and animals as ferritin, but storage of iron in special compounds has not been observed in algae (Boyer et al. 1987).

In addition to its role in cell growth, iron has several specific roles in cell metabolism. It is at the center of the cytochromes and ferredoxin, which transfer electrons in the respiratory chain and in photosynthesis. The importance of iron in electron transport lies in its ability to change valence between Fe^{2+} and Fe^{3+}, but it is also present in a number of oxidizing enzymes (such as catalase) in which it does not change valence (Bidwell 1979). Many cyanophytes can replace Fe-containing ferredoxin with non-Fe-containing flavodoxin, allowing N_2 fixation to proceed, even under periods of iron stress (Boyer et al. 1987). Although iron is not part of the chlorophyll molecule, it is required as a cofactor in the synthesis of chlorophylls. The chlorotic effect of iron deficiency in phytoplankton is well known (Boyer et al. 1987; Doucette & Harrison 1990). Many of the major enzymes of nitrogen metabolism are Fe-containing proteins (NR, NiR, GS, and nitrogenase), and thus nitrogen metabolism is extremely sensitive to iron stress.

In summary, Fe has a low solubility in seawater. Organisms do not respond to the total iron concentration, but rather to the "biologically available iron," the ratio of which is greatly affected by the level of chelators present. Production of the siderophore is induced along with a specific membrane uptake system under conditions of low iron availability. This ability to tie up iron as a siderophore-Fe complex that is available only to the cyanobacterium that produced the siderophore may play an important part in the establishment and domination of cyanobacterial populations. Cyanophytes can solubilize iron by producing a wide variety of highly specific siderophores that can actively transport the bound iron into the cell. The addition of humic substances (nonspecific chelators) increases phytoplankton growth, but it is difficult to determine whether the increased growth is due to increased solubility of the essential trace metals, such as iron, or to the decreased levels of inhibitory elements, such as copper. Virtually no work has been done on iron metabolism in seaweeds, and it is becoming increasingly clear that this important metal deserves considerable attention.

5.5.7 Trace elements

Uptake kinetics for iodine (I), zinc (Zn), cesium (Cs), strontium (Sr), cobalt (Co), molybdenum (Mo), and rubidium (Rb) have been studied (Gutknecht 1965; Penot & Videau 1975; Floc'h 1982; Amat & Srivastava 1985). Uptake of zinc by *Ascophyllum nodosum* (Skipnes et al. 1975) and uptake of cobalt by *Laurencia corallopsis* (Bunt 1970) appear to be by active transport. A simple exchange process involving intracellular polysaccharides is the main uptake mechanism for strontium (Skipnes et al. 1975), and in some algal species it is also the uptake mechanism for zinc (Gutknecht 1963, 1965).

The effects of Cu, Zn, Mn, and Co on the growth of gametophytes of *Macrocystis pyrifera* have been examined in chemically defined medium (Kuwabara 1982); the findings indicate that toxic concentrations of copper and zinc ions, together with cobalt and manganese deficiencies, may be among the factors controlling the growth of some marine macrophytes in deep seawater off southern California.

The principal roles of Mn, Cu, Zn, Se, Ni, and Mo are as enzyme cofactors (Table 5.1). Manganese plays a vital role in the oxygen-evolving system of photosynthesis and is a cofactor in several Krebs-cycle enzymes (Bidwell 1979). Copper is present in plastocyanin, one of the photosynthetic electron-transfer molecules, and is a cofactor in some enzyme reactions (Bidwell 1979). Urease, which catalyzes the hydrolysis of urea to ammonium in many microalgae, bacteria, and fungi, contains nickel (Hausinger 1987).

Zinc is an activator of several important dehydrogenases and is involved in protein-synthesis enzymes in higher plants. It is essential in algae, where it probably plays similar roles (O'Kelley 1974). McLachlan (1977) could not demonstrate a requirement for Zn, Cu, or Mn by embryos of *Fucus edentatus* in defined medium, but he concluded that these elements probably were present as contamination in sufficient concentrations to satisfy nutritional needs. Zinc, at an optimum concentration of 0.5 nM, was shown by Noda and Horiguchi (1972) to be required by *Porphyra tenera;* without it, chlorophyll and phycobilin productions were hindered, and the content of high-molecular-weight protein decreased.

Molybdenum is most important, in algae and other plants, in nitrate reduction and nitrogen fixation, where it is a component of the enzymes (NR and nitrogenase) involved in these processes (Howarth & Cole 1985). Nasr and Bekheet (1970) reported that addition of trace amounts of ammonium molybdate increased the dry weights of *Ulva lactuca, Dictyota dichotoma,* and *Pterocladia capillacea;* presumably much of the increase can be attributed to Mo stimulation of NR, because there was only a trace of ammonium in the culture medium. Like iron, molybdenum can participate in redox reactions because of its ability to change valence, in this case between Mo^{5+} and Mo^{6+}.

In general, marine plants tend to have lower amounts of Mo than freshwater plants, even though the concentration of dissolved Mo is about 20 times greater in seawater than in average fresh waters (Howarth & Cole 1985). Sulfate is 100 times more abundant in seawater than in fresh water, and the stereochemistry of Mo is similar to that of sulfate. Consequently, it has been shown that sulfate inhibits molybdate assimilation by phytoplankton, making Mo less available in seawater than in fresh water (Howarth & Cole 1985). Therefore, nitrogen fixation and nitrate assimilation may require a greater expenditure of energy in seawater than in fresh water.

Additions of selenium have been reported to increase the growth of *Fucus spiralis* and the red alga *Stylonema alsidii* (Fries 1982a); selenium at 0.01 μM was needed in artificial seawater for normal growth. Evidence that the Se requirement for growth of algae may be more widespread comes from Harrison et al. (1988),

who found that 15 out of 20 species of marine diatoms required Se for growth at concentrations of approximately 0.001 μM. The selenium-containing enzyme glutathione peroxidase occurs in the mitochondria and chloroplasts of marine phytoplankton, where it acts to detoxify injurious lipid peroxides and to maintain membrane integrity (Price & Harrison 1988). Vanadium, at a concentration of about 10 μg L^{-1}, is required for maximal growth of some macroalgae (Fries 1982b).

Long-distance translocation of ^{125}I was studied in *Laminaria saccharina,* and it followed a source-to-sink pattern (Amat & Srivastava 1985). The anion, I^-, was the only species of iodine transported toward the meristematic region at the blade–stipe junction. Although the exact role of iodine is still unknown, it has been shown to be stimulatory, and in some cases essential for growth and reproduction in some brown algae (Hsiao 1972).

5.5.8 Vitamins

The main roles for vitamins in algae appear to be as enzyme cofactors (Swift 1980). Thiamine, for example, as thiamine pyrophosphate, is a cofactor in the decarboxylation of pyruvic acid and other α-keto acids. Biotin is a cofactor in carboxylation reactions.

The cultivation of seaweeds under axenic conditions often results in aberrant morphology. This suggests that bacteria associated with the macrophyte or residing in the seawater may be producing a growth regulator. Possible growth regulators include lipophilic substances such as sterols (e.g., cholesterol and fucosterol). When these two sterols are exposed to UV irradiation, they can form the lipophilic vitamins D_3 and D_2, which have been found to influence growth in some higher plants (Buchala & Schmid 1979). Fries (1984) found a lipophilic growth regulator (suggested to be cholesterol and fucosterol) in the blades of *Fucus vesiculosus* and *Laminaria digitata* that stimulated the growth of axenically cultured *Fucus spiralis.* That suggestion was confirmed by large increases in fresh weight (up to 100%) in *Enteromorpha compressa* and *Fucus spiralis* when vitamins D_2 and D_3 at 10^{-8} to 10^{-7} M were added to the growth medium of these macroalgae. Fries (1984) suggested that intertidal macroalgae would be exposed to enough UV light to convert various sterols into vitamin D.

5.6 **Long-distance transport (translocation)**

Many large kelps are similar to vascular plants in that their movements of inorganic and organic compounds occur by translocation. The movement of organic compounds and the anatomical features of translocating tissues (basically, sieve elements that are in longitudinal files and form a three-dimensional interconnected system in the medulla in blades and stipes) were discussed in section 4.6. This section focuses on long-distance transport of inorganic ions.

Macrocystis sporophytes absorb nutrients from seawater and translocate photoassimilates from mature, nongrowing parts to support rapid growth, primarily from their basal (frond-producing) and apical (blade-producing) meristems. Analysis of the sieve-tube sap has revealed the presence of mannitol (65% dry weight), amino acids (15%), and inorganic ions (Fe, Mn, Co, Ca, Zn, Mo, I, Ni) (Manley 1983, 1984).

Various tracers (^{32}P, ^{86}Rb, ^{35}S, ^{99}Mo, ^{45}Ca, ^{36}Cl) have been used to show the movements of mineral elements in the thallus in *Laminaria digitata* (Floc'h 1982). Phosphorus, sulfur, and rubidium undergo pronounced long-distance transport, whereas chloride, molybdenum, and calcium do not appear to move. Nitrate was found in sieve-tube sap of *Macrocystis pyrifera,* but no translocation measurements were made (Manley 1983).

There is a high demand for phosphorus in growing or meristematic regions. Movement of ^{32}P in Fucales and Laminariales is consistently from the older tissues toward the younger, growing regions (Fig. 5.17). Therefore, a source-to-sink relationship exists, similar to that observed in vascular plants. In *Laminaria hyperborea* the older parts of the blade apparently serve as sources of phosphate for the meristematic regions, in much the same way as has been shown for carbon assimilation (Floc'h 1982). Evidence of phosphorus translocation is indirect; there is no significant difference in terms of ^{32}P uptake between young and old tissues, but the older, slowly growing tissues probably translocate unused phosphorus to new, actively growing tissues.

The midrib in many Fucales is the main pathway for mineral transport, as shown by the intensity of the labeling in autoradiographs (Fig. 5.17). Most of the translocation occurs through the medulla, with some secondary lateral transport from the medulla to the meristoderm (meristematic epidermis) in the stipe. Because ^{32}P has been found in the sieve-tube sap of *Macrocystis* and shows the same velocity and directionality of transport as ^{14}C, it most probably moves through the sieve elements (sec. 4.6).

Because phosphate translocation frequently occurs against a concentration gradient and is related to algal metabolism, the mechanism of transport is certainly not by simple diffusion. At present there is no general agreement on the mechanism of phosphate transport or on whether it is transported in inorganic form or organically bound. A few hours after inorganic phosphorus is taken up, it is incorporated into organic compounds, especially hexose monophosphates (Floc'h & Penot 1978; Floc'h 1982). This suggests that phosphorus may be translocated in an organically bound form, but it does not exclude the possibility that inorganic phosphorus may be translocated, followed by phosphorylation in the sieve-tube sap, as in higher plants (Bidwell 1979).

Nitrogen shows a pattern similar to that for phosphorus, with the highest demand for nitrogen in the meristematic regions. This demand cannot be met by uptake rates, and studies with *Laminaria digitata* have demonstrated that up to 70% of the nitrogen demand by the meristem is met by translocation of nitrogen assimilated by the mature blade, probably as organic nitrogen, such as amino acids (Davison & Stewart 1983).

Similarly, calculations based on measured iron-uptake rates, on the iron composition in the blade tissue, and on a description of frond growth have shown that iron-assimilation rates (tissue incorporation) for small, juvenile fronds of *Macrocystis pyrifera* can be accounted for only by imports of iron from mature fronds and that mature fronds take up an excess of iron (Manley 1984). Iron in the sieve-tube elements may be chelated with some organic compound; citrate is a known iron chelator in higher plants (Römheld 1987).

Recent studies have shown that translocation of ^{125}I in *Laminaria saccharina* follows a source-to-sink pattern. The anion I$^-$ was the only species transported, and the velocity ranged from 2 to 3.5 cm h^{-1} (Amat & Srivastava 1985).

5.7 Growth kinetics
5.7.1 Theory

The classic principles of microbial growth kinetics derived by Monod (1942) for growth limited by a single substrate are based on the assumption that the formation of new biomass bears a simple relationship to uptake of the substrate. Thus, the specific growth rate is related directly to the concentration of extracellular substrate, according to equation (1) in Table 5.8. Note that this equation, generally referred to as the Monod equation, is almost identical with the equation describing the relationship between uptake rate and nutrient concentration [Table 5.3, equation (4)]; both equations describe a rectangular hyperbola. In Monod's experiments, the growth-limiting substrate was glucose (carbon), which was metabolized almost immediately after uptake by bacterial cells. In that special case, growth was proportional to the external concentration of the limiting substrate, because glucose was not stored, and thus the yield (number of cells formed per unit of limiting substrate) remained constant.

In only a few cases in microalgae has the relationship between a steady-state growth rate and the external nutrient concentration been described by the Monod equation (Rhee 1980). Deviations from the equation appear to be due to increased cell mortality at low dilution rates and the ability of cells to store nutrients such as nitrogen and especially phosphorus. In many studies (Rhee 1980, 1982), external nutrient concentrations have been undetectable over a wide range of growth rates. In those cases, the growth rate was related to the intracellular nutrient concentration or the cell quota, q, according to equation (2) in Table 5.8 (the Droop equation). The subsistence cell quota, q_0, is the minimum concentration of the limiting nutrient per cell required before growth can proceed. When the specific

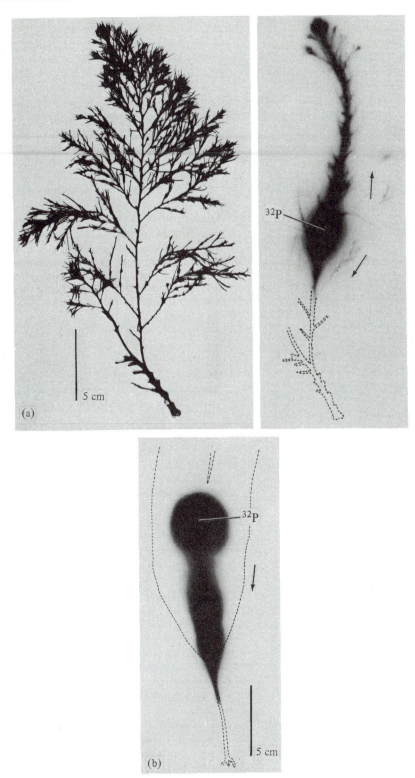

Figure 5.17. Translocation of phosphorus. Autoradiographs of brown algae labeled with ^{32}P for 3 h, showing point of uptake and direction of migration (arrows): (a) *Cystoseira baccata*, showing the whole thallus and the autoradiograph after isotope translocation; (b) autoradiograph of *Laminaria setchellii*. (From Floc'h 1982, with permission of Walter de Gruyter & Co.)

Table 5.8. *Algal growth equations*

$$(1) \quad \mu = \mu_m \frac{S}{K_s + S}$$

$$(2) \quad \mu = \mu'_m \left(1 - \frac{q_0}{q}\right)$$

$$(3) \quad \mu = \frac{100[\ln(N_t/N_0)]}{t}$$

Symbols: K_s, half-saturation constant for growth; N_0, initial biomass; N_t, biomass on day t; q, cell quota (amount of nutrient per cell); q_0, subsistence cell quota; S, substrate concentration; t, time in days; μ, specific growth rate; μ_m, true maximum specific growth rate; μ'_m, specific growth rate at infinite S.

growth rate (μ) is plotted against q, the Droop equation describes a rectangular hyperbola with a threshold q_0 and an asymptote equal to the maximum specific growth rate (μ'_m) (Fig. 5.18). In Droop's model, growth is empirically related to the total intracellular concentration of substrate. From a biochemical point of view, the growth rate is likely to be related to a certain intracellular pool (Rhee 1980). This internal nutrient pool is utilized during growth, but it is constantly replenished by concentration-dependent uptake of the external nutrient. At steady state, net uptake is the amount required to maintain equilibrium in the internal pool, and therefore the growth rate is related to the total intracellular concentration (i.e., the cell quota), assuming that the size of the internal nutrient pool is directly proportional to the cell quota.

Several assumptions or criteria are implied in each equation. The Monod equation is useful when carbon is the limiting nutrient, because the cell quota (reciprocal of cell yield) remains constant despite a varying growth rate. However, for other nutrients, especially phosphorus (Rhee 1980, 1982; Cembella et al. 1984), the cell quota increases sharply as the growth rate increases. This increase in cell quota allows growth to continue for several generations under conditions of phosphorus limitation, by mobilization of phosphorus storage reserves. Under these conditions, the growth rate is not related to the external phosphate concentration, but rather to the phosphorus cell quota. In this case, and similarly for nitrogen, the Droop equation is recommended for determining growth-kinetic parameters. Further comparisons between the Monod and Droop equations have been discussed by Goldman and McCarthy (1978), Cembella et al. (1984), and Droop (1983). The theory of growth kinetics described here has been developed from numerous experiments on bacteria and phytoplankton. No carefully executed experiments have been conducted on macroalgae.

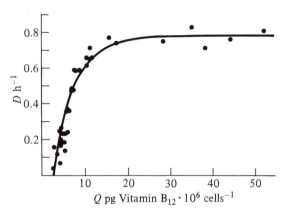

Figure 5.18. Relationship between dilution rate (D, which is equivalent to the steady-state growth rate) and cell quota (Q) for the phytoflagellate *Pavlova lutheri* in a vitamin B_{12}-limited chemostat. (From Droop 1968, *J. Mar. Biol. Assoc. U.K.* 48:689–733, with permission of Cambridge University Press.)

5.7.2 Measurement of growth kinetics

Various means are available for measuring the rates of growth for seaweeds (Chapman 1973a, Brinkhuis 1985). Nondestructive methods include measuring changes in wet weight or surface area or (in kelps) measuring the movement of holes punched in the base of the blade. Sampling to determine changes in dry weight, or changes in plant carbon or nitrogen, is destructive to the plant. On the basis of a time series of changes in any one of the foregoing parameters, the specific growth rate (μ) can be calculated as the percentage increase in the parameter per day. For example, the specific growth rate can be calculated from the daily increases in fresh weight according to equation (3) in Table 5.8, assuming steady-state exponential growth (DeBoer et al. 1978). For ecological or field purposes, nondestructive sampling is preferable, because the growth of a given plant can be followed over time and related to the limiting nutrient concentration in the ambient water.

The most accurate way to determine the relationship between external nutrient concentration and growth rate in seaweeds is to use continuous-flow cultures. In these cultures, an attempt is made to keep the external nutrient concentration constant by using a high dilution rate (flow rate/container volume) and a high container-volume/plant-biomass ratio. Then the growth rate is measured at a series of external nutrient concentrations when a steady state is achieved. During the approach to a steady state, a reasonably constant biomass must also be maintained by harvesting at frequent intervals or by increasing the inflowing nutrient concentration to compensate for the increase in biomass. In phytoplankton continuous cultures, this adjustment of biomass is achieved automatically, because cells are removed with

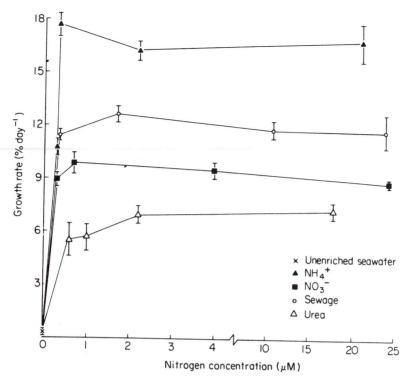

Figure 5.19. Growth rates (± 1 SE) for the red seaweed *Agardhiella subulata* as functions of residual nitrogen concentrations for various sources of nitrogen enrichment. (From DeBoer et al. 1978, with permission of *Journal of Phycology*.)

the outflowing medium. A steady state can also be approximated by using semicontinuous cultures, in which changes in nutrient concentration in the medium are minimized by making frequent changes in the medium. The time required to reach a steady state will depend primarily on the growth rate and culture conditions. DeBoer et al. (1978) have arbitrarily chosen the criterion that the biomass must increase 10-fold. For example, if an alga is growing at a rate of 10% per day, then a steady state will be reached in 23 days (DeBoer 1981). Similarly, many phytoplankton physiologists use the guideline of 10 generation times to reach a steady state.

The uptake rate is a good approximation of the growth rate when nutrients do not limit growth, or when steady-state growth occurs under conditions of nutrient limitation. However, nutrient-uptake rates may greatly exceed growth rates when nutrients are added to an algal culture growing under nutrient limitation. The half-saturation constant of uptake, K_s, determined during the transient conditions of short-term uptake experiments, may also be greater than the K_s value for growth.

5.7.3 Growth rates

In contrast to the numerous studies on the relationship between growth rate and ambient nutrient concentration in phytoplankton, only a few such studies have been conducted on seaweeds. DeBoer et al. (1978)

found that the nitrogen growth kinetics for two red algae, *Agardhiella subulata* and *Gracilaria foliifera*, followed typical growth-saturation curves (Fig. 5.19). The values of K_s ranged from 0.2 to 0.4 µM for various nitrogen sources, and growth rates were saturated at concentrations of NH_4^+ or NO_3^- as low as 1 µM. Similarly, the annual brown alga *Chordaria flagelliformis* had low K_s values (0.2–0.5 µM) for the three nitrogen substrates: nitrate, ammonium, and urea (Probyn & Chapman 1983). Those very low values were in contrast to the findings in other studies, where, for example, the growth rate was saturated at 10 µM nitrate for *Laminaria saccharina* (Chapman et al. 1978; Wheeler & Weidner 1983), between 6 and 15 µM nitrate for juvenile *Macrocystis pyrifera*, and at 30 µM nitrate or ammonium for the estuarine green alga *Cladophora* aff. *albida* (Gordon et al. 1981). These latter seaweeds would be poor competitors against phytoplankton, whose growth rate is saturated at 1 µM nitrogen or even less. Growth of juvenile sporophytes of *Macrocystis pyrifera* was found to be saturated at a PO_4^{3-} concentration greater than 1 µM (Manley & North 1984).

Although the growth rate is often related to the concentration of nutrients in the external medium, studies with phytoplankton have recently shown that the growth rate can be estimated more accurately from the concentrations of nutrients within cells (Droop 1968,

Table 5.9. *Representative maximum, critical, and minimum nitrogen and phosphorus tissue concentrations*

Species	Nutrient	Maximum (% dry wt)	Critical (% dry wt)	Minimum (% dry wt)	References
Phaeophyceae:					
Chordaria flagelliformis	NH_4^+		1.5	0.5	Probyn & Chapman (1983)
	NO_3^-		0.9	0.3	
Laminaria saccharina	NO_3^-		1.9	1.3	Chapman et al. (1978)
Macrocystis pyrifera	NO_3^-		^{a}	0.7	Wheeler & North (1980)
Macrocystis pyrifera	PO_4^{3-}		0.2		Manley & North (1984)
Pelvetiopsis limitata	NO_3^-		1.2–1.5		Fujita et al. (1989)
	NH_4^+		0.9		Fujita et al. (1989)
	Nitrogen	2.2	1.5	0.86	Wheeler & Björnsäter (1992)
	PO_4^{3-}	0.5		0.27	
Chlorophyceae:					
Chaetomorpha linum	Nitrogen	3.2	0.7	0.3	Lavery & McComb (1991b)
	PO_4^{3-}	0.23	0.04	0.01	Lavery & McComb (1991b)
Cladophora albida	Nitrogen		2.1	1.2	Gordon et al. (1981)
Codium fragile	NO_3^-	2.6	1.9	0.8	Hanisak (1979)
	PO_4^{3-}	0.48		0.28	
Enteromorpha intestinalis	Nitrogen	5.1	2.5	2.0	Björnsäter & Wheeler (1990)
	PO_4^{3-}	0.73		0.37	
Ulva fenestrata	PO_4^{3-}	0.6	0.5	0.3	Björnsäter & Wheeler (1990)
	Nitrogen	5.5	3.2	2.4	
Ulva rigida	NO_3^-		2.4		Fujita et al. (1989)
	NH_4^+		3.0		Fujita et al. (1989)
	Nitrogen	3.2	2.0	1.3	Lavery & McComb (1991b)
	PO_4^{3-}	0.06	0.025	0.02	Lavery & McComb (1991b)

^{a}Linear relationship between μ and Q over the experimental range of Q.

1973; Rhee 1980). This basic principle is related to the common agricultural practice of plant-tissue analysis, where the critical tissue concentration is that which just saturates growth. Higher or lower concentrations indicate nutrient reserves or deficiency, respectively. The technique has now been applied to aquatic vascular plants (Gerloff & Krombholz 1966) and seaweeds (Hanisak 1979, 1990; DeBoer 1981). Representative critical tissue concentrations of nitrogen for a number of macrophytes are given in Table 5.9, along with their minimum tissue concentrations. The growth of *Codium fragile* was found to be more directly related to the concentration of nitrogen present in its thallus than to the nitrogen concentration in the surrounding water (Fig. 5.20). Even though the values of internal nitrogen ranged from 0.9% to 4.8%, growth rates remained constant at nitrogen concentrations in excess of about 2%. Whereas tissue nitrogen generally reaches saturating values with increasing concentrations of external nitrogen, phosphorus does not saturate as readily, if at all, in three chlorophytes (Gordon et al. 1981; Björnsäter & Wheeler 1990) (Fig. 5.21). Therefore, the critical tissue phosphorus concentration is more difficult to determine accurately in some species. After 3 weeks of careful cul-

turing, tissue phosphorus concentrations in juvenile *Macrocystis pyrifera* were hyperbolically related to external PO_4^{3-} concentration, with tissue phosphorus ranging from 0.12% to 0.53% of dry weight for PO_4^{3-} concentrations from 0.3 to 6 μM. The critical tissue phosphorus was about 0.2% of dry weight (Manley & North 1984).

In fact, seaweeds do not have just one critical tissue concentration of nitrogen when other factors such as light are also considered. This is because light modifies the nitrogen requirement for maximal photosynthesis and growth by altering biochemical constituents (i.e., pigments, RuBisCO, and nitrogen reserves) that affect the nitrogen level (Lapointe & Duke 1984; Shivji 1985). For example, the nitrogen level for *Gracilaria tikvahiae* at 7% of surface irradiance averaged 25% more than that for plants grown at surface irradiance under similar nitrogen conditions, because of increases in pigments (e.g., phycoerythrin), tissue NO_3^-, and proteins. The ambient concentration of one nutrient may also affect the critical tissue concentration of another nutrient. Phosphorus-limited *Ulva* and *Enteromorpha* maintained high levels of tissue nitrogen, but nitrogen-limited algae experienced reductions or depletion of phosphorus

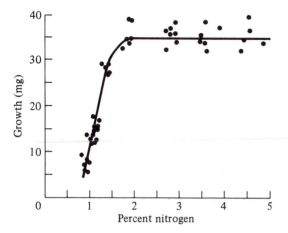

Figure 5.20. Relationship between growth and internal nitrogen concentration in *Codium fragile;* growth measured as increase in dry weight after 21 days. (From Hanisak 1979, with permission of Springer-Verlag, Berlin.)

(Björnsäter & Wheeler 1990). Therefore, because light and ambient nutrient concentrations vary seasonally, the critical nitrogen and phosphorus tissue concentrations also vary seasonally (Wheeler & Björnsäter 1992).

Monitoring of nutrient availabilities and limitations in the coastal waters of Denmark has been conducted using transplants of *Ceramium rubrum* (Lyngby 1990). Monthly monitoring of tissue nitrogen and phosphorus concentrations, followed by comparisons with the critical tissue concentrations, revealed that nitrogen limited growth, except in May, when growth was phosphorus-limited.

The macroalgal biomass has been monitored in the Peel-Harvey estuary in southwestern Australia for two decades (Lavery et al. 1991). Macroalgal blooms occurred suddenly in the late 1960s, and *Cladophora montagneana* dominated until 1979. A catastrophic event compounded by a series of unfavorable conditions resulted in the loss of *Cladophora* from the deep areas and its estuary-wide replacement by *Chaetomorpha,* which was more competitive in shallow areas. Periods of high nutrient concentrations favor *Ulva rigida* and *Enteromorpha intestinalis,* whereas *Chaetomorpha* resumes dominance during periods of low nutrients. As a result of a recent study of the nutritional ecophysiology of *Chaetomorpha* and *Ulva,* it is now possible to understand some of the physiological mechanisms that are responsible for their ecological distribution (Lavery & McComb 1991b). Critical tissue nitrogen and phosphorus concentrations have revealed that *Ulva* requires more nitrogen than *Chaetomorpha* (20 vs. 12 mg g_{dw}^{-1}), but less phosphorus (0.25 vs. 0.5 mg g_{dw}^{-1}). Even though *Ulva* requires almost twice as much nitrogen as *Chaetomorpha,* it takes up ammonium and nitrate at slower rates than does *Chaetomorpha.* Thus, *Ulva* frequently is

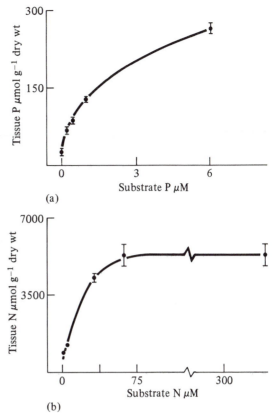

(a)

(b)

Figure 5.21. Concentrations of total nutrients in *Cladophora* tissues as functions of the nutrient concentrations supplied in complete medium: (a) phosphorus, supplied at 0–6 μM; nitrogen at 375 μM; (b) nitrogen, supplied at 0–375 μM; phosphorus at 12 μM. Each point is the mean (±SE) for three replicates. (From Gordon et al. 1981, with permission of Walter de Gruyter & Co.)

nitrogen-limited during the spring because of its high nitrogen requirements, its reduced ability to take up nitrogen, and its limited ability to store nutrients over winter. (Its ratio of maximum tissue nitrogen to minimum tissue nitrogen is 2.5; see Table 5.9.) *Chaetomorpha* persists over winter, allowing it to take advantage of elevated nutrient concentrations over the winter and store nutrients (its maximum : minimum for tissue nitrogen is 11, and its maximum : minimum ratio for tissue phosphorus is 23) that are used to support high growth rates in the spring and summer when nitrogen becomes limiting. In addition, dense accumulations (mats) of *Chaetomorpha* reduce the oxygen levels in the water overlying sediments, and that induces phosphorus release from the sediments (Lavery & McComb 1991a).

Growth of *G. tikvahiae* limited by both light and nitrogen yielded a parabolic function of the C : N ratio (Fig. 5.22), because light and nitrogen have opposite effects on the relationship between the C : N ratio

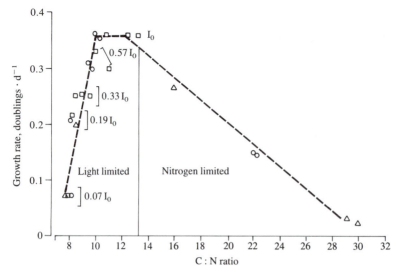

Figure 5.22. Growth rate for *Gracilaria tikvahiae* as a function of C:N ratio under light and nitrogen limitation. Data are from three studies by Lapointe and co-workers; all data were obtained by growing *G. tikvahiae* under identical conditions in outdoor, continuous-flow seawater cultures. (From Lapointe & Duke 1984, with permission of *Journal of Phycology*.)

and the growth rate (Lapointe & Duke 1984). Under light limitation, NO_3^- uptake is relatively rapid compared with rates of carbon fixation, and that results in a low C : N ratio and accumulation of nitrogen reserves such as phycoerythrin and tissue nitrate. Alternatively, under high light, with nitrogen limitation, carbon fixation is relatively rapid compared with the rates of NO_3^- uptake, and that results in a high C : N ratio and decreased phycoerythrin.

5.8 Effects of nutrient supply

Because nitrogen is the primary resource that limits seaweed growth, variations in seaweed growth rates should parallel variations in the nitrogen supply. The nitrogen-uptake capacity (V_{max}) is a direct function of the surface-area : volume (SA : V) ratio (Rosenberg & Ramus 1984), whereas the nitrogen-storage capacity varies approximately inversely with the SA : V ratio (Rosenberg & Ramus 1982; Duke et al. 1987). This suggests that the degree of coupling between seaweed growth rates and the nitrogen supply may also be a function of SA : V. High SA : V species, which have high V_{max} values, but low storage capacity, will have growth rates highly correlated with the nitrogen supply, whereas low SA : V species, with low V_{max} values, will store nitrogen and have growth rates relatively independent of the nitrogen availability (Rosenberg & Ramus 1982). For example, *Ulva curvata*, an opportunistic annual with a high SA : V ratio, is capable of utilizing transiently high ammonium concentrations (high V_{max}), and it is capable of high growth rates. In contrast, *Codium fragile*, a persistent perennial, cannot utilize ammonium pulses efficiently, has a slow growth

rate, and tends to integrate short-term variability by virtue of its low growth rate (Ramus & Venable 1987). Thus, a seaweed's "functional form" (Littler & Littler 1980) may determine its ability to buffer nutrient variability (Norton 1991). Thus, in general terms, opportunistic species often are annuals that feature rapid nutrient uptake and rapid growth potentials, high SA : V ratios, low nutrient-storage capabilities, and low levels of defense against herbivory (presumably because tissue loss is readily replaced via rapid nutrient uptake and growth), and they typically dominate eutrophic environments (Littler & Littler 1980). On the other hand, perennials tend to be slow-growing and can store large quantities of nutrients that can act as a buffer during times of nutrient shortage (Carpenter 1990).

Light and temperature limitations on growth tend to uncouple growth from nitrogen uptake (Duke et al. 1989). This may explain seaweeds' accumulation of nitrogen at low levels of light and temperature, because nitrogen uptake may be less limited by light and temperature than is growth. This, in turn, may partially compensate for the effects of reduced light and temperature on growth by increasing the pigment and enzyme levels (Duke et al. 1989).

5.8.1 Chemical composition

Extensive analysis of the chemical composition of marine plankton has revealed that the ratio relating carbon, nitrogen, and phosphorus is 106 : 16 : 1 (by atoms) (i.e., C : N = 7 : 1 and N : P = 16 : 1). This is commonly referred to as the Redfield ratio. Decomposition of this organic matter occurs according to the same ratio. However, Atkinson and Smith (1983) have recently shown that benthic marine macroalgae and seagrasses are much more depleted in phosphorus and less depleted in nitrogen, relative to carbon, than are phytoplankton. The median ratio C : N : P for seaweeds is about 550 : 30 : 1 (i.e., C : N = 18 : 1 and N : P = 30 : 1). An important ramification of these observations is that the amounts of nutrients required to support a particular level of net production are much lower for macroalgae than for phytoplankton. In addition, seaweeds, on average, should be less prone to phosphorus limitation with their N : P ratio of 30 : 1 than are phytoplankton, with an N : P ratio of 16 : 1. The high C : N : P ratios in seaweeds are thought to be due to their large amounts of structural and storage carbon, which vary taxonomically. Niell (1976) found higher C : N ratios in the Phaeophyceae than in either the Chlorophyceae or Rhodophyceae. The average carbohydrate and protein contents of seaweeds have been estimated at about 80% and 15%, respectively, of the ash-free dry weight (Atkinson & Smith 1983). In contrast, the average carbohydrate and protein contents of phytoplankton are 35% and 50%, respectively (Parsons et al. 1977).

In an extensive survey, Lapointe et al. (1992) found that macroalgae from a variety of carbonate-rich tropical waters were significantly depleted in phosphorus (relative to carbon and nitrogen), as compared with macroalgae from temperate waters (0.15% P vs. 0.07% P on a dry-weight basis). The mean tissue N : P ratio for tropical macrophytes was 43.4, compared with 14.9 for temperate forms – values similar to their mean ambient N : P ratios of 36 and 3, respectively. These data and observations of high alkaline phosphatase activities suggest that these tropical macrophytes tend to be phosphorus-limited, whereas temperate macrophytes tend to be nitrogen-limited.

Laboratory studies on two chlorophytes, *Ulva* and *Enteromorpha*, showed that nutrient supplies were related to nutrient compositions in tissues (Björnsäter & Wheeler 1990). Those researchers found that supply ratios high in nitrogen and low in phosphorus resulted in high tissue N : P ratios, and conversely supply ratios low in nitrogen and high in phosphorus yielded low tissue N : P ratios; supply ratios high in both nitrogen and phosphorus resulted in intermediate N : P tissue ratios. Wheeler and Björnsäter (1992) measured seasonal variations in tissue N : P in five seaweeds (range of N : P was 5–22) and compared them with seasonal variations in ambient nutrient concentrations and critical tissue nitrogen and phosphorus concentrations. They concluded that phosphorus limitation is more frequent than nitrogen in the macroalgae of the Pacific Northwest.

Munda and Gubensek (1976) described the total amino acid content in the acid hydrolysate for several brown, red, and green algae from Iceland. They noted that red algae generally had more total nitrogen than brown algae. The red algae showed greater variability in

amino acid composition, whereas the browns had a more uniform composition, with alanine, aspartate, and glutamate being the major amino acids (Rosell & Srivastava 1985).

Deviations from the Redfield ratio frequently are used to infer which nutrient is limiting the growth of phytoplankton (Goldman et al. 1979). Phytoplankton deprived of phosphorus during growth typically have N : P ratios greater than 30 : 1, whereas phytoplankton deprived of nitrogen during growth have N : P ratios less than 10 : 1. Similarly, C : N and C : P ratios are dependent on growth conditions. Recent studies have shown that seaweeds respond in a similar manner (Björnsäter & Wheeler 1990). C : N ratios generally are higher when plants are grown under nitrogen limitation, because of a decrease in proteins and an increase in carbohydrates. For example, in *Ulva lactuca*, the concentrations of β-alanine and asparagine can decrease 20-fold under nitrogen starvation (Nasr et al. 1968). *Chondrus crispus* has a higher carrageenan content in unenriched seawater than in nitrogen-enriched medium (Neish & Shacklock 1971), and a similar effect is seen in the agar content of *Gracilaria foliifera* (DeBoer 1979). Elevated transient uptake rates of ammonium have also been used to indicate nitrogen limitation in *Gracilaria foliifera* and *Agardhiella subulata* when C : N ratios have risen above 10 (D'Elia & DeBoer 1978).

The chemical composition of many temperate seaweeds varies seasonally, primarily because of the onset of nitrogen limitation in the coastal waters in summer. Wheeler and Srivastava (1984) found that tissue nitrate (as ethanol-soluble nitrate) and total nitrate paralleled the ambient nitrate levels and showed summer minima and winter maxima (from 0 to 70 μmol per gram fresh weight for nitrate, and from 0.9% to 2.9% of dry weight for total nitrogen) in *Macrocystis integrifolia*. In contrast, Wheeler and North (1981) found that neither nitrate nor ammonium accumulated in the tissue of *M. pyrifera;* free amino acids accounted for a major portion of the soluble nitrogen. Juvenile *M. pyrifera* sporophytes do not appear to store nitrogen (Wheeler & North 1980). Similarly, Chapman et al. (1978) have shown that for cultures of *Laminaria saccharina*, the internal concentrations of nitrate increase at substrate concentrations above 10 μM NO_3^- and reach concentrations several thousand times higher than that in the surrounding medium.

Higher-molecular-weight compounds may be involved in nitrogen accumulation and storage. The dipeptide L-citrullinyl-L-arginine can accumulate to high concentrations when *Chondrus crispus* is supplied with nitrate or ammonium at low temperatures (Laycock et al. 1981). Much of this reserve can be readily mobilized for growth of the plant when it encounters higher temperatures, increased irradiance, and low levels of external nitrogen, which are common during the late spring and summer months. Consequently, rapid growth rates, sustained by declining nitrogen reserves, persist well after the disappearance of ambient nitrogen. The accumulation of soluble nitrogen reserves appears to be optimized under conditions of low temperature and reduced light (Rosenberg & Ramus 1982). Under these conditions, the rate of accumulation exceeds the requirements for growth. Pigments, or more likely the proteins associated with them, may also play a secondary role in nitrogen storage (Perry et al. 1981; Smith et al. 1983). There have been several observations of marked decreases in pigment content under nitrogen deficiency (DeBoer 1981). Chlorophyll and phycoerythrin concentrations in *Gracilaria foliifera, Agardhiella subulata,* and *Ceramium rubrum* were strongly influenced by the concentrations of inorganic nitrogen in the culture medium (DeBoer & Ryther 1977).

Munda and Hudnik (1991) analyzed 23 common seaweeds from the intertidal and upper subtidal in the northern Adriatic for their levels of trace metals (Mn, Zn, Cu, Cd, Pb, Co, Ni). Their results indicated that the manganese concentration in algal tissue was the main feature distinguishing the various taxonomic groups, as well as in distinguishing seasonal and habitat variations.

5.8.2 Development, morphology, and reproduction

A consequence of multicellularity is a decreased SA : V ratio for the cells. Some seaweeds appear to respond to nutrient deficiency by the production of hyaline hairs from the thallus surface, akin to the production of root hairs in vascular plants. DeBoer (1981) and DeBoer and Whoriskey (1983) observed that in the presence of low nitrogen concentrations (e.g., $NH_4^+ < 0.5\,\mu M$) and moderate agitation, hair-cell formation was enhanced in *Hypnea musciformis, Gracilaria* species, *Agardhiella subulata,* and *Ceramium rubrum.* Ammonium concentrations greater than $20\,\mu M$ inhibited hair formation. Interestingly, cytoplasmic streaming occurred in hairs on the apical (meristematic) regions of the thalli, where nutrient-uptake rates are high, but not in hairs on the lower part of the thallus, where low uptake rates occur. Ammonium-uptake rates for *Ceramium rubrum* plants with hairs were approximately twice those for plants without hairs. DeBoer and Whoriskey (1983) suggested that these hairs may increase a plant's surface area and hence the number of nutrient-uptake sites. Increased development of apical hairs was observed in *Fucus spiralis* germlings grown in low nutrient concentrations (Schonbeck & Norton 1979c); these tufts could also be seen on germlings in the field, except when nutrients were at the highest concentrations. Further studies on hair formation in *Fucus* (Hurd et al. 1993) revealed that hair formation began in late February off the Isle of Man, even before phosphate was depleted from the seawater; hair production stopped in October. In the laboratory, hairs formed more rapidly under phosphorus limitation than they did in plants growing in nutrient-saturated me-

dium. Plants with hairs were capable of taking up phosphate two to three times faster than hairless *Fucus spiralis.* Therefore, the production of hairs may help macroalgae compete with phytoplankton and bacteria for the available supplies of phosphate. Phosphate may be taken up before the spring phytoplankton bloom and stored for use later in the spring and summer, when ambient phosphate will have been exhausted. The whorls of *Acetabularia* may also serve to increase the absorptive surface area (Gibor 1973; Adamich et al. 1975). Adamich et al. (1975) found an inverse relationship between whorl development and the concentration of nitrate in the medium.

Cladosiphon zosterae is a specific epiphyte on eelgrass, *Zostera marina.* Ammonium or urea in a culture medium will induce a hairless, compact morphology of the discoid stage of the epiphyte, as compared with the morphology produced in nitrate medium (Lockhart 1979). In the microscopic stages of the life history of the kelp *Lessonia nigrescens,* the level of irradiance and the nitrate and phosphate concentrations will influence the course of development of gametophytes and the attainment of fertility (Hoffman & Santelices 1982). The interacting effects are summarized in Figure 5.23. Under nitrogen limitation, the gametophytes do not survive; under phosphorus limitation, they are multicellular, but do not show the usual sexual differentiation.

Embryos of *Fucus edentatus* are morphologically distinct when they are grown under nitrogen deficiency (Fig. 5.24d). Generally, the nitrogen-deficient embryos were found to be smaller and tapered, with a single long primary rhizoid and reduced secondary rhizoid development, in contrast to those grown in complete medium (Fig. 5.24a) (McLachlan 1977). Growth of the embryos in phosphorus- or iron-deficient medium resulted in much smaller embryos, with normal morphology. Omission of bromine from the medium resulted in considerably reduced growth (Fig. 5.24b), and in a medium lacking boron, the embryos became moribund (Fig. 5.24c). The concentration of boron has also been reported to influence the development and reproduction of *Ulva lactuca* and *Dictyota dichotoma* (Nasr & Bekheet 1970). Iodine is required for vegetative growth and normal formation and maturation of plurilocular sporangia of *Ectocarpus siliculosus* (Woolrey & Lewin 1973) and for the development and reproduction of both crusts and blades of *Petalonia fascia* (Hsiao 1969).

Nutrient availability is known to influence reproduction in microalgae (Drebes 1977), and there have been a few reports for macroalgae. It was found that nitrogen depletion enhanced gamete formation in *Ulva fasciata,* whereas higher nitrogen concentrations favored vegetative growth and asexual reproduction (Mohsen et al. 1974). Abundant zoospore formation was observed in *Ulva lactuca* after ammonium was added to the medium (Nasr et al. 1968). Nitrogen defi-

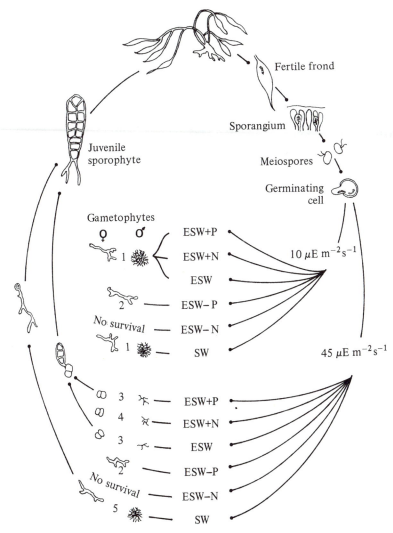

Figure 5.23. Life cycle of *Lessonia nigrescens*, showing the influences and interactions of light intensities and culture media on development and fertility. Gametophyte stages: 1, multicellular, vegetative; 2, multicellular, no sexual differentiation; 3, few cells, fertile; 4, few cells, vegetative; 5, multicellular, fertile. (From Hoffman & Santelices 1982, with permission of Elsevier Biomedical Press.)

ciency suppresses reproduction in many species (McLachlan 1982), but the species of nitrogen may also control reproduction, as in species of *Acetabularia* (Adamich et al. 1975). More information is needed on the effects of nutrients on reproduction, especially nutrients other than nitrogen.

5.8.3 Growth rate and distribution

The growth and productivity of seaweeds are controlled in part by environmental factors such as irradiance, temperature, nutrient availability, and water movement. Marked seasonal fluctuations in nutrient availabilities (especially nitrogen) occur, and they affect growth rates. This has been most thoroughly studied in the kelp *Laminaria longicruris* in Nova Scotia (Hatcher

et al. 1977; Gagné et al. 1982). In these kelp beds there are two main limiting factors: light and nitrogen availability. Nitrogen is present year-round at Centreville, in the southwestern part of the province, because of upwelling, but nitrogen is limiting for 8 months of the year at Boutlier's Point, St. Margaret's Bay (near Halifax). The interactions between light and nitrogen availabilities determine the seasonality of kelp growth. At Boutlier's Point, the plants grow mainly during the period of nitrogen availability, in winter and early spring. By also building up internal nitrogen reserves, they are able to prolong their rapid growth for at least 2 months and take advantage of improved light conditions during the spring. The bulky thallus is able to store substantial quantities of both inorganic and organic nitrogen during

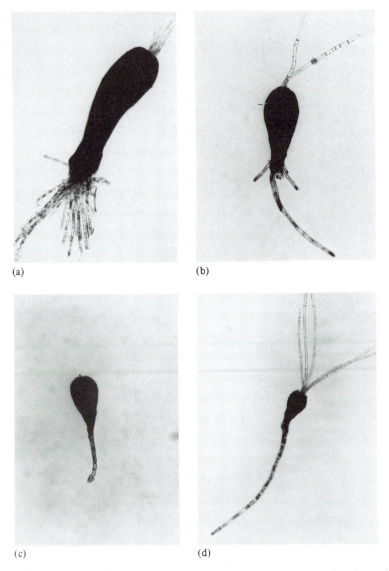

(a) (b)

(c) (d)

Figure 5.24. Fourteen-day-old embryos of *Fucus evanescens* grown in various media: (a) complete medium, (b) medium minus bromine, (c) medium minus boron, (d) medium minus nitrogen. (From McLachlan 1977, with permission of Blackwell Scientific Publications.)

late autumn and winter, to be used for growth in late spring and early summer, when nitrogen is becoming limiting but light conditions are improving (Fig. 5.25a,c). During the summer, when irradiance is high, plants at Boutlier's Point store carbohydrate as laminaran. These carbon reserves are then remobilized and used in conjunction with nitrogen (high ambient concentrations) to produce amino acids and proteins for growth in early winter. At Centreville, where nitrogen does not become limiting, plants do not build up laminaran reserves, and the kelp growth rate follows irradiance, being greatest in summer (Fig. 5.25b,d). Without carbohydrate reserves, their growth rate declines during late autumn and early winter as irradiance declines.

The seasonality of growth has been studied in other seaweeds, such as *Gracilaria foliifera* and *Ulva* species (Rosenberg & Ramus 1982). During late winter, both of these species accumulate substantial soluble-nitrogen reserves that become depleted during the spring–summer growth period. Species may survive during the summer by using intermittent peaks in nutrient concentrations as they occur. In contrast to the kelps, neither of these species starts the winter with a significant store of reserve carbohydrate. On the other hand, an annual, *Chordaria flagelliformis*, has been shown to maintain high growth rates during the summer even when the ambient nitrogen has become exhausted (Probyn & Chapman 1982). Measurements of K_s for

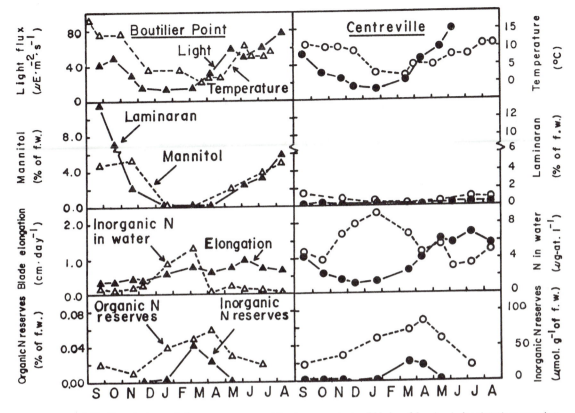

Figure 5.25. Growth, internal nitrogen reserves, and laminaran contents of blades of *Laminaria longicruris* at two sites in Nova Scotia with contrasting light and nitrogen environments. (From Gagné et al. 1982, with permission of Springer-Verlag, Berlin.)

Chordaria have indicated very low values (0.2–0.5 μM) for nitrate, ammonium, and urea, and therefore this annual effectively scavenges nitrogen from seawater (Probyn & Chapman 1983). The comparatively small intracellular nitrogen pool (0.1–0.4% of dry weight) indicates that in *Chordaria,* typical of many opportunistic species, newly absorbed nitrogen is directed into growth rather than storage.

Nitrogen-uptake rates for the kelp *Ecklonia maxima* and for phytoplankton were compared in an upwelling area off South Africa (Probyn & McQuaid 1985). The kelp took up nitrate and ammonium, but not urea, and nitrate uptake did not saturate at concentrations greater than 20 μM. Phytoplankton took up all three nitrogen forms, but preferred ammonium and urea. The fact that nitrate was the most abundant and most highly utilized nitrogen resource in this upwelling area indicates that most (80%) of the yearly kelp productivity is based on imported nitrogen (NO_3^-) rather than on recycled nitrogen (NH_4^+ and urea).

Another suggested source of nutrients, especially nitrate, is submarine discharge of ground water in coastal areas (Johannes 1980; Capone & Bautista 1985). Because the concentration of nitrate in groundwater is very high (50–120 μM), a small amount of discharge

could significantly enrich nitrogen-impoverished coastal waters in the summer. The availability of groundwater nitrate to *Sargassum filipendula* and *Enteromorpha intestinalis* has been confirmed by the induction of high nitrate reductase activity, as compared with beaches receiving much less groundwater (Maier & Pregnall 1990). Regenerated nutrients from the sediments could also be important sources of nitrogen enrichment. Martens et al. (1978) showed that in the top 1 m of sediments in Long Island Sound, ammonium concentrations may reach 10 mM, and those for reactive phosphate more than 1.0 mM. If the circulation of seawater is such as to entrain some fraction of the interstitial-sediment water into the surface layers, then nutrient enrichment of the water column will result from these regenerated nutrients. Smetacek et al. (1976) demonstrated this point in association with high-salinity seawater intrusions in the Kiel Bight. They showed that an intrusion of high-salinity water could displace low-salinity interstitial water and that the resulting inputs of ammonium, phosphate, and silicate to the water column raised their concentrations to approximately 10 times their concentrations prior to the intrusion. In Narragansett Bay, Rhode Island, up to 80% of the nutrients entering the bay at certain times of the year are estimated to come from the sediments (Nixon et al. 1976). Transport of nutrients into the water column is reported to occur as a result of two mechanisms whose effects are approximately equal in magnitude: (1) the activity of burrowing benthic organisms and (2) diffusion (Nixon et al. 1976).

Considerable attention has recently been devoted to macrophyte productivity in the tropics, particularly in coral-reef habitats. Coral reefs exhibit rates of primary productivity that are among the highest reported for any ecosystem (Lewis 1977). Algal turfs are responsible for the majority of whole-reef primary productivity. These turfs are densely packed assemblages of unicellular and filamentous algae from at least five algal Divisions and are characterized by diminutive canopy heights, generally less than 5 mm (Wanders 1977).

Members of the green-algal Order Bryopsidales (= Caulerpales) are important calcifying agents in tropical reefs, and they comprise two different life-form groups: (1) epilithic species with limited attachment structures and (2) psammophytic forms that have extensive subterranean rhizoidal systems (Fig. 1.1). Because the shallow-water habitats of the former have relatively low N : P ratios, as compared with the pore waters of the carbonate-rich sedimentary substrata in which the latter are anchored, Littler and Littler (1990) hypothesized that epilithic forms should tend to be nitrogen-limited, whereas psammophytic species should be phosphorus-limited. This hypothesis was subsequently confirmed by nutrient-addition bioassays (Littler et al. 1988; Littler & Littler 1990). The epilithic forms, *Halimeda opuntia*, *H. lacrimosa*, and *H. copiosa*, increased their photosynthetic rates when nitrogen was added, but not when phosphorus was added. In contrast, the psammophytic forms, *Udotea conglutinata*, *Halimeda monile*, *H. tuna*, and *H. simulans*, were more stimulated by phosphorus addition.

Productivity measurements during nutrient-addition assays for macroalgae growing off Belize demonstrated that the degrees and types of nutrient limitations on tropical macrophytes are also ecosystem-dependent (Lapointe et al. 1987). Two frondose algae, *Dictyota divaricata* and *Acanthophora spicifera*, growing on a barrier reef and at a detritus-rich mangrove island showed increased productivity following nitrogen and phosphorus additions for the reef-collected plants, but no response to the nutrient additions for the mangrove-collected plants.

Immediate enhancement of photosynthesis by coral-reef macrophytes (*Gracilaria* spp.) in response to ammonium enrichment has been reported off Guam (Nelson 1985); the respiration rate was not affected. Nelson (1985) suggested that these macrophytes can effectively exploit brief pulses of ammonium, such as those that may result from passing schools of coral-reef fish or invertebrates. In St. Croix, Virgin Islands, algal turfs grazed by the sea urchin *Diadema antillarum* were 2 to 10 times more productive per unit of chlorophyll *a*

than were turfs not grazed by sea urchins (Williams & Carpenter 1988). It was estimated that sea-urchin grazing supplied about 20% of the nitrogen required for algal-turf productivity. In some cases, tidal jets and nutrient upwelling on some reefs may also be important nutrient sources (Wolanski et al. 1988). Further evidence from Adey and Goertemiller (1987) indicates that, at least in the high-energy algal-turf environment, the physical forcing of primary production by waves and currents is likely to offset the limitations of low-nutrient conditions. This may explain why coral reefs are best developed in the trade-wind seas, where both local, wave-generated currents and equatorial currents carry large quantities of nutrients over their surfaces daily – albeit at low concentrations.

The physiological and growth parameters for *Macrocystis pyrifera* in southern California showed adverse effects from the low nutrient concentrations and high water temperatures (>20°C) associated with the 1982–83 El Niño event (Gerard 1984a). The kelp canopy of *M. pyrifera* and *Pleurophycus gardneri* deteriorated rapidly following the onset of high ambient surface temperatures, and the reductions in chlorophyll content, photosynthetic capacity, and growth rates for canopy fronds during that period were all attributable to depletion of internal nitrogen reserves and nitrogen starvation (Gerard 1984a; Germann 1988).

5.9 Synopsis

Seaweeds require various mineral ions and up to three vitamins for growth. Certain elements are essential for growth, and others may be taken up even though they apparently are not required. There are a few elements, such as nitrogen, phosphorus, iron, and possibly some trace metals (e.g., cobalt and manganese), whose concentrations in seawater may be low enough to limit growth rates of some seaweeds at certain times of the year because of their seasonal low concentrations in seawater. To date, nitrogen is the element whose insufficiency is most frequently observed to limit seaweed growth, but phosphorus may also be limiting in some cases.

Elements are taken up as ions (charged particles), which diffuse to the cell surface, through the cell wall, and thence to the plasmalemma. Some ions, especially cations, may not reach the plasmalemma because they become adsorbed to chemical components of the cell wall. Nonelectrolytes (uncharged particles) generally diffuse through the membranes at rates that are proportional to their solubilities in lipid and inversely proportional to their molecular sizes. Ions may pass through the plasmalemma passively, either by passive diffusion or by facilitated diffusion; however, they are usually transported by an active process requiring energy. For most ions, the relationship between an ion's uptake rate and the ion's concentration in the seawater can be described by a rectangular hyperbola, with the term V_{max}

denoting the maximum uptake rate, and K_s the concentration at which $V = \frac{1}{2}V_{max}$. Nutrient-uptake rates can be measured on the basis of isotope accumulation in the plant tissue or by monitoring the disappearance of the nutrient from the medium.

Various forms of dissolved inorganic nitrogen (nitrate, nitrite, ammonium) and dissolved organic nitrogen (urea, amino acids) are the main sources of nitrogen for seaweeds. Ammonium is generally taken up in preference to all other nitrogen sources. The uptake rates for nitrogen and other ions are influenced by the amount of light available, the temperature, the water motions, the level of desiccation, and the ionic form of the element. Biological factors that influence uptake include the type of tissue, the age of the plant, its nutritional past history, and interplant variability. After ions are taken up, some (e.g., N, P, S, Rb) may be translocated to other tissues within the plant.

After nitrate and ammonium are taken up, they usually are used to synthesize amino acids and proteins, although the ions may be stored in the cytoplasm and vacuoles. Nitrate is reduced intracellularly to ammonium via nitrate and nitrite reductases. Ammonium is incorporated into amino acids via two main pathways. In the glutamine synthetase pathway, glutamine is the first product. The second pathway, which is thought to be of secondary importance, is the glutamate dehydrogenase pathway, which results in formation of glutamic acid.

Seaweeds can take up phosphorus as orthophosphate ions or obtain phosphate from organic compounds through extracellular cleavage using the enzyme alkaline phosphatase. The most important role for phosphorus is in energy transfer through ATP and other high-energy compounds involved in photosynthesis and respiration.

Calcium ions are used in the maintenance of membranes and in cross-linking the cell-wall polysaccharides. Calcium carbonate is deposited in the walls of certain seaweeds. Magnesium is essential in chlorophylls and as a cofactor in many cellular reactions. Other important nutrients are K, S, Fe, Cu, Mn, Zn, and Co, as well as the vitamins B_{12}, thiamine, and biotin.

The relationships among external nutrient concentrations and growth rates can be described by a rectangular hyperbola in which the growth rate approaches a maximum at higher nutrient concentrations. An uptake rate will give a good approximation of the growth rate only when nutrients are not limiting growth or when steady-state growth is occurring under conditions of nutrient limitations. The half-saturation constants (K_s) for growth generally are much higher for seaweeds than for phytoplankton, indicating that phytoplankton have a higher affinity for nitrogen and that they probably can outcompete seaweeds when the nitrogen concentration is low. However, seaweeds, particularly kelps, can store large quantities of nutrients when the external concentrations of those nutrients are high. During periods of rapid growth, these cellular reserves can be utilized for up to 2 months after the nitrogen in the water is virtually depleted.

6

Temperature and salinity

6.1 Natural ranges of temperature and salinity

6.1.1 Open coastal waters

The surface temperatures of the oceans vary in two primary ways. First, they decrease toward higher latitudes, from about 28°C in the tropics to 0°C toward the poles, although this trend is markedly affected by ocean currents. Because of the California Current, for example, fairly uniform cool temperatures prevail in the seawater along much of the west coast of North America, even though land temperatures change considerably. Second, the seasonal changes in ocean temperatures are larger at midlatitudes. In the tropics and at the poles the annual temperature range is often less than 2°C (Kinne 1970), whereas in midlatitudes 5–10°C is common.

The salinity of open-ocean surface water is generally 34–37 parts per thousand (‰), though lower off the coasts of areas with great rainfall (e.g., the northwest coast of North America), and higher in subtropical areas with high rates of evaporation and low rainfall (Groen 1980). Certain seas have markedly higher or lower salinities: The Mediterranean, because there is a high rate of evaporation and little freshwater influx, has salinities of 38.4–39.0‰; the Baltic, essentially a gigantic estuary, is notably brackish, particularly at the surface, ranging from 10‰ near its mouth to 3‰ or less at the northern extreme. In coastal waters, especially those that are partially cut off from the ocean or are subject to heavy runoff, salinity is characteristically 28–30‰ or lower, even quite far along a coast from the mouth of a major river. Slight latitudinal trends in salinity are overshadowed in coastal areas by freshwater influx. In areas with marked seasonal differences in rainfall, or with winter snow periods followed by spring melt, the salinity, especially of surface water, may change dramatically over the year. In general, "marine and brackish waters are . . . almost infinitely variable

in the amplitude and frequency of their saline changes" (Russell 1987, p. 36).

Stratification is common in coastal waters, and especially in estuarine waters. Rather sharp temperature boundaries, called thermoclines, may develop between layers of water, especially in sheltered waters or where there is freshwater input (Fig. 6.1). Warm water flowing out of tropical lagoons at low tides does not readily mix with the cooler shelf waters, but forms a buoyancy front as it displaces the water on the shelf. The circulation associated with this displacement can cause plankton concentrations (slicks) (Wolanski & Hamner 1988). The salinity boundary is called a halocline (Fig. 6.1). Typically but not always, the surface water is warmer and less salty than the deeper water on which it floats. The combined effects of temperature and salinity on seawater density result in a density boundary, or pycnocline, which is a surprisingly strong barrier to mixing. Vertical mixing is thus slow unless wave action is vigorous. In places with moderate or large tidal amplitudes, such pycnoclines (which are often only a meter or two below the surface) will sweep portions of the shallow-subtidal and lower-intertidal zones, causing rapid temperature and salinity changes with each ebb and flood of the tide. An example of the importance of thermal stratification to seaweeds is found in two contrasting bays in Newfoundland (Hooper & South 1977). In Bonne Bay, a narrow fjord, a thermocline forms, warmer water stays near the surface, and Arctic species intolerant of temperatures above 5°C live in the deeper subtidal zone. In Placentia Bay, which is more exposed, there is more turbulence, but no thermocline and no Arctic algae, because the water that is too warm for them extends too far down. A quite different effect of a thermocline on algae is seen in the reduced animal-epiphyte settlement on the upper blades of *Macrocystis* due to restricted larval movement across a thermocline (Bernstein & Jung 1979).

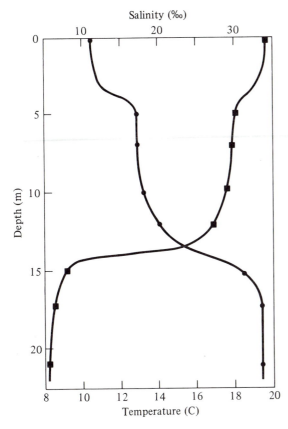

Salinity (‰)

Temperature (C)

Figure 6.1. Thermoclines and haloclines as seen in temperature and salinity profiles at a station between Copenhagen and Elsinore, Denmark. Rapid changes in temperature (squares) and salinity (circles) are seen at 4 m and 14 m below the surface, at the boundaries between three overlying water masses. (Modified from Friedrich 1969, with permission of Gebrüder Borntraeger.)

6.1.2 Estuaries and bays

The temperature and salinity ranges seen in coastal waters are even more pronounced in bays and estuaries. Shallow bays on Prince Edward Island, in the Gulf of St. Lawrence, freeze over in winter, but warm up to 22°C or more in summer. The annual range of salinity that *Cladophora* aff. *albida* can tolerate in Peel Inlet in Western Australia is 2–50‰ (Gordon et al. 1980). In general, as shown in Figure 6.2 and Table 6.1, the mean temperature will increase; the mean salinity will decrease; and the range of each will become greater with increasing distance up the estuary.

Salinities in river mouths depend on the proportions of river water and seawater; these proportions depend on the state of the tide and the state of the river, and they change both daily (Fig. 6.3) and seasonally. In addition to the gradient along the estuary, the salinity lower on the shore often is more variable than that higher on the shore, because the higher shore tends to be

covered only by the surface water (fresher water); whereas the lower shore is alternately under river water and saltwater (Anderson & Green 1980). Because of the slow rate of mixing of layers, even a small freshwater influx can have a pronounced, if very local, effect. That is, it can affect individuals as strongly as will a major influx, but it will affect fewer individuals. The course of a freshwater seep across the intertidal zone often can be seen as a bright green path, because *Enteromorpha*, but little else, will be growing in it.

A temperature–salinity (*T–S*) diagram is a useful way to describe the water climate of an estuary, and it can provide insight into the causes of distribution patterns that mean values conceal. Indeed, Druehl and Footit (1985) have argued that temperature and salinity should not be seen as separate factors affecting seaweed distribution. For instance, the mean annual temperatures and salinities for Nootka and Entrance Island, British Columbia, are nearly identical, but only Nootka supports growth of *Macrocystis integrifolia* (Druehl 1978). Nootka is on the outer coast of Vancouver Island, whereas Entrance Island is in the Strait of Georgia, essentially at the mouth of a large estuary fed by the Fraser River. The *T–S* diagram (Fig. 6.4) shows that for Nootka, when the temperature is high, the salinity is also high; for Entrance Island, when the temperature is high, the salinity is low. In the presence of high salinity, *M. integrifolia* is believed to be able to withstand higher temperatures, although the appropriate physiological tests have not yet been carried out.

6.1.3 Intertidal regions

The principal environmental feature of the intertidal zone is its regular exposure to atmospheric conditions. Its temperature regimes are thus much more complex than those of subtidal zones. Myriad microenvironments result from the many factors that affect the local temperature and temperatures of the resident organisms. Some factors, such as shading, affect the influx of heat to an organism, whereas others, such as evaporation, affect heat efflux. The major source of heat in the intertidal zone during ebb tide is direct solar radiation. Irradiance may be reduced because of shading by clouds, water, other algae, and shore topography (including overhangs, crevices, and the direction of slope). Small-scale topographic features also give shelter from breezes, and hence from evaporative cooling. The important temperature is that of the cytoplasm, but it is essentially unmeasurable. Two examples of actual algal temperatures recorded during emersion on hot days are given in Figure 6.5. *Endocladia muricata* is a stiff, tufty plant; the temperature at the interior of the clump, which is shaded and yet open to airflow, remains considerably lower than that of the air or the open rock surface (Fig. 6.5a) (Glynn 1965). *Porphyra fucicola*, on the other hand, is flattened against the rock surface like a little solar panel, and on a calm day, such as that

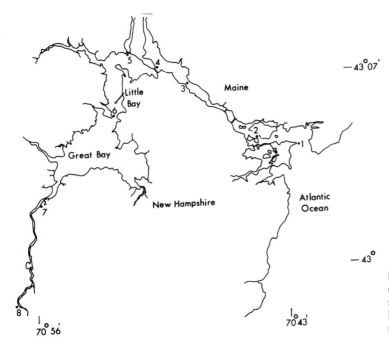

Figure 6.2. Map of Great Bay estuary system, New Hampshire, showing study locations (see Table 6.1). (From Fralick & Mathieson 1975, with permission of Springer-Verlag, Berlin.)

illustrated, it becomes much hotter than the air (Fig. 6.5b) (Biebl 1970). These graphs also show the sharp decreases in temperature as the tide covers the plants. Notice that the *Porphyra* thallus surface temperature fell from 33°C to 13°C in a matter of minutes after the water reached it.

　　Other variables that affect the temperatures of intertidal plants are the time of day at which low tide occurs and the extent of heating or cooling due to waves. Plants exposed by a low tide at dawn or dusk will suffer little heating by the sun, whereas if the low tide occurs in the middle of the day, heating (and also desiccation) can be extreme. In summer, the water may be cooler than the exposed rock and algae, as in Figure 6.5b. In winter, seawater frequently is warmer than the air, and so it can thaw algae that have been frozen during exposure to subzero air. However, if the rocks are cold enough, some of the water from the waves will freeze onto them, embedding the seaweeds in ice. The effects of coincident spring low tides and freezing air temperatures on *Laminaria digitata* have been documented by Todd and Lewis (1984) (Fig. 6.6a,b). Such defoliation of *Laminaria* can also be seen in the Bay of Fundy, Canada (C. S. Lobban, unpublished data). Wave splash can affect algal temperature by maintaining a supply of water for evaporation.

　　Littoral pools are subject to less extreme changes than are open rock surfaces. The longer the exposure and the higher the pool's surface-to-volume ratio, the greater the changes that will take place before the tide refloods the pool. Most of the factors that affect the temperatures of exposed rock pools also affect their salinity. Changes in tide-pool temperatures are more dras-

tic during the day and in the summer (Fig. 6.7a), because the chief source of heat is solar energy. However, the air temperature may warm or cool such pools (even freeze them). Water added to a pool as runoff, rain, or snow may heat or cool the pool. Tide pools experience little or no mixing and thus may easily become stratified. Temperature stratification that occurs during the day, in the absence of salinity stratification, usually breaks down at night.

　　Atmospheric variability also subjects seaweeds on open rock surfaces and in tide pools to frequent salinity fluctuations. Evaporation will cause an increase in the salinity of water in the surface film on seaweeds and, more slowly, in tidal pools. In contrast, rain, snow, and freshwater streams will cause a reduction in salinity. The community upset that can be caused by a flash flood in the intertidal zone has been documented by Littler and Littler (1987). Because fresh water floats on salt water, and because a long period of evaporation generally is necessary to effect any significant change in the salinity of pool water, salinity will change little in mid-intertidal and low-intertidal pools except during extremely hot days or following torrential downpours. High-intertidal pools, which are inundated infrequently or receive seawater only from wave splash, may become very brackish in rainy weather, or strongly hypersaline in hot, dry weather (Fig. 6.7b). Sharp increases in tide-pool salinities can come about as a result of freezing, because salts are initially excluded from the freezing layer and concentrated in the remaining liquid (Edelstein & McLachlan 1975). Seaweeds are found in pools with salinities from about 0.3 to 2.2 times that of normal (i.e., 35‰ seawater) – that is, about 10–77‰

Table 6.1. *Salinity and temperature gradients in a major estuary, the Great Bay estuary system in New Hampshire*

Study stations (nautical miles inland)	Temperature[a] (°C)	Salinity[a] (‰)
1. Jaffrey Point (0.0)	10.2	30.3
	23.0	32.5
	1.0	26.0
2. Pierce Island (2.0)	10.5	27.0
	21.5	32.0
	0.0	19.0
3. Newington Town Landing (5.5)	11.7	24.5
	24.0	31.0
	0.4	16.0
4. Dover Point (7.0)	12.5	23.7
	25.0	32.0
	0.0	15.0
5. Cedar Point (8.3)	13.6	22.1
	27.0	32.0
	0.0	6.0
6. Adams Point (10.7)	13.8	21.0
	28.0	31.5
	0.5	8.0
7. Chapman's Landing (15.2)	14.2	5.7
	26.5	22.0
	0.0	0.0
8. Exeter (19.0)	14.4	3.4
	28.1	20.0
	0.0	0.0

[a]First, second, and third values represent mean, maximum, and minimum values, respectively. *Source:* Fralick and Mathieson (1975), with permission of Springer-Verlag, Berlin.

(Gessner & Schramm 1971). The osmotic and other consequences of evaporative desiccation are taken up in section 6.4, following discussion of the biochemical, physiological, and ecological effects of temperature and salinity per se.

6.2 Temperature effects

Temperatures have fundamental effects on chemical-reaction rates. In turn, the metabolic pathways, each the sum of many chemical reactions, are affected by temperatures, but the interactions with other factors become more complex. Thus, we shall examine the effects of temperatures hierarchically, from the chemical level up to the population level.

6.2.1 Chemical-reaction rates

Temperatures have general effects on chemical-reaction rates that are embodied in the concept of

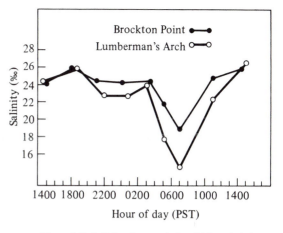

Figure 6.3. Salinity changes during 24-h periods in spring at two sites in Burrard Inlet, British Columbia. Values are averages for 3 days in separate years. (From Hsiao 1972, with permission of Simon Fraser University.)

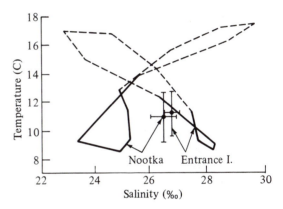

Figure 6.4. Temperature–salinity diagrams for two locations in British Columbia, one of which (Nootka) supports growth of *Macrocystis integrifolia*, the other (Entrance Island, Nanaimo) does not. The large crosses show means and standard deviations for annual temperatures and salinities. The outlines trace monthly temperature–salinity coordinates, with solid lines around winter conditions, and broken lines around summer conditions. The contrast in the two water climates is belied by their annual means. (From Druehl 1981, with permission of Blackwell Scientific Publications.)

Q_{10}, the ratio between the rate at a given temperature t and the rate at $t-10$. Typically, $Q_{10} \sim 2.0$; that is, the rate doubles over a 10°C increase in temperature, but rates can be much higher. The effect of temperature is greater on an uncatalyzed reaction than on one that is catalyzed (e.g., by an enzyme) (Raven & Geider 1988). In an enzyme-catalyzed reaction, with Michaelis-Menten kinetics,

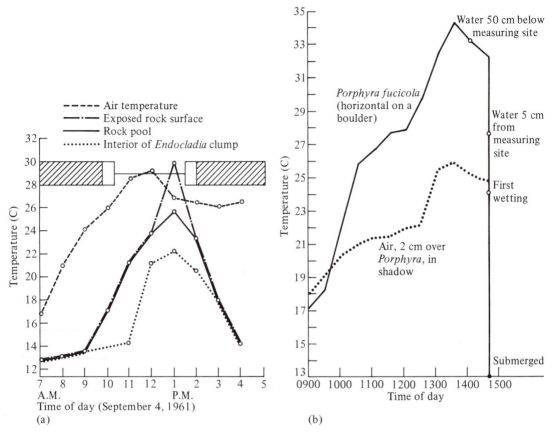

Figure 6.5. Temperature climates of two high-shore red algae. (a) Temperature observations of three microhabitats in the high-intertidal *Endocladia-Balanus* association at Monterey, California, as related to low-water exposure. Also shown is the air temperature at a nearby weather station during the observation period. The horizontal bar and line at the top of the graph show, for the level observed, the approximate durations of the submerged (cross-hatching), awash (clear), and exposed (line) periods. (b) *Porphyra fucicola* thallus temperature during ebb tide on a calm, sunny day. (Part a from Glynn 1965, with permission of the Zoological Museum, Amsterdam; b from Biebl 1970, with permission of Springer-Verlag, Berlin.)

$$v = V_{max}[S]/(K_m + [S])$$

both V_{max} and K_m are affected by temperature. Moreover, via temperature effects on membrane transport, *[S]* may also be influenced (Davison 1991).

If the rate of an enzyme-catalyzed reaction is measured as it varies with experimental changes in temperature over a broad range, often a peak will be found. This optimum temperature and the sharpness of the peak will also depend on pH and on the purity of the enzyme (in vitro). Thermal denaturation of the enzyme will occur above a critical temperature; on cooling, the enzyme may regain its active conformation, or it may be permanently damaged.

Cellular rates, as opposed in vitro rates, per unit of enzyme will depend on the amount of enzyme present, as well as on any seasonal changes in its kinetic parameters (Küppers & Weidner 1980; Davison & Davison 1987). Küppers and Weidner (1980) studied six *Laminaria hyperborea* enzymes from diverse metabolic pathways: RuBisCO, phosphoenolpyruvate carboxykinase (PEPCK); malate dehydrogenase (MDH), aspartate transaminase, glycerol phosphate dehydrogenase, and mannitol-1-phosphate dehydrogenase. RuBisCO is given as an example in Figure 6.8. Enzymes were extracted from the kelp, and their activities were measured under standard test conditions, which included a temperature of 25°C. Seasonal changes were found for all six enzymes, with peaks generally in February–April (Fig. 6.8a). In nature, of course, the temperature is not constant (nor as high as 25°C). By determining the effect of temperature on the in vitro activity of each enzyme and by recording seawater temperatures, Küppers and Weidner were able to calculate what the enzyme activities would have been in the living kelp on the shore. Again they found a seasonal change for each enzyme, but the peaks were in August (Fig. 6.8b). An enzyme's activity

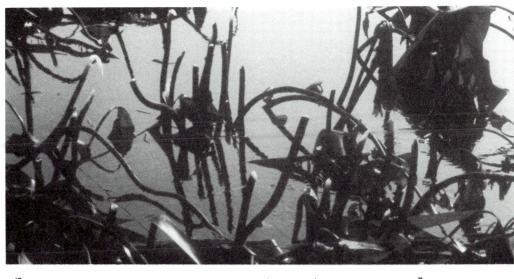

(a)

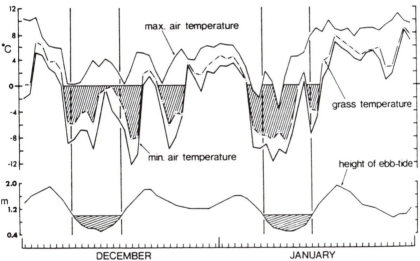

(b)

Figure 6.6. Exposure damage to kelp during cold weather. (a) Photograph of a *Laminaria* stand in Scotland showing the extent of mortality (blades and meristems missing). (b) Meteorological data for the period preceding the damage. Periods of subzero temperatures coincident with the lowest spring tides are highlighted. (From Todd & Lewis 1984, with permission of Springer-Verlag, Berlin.)

is the product of turnover rate (temperature-dependent) and enzyme concentration. The amount of each enzyme present will depend on nitrogen availability (Fig. 6.8c), with RuBisCO (and hence photosynthesis) most sensitive to nitrogen starvation (Wheeler & Weidner 1983). In summer, the kelp has high levels of enzyme activities because of smaller numbers of enzyme molecules with a higher turnover rate; in spring, during the period of nitrogen availability and a high growth rate, a high level of enzyme activity is achieved by an increase in the total amount of enzyme present, although the turnover rate is lower.

Davison and Davison (1987), studying *Laminaria saccharina,* found an inverse relation between temper-

ature and standard (20°C) activity for both RuBisCO and NADP-dependent glyceraldehyde-P dehydrogenase (GAPDH) (another Calvin-cycle enzyme). They also found an inverse relation between photosynthetic rate and temperature when measured at standard (15°C) temperature, but that relation disappeared when photosynthesis was measured at the growth temperature, indicating that the increase in Calvin-cycle enzymes compensates for low temperatures and makes *L. saccharina* photosynthesis virtually independent of temperature. However, enzymes in other pathways did not follow the same pattern, and those findings conflicted with some of Küppers and Weidner's data (e.g., for MDH and PEPCK), leaving many unanswered questions

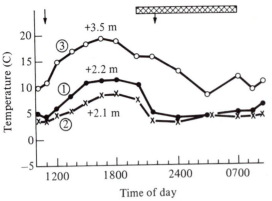

(a)

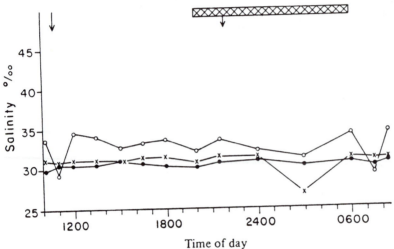

(b)

Figure 6.7. Temperature (a) and salinity (b) changes in three pools at different heights in the intertidal zone near Halifax, Nova Scotia, 8–9 May 1970. Times of high tide indicated by arrows, and darkness by the cross-hatched bar. Pools 1 and 2, between neap and spring higher high waters, were generally flushed twice daily, except for pool 1 during periods of calm or neap tides. Pool 3, 0.4 m above extreme high water, was washed only during severe storms and contained no perennial macroscopic algae (From Edelstein & McLachlan 1975, with permission of Springer-Verlag, Berlin.)

about the temperature acclimation of kelps at the biochemical level.

Temperature acclimation of RuBisCO in three Antarctic and two temperate diatoms was shown by Descolas-Gros and de Billy (1987) to involve changes in K_m. The K_m varied with temperature, and the minimum value for each species occurred at the culture temperature. In other words, the enzyme was most efficient at the acclimation temperature. The minimum values for K_m were the same in all species.

6.2.2 Metabolic rates

When rate-versus-temperature measurements are made for complex reactions, such as photosynthesis and respiration, the overall rate is a composite of all the individual reaction rates. If there is a rate-limiting reaction, it will not necessarily be the same one at all temperatures. The effects of a given temperature change will not be the same on all metabolic processes, because of differing temperature sensitivities of enzymes and the influences of other factors, including light, pH, and nutrients. In photosynthesis, for example, diffusion rates, carbonic anhydrase activity, and active transport of CO_2 and HCO_3^-, all affected by temperature, will determine

the supply of substrate to carbon-fixation pathways (Raven & Geider 1988; Davison 1991). In general, as Raven and Geider (1988) have pointed out, metabolic processes involve many enzymes and transport processes; so low temperature is likely to limit the overall rate via some particularly sensitive step. Algae can respond by altering the quantity or properties of the limiting component.

Temperature acclimation in enzymes (sec. 6.2.1) shows up as seasonal changes in the rates of photosynthesis and respiration versus temperature when performances of summer and winter plants are compared under otherwise identical conditions. For instance, Mathieson and Norall (1975a,b) showed that at a given irradiance, net photosynthesis for several algae was maximum at a lower temperature in winter specimens than in summer specimens (Fig. 6.9). More important, the rate of photosynthesis in cold water is higher in winter plants than in summer specimens; and summer plants can maintain near-peak photosynthesis through warmer temperatures than can winter plants.

Seaweeds in habitats where the temperature fluctuates seasonally may be better able to acclimate than are seaweeds from stable habitats. For instance, Dawes

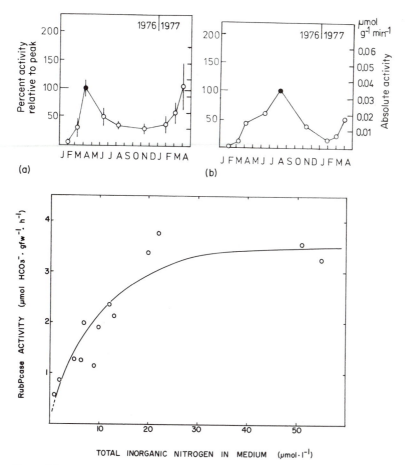

Figure 6.8. Activities of RuBisCO from *Laminaria* blades. (a) Seasonal changes in standard enzyme activity at 25°C. (b) Seasonal changes in temperature-adjusted enzyme activity. (In parts a and b, enzyme from *L. hyperborea.*) (c) Activity in juvenile sporophytes in culture as a function of nitrogen availability. (Parts a and b from Küppers & Weidner 1980, with permission of Springer-Verlag, Berlin; c from Wheeler & Weidner 1983, with permission of *Journal of Phycology.*)

(1989) compared two species of *Eucheuma*, one from Florida (temperature range 16–28°C) and one from the Philippines (temperature about 25°C). The Florida plants were able to acclimate, in stages, to 18°C, but the tropical specimens could not.

Different species, and even different populations of a given species, show diverse responses to temperature changes. Comparisons of macroalgal data from different sources can be facilitated by a mathematical model developed by Knoop and Bate (1990). Their equation relates the photosynthetic rate (P) at a given temperature (T) to the rate of increase at suboptimal temperatures (a) and the rate of decrease at supraoptimal temperatures (b) and to the maximum and minimum temperatures at which photosynthesis occurs (T_{max}, T_{min}). The curves in Figure 6.9 were calculated from Mathieson and Norall's (1975a,b) data by the equation

$$P = a(T - T_{min})\{1 - \exp[b(T - T_{max})]\}$$

Recent work by Davison and others has allowed us to look more closely into temperature–photosynthesis acclimation and to understand some of the factor interactions. The initial photochemical reactions are temperature-independent, but the enzymes of phosphorylation, electron transport, and plastoquinone diffusion are temperature-dependent. As a result, light-harvesting efficiency at subsaturating irradiances (initial slope, α, of a *P–I* curve; sec. 4.3.3) may vary with temperature (Davison 1991). Dark respiration generally increases with temperature. Thus, the amount of light necessary to reach compensation (I_c) increases with temperature. Such short-term effects are not easily related to the long-term responses of plants in the field, however, which are complicated by changes such as acclimation of the enzymes and increases in pigments in plants grown at higher temperatures. *Laminaria saccharina* grown at 15°C had more chlorophyll *a* (because of more PS-II reaction centers and possibly larger photosynthetic units) than did plants grown at 5°C. Because of

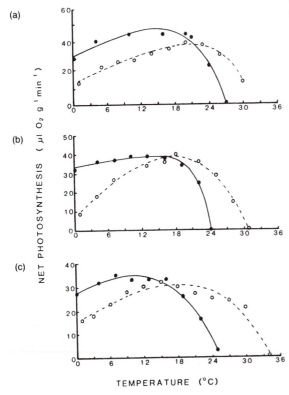

Figure 6.9. Actual data (winter, filled circles; summer, open circles) and model predictions (curves) of photosynthesis in three red algae: (a) *Callophyllis cristata*, (b) *Phycodrys rubens*, (c) *Chondrus crispus*. Data of Mathieson and Norall (1975a,b); curve fitting by Knoop and Bate (1990). (From Knoop & Bate 1990, with permission of Walter de Gruyter & Co.)

optimum temperature (i.e., that giving the maximum rate) has not been near that for the natural conditions. For instance, Fries (1966) found the optimum temperatures for growth of three red algae in axenic culture to be 20–25°C, whereas the water temperatures in their habitat, even in summer, rarely rose above 15°C. She speculated that a reason for the discrepancy may have been the absence of bacteria from the cultures: Seaweeds may use growth substances produced by their associated microflora, and marine bacteria grow best at lower temperatures. In other words, there is a physiological optimum for the alga alone, and there is an ecological optimum in nature, where it is interacting with bacteria and fungi.

Temperature optima vary among species and among strains, as well as between heteromorphic life-history stages (sec. 6.2.5). Changes with age of the thallus have also been noted. For example, the optimum for cultivated *Porphya yezoensis* drops from 20°C at the time of conchospore germination to 14–18°C for thalli 10–20 mm high, and still lower for larger thalli (Tseng 1981). Moreover, the temperature optimum for an alga under laboratory conditions, where temperature is kept constant, may appear narrower than it is in nature (even if other conditions are equal). A possible example of this is provided by a strain of freshwater green alga, *Scenedesmus*, which in the laboratory will not grow above 34°C, but in nature grows without inhibition at peak temperatures of 45°C (Soeder & Stengel 1974). Many flowering plants grow better when they face differing day and night temperatures, a phenomenon known as thermoperiodicity (Noggle & Fritz 1983). Although marine environments do not experience regular diurnal temperature fluctuations, the effects of fluctuating temperatures on seaweed growth and temperature tolerance need to be studied.

Norton (1977) has suggested that the very high growth rate of *Sargassum muticum* at high temperatures is what makes it so invasive in relatively warm waters. An example of such an advantage has been recorded by Kain (1969), involving *Saccorhiza polyschides* and *Laminaria* species. Gametophytes and young sporophytes of *S. polyschides* grow faster at 10°C and 17°C than do those of *Laminaria hyperborea*, *L. digitata*, or *L. saccharina* (Fig. 6.10), whereas growth of *Saccorhiza* at 5°C is slower than that of *L. hyperborea* and *L. saccharina*. Not only does *Saccorhiza* grow faster in warmer water, but also its cells are markedly larger (at all temperatures), and those in the stipe, especially, elongate greatly, giving the young sporophyte better access to light. At 5°C, *L. saccharina* is furthest ahead after 18 days; *L. hyperborea* does equally well at 10°C and 17°C; and *L. digitata* grows most slowly at all temperatures. *Saccorhiza* has a more southerly distribution than *Laminaria* species in Europe, which correlates with its poor growth in cold water. Correlations of the relative growth rates of the *Laminaria* species with their

this, when 5°C-grown plants were assayed at 15°C, I_c and I_k were lower than in 15°C-grown plants (Davison et al. 1991). Although I_c increased with incubation temperature in plants grown at both 5°C and 15°C, the effect was canceled out by acclimation effects; I_c and I_k values were similar for plants tested at their growth temperatures. This is important, given that many seaweeds grow in light-limited environments. These *L. saccharina* plants had lower Q_{10} when grown at 15°C (1.05 vs. 1.52); that is, increasing assay temperatures had greater effects on low-temperature plants. In general, Q_{10} tends to be greater in winter than in summer (Kremer 1981a). This change appears to be more pronounced in polar species than in temperate species (Drew 1977).

6.2.3 Growth optima

Numerous studies have investigated the effects of temperature on photosynthesis, respiration, and growth under otherwise uniform conditions. Not surprisingly, maximum rates often have been found to correlate with the temperature regime in an alga's habitat. However, there have been some reports of instances in which the

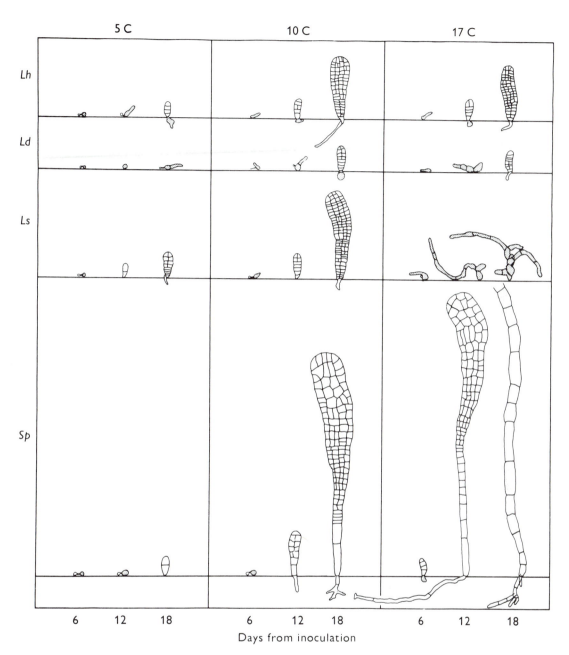

5 C   10 C   17 C

Lh

Ld

Ls

Sp

6   12   18   6   12   18   6   12   18

Days from inoculation

Figure 6.10. Growth and stipe elongation of juvenile kelp sporophytes after 6, 12, and 18 days in culture at 5°C, 10°C, or 17°C. *Lh, Laminaria hyperborea; Ld, L. digitata; Ls, L. saccharina; Sp, Saccorhiza polyschides.* (From Kain 1969, "The biology of *Laminaria hyperborea*," *J. Mar. Biol. Assoc. U.K.*, vol. 49, pp. 455–73, with permission of Cambridge University Press.)

distributions are not so clear. Kain has rationalized this as partly due to taxonomic difficulties with the species, making distribution records doubtful, but there may also be diverse ecological strains within species.

Within a given species there may be considerable genotypic variation in temperature tolerance and in optimum temperatures for growth (as for other responses) (Innes 1984, 1988; Gerard & Du Bois 1988). For in-

stance, high- and low-intertidal plants of *Enteromorpha linza* have different temperature responses (Innes 1988). Such variation may be great enough for geographically diverse populations to appear as distinct strains or races. Phenotypic variation within each population is also likely. However, among *Ectocarpus siliculosus*, the populations are not distinct from one another as races (or "ecotypes"), and Bolton (1983) has described their gradual changes as constituting an ecocline. This species is very widely distributed in eastern North America, from Texas to the high Arctic, but its range of genetic variability shifts gradually from population to population along a temperature gradient. Evidence for the genotypic basis of its variation was that the various geographic isolates has been maintained in culture at a uniform temperature of 20°C for several years.

Temperature limits can change, however, or the time of reproduction may change. *Ascophyllum nodosum* reproduces in late winter in Long Island Sound, but in summer in Greenland; gamete release begins at 6°C and ends at 15°C (Bacon & Vadas 1991). West (1972) concluded that strains of *Rhodochorton purpureum* are genetically selected for their temperature dependence of sporulation related to latitude. The very broadly distributed brown alga *Ectocarpus siliculosus* shows phenotypic variation within isolates and also genotypic variation between geographically diverse isolates, with respect not only to temperature optima but also to salinity tolerance (Russell & Bolton 1975). Even a more narrowly distributed species such as *Laminaria saccharina* shows genetic adaptation of temperature optima toward the edge of its range (Gerard & Du Bois 1988).

6.2.4 Temperature tolerance

A "stress factor" can be defined physiologically as a suboptimal or supraoptimal level of any environmental variable, particularly when the levels are far from optimal. "Strain" is the response of an organism (or material in general) to stress. In a mechanical analogy, stress is the force applied to stretch a rubber band; strain is the stretching of the rubber. Levitt (1972) defines two kinds of strain: elastic strain, which is completely reversible when the stress is removed; and plastic strain, which produces some permanent change in the organism. "Injury" can thus be defined as the result of plastic strain. Because the point at which injury begins can be difficult to determine, tolerance studies usually seek to determine the stress level at which 50% of a group of organisms (or cells) will be killed. Yet even the point of death is difficult to recognize. Although dead matter is physiologically inactive, cell processes tend to come slowly to a halt unless subjected to extreme conditions, and many organisms show remarkable ability to recover from strain or injury if returned to favorable conditions. For instance, Yarish et al. (1987) reported regeneration of purportedly heat-killed *Polyneura hilliae* after 4 weeks in posttest culture at 15°C.

Various physiological criteria have been used as measures of the extent of the stress on metabolism. Photosynthesis and respiration have been most commonly used, owing to the ease with which they can be measured, although a low metabolic rate does not necessarily indicate injury (plastic strain), as opposed to elastic strain. Hayden et al. (1972) developed an electrical-impedance technique for studying chilling, desiccation, and other forms of injury in plants, and MacDonald et al. (1974) used that method to assess injury to *Ascophyllum nodosum* and *Fucus vesiculosus*. Injury was indicated by a sudden increase in the slope of the impedance–cooling curve and by lower impedances after thawing than before freezing. *Ascophyllum* and *F. vesiculosus* tolerated chilling down to −20°C without any indication of injury, and *Ascophyllum* showed no injury by this criterion with up to 70% desiccation.

Experiments to test the tolerances of seaweeds and to compare the results with temperature data from distribution limits have been carried out for many years, with the major researchers being Biebl (e.g., Biebl 1970), van den Hoek and co-workers (e.g., Yarish et al. 1984; Cambridge et al. 1984) and Lüning (Lüning 1984; Lüning & Freshwater 1988). These authors have used widely different exposure times, ranging from 12 h (Biebl) to 2 months (Cambridge et al.). Not surprisingly, however, survival ability and regeneration capacity are decreased with longer exposure times (e.g., Yarish et al. 1987). Given that seaweeds have the potential to acclimate, the pretest conditions must also be taken into account (Yarish, et al. 1987). The rate of cooling is also an important, but often ignored, factor in plant responses (Minorsky 1989). Finally, the recovery time and conditions must be known and standardized. An interpretation of the findings must take into account not only water temperatures but also air temperatures if the algae are intertidal. Moreover, limitations in experimental material (age, stage, reproductive state, etc.) need to be recognized (Lüning & Freshwater 1988).

In general, temperate algae can tolerate cold water at least down to −1.5°C (seawater freezing point). Only 6 species out of 49 did not survive −1.5°C in tests by Lüning and Freshwater (1988) at Friday Harbor, Washington (where seawater never freezes). A few temperate algae, including *Endocladia muricata*, tolerated water as warm as 28°C, but none survived 30°C. Kelps showed the least heat resistance, being limited to 15–18°C. Some Arctic algae in Newfoundland have very low upper limits: *Papenfussiella callitricha* has a limit of 8°C (Hooper & South 1977), and the temperature range for *Phaeosiphoniella cryophila* is about −2°C to +5°C (Hooper et al. 1988). In the North Sea, Lüning (1984) found several seaweeds that would tolerate 30°C and one (*Protomonostroma undulatum*) that would tolerate only up to 10°C, with seasonal shifts of up to 5°C in some species (especially *Laminaria* spp. and *Desmarestia aculeata*).

Most seaweeds will be killed if they become frozen. However, the presence of solutes in water lowers its freezing point, and the high concentration of salts in cytoplasm provides some protection against freezing for intracellular water. Furthermore, the crystallization temperature typically is lower than the freezing point (Spaargaren 1984). Tissue water is not completely frozen until −35°C to −40°C. During progressive cooling, ice crystals form on the outsides of the cells first. This tends to draw water out of the protoplasts, causing dehydration, unless cooling is very rapid, in which case the protoplasts may freeze. Damage is also caused by mechanical disruption of cell components by ice-crystal formation (Bidwell 1979). Damage to the tonoplast will be especially injurious, because toxic materials stored in the vacuole can thus be released and poison the cell. Damage to *Chondrus crispus* frozen to −20°C was associated with increased plasma-membrane permeability and disruption of photosynthetic lamellae, with loss of pigments; but *Mastocarpus stellatus,* which lives a little higher on the shore, was not damaged by that temperature (Dudgeon et al. 1989).

Intertidal algae in cold temperate and polar regions must be resistant to subzero temperatures and are able to withstand a certain amount of freezing. In the Arctic, *Fucus vesiculosus* may survive for several months at −40°C (Gessner 1970). The ability of leafy *Porphyra* thalli to withstand −20°C has proved of great use to Japanese mariculturists, who store nets covered with young plants at that temperature, as insurance against loss of the crop on the nets in the sea (Miura 1975; Tseng 1981). The water content of the thalli has a pronounced effect on survival, however. Half-dried thalli are much more tolerant of freezing than are fully hydrated thalli, perhaps because there is less mechanical damage by intracellular ice. The freezing resistance of higher-plant parts, such as seeds, is also greater if the tissue has a low water content. *Porphyra pseudolinearis* males have the most extreme low-temperature tolerance known: 50% survived −70°C for 24 h (Terumoto 1964). Some species (e.g., *Chondrus crispus*) acclimate to freezing temperatures by physiological changes in the individual, whereas other species (e.g., *Mastocarpus stellatus)* are genetically adapted to be more resistant, and they do not acclimate (Dudgeon et al. 1990).

Whereas the effects of freezing are easy to explain, neither the damaging effects of chilling temperatures nor resistances to chilling are well understood for any plants. Damage may result from the low-temperature sensitivity of proteins in susceptible species (Gessner 1970; Graham & Patterson 1982), or it may result from photoinhibition when low temperatures limit electron transport and inhibit photon capture (Davison 1991).

Temperature-induced damage to thalli has been seen in both intertidal and subtidal plants. Schonbeck and Norton (1978, 1980a) described the temperature damage to the high-intertidal fucoids *Pelvetia canaliculata* and *Fucus spiralis* as consisting of reddish spots of decaying tissue developing some 10 days after thermal stress, along with narrowed apical growth and reduced rates of elongation and weight gain. High-temperature damage in both species was less severe at lower humidities (i.e., when the plants were drier). The plants re-

covered from this damage unless it was extreme. Adverse effects due to unusually high temperatures have been noted in populations of *Macrocystis pyrifera* in California, where warm water promotes black-rot disease (Andrews 1976), and in cultivated *Porphyra* in Japan (Tseng 1981).

Heat damage occurs through damage to PS-II, in turn controlled by the degree of saturation of thylakoid lipids (Davison 1991), and through the failure of one or a few thermolabile enzymes. The heat stability of an enzyme seems to be partly related to the temperature at which it was formed. Because there is constant turnover of protein molecules, the enzymic machinery may thus become gradually acclimated to changing temperatures. Moreover, recent evidence indicates that diverse organisms (including diatoms) respond to thermal shock by initiating synthesis of special heat-shock proteins that seem to protect against cellular damage (Schlesinger et al. 1982; Lai et al. 1988).

6.2.5 Temperature and geographic distribution

The water-temperature tolerances of different species of seaweeds are at least partly responsible for the patterns of the geographic distribution of adult plants (Fig. 6.11, Table 6.2), as reviewed by Lüning (1990). Indirect evidence regarding the importance of temperature for seaweed floras can be drawn from the effects of ocean currents, for the floral distributions do not have a strict latitudinal correlation (such a correlation might have been attributable to the amount of light available). The east and west coasts of South Africa are clear examples of the influences of warm and cool currents. The south-coast flora is a mixture of warm- and cold-water species that has come about as a result of the mixing of the warm water of the Agulhas Current and the cold water of the Benguela Current, along with some cold upwelling (Stephenson & Stephenson 1972). There are remarkable changes in the seaweed flora over a short distance around the Cape of Good Hope, with practically no change in latitude. Cluster analysis of the flora at the Cape by McQuaid and Branch (1984) has shown that temperature determines the species composition, whereas wave exposure determines the biomass for major species. Upwelling currents are not necessarily constant throughout the year, and so there can be marked changes in the flora related to the hydrographic "seasons," as off Baja California (Dawson 1951). The discovery of a species of *Laminaria* (a cold-water genus) off the coast of Brazil, only 22–23° south of the equator, is explained by the presence of cold upwelling at the considerable depths where these plants grow (Joly & de Oliveira 1967).

Work on temperature-defined zones of seaweed distributions dates back to the early part of this century. Setchell (1915) divided oceans into nine zones, which were defined by 5°C ranges of surface-water temperatures during the warmest month (drawn as isotherms on

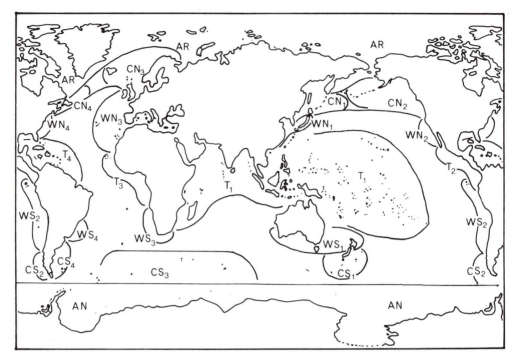

Figure 6.11. Phytogeographic regions for seaweeds (see Table 6.2). (From van den Hoek 1984, with permission of *Helgoländer Meeresuntersuchungen.*)

maps), except for the two polar zones, which covered 10°C ranges. Setchell later introduced the use of mean coldest-month temperatures (isocrymes) to define provinces within the broader zones.

Critics of Setchell, such as Hutchins (1947), argued that such a simple set of global temperature zones could not serve as a basis for biogeographic analysis. Hutchins pointed out that there are four critical temperatures: (1) the minimum for survival, which might determine the winter poleward boundary for a species; (2) the minimum for reproduction, controlling the summer poleward boundary; (3) the maximum for reproduction, controlling the winter equatorward boundary; (4) the maximum temperature for survival, determining the summer equatorward boundary. Van de Hoek (1982) added two more potential boundaries: those limiting growth poleward and equatorward. Thus, a species may be restricted by survival temperatures north and south, by reproduction temperatures north and south, or by one survival limit and one reproduction limit.

An evaluation of the experimental and phenological evidence led Breeman (1988) to the conclusion that temperature responses do account for most of the geographic boundaries of seaweeds. Restrictions are imposed by high or low survival limits for the hardiest life-history stage (often microthalli in heteromorphic life histories), temperature requirements for reproduction of any stage, and temperature limits on growth and

asexual propagation. The lethal limits for macrothalli and their growth/reproduction optima did not correlate with geographic limits for seaweeds that have heteromorphic life histories. Species with the same temperature tolerances may still have different geographic boundaries; for instance, rates of minimal growth will be higher in species susceptible to herbivores than in herbivore-resistant species (Breeman 1988). Added stresses will tend to decrease algal resistance to temperature extremes.

An example of temperature effects on the different life-history stages of algae can be seen in the geographic distribution of *Desmotrichum undulatum* (Punctariaceae), which has a northern limit determined by summer temperatures warm enough to permit development and fruiting of macrothalli. Its southern limit is set by winter temperatures too high for the development of macrothalli (in warmer water, the microthallus may exist without the macrothallus, but will not be noticed) (Rietema & van den Hoek 1981). The life cycle of this plant is as follows (Rhodes 1970; Rietema & van den Hoek 1981) (Fig. 6.12): Pluriseriate, strap-like macrothalli, present in nature during winter, produce plurilocular sporangia in culture at 21°C (i.e., under summer conditions); the zoospores germinate into creeping filamentous microthalli (Rhodes 1970). At 21°C these microthalli reproduce themselves via zoospores from plurilocular sporangia; however, at 6°C they form wide, erect filaments that subsequently develop into pluriser-

Table 6.2. Phytogeographic regions for seaweeds

Location in Fig. 6.11	Region	Approximate latitudinal extent
AR	Arctic Region	
CN$_1$	Cold temperate NW Pacific Region	55° N–40° N (Japan) and 35° N (Korea)
CN$_2$	Cold temperate NE Pacific Region	65° N–40° N
CN$_3$	Cold temperate NE Atlantic Region	70° N–55° N
CN$_4$	Cold temperate NW Atlantic Region	52°N–40° N
WN$_1$	Warm temperate NW Pacific Region	40° N–30° N (25° N in mainland China)
WN$_2$	Warm temperate NE Pacific Region	40° N–25° N
WN$_3$	Warm temperate NE Atlantic Region	55° N–20° N
WN$_4$	Warm temperate NW Atlantic Region	40° N–30° N
T$_1$	Tropical Indo-West Pacific Region (also includes Central Pacific islands: Micronesia, Polynesia, Hawaii)	30°(25°) N–30° S
T$_2$	Tropical E Pacific Region	25° N–5° S (0° near Galápagos Islands)
T$_3$	Tropical E Atlantic Region	20° N–10° S
T$_4$	Tropical W Atlantic Region	30° N–25° S
WS$_1$	Warm temperate SW Pacific Region	30° S–45° S
WS$_2$	Warm temperate SE Pacific Region (includes Juan Fernández Islands and Galápagos Islands)	5° S–45° S
WS$_3$	Warm temperate SE Atlantic Region (includes South Africa, Tristan da Cunha group)	10° S–45° S
WS$_4$	Warm temperate SW Atlantic Region	25° S–40° S
CS$_1$	Cold temperate SW Pacific Region (southern New Zealand, Stewart Island, Chatham Islands, Antipodes Islands, Macquarie Island, Bounty Islands, Campbell Island, Aukland Islands)	45° S–55° S
CS$_2$	Cold temperate SE Pacific Region	45° S–60° S
CS$_3$	Cold temperate SE Atlantic Region (includes Lindsay Island, Bouvet Island, Prince Edward Islands, Crozet Islands, Kerguelen Islands, McDonald Islands, Heard Island)	45° S–50° S
CS$_4$	Cold temperate SW Atlantic Region (SE America, Falkland Islands, South Georgia, S. Shetland Islands, S. Orkney Islands)	40° S–60° S
AN	Antarctic Region	

Source: van den Hoek (1984), with permission of Helgoländer Meeresuntersuchungen.

iate blades with uniseriate tips. No sporangia are seen in cultures at 6°C. Rietema and van den Hoek (1981) have elucidated some of the complexities of the development: Zoospores from the straplike thallus form microthalli under a wide range of temperatures (4–30°C). These microthalli produce erect (macro) thalli as follows: At high temperature, 20–30°C, growth of the macrothallus is rapid and brief, only a uniseriate filament being produced before zoosporogenesis; at 16–20°C the basal part of the erect filaments becomes pluriseriate; at still lower temperatures, and especially with long days, zoosporogenesis is long delayed, and the full straplike macrothallus is able to form.

However, temperature effects may be complicated by the effects of short-term extremes of temperature. Infrequent periods of unusual cold or heat may wipe out seaweed populations at the edges of their ranges (Fig. 6.6). Gessner (1970) has argued that such brief extremes, not the average warmest- or coldest-month temperatures, are the decisive circumstances that control algal distributions. Another criticism of the method of relating flora to temperature zones – whichever way they are defined – is that species distributions are tabulated using the remotest recorded finding of each species. In the most pessimistic view (Michanek 1979), "what such findings actually express

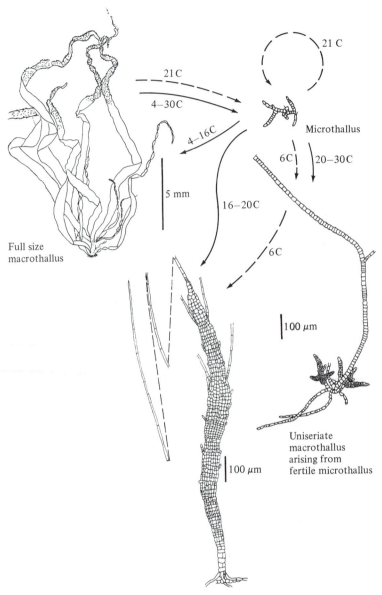

21 C

21 C

4–30C

4–16C

Microthallus

5 mm

6C   20–30C

16–20C

6C

Full size
macrothallus

100 µm

100 µm

Uniseriate
macrothallus
arising from
fertile microthallus

Macrothallus pluriseriate at base,
uniseriate at tip

Figure 6.12. Life history of *Desmotrichum undulatum*, showing effects of temperature adduced by Rhodes (1970) (broken arrows) and by Rietema and van den Hoek (1981) (solid arrows). (Redrawn from Rietema & van den Hoek 1981, from *Marine Ecology Press Series*, with permission of Inter-Research.)

is first how intensely an area has been investigated, secondly how rich it is in different biotopes and finally to what extent it contains enclaves with local conditions reminiscent of those prevailing in some other more or less distant region. . . . Such examples . . . represent exceptions, they are the products of local conditions, presumably occurring only in favored years, and observed quite accidentally.'' That last shortcoming was overcome by Bolton and Anderson (1990), who used seaweed community composition, rather than presence/

absence data, to demonstrate that, indeed, temperature is the controlling factor in the changes around the Cape of Good Hope.

The distribution of seaweeds is not only a reflection of each species' temperature responses but also a result of dispersal and vicariance (van den Hoek 1984). The presence of a species (or genus) in a given flora depends on its having arrived there at some earlier time (e.g., via ocean currents or in recent times, on ships' bottoms). The Japanese species *Sargassum muticum*, recently established in western North America and Europe as an invasive weed (Critchley et al. 1990), is an example. In the geologic past the shapes of the oceans were different – and the general climate changed too – so that species spreading was once possible between areas that are now isolated by land masses. The similarity between the Indo–West Pacific and the Western Atlantic/Caribbean seaweed floras probably is a result of their having populated the warm Tethys sea that once lay between Africa and Asia. In contrast, the cold temperate North Pacific, North Atlantic, and Southern oceans have long been separated by land or warm water and show little similarity (see van den Hoek's 1984 analysis of red-algal genera). Short-term climatic variations such as ENSO events (sec. 6.2.6) may play roles too, because the El Niño is stronger than normal currents in the area and follows different tracks (Richmond 1990).

6.2.6 El Niño

The significance of factor interactions is well illustrated in the attempts to understand the effects of the unusually severe ENSO (El Niño/southern oscillation) of 1982–3 in the eastern Pacific (Glynn 1988). ENSO events are large-scale climatic and hydrographic changes in the Pacific basic (Hansen 1990). Though normally occurring on the scale of seasonal changes in weather (the southern oscillation) and currents (El Niño), periodically these changes or shifts are large enough to cause unusual disturbances. In the eastern Pacific, storms and rain are more common and sea levels and water temperatures are higher than in the western Pacific, where sea levels are low (Fig. 6.13) and droughts are common. The 1982–3 ENSO was the most severe on record, with extensive and protracted increases in sea temperatures (2–5°C) and sea levels (0.2 m or more) in California (on the edge of the affected area). The center of the disturbance was off Peru, where the peak increase in water temperature was 8°C (Glynn 1988). Physical changes were detectable as far north as southern Alaska, though no effects on the biota could be attributed to El Niño that far north (Paine 1986). Extensive coral bleaching was reported in the tropical eastern Pacific (Panamá, Galápagos), and at least one species of coral endemic to Panamá apparently became extinct (Glynn & de Weerdt 1991). Mass mortalities of red and green algae in the Galápagos and kelp die-back in northern Chile were among the effects near the center of the El Niño. Low sea levels strongly affected corals (and presumably algae) on the reef flats of Pacific islands, which were exposed by tides half a meter or more below normal, where the range is only about 1 m (Fig. 6.13).

Much farther north, in southern California, the effects on algae were compounded (and their interpretations confounded) by severe winter storms in 1982 and 1983 that tore up plants. The storms had been preceded by cloudy skies in autumn that had mitigated the usual desiccating conditions of daytime low

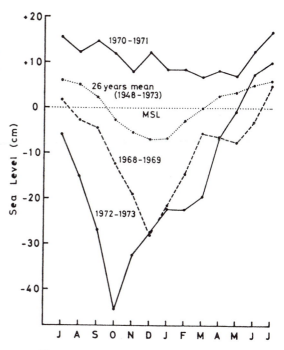

Figure 6.13. Sea-level changes on Guam (tidal range about 1 m) during two ENSO events (1968–9) and 1972–3) and a high-sea-level period in 1970–1, relative to the mean sea level for 1948–73. (From Yamaguchi 1975, with permission of *Micronesica*.)

tides (sec. 6.4) (Gunnill 1985; Tegner & Dayton 1987; Dayton & Tegner 1990). The storms, waves, and clouds were due to the ENSO, but even though Gunnill had quantitative data for several preceding years (Gunnill 1980a,b), he was able to extract only a few correlations with warm water per se (Gunnill 1985). Several perennial algae, such as *Pelvetia fastigiata*, recruited better during the warm-water period (as they had during a similar period in 1976); others, including kelps, showed decreased recruitment or survival, apparently because of the warm current.

Subtidal plants were also affected by nutrient shortages, because the thermoclines were deeper during El Niño, creating thicker layers of nutrient-depleted surface waters. Growth rates for *Macrocystis pyrifera* (Zimmerman & Robertson 1985) and *Pleurophycus gardneri* (Germann 1988) were reduced. Dean and Jacobsen (1986) were able to show directly that nutrient limitation, not high temperature, had caused the reduced growth of *Macrocystis*. Dayton and Tegner (1990) suggested that *Macrocystis* beds were particularly susceptible to El Niño because they are more closely involved in the pelagic system than are most benthic communities.

6.3 Biochemical and physiological effects of salinity

"Salinity" can be defined simply as grams of salts per kilogram of solution; yet this simple definition belies the physical, chemical, and biological complexities of this "factor." From the physical point of view, the complexity lies in the relationships that seawater density, light refraction, and electrical conductivity bear to salinity (and also to temperature) (Kalle 1971). The aspects of salinity that are of biological significance are ion concentrations, the density of seawater, and, especially, the osmotic pressure.

Salinity cannot be measured simply by evaporating the water and weighing the salts. Instead, related properties are measured. The international standard method is based on the chemical constancy of seawater: Chlorinity is determined (Parsons et al. 1984), and salinity is calculated from the relationship $S‰ = 1.8065Cl‰$ (Sharp & Culberson 1982). Salinity can also be determined (less accurately, but much more quickly) with a refractometer, which measures light refraction, or a hydrometer, which measures density. Just as an internal temperature cannot be measured, so there are important collections of water, such as that in intercellular spaces, that are inaccessible for measurement.

The most important effects of salinity are the osmotic consequences of the movement of water molecules along water-potential gradients and the flow of ions along electrochemical gradients. These processes take place simultaneously, and both are regulated in part by the semipermeable membranes that surround cells, chloroplasts, mitochondria, and vacuoles. Water movement is described in the next section, followed by an account of how cells cope with osmotic changes. As in the sections on temperature, we shall progress from the cellular level to the population level.

6.3.1 Water potential

To understand the physiological effects of salinity, one must understand the basic principles of water potential. (More detailed accounts can be found in plant-physiology textbooks; also see R. H. Reed 1990.) Movement of molecules requires free energy. Molecules may acquire free energy from a variety of sources as a result of changes in temperature, concentration, or pressure, as well as that due to gravity and other forces. The free energy of a substance is called its chemical potential, and the chemical potential of water is called water potential, denoted by the Greek letter ψ. With reference to a cell surrounded by a solution, there are several components to ψ. Matric potential, ψ_m, is a measure of the forces that bind water molecules to colloidal material (including proteins and cell walls); it is a minor component in submerged cells, but is important under desiccating conditions on the exposed seashore. Osmotic potential, ψ_π, is the potential of water to diffuse toward a solution. The osmotic potential of pure water is zero. Anything dissolved in water will lower its osmotic potential. The more particles there are in solution, the more negative the osmotic potential. Water flows down the potential gradient, that is, toward the

more negative ψ_π. Effectively, water movement results in a dilution of a more concentrated solution. The decrease in ψ_π is proportional to the number of particles dissolved, regardless of their size. Each dissociated ion of a salt counts as one particle, so that, ideally, a molar solution of sodium chloride has twice the osmotic potential of a molar solution of sucrose. (In practice, a small correction factor, the activity coefficient, must be included in the calculation.) The concentrations of solutions to be used for osmotic measurements are not given as molarity (moles per liter of solvent at 20°C), but as *molality* (moles per kilogram of solvent), because addition of solute molecules dilutes the solvent molecules. By using molality, we refer always to the same number of solvent molecules. Units are (milli)osmols per kilogram: One mole (1 mol) of undissociated solute in 1,000 g of water $= 1,000$ mosmol kg^{-1}. Seawater of 35‰ at 20°C has an osmolality of 1,050 mosmol kg^{-1}. (Reed et al. 1985; Kirst 1988).

As water flows into a plant cell, it pushes against the wall and creates a pressure. The tendency of water to move as a result of pressure is called the pressure potential, ψ_p. The pressure potential of water outside the cell is defined as zero at atmospheric pressure, but it will be positive for plants under water, owing to hydrostatic pressure. At equilibrium, when net water flow is zero,

$$\psi_{\pi e} + \psi_{pe} = \psi_{\pi i} + \psi_{pi}$$

(Bidwell 1979). The pressure potential is a property of the water, but as water presses against the cell wall, the cell wall reacts with an equal and opposite pressure, which is called turgor pressure ($P_i = -\psi_{pe}$). Note that turgor pressure is a property of the cell, not of the water. If the external pressure potential is negligible, turgor pressure at equilibrium will equal the difference between the osmotic potentials inside and outside the cell.

Frequently the overriding component of water potential is its osmotic (or chemical) potential, ψ_π, and inasmuch as pressures are the important components in osmotic adjustment to salinity changes, turgor pressure and osmotic pressure (π) are useful terms. These hydrostatic pressures are measured in pascals (Pa), the same metric units used for barometric pressure. At equilibrium,

$$\psi_e = \psi_i = P_i - \pi_i = -\pi_e$$

or

$$P_i = \pi_i - \pi_e$$

(R. H. Reed 1990). The osmotic pressure of fresh water is minimal, and the turgor pressure in cells is equivalent to the internal osmotic pressure ($P_i = \pi_i$). In seawater, where $\pi_e \approx 2.5$ MPa, cells must maintain a high internal solute concentration (ψ_i) to remain turgid.

What happens when a cell is placed in a solution with which it is not in equilibrium? [A solution of lower solute concentration (lower π_e or higher ψ_π), as compared with the cell, is called hypotonic; one of higher concentration, hypertonic.] The answer to this question is important in considering the effects of salinity changes on cells. If the cell is placed in hypertonic solution, water will flow rapidly out of the cell, because membranes are freely permeable to water. Turgor pressure will be reduced, and at first the cell will become flaccid. Then, as the cytoplasm and vacuole shrink further, the plasmalemma will tear away from the cell wall. The damage to the plasmalemma caused by this process, plasmolysis, is usually irreparable. There are some seaweeds, however, that can survive plasmolysis (Biebl 1962). If the cell is placed in hypotonic solution, water will enter the cell (and ions will leave), causing it to swell, and if the difference in osmotic potentials is great enough, it will burst. (Seaweed cells lack contractile vacuoles with which to expel water.) Again, such rupture will be fatal. During the first few minutes of submergence in distilled water, seaweed thalli rapidly lose ions from their "free space" (intercellular spaces and cell walls) as the solution in the free space comes to equilibrium with the medium (Gessner & Hammer 1968). Because the osmotic potential of seaweed cells is more negative than that of seawater, sometimes much more negative, its salinity must be greatly increased before seawater becomes hypertonic. The ability of seaweeds to tolerate high salinity (i.e., their ability to avoid plasmolysis) depends on the difference between the internal and external osmotic potentials and on the elasticity of the cell wall. (As long as the wall can collapse, the plasmalemma will not be torn away from it.) In the presence of reduced salinity, the turgor pressure will increase. Cells will expand as long as their walls are elastic, but because normal seawater is already hypotonic with respect to the inside of the cell, strain will increase with any reduction in salinity. The strength of the cell walls and the ability of the cells to make their internal osmotic potential less negative will determine their resistance to low salinity.

There is one further consideration. In systems containing charged particles, there is not only a chemical potential of the solutes but also an electrical potential. There is a tendency for the numbers of positive and negative charges to come to equilibrium. Gradients in these two potentials are not always in the same direction. The net passive movement of ions across a cell membrane will depend on the combined electrochemical gradient and will also be complicated by the fact that many molecules cannot freely cross the membrane. Moreover, cells actively import and export ions, across both the plasmalemma and the tonoplast. Many cells actively exclude Na^+, whereas Cl^- is not actively pumped or else is imported (Gutnecht & Dainty 1968; MacRobbie 1974). The resulting electrical imbalance is partly satisfied by uptake of nutritionally useful cations, such as Mg^{2+} and certain trace metals. Seawater has a very low water potential because of all the salts dissolved in it, but cells maintain even lower potentials (higher concentrations of particles), and the resulting turgor pressure is important for cell growth (Cosgrove 1981).

6.3.2 Cell volume and osmotic control

Control over cell volume clearly is vital to cells that have no walls (including seaweed gametes and spores; Russell 1987), but seaweeds, too, have been reported to alter their internal water potential in response to salinity changes; and these observations have led to the hypothesis that such seaweeds are able to regulate their cell volume. By increasing or decreasing $\psi_{\pi i}$ in response to $\psi_{\pi e}$, the cells can control their volume and turgor pressure. Some studies have reported findings to support this hypothesis, whereas others have indicated that, at least in some species, such changes in ion or metabolite concentrations do not affect cell volume. Moreover, some seaweeds do not regulate volume (*Porphyra*, discussed later), or only partly regulate turgor (e.g., *Bostrychia scorpioides;* Karsten & Kirst 1989).

Algal cells may alter their internal water potential by pumping inorganic ions in or out, or by interconversion of monomeric and polymeric metabolites (Hellebust 1976; Russell 1987) (Fig. 6.14). (Because membranes leak water freely in both directions, cells cannot change their water potential by pumping water molecules.) In terms of energy, a change in ion concentration is likely to be cheaper, but cytoplasmic enzymes and ribosomes cannot tolerate wide fluctuations in the ionic composition of the cytoplasm; so acclimation has to be on the basis of compatible solutes. For instance, inorganic salts such as KCl and KNO_3 have large inhibitory effects on enzymes from *Laminaria saccharina* and *Fucus vesiculosus* (Fig. 6.15). Ions are used in the vacuoles of these brown algae for osmotic adjustment and are important for short-term (0–30 min) acclimation in the cytoplasm (Borowitzka 1986). But in the cytoplasm and organelles such as mitochondria and chloroplasts, the water potential is controlled through the mannitol concentration (Davison & Reed 1985a,b; Reed et al. 1985). Similar results have been obtained for the red alga *Griffithsia monilis,* in which the compatible solute is digeneaside (Bisson & Kirst 1979). In the cortical cells of *Laminaria* there are large vacuoles, and the tissue contains more K^+, Na^+, and Cl^-, and less mannitol, than meristoderm tissue, in which cells have small vacuoles and relatively large proportions of cytoplasm (Davison & Reed 1985a).

In those algae in which each cell is largely filled by a vacuole, such as the siphonous green algae *Codium, Bryopsis,* and *Valonia,* changes in the turgor pressure of the vacuole will dominate the overall cell turgor changes, and inorganic salts will be accumulated (Raven 1976). Nevertheless, the water content of the cytoplasm is also important. Of course, any change in the water

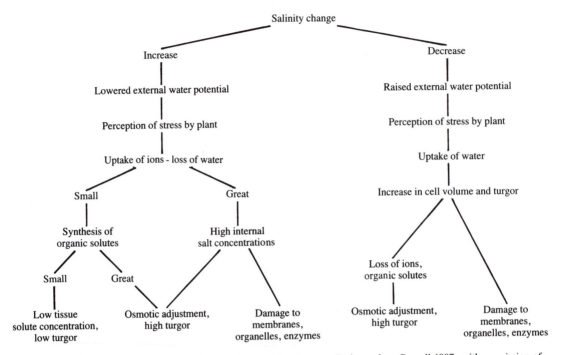

Figure 6.14. Short-term responses of seaweeds to salinity changes. (Redrawn from Russell 1987, with permission of Blackwell Scientific Publications.)

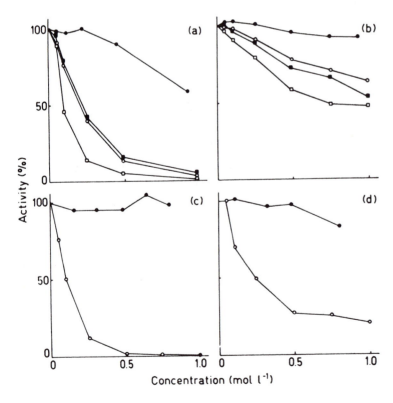

Figure 6.15. Effects of osmoticum concentration on kelp enzyme activities in vitro. The osmotica were mannitol (filled circles), KCl (open circles), NaCl (filled squares), and KNO_3 (open squares); the enzymes were (a) glyceraldehyde-3-phosphate dehydrogenase (GAPDH) (cytoplasmic), (b) glutamine synthetase, (c) RuBisCO (chloroplasts), and (d) phosphoenolpyruvate carboxykinase (PEPCK) (cytoplasmic). (From Davison & Reed 1985b, *Phycologia*, with permission of Blackwell Scientific Publications.)

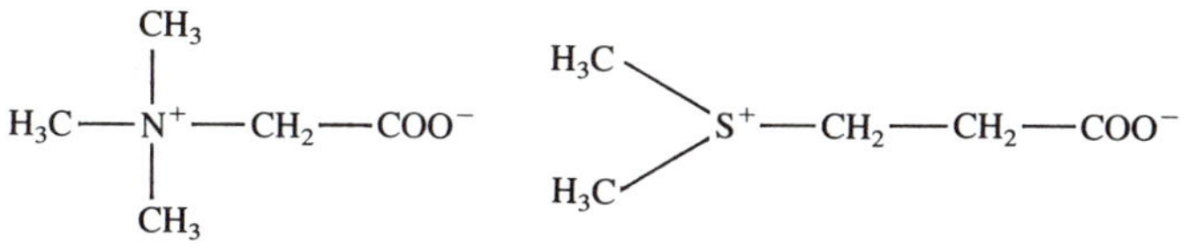

Figure 6.16. Two major osmotica in seaweeds: a quaternary ammonium compound (a) and a tertiary sulfonium compound (b). (After Blunden & Gordon 1986.)

(a) Glycinebetaine   (b) β-Dimethylsulfoniopropionate (DMSP)

potential of the vacuole will affect water movement between the vacuole and the cytoplasm, just as the water potential of seawater affects the cytoplasm from the other side.

Compatible solutes must be highly soluble, must have no net charge, and must be retained against their large concentration gradients across the plasmalemma and tonoplast (Borowitzka 1986; Blunden & Gordon 1986). Moreover, they should not be intermediates in any major biochemical pathway, but removed by two or three enzymatic steps so that there will be no conflict between osmotic adjustment and metabolism (Borowitzka 1986). Several compounds, such as sucrose, do not meet this criterion; mannitol, although thought of as a major photosynthetic product in brown algae, may in fact serve an osmotic role instead (Reed et al. 1985). However, compatible solutes do more than merely serve as particles in the osmotic equation; they interact with enzymes to stabilize them against conformation changes due to water loss (Borowitzka 1986).

Surprisingly small numbers of molecular types are used by the wide variety of organisms (from bacteria to higher plants and vertebrates) that must cope with osmotic stress. Mannitol, digeneaside, and sucrose are in the group of low-molecular-weight carbohydrates, in which the —OH groups are thought to blend well with the structure of cellular water; see Borowitzka (1986) for the biochemical theories. A second group includes amino acids, especially proline (e.g., in *Enteromorpha intestinalis;* Edwards et al. 1987). The third group comprises quaternary ammonium compounds and their tertiary sulfonium analogues (Blunden & Gordon 1986; Borowitzka 1986).

The presence of these compounds does not prove their use in osmotic adjustment, as Ritchie and Larkum (1987) have noted. However, several seaweeds have been shown to use betaines, especially glycinebetaine (Fig. 6.16a) in osmotic adjustment. The tertiary sulfonium compound β-dimethylsulfoniopropionate (DMSP) (Fig. 6.16b) is involved in the salinity responses of *Ulva lactuca* and is present in osmotically significant amounts in about a quarter of the other species tested, especially in Chlorophyceae (Reed 1983; Blunden & Gordon 1986). The importance of DMSP seems to be in long-term acclimation (Edwards et al. 1988). Betaines or their sulfonio analogues have been found in most seaweeds that have been screened (contrast blue-green algae, discussed later) and may even be

responsible for some of the antistress effects of commercial seaweed extracts used as foliar sprays (Blunden & Gordon 1986).

Blue-green algae, which have tremendous abilities to range from fresh waters through to marine waters, accumulate one of several compounds, depending on the halotolerance range (Reed et al. 1986; Borowitzka 1986). Study of their osmotic adjustment is easier because they are not compartmented like eukaryotes, and in particular have no vacuole. In general, blue-greens that grow in fresh water or soil accumulate sugars, sucrose or trehalose; marine strains accumulate glucosylglycerol; nonmarine hypersaline strains accumulate betaine. There are exceptions, however: The upper-intertidal species *Rivularia atra* accumulates trehalose in response to salinity stress (Reed & Stewart 1983). *Spirulina platensis* in seawater media accumulates only glucosylglycerol at 20°C, but also trehalose at 37°C (Warr et al. 1985a,b).

Seventeen seaweeds, including reds, greens, and browns, studied by Kirst and Bisson (1979) all maintained fairly constant turgor pressures over a wide range of external osmotic pressures by changing their internal concentrations of K^+, Na^+, and Cl^-, especially in the vacuole. The activity of an inwardly directed Cl^- pump regulated by external salinity seemed to be the principal means of turgor-pressure control in these species. In the studies of Kirst and Bisson the seaweeds were subjected to constant salinities, whereas in nature salinities will fluctuate either abruptly or according to a more or less sinusoidal curve. Dickson et al. (1982) studied the responses of *Ulva lactuca* to both kinds of fluctuations, and although there were some puzzles in their data, they were able to conclude the following: Changes in internal solute concentrations closely followed salinity fluctuations, reducing the changes in turgor pressure (Fig. 6.17a–d). Cellular K^+, Na^+, Cl^-, SO_4^{2-} and DMSP concentrations also closely followed salinity fluctuations, except that K^+ reentry into cells was slow in darkness (Fig. 6.17e,f). Mg^{2+} was not lost from cells in the presence of decreasing salinity.

The biochemical basis for changes in metabolite concentrations has been partly worked out by Kauss (1973) and Kauss et al. (1978) in the wall-less freshwater flagellate *Poterioochromonas malhamensis*, which responds to water-potential changes by adjusting its content of isofloridoside relative to its polymeric storage polysaccharide, chrysolaminaran (Fig. 6.18). The

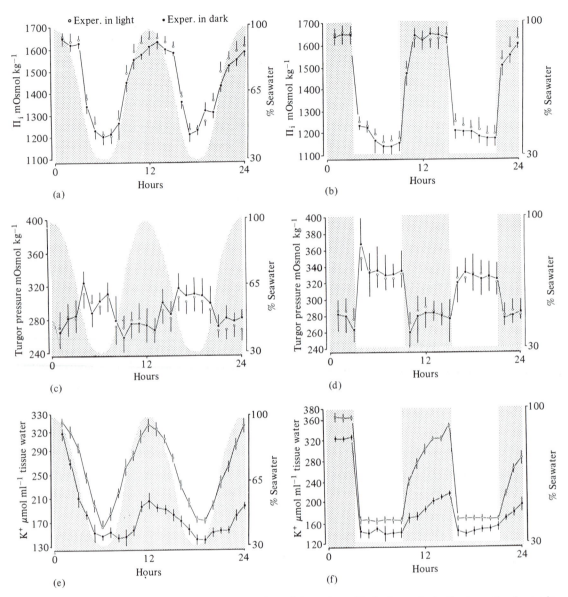

Figure 6.17. Responses of *Ulva lactuca* to fluctuating salinity regimes. The background stippling traces the changes in salinity; two regimes were used: sinusoidal changes (a, c, e) and abrupt changes (b, d, f). Changes in tissue osmolality (Π_i) (a, b), apparent turgor pressure (c, d), and K^+ concentration on a tissue-water basis (e, f) are given for experiments run in light (open circles) and in darkness (filled circles). (From Dickson et al. 1982, with permission of Springer-Verlag, Berlin.)

enzyme isofloridoside phosphate synthase exists as an inactive proenzyme as long as the cell is in a stable osmotic condition, but when the water potential fluctuates, another enzyme cleaves off part of the proenzyme, enabling it to bond galactose (in the form of UDP-galactose, from polymeric glucan) to glycerol phosphate (also from glucan), giving isofloridoside phosphate. When the external water potential decreases, the synthase drives the reaction toward isofloridoside; when the water potential increases, the reaction favors forma-

tion of the polymer. A change in the external water potential apparently is not sensed by the enzyme(s) affected; rather, there is some other sensor, possibly a membrane component, that responds to a pressure stimulus by producing some "controller" substance(s), possibly involving Ca^{2+} and the calcium-binding protein calmodulin (now isolated from one alga, a species of *Chlamydomonas;* Schleicher et al. 1984), and these in turn regulate the enzymes (Fig. 6.18) (Kauss & Thomson 1982).

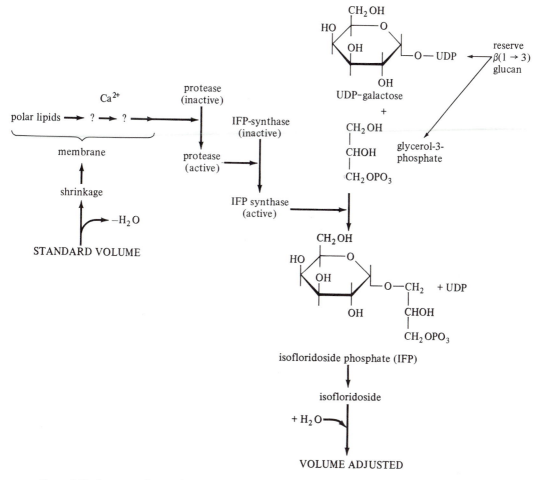

Figure 6.18. Sequence of events in osmotic adjustment by the wall-less flagellate *Poterioochromonas malhamensis*. (Redrawn from diagrams in Kauss 1978, and Kauss & Thomson 1982, with permission of Elsevier Biomedical Press and Pergamon Press, Ltd.)

Not all seaweeds rely completely on ion and metabolite concentrations as protection against salinity changes. Some also have morphological defenses and do change volume with salinity. *Porphyra purpurea*, in contrast to many seaweeds, has a nonrigid cell wall composed chiefly of mannan and xylan, rather than of cellulose. The cell-wall polymers are arranged as granules rather than as ordered microfibrils. This species does not regulate its cell volume or turgor pressure, and so salinity fluctuations cause water to flow in or out of the cells (Reed et al. 1980a). Nevertheless, it and other species of *Porphyra* do show changes in ionic composition (especially K^+ and Cl^-) and in metabolite concentrations in response to salinity changes (Reed et al. 1980b; Wiencke & Läuchli 1981). Similar phenomena occur in two species of *Enteromorpha*. *E. intestinalis* and *E. prolifera* are both euryhaline, and both exhibit changes in tissue water content and cell volume when the salinity is altered (Fig. 6.19) (Young et al. 1987a,b). Edwards et al. (1987) found that the

walls of estaurine *E. intestinalis* were thinner and hence stretchier than the walls of marine and rock-pool plants of the same species, allowing them to swell with the influx of water in low-salinity conditions. In the presence of high salinity, the cell volume decreased, although there were also increases in inorganic ions, sucrose, and proline (Edwards et al. 1987). Also, *Enteromorpha* and *Ulva* can be repeatedly plasmolyzed and deplasmolyzed without injury to the membranes (Ritchie & Larkum 1987).

6.3.3 Photosynthesis and growth

There tend to be optimum salinities for the processes of photosynthesis, respiration, and growth (Fig. 6.20), just as there are optimum temperatures. Numerous examples of salinity optima were given by Gessner and Schramm (1971), and the effects of hypersalinity were reviewed by Munns et al. (1983). A few studies have examined the interactions among salinity, temperature, and light (e.g., Lehnberg 1978). In Lehnberg's

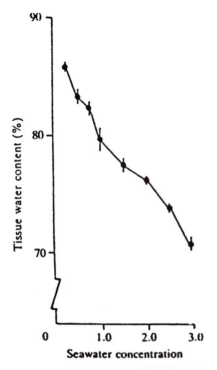

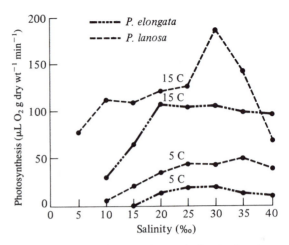

Figure 6.20. Apparent photosynthesis rates for estuarine *Polysiphonia lanosa* and *P. elongata* as functions of salinity at 5°C and 15°C, showing optima near full seawater salinity. (From Fralick & Mathieson 1975, with permission of Springer-Verlag, Berlin.)

Figure 6.19. Tissue water content [(wet weight − dry weight)/wet weight] × 100% for *Enteromorpha prolifera* as a function of salinity. Seawater concentration of 1.0 = 35‰ salinity. (From Young et al. 1987a, *Journal of Experimental Botany*, with permission of Oxford University Press.)

study of the red alga *Delesseria sanguinea* from the Baltic, the data on photosynthesis and respiration were presented as two- and three-dimensional maps (Fig. 6.21a,b) showing regions where one factor or another was limiting the metabolic rate. In Figure 6.21, the different responses of juvenile and adult thalli are evident; among other things, photosynthesis in adult thalli at low salinities is restricted to a very narrow temperature zone. Although these plants came from seawater of 15‰ salinity, their peak photosynthesis was at full seawater salinity. The same as with temperature effects, the rate of change can be important (Wiencke & Davenport 1987).

Several researchers have attempted to determine why diluted seawater causes a decline in photosynthesis. There is a sharp drop in the photosynthesis rate for several marine plants, including *Ulva lactuca*, when they are transferred from seawater to tap water, and a corresponding sharp return to normal when transferred back to seawater. This has been explained as an effect of carbon supply (CO_2 and HCO_3^-) (Hammer 1968; Gessner & Schramm 1971). Dawes and McIntosh (1981) explained why the rate of photosynthesis in the red alga *Bostrychia tenera* is temporarily greater in water of certain Florida estuaries than in either full seawater or seawater diluted with distilled water (Fig. 6.22). They found that the estuaries were all fed by spring water, the significant components of which were Ca^{2+} and HCO_3^-. Although *Bostrychia* will die if left too long in water of very low salinity, the increased photosynthesis in spring-water-diluted seawater enables the plants to survive short periods (a few days) of very low salinities better than in estuaries not fed by spring water. Calcium makes the plasmalemma less permeable to other ions, thus reducing the loss of ions that takes place (in addition to water influx) when cells are placed into dilute seawater (Gessner & Schramm 1971). Eppley and Cyrus (1960) found that lack of Ca^{2+} in fresh water resulted in loss of K^+ from *Porphyra perforata*. Yarish et al. (1980) found that Ca^{2+} and K^+ were limiting factors for photosynthesis in estuarine red algae. Although Ca^{2+} is unlikely to be absent from brackish water, its concentration may be too low in rainwater, which affects emersed intertidal algae. *Polysiphonia subtilissima* was recently found in fresh (but hard) water in spring-fed streams (Sheath & Cole 1990). The hardness of the water (Ca^{2+} concentration) may partly explain the survival of this species in fresh water, as it helps maintain *Bostrychia* photosynthesis. However, Yarish et al. (1979b) found that photosynthesis in *Bostrychia* and *Caloglossa* was reduced in the presence of low salinity, whereas that of *P. subtilissima* was not.

Gessner (1971) recorded the photosynthesis rates in seawater for two marine algae after pretreatment in distilled water of 1 M mannitol solution for various periods and found that in *Halymenia floresia* the mannitol effectively protected the photosynthetic apparatus from

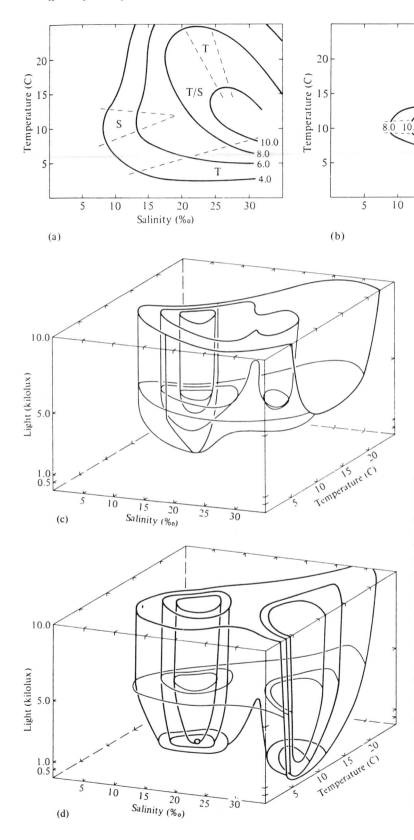

Figure 6.21. Temperature–salinity maps of photosynthesis rates for *Delesseria sanguinea* from the western Baltic Sea: (a) young thalli; (b) adult plants. The curves connect the locations of equal photosynthesis rates (in the same way that contours plot equal heights on a map); the rates plotted; in milligrams O_2 per gram dry weight per hour, are given as numbers on the curves. The maps are divided by dashed lines into regions showing where temperature (T), salinity (S), or both (T/S) are limiting. (c, d) Three-dimensional zones of photosynthetic ratio, PS/R, for young (c) and old (d) thalli. Ratios calculated from O_2 exchange. The contour lines are drawn for PS/R ratios of 20 (inner), 15, and 10 (outer). (From Lehnberg 1978, with permission of Walter deGruyter & Co.)

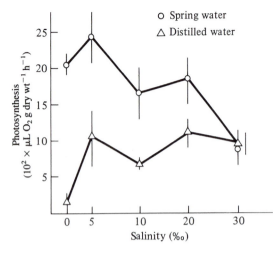

Figure 6.22. Photosynthesis rates for *Bostrychia tenella* at 28°C after 3 days in seawater diluted with various amounts of distilled water or spring water. (From Dawes & McIntosh 1981, with permission of Springer-Verlag, Berlin.)

low-salinity damage, whereas in *Dictyopteris polypodioides*, both osmotic-strength and ionic-composition changes affected photosynthesis. The reduction in photosynthesis may have been partly due to inhibition of individual enzymes, such as RuBisCO (Fig. 6.15), but it also could have been due to structural/mechanical effects. The effects of hypotonic medium on isolated chloroplasts of *Caulerpa* are disruption and loss of stromal proteins, such as glucose-6-phosphate dehydrogenase and glutamate dehydrogenase, which are not attached to membranes. RuBisCO, which is held in the pyrenoid, is not released by this treatment (Wright & Grant 1978).

Growth of seaweeds is reduced in high-salinity water not only because of cumulative enzyme effects but also because of reduced turgor pressure, which inhibits cell division (sec. 1.3.4). Lowered salinities often stunt the growth of seaweeds and have variable effects on branching (Norton et al. 1981). At the cellular level, Reed et al. (1980a) have noted that cell division in *Porphyra purpurea* is inhibited in concentrated seawater.

6.3.4 Tolerance and acclimation

If the frequencies of occurrence of various intensities of temperature and irradiance are plotted for the world, unimodal curves are obtained, whereas for salinities a bimodal curve is obtained, with one peak for freshwater habitats and one for marine; brackish-water habitats are relatively less common (Gessner & Schramm 1971). When organisms become acclimated to a new range of conditions, they generally lose the ability to perform as well under their previous conditions. This phenomenon has, in the course of evolution, resulted in two nearly separate groups of organisms: freshwater and marine. Very few species or even genera of eukaryotes

are able to cross the so-called salinity barrier; three that can are *Cladophora*, *Rhizoclonium*, and *Bangia*. One might suppose that brackish habitats, being mixtures of fresh water and seawater, might be populated by a mixture of freshwater and marine algae, but that is not the case. Brackish waters, even down to salinities of 10‰ or less, are populated by especially euryhaline marine algae such as *Fucus* and *Enteromorpha*, or typically brackish species such as *Rhizoclonium riparium* and *Vaucheria* species in salt marshes (Nienhuis 1987; Christensen 1988), and *Caloglossa* and *Bostrychia* in mangrove swamps (King 1990). Invasion of salt marshes by vascular plants takes place from the land, whereas the algae have invaded from the sea.

Intertidal seaweeds generally are able to tolerate seawater salinities of 10–100‰; subtidal algae are less tolerant, especially to increased salinities, generally withstanding 18–52‰ salinities (Biebl 1962; Gessner & Schramm 1971; Russell 1987). They must tolerate unpredictable changes in salinity during emersion. Estuarine subtidal plants experience more regular fluctuations.

Acclimation of a species to salinities higher or lower than normal (or other variables) may result from the development of genetically diverse populations (ecotypes) or through phenotypic change (without genetic change), either in individuals or through successive mitotic spore generations; or both processes may take place within one species (Yarish et al. 1979a). Genetic variation can be inferred (1) if populations of plants from different salinity regimes show different tolerances to salinity ranges and (2) if the progeny of laboratory-cultured plants show responses to salinity similar to those of the original isolates. Reed and Russell (1979), using regeneration of pieces of *Enteromorpha intestinalis*, studied salinity tolerances for populations from maritime pools 100 m from the top of the intertidal zone (salinity virtually zero), populations from high-intertidal pools, and populations from open intertidal rock. The range of experimental salinities was 0–136‰. Intertidal-zone plants showed the smallest range of salinity tolerance, with a peak at 34‰ (Fig. 6.23a). Plants from high-intertidal pools and maritime pools had broad salinity tolerances, with a plateau across 0–51‰ or more (Fig. 6.23b,c). These data are comparable to the findings of Young et al. (1987 a,b) for different populations of *E. intestinalis*, cited earlier. The tolerances to longer durations of exposure to various salinities were also broader for high-intertidal-pool populations than for plants from lower down the shore. The critical test to show that these variations were genetic, not phenotypic, was to culture swarmers from each population and then compare their salinity responses. The progeny showed responses similar to those of the parents, demonstrating that the variation was genetically maintained. (Contrast the experiments of Geesink and den Hartog, described later.) Bolton

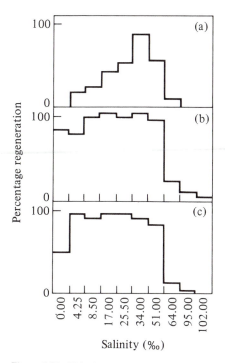

Percentage regeneration

Salinity (‰)

Figure 6.23. Rhizoid production by three populations
of *Enteromorpha intestinalis* in response to salinity:
(a) eulittoral-zone population; (b) littoral-fringe
plants; (c) plants from maritime pools influenced only
by sea spray. [From Reed & Russell 1979; reprinted
from *Estuarine and Coastal Marine Science*, vol. 8,
pp. 251–8, © Academic Press Inc. (London).]

(1979) reached a similar conclusion for *Pilayella littoralis*. Various populations from the head of an estuary to
the sea showed differing tolerances, especially to very
low salinities, even after culture for 2 months or more in
full seawater. Yarish et al. (1979a) concluded that their
two estuarine red algae showed ecotypic variation and
that some of the ecotypes also had some capacity for
phenotypic variation.

Phenotypic acclimation has been shown in *Bangia* by Geesink and den Hartog (den Hartog 1972b;
Geesink 1973); *B. fuscopurpurea* is a marine form, and
B. atropurpurea a freshwater form. The optimum salinity for growth of sporelings was that in which the spores
had formed. By transferring plants shortly before spore
formation into water of 10–20% lower salinity, den Hartog and Geesink were able to acclimate *B. fuscopurpurea*, over a number of generations, to completely
fresh water; *B. atropurpurea* could be acclimated to full
seawater in a similar way. (The authors concluded that
these are really a single species.) Reed (1985) showed
that freshwater *Bangia* rapidly produced osmotically
significant amounts of floridoside when grown in saline
media. Although *Bangia* grows in both fresh water and
seawater, it is rare in estuaries, occupying sites only
at the mouth and the head. Den Hartog surmised that

Bangia cannot tolerate the widely fluctuating salinities
occuring in mid-estuary. Geesink and den Hartog's experiments might seem to suggest that *Bangia* could have
invaded freshwater habitats by progressively moving up
estuaries, but the field observations suggest that it could
not. Den Hartog (1972b) suggested that probably the
freshwater populations came about when patches of marine habitats were cut off from the sea (e.g., by dikes or
by land uplift) and gradually became freshwater habitats. The Zuyder Zee in The Netherlands was closed
from the sea in 1932, becoming the Ysselmeer. After a
decade or more, *Bangia* was flourishing, whereas other
marine plants, lacking the ability to adapt to fresh water,
had died out.

The Baltic Sea gives interesting insights into salinity adaptations over longer times. The Baltic is connected to the North Sea by only a narrow channel. Its
postglacial history is known to have included several
major salinity shifts, with the most recent marine period
having started about 7,500 years ago (Russell 1987,
1988). Its salinity has been low (brackish) for about
3,000 years. Various fucoids and other marine algae
that occur in the Baltic also occur in the North Sea. Although the populations in the Baltic have consistently
lower salinity optima, their thallus organization differs
from that of their open-coast counterparts in diverse
ways. Some have small thalli but similar-sized cells;
some have smaller cells but not smaller thalli; some have
only certain cells reduced in size (Russell 1987, 1988).

Lowered salinity also promotes changes in the
chemical compositions of seaweeds, such as fucoids,
which have been studied in the field (Munda 1967) and
in the laboratory (Munda & Kremer 1977). Mannitol,
ash, and chloride contents declined, as did dry weight
as a percentage of fresh weight, whereas protein concentrations increased (Table 6.3). The decline in mannitol was attributed to decreased photosynthesis in the
presence of reduced salinities, but in view of the work
of Reed et al. (1985), that decline might better be interpreted as a direct response to increased water potential.

6.4 **Desiccation**

Seaweeds are essentially marine organisms, even
if they are out of the water more than half the time; their
exposure to the atmosphere is a stress, to which species
are more or less tolerant. Except under cool, very humid
conditions, loss of water from marine plants begins as
soon as they are emersed; in this way, desiccation is a
salinity stress. Removal of seaweeds from seawater also
deprives them of their source of nutrients, including
most of the inorganic carbon, although postdesiccation
enhancement of limiting-nutrient uptake has been reported for *Fucus distichus* (Thomas & Turpin 1980),
which would partly compensate for the intermittent
availability of nutrients. In a narrow sense, the term
"desiccation" is equivalent to "dehydration," but it
is used to encompass these nutrient changes as well,

Table 6.3. *Changes in chemical composition of* Fucus vesiculosus *(summer material) after 19 days of incubation in seawater of different salinities (control = 31.03‰)*

Salinity ‰	Dry weight as percentage of fresh weight	Percentage dry weight of:		
		Ash	Mannitol	Protein
31.03	17.45	22.60	14.57	15.67
20.94	15.70	22.50	6.61	15.31
15.53	13.21	18.20	5.81	15.54
10.67	12.50	16.20	5.38	16.31
5.15	10.80	15.90	3.18	17.74

Source: Munda and Kremer (1977), with permission of Springer-Verlag, Berlin.

because they normally occur together. However, on humid days these other stresses can occur in the absence of dehydration.

Desiccation is related to the surface-area-to-volume (SA : V) ratio. Small organisms are more susceptible. Sessile organisms can survive harsher conditions as they get larger, but they must be protected from desiccation when they are small (e.g., by growing in crevices, or by settling when low tides occur at night or in the early morning) (Denny et al. 1985).

Evaporation rates are affected by temperature; there tends to be more water loss during the day than at night, and more in summer than in winter (e.g., Mizuno 1984). In turn, evaporation relates to salinity through changes in the surface water potential, as we have seen. The seasonal changes in the timing of tides (sec. 2.2.1) are important, because daytime low tides are much more damaging during hot weather. In southern California, daytime lower low tides in October and November often coincide with dry offshore winds, further exacerbating evaporation stress and leading to algal die-off (Gunnill 1980b, 1985).

Another important factor in evaporation stress is the ratio of the area of the evaporating surface to the volume of the plant (Dromgoole 1980). A brief exposure to cool, humid air may hardly affect a seaweed, whereas prolonged exposure, particularly during hot summer days, may cause severe stress. The higher up on the shore a species grows, the longer it is exposed to desiccation. The small-scale habitat must be considered, because seaweeds may be partially protected by clumping and by the slope, orientation, and porosity of the shore. Whereas many intertidal seaweeds evidently tolerate desiccation, some largely avoid it by growing under overhangs or under other algae, in crevices, or in other wet areas. For example, *Corallina vancouveriensis* in parts of southern California receives protection where it grows among *Anthopleura elegantissima*, a colonial sea anemone that forms water retaining carpets (Taylor & Littler 1982). Saccate intertidal algae such as *Colpomenia peregrina* largely avoid desiccation stress by

retaining a reservoir of seawater inside the thallus (Vogel & Loudon 1985; Oates 1988). The nutrient levels and osmolality of these reservoirs are maintained by effective exchange of water when thalli are submerged.

Temperature can have an effect on growth form, which in turn can affect desiccation avoidance, as shown by Tanner (1986) for *Ulva californica*. Above 15°C in culture, and south of Point Conception, California, where water temperatures usually are above 15°C, *U. californica* forms densely tufted turfs. These hold water better than does the larger, more foliose growth form found in the cooler waters (and damper air) north of Point Conception, and formed in culture at 10°C. Abbott and Hollenberg (1976, p. 6) pointed out that intertidal red algae in southern California also tend to be shorter and less foliose.

What takes place during emersion? As soon as a seaweed is removed from water, its photosynthesis rate drops sharply, even before any desiccation takes place (e.g., Chapman 1966). This is because the inorganic-carbon supply is greatly restricted; a small amount of bicarbonate in the surface film of water on the plant is available for photosynthesis, but it is not quickly replenished. CO_2 must diffuse from the air into the water film and dissolve, but the concentration of CO_2 in air is about 10 times lower than the bicarbonate content of seawater. As desiccation begins, photosynthesis often declines even further, owing perhaps to the stress on the cells or to the evaporation of the surface film of water. In some species, photosynthesis actually increases again (Johnson et al. 1974; Brinkhuis et al. 1976; Quadir et al. 1979), although continued desiccation leads to another reduction. The reason for the increase seems to be that when the water film has evaporated, CO_2 from the air can penetrate more quickly into the cells; this explanation was given by Stocker and Holdheide (1937) and has yet to be confirmed or improved upon (Dring & Brown 1982).

If the relative humidity is experimentally maintained high enough to prevent desiccation, the photosynthesis rate may remain the same over long periods,

as found in *Fucus serratus* by Dring and Brown (1982).
This is evidence that emersion itself is not detrimental to
photosynthesis. A species of *Ulva* in Israel maintained a
constant photosynthesis rate over 0–20% water loss and
continued positive photosynthesis to about 35% water
loss. Beer and Eshel (1983) predicted that these plants
would maintain positive photosynthesis for some 90 min
after exposure in the morning, but for only 30 min at
midday. Only plants in the highest part of the population
were unable to maintain positive photosynthesis during
low tide. These presumably are saved from starvation
by more wetting at other times in the tidal sequence, or
through wave splash.

Many experiments have tested the recovery of
seaweeds from desiccation stress, often by measuring
rates of photosynthesis or respiration upon reimmersion;
clearly, many seaweeds are able to tolerate desiccation,
as detailed in a review of the older literature by Gessner
and Schramm (1971). Characteristically, these experiments consist in drying out a selection of intertidal
algae to measured degrees of water loss, then resub-
merging them in water and measuring the rates of gas
exchange at various times. Some representative results
are shown in Figure 6.24, where recoveries by *Fucus
vesiculosus* (mid-intertidal) and *Pelvitia canaliculata*
(high-intertidal) are compared. Not surprisingly, *Pelvetia* is able to withstand longer periods of desiccation.
The extent of desiccation that can be tolerated varies
among species. *Fucus* was shown to tolerate desiccation
to 25% of its original water content, with virtually complete recovery after some hours, whereas *Ulva* was
greatly impaired by such extreme desiccation (Quadir et
al. 1979). These experiments demonstrate that emersion
can be stressful. However, the ecological relevance of
these studies often has been limited because, as Chapman (1986) has noted, the exact conditions in their habitats were unknown and hence may or may not have
been tested in the experiments. Studies by Schonbeck
and Norton (1978, 1980b) and Dring and Brown (1982)
have correlated stress and recovery with the conditions
at the normal positions of the plants on the shore.

Dring and Brown (1982) interpreted the similar
straight-line relations between photosynthesis rates and
water losses (Fig. 6.25) to mean that the photosynthetic
apparatus of these high-shore British fucoids is not
more resistant to water loss than that of low-shore species. Nor is the rate of water loss significantly lower in
higher-intertidal species. What does differ is the ability
of the photosynthetic apparatus (and the cells in general) to *recover* from desiccation stress when resubmerged (Fig. 6.24). On the basis of their own data and
previous studies, Dring and Brown (1982) assessed
three hypotheses that might explain the effects of desiccation on intertidal plants and on zonation: (1) Species
from the upper shore are able to maintain active photosynthesis at lower tissue water contents than are species
from lower on the shore; this is refuted by the data in

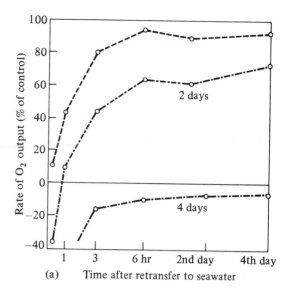

(a)   Time after retransfer to seawater

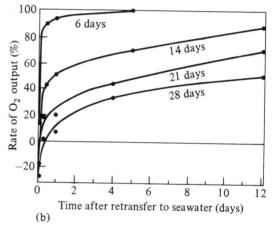

(b)

Figure 6.24. Recovery of photosynthesis in two intertidal fucoids, *Fucus vesiculosus* (a) and *Pelvetia
canaliculata* (b), following desiccation for several
days. The upper curve in (a) is the rate in a thallus
resubmerged immediately after reaching 10–12% of
the original water content. Photosynthesis rate, as
O_2 output, is expressed as a percentage of the rate
in un-dehydrated control plants. (From Gessner &
Schramm 1971).

Figure 6.25. (2) The rate of recovery of photosynthesis after a period of emersion is more rapid in species
from the upper shore; this hypothesis is also refuted by
the available data. (3) The recovery of photosynthesis
after a period of emersion is more complete in species
from the upper shore (this hypothesis was supported by
Dring & Brown). However, a more recent study by
Oates and Murray (1983) gave different results for two
California intertidal fucoids with a sharp boundary between them: *Hesperophycus* (the higher of the two)
showed greater photosynthesis rates in air than did
Pelvetia compressa f. *gracilis*, though both showed

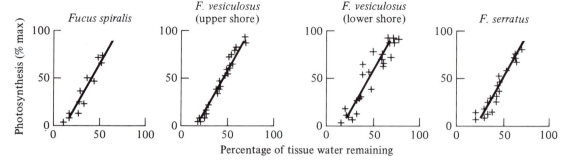

Figure 6.25. Photosynthesis during exposure to air as a function of water content for three fucoids from four heights in the intertidal zone. Linear-regression lines for the four populations are not significantly different. (From Dring & Brown 1982, with permission of Inter-Research.)

reduced rates compared with in-water photosynthesis. *Hesperophycus* contained more water when fully hydrated, and lost water more slowly when emersed. Thus Oates and Murray found that, contrary to Dring and Brown's conclusions, the differences in emersed net photosynthesis rates and rates of water loss would account for the boundary between these two species. Given the variation reported among different southern California kelp beds (sec. 1.1.2), the differences between the British and southern California fucoids should not be surprising.

There is little evidence yet on how higher-shore algae are able to recover more completely, but some other work has suggested that the explanation may lie in the biophysical properties of the photosynthetic apparatus. Red algae, such as *Porphyra*, have a mechanism to control the transfer of light energy from PS-II, which is connected to the phycobilisomes, to PS-I, which contains most of the chlorophyll *a*. Wiltens et al. (1978) found evidence that the steps in the photosynthetic sequence that are sensitive to desiccation are the transfer of electrons from PS-II to PS-I and the splitting of water (see Chapter 4). In a desiccation-tolerant plant, *Porphyra sanjuanensis*, rehydration first led to recovery of the intersystem electron-transfer process, and then recovery of the water-splitting process. In *P. perforata* under conditions of high light and desiccation, PS-II, which is the more sensitive to photooxidation, is protected by a cycling of electrons from the PS-II acceptor back to oxidants produced on the water-splitting side of PS-II. Light energy is thus dissipated as heat (Satoh & Fork 1983).

Some fucoids have an ability to "harden" to drought conditions as their exposure duration gradually increases (Schonbeck & Norton 1979a,b). The mechanism of hardening is not known, but two factors are potentially involved. First, an increase in dry matter (hence better water retention) probably causes the cells to collapse less severely and sustain less mechanical strain, especially to the plasmalemma. [This membrane, according to Levitt (1972), is the primary site of drought injury.] Second, there may be changes in the degree of saturation of membrane lipids (i.e., number of C=C double bonds). A saturated lipid layer offers the best protection against disruption by desiccation at high temperatures, whereas an unsaturated layer is most advantageous during dehydration caused by freezing, as shown by Levitt (1972), working with higher plants. The degree of saturation of fucoid lipids has been shown to be maximum in summer (Pham Quang & Laur 1976). Hardening and dehardening can take place rapidly, a few days of brief daily exposures being adequate to prepare upper-shore plants for the following prolonged exposure. Drought tolerance also depends on the temperature. *Pelvetia canaliculata* and *Fucus spiralis* survived much better at 9°C than at 25°C (Schonbeck & Norton 1980a).

Are there any species for which a periodic emersion is beneficial or even essential? The high-shore seaweed *Pelvetia canaliculata* is apparently one such species. It decays if submerged for more than 6 h out of every 12 h (Schonbeck & Norton 1978), perhaps because the endophytic fungus *Mycosphaerella* becomes parasitic in such conditions (Rugg & Norton 1987). Although the causative organisms of *P. canaliculata* rot have not been identified, the fungi of lichens can upset the delicate balance of symbiosis if conditions are good for fungal growth; moreover, other species of *Mycosphaerella* are known to be angiosperm pathogens. Rugg and Norton suggested that prolonged desiccation may be required to keep the fungus in check.

6.5 Salinity–temperature interactions and estuarine distribution

Although temperature and salinity often are studied as separate factors, they frequently change together, and factor interactions are common (Druehl 1981; Thomas et al. 1988). We have seen that salinity and temperature both affect water density and hence the development of pycnoclines, that the synergistic effects of salinity and temperature regulate *Macrocystis integrifolia* distribution, and that in some species par-

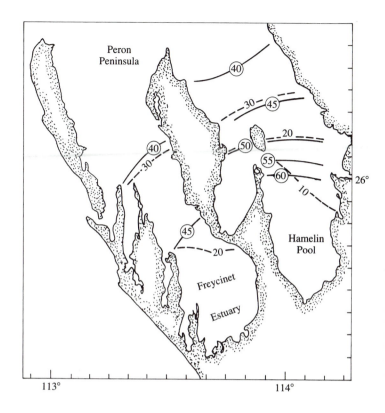

Figure 6.26. Map of Shark Bay, Western Australia, showing haloclines (solid lines, circled numbers, in ‰) and species-richness isoclines (dashed lines, numbers of species). The greater richness in Freycinet Estuary is thought to be due to greater water movement. (Modified from Kendrick et al. 1988, *Phycologia*, with permission of Blackwell Scientific Publication.)

tial desiccation allows greater tolerance to high or freezing temperatures, whereas lower temperatures can allow more desiccation tolerance. However, salinity–temperature interactions are most evident in estuaries, because the fluctuations in both factors tend to be large, and there are strong gradients in salinity and temperature along an estuary.

Many descriptions of algal distributions in estuaries, bay systems, and fjords have been carried out, such as those by Munda (1978) in Iceland, by Widdowson (1965) in British Columbia, by Silva (1979) in California, and by Mathieson et al. (1981) in New England. These studies tend to show that red and brown algae do not penetrate as far into estuaries as do green seaweeds (Gessner & Schramm 1971; Druehl 1981). Although in many such studies the distribution patterns have been attributed to salinity gradients, the physiological evidence of that is scarce. The distribution of *Phymatolithon calcareum* into the mouth of the Baltic Sea is restricted by low salinity, because the low Ca^{2+} concentration restricts calcification (King & Schramm 1982).

Polysiphonia species in the Great Bay estuary system in New Hampshire (Fig. 6.2, Table 6.1) are regulated by the temperature limits of the species, not by their salinity tolerances (Fralick & Mathieson 1975). The four species fall into two categories on the basis of their distributions and temperature optima: (1) Coldwater species, restricted to the outer coast and the mouth of the estuary, show peak photosynthesis rates at 21–24°C, but also show active photosynthesis at 5°C; these plants experience thermal injury (expressed as a sudden rise in respiration) at temperatures above 25°C. (2) Warmer-water species, which also have wider salinity tolerances, penetrate farther into the estuary; these plants show peak photosynthesis rates at 27–30°C, and little photosynthesis below 10°C. Thermal injury in the second group begins at 30°C.

Salinity gradients usually run from salt water to fresh water through highly variable and stressful brackish habitats. An interesting contrast is provided by Shark Bay, in subtropical Western Australia, where the salinity gradient starts at 36‰ and increases to 70‰ (Fig. 6.26; see also Table 6.1). Shark Bay experiences high evaporation rates and virtually no freshwater influx. It also contrasts with estuaries in a second way in that its salinity variation at any location is very small (ca. 2‰). The subtidal vegetation is dominated by seagrasses, especially *Amphibolis antarctica*, which supports a diverse epiphyte flora in up to 64‰ salinity (Harlin et al. 1985; Kendrick et al. 1988). The numbers of species decline in inverse proportion to salinity, although the relative abundances of individual species are not strongly correlated with salinity. Three major groups of algae have been identified: one that occurs or is abundant only below 50‰ salinity, a second group with high relative abundances only above 50‰ salinity, and a third group showing no correlation between abundances and salinity. The presence of a group of epiphytes most abundant above 50‰ salinity raises some interesting questions. Are their salinity optima near the salinity of seawater (Fig. 6.2c,d), or elevated? Is their abundance in the presence of high salinity the result of reduced competition from other species?

6.6 Synopsis

The surface temperatures of seawater vary with latitude and ocean currents. The annual range in the open ocean often is only about 5°C, but in the shallow waters of estuaries and bays it can be much greater. Moreover, intertidal seaweeds are exposed to atmospheric heating and cooling during low tide. Natural salinities in marine and brackish waters range from 10‰ to 70‰, but values of 25–35‰ are most common. Intertidal seaweeds may experience extreme salinity changes because of evaporation or rain/runoff.

Temperatures can have profound effects on seaweeds, owing ultimately to their effects on molecular structure and activity. Biochemical-reaction rates approximately double for every 10°C rise in temperature, but enzyme reactions show peak activities at certain optimum temperatures, above which any changes in tertiary or quaternary structures will inactivate and ultimately denature the enzymes. Photosynthesis, respiration, and growth, being sequences of enzyme reactions, also have optimum temperatures, but the effects of temperature are not uniform across all these processes. These optima vary between and within species. At these more complex levels, other environmental variables have larger effects and may overshadow the effects of temperature. Metabolic rates can become acclimated to gradually changing temperatures. Freezing kills many algae, especially if ice crystals form in the cells. However, many intertidal algae can withstand temperatures well below 0°C, especially if their cells are partially desiccated. On the other hand, tropical algae will be killed by low temperatures above 0°C.

In regions of extreme seasonal temperature changes, some seaweeds have life-history events cued by temperature (and also by photoperiod). Through their effects on the life histories of seaweeds and on the temperature ranges that seaweeds can tolerate, temperatures affect the geographic distributions of seaweeds, probably constituting the principal large-scale regulatory factor; salinity, wave action, and substratum configuration play important but local roles in phytogeography.

The components of salinity that are important to seaweed physiology are the total concentration of dissolved salts and the corresponding water potential, plus the availability of some specific ions, notably calcium and bicarbonate. The internal pressure in the cells of many seaweeds with rigid cell walls is regulated through active movement of ions across membranes or by interconversion of monomeric and polymeric compounds.

Desiccation stress on intertidal algae is partly a salinity stress, because as water evaporates from the plants, the salt concentration in the remaining water increases. During emersion, seaweeds are also deprived of bicarbonate for photosynthesis. Seaweeds higher on the shore generally are more drought-resistant than those nearer low water.

7

Water motion

The waters of the oceans are in constant motion. The causes of that motion are many, beginning with the great ocean currents, tidal currents, waves, and other forces, and ranging down to the small-scale circulation patterns caused by local density changes (Vogel 1981; Thurman 1988). Hydrodynamic force is a direct environmental factor, but water motion also affects other factors, including nutrient availability, light penetration, and temperature and salinity changes. The forces embodied in waves are difficult to comprehend, unless one has been dangerously close to them; because of the density of water, a wave or current exerts much more force than do the winds. "Imagine a human foraging for food and searching for a mate in a hurricane and you will have only an inkling of the physical constraints imposed on wave-swept life" (Patterson 1989, p. 1374). The energy amassed from a great expanse of air–ocean interactions is expended on the shoreline as waves break (Leigh et al. 1987). Equally difficult to visualize and measure are the microscopic layers of water next to plant surfaces where the plants' cells interact with water. Too much water motion imposes drag stresses on seaweeds; too little imposes diffusion stresses and impairs nutrient uptake (Wheeler 1988). Biomechanical studies of seaweed form and function are beginning to give some insight into this trade-off; see the reviews by Koehl (1984, 1986). Denny (1988) has provided a solid foundation in fluid mechanics for the study of marine organisms.

We begin this chapter at the microscopic level of gas and nutrient exchange and then examine waves and their effects on seaweeds and seaweed populations.

7.1 Water flow
7.1.1 Currents

Currents range in magnitude from the great ocean currents, of interest chiefly in regard to their temperature effects on phytogeography (sec. 6.2.5), to small-scale flows around and over surfaces. Currents attain extreme velocities only in narrow channels, where maximum speeds can surpass 2.5 m s^{-1} (5 knots) on a spring tide (Fig. 7.1). Steady currents of 0.5 m s^{-1} (1 knot) are generally considered strong, but that speed is low compared with the velocities briefly attained in breaking waves. The flow of a current along a shore is complicated by topography, as shown in Figure 7.1. Islands and reefs similarly affect flows and can concentrate planktonic organisms, eggs, and so forth, in slicks at fronts between eddies (Wolanski & Hamner 1988). Moreover, the velocity of a current decreases rapidly close to the seabed, owing to friction and eddy viscosity (the latter a property of turbulence). Thus, at the seabed, water motion will be due as much to turbulence as to the general flow of water.

7.1.2 Laminar flow and turbulent flow over surfaces

In the seaweed zone, currents change their directions frequently owing to surge and topography; special current meters are needed to measure flows, even at the macroscopic level (Forstner & Rützler 1970; Mathieson et al. 1977; Denny 1985a). To understand the exchange of gases and uptake of nutrients by seaweeds, we need to know something about water flow very close to the seaweed surface, for it is from the water layer immediately adjacent to it that a seaweed takes up nutrients (sec. 5.3). Current flows very near algal surfaces are not measurable even by "small-scale" probes, because they interfere with the flow, but the problem can be approached mathematically. The following brief, general account is based on the work of Streeter (1980), Hiscock (1983), Koehl (1986), Wheeler (1988), and Denny (1988).

The water in contact with a solid surface is stationary (Denny calls this the no-slip condition). Thus, there is a velocity gradient from the unimpeded current to the surface of the object (Fig. 7.2), forming a velocity boundary layer (defined as velocity less than 99% of the

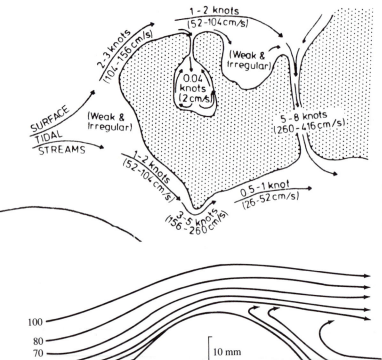

Figure 7.1. Some effects of coastal topography on the velocities of surface currents. (From Hiscock 1983, with permission of Oxford University Press.)

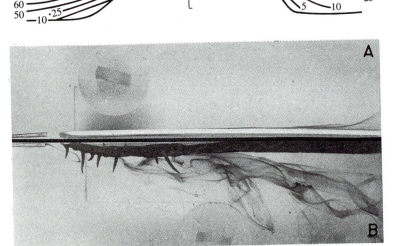

Figure 7.2. Water flow (velocities in cm s^{-1}) over a ridge on the seabed, illustrating laminar flow and an eddy on the downstream side. (From Hiscock 1983, with permission of Oxford University Press.)

Figure 7.3. Laminar and turbulent flow revealed by dye streams: (A) laminar flow over a flat plate; (B) turbulent flow over a *Macrocystis* blade. Dye was injected upstream (to the left) of the object into a current of velocity 20 mm s^{-1}. (From Wheeler 1980, with permission of Springer-Verlag, Berlin.)

free-stream velocity). The gradient is not linear, however: Within the lower 5% of the layer, the local velocity already reaches more than 50% of the free-stream velocity. Because the surface is effectively slowing down the water, there is a stress set up against the surface, called fluid (or surface) shear stress, τ_0. "Shear" refers to a force parallel to a surface, as opposed to compressive forces, which are applied perpendicularly. One can easily experience shear stress by holding a hand, parallel to the ground, out the window of a moving car.

When a current flows over an obstacle, eddies are set up at the edges, as shown on the right in Figure 7.2. If the obstacle is a smooth, flat plate, the water flow over the middle of the plate will be laminar at first. "Laminar flow" means smooth flow, which we can imagine as layers (laminae), with velocities decreasing toward the surface (Fig. 7.2.). Shear slows down the water flow next to the surface, and these flow layers drag against the current, causing the thickness of the laminar layer to increase [Fig. 7.3a; Table 7.1, equation (1)]; this happens regardless of whether the surface is smooth or rough. Some actual data are given in Figure 7.4. Sooner or later, depending on surface roughness, the laminar layer will become unstable and pass through a transition stage to turbulent flow. In turbulent flow there is vertical mixing as well as the horizontal flow. If

Table 7.1. *Equations defining various aspects of water flow and gas exchange*

(1) $\delta = 5\sqrt{\dfrac{x\mu}{\rho U}}$

(2) $\delta = \dfrac{\pi}{2}\sqrt{\dfrac{\nu W}{\pi}}$

(3) $\dfrac{\text{thickness of velocity boundary layer}}{\text{thickness of diffusion boundary layer}} = Sc^{1/3}$

(4) $Sc = \dfrac{\mu}{\rho_w D}$

(5) $T_l = 3Sc^{-1/3}\,Re_x^{-1/2}\,X$

(6) $T_t = 2.49Sc^{-1/4}\,Re_x^{-7/8}\,X$

(7) $Re_x = \dfrac{Ul}{\mu}$

(8) $J = -(D/T)(C_a - C_s)$

(9) $J = (C_a + K_s + rV_{max}) - [(C_a + K_s + rV_{max})^2 - 4r\,C_a V_{max}]^{1/2}/2r$

(10) $\sigma = \dfrac{F_{df}}{A_c} = BU^2 + B'l\dfrac{dU}{dt}$

Symbols: A_c, critical area; B and B', proportionality constants; C_a, ambient concentration; C_s, concentration at the cell surface or assimilation site (hence $C_a - C_s$ is the concentration gradient across the boundary layer); dU/dt, acceleration of water flow; D, diffusion coefficient of a particle; F_{df}, total force in direction of flow; J, flux of molecules across diffusion boundary layer; K_s, half-saturation constant of an enzyme; l, characteristic length of object; r, resistance of boundary layer; R, respiration; Re_x, Reynolds number, defined in equation (7); Sc, Schmidt number, defined in equation (4), and taken as constant by Wheeler (1980); T, T_l, T_t, thicknesses of diffusion boundary layers in general or for laminar or turbulent flow; U, current velocity; V_{max}, maximum velocity of enzyme reaction; W, wave period; x, distance along plate; X, half the mean length of object; μ, seawater dynamic viscosity; δ, velocity-boundary-layer thickness; ρ_w, seawater density; π, 3.14 . . . ; σ stress (force/area); ν, kinematic viscosity ($= \mu/\rho_w$).
Sources: Based on equations from Dromgoole (1978), Streeter (1980), Wheeler (1980, 1988), Denny et al. (1985), and Denny (1988).

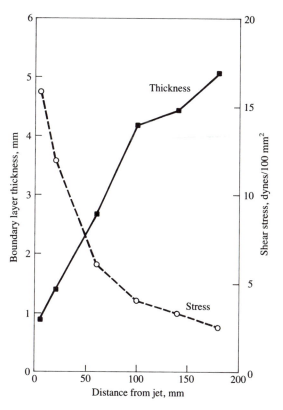

Figure 7.4. Thickness of the boundary layer and magnitude of the shear stress on a surface at various distances from the point at which a jet of water strikes it. (From Norton & Fetter 1981, "The settlement of *Sargassum muticum* propagules in stationary and flowing water," *J. Mar. Biol. Assoc. U.K.*, vol. 61, pp. 929–40, with permission of Cambridge University Press.)

the surface of the obstacle is rough, turbulence will develop more quickly (Fig. 7.3b). Plants can thus modify their hydrodynamic microhabitats by adjusting their morphologies (Wheeler 1988). The layer of motionless water that remains against the outer surface of an algal cell wall may be as thin as 5 μm for unicells or up to 150 μm for large algae, even in rapidly stirred water, according to estimates reviewed by Smith and Walker (1980). However, determining the thickness of a boundary layer in an oscillating flow (e.g., surge) is complicated by the fact that at a certain distance above the surface (depending on wave period), there is a velocity layer exactly out of phase with the oscillating flow. In these circumstances, equation (2) in Table 7.1 is preferred (Denny 1988).

In addition to the velocity gradient next to any surface in a current, there are, next to plant surfaces, gradients of nutrient concentrations. The thicknesses of these *diffusion* boundary layers will depend on the diffusion rates of the particular nutrients in water; they are not the same as the thickness of the velocity boundary layer. The thicknesses of the layers are related by the Schmidt number, Sc [Table 7.1, equations (3) and (4)]. For nutrients such as nitrate and phosphate in seawater, the diffusion boundary layers are about one-eighth the thickness of the velocity boundary layer (Charters in Neushul 1972).

A model of water flow and nutrient exchange over the surface of *Macrocystis pyrifera* blades was developed by Wheeler (1980). In that model, the thickness of the diffusion boundary layer is a function of the

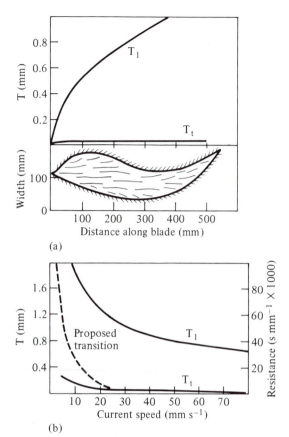

(a)

(b)

Figure 7.5. Theoretical curves for boundary-layer thicknesses (T) and diffusion resistances for a *Macrocystis pyrifera* lamina, based in part on equations (5) and (6) in Table 7.1. (a) Diffusion boundary layers over a kelp blade with water flow of 50 mm s^{-1} if there is laminar flow (T_l) or turbulent flow (T_t); below the graph is a drawing of the lamina. (b) Laminar (T_l) and turbulent (T_t) diffusion-boundary-layer thicknesses at a point 250 mm along the blade, plotted for different water velocities; boundary-layer resistance to the diffusion of HCO$_3^-$ is calculated on the right; dashed line is the estimated transition from a laminar to a turbulent boundary layer. (From Wheeler 1980, with permission of Springer-Verlag, Berlin.)

dimensions of the blade (length and area presented to the flow) and the velocity of the flow [Table 7.1, equations (5) and (6)]. Calculated thicknesses can be up to 1 mm if flow is laminar, but much thinner if flow is turbulent (Fig. 7.5a). The thickness of the layer will vary with the current velocity (Fig. 7.5b) [Table 7.1, equation (7)] and along the length of the lamina (Fig. 7.5a). Turbulence can develop over a *Macrocystis* blade at velocities as low as 10 mm s^{-1}. In nature, the current itself is very likely to be turbulent, enhancing mixing at the seaweed surface. At present, not much is known about how the edge of a seaweed thallus deflects, fans out, or focuses

water flow passing over it, nor about the turbulence set up by adjacent upcurrent plants (Norton et al. 1982).

Thalli composed of cylindrical branches, as is common in many of the smaller seaweeds, present quite a different set of hydrodynamic features: There the flow is over a series of narrow, closely spaced, rodlike surfaces. The effects of such a thallus on the water flow through it have been studied using *Gelidium nudifrons* (Anderson & Charters 1982): At all velocities, the thallus as a whole will damp large-scale turbulence in the water. At low velocities, the water leaving the thallus will exhibit smooth flow. Above a critical velocity, 60–120 mm s^{-1} (depending on the diameter and spacing of the branches), the branches will create microturbulence. Waves and currents may produce no effective turbulence around plants of this form, and so the microturbulence the plants create probably is very important to them for gas and nutrient exchange. If that is so, and because flow at low water velocities is smooth, one would expect *Gelidium nudifrons* to do poorly in areas where the water velocity is consistently below 120 mm s^{-1}. *G. nudifrons* offers a thallus that is relatively easy to study, because its fronds are stiff and smooth; plants that are soft and flexible (e.g., *Ectocarpus*) or are clothed with short lateral branches (e.g., *Platythamnion*) will present still other hydrodynamic characteristics.

7.1.3 Gas exchange and nutrient uptake

How does water flow affect gas exchange? The processes of photosynthesis, respiration, and growth in seaweeds are dependent on a flux of substrates (CO_2, O_2, nutrients). That flux is related to the thickness of the diffusion boundary layer by Fick's law [Table 7.1, equation (8)]. The boundary layer creates resistance. The overall resistance must include the resistances of the cell wall and the cytoplasm (which can change as organelles move), in addition to that of the boundary layer, but the boundary layer is by far the greatest contributor (Wheeler 1988). Thus, the thinner this layer, the smaller the resistance, and the easier it will be for fresh nutrient molecules to get to the seaweed surface.

Eventually, however, increasing current velocity will cease to restrict the overall reaction rate. Theoretical calculations, plus measurements of photosynthesis in *Macrocystis* blades, indicate that this velocity, U_p, is 36–60 mm s^{-1} (Wheeler 1980). For nutrient uptake, the velocity may be as low as 10 mm s^{-1} (C. Hurd, unpublished data). Its value will depend on the V_{max} of the reaction; for higher values of V_{max}, the flux of molecules must be greater. The effect of V_{max} on U_p shows that as the water velocity decreases, the influence of the boundary layer on V increases. Within this region, turbulence can be important, because the boundary layer will be much thinner, and the flux correspondingly greater when the flow is turbulent. Two features of the *Macrocystis* blade, its rugosity (wrinkled surface) and its marginal spines, affect water

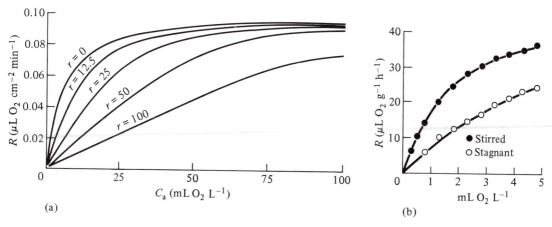

Figure 7.6. Calculated and measured effects of ambient oxygen concentration (C_a) on respiration (R). (a) Effects calculated from equation (9) in Table 7.1 for a range of resistances (r) from zero to 100 min per 10 mm. The highest value for r represents an increase in the effective boundary-layer thickness of 1,200 µm. Other values used in the equation: $R_m = 0.1$; for oxygen, $K_s = 0.50$ µL cm^{-3}. (b) Measured effects of oxygen concentration on respiration in primary axes of *Carpophyllum maschalocarpum* under stagnant and vigorously stirred conditions. (From Dromgoole 1978, *Aquatic Botany*, vol. 4, © 1978 Elsevier Science Publishers.)

flow; rugosity also increases its strength. However, both the rugosity and the number of spines in *M. integrifolia* will increase toward more exposed areas, where water velocities are high, and often greater than U_p (Wheeler 1980). Thus, in rough water, these anatomical features may serve to reduce drag rather than to increase turbulence (Wheeler 1980; Norton et al. 1982), but in areas of low and medium currents they may enhance nutrient uptake.

Given that water flow is so important in gas exchange, laboratory measurements of photosynthesis and respiration must be made in stirred chambers (Dromgoole 1978; Littler 1979). Dromgoole predicted the effects of various thicknesses of boundary layers on respiratory oxygen uptake at various concentrations of O_2, taking into account the enzymic nature of uptake and metabolism. His prediction was that when the boundary layer is thin – in vigorously stirred water – the curve for respiration versus ambient O_2 concentration will be hyperbolic, as are the curves for enzyme activity versus substrate concentration (Fig. 7.6a). This means that when the boundary layer is thin, the cells will be readily saturated with O_2. When the layer is thick, the shape of the curve will flatten, because greater O_2 concentrations will be needed to overcome boundary-layer resistance and achieve saturating concentrations at the cell surface. Experiments with *Carpophyllum maschalocarpum* (Sargassaceae) have borne out the predictions (Fig. 7.6b).

Stirring may not be enough, however. The productivity of coral-reef algal turfs was found to be 21% higher in an oscillating-flow chamber – which simulates wave surge (sec. 7.2.1) – than in a stirred chamber with the same water velocity (Carpenter et al. 1991). Whether this enhancement is peculiar to turfs or also ap-

plies to larger seaweeds remains to be seen. The hydrodynamics of turfs have yet to be analyzed.

Wheeler (1988) concluded that transport of nutrients often is restricted in low-velocity laminar flow (<0.2 m s^{-1}) and that algal size and productivity will depend on the plants' ability to create microturbulence at those speeds. He estimated the optimum range of water velocity to be 0.01–0.20 m s^{-1}, because in that range plants can get the maximum benefit from turbulent boundary layers and enhanced diffusion of nutrients, while having minimal drag effects.

7.2 Wave action

7.2.1 Physical nature of waves

Ocean waves are generated by wind blowing over the ocean surface. Certain characteristics of waves, such as height, period, and wavelength, will depend on the velocity and duration of the wind and the distance of open water over which the wind has blown. The last of these is referred to as the "fetch." The speed at which waves travel (or propogate) is called their "celerity," to distinguish it from any velocity the water itself may have (Denny 1988). Waves travel far beyond the storms that create them. As they travel, they lose energy, and because longer waves travel faster, they become sorted. Longer waves retain more of their energy because they spend less time in transit. Of course, the ocean surface rarely consists of a smooth progression of uniform swells – usually it is chaotic (Fig. 7.7), and its waves must be described statistically.

As waves approach the shore and interact with the seabed, their celerity will decrease. When the water depth becomes small compared with the wavelength, which is the case for swells near shore, the celerity is proportional to the square root of the water depth; the

(a)
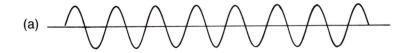

(b)
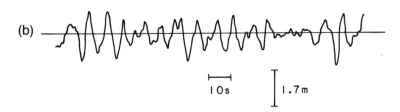

⊢——⊣ 10 s $\mathbb{I}$ 1.7 m

Figure 7.7. Sinusoidal and actual surface waves. (a) A monochromatic, sinusoidal wave train: wave height, length, and period are well defined. (b) A record of sea-surface elevations: wave height, length, and period vary. (From Denny 1988, with permission of the author.)

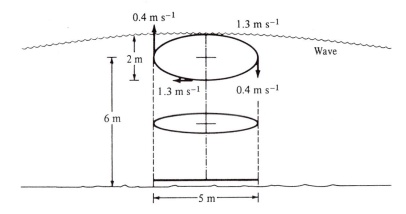

Figure 7.8. Waves and surge: the change, with depth, in the path of water motion from elliptical to horizontal. The values in this example were obtained at a site off Santa Catalina Island, California. A 5-m horizontal excursion of water was measured at a depth of 6 m when a 2-m wave moved past at a velocity of 8 m s^{-1}, with a wavelength of 120 m and a wave period of 15 s. Horizontal and vertical velocities in the upper elliptical path are indicated. (From Neushul 1972, with permission of the Japanese Society of Phycology.)

principle of conservation of energy leads to an increase in wave height for slower waves.

Waves tend to approach a beach nearly perpendicularly, whatever their direction was at sea. If a wave approaches obliquely, one end will be in shallower water, and its speed will decrease to less than that of the end that is in deeper water; as a result, the wave front will be curved and will tend to swing around parallel to the shore. The energies of waves tend to be concentrated on promontories and diminished in bays; hence headlands are more exposed to wave action. Some of the momentum of an oblique wave will drive water along the shore, creating a longshore current. Where opposing currents meet, water will flow offshore as a rip current. Both of these currents are important in the transport of spores.

The movement of water in the open ocean as a wave passes is circular (around an axis perpendicular to the wave direction) (Fig. 7.8). The radius of the circle decreases with depth, until there is no effect of the wave at all. In shallow water (less than about 30 m), the circular motion of water is progressively flattened, until at the seabed it is simply horizontal, back-and-forth mo-

tion, often referred to as "surge" (Fig. 7.8). On the surface, when the speed of the water at the top of its orbit exceeds the speed (celerity) of the wave, the water begins to spill over the front of the wave – the wave breaks. As the wave hits the shore, its energy is dissipated by friction. Water flow during the breaking of a wave on the shore consists of a jet of water coming down on the rock from the wave crest, plus a very strong surge parallel to the rock (Fig. 7.9 illustrates the hydrodynamic conditions in the surge component of a breaking wave). For "collapsing" waves, characteristic of gently sloping shores, the impact of the falling water will be relatively great compared with the surge; for "surging" waves, characteristic of steep shores, the surge component is greater.

In metric units, force [mass (kg) × acceleration (m s^{-2})] is measured in newtons (N = kg m s^{-2}); pressure [N m^{-2} = Pa (pascal)] is force per unit area. Mechanical energy is the energy available to do work, and it is measured in joules (J = N m). The available energy in an incompressible fluid such as seawater has three components (Denny 1988): (1) *Gravitational potential energy,* as in the falling water of a collapsing

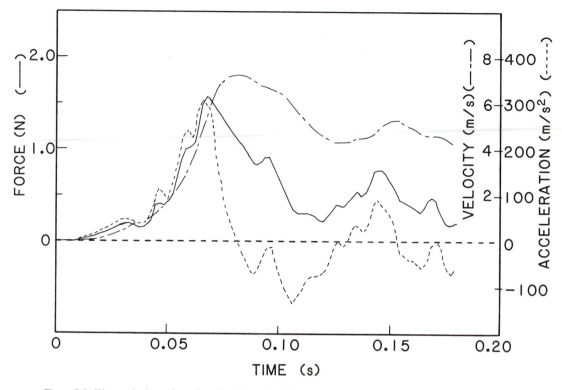

Figure 7.9. Water velocity and acceleration during the initial portion of a typical wave in the exposed intertidal zone on the coast of Washington, showing the hydrodynamic force they impose on an acorn barnacle, as measured in Denny's (1982) device. (From Denny et al. 1985, with permission of the Ecological Society of America.)

wave, is given by the mass of the water × its acceleration due to gravity × the height it can fall. (2) *Kinetic energy,* the amount of energy needed to accelerate a given mass of water to a given velocity (which will be lost when the water is stopped) is mass × (velocity)^2/2. (3) *Flow energy,* as in surge, is pressure × volume.

The pressures exerted by plunging breakers may reach 1.5×10^6 N m^{-2} (Denny et al. 1985). The pressures due to summer wave action usually are around 4×10^3 N m^{-2} (Palumbi 1984; Denny et al. 1985). The maximum water velocity at the moment a wave breaks may reach 14 m s^{-1} for a 9.75-m wave, according to estimates by Denny et al. (1985) (equivalent to a wind speed of 700 miles per hour on land; Wheeler 1988), and the acceleration as the wave hits a stationary object may reach 2,000 m s^{-2} (compare with the acceleration due to gravity, 10 m s^{-2}). Water velocity and acceleration – and hence the forces on organisms in it – are constantly varying (Fig. 7.9).

When winds are strong enough to cause whitecaps, the spilling water, as in a breaker, has forward movement and can drag the surface canopies of seaweed shoreward, although the actual force is uncertain because the air trapped in the water makes it less dense (Seymour et al. 1989). Such wind-generated waves

create a unique problem for *Macrocystis pyrifera:* Because the waves have wavelengths shorter than the lengths of the fronds in the canopy, the fronds may be pulled by more than one wave at a time. At the same time, the fronds damp the wave amplitude (Jackson & Winant 1983).

The destructiveness of large waves results largely from their direct hydrodynamic forces, but is increased by the fact that they can move rock particles, ranging from sand to cobbles. The most severe storm on record in southern California, January 1988, moved cobbles 200 mm in diameter that were situated as deep as 22 m – "such missiles are capable of extensive damage" (Seymour et al. 1989, p. 289).

The effects of wave action in broadening intertidal zones and altering community structures are well known (sec. 2.2.2, 2.2.3), but the measurement of wave action presents many practical difficulties, owing to the destructive capacity of waves and the complex flows involved (Denny 1988). The technique used will depend on the kind of information sought. Marine biologists have been principally interested in measuring surge, in connection with subtidal seaweeds (Charters et al. 1969; also see Foster et al. 1985), wave force (Jones & Demetropoulos 1968; Denny 1982, 1983,

1988; Palumbi 1984), and frequency of wetting (Druehl & Green 1970), in connection with intertidal organisms. Seymour et al. (1989) calculated drag forces on *Macrocystis* canopies from measurements of wave height and frequency. The instruments made by Charters et al. (1969), Jones and Demetropoulos (1968), and Palumbi (1984) were purely mechanical: Water pulled on a drogue, and a spring balance directly or indirectly measured the force applied, just as if a weight had been applied. There were serious drawbacks to that, however, which Denny (1988) enumerated. Denny's (1982, 1983, 1985a,b) own approach was to glue a replica of the organism to an instrument set flush with the rock surface, with the force on the organism being transmitted through a rod to an electronic or mechanical recording device.

Owing to the expense and the technical difficulties in direct measurements of wave action, less direct ways are frequently used. Relative wave action (or water motion in general) has been measured in terms of the erosion of plaster of Paris from "clod cards" (Doty 1971), plaster balls (Mathieson et al. 1977; Gerard & Mann 1979), and tethered concrete blocks (Craik 1980). Again, these methods have serious shortcomings (Denny 1985a).

Several authors have derived exposure indices from the factors that determine wave size. The simplest scale is the angle of open water (water unobstructed to some arbitrary distance from the site), as determined from a map (e.g., Baardseth 1970). This method relies on the relation of wave action to fetch, but does not take account of the other factors involved in generating or modifying wave action. A more sophisticated method (Thomas 1986) involves calculating an index based on the velocity, direction, and duration of the wind and the effective fetch; it can be modified for shore slope. Thus, most of the factors affecting wave action are included; the index can be derived from readily available hydrographic and weather data, and it is universally applicable. Nevertheless, all exposure indices are ultimately limited because of the circular argument that underlies them: The calibration or rationalization of the method is dependent on assumptions about the distribution of organisms, which is what the method is designed to explain.

Yet another approach is the biological-exposure scale, which has the advantage of integrating all the complex components of wave exposure, including the impact pressures of waves, the wetting effects of wind-driven spray, the presence of sedimentation in the intertidal zone, the mobility of loose stones, and the nutrient availability (Lewis 1964; Dalby 1980). Unfortunately, biological scales are applicable only locally, and they, like exposure indices, have the shortcoming of circularity. There is really no substitute for direct measurements. The kinds of measurements needed will depend on the nature of the problem to be solved; at the population level, the integrated effects of all components may be important, but more specific data will be needed for an understanding of the physical climate for individual organisms.

7.2.2 Form and function in relation to water motion

Solitary macroscopic organisms are chiefly affected by water flow outside the velocity boundary layer, whereas microscopic organisms and turfs are more strongly influenced by boundary-layer conditions. Aggregated organisms (e.g., kelp or seagrass beds) may experience much less severe hydrodynamic forces than solitary individuals, although Holbrook et al. (1991) argue that that is not so for the sea palm *Postelsia*. Water tends to flow over rather than through a group (Jackson & Winant 1983); the same is true of air movement around terrestrial forests. On most shores, organisms are crowded, but clumps can be treated mathematically as large organisms. The forces acting on an attached hemispherical organism or clump are the shear force in the direction of flow [F_{df}; Table 7.1, equation (10)] and lift force due to turbulence and to pressure differences as the flow is diverted around the object (consider a hydrofoil).

In tidal rapids, where there are currents, but little wave action, plants frequently grow to immense sizes if the current velocities are moderate, perhaps owing to the nutrient supply: Swift water maintains an ample supply of nutrients and sweeps away silt; yet there is not the buffeting associated with wave action. In very rapid currents, water motion causes large plants to flap about, and plants in such conditions often are as stunted as those that experience extreme wave exposure (Kitching & Ebling 1967; Norton et al. 1982; Mathieson et al. 1983).

The features of seaweed morphology by which plants might adapt to life in a current, as opposed to enduring wave action, are difficult to distinguish, but *Nereocystis* provides a model (Koehl & Alberte 1988). Like some other kelps, *Nereocystis* has narrow, smooth blades when it grows in rapidly moving water, but wider, undulate (ruffled) blades in sheltered water. Both blade types flap in the flow; in slow flow, that increases the turbulence and improves nutrient exchange. In high flow, ruffled blades tend to tear, whereas smooth blades tend to bundle together and present a more streamlined profile to the flow, but consequently experience more self-shading and probably less effective nutrient transport. Again we see the balance that must be maintained between diffusion stress and drag stress.

Wave action varies not only from place to place but also seasonally and unpredictably due to storms. Algae may grow large during periods of calm, only to be pruned back or destroyed in stormy periods. The algae in wave-exposed areas may consist of large kelps, short bushy or turfy algae, and crusts. Each has appropriate strategies for surviving wave force. Water motion affects plant morphology, just as plant morphology can affect the water forces on the plant.

Wave-swept organisms such as algae, corals, and gorgonians tend to be smaller than more sheltered individuals; many larger seaweeds either have narrower blades in more exposed habitats (Druehl 1978; Gerard & Mann 1979) or have their blades split into narrow strips (as in *Laminaria digitata* and *Lessonia*). These phenomena have been attributed to wave forces (Price 1978; Gerard & Mann 1979; Norton et al. 1982), but the extent to which such changes are due to wave exposure has been questioned by Russell (1978). Several examples will illustrate the problem of showing cause and effect rather than mere correlations (see also the discussion of *Macrocystis* blade rugosity and spines in sec. 7.1.3).

First, *Fucus vesiculosus* becomes progressively smaller and less vesiculate in more exposed habitats, but Russell (1978) has pointed out that if the plants do respond to wave action by becoming shorter, narrower, and less vesiculate, then there should be strong correlations between the morphological features, as well as between each feature and a scale of wave exposure. Russell's data show that only some of those expected correlations can be demonstrated. For example, vesicle number does not correlate with frond width or plant fresh weight.

Another problem in correlating form and function was encountered by Kraemer and Chapman (1991) in *Egregia menziesii:* Plants showed mechanical adaptations, but alginate compositions did not show the expected correlations with structural adaptations to wave exposure. Blades growing in high-energy environments had weaker alginate (less polyguluronate; sec. 4.5.2), which should have led to more flexible blades, and yet the blades were stiffer and stronger than those grown in calmer water. Finally, *Mastocarpus papillatus* shows considerable range in morphology (e.g., thallus thickness and degree of branching); yet the drag force on the plant is a function of the area of the thallus and is not strongly related to morphology (Carrington 1990). The diameter of the stipes does not increase with plant size, suggesting that thalli in wave-exposed habitats may be size-limited, regardless of their morphology.

Theoretical size limits imposed on rigid benthic organisms by wave forces were calculated by Denny et al. (1985). Their framework not only helps explain how wave forces affect seaweeds and invertebrates but also provides a basis for study of other potentially limiting factors. The model depends on the fact that of the two components of force imposed on marine organisms by waves, one, that due to acceleration, increases faster than the organism's structural strength as the organism grows, and thus potentially limits the organism's size. The model was worked out for rigid animals (urchins, limpets, barnacles, *Millepora* coral), but is applicable to some algae (e.g., branched, nonarticulated corallines such as *Lithophyllum moluccense*, character-

istic of high-energy tropical-reef crests) (Littler 1976). Drag force is proportional to the square of the velocity times the ratio between the area of the organism exposed to the flow and the critical attachment area; if this ratio remains constant during growth (i.e., isometric growth), the drag force will be proportional to the square of the velocity ($\Sigma = BU^2$). In other words, for *steady* flow, stress[*] is independent of the organism's size and cannot set a size limit for the organism. However, when the changing water velocity (i.e., acceleration) characteristic of breaking waves is taken into account, the size of the organism (as characteristic length, *l*) enters the equation [Table 7.1, equation (10)], and the larger the organism, the more it will be affected by acceleration forces.

Most seaweeds are flexible and are not well modeled by these equations. Moreover, the critical attachment area – usually the stipe or holdfast – tends to change in proportion to the projected area in algae, so drag forces will change as the plant ages. As Denny et al. (1985) have pointed out, flexibility is an important means of reducing drag and allowing growth to greater sizes. When tall, flexible organisms bend over and become streamlined, the area presented to the flow is greatly reduced, and therefore the drag force is greatly reduced, as is the acceleration reaction (see *Halosaccion glandiforme;* Vogel & Loudon 1985). [These are the two components of total force, F_{df}, in equation (10).] There will be reduced flow under a blade near the substrate (Koehl 1986), and if a plant is thin, it may be bent right over into the slower-moving water of the boundary layer. Stretching of the blade or stipe will also reduce the force on the holdfast. However, if an alga flaps in the flow, it can suffer more drag than it would if it were more rigid. Large kelps such as *Lessonia, Lessoniopsis, Eisenia*, and *Nereocystis*, and the kelplike fucoid *Durvillaea*, show several of these characteristics (Santelices & Ojeda 1984; Denny et al. 1985). Even apparently stiff plants such as *Corallina* and *Halimeda* have some flexibility owing to their noncalcified genicula.

Such means of reducing drag are remarkably effective, especially in the stipes of *Nereocystis luetkeana*, which become very narrow and elastic at the base (Koehl & Wainwright 1977, 1985). The stipe bases seem incongruously small compared with the holdfast, and yet they can withstand the strain imposed by most waves until they are damaged by grazing or abrasion (Koehl & Wainwright 1977). Two features of *Nereocystis* help the plant survive wave action. We can think of the plant as a sphere on the end of an elastic string. The maximum acceleration force occurs when a wave first strikes, but the wave force pulls on the stipe only when the plant has become fully extended in the direction of

---

[*] In an engineering sense, stress = force/area (Denny 1988).

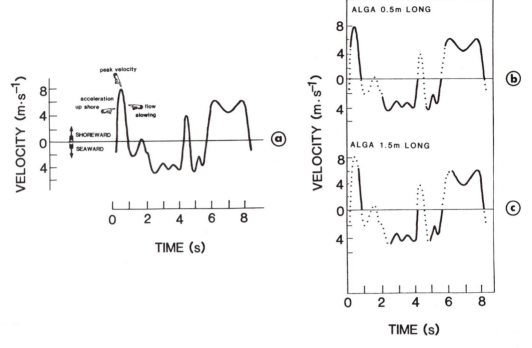

Figure 7.10. Water velocity and movement of a flexible seaweed. (a) Water velocity versus time 15 cm above the rock on an exposed shore. (b, c) The same velocity curve adapted to show times when a perfectly flexible seaweed would be moving with the water (dotted sections) and when the water would be moving over the seaweed (plant fully laid out in the direction of the flow), if the alga were (b) 0.5 m long or (c) 1.5 m long. (From Koehl 1986, with permission of Cambridge University Press.)

the surge. Assuming that the plant had become fully extended along the backflow of the preceding wave, it must travel some distance before it is again fully extended (Fig. 7.10). Second, it can stretch a long way before the force will have any chance of breaking the stipe. Extension and stretching take time – Denny (1988) estimates 12–13 s for a mature *Nereocystis* – and because the water travels shoreward for only half of a wave period, most waves have too short a period to harm the kelp: "If the stipe is long enough, the sphere never reaches the end of its tether before the water velocity changes direction" (Denny 1988, p. 245). Most seaweeds are much shorter than *Nereocystis* and thus will experience the pull of the wave force, but the delay still helps (Fig. 7.10).

The stiff stipes of *Eisenia arborea* and *Lessonia nigrescens* branch, and each branch has a bundle of blades arrayed at the tip. When the water is calm, this morphology holds the blades spread out for maximum light harvest. When there is a surge, the blades in each clump layer together, and the stipe bends, so that the whole thallus becomes streamlined (Charters et al. 1969; Neushul 1972; Koehl 1982, 1986). As with aggregated organisms, water tends to flow around rather than through the bundles of blades, and this, too, reduces drag.

Even though *Pterygophora* has a relatively stiff stipe for a kelp (six times more tensile strength than

Nereocystis), it is still very flexible (Biedka et al. 1987). Its resistance to breaking comes partly from different tensile and compressive properties. Stipe tissue is easier to compress than to stretch, so that in bending, the compression of the cells on the inside of the bend relieves part of the tension on the cells on the outside of the bend, and the stipe can bend farther before breaking (Biedka et al. 1987). In fact, *Pterygophora* stipes must be bent double and squeezed before they will snap, a situation that obviously never occurs in nature. *Pterygophora* stipes are very tough because the cortex is stiff, strong, and extensible. Yet the stipes are brittle, and surface flaws such as cracks or grazing marks will greatly reduce the mechanical strength of the stipes (Biedka et al. 1987). The effect of cracks is partly mitigated by the extensibility of the tissue, which rounds off the apex of the crack and makes it less liable to fracture further (Denny et al. 1989). Cracks may actually help the stipeless kelp *Hedophyllum sessile* in wave-exposed areas by allowing portions of the blade to be torn away by unusually strong waves, while saving the basal meristem (Armstrong 1987).

On the basis of his studies of wave stresses on individual organisms, Denny (1988) developed a "structural-exposure index." In contrast to the exposure indices described in section 7.2.1, this index is a property of the organism, not of its environment: It is the probability of being dislodged. The structural-exposure index incorporates all the factors that exert forces on the individual and the properties that enable it to resist. The index changes as the organism grows, as it is weakened by grazing, or even as it moves (a limpet is more prone to dislodgement when crawling). Aggregation and other behaviors may improve the prospect for an individual's survival. As with other indices, its value changes as the wave forces change.

Productivity is very high on wave-beaten shores, both on temperate rocky coasts and on tropical reefs. As noted earlier, the wave energy generated over some expanse of the ocean is expended at the shoreline. How does all this energy, which algae and animals cannot harness directly, help productivity? Leigh et al. (1987) have suggested four ways. First, wave action, by constantly moving algal fronds so that none is permanently shaded, maximizes the area available to trap light and permits a high frond-area index (ratio describing the area of algal surface per unit of rock surface), as compared with calmer shores or terrestrial forests. Second, water motion rapidly replenishes the nutrients in the boundary layer and maintains a high nutrient flux to the algal cells; this is a key component in tropical reefs, where algae maintain high levels of productivity in spite of very low nutrient concentrations (Adey & Goertemiller 1987). Third, waves hinder the grazing of herbivores such as urchins and fish, thus allowing seaweeds to invest less in chemical or structural defenses and channel more resources into growth. Finally, wave-induced whiplash can defend algae against herbivores and competitors (Santelices & Ojeda 1984). Often much of the productivity in exposed areas comes from animals, especially mussels. The high secondary productivity of mussels, in areas where primary productivity is relatively lower than on more sheltered shores, can be explained by the importation of productivity from the water column (McQuaid & Branch 1985).

Apart from the effects of wave force, water motion can cause morphogenetic responses, as found particularly in pseudoparenchymatous, siphonous green algae. The thalli of *Penicillus dumetosus* and *P. pyriformis*, species characteristic of moderately exposed areas, are rounded when grown in deeper water where wave action is slight, in shallow water if the directions of the waves change frequently, and in culture. But in shallow areas in which waves come predominantly from one direction, the stipe and bushy capitulum become flattened and oriented with the flat side perpendicular to the direction of the waves. (Notice that they thus maximize the area exposed to hydrodynamic forces.) Such orientation is also seen in some other genera, with blades that are always flat (e.g., *Udotea, Halimeda;* Friedmann & Roth 1977). Other species of *Penicillus,* characteristic of sheltered waters, do not show such flattening or thallus orientation.

A different response to water motion has been found in another siphonous green, *Codium fragile,* by Ramus (1972). Plants grown in culture will first

Figure 7.11. *Pterygophora* washed ashore with their substrate on the west coast of Vancouver Island testify to the force of moving water.

produce free filaments. These filaments will unite only if the culture is shaken. They bind into knots that become axis primordia, which then grow into the familiar multiaxial thallus. If shaking is discontinued, the filaments will start to grow apart again. The mechanisms of these various effects in *Penicillus* and *Codium* are unknown.

7.2.3 Wave action and other physical disturbances to populations

Water motion is a major cause of seaweed mortality at all stages of growth, perhaps especially for settling spores or zygotes (sec. 1.6.1). The windrows of seaweeds cast onto beaches testify to the power of waves to pull up seaweeds and animals, in some cases still attached to rocks (Fig. 7.11). Storms overturn boulders and move sand onto and off of beaches. All of these events are disturbances that destroy some organisms and create space for others. In many areas, storms constitute a regular seasonal phenomenon, although their intensities may vary markedly (e.g., Seymour et al. 1989). In some areas, cyclones (hurricanes or typhoons) create unpredictable disturbances. Earthquakes occasionally cause uplifts of intertidal communities (Nybakken 1969; Castilla 1988). In terms of scale, a disturbance can be categorized as a "disaster" or a "catastrophe" (Harper 1977): "Catastrophes" are infrequent, extreme events

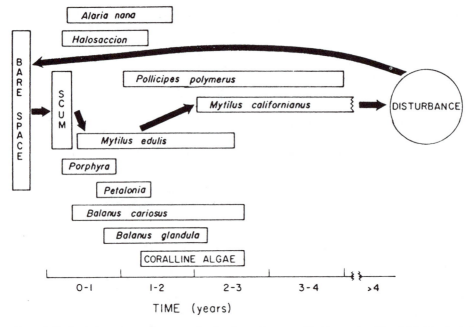

Figure 7.12. Patch dynamics on the exposed rocky shore of western Washington (see Fig. 3.10). The basic interactions and processes are indicated by the arrows. All other species and interrelationships, of which there are many, are of secondary importance. SCUM is an acronym for "successional community of unicellular and microalgae." (From Paine & Levin 1981, with permission of the Ecological Society of America.)

that destroy large portions of a population; "disasters" are more localized, but more frequent, events.

A 1980 hurricane had catastrophic effects on the corals in the reefs around Jamaica, as described by Woodley et al. (1981); another hurricane struck there in 1988. Hurricane Allen, in 1980, created waves up to 12 m high and devastated the shallow fore-reef communities, both by the direct force of the water and by the impact of entrained projectiles (broken coral, etc.), although damage was patchy. Randall and Eldredge (1977) gave a similar account of a typhoon in Guam. The changes in the algal communities were not assessed in either study, but they are likely to have been more complex than simply damage, because algal blooms occur on cleared substrate. (For instance, blooms of *Pseudobryopsis oahuensis* – an alga normally very rare on Guam – were seen there after major typhoons in 1976 and 1990; R. T. Tsuda & C. S. Lobban, unpublished observations.)

"Disasters," as defined earlier, are more interesting ecologically because in terms of scale and frequency they create an environmental variable to which organisms can respond. Their creation of clear spaces prevents competitively dominant organisms from permanently precluding other species. On exposed northeastern Pacific rocky shores, the mussel *Mytilus californianus* is a competitive dominant in the mid-intertidal zone (Paine & Levin 1981). When patches of mussels are ripped out, a "rotation" of algae and ani-

mals tends to give way to a restoration of the mussel bed (Fig. 7.12). ("The term *sequence* would imply too much order" in the process; Paine & Levin 1981, p. 174.) Within this community, one alga is particularly dependent on disturbance: the kelp *Postelsia palmaeformis* (Dayton 1973; Paine 1979, 1988). *Postelsia* lives in the mid-intertidal zone on only those headlands most severely exposed to wave action. Small sporophytes develop equally on bare rock, on animals (barnacles attached to mussels), and on other algae, but only those on bare rock persist to maturity (Paine 1988). Commonly, clumps of *Postelsia* and the barnacles/mussels to which they are attached are dislodged by waves. That, in conjunction with wave action directly on the mussels, creates bare space. According to Dayton's (1973) hypothesis, new *Postelsia* spores (from nearby plants) settle into the new bare patch and grow to maturity. However, according to L. D. Druehl and J. M. Green [unpublished data (but see Carefoot 1977)]; spores released in fall overwinter as gametophytes or very small sporophytes among smaller algae and under the mussels. In spring, those new sporophytes that are among small algae are able to grow up, whereas those under the mussels are inhibited by low levels of light unless the mussels are removed. Paine (1979) suggested that *Postelsia* patches are unable to persist on cleared spaces because the plants are annuals, because mussels encroach on the patches from the periphery, and because chiton grazing restricts the development of young plants. Yet

Postelsia also needs the mussels, because they can out-compete turflike *Corallina* and *Halosaccion*, whereas *Postelsia* cannot. The turfs are more resistant to grazing than is *Postelsia*, and more resistant to wave forces than are the mussels (Paine 1988).

Disturbances are also important to the mainte-nance of *Ecklonia radiata* kelp beds in Australia, but in that situation the kelp is the dominant organism in a sub-tidal habitat (Kirkman 1981; Kennelly 1987a,b). *Ecklo-nia* can maintain its dominance as long as disturbances are small, affecting only one or two plants. Shade-tolerant juveniles are present in the understory; when small clearings are created, these quickly grow up and replace the plants that have been lost. If the disturbance is catastrophic, removing a large part of the *Ecklonia* canopy, shade-intolerant species such as *Sargassum* or turf-forming algae will be able to outcompete *Ecklonia*. New sporophytes of the kelp that settle in these areas will remain as juveniles until the *Sargassum* canopy or the turf is removed by disaster or until kelps, encroach-ing from the edges, shade the turf too much.

Intertidal boulder fields are subjected to periodic disturbances in winter when storm waves overturn boul-ders. In southern California, small boulders in areas with short intervals between disturbances support only early-successional *Ulva*-barnacle communities. Large, infrequently disturbed boulders have a red turf domi-nated by *Gigartina canaliculata*. In the mid-intertidal zone, early-successional algae competitively inhibit a mid-successional association of red algae, which in turn competitively inhibit the late-successional *Gigartina* as-sociation (Sousa 1979). In its turn, the *Gigartina* turf outcompetes the kelp *Egregia laevigata*, which recruits only from spores and only at certain times of the year. The red algae expand vegetatively throughout all sea-sons via prostrate axes, encroaching on any space that becomes available. A 100-cm^2 clearing in the middle of a *G. canaliculata* bed was completely filled in 2 years. The turf traps sediment, which fills the spaces between the axes and prevents the settlement of other algal spores on the rock (Sousa 1979; Sousa et al. 1981). *E. laevi-gata* may settle if clearings become available at the right time, but by the time the kelp has matured (it lives only 8–15 months), the turf will have encroached all around, thus preventing the kelp from replacing itself.

Sand and sediment are major agents of distur-bance, associated with water movement. Wherever water-motion conditions are most suitable for algal-spore settling, they are also likely to be favorable for sediment settling. Spores that settle on sediment parti-cles are apt to be washed away before long, especially as they grow up into the faster-moving water layers. Also, if sediment settles on top of spores, the spores will be shaded and may be smothered. The interactions of sed-iment and water motion as they affected *Macrocystis pyrifera* spore settlement and gametophyte survival were studied by Devinny and Volse (1978). They found that even small amounts of sediment introduced before or along with the spores markedly reduced the percentage of spores able to settle and grow on glass slides. The in-teraction between spores and sediment is effectively one of competition for space. When the cultures in that study were shaken, either from the time that spores and sediment were added or starting 1 day later, survival was significantly reduced. Shaking also reduced sur-vival in the absence of sediment.

One might expect that scouring and burial of hab-itats by sand would prevent seaweed growth. Certainly, isolated rocks on a sandy beach and sandy areas in largely rocky shores have relatively few species of al-gae. These species tend to be robust, stress-tolerant pe-rennials such as *Sphacelaria radicans* and *Ahnfeltia plicata*, or opportunistic ephemerals such as *Chaeto-morpha linum*, *Enteromorpha* species, *Ectocarpus* spe-cies, and colonial diatoms (Daly & Mathieson 1977; Littler et al. 1983b). The opportunists are able to settle when scouring is at a minimum and the rocks are bare; they reproduce and disappear before scouring begins again. At the same time, sandy areas serve as refuges for these two groups of seaweeds (and animals) (Littler et al. 1983b). Many coralline communities in southern California are sand-stressed, but corallines resist sand scour and stabilize the sand (Stewart 1983, 1989). Sand scour causes saccate browns (*Colpomenia* and *Leathe-sia*) to be obligate epiphytes (especially on *Corallina*) in the mainland shore communities of southern California, whereas on the offshore islands they grow directly on rock (Oates 1989).

Sand movement on beaches typically is seasonal. Sand builds up in spring and is washed into the sub-tidal in autumn (Fig. 7.13). Tolerant seaweeds must sur-vive scouring and even months of burial. Such species include *Gymnogongrus linearis*, *Laminaria sinclairii*, *Phaeostrophion irregulare*, and *Ahnfeltia* species from the west coast of North America, and a *Polyides-Ahnfeltia* association and *Sphacelaria radicans* on the east coast of North America (Markham & Newroth 1972; Markham 1973; Sears & Wilce 1975; Daly & Mathieson 1977). The characteristics of these algae in-clude the following: tough, usually cylindrical, thalli, with thick cell walls; great ability to regenerate, or an asexual reproductive cycle functionally equivalent to re-generation (Norton et al. 1982); reproduction timed to occur when plants are uncovered; and physiological ad-aptations to withstand darkness, nutrient deprivation, anaerobic conditions, and H$_2$S. The nature of their physiological adaptations is at present unknown. Such algae have been called "psammophilic" (sand-loving), but as Littler et al. (1983b) have pointed out, that name implies that the algae do better (higher growth rates or reproductive output) in the presence of sand, whereas in reality they probably do worse, but are not as severely affected as their competitors. *Rhodomela larix* in the northeastern Pacific grows well on shores unaffected by

(a)   (b)

Figure 7.13. Habitat of *Laminaria sinclairii* on a sandy beach in Oregon (a) in April, with little sand present, and (b) in July, with rocks almost buried in sand. Arrows point to the same rock. (From Markham 1973, with permission of *Journal of Phycology*.)

sand, but it is much more abundant where sand accumulates (D'Antonio 1985a,b). It can survive long-term complete burial and sand scour, whereas its competitors, epiphytes, and herbivores are inhibited by sand in various ways. *Rhodomela* and other dense algal turfs tend to trap sand and silt, both on the rocky shore and on seagrass leaves. Perennial, sand-tolerant species may dominate the primary rock substratum, as *Corallina* does in the low-intertidal zone in southern California (Stewart 1982, 1983, 1989), but if they are not completely buried, they may in turn provide a substratum for epiphytic ephemerals, as mentioned earlier.

Some algae, especially the inhabitants of tropical lagoons, such as *Penicillus* and some *Halimeda* species, develop holdfasts in sand and silt. For these algae, sand is a necessary substratum, and the disturbances they face include burial and uprooting by surge and burrowing animals. For instance, *Caulerpa* species on a soft seabed in the Caribbean frequently were disturbed by conchs, ghost shrimp, rays, and so forth, which "uprooted plants, excavated holes which broke or undermined plants, trampled plants, or caused large-scale sediment redistributions" (Williams et al. 1985, p. 278). Stolons and upright shoots responded to burial by turning upward or forming new erect shoots, but their growth rates were reduced as compared with those of undisturbed plants. Stolon elongation involves very little increase in biomass, and so the cost of recovery from burial may not be great; chloroplasts move into the unburied parts of the siphonous plant.

7.3 Synopsis

A certain amount of water motion is necessary to maintain an effective supply of nutrients; beyond that, the force component of water motion becomes the more important. In the intertidal zone, a component of wave action is its wetting effect. Wave action is difficult to quantify, particularly if the integrated effects of all components are to be measured vis-à-vis the distribution of intertidal communities. The effects of wave forces may be best summarized as defining the probability of an individual plant being swept away, a value that accounts for the properties of the individual as well as the environment.

Water moves over plant surfaces as a current, with decreasing velocity closer to the surface owing to the shear created by the surface. Water flow may be laminar or turbulent, and the slow-moving velocity boundary layer is much thinner if flow is turbulent. Thus the rates of gas and mineral exchange are greater when flow is turbulent. Because of nutrient uptake by plants, there is also a diffusion boundary layer for each nutrient.

Seaweed morphology can be affected by water motion and can in turn affect the water flow over the plant. Most seaweeds survive wave action by being supple and stretchy, rather than tough or rigid. Their morphology must minimize both diffusion stress and drag stress.

Wave action plays a role in local geographic distributions of populations, both via the abilities of plants to compete successfully under drag stress and via disturbances of communities.

Water motion may involve sediment movement. Whereas sediment is generally deleterious to algae, some species are able to tolerate long periods of sand burial, and others gain a competitive advantage in sand-stressed areas.

8

Pollution

8.1 Introduction

Public concern over marine pollution has developed only relatively recently, because of several important events, such as the world's first major oil spill (106,000 metric tonnes) by the supertanker *Torrey Canyon,* which accelerated public concern in the early 1960s. This concern was renewed recently by the 11.2-million-gallon *Exxon Valdez* oil spill in Alaska (Leschine 1990; Maki 1991) and the largest spill in history in the Persian Gulf.

There is no precise definition of the term "pollution," but one possible general definition is a stress on the natural environment caused by human activities, resulting in unfavorable alteration of an ecosystem. Other definitions, referring to the introduction of a substance into the environment by humans, are more restrictive because they do not include thermal pollution. The term "unfavorable" in the definition involves human value judgments, and therefore it is common to see disagreement among scientists and politicians on whether or not certain events are examples of pollution (Rosenberg et al. 1981). The disagreement stems in part from the complexities of measuring pollutant effects over time scales ranging from minutes to decades and over at least five levels of biological organization, involving biochemical, physiological, population, community, and ecosystem structural changes (Hood et al. 1989) (Fig. 8.1).

The biochemical and physiological effects of a contaminant can result in reduced phenotypic fitness, as shown in Figure 8.2 (Bayne 1989). Pollution studies at the population level for many benthic invertebrates have focused on recruitment, mortality, size and age structure, and biomass and population production. It is likely that these parameters apply to seaweeds as well. Detecting pollution effects at the community level is more complex than detection at the cellular or population level because of the great variability in the time and space scales involved. To date, two approaches have been attempted: (1) Descriptive statistics. This involves describing the number of species detected per sample, the abundances of individual species, and the biomass, and then summarizing this information in the form of measures of diversity and species richness. (2) Patterns of species abundance and biomass. Houston (1979) described a general hypothesis of species diversity which states that under conditions of infrequent disturbances, competition between species will result in competitive displacement, whereby a few competitively superior species will dominate the community, and species diversity will be relatively low. If the community is subjected to disturbance by pollution, the competitive equilibrium will be disrupted, and species diversity will increase. At higher levels of disturbances, species will be eliminated from the community, and diversity will decline. Various systems to rank abundances and diversities have been proposed to measure community effects. One of the most common techniques is multivariate analysis. This procedure attempts to group co-occurring species into multivariate clusters based on a variety of organismal characters (Green 1979).

A new approach proposed by Warwick (1986) addresses a fundamental problem in measurements of community-level effects, namely, the requirement for reference samples. These are needed to control for temporal and spatial variability in order to be able to determine whether or not the observed changes are due to pollution. He proposed "internal controls," in which the different structural properties of a given faunal assemblage are expected to respond differently to the effects of pollution. He postulated that under stable, unpolluted conditions, the biomass of the community will be dominated by one or a few large species, and each species will be represented by relatively few individuals (Fig. 8.3A). In a grossly polluted situation, the community will become increasingly dominated

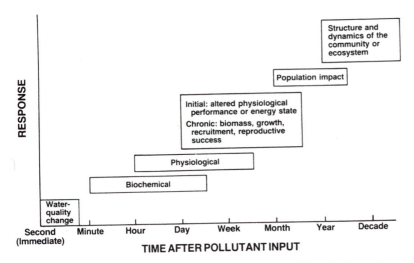

Figure 8.1. Hypothetical time-related sequence for the potential effects of pollutant input, observed at various levels of biological organization. (From Hood et al. 1989, with permission of E. W. Krieger Publishing Co.).

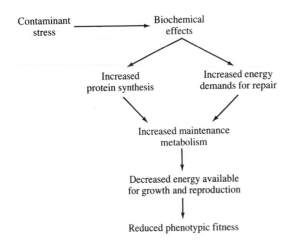

Figure 8.2. Reductions in physical fitness due to the biochemical and physiological effects of a contaminant. (From Bayne 1989, with permission of Hemisphere Publishing Corp.).

numerically by one or a few very small species, and a few large species will be present (Fig. 8.3C). Further research will be required to determine if these observations apply to seaweeds as well as to benthic invertebrates. It is clear that in order to determine statistically significant pollution effects at the community level, the field program must have the proper statistical design from the start. Advice on such matters is given by Green (1979) and White (1984), for example.

The adverse effects of pollutants on aquatic organisms generally are identified in terms of their acute and lethal impacts. Mortality can be readily recognized and quantified. The possible sublethal responses of an organism can be categorized according to the effects on the organism's (1) biochemistry/physiology, (2) mor-

phology, (3) behavior, and (4) genetics/reproduction. Physiologists and ecologists continue to debate about how to measure responses to sublethal concentrations of a pollutant, as well as whether or not laboratory bioassays give meaningful results (Underwood & Peterson 1988; Bayne 1989) and whether or not the responses observed in the laboratory can be extrapolated to the more natural and varied conditions in the oceans (White 1984). In reality, both laboratory and field measurements are necessary. In order to bridge the gap between the two areas, some laboratory facilities are scaled up and taken into the field to conduct experiments (Boyle 1985). For example, large plastic enclosures have been used to capture part of the water column in order to observe the effects of a pollutant on a community (Grice & Reeve 1982; Gray 1982).

The general stimulus–response relationship observed in biological assays of the effects of pollutants on aquatic organisms is shown schematically in Figure 8.4. The relationship frequently is nonlinear, with a threshold for the response. The dose–response relationship often is characterized by the parameter LC_{50}, the lethal concentration for 50% of the test organisms. Bioassays are particularly useful in assessing the toxicities of mixtures of substances such as pulp-mill effluents. At present we do not have an adequate basis in chemistry from which to estimate or anticipate how the biological consequences of chemical transformations, complexations, and interactions of contaminants will affect their toxicities, because only living systems can integrate the effects of those variables that are biologically important (Stebbing 1979). Bioassay techniques have been assessed, and the following three test organisms recommended: oyster larvae, sea-urchin larvae, and microalgae (Stebbing 1979). Stebbing has also recommended that manipulative techniques be used in conjunction

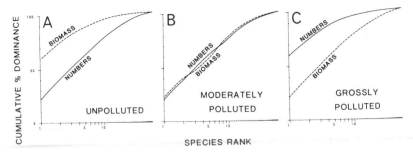

Figure 8.3. Hypothetical *k*-dominance curves for species biomass and numbers, showing unpolluted, moderately polluted, and grossly polluted conditions. (From Warwick 1986, with permission of Springer-Verlag, Berlin).

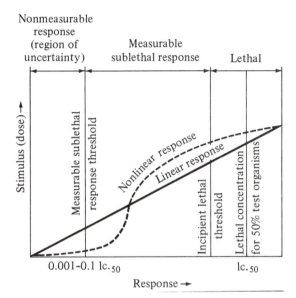

Figure 8.4. Hypothetical relation between concentrations of pollutant and responses of marine organisms, showing some significant points and regions on the curve. (From Waldichuk 1979, with permission of The Royal Society, London.)

with bioassays to aid in the categorization of contaminants (Table 8.1).

The Marine Pollution Subcommittee of the British National Committee on Oceanic Research has summarized past deficiencies and future needs in pollution research in the laboratory and field (Cole 1979b). First, in the laboratory, care must be taken in the experimental design and operation of bioassay tests. The physico-chemical aspects of the pollutant, such as solubility, adsorption, and chemical complexation and speciation, should be taken into account; because of such interactions, only a small portion of the pollutant may be biologically available (i.e., available to be taken up by an organism) (Burton 1979). The *total concentration* of a contaminant may give little indication of its toxicity. The choice of a test organism often is based on its ease of handling and culturing; frequently the most resistant

Table 8.1. *Manipulative techniques that might be used in conjunction with bioassays to aid in the identification of toxic contaminants.*

Manipulation	Technique
Removal of organics	Activated charcoal
Breakdown of organics	UV photooxidation
Removal of divalent metals	Ion-exchange resin
Binding of metals	Chelating agents (NTA, EDTA)
Removal of chlorinated hydrocarbons	Ion-exchange resin
Removal of PCBs and DDT	Membrane filtration

Source: Stebbing et al. (1980), with permission of Conseil International pour l'Exploration de la Mer.

stage (adult) is used for testing. A proper evaluation should take into account the organism's full life cycle, and certainly tests with the most sensitive stage should be conducted. If the findings are to be applicable to the survival of populations, the organism's response must ultimately be related to a healthy progression through the full life cycle, including successful reproduction. In many cases, several species should be tested (Cairns 1983). The recovery process should also be examined via experiments; the possibility that recovery may occur following damage by pollutants is often ignored and rarely assessed. The possibility of pollution-induced sexual reproduction leading to the development of resistance through genetic recombination certainly warrants further investigation. Short-term laboratory bioassays are particularly limited in terms of what they can determine, and both short- and long-term bioassays may be invalid in the absence of suspended particulate matter, which is known to profoundly influence the effects of many pollutants.

Because of the simplicity of laboratory experimental conditions in comparison with the complexity of the marine environment, there is a need for physiologists to make observations directly in the field. This could be done, for example, by attaching macrophytes to artificial substrates and taking them to polluted areas. Long-term monitoring of community structural changes in the field is required for a minimum of several years to take into account interannual variability and to assess effects at low pollutant concentrations. Remote sensing using infrared photography could be used to survey large areas of macrophytes rapidly and document major changes over several years. In the field, the ecologist must try to distinguish between the responses due to natural spatial and temporal variability and the effects attributable to pollution (Hawkins & Hartnoll 1983a; White 1984). Because often the bulk of most long-lived pollutants will end up in the sediments, the mechanisms and dynamics of uptake and release of pollutants from sediments and their transfer to biota need further investigation.

There are several categories of marine pollution. In this chapter, examples from five categories are discussed: thermal pollution; heavy metals, such as mercury, lead, cadmium, zinc, and copper; eutrophication (excessive nutrients, such as nitrogen or phosphorus); organic wastes, such as herbicides and pesticides; and hydrocarbons, particularly oil.

8.2 Thermal pollution

Some industries and most power plants use water for cooling, and they discharge their heated wastewater into the aquatic environment. The effects on macrophytes can be either deleterious or beneficial, depending on the geographic location, the season, and the species involved. Unlike higher plants, which may be exposed to a wide temperature range (ca. 50°C), seaweeds generally exist in a much narrower range (ca. 10–25°C). Whether or not macrophytes can survive the increased water temperature will depend on how close they are to their upper limit of temperature tolerance (sec. 6.2.4). This temperature tolerance is not constant for a species and may depend on other environmental factors, such as light, salinity, nutrients, and pollutants (Laws 1981). For example, during the summer in temperate shallow bays or tropical areas, the ambient water temperature may already be near the upper range of tolerable temperatures. Therefore, an increase of only a few degrees may quickly produce sublethal or lethal conditions for some species, with dramatic decreases in growth rates at temperatures above optimal. Extensive reductions in macrophytes have been noted in the areas of power-plant discharge around semitropical Biscayne Bay, Florida (Wood and Zieman 1969). On the other hand, in most temperate coastal areas with high rates of water exchange, localized increases in temperature usually will result in increases in growth rates and primary productivity. Many northern European countries now recommend that when a new power plant is to be built and is to be cooled by seawater, the temperature increase that will result after mixing not be allowed to exceed 2°C, and in the summer the temperature of the mixed water

not be allowed to rise above 26°C. In spite of such restrictions, the species composition of the affected area may change, even in the absence of thermal damage to individuals, because the increase in temperature may affect species competition. A change in algal-species composition may in turn affect the species compositions of herbivores and animals higher up the food chain. Other deleterious effects can occur if there is an abrupt plant shutdown, which can deliver a cold shock to many macrophytes. In addition, chlorine and copper are periodically introduced into such cooling-water systems to reduce fouling, creating a temporarily toxic environment in the vicinity of the discharge.

There is little published information on the direct effects of thermal effluents on macrophytes. The symptoms of thermal stress can include frond hardening, bleaching or darkening, and cell plasmolysis. Adult *Macrocystis* plants have shown substantial tissue deterioration in the presence of surface temperatures of 20°C or more for several weeks (North 1979). Three disorders have been observed: black rot, tumorlike swellings, and stipe rot. Black rot is a darkening of the blades that usually appears first at the tips, and then spreads toward the base. On the Atlantic coast of North America, a prolonged but intermittent thermal stress was reported to affect the growing (apical) tips of *Ascophyllum nodosum* (Fig. 8.5) (Vadas et al. 1978). Significant declines in percentage cover, biomass, growth, and survival were reported for *Ascophyllum*, but the basal sections of these plants survived, although weakly attached to the substrate. Apical meristems were not initiated in the second spring after the onset of the thermal discharge, but in later years the population recovered fully. However, the cover of *Fucus vesiculosus* decreased when the thermal discharge began, and it never reestablished.

The temperature curve for growth is the result of numerous metabolic interactions, some of which have different temperature optima (Cairns et al. 1975). Tolerance limits in the field are lower because of the interactions with other factors and the integration of long-term effects. For example, *Macrocystis pyrifera* displays a temperature optimum for photosynthesis between 20°C and 25°C, and yet plants begin to deteriorate when ocean temperatures reach 20°C (North 1979). Hence, field results are not always predictable from laboratory measurements.

Ocean Thermal Energy Conversion (OTEC) is a program for converting the solar energy stored in tropical waters into electrical energy by taking advantage of the temperature difference between the warm surface water and the colder water at a depth of 600 m. Surface water is pumped through a heat exchanger in order to evaporate a working fluid (Freon). The vapors then spin a turbine, and electricity is generated. The OTEC program has an effect on the environment opposite that of the traditional power plant, because colder water is pumped from depth to the surface. Two important environmental changes occur: Surface temperatures

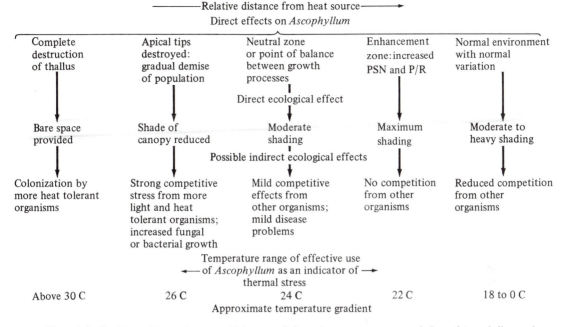

Figure 8.5. Graphic model showing potential impacts of elevated temperatures on populations of *Ascophyllum nodosum.* PSN, photosynthesis; P/R, primary productivity. (From Vadas et al. 1978, with permission of the Technical Information Center, U.S. Department of Energy).

Table 8.2. *Classification of elements according to toxicity and availability*

Noncritical				Toxic but very insoluble or very rare				Very toxic and relatively accessible			
Na	C	F		Ti	Ga	Hf	La	Be	As	Au	Co
K	P	Li		Zr	Os	W	Rh	Se	Hg	Ni	Te
Mg	Fe	Rb		Nb	Ir	Ta	Ru	Tl	Cu	Pd	Pb
Ca	S	Sr		Re	Ba			Zn	Ag	Sb	Sn
H	Cl	Al						Cd	Bi	Pt	
O	Br	Si	N								

Source: Wood (1974), *Science*, vol. 183, pp. 1048–52, copyright © 1974 by the American Association for the Advancement of Science.

are cooled from near 30°C to 20–25°C, and nutrient concentrations are increased as the nutrient-rich deep water mixes with the nutrient-impoverished surface water. Preliminary studies have shown that this process is beneficial: Two red algae, *Laurencia poitei* and *Gracilaria cervicornis,* and the brown *Sargassum fluitans* have all exhibited faster growth rates (Thorhaug & Marcus 1981).

8.3 **Heavy metals**

The term "heavy metal" generally has been used to describe those metals having atomic numbers higher than that of iron (59), or having densities greater than 5 g mL^{-1} (Sorentino 1979). From the standpoint of environmental pollution, metals may be classified into the following three groups: (1) noncritical; (2) toxic, but very insoluble or very rare; (3) very toxic and relatively accessible (Table 8.2) (Wood 1984). However, some heavy metals in category 3, such as manganese, iron, copper, and zinc, are essential micronutrients and frequently are referred to as trace metals (sec. 5.5.6, 5.5.7). They may limit algal growth if their concentrations are too low, but they can be toxic at higher concentrations; frequently the optimum concentration range for growth is narrow (Fig. 8.6). Other heavy metals in category 3, such as mercury and lead, are not required for growth, and they can become toxic to algae at very low concentrations (e.g., 10–50 μg L^{-1}).

Metals in minerals and rocks are generally harmless, becoming potentially toxic only when they

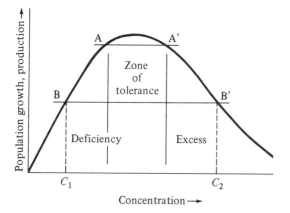

Figure 8.6. The oligodynamic-dose relation between population parameters and concentration of metabolite or pollutant. (From Perkins 1979, with permission of The Royal Society, London).

dissolve in water. They can enter the environment through natural weathering of rocks, leaching of soils and vegetation, and volcanic activity. Some of the highest mercury levels are found not in coastal waters but in the deep sea, near the mid-ocean ridges, deposited there by submarine volcanic activity. Therefore, in assessing marine pollution, a distinction must be made between natural sources and those due to human activities. Humans contribute metals to the environment during a variety of pursuits: mining and smelting ores, burning fossil fuels, disposing of industrial waste, and processing raw materials for manufacturing. Most of the metal load is transported by water in a dissolved or particulate state, and most of it reaches the oceans via rivers or land runoff. Also, rainwater carries significant amounts of cadmium, copper, zinc, and especially lead from the atmosphere to the oceans. These metals in the atmosphere come from the burning of fossil fuels. Metals in sediments may be reduced or oxidized, primarily by bacteria, and released into the overlying water.

Metals in an aquatic environment may exist in dissolved or particulate forms. They may be dissolved as free hydrated ions or as complex ions (chelated with inorganic ligands such as OH^-, Cl^-, or CO_3^{2-}, or they may be complexed with organic ligands such as amines, humic and fulvic acids, and proteins. Particulate forms may be found in a variety of situations: as colloids or aggregates (e.g., hydrated oxides); adsorbed onto particles; precipitated as metal coatings onto particles; incorporated into organic particles such as algae; held in the structural lattice in crystalline detrital particles (Beijer & Jerenlöv 1979). The physical and chemical forms of metals in seawater are controlled by environmental variables such as pH, redox potential, ionic strength, salinity, alkalinity, the presence of organic and particulate matter, and biological activity, as well as by the intrinsic properties of the metal. Changes in these variables can result in transformation of the metals'

chemical forms and can contribute to the availability, accumulation, and toxicity of these elements to aquatic organisms (Stokes 1983; Mance 1987).

In coastal waters, the concentrations of heavy metals decrease with distance from river mouths. This is the result not only of dilution but also of the salting-out process of high-molecular-weight fractions and flocculation of inorganic matter as salinity increases; metals may adsorb to these newly formed particles and sink to the sediments. On the other hand, some metals previously attached to particles in the river water may be displaced by chloride ions and become available for uptake by algae. Further information on the biogeochemistry of metals can be found in publications by Förstner and Wittmann (1979), Förstner (1980), and Laws (1981).

8.3.1 Uptake and accumulation

Metals are taken up both passively and actively by algae. Some, such as Pb and Sr, may be passively adsorbed by charged polysaccharides in the cell wall and intercellular matrix (sec. 4.5.2) (Morris and Bale 1975; Eide et al. 1980). Other metals (e.g., Zn, Cd) are taken up actively against large intracellular concentration gradients (Eide et al. 1980).

Macrophytes concentrate metal ions from seawater, and the variations in the concentrations of metals in the thallus often are taken to reflect the metal concentrations in the surrounding seawater. On that basis, macroalgae (especially the Phaeophyceae) have frequently been used as indicators of trace-metal pollution (Morris & Bale 1975; Phillips 1977, 1991). The rationale for using seaweeds as indicators of metal contamination has three main bases (Luoma et al. 1982): The first is that metal concentrations in solution often are near the limits of analytical detection and may be variable with time. Seaweeds concentrate metals from solution and integrate short-term temporal fluctuations in concentrations. Second, empirical methods for distinguishing the biologically available fraction of the total concentration of a dissolved metal have not been developed for natural systems. By definition, seaweeds will accumulate only those metals that are biologically available (assuming the degree of adsorption is slight). Finally, because plants do not ingest particulate-bound metals (as animals do), plants will accumulate metals only from solution.

Although the foregoing reasons once seemed to provide a sound basis for the use of seaweeds as indicators of metal pollution, such use probably has led to erroneous results, in view of what we now know. For example, light and nitrogen availabilities affected the rates of uptake of Fe, Mn, Zn, Cd, and Rb by *Ulva fasciata* in outdoor continuous cultures (Rice & Lapointe 1981). Light levels exerted greater overall effects than did nutrient levels. Concentrations of Cd and Rb decreased, and Mn increased, as the specific growth rate increased, implying metabolic regulation of these three

Table 8.3. *Concentrations of selected heavy metals in seawater and in brown algae and concentration factors*

Heavy metal	Concentration in seawater ($\mu g\ L^{-1}$)	*Fucus vesiculosus*		*Ascophyllum nosodum*	
		Concentration (ppm)	Concentration factor [a] ($\times 10^3$)	Concentration (ppm)	Concentration factor ($\times 10^3$)
Zn	11.3	116	10	149	13
Cu	1.4	9	6.4	12	8.6
Mn	5.3	103	19	21	3.9
Ni	1.2	8	6.8	5.5	4.6

[a]Concentration factor = ppm dried seaweed per microgram of dissolved metal per milliliter of seawater.
Source: Modified from Foster (1976), with permission of Applied Science Publishers.

metals. These results have now been formulated into a simple mass-transport model for metal uptake by macroalgae (Rice 1984). Algal nitrogen content controls the uptake of zinc and iron. Other investigators have found that the age of the fronds has an important bearing on the concentrations of Fe, Mn, Zn, and Cd accumulated, with older parts being more retentive (Bryan & Hummerstone 1973; Munda 1986; Higgins & Mackey 1987a). Other factors that can influence metal accumulation include the position of an alga on the shore (i.e., the length of time that the seaweed is immersed in seawater during the tidal cycle) (Bryan & Hummerstone 1973; Ferreira 1991), temperature (Munda 1986), salinity (Munda & Hudnik 1988), the season of the year (Burdon-Jones et al. 1982; Forsberg et al. 1988), and the presence of other pollutants in the surrounding water (Bryan 1969).

Therefore, it may be that seaweeds do *not* accurately reflect metal concentrations in the surrounding water. Several examples will illustrate this important point: A population of *Fucus* growing near a metal-polluted estuary showed the same seasonal pattern of changes in trace-metal concentrations observed in another population growing in unpolluted waters in the same region (Fuge & James 1973). Foster (1976) found that the concentrations of Cd, Pb, Cr, and Ni in *Fucus vesiculosus* and *Ascophyllum nodosum* from polluted areas were lower than those in control plants growing in unpolluted areas, even though the concentrations of these metals were higher in the water at the polluted sites. He suggested that the reason these two seaweeds did not reflect the ambient metal concentrations was that the high concentrations of dissolved Cu and Zn resulted in these elements being the predominant occupants of uptake binding sites to such an extent that accumulation of Cd, Pb, Cr, and Ni was inhibited. Likewise, Higgins and Mackey (1987a) found that pretreatment of kelp *Ecklonia radiata* with EDTA wash released 90% of their total Zn and Cd, 25% of their Cu, and 7% of their Fe, suggesting that large proportions of the Zn and Cd totals were associated with the apparent free

space (sec. 5.3.1). Zinc was found to be associated with extracellular polymers produced by epiphytic bacteria on *Gracilaria chilensis* (Holmes et al. 1991).

Because the amounts of metals found in seaweeds vary considerably, the concentration factors will vary also. The concentration factor is calculated as micrograms of the metal per gram of dry weight of seaweed divided by the concentration (in micrograms per milliliter) of the dissolved metal in seawater. Examples of concentration factors for four metals in two brown seaweeds are given in Table 8.3 (Rao & Tipnis 1967; Foster 1976). The concentration factors for these elements range from 10^3 to 10^4 (Ho 1987), though chromium reaches a factor of 10^6 (Saenko et al. 1976).

If macrophytes are to be good indicators of metal pollution, they must be able to release metal ions as well as take them up, especially if the metal concentrations in the water fluctuate over the long term. Several studies have examined metal release in situ by transplanting whole plants from polluted areas to unpolluted areas (Myklestad et al. 1978; Eide et al. 1980; Forsberg et al. 1988); zinc and cadmium showed some release from older tissue, whereas very little lead was released; mercury release was intermediate. Newly grown tissue attained the same heavy metal composition as the native plants, suggesting that only young portions should be used to assess ambient metal concentrations, because of the retention of metals by the older portions. Some authors have suggested that this may be advantageous in permitting the contamination history to be identified at any given location through analyses of different portions of the alga of different ages (Mykelstad et al. 1978; Eide et al. 1980).

Ho (1990) found that the cosmopolitan green alga *Ulva lactuca* was a good bioindicator for Cu, Zn, and Pb pollution associated with sewage in Hong Kong waters because of its high accumulation capacity. For similar reasons, Forsberg et al. (1988) found *Fucus vesiculosus* to be a good detector of metal pollution near Stockholm, and Say et al. (1990) recommended using *Enteromorpha* as a monitor for heavy metals in estuaries.

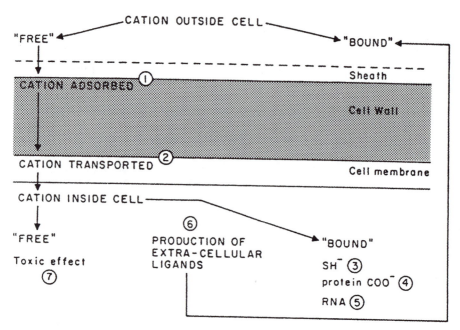

Figure 8.7. Routes of uptake for potentially toxic cations and possible sites (1–7) of tolerance mechanisms. (From Stokes 1983, with permission of Elsevier Biomedical Press.)

Despite all these optimistic assessments and recommendations for use of macroalgae as biomonitors, the problems inherent in this approach must not be forgotten; see the review by Phillips (1990). The primary problem is that the amount of a metal accumulated over a certain period of time usually is directly proportional to the growth rate of the macroalga (Rice & Lapointe 1981; Rice 1984). Therefore, any factor (nutrient limitation, salinity, temperature, turbidity, other pollutants, etc.) that tends to alter the growth rates of macroalgae will interfere with the accuracy of intersite comparisons of metal concentrations in bioindicator species (Phillips 1990). A further problem concerns the surface contamination of algae by metals associated with fine particulate material. This material is difficult to eliminate by washing prior to analysis, and it interferes with the quantification of metals actually present in the plant tissues (Bryan & Hummerstone 1973). This problem is particularly severe for metals that are closely associated with particulates in coastal waters, such as Fe, Pb, and Cr (Barnett & Ashcroft 1985).

Because of recent improvements in analytical techniques for measuring trace elements, direct measurements of trace-metal concentrations in the water are preferred to indirect measurements via bioaccumulation in seaweeds.

8.3.2 Mechanisms of tolerance to toxicity

Few studies on tolerance to metal toxicity have been conducted on seaweeds, and therefore some investigations that used microorganisms (Iverson & Brinckman 1978; Gadd & Griffiths 1978; Stokes 1983; Stauber & Florence 1987) and higher plants (Foy et al. 1978) will be discussed to illustrate basic principles (Fig. 8.7) that may apply to macrophytes.

Algae may produce extracellular compounds (Huntsman & Sunda 1980), compounds in the cell wall (Gekeler et al. 1988), or on the cell wall (Francke & Hillebrand 1980) that can bind to certain metal ions, rendering them nontoxic. Detoxification of metal ions in a culture medium or at the cell surface is referred to as an exclusion mechanism, because the metal ions do not cross the cell membrane, and consequently uptake of the metal is prevented (mechanism 1 in Fig. 8.7). This mechanism has been invoked to partially account for the high copper tolerance of some multicellular freshwater algae such as *Mougeotia*, *Microspora*, and *Hormidium*, where more than half the copper associated with cells is accumulated in the cell walls (Francke & Hillebrand 1980). Carrageenan from *Kappaphycus striatum* (Veroy et al. 1980) and fucoidan from *Ascophyllum nodosum* (Paskins-Hurlburt et al. 1976) have been shown to be capable of effectively binding heavy metals such as Cd, Pb, and Sr through an exchange mechanism. The metal-binding capacity of carrageenan was found to be correlated with its degree of sulfation. Hall et al. (1979) found that a tolerant strain of the ship-fouling alga *Ectocarpus siliculosus* accumulated less copper than did an intolerant strain. They observed that the tolerant strain did not always produce greater amounts of extracellular products; moreover, the extracellular products of the tolerant strain did not confer tolerance on the intolerant strain. Membrane and intracellular changes were believed to account for tolerance in the strain of *E. siliculosus*. Further studies on this strain revealed that it was also more tolerant of zinc and lead (Hall 1980).

Another exclusion mechanism, not yet observed for macrophytes, is adsorption or detoxification of a metal ion by surface-living microorganisms. Possibly, epiphytes such as diatoms or bacteria may take up and sequester the metal ion before it reaches the membrane surface of a seaweed.

In a thorough laboratory study, Stauber and Florence (1987) found that when trivalent metal ions (Al, Fe, and Cr), or divalent metals (Mn and Co) that could be oxidized by algae to trivalent species, were added to the culture medium, they protected two marine diatoms against Cu toxicity by forming a layer of hydrated metal (III) oxide around the cell. This layer was thought to adsorb Cu ions before they could penetrate the cell. A coating of hydrated metal oxide on the cells would be unlikely to form in synthetic growth medium containing strong chelators such as EDTA or nitrilotriacetate (NTA), because those ligands would complex trivalent metal ions. Manganese and iron, however, were unable to protect against the toxicity of lipid-soluble Cu complexes. This important study by Stauber and Florence (1987) establishes a mechanism for the previously reported antagonistic effects (Stauber & Florence 1985) of some metals, such as Mn, Fe, and Co, on Cu toxicity (i.e., an increase in Mn decreases Cu toxicity). Nickel, which has a chemistry similar to that of Co, cannot be oxidized in seawater to Ni(III) hydroxide, and therefore it is totally ineffective for protecting against Cu toxicity; similarly, Zn(II) is ineffective. The protective effects of Mn and Co are enhanced by their ability to scavenge damaging superoxide radicals and H_2O_2.

If no exclusion mechanism is operating, a metal ion can enter the cytoplasm. Several detoxification mechanisms are possible inside the cell (see mechanisms 3–6 in Fig. 8.7). Intracellular precipitation of copper has been observed within the vacuoles and nuclei of the freshwater green *Scenedesmus acutiformis* when grown at high copper concentrations (Silverberg et al. 1976), whereas in *Porphyra umbilicalis* the nucleus is the site for intracellular bound cadmium (McLean & Williamson 1977).

Once metals are inside the cell, they may, as a result of biological action, undergo changes in valence and/or conversion to organometallic compounds. Both processes can detoxify the metal if volatilization and removal of the metal results. Transformations involving changes in valence have been studies mainly with mercury. Several types of bacteria and yeast can reduce cationic mercury (Hg^{2+}) to the elemental state (Hg^0), which will volatilize from the medium (Gadd & Griffiths 1978). Certain metals (e.g., Hg, Pb, Cd, Sn) are transformed into organometallic compounds by methy-

lation. Although the products of methylation may be more toxic than the free ion, often they are volatile and can be released into the atmosphere.

Recently, Grill et al. (1987) discovered that when higher plants were exposed to a range of heavy metals, they synthesized a series of cysteine-rich peptides called phytochelatins that are functionally analogous to metallothioneins, which are produced by animals. The general structure for the set of peptides is (γ-Glu-Cys)$_n$-Gly, $n = 2$–11. That work was later extended to the microalga *Chlorella fusca*, which produced phytochelatins upon exposure to Cd, Pb, Zn, Ag, Cu, and Hg (Gekeler et al. 1988). Thus, it appears that algae and higher plants inactivate or detoxify heavy-metal ions in the cytoplasm by the synthesis of phytochelatins. It is interesting to note that animals respond to heavy-metal stress by induction of metallothionein-gene expression, whereas plants respond by induction of phytochelatin formation via enzymic polymerization of peptidic precursors. Thus, an evolutionary divergence in heavy-metal sequestration has occurred between plants and animals.

In addition to their physiological tolerances to heavy metals, algal resistance to metals is likely to be genetically controlled. Copper tolerance in the ship-fouling alga *Enteromorpha compressa* appears to be genetically determined, because progeny from the ship-fouling plants were also found to be very copper-tolerant, even though they had not previously been exposed to copper (Reed & Moffat 1983). The rate of accumulation of copper in ship-fouling thalli was equal to the rate for nonfouling thalli, suggesting that tolerance may be due primarily to internal detoxification rather than to an exclusion mechanism. This topic of genetic adaptation to heavy metals in a wide range of aquatic organisms was recently reviewed (Stokes 1983; Klerks & Weis 1987). It was concluded that microorganisms show the best evidence for the evolution of resistance and that this increased resistance to toxicants may have important implications for decisions regarding safe ambient levels of toxicants.

8.3.3 Effects of metals on algal metabolism

The order of metal toxicity to algae varies with the algal species and the experimental conditions, but generally the order is Hg > Cu > Cd > Ag > Pb > Zn (Rice et al. 1973; Rai et al. 1981).

Mercury, the most toxic metal, interacts with enzyme systems and inhibits their functions, especially enzymes with reactive sulfhydryl ($-SH$) groups (Van Assche & Clijsters 1990). The toxic effects of mercury on algae generally include (1) cessation of growth in extreme cases, (2) inhibition of photosynthesis, (3) reduction in chlorophyll content, and (4) increased cell permeability and loss of potassium ions from the cell (Rai et al. 1981). There have been two studies of the physiological effects of mercury on marine macroalgae. Hopkin and Kain (1978) studied how the different life-history stages of *Laminaria hyperborea* were affected and found that the growth of gametophytes was most sensitive. Respiration rates for sporophytes increased only at the highest concentrations of mercury. A study of the effects of Hg on increases in length for five intertidal Fucales showed that exposure to an average Hg concentration of 100–200 µg L^{-1} for 10 days gave a 50% reduction in growth rate (Strömgren 1980b). Even at 5–9 µg L^{-1}, a reduction in growth was seen in adults of *Fucus spiralis*.

Copper, even though it is an essential micronutrient, is the second most toxic metal, and copper sulfate has been used to control nuisance algae in fresh waters. Copper toxicity is dependent on the ionic activity (concentration of free Cu^{2+}), and not the total copper concentration (Sunda & Guillard 1976). However, some organic copper complexes (especially the lipid-soluble ones) are much more toxic than ionic Cu (Stauber & Florence 1987), because these lipid-soluble complexes can diffuse directly through the membrane into the cell.

The mechanism of Cu toxicity has been described by Stauber and Florence (1987) as a chronological sequence in microalgae. The initial Cu binding to the cell may be to carboxylic and amino residues in the membrane protein, rather than to thiol groups, because the Cu–algae stability constant is orders of magnitude lower than the thiol Cu-binding constant. At the cell membrane, Cu may interfere with cell permeability or the binding of essential metals. Following Cu transport into the cytosol, Cu may react with –SH enzyme groups and free thiols (e.g., glutathione), disrupting enzyme active sites and cell division. Cu may also exert its toxicity in subcellular organelles, interfering with mitochondrial electron transport, respiration, ATP production, and photosynthesis in the chloroplasts.

Toxicity effects have been shown to pass through several stages, depending on the Cu concentration (Sorentino 1979). First, copper affects the permeability of the plasmalemma, causing loss of K^+ from the cell and changes in cell volume. Next, Cu^{2+} may be transported to the cytoplasm and then to the chloroplasts, where it inhibits photosynthesis by uncoupling electron transport to $NADP^+$. As the ionic concentration increases, copper is bound to chloroplast membranes and other cell proteins, causing degradation of chlorophyll and other pigments. At still higher concentrations, copper produces irreversible damage to chloroplast lamellae, preventing photosynthesis and eventually causing death.

The effects of copper on macrophytes have been more extensively studied than have the effects of any other metal (because it is used in antifouling paints), although studies on its physiological effects are still lacking. Its toxic effects on phytoplankton have been summarized by Davies (1978, 1983), Hodson et al.

(1979), and Lewis and Cave (1982). Studies on *Laminaria hyperborea* showed that the effects of copper followed a pattern similar to that for mercury, in that gametophyte growth was more sensitive than sporophyte growth, but copper was less toxic than mercury (Hopkin & Kain 1978). In a similar study on *Laminaria saccharina*, Chung and Brinkhuis (1986) found that growth of sporophytes was the process most sensitive to Cu (>10 µg L^{-1}, followed by release of meiospores, development of gametophytes ($\geqq$50 µg L^{-1}, and settlement and germination of meiospores (500 µg L^{-1}). With Cu at 50 µg L^{-1}, sporophytes of *L. saccharina* showed abnormal growth patterns, haptera-like protuberances, giant cells, and abnormal branching patterns (Brinkhuis & Chung 1986). Bioassays using the kelp *Macrocystis pyrifera* revealed that reproduction in a long-term (16-day) test was three times more sensitive to Zn than was zoospore germination, but half as sensitive as germ-tube elongation (Anderson & Hunt 1988). Metals may inhibit reproduction by interfering with the ability of the sperm to find the egg, perhaps via a pheromone that is thought to be involved in this process (Maier & Müller 1986). Different developmental stages (spermatozoa, newly fertilized eggs, 3-week-old germlings, and adult plants) of *Fucus serratus* were tested by Scanlan and Wilkinson (1987) for their sensitivities to certain biocides to determine which stages could provide a basis for a toxicity test. They found that spermatozoa and newly fertilized eggs were the most sensitive stages. *Enteromorpha* rhizoid regeneration gave variable results, and the growth of *Ulva* discs was repeatable, but was not as sensitive as methods involving *Fucus* eggs. When the lowest concentration of copper that was toxic for sporophytes of *L. hyperborea* was compared with the concentrations of copper typically found in the ocean, the ratio was 3.3. This low ratio is in contrast to the ratios of 200 for mercury and up to 2,000 for cadmium. This suggests that if the copper concentrations in the ocean are increased by only a small factor, they will become toxic to this species. When the effects of copper concentrations on the growth of four intertidal fucoids were examined, copper was found to be somewhat more toxic than mercury and far more toxic than Zn, Pb, or Cd (Strömgren 1980a, b). A 50% reduction in growth took place at concentrations of 60–80 µg L^{-1} total copper, with *Pelvetia canaliculata* and *Fucus spiralis* being the most sensitive species. Fielding and Russell (1976) showed that species in mixed culture gave different responses to copper than when grown in unialgal culture. For example, *Ectocarpus* grew better in the presence of *Erythrotrichia* than by itself at the same copper concentration, for unknown reasons. These investigators warned that results from unialgal cultures may be misleading because they neglect possible species interactions (sec. 3.1.3).

Cadmium is a serious pollutant for plants and animals, particularly in coastal waters near industrial areas, where the concentrations may rise from the normal level of about 0.1 μg L^{-1} to several micrograms per liter. Some physiological investigations have been conducted on phytoplankton (Davies 1978; Simpson 1981), but there has been little research on macrophytes. Markham et al. (1980) reported on Cd uptake and its effects on growth, pigment content, and carbon assimilation in *Ulva lactuca* and *Laminaria saccharina*. The growth rate for sporophytes diminished to 50% of the control value when in a Cd concentration of 2,000 μg L^{-1} (but this is several orders of magnitude higher than the concentrations in most polluted areas). However, when plants exposed to different concentrations of cadmium for 6 days were measured for growth after a further 8 days in unpolluted seawater, the Cd concentration that caused a growth-rate reduction of 50% was 900 μg L^{-1}. Markham et al. (1980) concluded that the long-term effects are more serious than is immediately evident and that the exposure time is important in determining the extent of the effects. They found that Cd continued to accumulate over their 6-day experiment, with the slower-growing plants and regions of the thalli (e.g., stipe, holdfast) accumulating relatively more cadmium. At Cd concentrations greater than 2,300 μ L^{-1}, the blades showed a sharp loss of pigment in their distal regions. At sublethal concentrations (>2,000 μg L^{-1}), sharp reductions in photosynthesis and growth rates were observed. Enzymes that are involved in primary metabolism were extracted, and Cd was added to them in vitro (Kremer & Markham 1982). Generally, the activities of RuBP carboxylase, PEP carboxykinase, and mannitol-phosphate dehydrogenase were not affected. However, in vivo uptake and incorporation of ^{14}C-leucine were drastically reduced in Cd-treated plants. Kremer and Markham concluded from these observations that cadmium inhibits one or more steps in protein synthesis and thus leads to enzyme deficiencies and a series of secondary effects. In tests with five intertidal Fucales, Cd enhanced the growth of *Pelvetia canaliculata* and *Ascophyllum nodosum*, even at concentrations up to 1,000 μg L^{-1}. The reasons for the enhanced growth are unknown, but it seems possible that Cd displaced essential trace elements from particles, making them available to the plants, or else reduced the toxic effects of other metals.

There has been little research on the less toxic heavy metals, such as lead and zinc, in seaweeds. One preliminary report assessed the effects of lead on four small, finely branched red algae: *Platythamnion pectinatum*, *Platysiphonia decumbens*, *Pleonosporium squarrulosum*, and *Tiffaniella snyderae* (Stewart 1977). Significant reductions in growth occurred only at unrealistically high lead concentrations (10 mg L^{-1} as PbCl$_2$). Even though zinc is generally considered to be actively taken up by seaweeds (Skipnes et al. 1975) (actually entering the cell, as opposed to being adsorbed onto the surface), it has a relatively low toxic effect.

Strömgren (1979) found that Zn at 5–10 g L^{-1} was required for a 50% reduction in growth for five intertidal Fucales. In contrast, Cu and Hg toxicities occurred at 0.01 that concentration and Cd and Pb toxicities at 0.2 that concentration.

Some of the heavy metals in seawater are radioactive, and normal seawater has a radioactivity level of approximately 0.01 kBq L^{-1}. *Fucus* and *Porphyra* normally have radioactivities of 0.2–0.6 kBq kg^{-1} (wet weight), and the levels in molluscs and fish are 0.4–1 kBq kg^{-1} (Gerlach 1982). The levels of radioactivity may be even higher in certain areas because of fallout from the atmosphere or wastes from nuclear recycling plants. In the vicinity of Selafield (formerly Windscale), on the Irish Sea, the additional radioactivity emanating from the reprocessing plant results in contamination of *Porphyra*, especially with ^{106}Ru, with its levels of radioactivity ranging up to 12 kBq kg^{-1} (about 35 times normal). For residents of south Wales, "laver bread" made from *Porphyra* is a food specialty, and by regularly consuming it they can expose themselves to considerable radiation (up to about 20% of the permissible dosage of radioactivity) (Hetherington 1976).

8.3.4 Factors affecting metal toxicity

One of the most important of the factors that determine the biological availability of a metal is its physiochemical state (Langston 1990). Adsorption to particles in the water or complexation with dissolved organics generally will reduce toxicity of a metal. Because the form in which the metal exists often is difficult or even impossible to characterize, most studies have measured the total concentration of the metal, which does not correlate well with toxicity (Florence et al. 1984). This may explain why two studies examining the same *total* concentration of the metal in a particular alga may obtain quite different results.

Both the pH and the redox potential can have considerable effects on the availabilities and thus the toxicities of heavy metals (Peterson et al. 1984). At a low pH, metals generally exist as free cations, but at an alkaline pH, like that of seawater, they tend to precipitate as insoluble hydroxides, oxides, carbonates, or phosphates.

The interactions of salinity and temperature with toxicity are not always clear (Munda & Hudnik 1988). Usually the heavy-metal content of seawater is lower than that of fresh water. Munda (1984) found that the Zn, Mn, and Co accumulations in *Enteromorpha intestinalis* and *Scytosiphon simplicissimus* could be enhanced by decreasing the salinity. This could be associated with surface charge, because phytoplankton, and probably seaweeds, are negatively charged at low salinities. An increase in temperature has resulted in an increase in toxicity in some cases, but a reduction in other instances (Rai et al. 1981). Increased toxicity at higher temperature may be explainable by increases in the energy demand, which would result in enhanced respiration of the organism, but decreased toxicity at high temperature has not been satisfactorily explained (Förstner & Wittmann 1979).

High concentrations of certain nutrients, such as phosphorus, may reduce toxicity because of the formation of insoluble phosphates. It has been found that large additions of nitrate will reduce cadmium toxicity in a marine diatom, *Thalassiosira fluviatilis*, for unexplained reasons (Li 1978).

Algae growing in temperate coastal areas in summer may endure the double stress of nitrogen limitation and the presence of a pollutant. The possibility of increased sensitivity to the pollutant under this condition of double stress has not been extensively examined. In a study where the marine diatom *Skeletonema costatum* was exposed to mercury, Cloutier-Mantha and Harrison (1980) found that the nitrogen-limited cells grew as fast as nitrogen-saturated cells. Nevertheless, nitrogen-limited cells that had previously been exposed to Hg showed significantly reduced ability to take up NH_4^+ when it was added to the medium (i.e., had a higher K_s). Thus, mercury pollution may decrease the ability of a species to utilize the limiting nutrient during periods of seasonal nutrient limitation, thus decreasing its chances of surviving.

Algal extracellular products can reduce metal toxicity in laboratory cultures of phytoplankton when the culture density is high (Davies 1978, 1983; Lewis & Cave 1982). The importance of extracellular products in the natural environment is not clear, because dilution effects are considerable. However, abnormally high concentrations of dissolved organics from dispersed sewage occur near sewage outfalls and probably mitigate metal toxicity in those areas (sec. 8.6.1).

One interesting aspect of environmental research that has not received adequate consideration concerns the impacts of other pollutants on the toxicities of heavy metals. For example, the presence of (2,4-dichlorophenoxy)acetic acid (2,4-D) decreased the toxicity of Ni and Al in one marine phytoplankter tested, and Cu decreased the toxicity of the herbicide Paraquat to fresh water phytoplankters (Rai et al. 1981).

Heavy metals usually are discharged into the sea in combinations rather than singly. Therefore it is encouraging to see a few recent studies (e.g., Munda & Hudnik 1986) attempting to understand this more complex but realistic situation. Metal–metal antagonism has been observed. Selenium may relieve mercury toxicity (Rai et al. 1981), and manganese or iron may reduce copper toxicity in various microorganisms (Gadd & Griffiths 1978; Lewis & Cave 1982; Stauber & Florence 1985; Munda & Hudnik 1986). Significant antagonistic effects appeared with exposure to Cu + Zn and with Hg + Zn in measurements of increases in length of *Ascophyllum nodosum* (Strömgren 1980c). When two highly toxic metals such as Cu and Hg were added simultaneously, generally the toxic effects were additive. There are only a few examples of synergism between metals (i.e., where the total effect is greater than the sum of the effects of the individual metals). Munda and Hudnik (1986) observed that Mn and Co had synergistic effects on the growth of *Fucus vesiculosus*.

8.3.5 Ecological aspects

Given the high biomasses that are common in macroalgal ecosystems, large portions of the non-sediment-bound trace or heavy metals can be associated with the macroalgae, which act as substantial buffers of these elements. Because most of the macroalgal production enters the detrital pool, the decomposition of macrophyte detritus can play a significant role in the cycling of trace metals in coastal waters. Higgins and Mackey (1978b) found that detrital decomposition of the kelp *Ecklonia radiata* led to leaching of substantial amounts of trace metals and dissolved organic carbon (DOC). The DOC that was released contained organic ligands that were capable of forming strong complexes with Cu, Fe, and Zn. Polyphenolic compounds were important components of the decomposition exudate. These results suggest that this kelp plays a major role in regulating both the concentrations and speciations of heavy metals in near-shore environments.

The fact that metal concentrations in marine organisms typically are severalfold higher than the concentrations of the same metals in seawater has led to the suggestion that metals are accumulated in higher concentrations in higher trophic levels of the food chain because of biological magnification. Comparison of the concentrations of various metals in phytoplankton and zooplankton with those in seawater shows that the metal concentrations in the plankton are indeed about 1,000 times higher (Martin & Knauer 1973). However, only the concentrations of Cu, Zn, and Pb are substantially higher in zooplankton than in phytoplankton. For Mn, Ag, Cd, and Hg, the differences are small. Studies with mercury in anchovies and other animals have shown that bioaccumulation varies with the tissues sampled, with liver having the highest levels (Knauer & Martin 1972).

Support for the biological-magnification hypothesis in natural areas comes from the fact that the tuna and swordfish, both top-level carnivores, have mercury concentrations that are several orders of magnitude higher than those in phytoplankton (Laws 1981). Examples of bioaccumulation can also be found in heavily polluted areas. For example, coastal waters along the southern shore of the Bristol Channel, England, have very high concentrations of Cd, Zn, and Pb (Butterworth et al. 1972). Analysis of the seawater and a simple food chain showed that Cd and Zn were accumulated up the food chain. Table 8.4 shows that cadmium is found at relatively low levels in *Fucus* (the primary producer), at higher concentrations in the limpet *Patella* (herbivore), and in greatest concentrations in the carniv-

Table 8.4. *Cadmium concentrations in seawater, seaweeds, and shore animals at four collecting stations on the southern side of the Severn estuary and Bristol Channel*

Location	Distance from Avon mouth (km)	Seawater ($\mu g \, L^{-1}$)	*Fucus* ($mg \, kg^{-1}$)	*Patella* ($mg \, kg^{-1}$)	*Thais* ($mg \, kg^{-1}$)
Portishead	4	5.8	220	550	—[a]
Brean	25	2.0	50	200	425
Minehead	60	1.0	20	50	270
Lynmouth	80	0.5	30	50	65

[a] Not reported.

Source: Butterworth et al. (1972); reprinted by permission from *Marine Pollution Bulletin,* vol. 3, pp. 72–4, copyright © 1972 Macmillan Journals Limited.

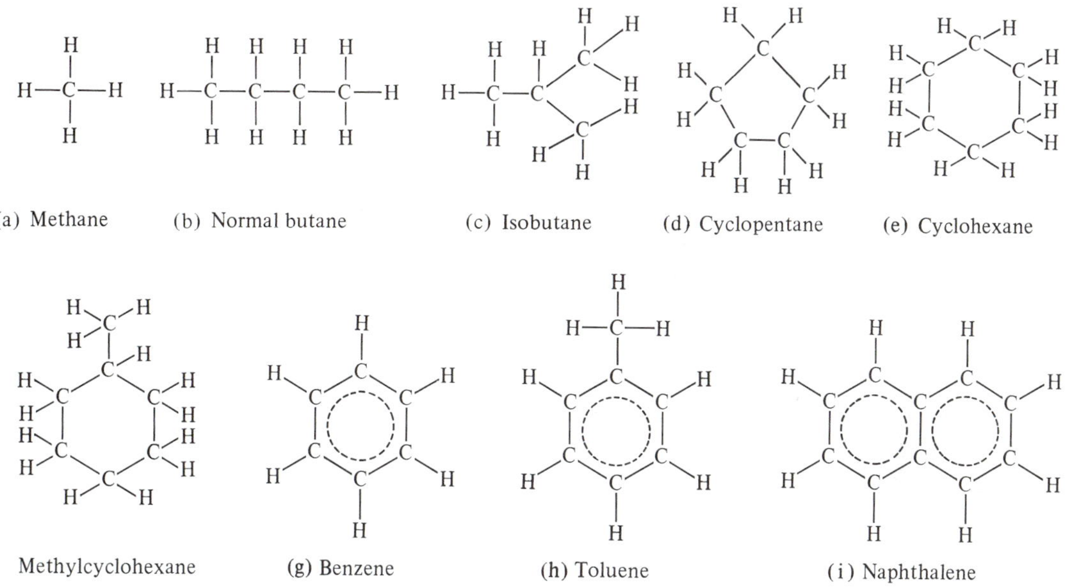

(a) Methane (b) Normal butane (c) Isobutane (d) Cyclopentane (e) Cyclohexane

Methylcyclohexane (g) Benzene (h) Toluene (i) Naphthalene

Figure 8.8. Some hydrocarbons from crude oil.

orous dog whelk *Thais*. The same pattern might also hold for other metals.

In the laboratory, the transfer of zinc and iron from *Fucus serratus* to *Littorina obtusata* did not result in accumulation in the snail, for unknown reasons (Young 1975). Similarly, when the abalone, a *Haliotis* species, was fed on a lead-treated brown alga, *Egregia laevigata*, little bioaccumulation occurred (Stewart & Schulz-Baldes 1976).

Although the elevated metal concentrations found in animals in some studies suggest biological magnification, it has been argued that the same effects might be produced by very different mechanisms (Mance 1987). On the basis of studies with DDT, Hamelink et al. (1971) suggested that elevated levels in animals may have resulted from direct uptake of the pollutant from the water and from differences in pollutant-exchange equilibria between water and different classes of organisms.

8.4 Oil

Petroleum, or crude oil, is an extremely complex mixture of hydrocarbons, with some additional compounds containing O, S, N, and metals such as Ni, V, Fe, and Cu (Preston 1988). The main components of oil may be classified into three broad categories (older names given in parentheses). The first category comprises the aliphatics (paraffins), which are straight-chain or branched hydrocarbons. They are composed mainly of alkanes (saturated; single bonds), whose general composition is C_nH_{2n+2} (Fig. 8.8a–c), with minor amounts of alkenes (unsaturated; carbon-to-carbon double bonds) and alkynes (unsaturated; carbon-to-carbon triple bonds). These aliphatic compounds make up about 20% of crude oil and are very common in gasoline and fuel oils. The presence of only saturated bonds makes reactions with alkanes difficult; hence they are very resistant to degradation. The more branched the

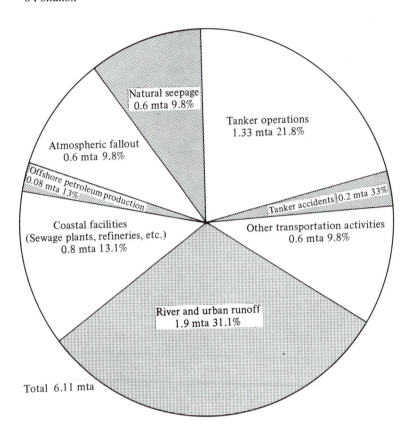

Figure 8.9. Sources of petroleum going into the oceans, millions of metric tonnes per annum (mta). (From Geyer 1980, with permission of Elsevier Science Publishers.)

molecule, the more difficult is the biodegradation. Low-molecular-weight ($>C_6$) alkanes are generally gases (e.g., methane, ethane, and propane), and high-molecular-weight aliphatics ($> C_{18}$) are solids (e.g., waxes). The lower the number of carbon atoms, the more volatile and more water-soluble the compound. They are relatively nontoxic. The second category comprises the cycloalkanes (cycloparaffins, alicyclic hydrocarbons, or naphthenes), which are similar to alkanes except that some or all of the carbon atoms are arranged in rings. These compounds have the general formula C_nH_{2n}) and account for about 50% of crude oil, the most prevalent being cyclopentane and cyclohexane (Fig. 8.8d–f). Frequently, alkyl groups (e.g., –CH_3) are substituted on the cycloalkane ring, forming compounds such as methylcyclohexane. Polycyclic naphthenes are very resistant to microbial degradation.

Aromatics compose the third major group of compounds; these contain one or more benzene rings, and the name comes from the pleasant aroma of these compounds. Aromatics commonly are found in crude oil or can be produced during refining; they include benzene, toluene, naphthalene, and phenol (Fig. 8.8g–i). Aromatics usually constitute less than 20% crude oil, but they are very toxic to plants and animals. They are readily degraded. Other hydrocarbons, such as alkenes, occur in crude oil in much smaller amounts. Alkenes (olefins) are unsaturated-chain compounds pos-

sessing double or triple bonds, but without the regular arrangement found in the benzene ring. Examples include ethylene and acetylene, which are produced during refining.

8.4.1 Inputs and fate of oil

The U.S. National Academy of Sciences periodically estimates the magnitudes of the spills and discharges of oil into the marine environment from various sources (Fig. 8.9), and those estimates have recently been reviewed by Preston (1988). Tanker accidents and oil-well blowouts, which make newspaper headlines, contribute only a small percentage of the total input, but they can be devastating in local areas. Most (53%) of the oil comes from the discharge of waste oil from industrial and municipal sources and from the routine operations of oil tankers, especially via shipping routes out of the Middle East (Fig. 8.10) (Preston 1988). Examination of the important source categories reveals that there probably will be no reduction in the input of oil into the ocean until there is a significant decline in the use of oil. The greatest inputs of oil occur in coastal areas, which often are the most biologically productive, rather than in the open ocean.

The main physical, chemical, and biological processes governing the fate of oil and weathering process for oil in the ocean are summarized in Figure 8.11 and Table 8.5, and they have been reviewed by Lee (1980)

Figure 8.10. Major oceanic transportation routes for petroleum. [From *Oil in the Sea* (1985), with permission of National Academy of Sciences.]

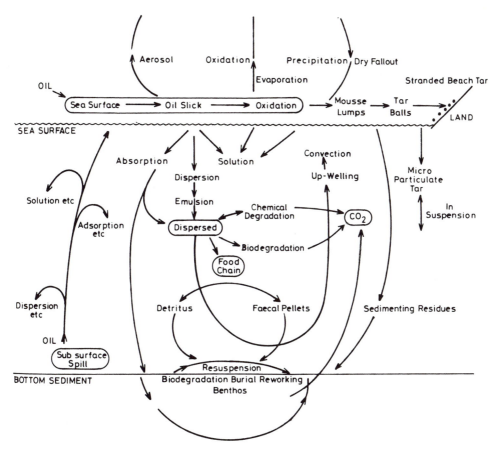

Figure 8.11. The weathering of a crude-oil slick at sea. (From Preston 1988, with permission of Academic Press.)

Table 8.5. *Pathways for the environmental fate of crude oil*

Pathway	Time scale (days)	Percentage of initial
Evaporation	1–10	25
Solution	1–10	5
Photochemical degradation	10–100	5
Biodegradation	50–500	30
Disintegration and sinking	100–1,000	15
Residue	>100	20
Total		100

Source: Reproduced with permission from Butler et al. (1976); © 1976 by the American Institute of Biological Sciences.

and Preston (1988). Weathering involves evaporation, dissolution, emulsification, dispersion, photooxidation, and biodegradation. The fate of the oil will depend on the type spilled and where it is spilled. The source of the crude oil will determine its unique characteristics (often denoted by place of origin, e.g., Nigerian or Kuwaiti crude oil). Many refined petroleum products are spilled, including gasoline, kerosene, fuel oils (no. 2, 3, 4, etc.) and lubricating oils.

Most oil spills immediately form a surface slick, a thin boundary layer between the seawater and the atmosphere. Light oils spread faster than heavy oils and may form films as thin as 0.1 μm. The effect of wind is to move oil at a speed of about 3% of the wind velocity, which is a speed similar to that for surface drift cards (Preston 1988). A knowledge of the film thickness, coupled with an estimate of the area of the slick, will give an indication of the total amount of oil involved. It is interesting to note that the characteristic rainbow-colored sheen of oil on water is indicative of a film thickness of only 0.3 μm. At that thickness, such a film could be produced over an area of 1 km^2 by only 350 L of oil (Preston 1988).

Many of the hydrocarbons are volatile and begin to evaporate immediately. After 24 h, half of the C_{14} compounds will have vaporized, but it will take 3 weeks to evaporate half of the hydrocarbons shorter than C_{17}. Vaporization continues slowly, leaving tarlike lumps; it is the most important natural factor in removing oil from the water surface. Refined products such as gasoline and kerosene may disappear almost completely, whereas viscous crudes may lose less than 25% by evaporation (Table 8.5) (Bishop 1983).

When the sea surface is agitated by wind, the oil may absorb water up to 50% of its weight and form brown masses called "chocolate mousse" (King 1984). Besides the water-in-oil emulsions, oil-in-water emulsions (dispersions) form, especially under the influence

of added chemicals (dispersants). Although emulsification and dispersion can give the impression that the oil has disappeared from the surface, it actually continues to exist as tiny droplets, and its potentially poisonous effects persist. However, the toxicity during such dispersion is reduced, because the lighter fractions such as the aromatics and aliphatics evaporate more quickly.

On a larger time scale, photochemical oxidation may contribute to the weathering of oil. Through the actions of atmospheric oxygen and solar radiation, the proportions of oxygenated compounds in the slick will increase. For example, aromatics and alkyl-substituted cycloalkanes tend to be oxidized more rapidly, forming soluble compounds and insoluble tars.

Microbial degradation begins to take place only after the oil at the surface has aged and lost some of its highly volatile, poisonous components by vaporization. At least 90 strains of marine bacteria and fungi and a few algae are capable of biodegrading some components of petroleum. No single microbial species is able to degrade all of the compounds in oil. Oil-decomposing bacteria increase in number slowly after an oil spill. In some cases, their growth is restricted because there is not enough nitrogen or phosphorus in the water, and it becomes necessary to supplement the low quantities normally present in oil. Moreover, many major oil spills occur in temperate waters in winter, when the temperature restricts bacterial growth rates. Nontoxic dispersants enhance biodegradation by greatly increasing the surface area of the oil (King 1984). Normal alkanes are the most easily degraded, whereas aromatics, cycloalkanes, and branched alkanes are more difficult (Preston 1988). Further aspects of microbial degradation of oil are discussed by Carlberg (1980), Stafford et al. (1982), and Gundlach et al. (1983) and reviewed by Preston (1988).

If an oil spill occurs near shore, and if the wind is in the right direction, beaching of the oil may occur. This oil may adhere to rocks, plants, and animals or may be worked into the sediments if a dispersant is used. Penetration into the interstitial system between sand grains will result in very slow degradation rates, often because of lack of oxygen in the interstitial water. As a result, oil may persist in sediments for years.

Tar-lump formation and sinking are the final stages of weathering. Oil may adsorb to particles, which will sink, or it may be consumed by filter-feeding plankton such as copepods and become incorporated into fecal pellets, which also will sink. Weathered oil may form lumps (usually about the size of peas), which can coalesce and become large enough to form a substratum for sedentary animals, such as gooseneck barnacles, again causing the lump to sink.

The *Amoco Cadiz* oil spill (223,000 metric tonnes) off the north coast of France was the largest (nearly twice the size of the *Torrey Canyon* spill) and the best-studied tanker spill in history. Evaporation and stranding on shore accounted for 60% of the oil spilled. After 3 years, most of the obvious effects had gone, but high hydrocarbon concentrations remained in estuaries and marshes that initially had received large amounts of oil (Gundlach et al. 1983).

8.4.2 Effects of oil on algal metabolism

The toxic effects of oil on algae fall into two categories: those associated with the coating of the organism and those due to uptake of hydrocarbons and the subsequent disruption of cellular metabolism. Coating reduces CO_2 diffusion and light penetration into the plant. Schramm (1972) observed that in *Porphyra umbilicalis, Fucus vesiculosus,* and *Laminaria digitata,* reductions in photosynthesis rates correlated with the thickness of the oil layer. He also found that during exposure to the air, the oil reduced desiccation of the blades, allowing photosynthesis to occur for longer than normal, but at a reduced rate. Severely oiled kelp fronds may break because of the weight of oil adhering to the fronds. The loss of thalli by this mechanism is associated primarily with the higher-molecular-weight, water-insoluble hydrocarbons (Nelson-Smith 1972).

The second category, disruption of cell metabolism, has been examined primarily by monitoring changes in photosynthesis, respiration, growth, pigment content, morphology, and ultrastructure. Bioassay investigations have revealed variable effects, depending on the physical and chemical properties of the oils and components tested (e.g., whole oils vs. refined products), the parameters being measured, and the test species employed. There are problems in interpreting many of the earlier published results (Vandermeulen & Ahern 1976; Vandermeulen 1987), especially if crude oil was used, because its composition was not known, and even today its composition cannot be easily determined. The major problems are the lack of details about the manner of preparation of the oil extract, how old the extract was before it was applied (hence what fraction of the volatiles had been lost), and the total or differential losses of hydrocarbons, especially during long-term experiments (Vandermeulen & Ahern 1976). For these reasons, many investigators have chosen to work with individual components of oil whose composition is at least understood and measurable, and then they extrapolate back to the original oil.

The penetration of oil will depend on the covering on the thallus. The brown algae, in particular, are thought to be largely protected from oil damage by the presence of the mucilaginous coating. This is assumed to be what "saved" the *Macrocystis* beds off Santa Barbara after an oil-well-blowout (Mitchell et al. 1970). However, it is important to note that some dispersants may damage this protective mucus layer. Oil adheres to dead seaweeds, and hence oil-coated seaweeds found on beaches may not necessarily have been killed as a result of the coating. The compounds that penetrate the

thallus most easily, and hence are most toxic, are the lower-molecular-weight, lipophilic compounds such as aromatics. The least toxic components, and the least water-soluble, are the long-chain alkanes. Intermediate in toxicity are the cycloalkanes, followed by olefins. The aromatics and other toxic hydrocarbons appear to exert their toxic effects by entering the lipophilic layer of the cell membrane, disrupting its spacing. As a result, the membrane ceases to properly control the transport of ions in and out of the cell.

Disruption of cellular metabolism usually has been measured through changes in the rates of photosynthesis or respiration. North et al. (1965) observed complete inhibition of photosynthesis in young blades of *Macrocystis* following 3 days of exposure to a 1% emulsion of diesel oil in seawater. In another study, as little as 10–100 ppm of unspecified fuel oils reduced photosynthesis by 50% during 4-day exposures (Clendenning & North 1960). More detailed studies conducted recently have shown that reductions in photosynthesis rates vary with the type of crude oil, its concentration, the length of exposure, the method of preparation of the oil–seawater mixture, the irradiance, and the algal species (Hsiao et al. 1978). *Cladophora stimpsonii, Ulva fenestrata,* and *Laminaria saccharina* showed photosynthetic inhibition by 7 ppm of Prudhoe Bay crude oil, while *Costaria costata* was unaffected (Shiels et al. 1973). The effects were most acute at high irradiance. In experiments with the green seaweed *Acrosiphonia sonderi,* a crude-oil extract inhibited photosynthesis during the first 4 h of incubation, presumably because of the toxicity of volatile aromatics such as benzene and naphthalene. In all the studies to date, the actual mechanism of inhibition has not been investigated. This is mainly because of the difficulty in separating the toxicity effect from the purely mechanical effects of the coating (smothering) of the thallus and the reduction in light reaching the plant. Bleaching is commonly observed among red algae and probably is caused by the breakdown of phycoerythrin by kerosene-related compounds. Lipid-soluble pigments such as chlorophylls may be leached out of cells by oil (O'Brien & Dixon 1976).

Several dioecious brown algae, including *Ectocarpus* and *Fucus,* secrete olefinic hydrocarbons into seawater as gamete attractants. The possibility that petroleum hydrocarbons could confound recognition of the attractant fucoserratene by *Fucus* spermatozoids was investigated by Derenbach and Gereck (1980). They found that a combination (rather than a single compound) of petroleum hydrocarbons attracted spermatozoids, but at concentrations about 100 times that at which fucoserratene is active.

Oil may interfere with sexual reproduction. The reproductive stages of *Fucus edentatus* and *Laminaria saccharina* are particularly sensitive to oil, especially during gamete or spore release (Steele & Hanisak

1979). Concentrations of Willamar crude or several fuel oil as low as 2 μg L^{-1} blocked fertilization in *Fucus*, apparently because of toxic effects on the sperm. *Laminaria* spores did not germinate above 20 μg L^{-1}. Male gametophytes were more sensitive to oils than were female gametophytes, because in both *Fucus* and *Laminaria* they are smaller than females and hence have a higher surface-to-volume ratio, possess fewer stored reserves, and respire at a higher rate, creating a greater energy demand. Sporophyte development in both species was inhibited at higher concentrations of 200 μg L^{-1}. Hopkin and Kain (1978) found the opposite response to phenol: Zoospores and gametophytes were more tolerant than sporophytes in *Laminaria hyperborea*.

The effects of oil contamination on algal respiration are not well known, because of a paucity of experimental evidence (Vandermeulen 1987). The respiration rate for *Laminaria hyperborea* was inhibited by phenol at 100 ppm immediately after addition of the pollutant (Hopkin & Kain 1978). To compare respiration and growth effects, Hopkin and Kain calculated the ratio between the minimum concentration of the pollutant that would reduce frond respiration and the minimum that would reduce sporophyte growth in culture; the ratios were phenol 1.3, Hg 100, Zn 500, and Cu 2,500. Because all ratios were greater than 1, growth of plants in culture was more sensitive than was that of tissue discs in the respirometer. Oil could interfere with respiration in a number of processes, such as gas diffusion, glycolysis, and oxidative phosphorylation. Mechanical blockage of gas diffusion is thought to be less pronounced for oxygen than for carbon dioxide (Schramm 1972). Other physiological mechanisms to explain the inhibition of respiration have not been examined in macrophytes, but some studies of higher plants have been reviewed by Baker (1970).

Inhibition of algal DNA and RNA activities has been reported following exposure to high concentrations of crude oil (Davavin et al. 1975). Exposures to emulsified oil–seawater mixtures (100–10,000 ppm) for 24 h resulted in decreased DNA in the red algae *Grateloupia dichotoma* and *Polysiphonia opaca* and significant reductions in DNA and RNA specific activities in the green alga *Ulva lactuca*. Given the fundamental importance of nucleic acids for reproduction and protein synthesis, this work provides a start in understanding the mechanisms whereby resistance to damage by oil may be conferred on tolerant species.

In a study in Norway, rocky-shore communities were kept in 50-m^3 concrete basins, and two commercial seaweeds were continuously exposed to diesel oil for 2 years (Bokn 1987). With diesel oil at 130 μg L^{-1} lengthwise growth for *Ascophyllum nodosum* and *Laminaria digitata* was significantly reduced, by about 50% over the 2-year period. At a lower diesel-oil concentration (30 μg L^{-1}) there was periodic inhibition of

growth, but no overall reduction in length. After 2 years of continuous exposure to oil, the plants completely recovered during the following oil-free growth season.

Straight-run gasoline, reformed gasoline, benzene, toluene, and *m*-xylene were shown to produce high acute toxicity in *Porphyra suborbiculata* and *Monostroma nitidum* at concentrations ranging from 1,000 to 10,000 mg L^{-1} (Tokuda 1987). Microscopic examination revealed that no cells were killed at concentrations of 100 mg L^{-1}. Kerosene was the least toxic substance tested.

8.4.3 Ecological aspects

Since the early 1960s, various attempts have been made to quantify the effects of large oil spills in different parts of the world on various flora and fauna. The conclusions drawn from these studies have varied considerably, ranging from minimal effects to severe damage. The assessments have varied depending on the ecosystem studied and the community or population observed. Gundlach and Hayes (1978) constructed an "oil-spill index" in which different ecosystems were ranked according to their vulnerability. Rocky exposed cliffs are the least vulnerable, whereas salt marshes and mangroves are extremely vulnerable (Table 8.6). Communities also have been ranked, with birds and benthic subtidal communities being most vulnerable, and plankton and benthic rocky intertidal communities only slightly vulnerable.

Another important factor in determining the magnitude of the ecological impact is the location of the spill relative to the shore. The impact is full-scale if the spill occurs close to the beach and is quickly washed ashore. The least impact occurs if the oil does not reach the shore for several days, giving time for many of the toxic volatile compounds to evaporate. Other factors that can affect the ecological impact are the type of oil, the amount spilled, the water temperature, the weather conditions, the prior exposure of the area to oil, the presence of other pollutants, and the type of remedial action (e.g., use of dispersants).

The ecological impacts of oil on several specific seaweed habitats can be examined against the general background presented earlier. On rocky shores there may be a slight, short-term impact, but no significant long-term effects on the macrophyte community have been observed (Nelson 1982; Gundlach et al. 1983). Rocky intertidal areas that have been cleansed with detergents after oil spills have shown recolonization rates comparable to the rates on control plots. The first macroalgae to recolonize the Cornwall shore after the *Torrey Canyon* spill in 1967 were *Ulva* and *Enteromorpha*. They quickly covered the entire area, because the herbivores that usually grazed on them (e.g., limpets and periwinkles) had been killed by the oil. The upper limit of distribution for *Laminaria digitata* and *Himanthalia elongata* was higher by as much as 2 m during the first

Table 8.6. *Expected impact of oil spills on marine-habitat types and cleanup recommendations*

Exposed rocky cliffs	In the presence of high-energy waves, oil-spill cleanup usually is unnecessary.
Exposed rocky platforms	Wave action causes rapid dissipation of oil, generally within weeks. In most cases, cleanup is not necessary.
Flat, fine-sand beaches	Because of close packing of the sediment, oil penetration is restricted. Oil usually forms a thin surface layer that can be efficiently scraped off. Cleanup should concentrate on the high-tide mark; lower beach levels are rapidly cleaned of oil by wave action.
Beaches with medium or coarse-grained sand	Oil forms thick oil–sediment layers and mixes down to 1 m deep with the sediment. Cleanup damages the beach and should concentrate on the high-water level.
Exposed tidal flats	Oil does not penetrate the compacted-sediment surface, but biological damage results. Cleanup is necessary only if oil contamination is heavy.
Mixed sand-and-gravel beaches	Oil penetration and burial occur rapidly; oil persists and has a long-term impact.
Gravel beaches	Oil penetrates deeply and is buried. Removal of oiled gravel is likely to cause future erosion of the beach.
Sheltered rocky coast	The lack of wave action enables oil to stick to rock surfaces and tidal pools. Severe biological damage results. Cleanup may cause more damage than if the oil is left untreated.
Sheltered tidal flats	Long-term biological damage occurs. Removal of the oil is nearly impossible without causing further damage. Cleanup is necessary only if the tidal flat is very heavily oiled.
Salt marshes and mangroves	Long-term deleterious effects occur. Oil may continue to exist for 10 years or more.

Source: From Gerlach (1982); reprinted with permission of Springer-Verlag, Berlin.

few years of succession (Freedman 1989). Limpets progressively recolonized, and within 7 years the distribution of seaweeds had returned to normal (Gerlach 1982). Similar observations were made on the Somerset coast of England, where the oil was reported not even to adhere to *Fucus spiralis*, and the percentage of cover of this alga increased from 50% to 100% after the spill (Crothers 1983). No significant effects of the *Amoco Cadiz* spill were observed for *Laminaria, Fucus,* or *Ascophyllum* (Gundlach et al. 1983). Some of the damage to corallines, such as loss of pigments, appears to have been partially or wholly due to the dispersant BP 1002 and its toxic aromatic solvent (Boney 1970). Several nontoxic dispersants such as Corexit are now available, and therefore toxic effects attributable to dispersants should no longer be a problem.

Modern dispersants are of two main types (Preston 1988). Hydrocarbon or conventional dispersants are based on hydrocarbon solvents; they contain about 20% surfactant and must be prediluted with seawater. Because of the large volumes required to treat even a moderate-size slick, these chemicals are more suitable for application from small ships. These dispersants require thorough mixing with the oil after application,

which can be achieved by a special towing device. The second group, the concentrates or self-mix dispersants, are alcohol- or glycol-based and usually contain higher concentrations of surfactant components. Typical dose rates are between 1:5 and 1:30 (dispersant:oil), and this makes them more suitable for aerial spraying. Usually the natural motion of the sea is enough to mix these dispersants, and therefore they are much more practical for large oil spills. With both types of dispersants, it is essential to apply the chemical as rapidly as possible (i.e., before mousse formation) for maximum effectiveness. Unfortunately, this traps the more volatile and toxic components that normally would evaporate. Light fuel oils such as gasoline should be left to evaporate, and heavy fuel oils and mousses are not amenable to dispersion.

In cases in which dispersants were not used to aid in the oil cleanup, algal growth generally was less affected (Foster et al. 1990). In the case of the San Francisco Bay oil spill of 1971, caused by the collision of two tankers carrying Bunker C fuel oil, pre-spill algal densities were restored in 2 years. The upper-shore algae, *Endocladia muricata* and *Gigartina cristata*, became coated with oil, but their growth in the following summer appeared to be normal. Other algae, such as *Halosaccion glandiforme, Enteromorpha intestinalis, Urospora penicilliformis,* and *Ralfsia pacifica,* were more dense than normal, possibly because of a reduction in grazers.

In cases in which weathering processes have time to eliminate the more volatile toxic components before the oil reaches the shore, the effects of even a heavy oil deposition on intertidal flora appear to be largely physical, with injury due to smothering and adsorption of oil. Most seriously affected by oil coating are species that grow between neap and spring high-tide marks, especially those algae near spring high tide, where oil may be stranded for a long time. Many high-intertidal species of Rhodophyceae and Phaeophyceae become oleophilic as their surfaces dry out (O'Brien & Dixon 1976). This strong adsorptive capacity for oil has been documented for *Ascophyllum nodosum, Fucus* species, *Pelvetia canaliculata, Mastocarpus stellatus* and *Gelidium crinale.* Such algae can become severely overweighted by adsorbed oil and subject to breakage by waves. For algae with annual basal regrowth, loss of distal blades may be no more debilitating to a plant than losses during a winter storm (Nelson-Smith 1972). However, the loss of too many photosynthetic blades during the growing season, when metabolic products are stored, could impair a plant's regenerative ability (O'Brien & Dickson 1976).

No clear patterns emerge from the relationships between systematics and the susceptibilities of intertidal algae to oil. Several studies have indicated that Cyanophyceae are particularly resistant to oil (O'Brien & Dixon 1976). Species of Chlorophyceae, in particular, have a remarkable ability to invade areas where other species have been eliminated. The spread of green algae often is due to the die-off of herbivores, which are more susceptible to oil damage than are algae. Early observations suggested that filamentous red algae and corallines were most susceptible to oil and oil–emulsifier blends, possibly because of the destruction of phycoerythrin (Nelson-Smith 1972), but this suggestion will require further confirmation.

Salt marshes (Sanders et al. 1980) and coral reefs (Loya & Rinkevich 1980) are the habitats most severely affected by oil spills. Following the death of corals, rapid colonization of algae on the skeletons of the dead corals may be enhanced by oil pollution. In an oil-polluted reef at Eilat (Red Sea), up to 50% of the surfaces of the coral *Stylophora pistillata* were covered by the brown algae *Lobophora variegata* (Loya & Rinkevich 1980).

8.5 Synthetic organic chemicals
8.5.1 Herbicides
Among herbicides, phenoxycarboxylic acid derivatives (2,4-D, 2,4,5-T, 2,4,5-TB, etc.) outweigh all other compounds in tonnage produced. These weed killers are used in high concentrations (mg L^{-1}) in freshwater streams and ponds to control nuisance vascular plants such as *Elodea.* Two other compounds, paraquat and diquat, are widely used because both disappear rapidly from the water and do not appear to be released from the sediments where they tend to concentrate (Duursma & Marchand 1974; Hurlbert 1975). After a herbicide treatment, the growth of freshwater phytoplankton may be temporarily depressed and then flourish, usually because of the nutrients released from the macrophyte kill and subsequent decomposition (Kohler & Labus 1983).

Although herbicides have not been directly used in the marine environment, they can enter estuarine areas through river discharge or runoff. Studies of herbicide effects on marine algae have been conducted primarily in the laboratory. Sporeling growth for five species of red macroalgae, *Pterothamnion plumula, Plumaria elegans, Callithamnion tetricum, Nemalion multifidum,* and *Brongniartella byssoides,* was inhibited by 3-amino-1,2,4-triazole (3AT or amitrole) at about 10 mg L^{-1} (Boney 1963). Short-term immersions in culture medium containing 3AT reduced the growth of sporelings, whereas protracted contact with 3AT resulted in chlorosis. Boney also found that growth inhibition was more pronounced for sporelings of intertidal algae than for those of sublittoral species.

Paraquat and 3AT were also tested for their effects on the settlement, germination, and growth of *Enteromorpha* (Moss & Woodhead 1975). Zygotes were able to develop into filaments in the presence of paraquat at 7 mg L^{-1}, but germination was deferred at higher concentrations. Increased resistance of zygotes was observed when they settled in clumps on the substratum. Green filaments of *Enteromorpha* were more susceptible than ungerminated zygotes. *Enteromorpha* was more sensitive to 3AT than to paraquat. Atrazine at 0.01 mg L^{-1} was lethal to young sporophytes of *Laminaria hyperborea* (Hopkin & Kain 1978). Three other herbicides tested by Hopkin and Kain, 2,4-D, dalapon, and (4-chloro-*o*-toloxy) acetic acid (MCPA), were nontoxic to the sporophytes at the highest concentration tested (>100 mg L^{-1}). In order to compare the effects of atrazine and the toxicities of metals, Hopkin and Kain calculated, on a molarity basis, the minimum concentrations that would have detrimental effects (Table 8.7); atrazine was found to be even more toxic than Cu or Hg.

The growth and photosynthesis of marine phytoplankton can be adversely affected by herbicide concentrations between 10 and 500 mg L^{-1}, depending on the compound (Duursma & Marchand 1974). Phytoplankton are most sensitive to the triazine group of herbicides, and least sensitive to diquat and paraquat (Kohler & Labus 1983).

8.5.2 Insecticides
On the basis of their different chemical natures, two general categories of insecticides are delimited: the

Table 8.7. *Molarity of pollutants that cause detectable effects on Laminaria hyperborea*

Pollutant	Molarity causing minimum detectable effect ($\times 10^{-6}$)
Atrazine	46
Copper	79
Mercury	250
Cadmium	890
DOBS 055	2,800
SDBS	2,900
Zinc	7,700

Source: Reprinted with permission from Hopkins and Kain (1978), *Estuarine and Coastal Marine Science*, vol. 7, pp. 531–53, copyright © 1978 Academic Press Inc. (London) Ltd.

organic-phosphate compounds (e.g., parathion), which are the more degradable; and the chlorinated hydrocarbons (e.g., DDT, dieldrin, endrin), which are the more persistent. In studying these two categories, Ukeles (1962) found that organochlorine insecticides were more toxic to five species of phytoplankton.

DDT came into use as a pesticide in 1942 and was used extensively throughout the world until 1972, when its use in the United States was banned except under special circumstances. Although the impacts of insecticides on freshwater communities have been demonstrated (e.g., Rudd 1964; Preston 1988), no effects on natural marine macrophyte communities have been documented. In laboratory experiments on six macrophytes, Ramachandran et al. (1984) found that at concentrations of 50 μg L^{-1}, organochlorine pesticides inhibited photosynthesis and respiration to a greater extent than did organophosphorus compounds. DDT and lindane were the most toxic. These pesticides are highly soluble in lipids, and therefore the lipid layers of the outer cellular membrane are prone to pesticide interactions.

The concentration of DDT and its breakdown products in the open ocean is about 0.002 μg L^{-1} (Duursma & Marchand 1974). Reduction of phytoplankton photosynthesis occurs at about 10 μg L^{-1} or higher, depending on the species (Wurster 1968), and photosynthesis and respiration were found to be inhibited in six macrophytes at 50 μg L^{-1} (Ramachandran et al. 1984). To achieve such high concentrations in tests, DDT must first be dissolved in ethanol, because its solubility in seawater is only 1.2 μg L^{-1}. Therefore, DDT seems unlikely to affect phytoplankton in nature, and it also may have virtually no effect on macrophytes, although more experiments are needed to test this (Duursma & Marchand 1974; Hurlbert 1975; Laws 1981). Estimates indicate that zooplankton are 100 more times more sensitive

and fish 10 times more sensitive than phytoplankton. In the freshwater environment, it was shown that insecticide treatment reduced benthic herbivore populations, and that was followed by increases in benthic filamentous algae such as *Zygnema* and *Mougeotia* (Hurlbert 1975). Epiphytes such as benthic diatoms also increased because of the susceptibility of the herbivores to DDT.

The rate of degradation of DDT in the ocean varies from more than a few days to several months (Reutergårdh 1980; Laws 1981). However, its degradation components DDD and DDE may be as toxic as DDT itself (Preston 1988). Much of the DDT in the ocean may exist inside or adsorbed onto plankton or other particulates. It is especially concentrated in the fatty tissues of animals, where its half-life may be considerably longer. The best-documented effect of DDT is that it decreases the thicknesses of birds' eggshells, but the issue whether or not DDT exhibits food-chain magnification or amplification remains controversial (Laws 1981: Preston 1988).

8.5.3 Industrial chemicals: PCBs

Polychlorinated biphenyls (PCBs) are complex mixtures of chlorine-substituted biphenyls. They have been marketed under the trade names Aroclor 1242 and 1254, with the last two digits signifying the average percentage chlorine by weight in the mixture. PCBs are exceptionally stable compounds (destruction by burning requires temperatures >1,300°C), and they are toxic to most organisms. For these reasons and others, PCBs are no longer manufactured in the United States. Although they are being phased out, a few specialized uses still remain; they are still used as dielectric fluids in capacitors, and plasticizers in waxes, in transformer fluids and hydraulic fluids, in lubricants, and in heat-transfer fluids.

PCB concentrations in the upper 200 m of the North Atlantic Ocean have been reported to about 20 ng L^{-1}, but near industrial areas they may be as high as 320 ng L^{-1} (Peakall 1975). Concern over these concentrations is warranted, because at 100 ng L^{-1}, PCBs have been shown to affect the growth of phytoplankton communities in continuous cultures (Fisher et al. 1974). Furthermore, Fisher and Wurster (1973) showed that phytoplankton living at suboptimal temperatures were even more sensitive to PCBs. Further studies showed that there was no loss of photosynthesis per unit of chlorophyll *a*, but nevertheless carbon fixation was reduced because chlorophyll *a* per cell was reduced. Generally, the higher the degree of chlorination, the higher the toxicity. The estimates of the toxicities due to the concentrations of PCBs in our waters have become more alarming because of recent findings that the large amounts of PCBs adsorbed onto particles can be taken up by phytoplankton when they make contact with the particles (Harding & Phillips 1978). It has been shown that PCBs initially associated with microparticulates are rapidly transferred to four species of marine diatoms.

The transferred PCBs can inhibit photosynthesis at a site on the electron-transport chain, close to PS-II (Sinclair et al. 1977). This research demonstrates that particle-bound PCBs are of great biological importance, especially in coastal and estuarine areas, where suspended particulate loads can be very high.

The effects of PCBs on natural phytoplankton communities have been studied outdoors in large controlled experimental enclosures or mesocosms (Iseki et al. 1981). Following the addition of PCBs at 50 $\mu g\ L^{-1}$ (more than 200 times the natural concentrations) to those enclosures, the primary productivity was reduced by 30%, the settling velocity of the particulate matter was accelerated, the zooplankton standing stocks were reduced, and the decomposition activity of sedimented matter by bacteria was reduced by 90%.

The effects of PCBs on invertebrates and vertebrates, especially birds, have been relatively well studied (Duursma & Marchand 1974; Peakall 1975; Reutergårdh 1980; Preston 1988). There has been only one study of the effects of PCBs on marine macrophytes. The relatively simple PCB, 4,4'-dichlorophenyl$_n$ (DCB), inhibited growth, gametogenesis, and sporophyte recruitment in *Macrocystis pyrifera* at a concentration of 5 $\mu g\ L^{-1}$ (James et al. 1987). Further tests with a PCB containing three times the number of chlorine atoms found in 4,4'-DCB showed only a twofold increase in toxicity. Total PCB concentrations in sewage effluents in southern California during 1980–1 ranged from 0.03 to 1.55 ppb. Following discharge, those effluents are diluted 10-fold to 100-fold by diffusion systems, so it is unlikely that PCBs at these sewage outfalls will be toxic to kelp growth.

8.5.4 Antifouling compounds: triphenyltin

Since 1970, organotins, particularly the trialkyltin compounds such as triphenyltin (TPT) and tributyltin (TBT), have been widely used as biocides in antifouling compositions for boat hulls and fish-farming gear. Although TBT eventually degrades in the environment, effects on nontarget organisms have recently been recognized at lower levels than were previously anticipated (Langston 1990). Shell abnormalities and reduced growth and recruitment in oysters sampled near marinas were the first indication of the TBT problem. Subsequently, effects have been demonstrated in a number of marine and estuarine species (Langston 1990). Although TBT and TPT are highly effective against *Enteromorpha* species, they are less effective in controlling *Ectocarpus* and the microfouling film that generally precedes settlement by macroalgae (Millner & Evans 1981). This microfouling film generally comprises bacteria, benthic diatoms, and filaments of green algae, such as *Ulothrix pseudoflacca*.

Although a number of organic and inorganic tin compounds have been tested, the trialkyltin compounds have been found to be the most toxic to fouling algae (Wong et al. 1982). For that reason, and the economic aspects associated with fouling, a number of extensive physiological and biochemical studies have been conducted (Evans & Christie 1970; Millner & Evans 1980, 1981).

The photosynthetic apparatus of zoospores and the vegetative tissues of *Ulothrix* was found to be relatively insensitive to triphenyltin, compared with those of *Enteromorpha intestinalis* (Millner & Evans 1980). However, respiration rates in the zoospores and vegetative tissues of both species were equally affected. The fact that respiration is more affected than photosynthesis suggests that the specific binding site might be in the mitochondria. Trialkyltins act as energy-transfer inhibitors in animal mitochondria (Gould 1976). There are questions remaining to be answered: For example, why is *Ulothrix* more resistant to organotins than is *Enteromorpha*, even though *Ulothrix* takes up organotins more rapidly? Laboratory studies with marine phytoplankton have revealed that for the ubiquitous marine diatom *Skeletonema costatum*, virtually no growth occurs in medium containing organotins at 0.1 $\mu g\ L^{-1}$ (Beaumont & Newman 1986). Because TBT has been reported to range from 0.1 to 2 $\mu g\ L^{-1}$ in estuaries and especially marinas, these higher concentrations may be inhibiting primary productivity of microalgae and macroalgae (Hall & Pinkley 1984).

8.6 Complex wastes and eutrophication
8.6.1 Eutrophication

Sewage is classified as a complex waste because it contains inorganic nutrients (N and P in particular), organics, chlorine (from chlorination), and some heavy metals. In this section the focus will be on the inorganic nutrients. Sewage usually is delivered to a body of water by means of a pipe, often with holes in it to disperse the sewage over a wider area. Before the sewage is released, several treatments are possible. In primary treatment, the sewage is screened to remove large particulates and then is passed to settling chambers, where particles settle out. In secondary treatment, the remaining liquid sewage is put into another tank, where it is aerated to encourage bacterial growth and aerobic oxidation of the dissolved organics. This process removes a large portion of the organics from the sewage. In tertiary treatment, nutrients such as N and P are removed by chemical treatment (e.g., precipitation of phosphate by alum) or biological treatment (e.g., growing phytoplankton to remove nutrients) (Ryther et al. 1972). Because it is so expensive, tertiary treatment is not widely used. On a worldwide basis, more than 90% of the sewage from coastal areas enters the sea completely untreated (Cole 1979a). In developed countries, where most sewage is treated, the sludge that settles out in primary treatment is loaded onto ships and dumped farther out to sea.

The dumping of nutritive organic wastes into coastal areas with low rates of water exchange may stimulate the growth of algae to the point that excessive amounts of phytoplankton and/or macrophytes in the water may create biological, aesthetic, or recreational problems. Above-normal plant growth and biomass production in response to added nutrients is termed "eutrophication." The water body affected is said to be eutrophic or hypertrophic (Gerlach 1982); the increased nutrient input may have resulted from land runoff, river inflow, or sewage discharge. The latter source is the most important and the least studied. When eutrophication is extensive, the large volume of algal biomass (both phytoplankton and macrophytes) soon begins to decay and seriously depletes the oxygen concentration needed by animals. Where the rate of water exchange in a bay is extremely low, this high biological oxygen demand (BOD) can result in an extensive fish kill.

Most studies that have examined the responses of macrophytes to eutrophication have been concerned with sewage outfalls, probably because it is convenient to study them. A sewage outfall affects a small, defined area, with a gradient in nutrient concentration away from the outfall. Generally, such studies have been from an ecological point of view, examining changes in community structure and diversity (Borowitzka 1972; Munda 1974). The best-studied sewage outfall is that on San Clemente Island, off southern California. This is a low-volume outfall, producing only 95,000 L of untreated domestic sewage each day. Littler and Murray (1975) found 17 fewer species of macrophytes and less cover near the outfall than in a nearby control area. The outfall flora was less diverse and showed a reduction in community stratification (spatial heterogeneity) because of the absence of *Egregia laevigata*, *Halidrys dioica*, *Sargassum agardhianum*, and the seagrass *Phyllospadix torreyi*. These had been replaced in the mid-intertidal near the outfall by a low turf of cyanobacteria, *Ulva californica*, *Gelidium pusillum*, and small *Pterocladia capillacea*, and in the lower intertidal by *Serpulorbis squamigerus* covered with *Corallina officinalis* var. *chilensis*. Littler and Murray suggested that sewage favors rapid colonizers and more sewage-tolerant organisms. Macrophytes near the outfall exhibited relatively higher net productivities, smaller growth forms, and simpler and shorter life histories, and most were components of early-successional stages.

Further studies were undertaken at that site to determine experimentally whether or not algal communities that are characteristic of sewage-stressed habitats showed high resilience. The measure was their ability to recover quickly after removal of all biota from some quadrants on the rocky shore. Murray and Littler (1978) found that cyanobacteria, filamentous Ectocarpaceae, and colonial diatoms were the dominant forms during the early-successional stages in the cleared areas in both the sewage and control plots. The outfall plots showed rapid recovery by algae such as *Ulva californica*, *Gelidium pusillum*, and *Pseudolithoderma nigrum*, which

have a capacity for rapid recruitment. The algal communities in the unpolluted (control) denuded areas did not fully recover, even after 30 months.

The studies by Littler and colleagues (Murray and Littler 1978; Kindig & Littler 1980) on the impact of sewage on macrophytes provide an excellent example of the combination of field studies and laboratory studies; they progressed from a community field study to an experimental manipulation (denuded plots) in the field and then to studies of the environmental physiology of important species in the laboratory. Kindig and Littler (1980) studied the responses of 10 macrophytes to various sewage effluents (untreated, primary, secondary, and secondary-chlorinated) during long-term culture studies in the laboratory. *Bossiella orbigniana* and *Corallina officinalis* var. *chilensis* exhibited increased photosynthesis rates when exposed to primary-treated sewage, and in long-term cultures their growth was enhanced. Chlorination of effluent produced only a short-term reduction in growth for the first week of culturing. Three populations of *C. officinalis* with differing pollution histories (preexposure to pollution) showed tolerances to sewage corresponding to the extent of their prior exposures. This finding indicates that this species may be able to adapt physiologically to sewage stress and suggests that considerable caution must be exercised in the selection of benthic algae as biological indicators of pollution (Burrows 1971). The studies by Littler and colleagues confirm earlier reports that coralline algae are extremely tolerant of high concentrations of sewage (Dawson 1959). Downstream and inshore from a domestic sewage outfall in Laguna Beach, California, Dawson (1959) observed that 90% of the algal biomass was composed of the corallines *Bossiella* and *Corallina*.

Excessive growth of green seaweeds in response to sewage effluents is becoming an increasingly common phenomenon in sheltered marine bays (Perkins & Abbott 1972; Reise 1983; Montgomery et al. 1985; Soulsby et al. 1985). An overabundance of *Enteromorpha* species on the tidal flats of the Wadden Zee during the summer was attributed to eutrophication by adjacent sewage effluents (Reise 1983). The mats were first composed primarily of *Enteromorpha*, but later other algae such as species of *Ulva*, *Cladophora*, *Chaetomorpha*, and *Porphyra* appeared as secondary components. Mats of these algae cover wide areas of sheltered flats, and in sandy flats the strands of *Enteromorpha* become anchored in the feeding tunnels of the abundant polychaete *Arenicola marina*, enabling the algae to resist displacement by tidal currents. Storms are able to dislocate the algal mats (5–20 cm deep), and usually the sand flats are covered for no more than 1 month. The sediments under the mats become anoxic. This condition is tolerated by polychaetes, but the more sensitive *Turbellaria* decreases in abundance and species richness. Even in tropical areas (off the coast of India) it was found that domestic sewage stimulated greens (*Ulva* and *Entero-morpha*) much more than brown seaweeds (Tewari & Joshi 1988).

In two intertidal basins in southern England, large crops of *Ulva* and *Enteromorpha* develop each summer. Studies were initiated to help predict whether or not increased discharge of sewage effluent in the future would lead to increased macroalgal mats in intertidal areas (Lowthion et al. 1985; Montgomery et al. 1985; Soulsby et al. 1985). A 5-year study indicated that the habitat would be unsuitable to support a standing crop much in excess of the current biomass, and therefore an increase in sewage-derived nutrients would not lead to increased macroalgal biomass (Soulsby et al. 1985). However, the opposite conclusion was reached for a lagoon in Tasmania, because the water circulation was more restricted than in the case in southern England (Buttermore 1977).

The distribution of littoral algae in the inner part of Oslofjord in Norway has been studied over the past 40 years, and *Ascophyllum nodosum* had been observed to be a dominant alga in the area before 1940. More than 20 years ago there occurred a large increase in the sewage load. Many species, such as *Rhodochorton purpureum*, *Phyllophora truncata*, *Spermothamnion repens*, and *Ascophyllum nodosum*, have disappeared or become rare (Rueness 1973). Rueness cleared plots in the inner part of Oslofjord near the sewage outfall and in a control area to observe recolonization. In addition, rocks from the control area to which *A. nodosum* had become attached were transplanted to the sewage-stressed area. Regrowth was much faster in the inner fjord than in the control area. In the inner fjord, the dominant recolonizing species was *Enteromorpha compressa*, followed by *Fucus spiralis*. No *Ascophyllum* germlings were observed. In the cleared control plots, regrowth proceeded more slowly, and green algae were less predominant. The number of species that recolonized was also greater, and after 6 months the regrowth was primarily dominated by a dense stand of *Porphyra purpurea* in the cleared control area. The *Ascophyllum* transplants into the sewage-stressed area were heavily infested with epiphytes and frequently were overgrown by *Enteromorpha* species, *Ulva lactuca*, *Ceramium capillaceum*, and small mussels (*Mytilus*). Rueness concluded that the increased competition for substrate and the shading effect of the *Enteromorpha* carpet reduced the chances of the *Ascophyllum* germlings becoming established near the sewage outfall.

Although nutrient inputs to the Baltic Sea have increased since the end of the nineteenth century, there still is little evidence of a general eutrophication of the Baltic Sea (outside locally polluted areas). A revisit after 40 years to some well-documented diving stations in the northern Baltic Sea revealed that the lower limit for growth of *Fucus vesiculosus* had moved upward, from 11.5 m in 1943–4 to 8.5 m in 1984 (Kautsky et al. 1986). Currently, the deepest specimens, at 8.5 m have the same dwarfed appearance as those found at 11.5 m in the 1940s. The decrease in *F. vesiculosus* coverage with depth toward the lower limit was described by an exponentially decreasing light-attenuation curve. The change in depth penetration was thought to be due to the decreased water transparency, as a result of 40–50% increases in summer values for chlorophyll *a* and nutrients in the offshore surface water of the Baltic Sea since the 1940s. The main nitrogen inputs to the Baltic are from rivers (use of fertilizers in agriculture, not sewage effluent) and rainfall (nitrogen oxides from the combustion of fossil fuels) (Rosenberg 1985).

The green alga *Cladophora* cf. *albida* is an acknowledged symptom of increased eutrophication in Peel Inlet in Western Australia (Gordon et al. 1981). Rivers flowing into the inlet provide large inputs of nitrogen and phosphorus, presumably from agricultural runoff and sewage. This increase in nutrients has resulted in the formation of thick algal beds (10–100 mm deep) that accumulate in shallow waters and decompose. The resulting deterioration of the previously clean beaches is a concern for recreational usage, and the commercial fishery may be threatened.

Even in oligotrophic areas of the oceans and coral-reef communities, significant effects from sewage outfall have been observed (Laws 1981; Pastorok & Bilyard 1985). Kaneohe Bay is a subtropical embayment in the Hawaiian Islands. By 1972, about 4×10^6 L of sewage were being emptied into it each day. That sewage affected the coral-reef community in two ways. First, a reduction in water clarity was caused by increased phytoplankton growth. That reduced the amount of light available for the symbiotic zooxanthellae living in the hermatypic corals and thus resulted in reduced coral growth. Second, the sewage discharge stimulated the growth of the green alga *Dictyosphaeria cavernosa*, commonly known as the bubble alga, which usually establishes itself within a coral head at the base of the frond and then grows outward, eventually enveloping the coral head and killing the coral. That alga is not abundant beyond the sewage-stressed area and therefore appears to have spread in response to the elevated nutrient concentrations. Eventually the sewage was diverted from the bay, and the recovery of the community has been documented (Laws & Redalje 1982). Although the inorganic-nutrient concentrations have reverted to their pre-sewage levels, the system took some time to fully stabilize, because of the slow release of nutrients from plankton that had sunk and accumulated in the sediment during the sewage-discharge period.

The productivity of *Cladophora prolifera* is limited by both nitrogen and phosphorus in Bermuda's shallow, oligotrophic inshore surface waters (Lapointe & O'Connell 1989). Seepage of nitrogen-rich ground-

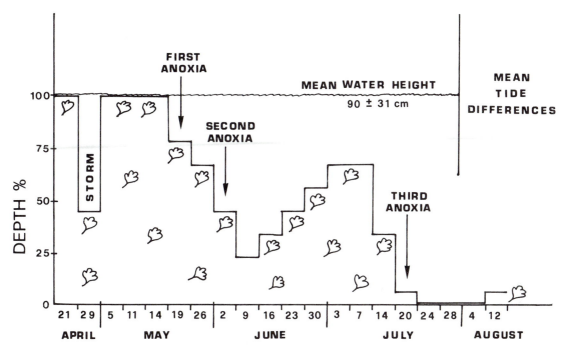

Figure 8.12. Vertical distribution of the macroalgal biomass. (From Sfriso et al. 1987, with permission of Elsevier Applied Science Publishers.)

water, combined with a high alkaline phosphatase capacity, accounts for this cumulative increase in biomass. The proliferation of *Cladophora* in Bermuda's inshore waters over the past 20 years exemplifies the dramatic ecological changes that occur when oligotrophic marine ecosystems are impacted by nutrient-enriched groundwaters.

The interplay between macroalgal growth and the triggering of phytoplankton blooms was clearly demonstrated in the hypertrophic Venice Lagoon (Sfriso et al. 1987). Under aerobic conditions, nutrients were taken up primarily during the spring and summer periods. When there was an imbalance between production and consumption of oxygen, anoxic conditions occurred (Fig. 8.12), and large amounts of nutrients were released by the decomposition of macroalgae. The nutrients were released to the sediments and the water; the latter nutrient source triggered a phytoplankton bloom (chlorophyll *a* increased from 5 to 100 mg m^{-3}).

The ammonium concentrations in discharged sewage can be very high (up to 2,200 μM). However, the maximum value for ammonium found in the surface waters over the White's Point sewage outfall, off Los Angeles, was 35 μM, and more frequently the concentrations were 5–10 μM. That represents a dilution of about 100-fold compared with the discharged concentration. Ammonium concentrations of 10–30 μM are not toxic to phytoplankton, but Thomas et al. (1980) found that at concentrations of 200 μM, the growth of two di-

noflagellates was inhibited, whereas three diatoms were not inhibited. Macrophytes from the Chlorophyceae seem to be more tolerant of sewage toxicity than are many phytoplankton species. *Enteromorpha linza* grew well in full-strength sewage effluent, even though the ammonium concentration was 500 μM (Chan et al. 1982). *Enteromorpha compressa* appears to be more sensitive, showing inhibition of photosynthesis when exposed to NH_4^+ at about 75 μM. The germination rates for the zygotes of three species of *Sargassum* were 50% and zero for secondary effluent with and without ammonium (3.5 mM), respectively, but toxicity was reduced by 50% when the pH of the medium was less than 7 (Ogawa 1984); residual chlorine concentrations greater than 3 mg L^{-1}, and an anionic surfactant, were also toxic. The inhibition of growth that has been observed near sewage outfalls probably is not due to an excessively high ammonium level per se, but rather to a combination of high ammonium concentrations and other inhibiting factors, such as heavy metals (Hershelman et al. 1981), chlorinated compounds such as chloramine in chlorinated sewage (MacIsaac et al. 1979; Thomas et al. 1980), and surfactants.

In addition to the direct inhibitory effects of sewage on macrophytes, secondary effects may account for macrophyte decline in progressively eutrophicated fresh waters. There is recent evidence that some decreases in macrophyte numbers have been due to increased growth and shading by epiphytes and filamentous algae associated with beds of (vascular) macrophytes, as well as increased turbidity of surface layers because of phytoplankton growth (Phillips et al. 1978).

Other components of concern in the sewage problem are detergents (which can cause oxygen depletion because of the organic load) and sewage sludge and its disposal. Because pollutants such as heavy metals and possibly PCBs are greatly concentrated in sludge, it is generally dumped farther out to sea, but in the United States, ocean dumping is being banned. Detergents and surfactants in sewage are also considered to be pollutants. Anionic detergents account for the bulk of the detergents in household sewage, and three of these, sodium lauryl ether sulfate, sodium dodecylbenzenesulfonate (SDBS), and DOBS 055, were tested on *Laminaria hyperborea* (Hopkin & Kain 1978). SDBS and DOBS 055 reduced the growth of sporophytes at concentrations of 1–10 mg L^{-1}, and gametophyte germination was also inhibited by SDBS. The toxicity of anionic detergents is intermediate between that of nonionic detergents (the least toxic) and that of cationic detergents, based on tests with phytoplankton (Duursma & Marchand 1974; Kohler & Labus 1983). The toxicity of detergents and surfactants is attributed to disruption of cellular and intracellular membranes. Indeed, detergents such as Triton X-100 are used in physiological and biochemical research to help extract cell components.

The effects of chlorine on algal photosynthesis have been documented in the laboratory, but no effects of chlorinated wastewater on algae have been observed in the field. Chlorine is highly reactive and rapidly forms a number of compounds. It can form highly toxic chlorinated organic compounds, initiate the production of the strong biocide hypobromite, and react with ammonium to produce chloramines, which are particularly toxic to larval zooplankton (Bishop 1983). At 10 ppm, chlorine irreversibly inhibits the photosynthetic activity of phytoplankton (Eppley et al. 1976). However, field experiments have shown that there is no evidence of deleterious effects of chlorine on phytoplankton photosynthesis in waters receiving chlorinated sewage wastes off southern California (Thomas et al. 1974). This probably is attributable to the jet diffusion system that is used, which provides immediate dilution of more than 100-fold. Tests of chlorine toxicity to marine macrophytes have not been conducted.

Disposal of industrial waste into the oceans is being phased out in the United States because the wastes contain many compounds that are extremely toxic. All such dumping had been scheduled to cease by the end of 1981, but that deadline was postponed indefinitely because of lack of suitable disposal alternatives. Elevated concentrations of trace metals in surficial sediments near the Los Angeles County outfall have resulted in contamination factors (median outfall/median baseline) greater than 20 for Ag, Cd, Cu, and Hg (Hershelman et al. 1981). These elevated metal concentrations could have significant effects on small macrophytes that remain close to the sediment/substrate.

8.6.2 Pulp-mill effluent

Different wood-processing systems have different wastes, depending on the quality of the final product. Two methods of making pulp from coniferous trees, used in Canada and the United States, are the kraft and sulfite processes. In the kraft process, wood chips are initially digested in an alkaline solution of sodium sulfide and sodium hydroxide. This is the cleaner process, because most of the digesting chemicals are recovered before the effluent is discharged. In the sulfite process, digestion occurs with an acidic calcium bisulfite solution, and much less of the digestive solution is reclaimed. Wood-processing industries require large quantities of water (200,000 L per metric ton of pulp) and therefore release large quantities of effluent containing such toxic compounds as hydrogen sulfide, methyl mercaptans (giving most of the smell), resins, fatty acid soaps, and sodium thiosulfate (Carefoot 1977). In addition, the effluent contains large amounts of waste organic matter such as lignins, which color the water brown, and wood fibers, which blanket the sediments in the area of the discharge, creating a high BOD and possibly anaerobic conditions. Both the lignins and fibers severely reduce light penetration into the water.

There has been only one study of the effects of pulp-mill effluent on seaweeds (Hellenbrand 1978): Under normal field conditions, *Chondrus crispus*, *Ascophyllum nodosum*, and *Fucus vesiculosus* were not adversely affected by treated kraft-mill effluent. In fact, productivity increased for all seaweeds, probably because of the nutrients in the effluent.

Laboratory experiments on the effects of six different pulp-mill effluents on marine phytoplankton were conducted by Stockner and Costello (1976). They found that some species required a preadaption period before the cultures resumed exponential growth in relatively high concentrations (20–30%) of the kraft-mill effluent. A green flagellate, *Dunaliella tertiolecta*, exhibited exponential growth even in 90% kraft effluent, which was the most toxic effluent of the six types tested. Their results suggest that in marine waters that receive effluent without a drastic pH change, phytoplankton may not be seriously affected except when effluent concentrations exceed 30–40%. If the area receives effluent from a sulfite process, then lower concentrations (ca. 10%) may produce some inhibition of growth. In actual field experiments, Stockner and Cliff (1976) found that light attentuation by the effluent, especially in the 400–500-nm region, was the major cause of the reductions in daily rates of primary production. The tea-colored effluent would also be expected to reduce light availability and primary productivity for some macrophytes in the area, but that has not been tested.

The storage of logs in booms while they are waiting to be processed through the pulp mill may destroy local macrophyte beds, primarily because they prevent light prevention.

8.7 **Synopsis**

Pollution includes human additions of deleterious materials and energy into the environment. The effects of pollutants on macrophytes can be lethal (acute) or sublethal, and the effects are assessed with bioassay experiments, which should be conducted both in the laboratory and in the field. The physicochemical aspects, such as solubility, adsorption and chemical complexation, and speciation, are extremely important in quantifying the effects of a pollutant. On the other hand, the total concentration of a contaminant may give little indication of its toxicity. The choice of bioassay organism, its life-history stage, and its potential for long-term recovery are also important in pollution assessments. There are obvious limitations to laboratory bioassays, especially because they do not contain nature's suspended particulates, which are known to radically reduce the toxicity of pollutants such as heavy metals through adsorption in estuarine areas.

Thermal pollution, originating primarily from the cooling water discharged from power plants, can be stimulatory if the water temperature does not rise above the optimal temperature for growth of a species. Thermal stress on seaweeds has occurred in areas off southern California, and the symptoms include frond hardening, bleaching or darkening, and cellular plasmolysis in *Macrocystis*.

Increased supplies of nutrients, especially nitrogen, near sewage outfalls generally have resulted in changes in community structure and diversity and have increased epiphytism on macrophytes. Macrophytes near the outfall tend to show relatively higher net primary productivities, smaller growth forms, and simpler and shorter life histories; most are components of early-successional stages.

The degree of heavy-metal toxicity is influenced by the type of metal ion, the amount of particulates in the water, and the algal species. Generally, the order of metal toxicities for seaweeds is $Hg > Cu > Cd > Ag > Pb > Zn$. Because metal toxicity usually occurs only when the metal exists as a free ion, adsorption of the ions onto particles maybe a very significant detoxification process in some environments. Macrophytes show several mechanisms to detoxify the metals or to increase their tolerances. Extracellularly, metals may be detoxified by binding to algal extracellular products. Exclusion of the metal ion may occur at the cell wall via binding to cell-wall polysaccharides, or at the cell membrane via changes in the transport properties of the membrane. Intracellularly, metals may undergo changes in valence or may be converted into nontoxic organometallic compounds. Intracellular precipitation within vacuoles and nuclei has been observed for copper in phytoplankton. If significant detoxification does not occur, the metal ion may inhibit the functioning of algal enzyme systems, eliciting the following responses: cessation of growth, inhibition of photosynthesis, reduction of chlorophyll content, an increase in cell permeability, and loss of K^+ from the cell. Macrophytes tend to concentrate many heavy metals to several orders of magnitude above ambient seawater concentrations. The tendency for further bioaccumulation along the food chain is less clear because of the variations among seaweeds and animals, and it would also be dependent on the type of pollutant and even the kind of metal.

Generally, trace metals never constitute a threat to the marine environment other than in estuarine or hydrodynamically restricted areas. The difference between the natural concentration and that at which acute effects are observable is normally several orders of magnitude. This is reflected in the government's water-quality criteria, in which the allowable concentrations are considerably higher than any ever found in normal circumstances. It is therefore the more insidious sublethal effects that are most likely to be encountered, and they can occur at concentrations more than an order of magnitude lower than the concentrations that will produce acute effects, which were determined in earlier studies using LC_{50} tests. The trend toward increasingly more sensitive indices may eventually enable us to detect effects at even lower levels; results from biochemical studies, relating to the induction and saturation of detoxification mechanisms, have been promising. Sublethal effects have been demonstrated under laboratory conditions, but they have rarely been identified under natural field conditions. This is not surprising, considering the complexity of the different environmental stresses to which marine organisms are subjected. Sublethal and acute toxicities are critically dependent upon the stage of development of an organism. Reproduction and early developmental stages generally are the most vulnerable. Unfortunately, the life-cycle studies that are needed to examine the sublethal effects as a result of prolonged exposure to a contaminant will be complex and expensive. Nevertheless, it is undoubtedly by such studies that the real effects of trace-metal contamination will be revealed.

Petroleum is an extremely complex mixture of hydrocarbons, including alkanes, cycloalkanes, and aromatics. Oil can reduce photosynthesis and growth in macrophytes by preventing gas exchange, disrupting chloroplast membranes, destroying chlorophyll, and altering cell permeability. In some cases, penetration of oil is reduced by the mucilaginous coating, especially on some brown seaweeds. The components that penetrate the thallus most easily and hence are the most toxic are the lower-molecular-weight, volatile, lipophilic compounds, including the aromatics. Alkanes are least toxic. In the laboratory, the concentrations at which oil will be toxic will depend on the type of oil, how the extract was prepared, and when it was used, as well as on the water temperature and the presence of other pollutants or dispersants. Additional factors in the field that can influence oil toxicity include the proximity of the spill to the shore and weather conditions, especially wind. Rocky intertidal areas suffer slight, short-term harm from oil spills, whereas the impacts on salt marshes and coral reefs are severe and longer-term. Weathering of oil occurs by a number of processes, the most important of which is evaporation of the most toxic compounds. Herbivores often are more susceptible to oil than macrophytes, and often an increase in ephemeral algal biomass is a response to the reduced grazing pressure.

The effects of synthetic organic chemicals, such as insecticides, herbicides, industrial chemicals, and antifouling compounds, on macrophytes have received little attention. Likewise, complex wastes such as pulp-mill effluent and domestic sewage have been largely ignored.

Most of the ecosystems that have been studied to date have shown remarkable abilities to recover when the source of the pollutant has been removed. Most of the effects that have been discussed have been local and confined to coastal areas, where point sources of pollutants predominate. In many cases, the animals were found to be more sensitive than the macrophytes, resulting in a decrease in grazing and an increase in some species of seaweeds. Other changes were at the community-structure level, where the pollutant rendered one species less competitive than another.

Although the temptation to generalize about pollutant effects may be great, extreme caution is warranted in view of the large number of environmental and physiological factors that influence toxicity, notably the wide range of tolerances displayed by different organisms. In addition, indirect effects caused by the elimination of sensitive species could have far greater significance for marine communities than is indicated in toxicity studies with single species. Consequently, the incidents of pollution described in this chapter merely serve to highlight the types of changes that can occur at contaminated sites and do not necessarily signify universally applicable responses.

9
Seaweed mariculture

9.1 Introduction

Mariculture, or marine agronomy (Doty 1977), distinct from simple harvesting of wild stocks, is the cultivation of the sea. It involves large-scale cultivation of commercially useful organisms, including seaweeds. In Japan, China, and other Asian countries, where seaweeds have long composed an important part of the human diet, seaweed farming is a major business (Table 9.1). In other regions of the world, where the primary uses of seaweeds are as animal fodder, fertilizers, or sources of phycocolloids, wild stocks usually are harvested (Hoppe & Schmid 1969) and managed (e.g., some habitat improvement). In recent years, seaweeds have also been considered as potential solar-energy converters, to provide biomass as a source of nutrients and energy for methane-producing bacteria.

Mariculture depends on improving the conditions found in the sea, improving the plant material, or creating artificial environments, which can provide optimum conditions for growth of the plant. Thus, just as agriculture depends on vascular-plant ecology and physiology for a basic understanding of the crops, successful mariculture depends on an extensive basic knowledge of the biology and physiology of the seaweeds under cultivation and how factors important to seaweed growth can be manipulated to improve yields.

Ancient records show that people collected seaweeds for food as long ago as 2500 B.P. in China (Tseng 1981) and 1500 B.P. in Europe (Levring 1977). In the past 300 years, and particularly in the past 50 years, the practice has grown and changed, first in Japan and then in China, from the process of simply harvesting the wild stands to the processes of selecting, breeding, and cultivating certain species.

As part of the human diet, seaweeds provide protein, vitamins, and minerals (especially iodine from kelp). In addition, commercially important phycocolloids – agars, carrageenans, and alginates – are ex-tracted from red and brown algae. Agars obtained from *Gelidium* are used extensively in microbiology and tissue culture for solidifying growth media, and more recently in electrophoretic gels. The agar from *Gracilaria* is used mainly in foods. Carrageenans, chiefly from *Eucheuma* and *Chondrus*, are widely used as thickeners in dairy products. Alginates, from *Macrocystis* and *Laminaria*, are also used as thickeners in a multitude of products ranging from salad dressings to oil-drilling fluids to the coatings in paper manufacture (Chapman & Chapman 1980; Waaland 1981).

Some 400–500 species of seaweeds are collected for food, fodder, or chemicals, but fewer than 20 species in 11 genera are commercially cultivated (Michanek 1978; Tseng 1981; van der Meer 1983; Shokita et al. 1991). Four major crop-plant genera in Asia are the red algae *Eucheuma* and *Porphyra* and the brown algae *Laminaria* and *Undaria*. In the following sections, the mariculture practices used in growing these seaweeds are considered as examples of the application of the ecological and physiological principles described in the preceding chapters.

9.2 *Porphyra* mariculture

Porphyra is used extensively for food, and it is known as *nori* in Japan, *zicai* in China, and "purple laver" in Great Britain. *Porphyra* is one of the most extensively eaten seaweeds by coastal peoples in southeast Asia and the Pacific Ocean basin (Abbott 1988). In New Zealand, it is relished by the Maoris. *Porphyra* has high contents of digestible protein (20–25% wet weight) and free amino acids (especially glutamic acid, glycine, and alanine), which are responsible for its specific taste. The vitamin C content of *Porphyra* is similar to that in lemons, and it is also rich in the B vitamins. It is an excellent source of iodine and other trace elements. The retail value of *Porphyra* produced in Japan was about $1 billion (U.S.) in 1986. That makes the *Porphyra*

Table 9.1. *World seaweed production (metric tonnes wet weight), 1987*

	Brown algae	Red algae	Green algae	Other seaweeds	Total
China	1,073,400	122,900	—[a]	—	1,196,300
Japan	294,500	330,500	1,300	37,800	664,100
Korea	339,300	93,100	11,200	13,000	456,600
Philippines	—	222,000	—	—	222,000
Norway	174,100	—	—	—	174,100
Chile	33,532	83,643	—	—	117,175
World total	2,160,700	973,000	12,600	97,200	3,243,400

[a]Insignificant production.
Source: After *Fisheries Journal*, no. 31, based on FAO statistics.

industry in Japan, which produces about 60% of the worldwide *Porphyra* total, the world's highest-valued near-shore fishery (Mumford & Miura 1988). *Porphyra* is the most highly domesticated marine alga, which reflects the relatively more advanced state of our understanding of the biology of this genus (Miura 1975; Tseng 1981; Mumford & Miura 1988; Shokita et al. 1991).

Porphyra species are primarily intertidal, occurring mainly in temperate areas, but also in subtropical and sub-Arctic regions. There are more than 100 species worldwide, and many are difficult to distinguish from one another. In China and Japan, at least seven species are used in commercial cultivation, but *P. yezoensis*, *P. tenera*, and *P. haitanensis* account for more than 90% of the total production (Tseng 1981). In the Pacific Northwest in the United States, 5 of the 17 or more native species have been identified as having the qualities that can yield high-quality nori; they are *P. fallax* (as *P. perforata*), *P. abbottiae*, *P. torta*, *P. pseudolanceolata*, and *P. nereocystis* (Waaland et al. 1986; Mumford & Miura 1988).

9.2.1 Biology

The life cycle of *Porphyra* involves a heteromorphic alternation of generations (see Fig. 1.31). It is the foliose gametophyte that is eaten. The blade can be yellow, olive, pink, or purple, 1 or 2 cells thick and over 1 m in length. The blade can reproduce asexually in some species by means of monospores or aplanospores. In the commercially cultivated species, sexual reproduction occurs under the stimuli of increasing day length and rising temperatures (sec. 1.5.3). Male gametes are released from the spermatangium and fuse with the female cell (carpogonium). Following fertilization, division of the carpogonium is mitotic, forming packets of diploid carpospores. The released carpospores develop into the conchocelis phase (the diploid sporophyte consisting of microscopic filaments), which in the wild will bore into shells, where it grows vegetatively. The conchocelis filaments can reproduce asexually. In the pres-

ence of decreasing day length and falling temperatures, terminal cells of the conchocelis phase produce conchospores inside conchosporangia. Meiosis occurs during the germination of the conchospore, producing the macroscopic gametophyte. This life cycle can vary among *Porphyra* species (Cole & Conway 1980).

The great success of the nori industry is due to its application of what has been learned in studies of the life history of *Porphyra*. Until 1949, when Drew (1949) showed that the genus *Conchocelis* is a stage in the *Porphyra* life cycle, fishermen did not know where the spores came from, nor that the habitat of the conchocelis was quite different from that of the crop. Drew's revelation transformed the nori industry, allowing indoor mass cultivation of the filaments in sterilized oyster shells and the "seeding" of conchospores directly onto nets for outplanting in the sea. All Japanese species investigated thus far can produce the conchocelis phase. This phase can be maintained for long periods of time in free culture, and it grows vegetatively under a wide range of temperatures, irradiances, and photoperiods. It probably is a perennial, persistent stage in the life histories of many *Porphyra* species in nature as well.

9.2.2 Cultivation

The farming practices developed in Japan and China for *Porphyra* illustrate the basic principles of seaweed mariculture for food. Cultivation began in Tokyo Bay some 300 years ago and remained there until the early nineteenth century, when the practice gradually spread to other areas of Japan (Okazaki 1971; Tseng 1981). Enhancement of wild stocks was originally achieved by pushing tree branches or bamboo shoots into the mud on the bottom of the bay, or by clearing rock surfaces, so that when conchospores were liberated in early autumn they would have space for attachment and growth. Later, horizontal nets were strung between poles. The nets were more easily transported from the collecting grounds to the cultivation areas.

A flow diagram summarizes the modern production of nori (Fig. 9.1): Mass culture of shells inoculated

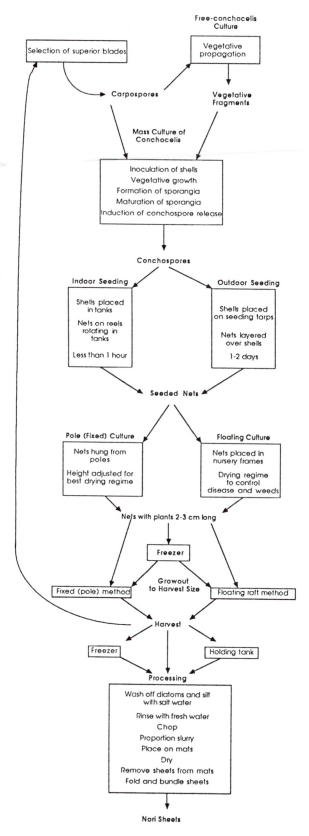

with conchocelis takes place in tanks in greenhouses (Miura 1975; Tseng 1981). In February or March, *Porphyra* thalli are induced to release carpospores by being dried overnight and then reimmersed in seawater for 4–5 h. Between 15 and 150 g (wet weight) of *Porphyra* blades, depending on the species, are sufficient for coverage of about 3 m^2 of shells. Sterile oyster or scallop shells, or artificial substrata treated with calcite granules, are placed in seawater tanks with the fertile *Porphyra* blades, or sprinkled with a suspension of carpospores. It is also possible to take conchocelis filaments grown in vitro, fragment them, and sprinkle them onto shells. The best conditions for conchocelis filaments to bore into the shells include a temperature of 10–15°C and bright, but not direct, sunlight. Good growth of conchocelis requires bright daylight and abundant nutrients. Nitrogen, phosphorus, and potassium are added, and the water in the culture tanks is stirred to improve gas exchange and nutrient uptake. During the early summer the water temperature increases from less than 15°C to more than 25°C. Midday irradiance is kept to about 55 $\mu E\ m^{-2}\ s^{-1}$ by using screens.

From early July to late August–September, the water temperature rises from about 22°C to 28–30°C and then gradually decreases. This is a critical time for formation of conchosporangia and maximum conchospore production, which are dependent on temperature and photoperiod. Light is manipulated so as to promote accumulation of reserves and to delay sporulation (Kurogi & Akiyama 1966). Irradiance is first reduced to about one-quarter (ca. 15 $\mu E\ m^{-2}\ s^{-1}$) in early to middle July and held there as the temperature continues to rise. In September, when the ambient temperature has fallen to about 23°C, sporulation is encouraged by artificially reducing the photoperiod to 8–10 h per day and dropping the tank temperature to 17–18°C. Conchospores are collected on nets either by running nets through the indoor tanks containing the shells or by placing shells under nets in the field (Fig. 9.2). Conchospore adherence and germination require brighter light, 50 $\mu E\ m^{-2}\ s^{-1}$ or more, and usually germination is carried out in the sea.

Successful seeding requires settlement of two to five spores per square millimeter. This density is achieved in 8–10 min in tank seeding, but it can take 1–5 days in the field. The nets are then attached to poles or rafts in the field for nursery cultivation. When the plants reach 2–3 cm in length, they can be left to grow further, or the nets can be rolled up and frozen for up to 6 months or more.

Several methods are used for suspending nets, depending on the depth of the water and the tidal amplitude (Fig. 9.3). In shallow areas, the nets are suspended

Figure 9.1. Flow diagram for the production of *hoshinovi* (sheets of *Porphyra*) as practiced in Japan. (From Mumford & Miura 1988, with permission of Cambridge University Press.)

Figure 9.2. Outdoor seeding of netting with *Porphyra* conchospores. The conchocelis-phase-bearing shells are placed on a semifloating tarp. Up to 50 nets are spread over the shells; the spores float up and attach to the netting. (From Mumford & Miura 1988, with permission of Cambridge University Press.)

from fixed poles, so that the plants are regularly exposed to the atmosphere (Fig. 9.3a). If the tidal range is greater than about 2 m, the nets are attached to poles so that they rest just above the bottom at low tide, but float as the tide rises (Fig. 9.3c). This avoids too much shading by the water column. In deep water, *Porphyra* is grown on nets attached to floating rafts near the surface. Intertidal pole cultivation often is preferred because it ensures periodic exposure of the proper duration, which helps to reduce the incidence of disease and the growth of competitive (weed) species, especially epiphytic diatoms (Tseng 1981).

As discussed in section 6.2.3, the optimum temperature for growth decreases as the thallus ages. Thus, the timing of outplanting is important. Delay will slow the growth of the germlings and result in a later initial harvest. Seawater is considered infertile for nori growth if the $NH_4^+ + NO_3^-$ concentration is less than 3 μM. The best-quality nori is obtained when the nitrogen concentration is greater than 15 μM. Fertilizer, such as ammonium sulfate, is applied as a spray over the beds or is allowed to diffuse from bottles hung on the support poles. The best-quality plants are those harvested from

October to December under normal growth conditions. Harvesting is done every 5–10 days, and each net is harvested three or more times. Automated harvesters are used (Mumford & Miura 1988). After the first net has been harvested several times, it is replaced by another net brought from the freezer. This process is repeated at least three or four times until the growing season ends, usually in January or February, because of decreasing quality (Okazaki 1971).

The discovery that *Porphyra* germlings could survive deep freezing added a new dimension to nori farming. If thalli are allowed to dry to between 20% and 30% of their initial moisture content, and then are frozen and stored at −20°C, they can resume normal growth as much as a year later. This practice can be used to extend the useful harvest period into March or April, and it also serves as insurance against failure of the early crop. Freezing the nets allows the farmer to produce more nursery sets than he has grow-out areas and permits flexibility in the control of disease and fouling.

The harvested thalli are thoroughly washed in seawater, and all epiphytes and dead tissues are removed. The thalli are chopped and made into a freshwater slurry, then spread over screens and dried. Finished nori sheets are approximately 200 × 180 mm. One net, 18 × 1.2 m, with a stretched mesh size of 300 mm, will produce between 300 and 2,000 sheets of dried nori (Miura 1975). The sheets are made by a large machine that produces up to 4,500 sheets per hour (Mumford & Miura 1988). The finished product (in Japanese, *hoshi-nori*) can be eaten directly in sauces, soups, salads, and sushi or can undergo secondary processing. This involves toasting the nori sheets to produce *yaki-nori* for the sushi trade, or else toasting and seasoning the sheets and cutting them into smaller pieces to be sold as *ajitsuki-nori*.

9.2.3 Problems in *Porphyra* culture

Just as there are problems with weeds in land agriculture, there are similar problems in *Porphyra* cultivation. In general, there are two kinds of weedy algae: green algae (usually *Monostroma*, *Enteromorpha*, and *Urospora*) and diatoms (most often *Licmophora*). Attachment of the spores of these algae is prevented in three ways: (1) Monospore production by young thalli increases seedling density, so that little space is left for weed spores to attach. The production nets are handled carefully so that the germlings will not be scraped off and leave free space. (2) Only the distal ends of thalli are harvested, so the nets remain densely covered with nori. (3) If weed algae do establish on the nets, they can be killed off by exposing the net to the air for hours or even days, because weed algae are more susceptible to desiccation than is *Porphyra* (Mumford & Miura 1988).

Another problem is grazing of *Porphyra* by herbivorous fish. If the problem is severe, special nets must be used to protect the crop (Tseng 1981). *Porphyra* cul-

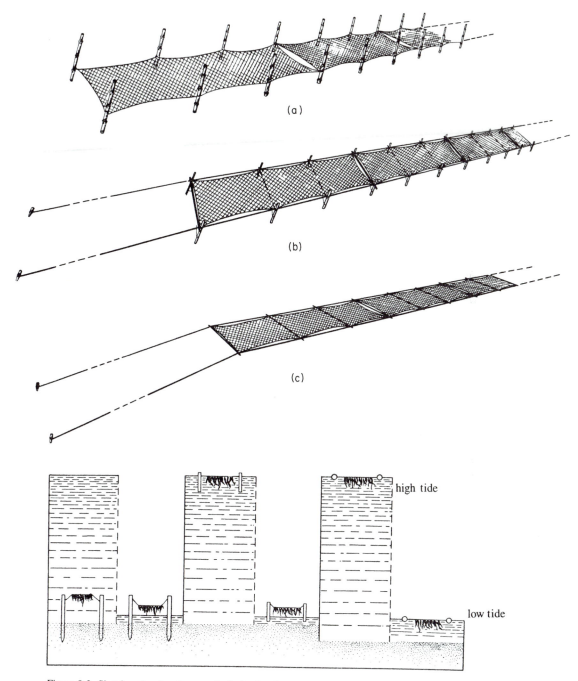

Figure 9.3. Sketches showing three methods for *Porphyra* cultivation: (a) fixed type of the pillar method: (b) semifloating method: (c) floating method. In the diagrams below, for each of the methods, the left side represents the position of the net during high tide, and the right side represents low tide. (From Tseng 1981, with permission of Blackwell Scientific Publications.)

tivation is also susceptible to environmental factors that induce physiological stress. In Japan, very dense nets in protected bays can be damaged by a sudden rise in temperature. This often occurs in November, when the seawater is very calm. Such a rise in temperature will kill

some of the young nori, and the nets will become partially denuded, sometimes in large patches, resulting in low production.

Disease remains one of the biggest threats to nori producers. Most of the diseases are fungal (Andrews

1976). *Porphyra* is susceptible to "red rot" or "red wasting disease," caused by an oomycete fungus, *Pythium*, and to "chytrid blight," caused by a fungus, *Olpidiopsis*. In Japan, farmers have resisted the use of chemicals on this food crop, and many diseases are controlled by drying, freezing, and maintaining healthy plants by good cultivation practices and other innovative methods. For example, a green color mutant has been found to have high resistance to the red-rot disease that can attack and quickly devastate *Porphyra*. This resistance may be due to the fact that the green mutant has relatively thick cell walls, and there seems to be an inverse relationship between susceptibility and cell-wall thickness. Unfortunately, for decades the strain selection in *Porphyra* has been for thin cell walls (better taste and texture). It is doubtful that the green form will be acceptable to consumers, because of its color and its thick walls.

A new bacterial disease has appeared in southern Japan; it leads to what is called *suminori* or charcoal nori. The plants appear normal to the naked eye, but the disease can be detected microscopically. If the diseased material is processed, the sheets will be gray, lusterless, and worthless. Thus far there is no means to control the disease. "Green-spot disease" is caused by pathogenic species of the bacteria *Vibrio* and *Pseudomonas*.

Each commercially cultivated seaweed is susceptible to one or more diseases, some due to pathogens, others to adverse physical conditions (Andrews 1976; Tseng 1981). "Crown-gall disease" results in tumorous growths, possibly from carcinogenic substances in sewage; this disease can be fatal (Tseng 1981). In Japan, fog has been known to cause serious damage to the nori crops, which are exposed to the atmosphere part of each day. Sulfites in polluted air account for part of the cause, because they form sulfurous acid when they dissolve in water. Plants in greenhouses can be affected by H_2S coming from sulfate-reducing bacteria in the water pipes. A disease need not kill the plants to destroy or reduce their value as a crop.

Recently in Japan there has been concern that the use of highly selected strains can lead to genetic uniformity and hence to increased potential for crop failure through either disease or unfavorable conditions affecting all plants similarly. As a safeguard, nets are now seeded with several strains. It has also been suggested that "gene banks" of conchocelis cultures be established to maintain the genetic diversity that is being threatened by widespread use of only a few cultivars (Mumford & Miura 1988).

9.2.4 Future trends

Recent scientific advances in genetics, physiology, and biochemistry are quickly being applied in new production techniques (Mumford & Miura 1988). With the advent of research with color mutants (Miura 1985) and the discovery of the germinating conchospore as the site of meiosis, great progress should be made in the near future on the genetics of *Porphyra*. At the present time, however, strain selection is the major area of progress. In the past 15 years, strains have been selected for a long, narrow shape, late maturation (Miura 1984), and monospore production. Narrow plants give a greater yield on nets. When plants become reproductive, their rapid growth is partially offset by erosion of the margin. Secondary settlement of monospores on nets can help overcome an otherwise insufficient initial seeding. In addition, nets can be seeded entirely with monospores, which lowers the costs for conchocelis production per net. Monospores also grow and mature faster and can be used as a primary seed source (Li 1984).

Vegetative propagation of the gametophyte via protoplasts could solve two problems. By elimination of the conchocelis phase, production costs could be lowered, and genetic diversity eliminated. By propagation of vegetative cells from the blade in tissue culture, much greater control can be maintained over desirable plant genotypes. Protoplast production and fusion techniques (Polne-Fuller & Gibor 1984; Chen 1986; Fujita & Migata 1986) have been successful for a number of *Porphyra* species (Polne-Fuller & Gibor 1990). Waaland et al. (1986) have been able to mass-produce the conchocelis in free-living culture, but that does not seem economically feasible on a large scale. The ability to isolate and regenerate viable vegetative cells allows induction and selection of desired mutations, as well as vegetative cloning of specific isolates. Another benefit of single-cell and protoplast technology is the ability to bypass sexuality and thus maintain a pure gene pool.

Traditionally, the typical Japanese nori farmer was largely self-sufficient. The trend is now toward specialization. Now a farmer can buy conchocelis shells from a large firm that grows only conchocelis, or he can buy nets that are already seeded. Lastly, he can sell the raw *Porphyra* to a processor, rather than processing it himself or participating in a cooperative.

Nori production in Japan increased steadily through the 1970s as new sites came into use, new production techniques became available, and a strong market persisted. The record production of 10 billion sheets in 1974, however, greatly exceeded the market demand for the product, and as a result the price declined sharply. Repeated overproduction led to voluntary production restraints. The problem seems to be an excess of low- and medium-quality nori because of the greater production from floating-raft-style cultivation. This low quality is due to the lack of regular daily emersion when rafts are used.

Nori consumption has been expanding worldwide, particularly in North America, over the past decade. This is due in part to Japanese marketing efforts and increasing consumption of Japanese cuisine.

Besides facing competition from Korea and China, Japan may lose some of the rapidly increasing

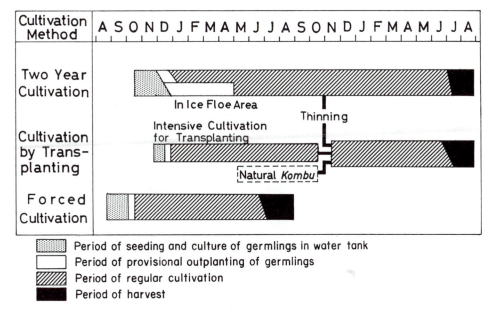

Figure 9.4. Major seasonal events in the 2-year cultivation method, the forced-cultivation method, and the transplanted-seedstock method of rearing edible kelp. (From Kawashima 1984, with permission of the Japanese Society of Phycology.)

American market. In the 1980s, a very small *Porphyra*-cultivation industry in the state of Washington and in British Columbia, Canada, emerged after 10 years of development (Mumford 1987, 1990). The cultivation technology has been transferred and modified from Japan and Korea, and both local and Japanese species are being used. The early results indicate that such cultivation is biologically feasible and could be economically viable; the products are of high quality. At present, the development of the nori industry in the Pacific Northwest has stalled. The constraints on future development are institutional – especially obtaining the necessary permits for use of water areas and finding adequate financing.

9.3 *Laminaria* mariculture

Edible *Laminaria* was collected in northern Japan as early as the eighth century, and some wild harvest was even exported to China (Tseng 1981, 1987b). Kelp cultivation was initiated in China and Japan in the early 1950s (Kawashima 1984), and today cultivation also occurs in Korea and Russia (Druehl 1988). China dominates the harvest of *Laminaria*, with 1.5 million metric tonnes (wet weight) out of a worldwide harvest of 2 million tonnes. Japan produces only 50,000 metric tonnes (Brinkhuis et al. 1987; Tseng & Fei 1987). In Japan, *Laminaria* is known as *kombu*, and in China it is called *haidai*.

Laminaria is a temperate seaweed that grows best at 8–16°C and lives in the low intertidal and upper subtidal. The harvestable sporophyte alternates with microscopic gametophytes (see Figs. 1.27 and 1.31).

Laminaria japonica is the main species that is cultivated. In North America and Europe, *L. saccharina* and *L. groenlandica* have been investigated in pilot projects (Druehl et al. 1988; Kain et al. 1990).

9.3.1 Cultivation

Seedstock is produced from meiospores released from the sori of wild or cultivated sporophytes. The sori are cleaned by vigorous wiping or by brief immersion in bleach and are left in a cool, dark place for up to 24 h. Spore release usually occurs within 1 h after reimmersion. The zoospores attach to a substratum within 24 h and develop into gametophytes. Release of gametes, fertilization, and growth into sporophytes 4–6 mm long require 45–60 days. In the original 2-year cultivation method, seedstock was produced in late autumn, when the sporophytes produce their sori. Seedstock was available for outplanting from December to February, and the crop was ready to harvest in 20 months (Kawashima 1984) (Fig. 9.4). Hasegawa (1971) developed the forced-cultivation method of seedstock production. In that method, seedstock was produced in the summer, because sporophytes that spent 3 months in the autumn in the field prior to their second growth season behaved as second-year plants. This method saves 1 year (Fig. 9.4). The natural cycle has been manipulated even further. By depriving the gametophyte stage of blue light, gametogenesis can be delayed, and sporelings will be produced throughout the year (Lüning & Dring 1972; Druehl et al. 1988).

Rearing of seedstock usually is carried out on horizontal or vertical strings. The seedstock is placed in

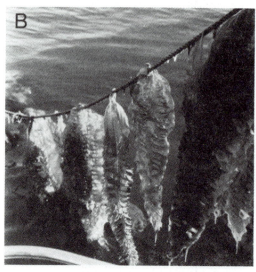

Figure 9.5. (A) Seedlings of *Laminaria groenlandica* as they appear at the time of planting in the sea. Plants approximately 4 mm long. (B) Clusters of *L. groenlandica* after 2 months of cultivation in the sea. Plants approximately 60 cm long. (From Druehl et al. 1988.)

sheltered waters for 7–10 days, where weaker plants are culled from the string by water motion. The string with the seedstock on it is then cut into small pieces and either inserted into the warp of the culture rope or attached by string or tape (Fig. 9.5). The ropes are checked every few months to thin the plant densities and to remove trapped debris and fouling organisms. The plants are kept about 5 m below the surface in the winter to avoid winter-storm waves, and 2 m below in the spring and summer to get more light for growth. Harvesting takes place in the summer (Figs. 9.5 and 9.6). Lengthwise growth occurs by production of new tissue in the meristematic zone between the stipe and the base of the blade. Natural shedding of the older distal part of the blade is common in *Laminaria* in late spring. This loss amounts to as much as 25% of the total harvest. In China, tip-cutting is employed between April and May of the second year to eliminate this loss of harvestable material (Tseng 1987b). In northern China, the kelp farms are fertilized with nitrogen in the summer to overcome the nitrogen limitation in those waters.

Among the more important diseases encountered in *Laminaria* cultivation, three are caused by adverse conditions (Tseng 1987b): green-rot disease (irradiance is too low), white-rot disease (irradiance is too high, and nitrogen too low) and blister disease (effects of fresh water from rainfall). There are two pathogenic diseases: Malformation disease of summer sporelings is characterized by abnormal cell division due to hydrogen sulfide produced by sulfate-reducing bacteria. Swollen-stipe, twisted-frond disease has appeared sporadically in China, and the organism causing it is unknown (Tseng 1987b).

The productivities of rope-cultured plants are most easily compared as wet weight per unit length of culture rope. Values range from 5 to 28 kg m^{-1}. On an areal basis, production ranges from 50 to 130 metric tonnes per hectare (Druehl 1988).

9.3.2 Utilization and future prospects

Laminaria is an excellent source of iodine, and it has been used in China as a dietary iodine supplement to prevent goiter (Brinkhuis et al. 1987). Kelps have been used for medicinal and food purposes for over 1,500 years in China, Japan, and Korea (Tseng 1984, 1987a). *Laminaria* has been the main source for the alginate industry for some time, especially in Europe (Kain & Dawes 1987).

Most of the *Laminaria* is dried and eaten directly in soups, salads, and tea, or used to make secondary products with various seasonings (e.g., sugar, salt, soy sauce) (Nisizawa et al. 1987). There are several grades of *kombu*, the highest of which can fetch $15 per kilogram dry weight (Brinkhuis et al. 1987).

The Chinese have conducted genetic studies of *Laminaria* and have employed methods of continuous inbreeding and selection in developing new strains. They have developed strains that are more tolerant of high temperatures and have high growth rates and high iodine contents (Tseng 1987b). Currently, microscopic gametophytes, produced from single spores and kept asexual by withholding blue light (Lüning & Dring 1972), can be cloned by fragmentation to provide populations of genetically identical gametophytes. These gametophytes, when crossed, produce genetically identical sporophytes. Using this system, superior strains

Figure 9.6. *Kombu* harvest in Japan. (Courtesy of and © 1991, Yamaha Motor Corp.)

can be maintained for long periods. The ability to clone the sporophyte directly would provide an optimal system to maintain a superior strain, but that has not been achieved yet.

The most feasible way to increase production of kelp is through cultivation, because wild harvests have not increased over the past 20 years (Druehl 1988). China is considering offshore cultivation. Even though the Japanese rely primarily on *Undaria* as an edible kelp, the cultivation of *Laminaria* is becoming increasingly important in meeting market demands.

9.4 *Undaria* mariculture

All kelps are more or less edible, but the two genera that are most important economically are *Undaria* and *Laminaria*.

Undaria has been a foodstuff of great importance and high value in Japan since about A.D. 700 (Nisizawa et al. 1987). At the beginning of this century the demand for *Undaria* exceeded its production from the wild. Cultivation began in Japan and was followed later by China. In the 1970s Korea began cultivation, and today it is the largest producer. The annual production of *Undaria* is about 470,000 tonnes of fresh crop (Japan, 100,000 t; Korea, between 290,000 and 333,000 t; China >35,000 t) (Perez et al. 1988). Japan consumes about four times as much as Korea (40,000 vs. 10,000 t per year, dry weight) (Boude et al. 1988), and it imports *Undaria* from Korea (up to one-third of its production). *Undaria* is becoming more popular in North America and Europe, but that market is still insignificant compared with Asia.

There are three species of *Undaria*, *U. pinnatifida*, *U. undarioides*, and *U. peterseniana*, but *U. pinnatifida* is the main species cultivated. It grows on rocks and reefs at 1–8 m below low tide, in places facing the open sea and along coasts that have warm currents, typical of southern Japan (Saito 1975; Kafuku & Ikenoue 1983). It is an annual that grows 1–2 m long.

In Japan, *U. pinnatifida* (also called *wakame*) is more important than *Laminaria japonica* in terms of the

value of the harvest, the amount produced, and its epicurean quality. For many years, small increases in the natural production were achieved by depositing stones on the bottom and blasting rocky reefs to increase the area for attachment. However, by 1968, production by cultivation exceeded the natural harvest.

9.4.1 Cultivation

Typically, twine or rope is immersed in seawater tanks with fertile sporophytes during April and May, when the water temperature is 17–20°C (Mathieson 1986). Zoospores produced from sporophytes, which previously had been dried for several hours, are allowed to settle until about 100 spores per centimeter of twine have become attached; higher densities would enhance opportunities for fungal and bacterial diseases. The seedling lines are then lashed to frames submerged in seawater tanks until September or November (Saito 1975). When the ambient temperature falls below 15°C, the seedling lines are wound around heavy-gauge ropes and positioned on horizontally floating bamboo poles at intervals of 1–6 m. Because it is an annual and grows quickly, young kelp sporophytes can be harvested by late winter. On shore, the midrib is cut out, and the thalli are put up on lines for drying. The semidried product is kneaded and then dried completely and processed into different types of *wakame* (Levring et al. 1969).

Undaria is often intermixed with *Laminaria* on floating rafts in Japan (Tseng 1981). Because *Undaria* has an earlier, shorter growing season than *Laminaria*, it is harvested in March and does not interfere with the growth of *Laminaria*, which is harvested in June. This provides the farmers with two crops per year. Because the Chinese prefer *Laminaria* to *Undaria*, many farmers in China do not mix the two species.

Undaria was accidentally introduced into France in 1971 (Perez et al. 1988). The high demand for *Undaria* in Japan has encouraged the French to explore its cultivation potential. The French use another seeding method, which allows them to obtain a great quantity of "seeds" at the gametophyte stage at any time of the year. The gametophytes are kept in the laboratory as vegetative stock, from which a suspension is produced and sprayed onto collectors (lines of thread on a square frame). The collectors are hung under a bank of neon lights in tanks of seawater until the young sporophytes are 3–5 mm, large enough to survive in the open sea (Perez et al. 1988). Then the thread is unrolled from the frame, wound around a rope, and placed at a depth of 1 m in the sea.

9.4.2 *Undaria* as food

Undaria is processed into a variety of food products. Raw *wakame* fronds, with the midrib removed, are diced to produce *subasuki wakame*. The main problem with this product is that the fronds often fade and soften during storage because of the activities of enzymes such as chlorophyllase and alginate lyase (Watanabe & Nisizawa 1984).

   Haibashi wakame was developed to remedy the problem of the softening and fading of the fronds: Fresh *wakame* is mixed with the ashes of straw, wood, or briquets in a rotary mixer. The mixture is spread on a sandy beach and dried in the sun for 2–3 days. Then it is packed in a plastic bag and kept in the dark. Later it is washed with seawater and then fresh water to remove the adhering ash and salt. The midrib is removed, along with faded fronds, and the remaining tissue is dried. This treatment preserves the deep greenish brown color of the alga and makes the product more elastic and flavorful for chewing. The water-extractable calcium in the ash aids in the retention of elasticity. The alkalinity of the ash inactivates alginate lyase, prevents the degradation of alginate, and leads to the formation of insoluble calcium alginate. This prevents the frond from softening (Watanake & Nisizawa 1984).

   The main *wakame* product is salted *wakame*, which is baled: Fresh *wakame* is heated in 80°C water for 1 min and then quickly cooled. Thirty kilograms of salt are added per 100 kg of raw seaweed; they are mixed and stored for 24 h and then placed in a net bag to remove excess water. The fronds are stored at −10°C and packaged in plastic bags for sale (Nisizawa et al. 1987).

   The consumption of cut *wakame* is increasing rapidly because of its use in instant foods such as soups and noodles. It is processed from boiled and salted *wakame*. The excess salt is removed, and the frond is cut into small pieces, then dried and marketed as chips.

   Several North American substitutes for *wakame* have been developed using *Alaria* and *Nereocystis*.

9.4.3 Future trends

   There have been some recent improvements in the cultivation of *Undaria*. Enhancement of the growth and development of the *Undaria* thalli has been achieved by addition of squid-liver protein to the gametophytes (Yaba et al. 1984). Also, it has been found that totipotency can be induced in *Undaria*. Therefore, somatic-cell clones can be subcultured, and tissue culture can be used to breed varieties, fix heterosis, and preserve good genotypes (Yan 1984).

   Another technique to improve the productivity of the *Undaria* species has been to create hybrids. For example, the hybrid between *U. pinnatifida* and *U. undarioides* grows better at high water temperatures than do the parents, and it has a thicker, heavier, more shallowly lobed blade than *U. pinnatifida* (Saito 1975).

9.5 *Eucheuma* and *Kappaphycus* mariculture

   The world's sources of carrageenans include eight genera of red algae, which yielded 43,000 tonnes of commercial dry seaweeds in 1984 (Lewis et al. 1988). Three phycocolloid-producing genera account for over 75% of the commercial effort in the tropics: *Eucheuma*, *Kappaphycus*, and *Hypnea* (Dawes 1987, 1990). Approximately half of the world's carrageenophytes are produced in the Philippines and Indonesia, where several species of *Eucheuma* and *Kappaphycus* are cultivated on a large scale. Much of the remaining supply comes from natural harvests of *Chondrus* in Europe, Chile, and eastern Canada.

   Eucheuma and *Kappaphycus* are subtropical and tropical red algae, and they occur mainly between 20° N and 20° S latitudes. Tolerance to aerial exposure governs their upward distributional limits (Doty 1986). They grow best in open-ocean waters that feature high levels of water motion.

   Eucheuma and *Kappaphycus* are valuable commercially because they produce gelling carrageenan (either κ or ι) in both haploid and diploid stages. This contrasts with other carrageenan producers, such as *Iridaea* and *Gigartina*, in which the gametophyte usually produces κ-carrageenan, and the sporophyte usually produces weakly gelling λ-carrageenan (Dawes 1987).

   The world harvest of *Eucheuma* and *Kappaphycus* has dwindled, and now over 95% of the crop is farmed. Several species have been farmed in the Philippines, but because *Kappaphycus alvarezii* (commonly known as *cottonii*) is so much easier to grow, it has completely replaced the other species such as *Kappaphycus striatum*, formerly the dominant cultivated species. Other commercial species are *Eucheuma denticulatum* (common name, spinosum) and *E. gelatinum* (common name, gelatinae). *E. gelatinum* is grown in China, where it is reported to be less affected by storms and cyclones than is *K. alvarezii* (Pringle et al. 1989). In the tropical Western Atlantic, *Eucheuma isiforme* is being explored for cultivation (Dawes 1990). *K. alvarezii* produces ι-carrageenan, whereas *E. denticulatum* and *E. gelatinum* produce κ-carrageenan. In the following discussion, "*Eucheuma*" will be used to include *Kappaphycus alvarezii* and *K. striatum*, which formerly were recognized as species of *Eucheuma* (Doty 1988).

9.5.1 Biology

   The life history of "*Eucheuma*" is the triphasic scheme common for red algae (see sec. 1.5). A diploid tetrasporophyte phase produces haploid nonmotile meiospores called tetraspores. The tetraspores usually produce separate haploid male and female gametophytes. A diploid carposporophyte develops in situ on the female gametophyte after fertilization. The carposporophyte releases diploid carpospores, which initiate the tetrasporophyte stage again. In contrast to *Porphyra*, the phycocolloid-producing red algae, such as "*Eucheuma*," have an alternation of isomorphic macroscopic gametophytic and sporophytic generations.

9.5.2 Cultivation

   "*Eucheuma*" farming on a commercial scale began in the early 1970s in the Philippines. Because the

Filipinos are now the world's largest producers, the farming methods described are the methods used in that area (Doty & Alvarez 1975; Ricohermoso & Deveau 1979; Trono & Ganzon-Fortes 1989).

Farm sites must have the following characteristics: salinities greater than 30‰; good water movement, but without large waves; clear water at 25–30°C; and a coarse-sand bottom to retain mangrove poles and keep plants from being covered with silt. Usually a site is tested on a small scale, and if the test plants double in size in 30 days or if daily growth rates are 3–5%, then the site is considered a good one. The bottom is cleared of seagrasses, seaweeds, large stones, corals, and grazers, such as sea urchins. Plant growth will decrease if plants become desiccated at low tide.

Bottom- and floating-monoline methods are used. In the first method, heavy nylon-monofilament line is stretched between stakes 10 m apart and 0.3–0.5 m above the bottom. The floating-monoline-bamboo method is similar except that the whole line structure floats because of the bamboo, and it is tied to stakes on the bottom. This method is better for deeper sites and sites with very irregular bottom contours.

Thalli are cut into small (~200-g) pieces (called "seedlings") and tied to the monofilament line with soft plastic twine at 20–30-cm intervals. During the growing period, invading plants and animals (especially sea urchins) are removed by hand.

Plants are harvested when they reach 1 kg or larger. This takes about 2–3 months, and therefore a farmer can obtain up to five harvests per year. A plant can be harvested whole or pruned back to seedling size. The best plants sometimes are kept as seedlings for the next crop. The plants are washed, air-dried for 3 days, washed again, and redried. This washing removes dead epiphytes and salt.

This type of "*Eucheuma*" farming involves low capital costs and is simple to set up; it provides families with extra income. It is labor-intensive, because there is no automation. The dried plant is sold for $700 (U.S.) per tonne (McHugh 1990), and the average farmer realizes an annual net income of over $1,000 per hectare.

Fouling of "*Eucheuma*" plants by other seaweeds and grazing by starfish and urchins can be controlled by regular maintenance. A disease known as *ice-ice* can form a white powdery growth over the thallus and causes loss of pigment and gradual consumption and fragmentation of the plant (Uyenco et al. 1981). It can be caused by unfavorable ecological conditions and/or pathogenic microorganisms, but further research will be required to clarify the precise cause. At present, there is no control for it.

9.5.3 Production, uses, and future prospects

Live or freshly dried seaweed can be immersed in a weak alkali solution to give a "stabilized" raw material. This treatment impedes chemical or biological degradation of the seaweed. NaOH or KOH is added to seawater to give a 0.5–1.5-N solution. The plants are soaked for 3 h, dried in the sun to less than 30% moisture, and then baled and exported.

Chips are made by heating seaweed for 2 h at 85°C in a 2-N KOH solution. This increases the gel strength of the carrageenan. The plants are washed to remove the alkali, chopped wet, and then dried to form chips. The chips can be ground to produce seaweed flour. The flour is used in products not intended for human consumption, such as gelation of pet foods and air fresheners and stabilization of industrial slurries. Seaweed flour has 9–15% fiber content, which causes its gels and solutions to be distinctly cloudy and grainy, whereas alcohol-precipitated carrageenan solutions are clear. However, some manufacturers insist on trying to use seaweed flour rather than extracted carrageenan for human foods because it is cheaper (Adams & Foscarini 1990).

Estimates of growth rates vary from 0.5% to 5% per day. The production of new strains by traditional hybridization has not been carried out because there is a lack of recognizable, genetically different strains and males in this genus. Sexual reproduction and the rearing of crops from spores or very small vegetative propagules do not seem practical (Doty 1986), and therefore the practice of macrovegetative reproduction is likely to continue.

Quality control has emerged as a recent problem for farmers. The quality of the carrageenan has been decreasing recently because of hybridization with native plants in the farming areas in the Philippines. In the future, carrageenan may be used to produce nearly fat-free hamburgers. Most of the market is for ι- and κ-carrageenans, which form strong gels. The water-extractable ι- and κ-carrageenans are used in instant foods that come in powdered form, in chocolate milk, and in toothpaste; λ-carrageenan forms a highly viscous, nongelling, polyanionic hydrocolloid, and it is well suited for instant-mix food products.

9.6 **Other seaweeds**

Agar-producing genera of interest include *Gracilaria* and *Gelidium* (Mathieson 1986). Prior to 1942, *Gelidium* was the primary source of agar, and it was prized for its high-quality agar (high gelling strength and low sulfation content). Because of its growth, the natural beds were overharvested, and interest turned to fast-growing *Gracilaria* species (Hanisak 1987). Roughly 35% of the world's agar production comes from *Gelidium*, and bacteriological-grade agar is obtained exclusively from this genus (Santelices 1990a). Attempts to increase the production from the natural beds by increasing the areas of the rock substrates, farming it with ropes and rafts, and even cultivating free-floating plants in onshore tanks have met with some success (Melo et al. 1990). Agar, particularly

from *Gelidium,* can be fractionated into two components: agarose (the gel-forming component) and agar-
opectin (a mixture of variously sulfated molecules).
Agarose either is not sulfated or is low in sulfate. It is
used in electrophoresis and other specialized laboratory
procedures, and some forms cost $3,600 (U.S.) per ki-
logram (Rogers & Gallon 1988).

Gracilaria is the main source of food-grade agar.
It produces a lower-quality agar because of higher sulfate content. It is harvested from the wild in more than
20 countries, but attempts are being made to farm it and
grow it in land-based tanks. *Gracilaria* grows quickly
in tanks, and so adequate supplies of nutrients (NO_3^-,
PO_4^{3-}, CO_2) and light are essential. Pond culture is used
in Taiwan: Wild plants are collected and broken into 10-
cm-long pieces. Plants grow to 10 times their original
mass in 3 months. Annual yields are estimated to be 24
dry tonnes per hectare. Raft culture has been employed
in China (Ren et al. 1984). In St. Lucia, *Gracilaria
debilis* and *G. domingensis* are planted together in
ropes, similar to the cultivation of "*Eucheuma*" (Smith
et al. 1984). The faster-growing *G. domingensis* sweeps
G. debilis clean of most epiphytes. Many species of
Gracilaria are delicate and brittle, but these two species
are more robust and are amenable to rope culture. The
search for new strains continues, especially for a strain
that will produce a strong-gelling agar. Normally, harvested agarophytes must be treated with alkali to remove
6-sulfate from L-galactose so that the agar will have adequate gel strength.

Several other kelps are cultivated or harvested
from nature for food and alginates (Mathieson 1986).
Wild *Macrocystis pyrifera* is harvested off the west
coast of the United States (North et al. 1986; North
1987). Small-scale cultivation of *Laminaria saccharina*
and *L. longicruris* is occurring in eastern Canada (Chapman 1987), the eastern United States (Yarish et al.
1990), and Europe (Kain & Dawes 1987). The Europeans are also experimenting with *Alaria esculenta*
(Kain & Dawes 1987).

Other brown seaweeds include *Ascophyllum nodosum,* harvested for its alginate content in eastern
North America (Sharp 1987) and northern Europe (Kain
& Dawes 1987), and *Hizikia fusiformis* (Sargassaceae),
which is grown for food in Japan (Nisizawa et al. 1987).

The Japanese eat *aonori,* which is a mixture of
sea lettuce (*Ulva*), green laver (*Enteromorpha*), and *hi-
toegusa* (*Monostroma latissimum*). *Monostroma* is
grown commercially and is an important component of
tsukudani, a green paste made by boiling down this seaweed mixture in soy sauce (Nisizawa et al. 1987).

9.7 Domestication of seaweeds: application of ecology and physiology

Only four genera, *Porphyra,* "*Eucheuma,*" *Laminaria,* and *Undaria,* can be regarded as true marine
crops, where the amounts harvested from cultivated

sources exceed those harvested from wild populations.
Research and experimentation on how to grow these
plants most profitably have been ongoing for more than
40 years. Domestication of a lesser-known seaweed,
Chondrus crispus (Irish moss), has taken place in the
past two decades, and it will be used as a case study to
demonstrate how fundamental knowledge of ecology,
physiology, and genetics is applied to the cultivation of
a new species.

The North Atlantic red alga *Chondrus crispus*
was initially the main source of carrageenan for industry. Before 1975, *Chondrus* supplied 75% of all carrageenan requirements for industry, but by 1985 the figure
was less than 25%. That decrease was primarily due to
the farming of "*Eucheuma*" species in the Philippines
(Pringle & Mathieson 1987). Because no more *Chondrus* could be harvested from wild stocks, cultivation in
tanks on land was explored in a 3.4-hectare site in Nova
Scotia, Canada (Surette 1988).

Initially the intention was to cultivate *Chondrus
crispus* as a high-grade source of κ-carrageenan (Neish
& Fox 1971), but less expensive production of
κ-carrageenan from "*Eucheuma*" in the Philippines
made it imperative to develop *Chondrus* as a source of
λ-carrageenan. It was soon discovered that λ-
carrageenan was produced by *Chondrus* tetraspro-
phytes, and κ-carrageenan by gametophytes (Chen et al.
1973; McCandless et al. 1973). Thus, for the first time,
a commercial source of pure λ-carrageenan could be
made available. For the next two decades, researchers
worked toward the goal of growing *Chondrus* in land-
based tanks in a temperature environment where the
availability of light and the temperature usually limited
growth for much of the year. Could cold-water seaweed
aquaculture make a profit?

The scientific bases for designing an efficient,
profitable cultivation system have come from research
into the strains, the sites, and the environmental factors affecting productivity (Pringle et al. 1989). The
product to be marketed dictates the choice of species.
To make the operation profitable, considerable effort
should be devoted to strain selection, (i.e., choosing an
individual that will grow the fastest and/or produce the
most product for the environmental conditions provided). From an initial screening of several hundred
plants, the T4 clone of *Chondrus* was chosen as the best
clone (Neish & Fox 1971). A decade later, further
screening revealed other clones of gametophytes that
had even higher growth rates than T4 (Cheney et al.
1981). The encroaching competition from "*Eucheuma*"
for the production of κ-carrageenan initiated a screening
program for high-growth-rate sporophytes that would
produce λ-carrageenan. The sporophyte BH-D was superior to T4, and it was adopted for commercial production (Craigie & Shacklock 1989).

Several years may be needed to scale up from one
frond (*Chondrus* can be propagated vegetatively by

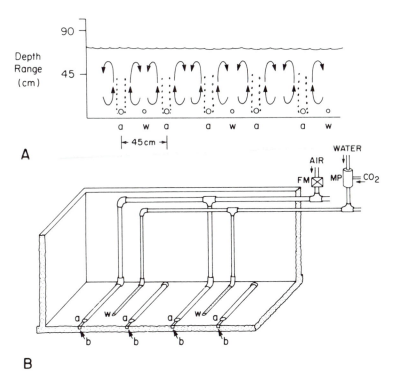

Figure 9.7. Tank design for *Chondrus crispus* cultivation: (A) section; (B) plan. Key: a, air pipe; b, batten for attaching air pipe; FM, flowmeter; MP, mixing pump; w, water pipe. Arrows show directions of water circulation induced by air bubbles. (From Bidwell et al. 1985, with permission of Kluwer Academic Publishers.)

fragmentation) to obtain enough biomass to stock tanks covering several hectares. This is another reason for starting the screening program early. The screening program should continue and be used to help solve other problems as they arise (e.g., selecting strains that will be less prone to epiphytes and disease, or a different-shaped thallus to capture more light).

Site selection is also an extremely important choice, because major factors such as light availability, water temperature, and water quality are involved. Choosing to locate on the east coast of Canada in a temperature climate meant that plant growth would be limited by light and temperature for much of the year (Craigie 1990b). Despite those handicaps, the project was started because of the decline in the wild harvest of *Chondrus*.

After selection of a strain and a site, the next challenge was to conduct experiments on how to overwinter *Chondrus* in tanks (Craigie 1990b). Craigie and Shacklock (1989) found that *Chondrus* could survive the low temperatures ($<1.9°C$) and high salinity (58‰) encountered when up to half the water in the tanks froze if the stocking densities were less than 10 kg m^{-3} and the water was enriched with nutrients.

The keys to an economically successful cultivation system using tanks are the engineering considerations and the production costs, usually not the biology. Other crucial factors in the design include selection of a system that will be sustainable, that will be easy to harvest, clean, and maintain, and that will remain relatively free of epiphyte contamination and disease. In North America and Europe, where labor is expensive, such projects are necessarily capital-intensive.

The engineering and design processes described next are from the early published reports, but because this production system became private, as Acadian Sea Plants (Surette 1988), modifications to the system have been kept secret. Mixing and water exchange were major factors in attaining successful *Chondrus* production. Initially, paddle wheels were used for mixing, but the use of air bubbles proved to be trouble-free and 10 times more energy-efficient (Bidwell et al. 1985). Agitation with air serves several important functions: (1) it suspends the alga, (2) it reduces boundary layers, thus facilitating nutrient/metabolite movements, (3) it breaks up larger plants, (4) it rubs plants together and aids in epiphyte removal, (5) it suspends debris to be flushed away, and (6) it minimizes oxygen supersaturation (Craigie & Shacklock 1989). Initially, air was forced through sparagers or bubblers from holes in the floors of the tanks, but that tended to create "dead" areas, where the plants stagnated. The final design was to lay parallel air pipes across the bottom of the tank with 1-mm holes drilled at 50-mm intervals (Fig. 9.7). The operating principle of the tank design was to create circulating cells of water produced by rows of rising air bubbles. The bubbles keep the water and plants in constant motion, and the plants remain evenly distributed across the tank. These circulation cells represent modules that can be expanded as culture facilities expand. Individual plants are exposed to surface light and then plunged to the bottom of the tank (complete darkness). This is an

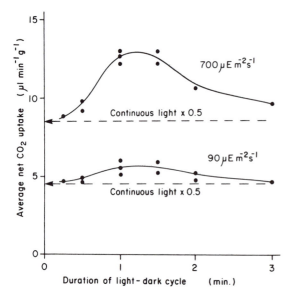

Figure 9.8. Photosynthesis by *Chondrus crispus* plants from a natural population in St. Margaret's Bay, Nova Scotia, in intermittent light of different intensities. Light cycles consisted of equal periods of light and darkness. Irradiance was provided by UHO fluorescent lamps. Temperature at high irradiance, 20–21°C; at low irradiance, 18–19°C. Replication (n = 1–3) indicated by points on graphs. (From Bidwell et al. 1985, with permission of Kluwer Academic Publishers.)

advantage, because photosynthesis under flashes of very high illumination is about 40% more efficient than that in continuous light (Bidwell et al. 1985); the optimum duration is about 1 min (Fig. 9.8). Therefore, the tank depth and the spacing of air pipes and airflows were selected to give the plants this optimum light flash of about 1 min.

Another critical factor in tank design is the water depth. Through experimentation with depths ranging from 30 to 90 cm, it was found that 90 cm gave the highest production, measured as grams dry weight per square meter per week.

Similarly, tests with inoculum densities ranging from 5 to 12 kg m^{-2} revealed that plants at 12 kg m^{-2} produced the most biomass and carrageenan (g m^{-2} wk^{-1}). A beneficial side effect of increasing the plant density was that epiphyte growth was dramatically inhibited at high inoculum densities, to the extent that chemical control of epiphytes was seldom required.

Water flushing has three purposes: (1) to add nutrients and carbon for photosynthesis, (2) to remove waste products, and (3) to help control the temperature. Temperature control is vital for sustained yield. Above 18°C, photosynthesis declines rapidly, and carrageenan yields decline (Bidwell et al. 1984). Flushing rates of three to seven tank volumes of new seawater per week

help maintain temperature and adequate water quality (Craigie 1990b).

At high plant densities, at which all the light in the tank is absorbed, dissolved inorganic carbon (DIC) is used up, and photosynthesis can be reduced by 50%, even under high flushing rates and strong aeration (Bidwell et al. 1985). Because the CO_2 concentration in air is only 3%, and it dissolves very slowly, pure CO_2 is added with special diffusers to make certain that most of the CO_2 dissolves in the water. The pH value serves as a useful indicator of DIC concentration, because as photosynthesis increases, DIC decreases, and pH increases from the normal 8.2 value to over 9. Therefore, a pH-stat can be installed to monitor pH and control the delivery of CO_2.

Nitrogen and phosphorus must be added, in addition to carbon. However, the additions of nutrients that give maximum growth do not give the maximum yield of carrageenan (Table 9.2) (Neish et al. 1977). This inverse relationship between ambient nutrient concentrations and carrageenan content is referred to as the "Neish effect." Phosphorus fertilization had no effect on the yield of carrageenan, whereas low levels of nitrogen fertilization produced the most carrageenan. Nitrogen depletion in batch cultures becomes evident in 3–5 days, whereas phosphorus deficiency requires 1 week to reduce the growth rate. The nitrogen content of *Chondrus* varies from 0.8% to 4.5%, whereas the phosphorus content is about 0.4% of dry weight. Therefore, N and P are added in a ratio of 10:1 as a mixture of two agricultural fertilizers: NH_4NO_3 and $(NH_4)_2HPO_4$. These are added in pulses, usually at night, when flushing is stopped, so that the plants will have time to take up the pulses. Pulse fertilization is beneficial in minimizing epiphyte contamination because it takes advantage of the nutrient-storage capabilities of *Chondrus*. Phosphorus storage is thought to occur in an unknown organic form (Chopin 1986), whereas nitrogen is stored mainly as a dipeptide: citrullinylarginine (Laycock et al. 1981). Attempts to induce nutrient deficiencies of Co, Cu, Fe, Mn, and Zn in actively growing *Chondrus* cultures were unsuccessful, except for iron (Craigie 1990b), which slowed growth considerably. Therefore, during intensive seaweed cultivation it may be necessary to add Fe, in addition to N and P.

The net productivity of *Chondrus* in tank culture is a linear function of irradiance over temperatures of 10–20°C when nutrients are not limiting (Craigie 1990b). The net dry-weight production over a 365-day period was 6.3 kg m^{-2} (Craigie & Shacklock 1989). The overall annual conversion of solar radiation was 1.76%, which is high for plants in temperate regions. This system made a profit for 2 years before the sole purchaser of the crop decided to buy cheaper, lower-quality carrageenan from the Philippines. However, the system is adaptable, and it has been successfully tested for the cultivation of several other potentially

Table 9.2. *Average dry-weight production and carrageenan production for* Chondrus crispus *in tank culture during January–September (Nova Scotia)*

Fertilizer level	Dry-weight production (g m^{-2} wk^{-1})	Carrageenan content (%)	Carrageenan production (g m^{-2} wk^{-1})
No addition	112	35.1	39
Low	120	38.3	46
Medium	129	34.5	45
High	130	32.6	42

Source: Bidwell et al. (1985); reprinted from *Botanica Marina* with permission of Walter de Gruyter & Co.

more valuable species of seaweeds. The first on-land, temperate-climate seaweed farm has shown that although it is feasible biologically and engineeringwise to grow seaweeds in temperate areas, it may not be economically feasible unless the price of the particular product is very high.

9.8 Seaweed biotechnology: current status and future prospects

The current utilization of seaweeds can be divided roughly into industrial use of phycocolloids in the developed countries and human consumption of seaweeds in the Far East. Table 9.3 gives an outline of the major uses of seaweeds today. Although seaweeds will remain important as food supplements in the Far East, the most notable trend over the past three or four decades has been the significant growth in the output of the colloids: agar, carrageenan, and alginate (McLachlan 1985). Aside from this increasing demand for vegetable gums, any future large-scale exploitation of marine algae will require new algal products or genetically improved seaweeds that will make production of existing products economical. Future products from algae may include methane, alcohols, fatty acids, and esters, as well as resynthesis into carbohydrates (Indergaard 1983; McHugh 1987) (Fig. 9.9). There has been extensive interest in obtaining energy from seaweeds in the United States. *Gracilaria* has been studied in Florida (Hanisak 1987), *Laminaria* in the northeastern United States (Brinkhuis et al. 1987), and *Macrocystis* off California (North 1987).

Progress in the tissue culture and genetic engineering of plants is now making possible (1) large-scale, rapid propagation of genetically uniform plants from superior stock, (2) the selection of improved varieties using somaclonal variation techniques, (3) the development of new hybrids between different cultivars and species by protoplast fusion and cell-culture techniques, and (4) the use of recombinant DNA technology to introduce new genetic material into plant cells (Evans & Butler 1988; van der Meer 1988). Not surprisingly, advances in macroalgal biotechnology lag far behind research with higher plants because of a low level of interest in seaweeds and their commercial products and because there are substantial gaps in our knowledge of the biochemistry and physiology of these diverse plants. An increasing contribution from molecular-genetic techniques is to be expected, but for the near future, classical plant breeding will continue to make the largest contribution (van der Meer 1988).

The starting point for applying new developments in technology to seaweeds is successful isolation of viable protoplasts from diploid and haploid tissues under axenic conditions. Protoplast fusion offers a unique opportunity to produce new hybrids between related but sexually incompatible or sterile species. These somatic hybrids should have a combined genome and a novel cytoplasmic combination. Vegetative tissues of some seaweeds have been dissociated into single cells that can be regenerated into complete plants (Polne-Fuller & Gibor 1987), and protoplasts have been isolated from several seaweeds. There has been some success in fusing protoplasts, but regeneration from these protoplasts has not been achieved (Evans & Butler 1988). The formation of callus (undifferentiated cell masses on tissue explants) can be used as a source of vegetative cells or callus that can be induced to differentiate into whole plants. This has been achieved in several seaweeds (Evans & Butler 1988). It is difficult to say whether or not it will be possible to produce phycocolloids such as agar and carrageenan cost-effectively by these new technologies, because it is not yet known to what extent improved quality or yield can be achieved. On the other hand, somatic hybridization for producing strains resistant to fungal diseases, for example, is possible (Evans & Butler 1988). Increased research efforts must therefore be devoted to perfecting the underlying biotechnological techniques (protoplast isolation, somatic fusion, callus formation, and plant regeneration) to realize the potential offered by seaweeds.

9.9 Synopsis

Seaweed mariculture involves large-scale cultivation of seaweeds in the sea or in land-based tanks.

Table 9.3. *Estimates of global use of seaweeds (production values estimated for mid-1980s)*

Product/species	Product (tonne yr^{-1})	Algal consumption (tonne yr^{-1}, wet weight)
Industrial uses		
Alginates (ca. 230 × 10^6 US$/yr):		
Macrocystis pyrifera, Laminaria spp.,		
A. nodosum, Durvillaea sp., *Lessonia* sp.	25,000	ca. 500,000
Agar (ca. 160 × 10^6 US$/yr):		
Gelidium spp., *Gracilaria* spp., *Gelidiella* sp.,		
Pterocladia sp.	11,000	ca. 180,000
Carrageenans (ca. 100 × 10^6 US$/yr):		
Eucheuma spp., *Chondrus crispus, Gigartina* spp.,		
Furcellaria lumbricalis, Mastocarpus sp.		
Gymnogongrus sp., *Hypnea* sp.	15,500	ca. 250,000
Seaweed meal (ca. 5 × 10^6 US$/yr):		
Ascophyllum nodosum	10,000	ca. 50,000
Maërl (ca. 10 × 10^6 US$/yr):		
Lithothamnion calcareum	510,000	ca. 550,000
Seaweed extracts (ca. 5 × 10^6 US$/yr):		
A. nodosum, Ecklonia maxima, Laminaria sp.,		
Fucus sp.	1,000	ca. 10,000
Total: ≈1,540,000 tonnes wet weight		
Nutritional uses		
Nori (>1,800 × 10^6 US$/yr):		
Porphyra spp.	ca. 40,000	ca. 400,000
Wakame (>600 × 10^6 US$/yr):		
Undaria spp.	ca. 20,000	ca. 300,000
Kombu (>>600 × 10^6 US$/yr):		
Laminaria spp.	ca. 250,000	ca. 1,000,000
Total: ≈1,700,000 tonnes wet weight		

Source: From Indergaard and Jensen (1991), with permission of M. Indergaard.

It is distinct from simple harvesting of wild stock. The major seaweed-farming countries are China, Japan, Korea, the Philippines, and Indonesia. Seaweeds frequently are grown for human food, and they provide protein, vitamins, and minerals (especially iodine). The commercially important phycocolloids are agars and carrageenans from red algae and alginates from brown algae. The four major genera that are cultivated in Asia are the red algae "*Eucheuma*" and *Porphyra* and the brown algae *Laminaria* and *Undaria*. Cultivation has two major components: seedstock production and field cultivation of the seedstock to a harvestable product.

In the laboratory, during February or March, *Porphyra* thalli are induced to release carpospores, which adhere to shells and germinate into a small microscopic filament called the conchocelis. The conchocelis produces conchospores, which are collected on nets that are then taken to the field, where the spores germinate and grow into the harvestable thallus. Har-

vested thalli are chopped and made into a slurry, spread over screens, dried, and then pressed into nori sheets. The problems surrounding *Porphyra* culture include fouling of nets by fast-growing weed species, predation by herbivorous fish, and threats from several diseases.

The two most economically important edible kelps are *Undaria* (mainly from Korea) and *Laminaria* (mainly from China). The kelps have heteromorphic life cycles, and the microscopic stage is reared and manipulated in the laboratory. The gametophytic seedstock is attached to rope on a floating-raft system. In Japan, *Undaria* is harvested by hand, dried, and made into different types of *wakame*. Most of the *Laminaria* is dried and eaten directly in soups, salads, and tea, or is used to make secondary products with various seasonings.

Half of the world's carrageenan is produced in the Philippines and Indonesia, where several species of *Eucheuma* and *Kappaphycus* are cultivated on a large scale. Farming of these seaweeds began in the 1970s,

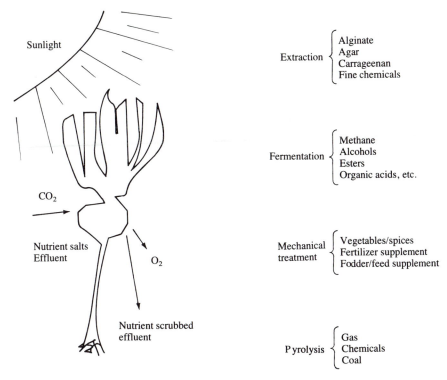

Figure 9.9. Current and potential future uses of seaweeds. (After Indergaard & Jensen 1991, courtesy of M. Indergaard.)

and today it is one of the most notable success stories of the seaweed-farming industry. Floating-monoline methods are used.

The domestication of a seaweed involves the application of ecology, physiology, and genetics to provide a scientific basis for seaweed aquaculture. The domestication of a κ- and λ-carrageenan producer, *Chondrus crispus*, has been well documented over the past two decades. The protocol for domestication involves species selection (i.e., product selection and market survey), site selection, and strain selection (i.e., choosing an individual that will grow the fast-est and/or produce the most product in the environmental conditions provided). Tank design and water circulation are also critical. Physiological experiments have provided fundamental knowledge used in growing the species efficiently. This land-based system represents the first attempt at large-scale (3.4 hectares), cold-water seaweed aquaculture. Although it is suffering from competition from less expensive carrageenan produced from "*Eucheuma*" in the Philippines, the tank system is adaptable, and it has been successfully tested for the cultivation of several other potentially more valuable species of seaweeds.

Appendix: Taxonomic classification of algae mentioned in the text

Paul C. Silva and Richard L. Moe

Note: Synonyms and life-history stages are listed under accepted names. Superscript numbers refer to recent nomenclature changes added in proof, which are listed by number on p. 307.

Cyanophyceae
  Chroococcales
    Hydrococcaceae
      Pleurocapsa violacea Weber-van Bosse
      Radaisia
  Oscillatoriales
    Oscillatoriaceae
      Blennothrix lyngbyacea (Kützing) Anagnostidis
        & Komárek
          syn: *Microcoleus lyngbyaceus* (Kützing)
            P. Crouan & H. Crouan
      Oscillatoria
      Plectonema
    Phormidiaceae
      Spirulina platensis (Nordstedt) Geitler[1]
      Trichodesmium
    Schizotrichaceae
      Schizothrix
  Nostocales
    Microchaetaceae
      Tolypothrix
    Nostocaceae
      Anabaina
    Rivulariaceae
      Calothrix crustacea Thuret
      Rivularia atra Roth
    Scytonemataceae
      Scytonema

Glaucocystophyceae
  Cyanophorales
    Cyanophoraceae
      Cyanophora paradoxa Korshikov

Rhodophyceae
  Porphyridiales
    Porphyridiaceae
      Porphyridium purpureum (Bory de Saint-
        Vincent) Drew & Ross

      Rhodella maculata L. V. Evans
      Stylonema alsidii (Zanardini) Drew
         syn: *Goniotrichum alsidii* (Zanardini)
          Howe
  Erythropeltidales
    Erythrotrichiaceae
      Erythrotrichia carnea (Dillwyn) J. Agardh
      Smithora naiadum (C. L. Anderson) Hollenberg
  Bangiales
    Bangiaceae
      Bangia atropurpurea (Roth) C. Agardh
         syn: *Bangia fuscopurpurea* (Dillwyn)
          Lyngbye
      Porphyra abbottiae Krishnamurthy
      Porphyra fallax S. C. Lindstrom & K. M. Cole
      Porphyra fucicola Krishnamurthy
      Porphyra haitanensis T. J. Chang & B. F. Zheng
      Porphyra lanceolata (Setchell & Hus)
        G. M. Smith
      Porphyra miniata (C. Agardh) C. Agardh
      Porphyra nereocystis C. L. Anderson
      Porphyra perforata J. Agardh
      Porphyra pseudolanceolata Krishnamurthy
      Porphyra pseudolinearis Ueda
      Porphyra purpurea (Roth) C. Agardh
      Porphyra sanjuanensis Krishnamurthy
      Porphyra suborbiculata Kjellman
      Porphyra tenera Kjellman
      Porphyra torta Krishnamurthy
      Porphyra umbilicalis (Linnaeus) Kützing
      Porphyra yezoensis Ueda
  Acrochaetiales
    Acrochaetiaceae
      Rhodochorton purpureum (Lightfoot)
        Rosenvinge
  Palmariales
    Rhodophysemataceae
      Rhodophysema elegans (P. Crouan & H. Crouan
        ex J. Agardh) Dixon

Palmariaceae
Devaleraea ramentaceum (Linnaeus) Guiry
Halosaccion glandiforme (S. G. Gmelin)
Ruprecht
Palmaria palmata (Linnaeus) Kuntze
Nemaliales
Galaxauraceae
Tricleocarpa oblongata (Ellis & Solander)
Huisman & Borowitzka
syn: *Galaxaura oblongata* (Ellis & Solander) Lamouroux
Nemaliaceae
Nemalion helminthoides (Velley) Batters
syn: *Nemalion multifidum* (Weber & Mohr) Chauvin
Trichogloea
Liagoraceae
Liagora tetrasporifera Børgesen
Ahnfeltiales
Ahnfeltiaceae
Ahnfeltia plicata (Hudson) E. Fries
sporophyte: *Porphyrodiscus simulans* Batters
Gelidiales
Gelidiaceae
Gelidiella[2]
Gelidium crinale (Turner) Gaillon
Gelidium nudifrons Gardner
Gelidium pusillum (Stackhouse) Le Jolis
Pterocladia capillacea (S. G. Gmelin) Bornet
Gracilariales
Gracilariaceae
Gracilaria cervicornis (Turner) J. Agardh
syn: *Gracilaria ferox* J. Agardh
Gracilaria chilensis Bird, McLachlan, & Oliveira
syn: *Gracilaria sordida* W. A. Nelson
Gracilaria conferta (Montagne) Montagne
Gracilaria debilis (Forsskål) Børgesen
Gracilaria domingensis (Kützing) Sonder ex Dickie
Gracilaria foliifera (Forsskål) Børgesen
Gracilaria pacifica Abbott
Gracilaria secundata Harvey
Gracilaria tikvahiae McLachlan
Gracliaria verrucosa (Hudson) Papenfuss
Gracilariophila
Gracilariopsis lemaneiformis (Bory de Saint-Vincent) Dawson, Acleto, & Foldvik
syn: *Gracilaria lemaneiformis* (Bory de Saint-Vincent) Greville
Pterocladiophilaceae
Holmsella pachyderma (Reinsch) Sturch
Bonnemaisoniales
Bonnemaisoniaceae
Asparagopsis taxiformis (Delile) Trevisan
Bonnemaisonia hamifera Hariot
Cryptonemiales
Choreocolacaceae
Choreocolax polysiphoniae Reinsch
Harveyella mirabilis (Reinsch) Schmitz & Reinke
Cryptonemiaceae
Grateloupia dichotoma J. Agardh

Halymenia floresia (Clemente) C. Agardh
Prionitis lanceolata (Harvey) Harvey
Thamnoclonium dichotomum (J. Agardh) J. Agardh
Dumontiaceae
Acrosymphyton purpuriferum (J. Agardh) Sjöstedt
Dumontia incrassata (O. F. Müller) Lamouroux
syn: *Dumontia contorta* (S. G. Gmelin) Ruprecht
Neodilsea integra (Rosenvinge) Zinova
Weeksia fryeana Setchell
Endocladiaceae
Endocladia muricata (Endlicher) J. Agardh
Gloiopeltis
Gloiosiphoniaceae
Gloiosiphonia capillaris (Hudson) Berkeley
Kallymeniaceae
Callophyllis cristata (Linnaeus ex Turner) Kützing
syn: *Euthora cristata* (Linnaeus ex Turner) J. Agardh
Hommersandia maximicarpa Hansen & Lindstrom
Kallymenia
Peyssonneliaceae
Peyssonnelia
Weeksiaceae
Constantinea
Hildenbrandiales
Hildenbrandiaceae
Apophlaea sinclairii Hooker f. & Harvey
Hildenbrandia
Corallinales
Corallinaceae
Amphiroa anceps (Lamarck) Decaisne
Amphiroa foliacea Lamouroux
Bossiella orbigniana (Decaisne) P. C. Silva
Cheilosporum proliferum (Lamouroux) Hariot
Clathromorphum circumscriptum (Strömfelt) Foslie
Clathromorphum compactum (Kjellman) Foslie
Corallina elongata Ellis & Solander
Corallina officinalis Linnaeus
Corallina officinalis var. *chilensis* (Decaisne) Kützing
Corallina vancouveriensis Yendo
Leptophytum foecundum (Kjellman) Adey
Leptophytum laeve Adey
Lithophyllum congestum (Foslie) Foslie
Lithophyllum impressum Foslie
Lithophyllum moluccense (Foslie) Foslie
Lithothamnion australe Foslie
Lithothamnion calcareum (Pallas) Lemoine
Lithothamnion phymatodeum Foslie
Melobesia
Metagoniolithon radiatum (Lamarck) Ducker
Phymatolithon calcareum (Pallas) Adey & McKibbin
Porolithon pachydermum (Foslie) Foslie
Pseudolithophyllum muricatum (Foslie) Steneck & Paine

Pseudolithophyllum whidbeyense (Foslie)
    Steneck & Paine
Serraticardia maxima (Yendo) P. C. Silva
Gigartinales
   Calosiphoniaceae
    Calosiphonia vermicularis (J. Agardh) Schmitz
   Caulacanthaceae
    Catenella caespitosa (Withering) L. Irvine
     syn: *Catenella opuntia* (Goodenough &
      Woodward) Greville
   Cruoriaceae
    Cruoria arctica Schmitz
   Furcellariaceae
    Furcellaria lumbricalis (Hudson) Lamouroux
   Gigartinaceae
    Chondrus crispus Stackhouse
    Gigartina canaliculata Harvey[3]
    Gigartina cristata (Setchell) Setchell &
     Gardner[3]
    Gigartina exasperata Harvey & Bailey
    Iridaea boryana (Setchell & Gardner)
     Skottsberg
    Iridaea splendens (Setchell & Gardner)
     Papenfuss[3]
     syn: *Iridaea cordata* auct.
   Haemeschariaceae
    Haemescharia polygyna Kjellman
   Hypneaceae
    Hypnea musciformis (Wulfen) Lamouroux
    Hypnea spinella (C. Agardh) Kützing
   Nemastomataceae
    Platoma bairdii (Farlow) Kuckuck
   Petrocelidaceae
    Mastocarpus papillatus (C. Agardh) Kützing
     syn: *Gigartina papillata* (C. Agardh) J.
      Agardh[3]
    Mastocarpus stellatus (Stackhouse) Guiry
     syn: *Gigartina stellata* (Stackhouse)
      Batters
   Phyllophoraceae
    Ahnfeltiopsis furcellata (C. Agardh) P. C. Silva
     & DeCew
     syn: *Gymnogongrus furcellatus*
      (C. Agardh) J. Agardh
    Ahnfeltiopsis linearis (C. Agardh) P. C. Silva
     & DeCew
     syn: *Gymnogongrus linearis* (C. Agardh)
      J. Agardh
    Phyllophora sicula (Kützing) Guiry & L.
     Irvine
     syn: *Phyllophora palmettoides* J. Agardh
    Phyllophora traillii Holmes & Batters
     sporophyte: *Erythrodermis allenii* Batters
    Phyllophora truncata (Pallas) Zinova
     syn: *Phyllophora brodiei* (Turner)
      J. Agardh
    Schottera nicaeensis (Lamouroux ex Duby)
     Guiry & Hollenberg
   Plocamiaceae
    Plocamiocolax pulvinata Setchell
    Plocamium cartilagineum (Linnaeus) Dixon
   Polyidaceae
    Polyides

Solieriaceae
   Agardhiella subulata (C. Agardh) Kraft &
    Wynne
    syn: *Neoagardhiella baileyi* (Harvey ex
     Kützing) Wynne & W. R. Taylor
   Eucheuma denticulatum (N. Burman) Collins
    & Hervey
   Eucheuma gelatinum (Esper) J. Agardh
   Eucheuma isiforme (C. Agardh) J. Agardh
   Kappaphycus alvarezii (Doty) Doty
    syn: *Eucheuma alvarezii* Doty
   Kappaphycus striatum (Schmitz) Doty
    syn: *Eucheuma striatum* Schmitz
   Sarcodiotheca gaudichaudii (Montagne)
    Gabrielson
   Solieria tenera (J. Agardh) Wynne & W. R.
    Taylor
   Turnerella pennyi (Harvey) Schmitz
Rhodymeniales
   Champiaceae
    Champia parvula (C. Agardh) Harvey
    Gastroclonium subarticulatum (Turner) Kützing
     syn: *Gastroclonium coulteri* (Harvey) Kylin
   Rhodymeniaceae
    Ceratodictyon spongiosum Zanardini
    Maripelta rotata (Dawson) Dawson
    Rhodymenia californica Kylin
Ceramiales
   Ceramiaceae
    Aglaothamnion cordatum (Børgesen)
     Feldmann-Mazoyer
     syn: *Callithamnion cordatum* Børgesen
    Aglaothamnion neglectum Feldmann-Mazoyer
    Anotrichium tenue (C. Agardh) Nägeli
    Antithamnion defectum Kylin[4]
    Callithamnion tetricum (Dillwyn) S. F. Gray
    Centroceras clavulatum (C. Agardh) Montagne
    Ceramium capillaceum Meneghini[5]
     syn: *Ceramium strictum* (Kützing)
      Rabenhorst
    Ceramium rubrum (Hudson) C. Agardh
    Griffithsia monilis Harvey
    Griffithsia pacifica Kylin
    Microcladia californica Farlow
    Microcladia coulteri Harvey
    Platythamnion pectinatum Kylin
    Pleonosporium squarrulosum (Harvey) Abbott
    Plumaria elegans (Bonnemaison) Schmitz
    Pterothamnion plumula (Ellis) Nägeli
     syn: *Antithamnion plumula* (Ellis) Thuret
    Ptilota serrata Kützing
    Scagelia pylaisaei (Montagne) Wynne
     syn: *Antithamnion boreale* (Gobi)
      Kjellman
    Spermothamnion repens (Dillwyn) Rosenvinge
    Spyridia hypnoides (Bory de Saint-Vincent)
     Papenfuss
     syn: *Spyridia aculeata* (C. Agardh ex
      Decaisne) Kützing
    Tiffaniella snyderae (Farlow) Abbott
   Dasyaceae
    Dasya
   Delesseriaceae

Caloglossa
Cryptopleura
Delesseria sanguinea (Turner) Lamouroux
Phycodrys rubens Batters
Platysiphonia decumbens Wynne
Polyneura hilliae (Greville) Kylin[6]
Rhodomelaceae
  Acanthophora spicifera (Vahl) Børgesen
  Acrocystis groenlandica Rosenvinge
  Amansia
  Bostrychia scorpioides (Hudson) Montagne
  Bostrychia tenella (Lamouroux) J. Agardh
    syn: *Bostrychia binderi* Harvey
  Brongniartella byssoides (Goodenough &
    Woodward) Schmitz
  Herposiphonia
  Janczewskia gardneri Setchell & Guernsey
  Laurencia corallopsis (Montagne) Howe
  Laurencia pacifica Kylin
  Laurencia poitei (Lamouroux) Howe
  Laurencia spectabilis Postels & Ruprecht
  Polysiphonia arctica J. Agardh
  Polysiphonia elongata (Hudson) Sprengel
  Polysiphonia harveyi J. W. Bailey
  Polysiphonia lanosa (Linnaeus) Tandy
  Polysiphonia opaca (C. Agardh) Moris &
    DeNotaris
  Polysiphonia scopulorum Harvey
  Polysiphonia subtilissima Montagne
  Rhodomela larix (Turner) C. Agardh
  Stictosiphonia

Phaeophyceae
Ectocarpales
  Ectocarpaceae
    Ectocarpus fasciculatus Harvey
    Ectocarpus siliculosus (Dillwyn) Lyngbye
    Hincksia mitchelliae (Harvey) P. C. Silva
    Hincksia rallsiae (Vickers) P. C. Silva
    Pilayella littoralis (Linnaeus) Kjellman
    Streblonema
Ralfsiales
  Lithodermataceae
    Pseudolithoderma nigrum Hollenberg
    Sorapion simulans Kuckuck
  Ralfsiaceae
    Jonssonia pulvinata S. Lund
    Ralfsia expansa (J. Agardh) J. Agardh
    Ralfsia ovata Rosenvinge
    Ralfsia pacifica Hollenberg
    Stragularia spongiocarpa (Batters) Hamel
      syn: *Ralfsia spongiocarpa* Batters
Sphacelariales
  Sphacelariaceae
    Sphacelaria arctica Harvey
    Sphacelaria cirrosa (Roth) C. Agardh
    Sphacelaria plumosa Lyngbye
    Sphacelaria plumula Zanardini
    Sphacelaria radicans (Dillwyn) Harvey
    Sphacelaria rigidula Kützing
      syn: *Sphacelaria furcigera* Kützing
    Sphacelaria tribuloides Meneghini
Dictyotales

Dictyotaceae
  Dictyopteris polypodioides (De Candolle)
    Lamouroux
    syn: *Dictyopteris membranacea* (Stack-
    house) Batters
  Dictyota dichotoma (Hudson) Lamouroux
  Dictyota diemensis Sonder ex Kützing
  Dictyota divaricata (J. Agardh) J. Agardh
  Lobophora variegata (Lamouroux) Womersley
  Padina jamaicensis (Collins) Papenfuss
  Spatoglossum pacificum Yendo
  Stypopodium zonale (Lamouroux) Papenfuss
  Zonaria angustata (Sonder) Papenfuss
Cutleriales
  Cutleriaceae
    Cutleria multifida (J. E. Smith) Greville
Syringodermatales
  Syringodermataceae
    Syringoderma floridana E. C. Henry
Tilopteridales
  Tilopteridaceae
    Haplospora globosa Kjellman
    Phaeosiphoniella cryophila Hooper, E. C.
    Henry, & Kuhlenkamp
Chordariales
  Chordariaceae
    Analipus japonicus (Harvey) Wynne
    Cladosiphon zosterae (J. Agardh) Kylin
    Chordaria flagelliformis (O. F. Müller)
    C. Agardh
    Papenfussiella callitricha (Rosenvinge) Kylin
  Corynophlaeaceae
    Leathesia difformis (Linnaeus) J. E. Areschoug
  Elachistaceae
    Elachista stellaris J. E. Areschoug
  Splachnidiaceae
    Splachnidium
Scytosiphonales
  Chnoosporaceae
    Chnoospora
  Scytosiphonaceae
    Colpomenia peregrina Sauvageau
    Petalonia fascia (O. F. Müller) Kuntze
    Scytosiphon simplicissimus (Clemente)
    Cremades
      syn: *Scytosiphon lomentaria* (Lyngbye)
    Link
    Scytosiphon simplicissimus var. *complanatus*
    (Rosenvinge) Cremades
      syn: *Scytosiphon lomentaria* var. *com-*
    planatus Rosenvinge
Dictyosiphonales
  Coelocladiaceae
    Coelocladia arctica Rosenvinge
  Dictyosiphonaceae
    Dictyosiphon foeniculaceus (Hudson) Greville
  Punctariaceae
    Adenocystis utricularis (Bory de Saint-Vincent)
    Skottsberg
    Desmotrichum undulatum (J. Agardh) Reinke
    Phaeostrophion irregulare Setchell & Gardner
    Platysiphon verticillatus Wilce
    Punctaria glacialis Rosenvinge

Striariaceae
Stictyosiphon tortilis (Ruprecht) Reinke
Desmarestiales
Desmarestiaceae
Desmarestia aculeata (Linnaeus) Lamouroux
Desmarestia firma (C. Agardh) Skottsberg
Desmarestia viridis (O. F. Müller) Lamouroux
Himantothallus grandifolius (A. Gepp &
E. Gepp) Zinova
Laminariales
Alariaceae
Alaria crassifolia Kjellman
Alaria esculenta (Linnaeus) Greville
Ecklonia maxima (Osbeck) Papenfuss
Ecklonia radiata (C. Agardh) J. Agardh
Egregia laevigata Setchell
Egregia menziesii (Turner) J. E. Areschoug
Eisenia arborea J. E. Areschoug
Pterygophora californica Ruprecht
Undaria peterseniana (Kjellman) Okamura
Undaria pinnatifida (Harvey) Suringar
Undaria undarioides (Yendo) Okamura
Chordaceae
Chorda tomentosa Lyngbye
Laminariaceae
Agarum clathratum Dumortier
syn: *Agarum cribrosum* Bory de
Saint-Vincent
Agarum fimbriatum Harvey
Costaria costata (C. Agardh) Saunders
Hedophyllum sessile (C. Agardh) Setchell
Laminaria angustata Kjellman
Laminaria digitata (Hudson) Lamouroux
Laminaria farlowii Setchell
Laminaria groenlandica Rosenvinge
Laminaria hyperborea (Gunnerus) Foslie
Laminaria japonica J. E. Areschoug
Laminaria longicruris De la Pylaie
Laminaria pallida Greville
Laminaria saccharina (Linnaeus) Lamouroux
Laminaria setchellii P. C. Silva
Laminaria sinclairii (Hooker f. & Harvey) Far-
low, Anderson, & Eaton
Laminaria solidungula J. Agardh
Pleurophycus gardneri Setchell & Saunders
Saccorhiza dermatodea (De la Pylaie) J. E.
Areschoug
Saccorhiza polyschides (Lightfoot) Batters
Lessoniaceae
Lessonia nigrescens Bory de Saint-Vincent
Lessoniopsis
Macrocystis angustifolia Bory de Saint-Vincent
Macrocystis integrifolia Bory de Saint-Vincent
Macrocystis pyrifera (Linnaeus) C. Agardh
Nereocystis luetkeana (Mertens) Postels &
Ruprecht
Pelagophycus porra (Leman) Setchell
Postelsia palmaeformis Ruprecht
Fucales
Cystoseiraceae
Acrocarpia paniculata (Turner) J. E. Areschoug
Cystophora
Cystoseira baccata (S. G. Gmelin) P. C. Silva

Cystoseira osmundacea (Turner) C. Agardh
Cystoseira stricta (Montagne) Sauvageau
Halidrys dioica Gardner
Halidrys siliquosa (Linnaeus) Lyngbye
Fucaceae
Ascophyllum nodosum (Linnaeus) Le Jolis
Fucus edentatus De la Pylaie
Fucus evanescens C. Agardh
Fucus gardneri P. C.Silva
syn: *Fucus distichus* subsp.
edentatus (De la Pylaie) Powell (Pacific
coast populations only)
Fucus serratus Linnaeus
Fucus spiralis Linnaeus
Fucus vesiculosus Linnaeus
Hesperophycus
Pelvetia canaliculata (Linnaeus) Decaisne &
Thuret
Pelvetia compressa (J. Agardh) De Toni
syn: *Pelvetia fastigiata* (J. Agardh) De Toni
Pelvetia compressa f. *gracilis* (Setchell &
Gardner) P. C. Silva
syn: *Pelvetia fastigiata* f. *gracilis* Setchell
& Gardner
Pelvetiopsis limitata (Setchell) Gardner
Himanthaliaceae
Himanthalia elongata (Linnaeus) S. F. Gray
Hormosiraceae
Hormosira banksii (Turner) Decaisne
Sargassaceae
Carpophyllum maschalocarpum (Turner)
Greville
Hizikia fusiformis (Harvey) Okamura
Sargassum agardhianum Farlow
Sargassum filipendula C. Agardh
Sargassum fluitans Børgesen (Børgesen)
Sargassum muticum (Yendo) Fensholt
Sargassum natans (Linnaeus) Gaillon
Turbinaria
Durvillaeales
Durvillaeaceae
Durvillaea potatorum (Labillardière)
J. E. Areschoug
Genus incertae sedis
Pilinia

Bacillariophyceae
Naviculales
Bacillariaceae
Cylindrotheca
Fragilariaceae
Licmophora
Naviculaceae
Berkeleya rutilans (Roth) Grunow
Thalassiosirales
Thalassiosiraceae
Skeletonema costatum (Greville) Cleve
Thalassiosira fluviatilis Hustedt

Chrysophyceae
Ochromonadales
Dinobryaceae
Poterioochromonas malhamensis (E. G. Prings-
heim) L. S. Péterfi

Sarcinochrysidales
Chrysomeridaceae
Chrysonephos lewisii (W. R. Taylor) W. R.
Taylor

Prymnesiophyceae
Isochrysidales
Hymenomonadaceae
Pleurochrysis
syn: *Cricosphaera*
Prinsiaceae
Emiliana
Pavlovales
Pavlovaceae
Pavlova lutheri (Droop) J. C. Green
syn: *Monochrysis lutheri* Droop
Prymnesiales
Phaeocystaceae
Phaeocystis pouchetii (Hariot) Lagerheim

Raphidophyceae
Raphidomonadales
Vacuolariaceae
Olisthodiscus luteus N. Carter

Euglenophyceae
Euglenales
Euglenaceae
Euglena

Chlorophyceae
Chlorococcales
Scenedesmaceae
Scenedesmus fuscus (Shihara & Krauss)
Hegewald
syn: *Chlorella fusca* Shihara & Krauss
Chlorosarcinales
Collinsiellaceae
Collinsiella
Tetrasporales
Palmellopsidaceae
Palmoclathrus
Volvocales
Chlamydomonadaceae
Chlamydomonas
Dunaliellaceae
Dunaliella tertiolecta Butcher
Chaetophorales
Chaetophoraceae
Uronema curvatum Printz
Ctenocladales
Ulvellaceae
Acrochaete operculata Correa & R. Nielsen
Entocladia
Ulotrichales
Ulotrichaceae
Ulothrix flacca (Dillwyn) Thuret[7]
Ulothrix pseudoflacca Wille
Ulvales
Monostromataceae
Blidingia minima var. *vexata* (Setchell & Gardner) J. Norris
Monostroma grevillei (Thuret) Wittrock
Monostroma latissimum Wittrock

Monostroma nitidum Wittrock
Protomonostroma undulatum (Wittrock)
Vinogradova
syn: *Monostroma undulatum* Wittrock
Ulvaceae
Enteromorpha compressa (Linnaeus) Nees
Enteromorpha intestinalis (Linnaeus) Nees
Enteromorpha linza (Linnaeus) J. Agardh
Enteromorpha procera Ahlner[8]
syn: *Enteromorpha ahlneriana* Bliding
Enteromorpha prolifera (O. F. Müller)
J. Agardh[8]
Enteromorpha torta (Mertens) Reinbold[8]
Gayralia oxysperma (Kützing) Vinogradova
syn: *Monostroma oxyspermum* (Kützing)
Doty
Percursaria percursa (C. Agardh) Rosenvinge
Ulva californica Wille
Ulva curvata (Kützing) De Toni
Ulva fasciata Delile
Ulva fenestrata Postels & Ruprecht
Ulva japonica (Holmes) Papenfuss
Ulva lactuca Linnaeus
Ulva mutabilis Föyn
Ulva olivascens Dangeard
Ulva pertusa Kjellman
Ulva rigida C. Agardh
Prasiolales
Prasiolaceae
Prasiola borealis Reed
Prasiola stipitata Suhr ex Jessen
Acrosiphoniales
Acrosiphoniaceae
Acrosiphonia arcta (Dillwyn) Gain
syn: *Spongomorpha arcta* (Dillwyn)
Kützing
Acrosiphonia sonderi (Kützing) Kornmann
Cladophorales
Anadyomenaceae
Anadyomene stellata (Wulfen) C. Agardh
Cladophoraceae
Chaetomorpha aerea (Dillwyn) Kützing
Chaetomorpha coliformis (Montagne) Kützing
syn: *Chaetomorpha darwinii* (Hooker f. &
Harvey) Kützing
Chaetomorpha linum (O. F. Müller) Kützing
Chaetomorpha melagonium (Weber & Mohr)
Kützing
Cladophora albida (Nees) Kützing
Cladophora glomerata (Linnaeus) Kützing
Cladophora montagneana Kützing
Cladophora prolifera (Roth) Kützing
Cladophora rupestris (Linnaeus) Kützing
Cladophora stimpsonii Harvey
Rhizoclonium riparium (Roth) Harvey
Urospora penicilliformis (Roth) J. E.
Areschoug
Siphonocladales
Siphonocladaceae
Boergesenia forbesii (Harvey) J. Feldmann
Boodlea coacta (Dickie) G. Murray & De Toni
Siphonocladus
Ventricaria

Valoniaceae
  Dictyosphaeria cavernosa (Forsskål)
    Børgesen
  Ernodesmis
  Valonia
  Valoniopsis pachynema (Martens) Børgesen
Bryopsidales
  Bryopsidaceae
    Bryopsis hypnoides Lamouroux
    Bryopsis maxima Okamura
    Bryopsis plumosa (Hudson) C. Agardh
    Bryopsidella
    Derbesia marina (Lyngbye) Solier
    Derbesia tenuissima (Moris & De Notaris)
      P. Crouan & H. Crouan
    Halicystis
    Pseudobryopsis oahuensis Egerod
  Caulerpaceae
    Caulerpa cactoides (Turner) C. Agardh
    Caulerpa cupressoides (Vahl) C. Agardh
    Caulerpa paspaloides (Bory de Saint-Vincent)
      Greville
    Caulerpa prolifera (Forsskål) Lamouroux
    Caulerpa racemosa (Forsskål) J. Agardh
    Caulerpa simpliciuscula (R. Brown ex Turner)
      C. Agardh
    Caulerpa taxifolia (Vahl) C. Agardh
  Codiaceae
    Codium decorticatum (Woodward) Howe
    Codium dimorphum Svedelius
    Codium fragile (Suringar) Hariot
    Codium fragile subsp. *tomentosoides* (van
      Goor) P. C. Silva
    Codium intertextum Collins & Hervey

Codium isthmocladum Vickers
Codium magnum Dawson
Halimedaceae
  Halimeda copiosa Goreau & Graham
  Halimeda incrassata (Ellis) Lamouroux
  Halimeda lacrimosa Howe
  Halimeda macroloba Decaisne
  Halimeda monile (Ellis & Solander)
    Lamouroux
  Halimeda opuntia (Linnaeus) Lamouroux
  Halimeda simulans Howe
  Halimeda tuna (Ellis & Solander) Lamouroux
Ostreobiaceae
  Ostreobium
Udoteaceae
  Avrainvillea gardineri A. Gepp & E. Gepp
  Chlorodesmis
  Penicillus capitatus Lamarck
  Penicillus dumetosus (Lamouroux) Blainville
  Penicillus pyriformis A. Gepp & E. Gepp
  Rhipiliopsis
    syn: *Johnson-sea-linkia*
  Udotea conglutinata (Solander & Ellis)
    Lamouroux
  Udotea flabellum (Ellis & Solander) Howe
  Udotea indica A. Gepp & E. Gepp
Dasycladales
  Dasycladaeae
    Acetabularia acetabulum (Linnaeus) P. C. Silva
    Acetabularia calyculus Lamouroux
    Acetabularia crenulata Lamouroux
    Batophora
    Cymopolia barbata (Linnaeus) Lamouroux
    Neomeris

Note added in proof: Recent nomenclature changes

1. *Arthrospira platensis* (Nordstedt) Gomont
   syn: *Spirulina platensis* (Nordstedt) Geitler

2. *Gelidiella* has been transferred to a new family,
   Gelidiellaceae, within Gelidiales.

3. *Iridaea cordata* auct. is now considered synony-
   mous with *Mazzaella lilacina*, not with *I. splendens*. *Gigartina cristata* is included in synonymy
   with *Mastcarpus papillatus* (Petrocelidaceae). The
   list for Gigartinaceae should now read:
   Gigartinaceae
     Chondracanthus canaliculatus (Harvey) Guiry
       syn: *Gigartina canaliculata* Harvey
     Chondrus crispus Stackhouse
     Gigartina exasperata Harvey & Bailey
     Iridaea boryana (Setchell & Gardner)
       Skottsberg
     Mazzaella lilacina (Postels & Ruprecht)
       Leister
         syn: *Iridaea cordata* auct.

4. *Antithamnion densum* (Suhr) Howe
   syn: *Antithamnion defectum*
     Kylin

5. Now *Ceramium deslongchampii* Chauvin
   ex Duby.

6. *Polyneura bonnemaisonii* (C. Agardh) Maggs &
   Hommersand
     syn: *Polyneura hilliae* (Greville) Kylin

7. *Ulothrix flacca* (Dillwyn) Thuret
   syn: *Ulothrix pseudoflacca* Wille

8. *Enteromorpha ahlneriana* and *E. torta* now
   synonyms of *E. prolifera*:
   Enteromorpha prolifera (O. F. Müller)
     J. Agardh
       syn: *Enteromorpha ahlneriana* Bliding
       syn: *Entermorpha torta* (Mertens)
         Reinbold

References

Section numbers for text citation of references are provided at the end of each reference.

Abbott, I. A. (1988). Food and food products from seaweeds. In C. A. Lembi & J. R. Waaland (eds.), *Algae and Human Affairs* (pp. 135–48). Cambridge University Press. [9.2, 9.2.4]

Abbott, I. A. & G. J. Hollenberg (1976). *Marine Algae of California*. Stanford University Press. [6.4]

Adamich, M., A. Gibor, & B. M. Sweeney (1975). Effects of low nitrogen levels and various nitrogen sources on growth and whorl development in *Acetabularia* (Chlorophyta). *J. Phycol.* 11:364–7. [5.8.2]

Adams, T., & R. Foscarini (eds.) (1990). *Proceedings of the Regional Workshop on Seaweed Culture and Marketing* (Suva, Fiji, 14–17 November 1989). South Pacific Aquacult. Devel. Project, FAO, UN, 86 pp. [9.5.3]

Adey, W. H. (1970). The effects of light and temperature on growth rates in boreal-subartic crustose corallines. *J. Phycol.* 6:269–76. [6.2.2]

Adey, W. H. (1973). Temperature control of reproduction and productivity in a subarctic coralline alga. *Phycologia* 12: 111–18. [3.2.2]

Adey, W. H. (1991). *Dynamic Aquaria*. New York: Academic Press. [Fig. 1.15b]

Adey, W. H., & T. Goertemiller (1987). Coral reef algal turfs: master producers in nutrient poor seas. *Phycologia* 26:374–86. [3.2.1, 3.2.2, 5.8.3, 7.2.2]

Adey, W. H., & J. M. Vassar (1975). Colonization, succession and growth rates of tropical crustose coralline algae (Rhodophyta, Cryptonemiales). *Phycologia* 14:55–69. [3.2.2]

Alderdice, D. F. (1972). Factor combinations. Responses of marine poikilotherms to environmental factors acting in concert. In O. Kinne (ed.), *Marine Ecology* (vol. 1, pt. 3, pp. 1659–722). New York: Wiley. [1.1.3]

Algarra, P., & F. X. Neill (1987). Structural adaptations to light reception in two morphotypes of *Corallina elongata* Ellis & Soland. *P.S.Z.N.I. Mar. Ecol.* 8:253–61. [4.3.2]

Algarra, P., J. C. Thomas, & A. Mousseau (1990). Phycobilisome heterogeneity in the red alga *Porphyra umbilicalis*. *Plant Physiol.* 92:570–6. [4.3.1]

Allen, T. F. H. (1977). Scale in microscopic algal ecology: a neglected dimension. *Phycologia* 16:253–7. [1.1.3]

Al-ogily, S. M., & E. W. Knight-Jones (1977). Anti-fouling role of antibiotics produced by marine algae and bryozoans. *Nature* 265:728–9. [3.1.2]

Amat, M. A., & J.-P. Braud (1990). Ammonium uptake by *Chondrus crispus* Stackhouse (Gigartinales; Rhodophyta) in culture. *Hydrobiologia* 204/205:467–71. [5.5.1]

Amat, M. A., & L. M. Srivastava (1985). Translocation of iodine in *Laminaria saccharina* (Phaeophyta). *J. Phycol.* 21:330–3. [5.5.7, 5.6]

Amsler, C. D., & M. Neushul (1989a). Diel periodicity of spore release from the kelp *Nereocystis luetkeana* (Mertens) Postels et Ruprecht. *J. Exp. Mar. Biol. Ecol.* 134:117–27. [1.6.1]

Amsler, C. D. & M. Neushul. (1989b). Chemotactic effects of nutrients on spores of the kelps *Macrocystis pyrifera* and *Pterygophora californica*. *Mar. Biol.* 102:557–64. [1.6.1]

Amsler, C. D., & M. Neushul (1990). Nutrient stimulation of spore settlement in the kelps *Pterygophora californica* and *Macrocystis pyrifera*. *Mar. Biol.* 107:297–304. [1.6.1]

Amsler, C. D., & R. B. Searles (1980). Vertical distribution of seaweed spores in a water column offshore of North Carolina. *J. Phycol.* 16:617–19. [1.1.1, 1.6.1]

Anderson, B. S., & J. W. Hunt (1988). Bioassay methods of evaluating the toxicity of heavy metals, biocides and sewage effluent using microscopic stages of giant kelp *Macrocystis pyrifera* (Agardh): a preliminary report. *Mar. Environ. Res.* 26:113–34. [8.3.3]

Anderson, E. K., & W. J. North (1966). *In situ* studies of spore production and dispersal in the giant kelp, *Macrocystis pyrifera*. In E. G. Young & J. L. McLachlan (eds.), *Proc. Intl. Seaweed Symp.* 5:73–86. [2.5.2]

Anderson, F. E., & J. Green (1980). Estuaries. *The New Encyclopaedia Britannica, Macropaedia* (vol. 6, pp. 968–76). [6.1.2]

Anderson, L. M. (1984). Iron reduction by juvenile *Macrocystis pyrifera* (L.) C. Agardh. *Hydrobiologia* 116/117:493–7. [5.5.6]

Anderson, M. A., & F. M. M. Morel (1980). Uptake of Fe(II) by a diatom in oxic culture medium. *Mar. Biol. Lett.* 1:263–8. [5.5.6]

Anderson, M. A., & F. M. M. Morel (1982). The influence of aqueous iron chemistry on the uptake of iron by the coastal diatom *Thalassiosira weissflogii*. *Limnol. Oceanogr.* 27:789–813. [5.5.6]

Anderson, R. J., & J. J. Bolton (1989). Growth and fertility, in relation to temperature and photoperiod, in South African *Desmarestia firma* (Phaeophyceae). *Bot. Mar.* 32:149–58. [1.5.3]

Anderson, S. M., & A. C. Charters (1982). A fluid dynamics study of seawater flow through *Gelidium nudifrons*. *Limnol. Oceanogr.* 27:399–412. [7.1.2]

Andrews, J. H. (1976). The pathology of marine algae. *Biol. Rev.* 51:211–53. [3.3.2, 6.3.3, 9.2.3]

Antia, N. J., P. J. Harrison, & L. Oliveira (1991). The role of dissolved organic nitrogen in phytoplankton nutrition, cell biology and ecology. *Phycologia* 30:1–89. [5.5.1]

Archambault, D., & E. Bourget (1983). Importance du régime de dénudation sur la structure et la succession des communautés intertidales de substrat rocheux en milieu subarctique. *Can. J. Fish. Aquat. Sci.* 40:1278–92. [2.2.2]

Armstrong, S. L. (1987). Mechanical properties of the tissues of the brown alga *Hedophyllum sessile* (C. Ag.) Setchell: variability with habitat. *J. Exp. Mar. Biol. Ecol.* 114:143–51. [7.2.2]

Arndt, E. A. (1989). Ecological, physiological and historical aspects of brackish water fauna distribution. In J. S. Ryland & P. A. Tyler (eds.), *Reproduction, Genetics and Distribution of Marine Organisms* (pp. 327–38). 23rd European Marine Biology Symposium. Fredensborg: Olsen & Olsen. [2.4.1]

Arnold, K. E., & S. L. Manley (1985). Carbon allocation in *Macrocystis pyrifera* (Phaeophyta): intrinsic variability in photosynthesis and respiration. *J. Phycol.* 21:154–67. [1.7.2, 4.7.1]

Arnold, K. E., & S. N. Murray (1980). Relationships between irradiance and photosynthesis for marine benthic green algae (Chlorophyta) of differing morphologies. *J. Exp. Mar. Biol. Ecol.* 43:183–92. [4.3.2]

Arnon, D. I., & P. R. Stout (1939). The essentiality of certain elements in minute quantity for plants, with special reference to copper. *Plant Physiol.* 14:371–5. [5.1.1]

Asare, S. O., & M. M. Harlin (1983). Seasonal fluctuations in tissue nitrogen for five species of perennial macroalgae in Rhode Island Sound. *J. Phycol.* 19:254–7. [5.5.1]

Atkins, G. L., & I. A. Nimmo (1980). Current trends in the estimation of Michaelis-Menten parameters. *Anal. Biochem.* 104:1–9. [5.4]

Atkinson, M. J., & R. W. Grigg (1984). Model of a coral reef ecosystem. II. Gross and net benthic primary production at French Frigate Shoals, Hawaii. *Coral Reefs* 3:13–22. [4.7.1, 4.7.6]

Atkinson, M. J., & S. V. Smith (1983). C:N:P ratios of benthic marine plants. *Limnol. Ocenaogr.* 28:568–74. [5.5.2, 5.8.1]

Augier, H. (1978). Les hormones des algues. Etat actuel des connaissances. VII. Applications, conclusion, bibliographie. *Bot. Mar.* 21:175–97. [1.7.3]

Austin, A. P., & J. D. Pringle (1969). Periodicity of mitosis in red algae. *Proc. Intl. Seaweed Symp.* 6:41–52. [1.3.5]

Azanza-Corrales, R., & C. J. Dawes (1989). Wound healing in cultured *Eucheuma alvarezii* var. *tambalang* Doty. *Bot. Mar.* 32:229–34. [1.7.4]

Azov, Y. (1986). Seasonal patterns of phytoplankton productivity and abundance in nearshore oligotrophic waters of the Levant Basin (Mediterranean). *J. Plankton Res.* 8:41–53. [5.5.2]

Baardseth, E. (1970). *A Square-Scanning, Two-Stage Sampling Method of Estimating Seaweed Quantities.* Norwegian Institute of Seaweed Research, Report 33, 41 pp. [7.2.1]

Bacon, L. C., & R. L. Vadas (1991). A model for gamete release in *Ascophyllum nodosum* (Phaeophyta). *J. Phycol.* 27:166–73. [6.2.3]

Baker, J. M. (1970). The effects of oils on plants. *Environ. Pollut.* 1:27–44. [8.4.2]

Bakus, G. J. (1981). Chemical defense mechanisms on the Great Barrier Reef, Australia. *Science* 211:497–9. [3.3.1]

Bakus, G. J., N. M. Targett, & B. Schule (1986). Chemical ecology of marine organisms: an overview. *J. Chem. Ecol.* 12:951–87. [3.2.3]

Banner, A. H. (1974). Kaneohe Bay, Hawaii: urban pollution and a coral reef ecosystem. *Proc. 2nd. Intl. Symp. Coral Reefs, Brisbane, Australia* 2:685–702. [2.1.2]

Bannister, P. (1976). *Introduction to Physiological Plant Ecology.* Oxford: Blackwell Scientific, 273 pp. [Fig. 3.4]

Bannwarth, H. (1988). Nucleo-cytoplasmic interaction in cultured *Acetabularia*. In C. S. Lobban, D. J. Chapman, & B. P. Kremer (eds.), *Experimental Phycology: A Laboratory Manual* (pp. 260–71). Cambridge University Press. [1.4.2]

Barber, R. T., & J. H. Ryther (1969). Organic chelators: factors affecting primary production in the Cromwell Current upwelling. *J. Exp. Mar. Biol. Ecol.* 3:191–9. [5.2]

Barber, R. T., R. C. Dugdale, J. J. MacIsaac, & R. L. Smith (1971). Variations in phytoplankton growth associated with the source and conditioning of upwelling water. *Investig. Pesq.* 35:171–99. [5.2]

Barnes, D. J., & B. E. Chalker (1990). Calcification and photosynthesis in reef-building corals and algae. In Z. Dubinsky (ed.), *Ecosystems of the World* (vol. 25, pp. 109–31). Amsterdam: Elsevier. [5.5.3]

Bauernfeind, J. C., & E. DeRitter (1970). *Handbook of Biochemistry*, 2nd. ed. Boca Raton, FL: CRC Press. [Fig. 5.1]

Barnett, B. E., & C. R. Ashcroft (1985). Heavy metals in *Fucus vesiculosus* in the Humber Estuary. *Environ. Pollut.* B-9:193–201. [8.3.1]

Bauman, R. W., Jr., & B. R. Jones (1986). Electrophysiological investigations of the red alga *Griffithsia pacifica* Kyl. *J. Phycol.* 22:49–56. [1.3.5]

Bayne, B. L. (1989). The biological effects of marine pollutants. In J. Albaigés (ed.), *Marine Pollution* (pp. 131–51). New York: Hemisphere. [8.1]

Beaumont, A. R., & P. B. Newman (1986). Low levels of tributyl tin reduce growth of marine micro-algae. *Mar. Pollut. Bull.* 17:457–61. [8.5.4]

Becker, W. M., & D. W. Deamer (1991). *The World of the Cell*, 2nd ed. New York: Benjamin/Cummings, 886 pp. [Fig. 5.3b]

Beer, S. & A. Eshel (1983). Photosynthesis of *Ulva* sp. I. Effects of desiccation when exposed to air. *J. Exp. Mar. Biol. Ecol.* 70:91–7. [4.4.1, 6.4]

Beer, S., & B. Shragge (1987). Photosynthetic carbon metabolism in *Enteromorpha compressa* (Chlorophyta). *J. Phycol.* 23:580–4. [4.4.1]

Beevers, L. (1976). *Nitrogen Metabolism in Plants.* London: Edward Arnold, 339 pp. [5.5.1]

Beijer, K., & A. Jerenlöv (1979). Sources, transport and transformation of metals in the environment. In L. Friberg,

G. F. Nordberg, & V. B. Vouk (eds.), *Handbook on the Toxicology of Metals* (pp. 47–63). Amsterdam: Elsevier/North Holland. [8.3]

Bekheet, I. A., K. M. Kandil, & N. Z. Shaban (1984). Studies on urease extracted from *Ulva lactuca*. *Hydrobiologia* 116/117:580–3. [5.5.1]

Belliveau, D. J., D. J. Garbary, & J. L. McLachlan (1990). Effects of fluorescent brighteners on growth and morphology of the red alga *Antithamnion kylinii*. *Stain Technol.* 65:303–11. [1.3.4]

Bender, M. E., G. P. Klinkhammer, & D. W. Spencer (1977). Manganese in seawater and the marine manganese balance. *Deep Sea Res.* 24:799–812. [5.2]

Benson, E. E., J. C. Rutter, & A. H. Cobb (1983). Seasonal variation in frond morphology and chloroplast physiology of the intertidal alga *Codium fragile* (Suringar) Hariot. *New Phytol.* 95:569–80. [1.7.2]

Bernstein, B. B., & N. Jung (1979). Selective pressures and coevolution in a kelp canopy community in southern California. *Ecol. Monogr.* 49:335–55. [3.1.2, 6.1.1]

Beutlich, A., B. Borstelmann, R. Reddemann, K. Speckenbach, & R. Schnetter (1990). Notes on the life histories of *Boergesenia* and *Valonia* (Siphonocladales, Chlorophyta). *Hydrobiologia* 204/205:425–34. [1.5.2]

Bhattacharya, D., & L. D. Druehl (1988). Phylogenetic comparison of the small-subunit ribosomal DNA sequence of *Costaria costata* (Phaeophyta) with those of other algae, vascular plants and oomycetes. *J. Phycol.* 24:539–43. [1.4.1]

Bhattacharya, D., D. L. Baillie, & L. D. Druehl (1990). Population analysis of the kelp *Costaria costata* (Phaeophyta) using a polymorphic ribosomal DNA probe. *Plant Syst. Evol.* 170:177–91. [1.4.1]

Bidwell, R. G. S. (1979). *Plant Physiology*, 2nd. ed. New York: Macmillan. [5.5.1–6, 5.6, 6.2.4, 6.3.1, 6.4.1]

Bidwell, R. G. S., & J. McLachlan (1985). Carbon nutrition of seaweeds: photosynthesis, photorespiration and respiration. *J. Exp. Mar. Biol. Ecol.* 86:15–46. [4.4.1, 4.4.3, 4.4.4, 4.7.1]

Bidwell, R. G. S., N. D. H. Lloyd, & J. McLachlan (1984). The performance of *Chondrus crispus* (Irish moss) in laboratory simulation of environments in different locations. *Proc. Intl. Seaweed Symp.* 11:292–4. [9.7]

Bidwell, R. G. S., J. McLachlan, & N. D. H. Lloyd (1985). Tank cultivation of Irish moss, *Chondrus crispus* Stackh. *Bot. Mar.* 28:87–97. [9.6, 9.7]

Biebl, R. (1962). Seaweeds. In R. A. Lewin (ed.), *Physiology and Biochemistry of the Algae* (pp. 799–815). New York: Academic Press. [6.4.1, 6.4.4]

Biebl, R. (1970). Vergleichende Untersuchungen zur Temperaturresistenz von Meeresalgen entlang der pazifischen Küste Nordamerikas. *Protoplasma* 69:61–83. [6.1.3, 6.2.4]

Biedka, R. F., J. M. Gosline, & R. E. De Wreede (1987). Biomechanical analysis of wave-induced mortality in the marine alga *Pterygophora californica*. *Mar. Ecol. Prog. Ser.* 36:163–70. [7.2.2]

Biggins, J., & D. Bruce (1989). Regulation of excitation energy transfer in organisms containing phycobilins. *Photosynth. Res.* 20:1–34. [4.3.1]

Bingham, S., & J. A. Schiff (1979). Conditions for attachment and development of single cells released from mechanically-disrupted thalli of *Prasiola stipitata* Suhr. *Biol. Bull.* 156:257–71. [1.6.1, 1.7.1]

Bird, K. T. (1976). Simultaneous assimilation of ammonium and nitrate by *Gelidium nudifrons* (Gelidiales: Rhodophyta). *J. Phycol.* 12:238–41. [5.4.2]

Bird, K. T., C. J. Dawes, & J. T. Romeo (1980). Patterns of non-photosynthetic carbon fixation in dark held, respiring thalli of *Gracilaria verrucosa*. *Z. Pflanzenphysiol.* 98:359–64. [5.5.1]

Bird, K. T., C. Habig, & T. DeBusk (1982). Nitrogen allocation and storage patterns in *Gracilaria tikvahiae* (Rhodophyta). *J. Phycol.* 18:344–8. [5.5.1]

Bisalputra, T. (1974). Plastids. In W. D. P. Stewart (ed.), *Algal Physiology and Biochemistry* (pp. 124–60). Oxford: Blackwell Scientific. [1.3]

Bishop, P. L. (1983). *Marine Pollution and Its Control.* New York: McGraw-Hill, 357 pp. [8.4.1, 8.6.1]

Bisson, M. A., & G. O. Kirst (1979). Osmotic adaptation in the marine alga *Griffithsia monilis* (Rhodophyceae): the role of ions and organic compounds. *Austr. J. Plant Physiol.* 6:523–38. [6.3.2]

Björnsäter, B. R., & P. A. Wheeler (1990). Effect of nitrogen and phosphorus supply on growth and tissue composition of *Ulva fenestrata* and *Enteromorpha intestinalis* (Ulvales, Chlorophyta). *J. Phycol.* 26:603–11. [5.7.3, 5.8.1]

Black, W. A. P. (1949). Seasonal variation in chemical composition of some of the littoral seaweeds common to Scotland. Part II. *Fucus serratus, Fucus vesiculosus, Fucus spiralis* and *Pelvetia canaliculata*. *J. Soc. Chem. Ind.* 68:183–9. [4.5.1]

Black, W. A. P. (1950). The seasonal variation in weight and chemical composition of the common British Laminariaceae. *J. Mar. Biol. Assoc. U.K.* 29:45–72. [4.5.1]

Blackburn, S. I., & P. A. Tyler (1987). On the nature of eclectic species – a tiered approach to genetic compatability in the desmid *Micrasterias thomasiana*. *Br. Phycol. J.* 22:277–98. [1.1.3]

Bliding, C. (1963). A critical survey of European taxa in Ulvales. I. *Capsosiphon, Percursaria, Blidingia, Enteromorpha. Opera Bot.* 8(3):1–60. [2.4.1]

Blunden, G., & S. M. Gordon (1986). Betaines and their sulphonio analogues in marine algae. *Prog. Phycol. Res.* 4:39–80. [6.3.2]

Bogorad, L. (1975). Phycobiliproteins and complementary chromatic adaptation. *Annu. Rev. Plant Physiol.* 26:369–401. [4.3.4]

Bokn, T. (1987). Effects of diesel oil and subsequent recovery of commercial benthic algae. *Hydrobiologia* 151/152:277–84. [8.4.2]

Bold, H. C., & M. J. Wynne (1985). *Introduction to the Algae: Structure and Reproduction*, 2nd ed. Englewood Cliffs, NJ: Prentice-Hall, 706 pp. [1.3.5, 1.5.2, 1.5.4, 1.7.2, 3.3.1, 3.3.2]

Bold, H. C., C. J. Alexopoulos, & T. Delevoryas (1980). *Morphology of Plants and Fungi*, 4th ed. New York: Harper & Row, 819 pp. [1.7.1]

Bolton, J. J. (1979). Estuarine adaptation in populations of *Pilayella littoralis* (L.) Kjellm. (Phaeophyta, Ectocarpales). *Estu. Cstl. Mar. Sci.* 9:273–80. [6.3.4]

Bolton, J. J. (1981). Community analysis of vertical zonation patterns on a Newfoundland rocky shore. *Aquat. Bot.* 10:299–316. [2.5.1]

Bolton, J. J. (1983). Ecoclinal variation in *Ectocarpus siliculosus* (Phaeophyceae) with respect to temperature growth optima and survival limits. *Mar. Biol.* 73:131–8. [6.2.3]

Bolton, J. J., & R. J. Anderson (1990). Correlation between intertidal seaweed community composition and sea water temperature patterns on a geographical scale. *Bot. Mar.* 33:447–57. [6.2.5]

Bolwell, G. P., J. A. Callow, M. E. Callow, & L. V. Evans (1979). Fertilization in brown algae. II. Evidence for lectin-sensitive complementary receptors involved in gamete recognition in *Fucus serratus. J. Cell Sci.* 36:19–30. [1.5.4]

Bolwell, G. P., J. A. Callow, & L. V. Evans (1980). Fertilization in brown algae. III. Preliminary characterization of putative gamete receptors from eggs and sperm of *Fucus serratus. J. Cell Sci.* 43:209–24. [1.5.4]

Boney, A. D. (1963). The effects of 3-amino-1,2,4-triazole on the growth of sporelings of marine red algae. *J. Mar. Biol. Assoc. U.K.* 43:643–52. [8.5.1]

Boney, A. D. (1970). Toxicity studies with an oil-spill emulsifier and the green alga *Prasinocladus marinus. J. Mar. Biol. Assoc. U.K.* 50:461–73. [8.4.2]

Bonotto, S. (1988). Recent progress in research on *Acetabularia* and related Dasycladales. *Prog. Phycol. Res.* 6:59–235. [1.3.2, 1.3.3, 1.4.1, 1.4.2, 1.5.2]

Borowitzka, L. M. (1986). Osmoregulation in blue-green algae. *Prog. Phycol. Res.* 4:243–56. [6.3.2]

Borowitzka, M. A. (1972). Intertidal algal species diversity and the effect of pollution. *Austr. J. Mar. Freshwater Res.* 23:73–84. [8.6.1]

Borowitzka, M. A. (1977). Algal calcification. *Oceanogr. Mar. Biol. Annu. Rev.* 15:189–223. [5.5.2]

Borowitzka, M. A. (1979). Calcium exchange and the measurement of calcification rates in the calcareous coralline red alga *Amphiroa foliacea. Mar. Biol.* 50:339–47. [5.5.2]

Borowitzka, M. A. (1982). Mechanisms in algal calcification. *Prog. Phycol. Res.* 1:137–77. [4.4.1, 5.5.2]

Borowitzka, M. A. (1987). Calcification in algae: mechanisms and the role of metabolism. *CRC Crit. Rev. Plant Sci.* 6:1–45. [5.5.3]

Borowitzka, M. A. (1989). Carbonate calcification in algae – initiation and control. In S. Mann, J. Webb, & R. J. P. Williams (eds.), *Biomineralization – Chemical and Biochemical Perspectives* (pp. 63–94). Weinheim, Germany: VCH Publications. [5.5.3]

Borstlap, A. C. (1981). Invalidity of the multiphasic concept of ion absorption in plants. *Plant Cell Environ.* 4:189–95. [5.4]

Borum, J., H. Kaas, & S. Wium-Andersen (1984). Biomass variation and autotrophic production of an epiphyte-macrophyte community in a coastal Danish area: II. Epiphyte species composition, biomass and production. *Ophelia* 23:165–79. [2.4.2]

Bot, P. V. M., W. T. Stam, S. A. Boele-Bos, C. van den Hoek, & W. van Delden (1989a). Biogeographic and phylogenetic studies on three North Atlantic species of *Cladophora* (Cladophorales, Chlorophyta) using DNA-DNA hybridization. *Phycologia* 28:159–68. [1.4.1]

Bot, P. V. M., R. W. Holton, W. T. Stam, & C. van den Hoek (1989b). Molecular divergence between North Atlantic and Indo-West Pacific *Cladophora albida* (Cladophorales: Chlorophyta) isolates as indicated by DNA-DNA hybridization. *Mar. Biol.* 102:307–13. [1.4.1]

Boude, J. P., Y. Negro, A. Fauchoux, & B. Simon (1988). Food algae market in the world in 1987. Position of *Undaria pinnatifida. ASFA Aquacult. Abst.* 6:157. [9.4]

Bowes, G. (1985). Pathways of CO_2 fixation by aquatic organisms. In W. J. Lucas & J. A. Berry (eds.), *Inorganic Carbon Uptake by Aquatic Photosynthetic Organisms* (pp. 187–210). Bethesda, Md.: American Society of Plant Physiologists. [4.4.4]

Box, G. E. P., W. G. Hunter, & J. S. Hunter (1978). *Statistics for Experimenters: An Introduction to Design, Data Analysis and Model Building.* New York: Wiley, 653 pp. [1.1.2]

Boyce, M. S. (1984). Restitution of *r-* and *K*-selection as a model of density-dependent natural selection. *Annu. Rev. Ecol. Syst.* 15:427–47. [2.5.2]

Boyen, C., B. Kloareg, & V. Vreeland (1988). Comparison of protoplast wall regeneration and native wall deposition in zygotes of *Fucus distichus* by cell wall labelling with monoclonal antibodies. *Plant Physiol. Biochem.* 26:653–9. [1.3.1, 1.6.1]

Boyer, G. L., A. H. Gilliam, & C. Trick (1987). Iron chelation and uptake. In P. Fay & C. Van Baalen (eds.), *The Cyanobacteria* (pp. 415–36). Amsterdam: Elsevier. [5.5.6]

Boyle, E. A. (1979). Copper in natural waters. In J. O. Nriagu (ed.), *Copper in the Environment* (pt. I, pp. 77–88). New York: Wiley. [5.2]

Boyle, T. P. (1985). *Validation and Predictability of Laboratory Methods for Assessing the Fate and Effects of Contaminants in Aquatic Ecosystems.* ASTM STP 865. Philadelphia: American Society for Testing and Materials. [8.1]

Bradley, P. M. (1991). Plant hormones do have a role in controlling growth and development of algae. *J. Phycol.* 27:317–21. [1.7.2, 1.7.3]

Branch, G. M. (1975). Mechanisms reducing intraspecific competition in *Patella* spp.: migration, differentiation and territorial behaviour. *J. Animal Ecol.* 44:575–600. [3.2.2]

Branch, G. M., & C. L. Griffiths (1988). The Benguela ecosystem. Part V. The coastal zone. *Oceanogr. Mar. Bio. Annu. Rev.* 26:395–486. [4.7.6]

Brand, L. E., W. G. Sunda, & R. R. L. Guillard (1983). Limitation of marine phytoplankton reproductive rates by zinc, manganese and iron. *Limnol. Oceanogr.* 28:1182–98. [5.5.6]

Brand, L. E., W. G. Sunda, & R. R. L. Guillard (1986). Reduction of marine phytoplankton reproduction rates by copper and cadmium. *J. Exp. Mar. Biol. Ecol.* 96:225–50. [5.2]

Bråten, T. (1983). Autoradiographic evidence for the rapid disintegration of one chloroplast in the zygote of the green alga *Ulva mutabilis. J. Cell Sci.* 12:385–9. [1.3.2]

Brattström, H. (1980). Rocky-shore zonation in the Santa Marta area, Colombia. *Sarsia* 65:163–226. [2.2.2]

Brattström, H. (1985). Rocky-shore zonation on the Atlantic coast of Panama. *Sarsia.* 70:179–216. [2.2.2]

Brawley, S. H. (1987). A sodium-dependent, fast block to polyspermy occurs in eggs of fucoid algae. *Devel. Biol.* 124:390–7. [1.5.4]

Brawley, S. H., & W. H. Adey (1977). Territorial behavior of three-spot damselfish (*Eupomacentrus planifrons*) increases reef algal biomass and productivity. *Env. Biol. Fish.* 2:45–51. [3.2.2]

Brawley, S. H., & W. H. Adey (1981). The effect of micrograzers on algal community structure in a coral reef microcosm. *Mar. Biol.* 61:167–77. [3.2.1]

Brawley, S. H., & X. G. Fei (1987). Studies of mesoherbivory in aquaria and in an unbarricaded mariculture farm on the Chinese coast. *J. Phycol.* 23:614–23. [3.2.1]

Brawley, S. H., & R. Wetherbee (1981). Cytology and ultrastructure. In C. S. Lobban & M. J. Wynne (eds.), *The Bi-*

ology of Seaweeds (pp. 248–99). Oxford: Blackwell Scientific. [1.3, 1.3.2, 1.3.5, 4.6.]

Brechignac, F., M. Andre, & G. Gerbaud (1986). Preferential photosynthetic uptake of exogenous HCO_3^- in the marine macroalga *Chondrus crispus*. *Plant Physiol*. 80:1059–62. [4.4.1]

Breeman, A. M. (1988). Relative importance of temperature and other factors in determining geographic boundaries of seaweeds: experimental and phenological evidence. *Helgol. Meeresunters*. 42:199–241. [6.2.5]

Breeman, A. M., & M. D. Guiry (1989). Tidal influences on the photoperiodic induction of tetrasporogenesis in *Bonnemaisonia hamifera* (Rhodophyta). *Mar. Biol*. 102:5–14. [1.5.3]

Breeman, A. M., & A. ten Hoopen (1987). The mechanism of daylength perception in the red alga *Acrosymphyton purpuriferum*. *J. Phycol*. 23:36–42. [1.5.3]

Breeman, A. M., S. Bos, S. van Essen, & L. L. van Mulekom (1984). Light–dark regimes in the intertidal zone and tetrasporangial periodicity in the red alga *Rhodochorton purpureum* (Lightf.) Rosenv. *Helgol. Meeresunters*. 37:365–87. [1.5.3]

Breeman, A. M., E. J. S. Meulenhoff, & M. D. Guiry (1988). Life history regulation and phenology of the red alga *Bonnemaisonia hamifera*. *Helgol. Meeresunters*. 42:535–51. [1.5.3]

Brinkhuis, B. H. (1985). Growth patterns and rates. In M. M Littler & D. S. Littler (eds.), *Handbook of Phycological Methods: Ecological Field Methods: Macroalgae* (pp. 461–77). Cambridge University Press. [5.7.2]

Brinkhuis, B. H., & I. K. Chung (1986). The effects of copper on the fine structure of the kelp *Laminaria saccharina* (L.) *Mar. Environ. Res*. 19:205–23. [8.3.3]

Brinkhuis, B. H., N. R. Tempel, & R. F. Jones (1976). Photosynthesis and respiration of exposed salt-marsh fucoids. *Mar. Biol*. 34:349–59. [6.4]

Brinkhuis, B. H., H. G. Levine, C. G. Schlenk, & S. Tobin (1987). *Laminaria* cultivation in the Far East and North America. In K. T. Bird & P. H. Benson (eds.), *Seaweed Cultivation for Renewable Resources* (pp. 107–46). Amsterdam: Elsevier. [9.3, 9.3.2, 9.8]

Brinkhuis, B. H., R. Li, C. Wu, & X. Jiang (1989). Nitrite uptake transients and consequences for in vivo algal nitrate reductase assays. *J. Phycol*. 25:539–45. [5.5.1]

Britz, S. J., & W. R. Briggs (1976). Circadian rhythms of chloroplast orientation and photosynthetic capacity in *Ulva*. *Plant Physiol*. 58:22–7. [4.7.2]

Broadwater, S. T., & J. Scott (1982). Ultrastructure of early development in the female reproductive system of *Polysiphonia harveyi* Bailey (Ceramiales, Rhodophyta). *J. Phycol*. 18:427–41. [1.5.4]

Brostoff, W. N. (1988). Seaweed community structure and productivity: the role of mesograzers. *Proc. 6th Intl. Coral Reef Congr*. 2:1–5. [3.2.1]

Brown, J. C. (1978). Mechanism of iron uptake by plants. *Plant Cell Environ*. 1:249–57. [5.5.6]

Brown, V., S. C. Ducker, & K. S. Rowan (1977). The effect of orthophosphate concentrations on the growth of the coralline algae (Rhodophyta). *Phycologia* 16:125–31. [5.5.2]

Brownlee, C. (1990). Light and development: cellular and molecular aspects of photomorphogenesis in brown algae. In P. J. Herring, A. K. Campbell, M. Whitfield, & L. Maddock (eds.), *Light and Life in the Sea* (pp. 115–26). Cambridge University Press. [1.6.2]

Bruland, K. W. (1983). Trace elements in sea-water. In J. P. Riley (ed.), *Chemical Oceanography* (vol. 8, pp. 157–219). London: Academic Press. [5.2]

Bryan, G. W. (1969). The absorption of zinc and other metals by the brown seaweed *Laminaria digitata*. *J. Mar. Biol. Assoc. U.K*. 49:225–43. [8.3.1]

Bryan, G. W., & L. G. Hummerstone (1973). Brown seaweed as an indicator of heavy metals in estuaries in south-west England. *J. Mar. Biol. Assoc. U.K*. 53:705–20. [8.3.1]

Bryhni, E. (1978). Quantitative differences between polysaccharide compositions in normal differentiated *Ulva mutabilis* and the undifferentiated mutant *lumpy*. *Phycologia*. 17:119–24. [1.3.4]

Brylinski, M. (1977). Release of dissolved organic matter by marine macrophytes. *Mar. Biol*. 39:213–20. [4.7.3]

Buchala, A. J., & A. Schmid (1979). Vitamin D and its analogues as a new class of plant growth substances affecting rhizogenesis. *Nature* 280:230–1. [5.5.7]

Buggeln, R. G. (1974). Negative phototropism of the haptera of *Alaria esculenta* (Laminariales). *J. Phycol*. 10:80–2. [1.7.2]

Buggeln, R. G. (1976). Auxin, an endogenous regulator of growth in algae? *J. Phycol*. 12:355–8. [1.7.3]

Buggeln, R. G. (1978). Physiological investigations on *Alaria esculenta* (Laminariales, Phaeophyceae) IV. Inorganic and organic nitrogen in the blade. *J. Phycol*. 14:156–60. [5.5.1]

Buggeln, R. G. (1981). Morphogenesis and growth substances. In C. S. Lobban & M. J. Wynne (eds.), *The Biology of Seaweeds* (pp. 627–60). Oxford: Blackwell Scientific. [1.7.2, 1.7.3]

Buggeln, R. G. (1983). Photoassimilate translocation in brown algae. *Prog. Phycol. Res*. 2:283–332. [4.6]

Buggeln, R. G., D. S. Fensom, & C. J. Emerson (1985). Translocation of ^{11}C-photoassimilate in the blade of *Macrocystis pyrifera* (Phaeophyceae). *J. Phycol*. 21:35–40. [4.6]

Bunt, J. S. (1970). Uptake of cobalt and vitamin B_{12} by tropical macroalgae. *J. Phycol*. 6:339–43. [5.5.6]

Burdon-Jones, C., G. R. W. Denton, G. B. Jones, & K. A. McPhie (1982). Regional and seasonal variations of trace metals in tropical Phaeophyceae from North Queensland. *Mar. Environ. Res*. 7:13–30. [8.3.1]

Burgman, M. A., & V. A. Gerard (1990). A stage-structured, stochastic population model for the giant kelp, *Macrocystis pyrifera*. *Mar. Biol*. 105:15–23. [2.5.2, 4.7.4]

Burris, J. E. (1977). Photosynthesis, photorespiration, and dark respiration in eight species of algae. *Mar. Biol*. 39:371–9. [4.4.2]

Burrows, E. M. (1971). Assessment of pollution effects by the use of algae. *Proc. R. Soc. Lond*. 177:295–306. [8.6.1]

Burrows, E. M., & S. M. Lodge (1951). Autecology and the species problem in *Fucus*. *J. Mar. Biol. Assoc., U. K*. 30:161–76. [2.1.1]

Burton, J. D. (1979). Physico-chemical limitations in experimental investigations. *Phil. Trans. R. Soc. Lond*. B286:443–56. [8.1]

Buschmann, A. H., & A. Bravo (1990). Intertidal amphipods as potential dispersal agents of carpospores of *Iridaea laminarioides* (Gigartinales, Rhodophyta). *J. Phycol*. 26:417–20. [3.2.1]

Butler, D. M., K. Ostgaard, C. Boyen, L. V. Evans, A. Jensen, & B. Kloareg (1989). Isolation conditions for high yields of protoplasts from *Laminaria saccharaina* and *L. digitata* (Phaeophyceae). *J. Exp. Bot.* 40:1237–46. [1.5.2, 1.7.1]

Butler, J. N., B. F. Morris, & T. D. Sleeter (1976). The fate of petroleum in the open ocean. In *Sources, Effects, and Sinks of Hydrocarbons in the Aquatic Environment* (pp. 287–97). Washington, D.C.: American Institute of Biological Sciences. [Table 8.5]

Buttermore, R. E. (1977). Eutrophication of an impounded estuarine lagoon. *Mar. Pollut. Bull.* 8:13–15. [8.6.1]

Butterworth, J., P. Lester, & G. Nickless (1972). Distribution of heavy metals in the Severn estuary. *Mar. Pollut. Bull.* 3:72–4. [8.3.5]

Byrne, R. H., & D. R. Kester (1976). Solubility of hydrous ferric oxide and iron speciation in seawater. *Mar. Chem.* 4:255–74. [5.2]

Cabioch, J. (1988). Morphogenesis and generic concepts in coralline algae – a reappraisal. *Helgol. Meeresunters.* 42: 493–509. [1.2.1]

Cabioch, J., & G. Giraud (1986). Structural aspects of biomineralization in the coralline algae (calcified Rhodophyceae). In B. S. C. Leadbeater & R. Riding (eds.), *Biomineralization in Lower Plants and Animals* (pp. 141–56). Oxford: Clarendon Press. [5.5.3]

Caccamese, S., R. M. Toscano, G. Furani, & M. Cormaci. (1985). Antimicrobial activities of red and brown algae from southern Italy coast. *Bot. Mar.* 28:505–7. [3.1.2]

Cairns, J., Jr. (1983). Are single species toxicity tests alone adequate for estimating environmental hazard? *Hydrobiologia* 100:47–57. [8.1]

Cairns, J., Jr., A. G. Heath, & B. C. Parker (1975). Temperature influence on chemical toxicity to aquatic organisms. *J. Water Pollut. Control Fed.* 47:267–80. [8.2]

Callow, J. A., M. E. Callow, & L. V. Evans (1979). Nutritional studies on the parasitic red alga *Choreocolax polysiponiae*. *New Phytol.* 83:451–62. [3.3.2]

Callow, M E. (1989). Algal attachment to non-toxic antifouling coatings. [abstract]. *Br. Phycol. J.* 24:300. [1.6.1]

Callow, M. E., & L. V. Evans (1974). Studies on the shipfouling alga *Enteromorpha*. III. Cytochemistry and autoradiography of adhesive production. *Protoplasma* 80:15–27. [1.6.1]

Callow, M. E., & L. V. Evans, G. P. Bolwell, & J. A. Callow (1978). Fertilization in brown algae. I. SEM and other observations on *Fucus serratus*. *J. Cell. Sci.* 32:45–54. [1.5.4]

Calvo, M. A., F. J. Cabanes, & L. Abarca (1986). Antifungal activities of some Mediterranean algae. *Mycopathologia* 93:61–3. [3.1.2]

Cambridge, M., A. M. Breeman, & C. van den Hoek (1984). Temperature responses of some North Atlantic *Cladophora* species (Chlorophyceae) in relation to their geographic distribution. *Helgol. Meeresunters.* 38:349–63. [6.2.3]

Campbell, G. S. (1977). *An Introduction to Environmental Biophysics*. Berlin: Springer Verlag. 159 pp. [7.1.2]

Campbell, J. W., & T. Aarup (1989). Photosynthetically available radiation at high latitudes. *Limnol. Oceanogr.* 34:1490–9. [4.2.2]

Cannon, M. I. (1989). Cell division patterns and diurnal cycle of *Porphyra abbottae* (Rhodophyta, Bangiales) carpogonial and carpospore development. *J. Phycol.* 25:612–15. [1.3.5, 1.5.4]

Capone, D. G., & M. F. Bautista (1985). A groundwater source of nitrate in nearshore marine sediments. *Nature* 313:214–16. [5.8.3]

Carefoot, T. H. (1977). *Pacific Seashores. A Guide to Intertidal Ecology.* Vancouver, B.C: J. J. Douglas, 208 pp. [2.2.1, 7.2.3, 8.6.2]

Carefoot, T. H. (1982). Phagostimulatory properties of various chemical compounds to sea hares (*Aplysia kurodai* and *A. dactylomela*). *Mar. Biol.* 68:207–15. [3.2.2]

Cariello, L., & L. Zanetti (1979a). Effect of *Posidonia oceanica* extracts on the growth of *Staphylococcus aureus*. *Bot. Mar.* 22:129–31. [2.4.2]

Cariello, L., & L. Zanetti (1979b). Distribution of chicoric acid during leaf development in *Posidonia oceanica*. *Bot. Mar.* 22:359–60. [2.4.2]

Carlberg, S. R. (1980). Oil pollution of the marine environment – with emphasis on estuarine studies. In E. Olausson & I. Cato (eds.), *Chemistry and Biogeochemistry of Estuaries* (pp. 367–402). New York: Wiley. [8.4.1]

Carlson, D. J., & M. L. Carlson (1984). Reassessment of exudation by fucoid macroalgae. *Limnol Oceanogr.* 29:1077–87. [4.7.3]

Carpenter, E. J. (1972). Nitrogen fixation by a blue-green epiphyte on pelagic *Sargassum*. *Science* 178:1207–9. [5.5.2]

Carpenter, E. J., & J. S. Lively (1980). Review of estimates of algal growth using ^{14}C tracer techniques. In P. G. Falkowski (ed.), *Primary Production in the Sea* (pp. 161–78). New York: Plenum Press. [4.7.1]

Carpenter, R. C. (1986). Partitioning herbivory and its effects on coral reef algal communities. *Ecol. Monogr.* 56:345–63. [3.2.1]

Carpenter, R. C. (1988). Mass mortality of a Carribean sea urchin: immediate effects on community metabolism and other herbivores. *Proc. Natl. Acad. Sci. USA* 85:511–14. [3.2.1]

Carpenter, R. C. (1990). Competition among marine macroalgae: a physiological perspective. *J. Phycol.* 26:6–12. [2.1.1, 3.0, 3.1, 5.8]

Carpenter, R. C., J. M. Hackney, & W. H. Adey (1991). Measurements of primary productivity and nitrogenase activity of coral reef algae in a chamber incorporating oscillatory flow. *Limnol. Oceanogr.* 36: 40–9. [7.1.3]

Carrington, E. (1990). Drag and dislodgement of an intertidal macroalga: consequences of morphological variation in *Mastocarpus papillatus* Kützing. *J. Exp. Mar. Biol. Ecol.* 139:185–200. [7.2.2]

Casola, E., M. Scardi, L. Mazzella, & E. Fresi (1987). Structure of the epiphytic community of *Posidonia oceanica* leaves in a shallow meadow. *P.S.Z.N.I. Mar. Ecol.* 8:285–96. [2.4.2]

Castilla, J. C. (1988). Earthquake-caused coastal uplift and its effects on rocky intertidal kelp communities. *Science* 242:440–3. [7.2.3]

Castilla, J. C., & R. H. Bustamente (1989). Human exclusion from rocky intertidal of Las Cruces, central Chile: effects on *Durvillaea antarctica* (Phaeophyta, Durvillaeales). *Mar. Ecol. Prog. Ser.* 50:203–14. [3.2.1]

Cattolico, R. A., & S. Loiseaux-de Goër (1989). Analysis of chloroplast evolution and phylogeny: a molecular approach. In J. C. Green, B. S. C. Leadbeater, & W. I. Diver (eds.),

The Chromophyte Algae: Problems and Perspectives (pp. 85–100). Oxford: Clarendon Press. [1.3.2, 1.4.1]

Cavalier-Smith, T. (1986). The kingdom Chromista: origin and sysematics. *Prog. Phycol. Res.* 4:309–47. [1.1.1, 1.3.3]

Cembella, A., N. J. Antia, & P. J. Harrison (1984). The utilization of inorganic and organic phosphorus compounds as nutrients by eukaryotic microalgae: a multidisciplinary perspective. *CRC Crit. Rev. Microbiol.* 10:317–91. [5.2, 5.5.2]

Chamberlain, A. H. L., J. Gorham, D. F. Kane, & S. A. Lewey (1979). Laboratory growth studies on *Sargassum muticum* (Yendo) Fensholt. II. Apical dominance. *Bot. Mar.* 22:11–19. [1.7.2, 1.7.3]

Chamberlain, Y. M. (1984). Spore size and germination in *Fosliella, Pneophyllum* and *Melobesia* (Rhodophyta, Corallinaceae). *Phycologia* 23:433–42. [1.6.2]

Chan, K., P. K. Wong, & S. L. Ng (1982). Growth of *Enteromorpha linza* in sewage effluent and sewage effluent–seawater mixtures. *Hydrobiologia* 97:9–13 [8.6.1]

Chapman, A. R. O. (1973a). Methods for macroscopic algae. In J. R. Stein (ed.), *Handbook of Phycological Methods: Culture Methods and Growth Measurements* (pp. 87–104). Cambridge University Press. [1.5.3, 5.7.2]

Chapman, A. R. O. (1973b). A critique of prevailing attitudes towards the control of seaweed zonation on the sea shore. *Bot. Mar.* 16:80–2. [2.2.3]

Chapman, A. R. O. (1984). Reproduction, recruitment and mortality in two species of *Laminaria* in south-west Nova Scotia. *J. Exp. Mar. Biol. Ecol.* 78:99–108. [2.5.2]

Chapman, A. R. O. (1985). Demography. In M. M. Littler & D. S. Littler (eds.), *Handbook of Phycological Methods: Ecological Field Methods: Macroalgae* (pp. 251–68). Cambridge University Press. [2.5.1, 2.5.2]

Chapman, A. R. O. (1986). Population and community ecology of seaweeds. In J. H. S. Blaxter & A. J. Southward (eds.), *Advances in Marine Biology* (vol. 23, pp. 1–161). London: Academic Press. [2.2.2, 2.5.2, 3.2.1, 6.4]

Chapman, A. R. O. (1987). The wild harvest and culture of *Laminaria longicruris* in eastern Canada. In M. S. Doty, J. F. Caddy, & B. Santelices (eds.), *Case Studies of Seven Commercial Seaweed Resources* (pp. 193–237). FAO Fish. Tech. Pap. 281. [9.6]

Chapman, A. R. O., & E. M. Burrows (1970). Experimental investigations into the controlling effects of light conditions on the development and growth of *Desmarestia aculeata* (L.) Lamour. *Phycologia* 9:103–8. [1.5.3]

Chapman, A. R. O., & J. S. Craigie (1977). Seasonal growth in *Laminaria longicruris:* relations with dissolved inorganic nutrients and internal reserves of nitrogen. *Mar. Biol.* 40:197–205. [4.5.1, 5.5.1]

Chapman, A. R. O., & C. L. Goudey (1983). Demographic study of the macrothallus of *Leathesia difformis* (Phaeophyta) in Nova Scotia. *Can J. Bot.* 61:319–23. [2.5.2]

Chapman, A. R. O., & J. E. Lindley (1980). Seasonal growth of *Laminaria solidungula* in the Canadian High Arctic in relation to irradiance and dissolved nutrient concentrations. *Mar. Biol.* 57:1–5. [4.2.2]

Chapman, A. R. O., J. W. Markham, & K. Lüning (1978). Effects of nitrate concentration on the growth and physiology of *Laminaria saccharina* (Phaeophyta) in culture. *J. Phycol.* 14:195–98. [5.7.3, 5.8.1]

Chapman, D. J., & P. J. Harrison (1988). Nitrogen metabolism and the measurement of nitrate reductase. In C. S. Lobban, D. J. Chapman, & B. P. Kremer (eds.), *Experimental Phy-*

cology: A Laboratory Manual (pp. 196–202). Cambridge University Press. [5.5.1]

Chapman, M. S., S. W. Suh, P. M. G. Curmi, D. Cascio, W. W. Smith, & D. S. Eisenberg (1988). Tertiary structure of plant RuBisCO: domains and their contacts. *Science* 241:71–4. [4.1, 4.4.2]

Chapman, V. J. (1966). The physiological ecology of some New Zealand seaweeds. *Proc. Intl. Seaweed Symp.* 5:29–54. [6.4]

Chapman, V. J., & D. J. Chapman (1980). *Seaweeds and Their Uses,* 3rd ed. London: Chapman & Hall, 334 pp. [9.1]

Charters, A. C., M. Neushul, & C. Barilotti (1969). The functional morphology of *Eisenia arborea. Proc. Intl. Seaweed Symp.* 6:89–105. [7.2.1, 7.2.2]

Charters, A. C., M. Neushul, & D. Coon (1973). The effect of water motion on algal spore adhesion. *Limnol. Oceanogr.* 18:884–96. [7.1.2]

Chen, L. C.-M. (1986). Cell development of *Porphyra miniata* (Rhodophyceae) under axenic culture. *Bot. Mar.* 29:435–9. [9.2.4]

Chen, L. C.-M., T. Edelstein, E. Ogata, & J. McLachlan (1970). The life history of *Porphyra miniata. Can. J. Bot.* 48:385–9. [1.5.3]

Chen, L. C.-M., J. McLachlan, A. C. Neish, & P. Shacklock (1973). The ratios of kappa- to lambda-carrageenan in nuclear phases of the rhodophycean algae *Chondrus crispus* and *Gigartina stellata. J. Mar. Biol. Assoc. U.K.* 53:11–16. [9.7]

Cheney, D. P. (1982). The determining effects of snail herbivore density on intertidal algal recruitment and composition. Abstr. 1st Intl. Phycol. Congr., p. a8 [3.2.2]

Cheney, D., A. Mathieson, & D. Schubert (1981). The application of genetic improvement techniques to seaweed cultivation: I. Strain selection in the carrageenophyte *Chondrus crispus. Proc. Intl. Seaweed Symp.* 10:559–76. [9.7]

Cheshire, A. C., & N. D. Hallam (1985). The environmental role of alginates in *Durvillaea potatorum* (Fucales, Phaeophyta). *Phycologia* 24:147–53. [4.5.2]

Cheshire, A. C., & N. D. Hallam (1989). Methods for assessing the age composition of native stands of subtidal macroalgae: a case study on *Durvillaea potatorum. Bot. Mar.* 32:199–204. [2.5.2]

Chida, Y., & K. Ueda (1986). Mitochondrial number and form change during autospore formation in *Chlorococcum infusionum* (Schrank) Meneghnini (Chlorococcales, Chlorophyta). *Phycologia* 25:503–9. [1.3]

Chopin, T. (1986). *The Red Alga Chondrus crispus Stackhouse (Irish Moss) and Carrageenans – a Review.* Can. Tech. Rept. Fish. Aquat. Sci. no. 1514, 69 pp. [9.7]

Chopin, T., A. Hourmat, J.-Y. Floc'h, & M. Penot (1990). Seasonal variations in the red alga *Chondrus crispus* on the Atlantic French coast. II. Relations with phosphorus concentration in seawater and intertidal phosphorylated fractions. *Can. J. Bot.* 68:512–17. [5.5.2]

Christensen, T. (1988). Salinity preferences of twenty species of *Vaucheria* (Tribophyceae). *J. Mar. Biol. Assoc. U.K.* 68:531–45. [6.3.4]

Christie, A. O., & M. Shaw (1968). Settlement experiments with zoospores of *Enteromorpha intestinalis* (L.) Link. *Br. Phycol. Bull.* 3:529–534. [1.6.1]

Chu, H., & B. T. Wu (1983). Studies on the lime-boring algae from China. I. The lime-boring algae collected in the Xisha Islands. *Abstr. Intl. Seaweed Symp., 11,* p. 46 [2.1.2]

Chung, I. K., & B. Brinkhuis (1986). Copper effects in early stages of the kelp, *Laminaria saccharina*. *Mar. Poll. Bull.* 17:213–18. [8.3.3]

Cinelli, F., M. Cormaci, G. Furnari, & L. Mazzella (1984). Epiphytic macroflora of *Posidonia oceanica* (L.) Delile leaves around the Island of Ischia (Gulf of Naples). In C. F. Boudouresque, A. Jeudy de Grissac, & J. Olivier (eds.), *International Workshop on Posidonia oceanica Beds* (pp. 91–9). Paris: G.I.S. Posidonie Publ. [2.4.2]

Clayton, M. N. (1981). Correlated studies on seasonal changes in the sexuality, growth rate and longevity of complanate *Scytosiphon* (Scytosiphonaceae, Phaeophyta) from southern Australia growing *in situ*. *J. Exp. Mar. Biol. Ecol.* 51:87–96. [1.5.1, 2.5.2]

Clayton, M. N. (1984). An electron microscope study of gamete release and settling in the complanate form of *Scytosiphon* (Scytosiphonaceae, Phaeophyta). *J. Phycol.* 20:276–85. [1.6.1]

Clayton, M. N. (1988). Evolution and life histories of brown algae. *Bot. Mar.* 31:379–87. [1.5.1]

Clayton, M. N., & G. W. Beakes (1983). Effects of fixatives on the ultrastructure of physodes in vegetative cells of *Scytosiphon lomentaria* (Scytosiphonaceae, Phaeophyta). *J. Phycol.* 19:4–16. [1.3]

Clayton, M. N., & C. M. Shankly (1987). The apical meristem of *Splachnidium rugosum* (Phaeophyta.) *J. Phycol.* 23:296–307. [1.7.2]

Clayton, M. N., N. D. Hallam, S. E. Luff, & T. Diggins (1985). Cytology of the apex, thallus development and reproductive structures of *Hormosira banksii* (Fucales, Phaeophyta). *Phycologia* 24:181–90. [1.7.2]

Clendenning, K. A., & W. J. North (1960). Effects of wastes on the giant kelp, *Macrocystis pyrifera*. *Proc. Intl. Conf. Waste Disposal Mar. Environ.* 1:82–91. [8.4.1]

Cloutier-Mantha, L., & P. J. Harrison (1980). Effects of sublethal concentrations of mercuric chloride on ammonium limited *Skeletonema costatum*. *Mar. Biol.* 56:219–31. [8.3.4]

Coale, K. H., & K. W. Bruland (1988). Copper complexation in the Northeast Pacific. *Limnol. Oceanogr.* 33:1084–101. [5.2]

Cohen-Bazire, G., & D. A. Bryant (1982). Phycobilisomes: composition and structure. In N. G. Carr & B. A. Whitton (eds.), *The Biology of Cyanobacteria* (pp. 143–90). Oxford: Blackwell Scientific. [4.3.1]

Cole, H. A. (1979a). Pollution of the sea and its effects. *Proc. R. Soc. Lond.* B205:17–30. [8.6.1]

Cole, H. A. (1979b). Summing-up: deficiencies and future needs. *Phil. Trans. R. Soc. Lond.* B286:625–33. [8.1]

Cole, K., & E. Conway (1980). Studies in the Bangiaceae: reproductive modes. *Bot. Mar.* 23:545–53. [9.2.1]

Coleman, A. W. (1985). Diversity of plastid DNA configuration among classes of eukaryotic algae. *J. Phycol.* 21:1–16. [1.3.2]

Coley, P. D., J. P. Bryant, & F. S. Chapin II (1985). Resource availability and plant antiherbivore defense. *Science* 230: 895–9. [3.2.3]

Colijn, F., & C. van den Hoek (1971). The life-history of *Sphacelaria furcigera* Kütz. (Phaeophyceae). II. The influence of daylength and temperature on sexual and vegetative reproduction. *Nova Hedwigia* 21:901–22. [1.5.3]

Colman, J. (1933). The nature of the intertidal zonation of plants and animals. *J. Mar. Biol. Assoc. U.K.* 18:435–76. [2.2.3]

Connell, J. H., & R. O. Slatyer (1977). Mechanisms of succession in natural communities and their role in community stability and organization. *Am. Nat.* 111:1119–44. [2.4.1, 2.5.1]

Conover, J. T., & J. M. Sieburth (1964). Effect of *Sargassum* distribution on its epibiota and antibacterial activity. *Bot. Mar.* 6:147–157. [3.1.2]

Conway, H. L., P. J. Harrison, & C. O. Davis (1976). Marine diatoms grown in chemostats under silicate or ammonium limitation. II. Transient response of *Skeletonema costatum* to a single addition of the limiting nutrient. *Mar. Biol.* 35:187–9. [5.4.1]

Cook, C. M., & B. Colman (1987). Some characteristics of photosynthetic inorganic carbon uptake of a marine macrophytic red alga. *Plant Cell Environ.* 10:275–8. [4.4.1]

Coomans, R. J., & M. H. Hommersand (1990). Vegetative growth and organization. In K. M. Cole & R. G. Sheath (eds.), *Biology of the Red Algae* (pp. 275–304). Cambridge University Press. [1.2.1, 1.7.2]

Coon, D. A., M. Neushul, & A. C. Charters (1972). The settling behavior of marine algal spores. *Proc. Intl. Seaweed Symp.* 7:237–42. [1.6.1, 7.1.2]

Cornish-Bowden, A. (1979). *Fundamentals of Enzyme Kinetics*. London: Butterworth, 203 pp. [5.4]

Correa, J. A., & J. McLachlan (1991). Endophytic algae of *Chondrus crispus* (Rhodophyta). III. Host specificity. *J. Phycol.* 27:448–59. [3.1.2, 3.3.2]

Correa, J., I. Novaczek, & J. McLachlan (1986). Effect of temperature and daylength on morphogenesis of *Scytosiphon lomentaria* (Scytosiphonales, Phaeophyta) from eastern Canada. *Phycologia* 25: 469–75. [1.5.3]

Cortel-Breeman, A. M., & A. ten Hoopen (1978). The short-day response in *Acrosymphyton purpuriferum* (J. Ag.) Sjöst. (Rhodophyceae, Cryptonemiales). *Phycologia* 17:125–32. [1.5.3]

Corzo, A., & F. X. Niell (1991). Determination of nitrate reductase activity in *Ulva rigida* C. Agardh by the in situ method. *J. Exp. Mar. Biol. Ecol.* 146:181–91. [5.5.1]

Cosgrove, D. J. (1981). Analysis of the dynamic and steady-state responses of growth rate and turgor pressure to changes in cell parameters. *Plant Physiol.* 68:1439–46. [6.3.1]

Cosson, J. (1977). Action de l'éclairement sur la morphogénèse des gamétophytes de *Laminaria digitata* (L.) Lam. (Phaeophycée, Laminariale). *Bull. Soc. Phycol. Fr.* 22:19–26. [1.5.3]

Coughlan, S. (1977). Sulphate uptake in *Fucus serratus*. *J. Exp. Bot.* 28:1207–15. [5.5.5]

Coughlan, S., & L. V. Evans (1978). Isolation and characterization of Golgi bodies from vegetative tissue of the brown alga *Fucus serratus*. *J. Exp. Bot.* 29:55–68. [4.5.3]

Court, G. J. (1980). Photosynthesis and translocation studies of *Laurencia spectabilis* and its symbiont *Janczewskia gardneri* (Rhodophyceae). *J. Phycol.* 16:270–9. [3.3.2, 4.4.1]

Cousens, R. (1981). The population biology of *Ascophyllum nodosum* (L.) Le Jolis. Ph.D. dissertation, Dalhousie University, Halifax, Nova Scotia. [2.5.2]

Cousens, R., & M. J. Hutchings (1983). The relationship between density and mean frond weight in monospecific seaweed stands. *Nature* 301:240–1. [3.1.3]

Coutinho, R., & Y. Yoneshigue (1990). Diurnal variation in photosynthesis vs. irradiance curves from ''sun'' and ''shade'' plants of *Pterocladia capillacea* (Gmelin) Bornet et

Thuret (Gelidiaciaceae [*sic*]: Rhodophyta) from Cabo Frio, Rio de Janeiro, Brazil. *J. Exp. Mar. Biol. Ecol.* 118:217–28. [4.3.2]

Cox, P. A., & P. B. Tomlinson (1988). Pollination ecology of a seagrass, *Thalassia testudinum* (Hydrocharitaceae), in St. Croix. *Am. J. Bot.* 75:958–65. [1.5.4]

Craigie, J. S. (1974). Storage products. In W. D. P. Stewart (ed.), *Algal Physiology and Biochemistry* (pp. 206–35). Oxford: Blackwell Scientific. [4.4.2, 4.5.1]

Craigie, J. S. (1990a). Cell walls. In K. M. Cole & R. G. Sheath (eds.), *Biology of the Red Algae* (pp. 221–57). Cambridge University Pres. [1.3.1, 4.5.2]

Craigie, J. S. (1990b). Irish moss cultivation: some reflections. In C. Yarish, C. A. Penniman, & P. Van Patten (eds.), *Economically Important Plants of the Atlantic: Their Biology and Cultivation* (pp. 37–52). Groton: Connecticut Sea Grant College Program. [9.7]

Craigie, J. S., & J. McLachlan (1964). Excretion of coloured ultraviolet-absorbing substances by marine algae. *Can. J. Bot.* 42:23–33. [1.3]

Craigie, J. S., & P. F. Shacklock (1989). Culture of Irish moss. In A. D. Boghen (ed.), *Cold Water Aquaculture in Atlantic Canada* (pp. 243–70). Moncton, New Brunswick: Can. Inst. Res. Region. Develop, University of Moncton. [9.7]

Craigie, J. S., E. R. Morris, D. A. Rees, & D. Thom (1984). Alginate block structure in Phaeophyceae from Nova Scotia: variation with species, environment and tissue-type. *Carbohydr. Polymers* 4:237–52. [1.3.1]

Craik, G. J. S. (1980). Simple method for measuring relative scouring of intertidal areas. *Mar. Biol.* 59:257–60. [7.2.1]

Crayton, M. A., E. Wilson, & R. S. Quatrano (1974). Sulfation of fucoidan in *Fucus* embryos. II. Separation from initiation of polar growth. *Devel. Biol.* 39:134–7. [1.6.1]

Critchley, A. T., P. H. Nienhuis, & K. Verschuure (1987). Presence and development of populations of the brown alga *Sargassum muticum* in the SW Netherlands. *Hydrobiologia* 151/152:245–55. [2.4.1]

Critchley, A. T., W. F. Farnham, T. Yoshida, & T. A. Norton (1990). A bibliography of the invasive alga *Sargassum muticum* (Yendo) Fensholt (Fucales: Sargassaceae). *Bot. Mar.* 33:551–62. [6.2.5]

Cross, F. A., & W. G. Sunda (1977). Relationship between bioavailability of trace metals and geochemical processes in estuaries. In M. L. Wiley (ed.), *Estuarine Interactions* (pp. 429–42). New York: Academic Press. [5.2]

Crothers, J. H. (1983). Field experiments on the effects of crude oil and dispersant on the common animals and plants of rocky seashores. *Mar. Environ. Res.* 8:215–39. [8.4.2]

Cunningham, E. M., & M. D. Guiry (1989). A circadian rhythm in the long-day photoperiodic induction of erect axis development in the marine red alga *Nemalion helminthoides*. *J. Phycol.* 25:705–12. [1.5.3]

Dalby, D. H. (1980). Monitoring and exposure scales. In J. H. Price, D. E. G. Irvine, & W. F. Farnham (eds.), *The Shore Environment. Vol. 1: Methods* (pp. 117–36). New York: Academic Press. [7.2.1]

Daly, M. A., & A. C Mathieson (1977). The effects of sand movement on intertidal seaweeds and selected invertebrates at Bound Rock, New Hampshire, USA. *Mar. Biol.* 43:45–55. [7.2.3]

D'Antonio, C. M. (1985a). Role of sand in the domination of hard substrata by the intertidal alga *Rhodomela larix*. *Mar. Ecol. Prog. Ser.* 27:263–75. [7.2.3]

D'Antonio, C. (1985b). Epiphytes on the rocky intertidal alga *Rhodomela larix* (Turner) C. Agardh: negative effects on the host and food for herbivores? *J. Exp. Mar. Biol. Ecol.* 86:197–218. [3.2.1, 7.2.3]

Davavin, I. A., O. G. Mironov, & I. M. Tsimbal (1975). Influence of oil on nucleic acids of algae. *Mar. Pollut. Bull.* 6:13–15. [8.4.2]

Davies, A. G. (1978). Pollution studies with marine plankton. Part II. Heavy metals. *Adv. Mar. Biol.* 15:381–508. [8.3.3, 8.3.4]

Davies, A. G. (1983). The effects of heavy metals upon natural marine phytoplankton populations. *Prog. Phycol. Res.* 2:113–45. [8.3.3, 8.3.4]

Davies, P. J., & J. F. Marshall (1985). *Halimeda* bioherms – low energy reefs, Northern Great Barrier Reef. *Proc. 5th Intl. Coral Reef Congr., Tahiti.* 5:1-7. [2.1.2]

Davison, I. R. (1991). Environmental effects on algal photosynthesis: temperature. *J. Phycol.* 27:2–8. [6.2.1, 6.2.2, 6.2.3, 6.2.4]

Davison, I. R., & J. O. Davison (1987). The effect of growth temperature on enzyme activities in the brown alga *Laminaria saccharina*. *Br. Phycol. J.* 22:77–87. [6.2.1]

Davison, I. R., & R. H. Reed (1985a). Osmotic adjustment in *Laminaria digitata* (Phaeophyta) with particular reference to seasonal changes in internal solute concentrations. *J. Phycol.* 21:41–50. [6.3.2]

Davison, I. R., & R. H. Reed (1985b). The physiological significance of mannitol accumulation in brown algae: the role of mannitol as a compatible cytoplasmic solute. *Phycologia* 24:449–57. [6.3.2]

Davison, I. R., & W. D. P. Stewart (1983). Occurrence and significance of nitrogen transport in the brown alga *Laminaria digitata*. *Mar. Biol.* 77:107–12. [5.4.2, 5.6]

Davison, I. R., & W. D. P. Stewart (1984). Studies on nitrate reductase activity in *Laminaria digitata* (Huds.) Lamour. I. Longitudinal and transverse profiles of nitrate reductase activity within the thallus. *J. Exp. Mar. Biol. Ecol.* 74:201–10. [5.4.2, 5.5.1]

Davison, I. R., M. Andrews, & W. D. P. Stewart (1984). Regulation of growth in *Laminaria digitata*: use of *in vivo* nitrate reductase activities as an indicator of nitrogen limitation in field populations of *Laminaria* spp. *Mar. Biol.* 84:207–17. [5.5.1]

Davison, I. R., R. M. Greene, & E. J. Podolak (1991). Temperature acclimation of respiration and photosynthesis in the brown alga *Laminaria saccharina*. *Mar. Biol.* 110:449–54. [4.3.3, 6.2.2]

Dawes, C. J. (1979). Physiological and biochemical comparisons of *Eucheuma* spp. (Florideophyceae) yielding *iota*-carrageenan. *Proc. Intl. Seaweed Symp.* 9:188–207. [4.5.2]

Dawes, C. J. (1981). *Marine Botany*. New York: Wiley-Interscience, 628 pp. [1.1.1]

Dawes, C. J. (1985). Macroalgae of the Tampa Bay Estuarine System. In S. F. Treat, J. L. Simon, R. R. Lewis III, & R. L. Whitman, Jr. (eds.), *Proceedings Tampa Bay Area Scientific Information Symposium* (May 1982) (pp. 184–209). Minneapolis, MN: Burgess Pubishing. [2.4.2]

Dawes, C. J. (1987). The biology of commercially important tropical marine algae. In K. Bird & P. H. Benson (eds.),

Seaweed Cultivation for Renewable Resources (pp. 155–90). Amsterdam: Elsevier. [9.5]

Dawes, C. J. (1989). Temperature acclimation in cultured *Eucheuma isiforme* from Florida and *E. alvarezii* from the Phillipines. *J. Appl. Phycol.* 1:59–65. [6.2.2]

Dawes, C. J. (1990). Observations on the physiological ecology of *Eucheuma isiforme* from the western Atlantic: a review. In C. Yarish, C. A. Penniman, & P. Van Patten (eds.), *Economically Important Plants of the Atlantic: Their Biology and Cultivation* (pp. 27–35). Groton: Connecticut Sea Grant College Program. [9.5]

Dawes, C. J., & C. Barilotti (1969). Cytoplasmic organization and rhythmic streaming in growing blades of *Caulerpa prolifera*. *Am. J. Bot.* 56:8–15. [1.7.2]

Dawes C. J., & C. A. Lohr (1978). Cytoplasmic organization and endosymbiotic bacteria in the growing points of *Caulerpa prolifera*. *Rev. Algol. N.S.* 13:309–14. [1.3.2, 1.7.2]

Dawes, C. J., & R. P. McIntosh (1981). The effect of organic material and inorganic ions on the photosynthetic rate of the red alga *Bostrychia binderi* from a Florida estuary. *Mar. Biol.* 64:213–18. [6.3.3]

Dawes, C. J., J. M. Lawrence, D. P. Cheney, & A. C. Mathieson (1974). Ecological studies of Floridian *Eucheuma* (Rhodophyta, Gigartinales). III. Seasonal variation of carrageenan, total carbohydrate, protein and lipid. *Bull. Mar. Sci.* 24:286–99. [4.5.1]

Dawes, C. J., N. F. Stanley, & D. J. Stanicoff (1977). Seasonal and reproductive aspects of plant chemistry, and *iota*-carrageenan from Florida *Eucheuma* (Rhodophyta, Gigartinales). *Bot. Mar.* 20:137–147. [4.5.1]

Dawson, E. Y. (1951). A further study of upwelling and associated vegetation along Pacific Baja California, Mexico. *J. Mar. Res.* 10:39–58. [6.2.5]

Dawson, E. Y. (1959). A preliminary report on the benthic marine flora of southern California. In *Oceanographic Survey of the Continental Shelf Area of Southern California* (pp. 169–264). Sacramento: Publications of the California State Water Quality Control Board, no. 20. [8.6.1]

Dawson, E. Y. (1966). *Marine Botany. An Introduction.* New York: Holt, Rinehart & Winston. [1.2.1]

Dayton, P. K. (1973). Dispersion, dispersal, and persistence of the annual intertidal alga, *Postelsia palmaeformis* Ruprecht. *Ecology* 54:433–8. [2.1.1, 7.2.3]

Dayton, P. K. (1985). Ecology of kelp communities. *Annu. Rev. Ecol. Syst.* 16:215–45. [2.1.3]

Dayton, P. K., & J. S. Oliver (1980). An evaluation of experimental analyses of population and community patterns in benthic marine environments. In K. R. Tenore & B. C. Coull (eds.), *Marine Benthic Dynamics* (pp. 93–120). Belle W. Baruch Library of Marine Science Publication no. 11. Columbia: University of South Carolina Press. [2.5.1]

Dayton, P. K., & M. J. Tegner (1990). Bottoms beneath troubled waters: benthic impacts of the 1982–1984 El Niño in the temperate zone. In P. W. Glynn (ed.), *Global Ecological Consequences of the 1982–83 El Niño–Southern Oscillation* (pp. 433–65). Amsterdam: Elsevier. [6.2.6]

Dean, T. A., & F. R. Jacobsen (1986). Nutrient-limited growth of juvenile kelp, *Macrocystis pyrifera*, during the 1982–1983 "El Niño" in southern California (U.S.A). *Mar. Biol.* 90:597–602. [6.2.6]

DeBoer, J. A. (1979). Effects of nitrogen enrichment on growth rate and phycocolloid content in *Gracilaria foliifera*

and *Neoagardhiella baileyi* (Florideophyceae). *Proc. Intl. Seaweed Symp.* 9:263–71. [5.8.1]

DeBoer, J. A. (1981). Nutrients. In C. S. Lobban & M. J. Wynne (eds.), *The Biology of Seaweeds* (pp. 356–91). Oxford: Blackwell Scientific. [1.5.3, 1.7.2, 5.1.1, 5.1.2, 5.4.2, 5.5.1, 5.5.2, 5.7.2, 5.7.3, 5.8.1, 5.8.2]

DeBoer, J. A., & J. H. Ryther (1977). Potential yields from a waste-recycling algal mariculture system. In R. Krauss (ed.), *The Marine Plant Biomass of the Pacific Northwest Coast* (pp. 231–49). Corvallis: Oregon State University Press. [5.1.3, 5.8.1]

DeBoer, J. A., & F. G. Whoriskey (1983). Production and role of hyaline hairs of *Ceramium rubrum*. *Mar. Biol.* 77:229–34. [5.4.2, 5.8.2]

DeBoer, J. A., H. J. Guigli, T. L. Israel, & C. F. D'Elia (1978). Nutritional studies of two red algae. I. Growth rate as a function of nitrogen source and concentration. *J. Phycol.* 14:261–6. [5.5.1, 5.7.2, 5.7.3]

de Lestang-Brémond, G., & M. Quillet (1981). The turnover of sulphates on the lambda-carrageenan of the cell-walls of the red seaweed Gigartinale: *Catenella opuntia* (Grev.). *Proc. Intl. Seaweed Symp.* 10:449–54. [4.5.3]

D'Elia, C. F., & J. A. DeBoer (1978). Nutritional studies of two red algae. II. Kinetics of ammonium and nitrate uptake. *J. Phycol.* 14:266–72. [5.4, 5.4.2, 5.8.1]

DeManche, J. M., H. C. Curl, D. W. Lundy, & P. L. Donaghay (1979). The rapid response of the marine diatom *Skeletonema costatum* to changes in external and internal nutrient concentration. *Mar. Biol.* 53:323–33. [5.5.1]

den Hartog, C. (1972a). Substratum. Multicellular plants. In O. Kinne (ed.), *Marine Ecology* (vol. I, pt. 3, pp. 1277–89). New York: Wiley [3.1.2]

den Hartog, C. (1972b). The effect of the salinity tolerance of algae on their distribution, as exemplified by *Bangia*. *Proc. Intl. Seaweed Symp.* 7:274–6. [6.3.4]

Denley, E. J., & P. K. Dayton (1985). Competition among macroalgae. In M. M. Littler & D. S. Littler (eds.), *Handbook of Phycological Methods: Ecological Field Methods: Macroaglae* (pp. 511–30). Cambridge University Press. [3.0, 3.1]

Denny, M. W. (1982). Forces on intertidal organisms due to breaking ocean waves: design and application of a telemetry system. *Limnol. Oceanogr.* 27:178–83. [7.2.1]

Denny, M. W. (1983). A simple device for recording maximum force exerted on intertidal organisms. *Limnol. Oceanogr.* 28:1269–74. [7.2.1]

Denny, M. W. (1985a). Water motion. In M. M. Littler & D. S. Littler (eds) *Handbook of Phycological Methods: Ecological Field Methods: Macroalgae* (pp. 7–32). Cambridge University Press. [7.1.2, 7.2.1]

Denny, M. W. (1985b). Wave forces on intertidal organisms: a case study. *Limnol. Oceanogr.* 30:1171–87. [7.2.1]

Denny, M. W. (1988). *Biology and the Mechanics of the Wave-Swept Environment*. Princeton University Press. [2.2.1, 7.0, 7.1.2, 7.2.1, 7.2.2]

Denny, M. W., T. L. Daniel, & M. A. R. Koehl (1985). Mechanical limits to size in wave-swept organisms. *Ecol. Monogr.* 55:69–102. [6.4, 7.2.1, 7.2.2]

Denny, M. [W.], V. Brown, E. Carrington, G. Kraemer, & A. Miller (1989). Fracture mechanics and the survival of wave-swept macroalgae. *J. Exp. Mar. Biol. Ecol.* 127:211–28. [7.2.2]

Derenbach, J. B., & M. V. Gereck (1980). Interference of petroleum hydrocarbons with the sex pheromone reaction of *Fucus vesiculosus* (L.) *J. Exp. Mar. Biol. Ecol.* 44:61–5. [8.4.2]

de Ruyter van Steveninck, E. D., & R. P. M. Bak (1986). Changes in abundance of coral-reef bottom components related to mass mortaility of the sea urchin *Diadema antillarum*. *Mar. Ecol. Prog. Ser.* 34:87–94. [3.2.1]

de Ruyter van Steveninck, E. D., L. L. van Mulekom, & A. M. Breeman (1988). Growth inhibition of *Lobophora variegata* (Lamouroux) Womersley by scleractinian corals. *J. Exp. Mar. Biol. Ecol.* 115:169–78. [3.1.2]

Descolas-Gros, C., & G. de Billy (1987). Temperature adaptation of RuBP carboxylase: kinetic properties in marine Antarctic diatoms. *J. Exp. Mar. Biol. Ecol.* 108:147–58. [6.2.1]

Dethier, M. N. (1982). Pattern and process in tidepool algae: factors influencing seasonality and distribution. *Bot. Mar.* 25:55–66. [2.2.1, 3.2.1, 3.2.2]

Dethier, M. N. (1984). Disturbance and recovery in intertidal pools: maintenance of mosaic patterns. *Ecol. Monogr.* 54:99–118. [2.3.2]

De Vecchi, L., & M. Grilli Caiola (1986). An ultrastructural and cytochemical study of *Anabaena* sp. (Cyanophyceae) envelopes. *Phycologia* 25:415–22. [1.3.1]

Devinny, J. S., & L. A. Volse (1978). Effects of sediments on the development of *Macrocystis pyrifera* gametophytes. *Mar. Biol.* 48:343–8. [7.2.3]

DeWreede, R. E. (1985). Destructive (harvest) sampling. In M. M. Littler & D. S. Littler (eds.), *Handbook of Phycological Methods: Ecological Field Methods: Macroalgae* (pp. 147–60). Cambridge University Press. [2.5.1]

Deysher, L., & T. A. Norton (1982). Dispersal and colonization in *Sargassum muticum* (Yendo) Fensholt. *J. Exp. Mar. Biol. Ecol.* 56:179–95. [1.6.1, 2.1.1]

Dickson, D. M., R. G. Wyn-Jones, & J. Davenport (1982). Osmotic adaptation in *Ulva lactuca* under fluctuating salinity regimes. *Planta* 155:409–15. [6.3.2, 6.4.2]

Dickson, L. G., & J. R. Waaland (1985). *Porphyra nereocystis*: a dual-daylength seaweed. *Planta* 165:548–53. [1.5.3]

Dieckmann, G. S. (1980). Aspects of the ecology of *Laminaria pallida* (Grev.) J. Ag. off the Cape Pennisula (South Africa). I. Seasonal growth. *Bot. Mar.* 23:579–85. [4.7.4]

Digby, P. S. B. (1977). Photosynthesis and respiration in the coralline algae *Clathromorphum circumscriptum* and *Corallina officinalis* and the metabolic basis of calcification. *J. Mar. Biol. Assoc. U.K.* 57:1111–24. [5.5.3]

Dillon, P. S., J. S. Maki, & R. Mitchell (1989). Adhesion of *Enteromorpha* swarmers to microbial films. *Micro. Ecol.* 17:39–47. [1.6.1]

Diouris, M., & J. Y. Floc'h (1984). Long-distance transport of ^{14}C-labelled assimilates in the Fucales: directionality, pathway and velocity. *Mar. Biol.* 78:199–204. [4.6.]

Dipierro, S., C. Perrone, & A. P. Felicini (1977). *In vitro* nitrate reductase assay in *Petroglossum nicaeense* (Duby) Schotter (Rhodophyta, Phyllophoraceae). *Phycologia* 16: 179–82. [5.5.1]

Dixon, P. S. (1973). *Biology of the Rhodophyta.* Edinburgh: Oliver & Boyd, 285 pp. [1.5.4, 1.6.2, 1.7.2]

Dixon, P. S., & W. N. Richardson (1970). Growth and reproduction in red algae in relation to light and dark cycles. *Ann N.Y. Acad. Sci.* 175:764–77. [1.6.2]

Done, T. J., P. K. Dayton, A. E. Dayton, & R. Steger (1991). Regional and local variability in recovery of shallow coral communities: Moorea, French Polynesia and central Great Barrier Reef. *Coral Reefs* 9:183–92. [Fig. 3.11]

Dortch, Q. (1982). Effect of growth conditions on accumulation of internal nitrate, ammonium, and protein in three marine diatoms. *J. Exp. Mar. Biol. Ecol.* 61:243–64. [5.5.1]

Dortch, Q. (1990). The interaction between ammonium and nitrate uptake in phytoplankton. *Mar. Ecol. Prog. Ser.* 61:183–201. [5.4.2]

Doty, M. S. (1946). Critical tide factors that are correlated with the vertical distribution of marine algae and other organisms along the Pacific Coast. *Ecology* 27:315–28. [2.2.3]

Doty, M. S. (1971). Measurement of water movement in reference to benthic algal growth. *Bot. Mar.* 14:32–5. [7.2.1]

Doty, M. S. (1977). *Eucheuma* – current marine agronomy. In R. W. Krauss (ed.), *The Marine Plant Biomass of the Pacific Northwest Coast. A Potential Economic Resource* (pp. 203–14). Corvallis: Oregon State University Press. [9.1]

Doty, M. S. (1986). The production and use of *Eucheuma*. In M. S. Doty, J. F. Caddy, & B. Santelices (eds.), *Case Studies of Seven Commercial Seaweed Resources* (pp. 123–61). FAO Fish. Tech. Pap 281. [9.5, 9.5.3]

Doty, M. S. (1988). A tribe of commercial seaweeds related to *Eucheuma* (Solieriaceae, Gigartinales). In I. A. Abbott (ed.), *Taxonomy of Economic Seaweeds: With Reference to Some Pacific and Caribbean Species* (vol. 2, pp. 157–207). Sacramento: California Sea Grant Program. [9.5]

Doty, M. S., & V. B. Alvarez (1975). Status, problems, advances and economics of *Eucheuma* farms. *Mar. Technol. Soc. J.* 9:30–5. [9.5.2]

Doty, M. S., W. J. Gilbert, & I. A. Abbott (1974). Hawaiian marine algae from seaward of the algal ridge. *Phycologia* 13:345–57. [4.3.4]

Douce, R., & M. Neuburger (1989). The uniqueness of plant mitochondria. *Annu. Rev. Plant Physiol. Plant Molec. Biol.* 40:371–414. [1.3]

Doucette, G. J., & P. J. Harrison (1990). Some effects of iron and nitrogen stress on the red tide dinoflagellate *Gymnodinium sanguineum. Mar. Ecol. Prog. Ser.* 62:293–306. [5.5.6]

Dowd, J. E., & D. S. Riggs (1965). A comparison of estimates of Michaelis-Menten kinetic constants from various linear transformations. *J. Biol. Chem.* 240:863–9. [5.4]

Drebes, G. (1977). Sexuality. In D. Werner (ed.), *The Biology of Diatoms* (pp. 250–83). Oxford: Blackwell Scientific. [5.8.2]

Drew, E. A. (1977). The physiology of photosynthesis and respiration in some Antarctic marine algae. *Br. Antarct. Survey Bull.* 46:59–76. [6.2.2, 6.6.4]

Drew, E. A. (1983). Light. In R. Earll & D. G. Erwin (eds.), *Sublittoral Ecology. The Ecology of the Shallow Sublittoral Benthos* (pp. 10–57). Oxford: Clarendon Press. [4.2.1, 4.2.2]

Drew, E. A., & K. M. Abel (1990). Studies on *Halimeda*. III. A daily cycle of chloroplast migration within segments. *Bot. Mar.* 33:31–45. [1.3.2, 3.2.3]

Drew, K. M. (1949). Conchocelis phase in the life history of *Porphyra umbilicalis* (L.) Kütz. *Nature* 164:748 [1.5.2, 9.2.1]

Drews, G., & J. Weckesser (1982). Function, structure and composition of cell walls and external layers. In N. G. Carr & B. A. Whitton (eds.), *The Biology of Cyanobacteria* (pp. 333–57). Oxford: Blackwell Scientific. [1.3.1, 1.3.5]

Dring, M. J. (1967). Phytochrome in red alga, *Porphyra tenera*. *Nature* 215:1411–12. [1.5.3]

Dring, M. J. (1974). Reproduction. In W. D. P. Stewart (ed.), *Algal Physiology and Biochemistry* (pp. 814–37). Oxford: Blackwell Scientific. [1.5.3]

Dring, M. J. (1981). Photosynthesis and development of marine macrophytes in natural light spectra. In H. Smith (ed.), *Plants and the Daylight Spectrum* (pp. 297–314). London: Academic Press. [4.3.4]

Dring, M. J. (1982). *The Biology of Marine Plants*. London: Arnold. [1.1.1]

Dring, M. J. (1984a). Photoperiodism and phycology. *Prog. Phycol. Res.* 3:159–92. [1.5.3]

Dring, M. J. (1984b). Blue light effects in marine macroalgae. In H. Senger (ed.), *Blue Light Effects in Biological Systems* (pp. 509–16). Berlin: Springer-Verlag. [1.7.2]

Dring, M. J. (1987). Light climate in intertidal and subtidal zones in relation to photosynthesis and growth of benthic algae: a theoretical model. In R. M. M. Crawford (ed.), *Plant Life in Aquatic and Amphibious Habitats* (pp. 23–34). Oxford: Blackwell Scientific. [4.2.2]

Dring, M. J. (1988). Photocontrol of development in algae. *Annu. Rev. Plant Physiol. Plant Molec. Biol.* 39:157–74. [1.5.3]

Dring, M. J. (1989). Stimulation of light-saturated photosynthesis in *Laminaria* (Phaeophyta) by blue light. *J. Phycol.* 25:254–8. [4.3.3, 4.3.4]

Dring, M. J. (1990). Light harvesting and pigment composition in marine phytoplankton and macroalgae. In P. J. Herring, A. K. Campbell, M. Whitfield, & L. Maddock (eds.), *Light and Life in the Sea* (pp. 89–103). Cambridge University Press. [4.3.1, 4.3.4]

Dring, M. J., & F. A. Brown (1982). Photosynthesis of intertidal brown algae during and after periods of emersion: a renewed search for physiological causes of zonation. *Mar. Ecol. Prog. Ser.* 8:301–8. [6.4]

Dring, M. J., & K. Lüning (1975). A photoperiodic response mediated by blue light in the brown alga *Scytosiphon lomentaria*. *Planta* 125:25–32. [1.5.3]

Dring, M. J., & K. Lüning (1983). Photomorphogenesis in marine macroalgae. In W. Shropshire, Jr., & H. Mohr (eds.), *Encyclopaedia of Plant Physiology. Vol. 16B: Photomorphogenesis* (pp. 545–68). Berlin: Springer-Verlag. [Fig. 1.31]

Dring, M. J., & J. A. West (1983). Photoperiodic control of tetrasporangium formation in the red alga *Rhodochorton purpureum*. *Planta* 159:143–50. [1.5.3]

Dromgoole, F. I. (1978). The effects of oxygen on dark respiration and apparent photosynthesis of marine macroalgae. *Aquat. Bot.* 4:281–97. [6.6.4, 7.1.2, 7.1.3]

Dromgoole, F. I. (1980). Desiccation resistance of intertidal and subtidal algae. *Bot. Mar.* 23:149–59. [6.4]

Dromgoole, F. I. (1981). Form and function of the pneumatocysts of marine algae. I. Variations in the pressure and composition of internal gases. *Bot. Mar.* 24:257–66. [1.7.2]

Dromgoole, F. I. (1982). The buoyant properties of *Codium*. *Bot. Mar.* 25:391–7. [1.1.1]

Dromgoole, F. I. (1990). Gas-filled structures, buoyancy and support in marine macro-algae. *Prog. Phycol. Res.* 7:169–211. [1.7.2]

Droop, M. R. (1968). Vitamin B_{12} and marine ecology. IV. The kinetics of uptake, growth and inhibition in *Monochrysis lutheri*. *J. Mar. Biol. Assoc. U.K.* 48:689–733. [5.7.3]

Droop, M. R. (1973). Some thoughts on nutrient limitation in algae. *J. Phycol.* 9:264–72. [5.1.3]

Droop, M. R. (1974). The nutrient status of algal cells in continuous culture. *J. Mar. Biol. Assoc. U.K.* 54:825–55. [5.1.3]

Droop, M. R. (1983). 25 years of algal growth kinetics. A personal view. *Bot. Mar.* 26:99–112. [5.7.1]

Druehl, L. D. (1978). The distribution of *Macrocystis integrifolia* in British Columbia as related to environmental parameters. *Can. J. Bot.* 56:69–79. [6.1.2]

Druehl, L. D. (1981). Geographic distribution. In C. S. Lobban & M. J. Wynne (eds.), *The Biology of Seaweeds* (pp. 306–25). Oxford: Blackwell Scientific. [6.5, Fig. 6.4]

Druehl, L. D. (1988). Cultivated edible kelp. In C. A. Lembi & J. R. Waaland (eds.), *Algae and Human Affairs* (pp. 119–34). Cambridge University Press. [9.3, 9.3.1, 9.3.2]

Druehl, L. D., & R. G. Footit (1985). Biogeographical analysis. In M. M. Littler & D. S. Littler (eds.) *Handbook of Phycological Methods: Ecological Field Methods* (pp. 315–25). Cambridge University Press. [6.1.2]

Druehl, L. D., & J. M. Green (1970). A submersion-emersion sensor for intertidal biological studies. *J. Fish. Res. Bd. Can.* 27:401–3. [7.2.1]

Druehl, L. D., & J. M. Green (1982). Vertical distribution of intertidal seaweeds as related to patterns of submersion and emersion. *Mar. Ecol. Prog. Ser.* 9:163–70. [2.2.3]

Druehl, L. D., R. Baird, A. Lindwall, K. E. Lloyd, & S. Pakula (1988). Longline cultivation of some Laminariaceae in British Columbia. *Aquacult. Fish. Management* 19:253–63. [9.3, 9.3.1]

Dube, M. A., & E. Ball (1971). *Desmarestia* sp. associated with the seapen *Ptilosarcus gurneyi* (Gray). *J. Phycol.* 7:218–20. [2.4]

Ducreux, G. (1984). Experimental modification of the morphogenetic behavior of the isolated sub-apical cell of the apex of *Sphacelaria cirrosa* (Phaeophyceae). *J. Phycol.* 20:447–54. [1.7.1]

Ducreux, G., & B. Kloareg (1988). Plant regeneration from protoplasts of *Sphacelaria* (Phaeophyceae). *Planta* 174:25–9. [1.7.1]

Dudgeon, S. R., I. R. Davison, & R. L. Vadas (1989). Effect of freezing on photosynthesis of intertidal macroalgae: relative tolerance of *Chondrus crispus* and *Mastocarpus stellatus* (Rhodophyta). *Mar. Biol.* 101:107–14. [3.1.1, 6.3.4]

Dudgeon, S. R., I. R. Davison, & R. L. Vadas (1990). Freezing tolerance in the intertidal red algae *Chondrus crispus* and *Mastocarpus stellatus*: relative importance of acclimation and adaptation. *Mar. Biol.* 106:427–36. [6.3.4]

Duffy, J. E., & M. E. Hay (1990). Seaweed adaptations to herbivory. *BioScience* 40:368–75. [3.2.2]

Duggins, D. O. (1980). Kelp beds and sea otters: an experimental approach. *Ecology* 61:447–53. [3.2.1]

Duggins, D. O., C. A. Simenstad, & J. A. Estes (1990). Magnification of secondary production by kelp detritus in coastal marine ecosystems. *Science* 245:170–3. [4.7.6]

Duke, C. S., R. W. Litaker, & J. Ramus (1987). Seasonal variation in RuBPCase activity and N allocation in the chlorophyte seaweeds *Ulva curvata* (Kütz.) de Toni and *Codium decorticatum* (Woodw.) Howe. *J. Exp. Mar. Biol. Ecol.* 112:145–64. [5.8]

Duke, C. S., W. Litaker, & J. Ramus (1989). Effects of temperature, nitrogen supply, and tissue nitrogen on ammonium uptake rates of the chlorophyte seaweeds *Ulva curvata* and *Codium decorticatum*. *J. Phycol.* 25:113–20. [5.8]

Duncan, M. J., & R. E. Foreman (1980). Phytochrome-mediated stipe elongation in the kelp *Nereocystis* (Phaeophyceae). *J. Phycol.* 16:138–42. [1.7.2]

Dunton, K. H., & D. M. Schell (1986). Seasonal carbon budget and growth of *Laminaria solidungula* in the Alaskan High Arctic. *Mar. Ecol. Prog. Ser.* 31:57–66. [4.2.2]

Durako, M. J., & C. J. Dawes (1980). A comparative seasonal study of two populations of *Hypnea musciformis* from the east and west coasts of Florida. U.S.A. II. Photosynthetic and respiratory rates. *Mar. Biol.* 59:157–62. [4.3.2, 4.5.1]

Duursma, E. K., & M. Marchand (1974). Aspects of organic marine pollution. *Oceanogr. Mar. Biol. Annu. Rev.* 12:315–431. [8.5.1, 8.5.2, 8.5.3, 8.6.1]

Easton, W. H., & E. A. Olson (1976). Radiocarbon profile of Hanauma Reef, Oahu, Hawaii: reply. *Geol. Soc. Am. Bull.* 87:711–19. [2.1.2]

Ebeling, A. W., D. R. Laur, & R. J. Rowley (1985). Severe storm disturbances and the reversal of community structure in a southern California kelp forest. *Mar. Biol.* 84. 287–94. [2.5.2]

Eckhardt, R., R. Schnetter, & G. Seibold (1986). Nuclear behaviour during the life cycle of *Derbesia* (Chlorophyceae). *Br. Phycol. J.* 21:287–95. [1.5.2]

Edelstein, T., & J. McLachlan (1975). Autecology of *Fucus distichus* ssp. *distichus* (Phaeophyceae: Fucales) in Nova Scotia, Canada. *Mar. Biol.* 30:305–24. [6.1.3]

Edwards, D. M., R. H. Reed, J. A. Chudek, R. Foster, & W. D. P. Stewart (1987). Organic solute accumulation in osmotically-stressed *Enteromorpha intestinalis*. *Mar. Biol.* 95:583–92. [6.3.2]

Edwards, D. M., R. H. Reed, & W. D. P. Stewart (1988). Osmoacclimation in *Enteromorpha intestinalis*: long-term effects of osmotic stress on organic solute accumulation. *Mar. Biol.* 98:467–76. [6.3.2]

Edwards, P. (1977). An investigation of the vertical distribution of selected benthic marine algae with a tide-simulating apparatus. *J. Phycol.* 13:62–8. [2.1.1, 6.5]

Eide, I., S. Myklestad, & S. Melson (1980). Long-term uptake and release of heavy metals by *Ascophyllum nodosum* (L.) Le Jol. (Phaeophyceae) *in situ*. *Environ. Pollut.* 23:19–28. [8.3.1]

Elner, R. W., & R. L. Vadas, Sr. (1990). Inference in ecology: the sea urchin phenomenon in the northwest Atlantic. *Am. Nat.* 136:108–25. [3.2.1]

Emerson, C. J., R. G. Buggeln, & A. K. Bal (1982). Translocation in *Saccorhiza dermatodea* (Laminariales, Phaeophyceae): anatomy and physiology. *Can. J. Bot.* 60:2164–84. [4.6]

Engelmann, T. W. (1883). Farbe und Assimilation. *Bot. Zeit.* 41:1–13. [4.3.4]

Engelmann, T. W. (1884). Untersuchungen über die quantitativen Beziehungen zwischen Absorption des Lichtes und Assimilation in Pflanzenzellen. *Bot. Zeit.* 42:81–93. [4.3.4]

Enright, C. T. (1979). Competitive interaction between *Chondrus crispus* (Florideophyceae) and *Ulva lactuca* (Chlorophyceae) in *Chondrus* aquaculture. *Proc. Intl. Seaweed Symp.* 9:209–18. [3.1.3]

Eppley, R. W. (1958). Sodium exclusion and potassium retention by the marine red alga *Porphyra perforata*. *J. Gen. Physiol.* 41:901–11. [5.5.2]

Eppley, R. W. (1978). Nitrate reductase in marine phytoplankton. In J. A. Hellebust & J. S. Craigie (eds.), *Handbook of Phycological Methods. Physiological and Biochemical Methods* (pp. 217–23). Cambridge University Press. [5.5.1]

Eppley, R. W., & L. R. Blinks (1957). Cell space and apparent free space in the red alga, *Porphyra perforata*. *Plant Physiol.* 32:63–4. [5.3.1]

Eppley, R. W., & B. S. Cyrus (1960). Cation regulation and survival of the red alga *Porphyra perforata* in diluted and concentrated seawater. *Biol. Bull.* 118:55–65. [6.3.3]

Eppley, R. W., E. H. Renger, & P. M. Williams (1976). Chlorine reactions with seawater constitutents and the inhibition of photosynthesis of natural marine phytoplankton. *Estu. Cstl. Mar. Sci.* 4:147–61. [8.6.1]

Epstein, E. (1972). *Mineral Nutrition of Plants: Principles and Perspectives*. New York: Wiley. [5.4]

Estep, M. F., F. R. Tabita, & C. van Baalen (1978). Purification of ribulose 1,5-bisphosphate carboxylase and carbon isotope fractionation by whole cells and carboxylase from *Cylindrotheca* sp. (Bacillariophyceae). *J. Phycol.* 14:183–8. [4.4.2]

Estes, J. A., & P. D. Steinberg (1988). Predation, herbivory, and kelp evolution. *Paleobiology* 14:19–36. [2.1.3]

Evans, G. C. (1972). *The Quantitative Analysis of Plant Growth*. Oxford: Blackwell Scientific, 734 pp. [1.12, 1.5.3]

Evans, L. V. (1974). Cytoplasmic organelles. In W. D. P. Stewart (ed.), *Algal Physiology and Biochemistry* (pp. 86–123). Oxford: Blackwell Scientific. [1.3]

Evans, L. V., & D. M. Butler (1988). Seaweed biotechnology—current status and future prospects. In L. J. Rogers & J. R. Gallon (eds.) *Biochemistry of the Algae and Cyanobacteria* (pp. 335–50). Oxford: Clarendon Press. [9.8]

Evans, L. V., & A. O. Christie (1970). Studies on the shipfouling alga *Enteromorpha*. I. Aspects of the fine-structure and biochemistry of swimming and newly settled zoospores. *Ann. Bot.* 34:451–66. [1.6.1, 8.5.4]

Evans, L. V., & A. J. Trewavas (1991). Is algal development controlled by plant growth substances? *J. Phycol.* 27:322–6. [1.7.3]

Evans, L. V., J. A. Callow, & M. E. Callow (1973). Structural and physiological studies on the parasitic red alga *Holmsella*. *New Phytol.* 72:393–402. [3.3.2]

Evans, L. V., J. A. Callow, & M. E. Callow (1978). Parasitic red algae: an appraisal. In D. E. G. Irvine & J. H. Price (eds.), *Modern Approaches to the Taxonomy of Red and Brown Algae* (pp. 87–109). New York: Academic Press. [3.3.2]

Evans, L. V., J. A. Callow, & M. E. Callow (1982). The biology and biochemistry of reproduction and early development in *Fucus*. *Prog. Phycol. Res.* 1:67–110. [1.5.4, 1.6.2]

Fagerberg, W. R., & C. J. Dawes (1977). Studies on *Sargassum*. II. Quantitative ultrastructural changes in differentiated stipe cells during wound regeneration and regrowth. *Protoplasma* 92:211–27. [1.7.4]

Fagerberg, W. R., R. Moon, & E. Truby (1979). Studies on *Sargassum*. III. A quantitative ultrastructural and correlated physiological study of the blade and stipe organs of *S. filipendula*. *Protoplasma* 99:247–61. [1.1.1, 1.7.2]

Fain, S. R., L. D. Druehl, & D. L. Baillie (1988). Repeat and single copy sequences are differentially conserved in the evolution of kelp chloroplast DNA. *J. Phycoll.* 24:292–302. [1.4.1]

Falkowski, P. G., & J. LaRoche (1991). Acclimation to spectral irradiance in algae. *J. Phycol.* 27:8–14. [4.3.3]

Falkowski, P. G., & R. B. Rivkin (1976). The role of glutamine synthetase in the incorporation of ammonium in *Skeletonema costatum* (Bacillariophyceae). *J. Phycol.* 12:448–50. [5.5.1]

Farnham, W. F. (1980). Studies on aliens in the marine flora of southern England. In J. H. Price, D. E. G. Irvine, & W. F. Farnham (eds.), *The Shore Environment. 2: Ecosystems* (pp. 875–914). London: Academic Press. [2.4.1.]

Fawley, M. W., C. A. Douglas, K. D. Stewart, & K. R. Mattox (1990). Light-harvesting pigment-protein complexes of the Ulvophyceae (Chlorophyta): characterization and phylogenetic significance. *J. Phycol.* 26:186–95. [4.3.1]

Fenical, W. (1975). Halogenation in the Rhodophyta. A review. *J. Phycol.* 11:245–59. [3.1.2]

Ferguson, R. L., G. W. Thayer, & T. R. Rice (1980). Marine primary producers. In F. J. Vernberg & W. B. Vernberg (eds.), *Functional Adaptations of Marine Organisms* (pp. 9–69). New York: Academic Press. [1.1.1]

Ferreira, J. G. (1991). Factors governing mercury accumulation in three species of marine microalgae. *Aquat. Bot.* 39:335–43. [8.3.1]

Ferreira, J. G., & L. Ramos (1989). A model for the estimation of annual production rates of macrophyte algae. *Aquat. Bot.* 33: 53–70. [4.7, 4.7.4]

Fetter, R., & M. Neushul (1981). Studies in developing and released spermatia in the red alga *Tiffaniella snyderae* (Rhodophyta). *J. Phycol.* 17:141–59. [1.5.3]

Field, J. G., N. G. Jarman, G. S. Dieckmann, C. L. Griffiths, B. Velimirov, & P. Zoutendyk (1977). Sun, waves, seaweed and lobsters: the dynamics of a west coast kelp bed. *S. Afr. J. Sci.* 73:7–10. [4.7.6]

Field, J. G., C. L. Griffiths, N. Jarman, P. Zoutendyk, B. Velmirov, & A. Bowes (1980). Variation in structure and biomass of kelp communities along the south-west Cape coast. *Trans. R. Soc. S. Afr.* 44:145–203. [2.3.2]

Fielding, A. H., & G. Russell (1976). The effect of copper on competition between marine algae. *J. Appl. Ecol.* 13:871–6. [8.3.3]

Fillion-Myklebust, C., & T. A. Norton (1981). Epidermis shedding in the brown seaweed *Ascophyllum nodosum* (L.) Le Jolis, and its ecological significance. *Mar. Biol. Lett.* 2:45–51. [3.1.2]

Finckh, A. E. (1904). Biology of reef-forming organisms at Funafuti Atoll, Section VI. In *The Atoll of Funafuti* (pp. 125–50). Report of the Coral Reef Committee, Royal Society, London. [2.1.2]

Finden, D. A. S., E. Tippings, G. H. M. Jaworski, & C. S. Reynolds (1984). Light-induced reduction of natural iron (III) oxide and its relevance to phytoplankton. *Nature* 309:783–84. [5.5.6]

Fischer, N. S., & C. F. Wurster (1973). Individual and combined effects of temperature and polychlorinated biphenyls on the growth of three species of phytoplankton. *Environ. Pollut.* 5:205–12. [8.5.3]

Fisher, N. S., E. J. Carpenter, C. C. Remsen, & C. F. Wurster (1974). Effects of PCB on interspecific competition in natural and gnotobiotic phytoplankton communities in continuous and batch cultures. *Microb. Ecol.* 1:39–50. [8.5.3]

Fitter, A. H., & R. K. M. Hay (1981). *Environmental Physiology of Plants.* New York: Academic Press, 355 pp. [6.2.2]

Fjeld, A. (1972). Genetic control of cellular differentiation in *Ulva mutabilis.* Gene effects in early development. *Devel. Biol.* 28:326–43. [1.7.1]

Fjeld, A., & A. Løvlie (1976). Genetics of multicellular marine algae. In R. A. Lewin (ed.), *The Genetics of Algae* (pp. 219–35). Oxford, Blackwell Scientific. [1.4.1, 1.7.1]

Fleischmann, E. M. (1989). The measurement and penetration of ultraviolet radiation into tropical marine water. *Limnol. Oceanogr.* 34:1623–9. [4.2.2]

Fletcher, R. L. (1975). Heteroantagonism observed in mixed algal cultures. *Nature* 253:534–5. [3.1.2]

Fletcher, R. L., R. E. Baier, & M. S. Fornalik (1985). The effects of surface energy on germling development of some marine macroalgae (abstract). *Br. Phycol. J.* 20:184–5. [1.6.1]

Floc'h, J.-Y. (1982). Uptake of inorganic ions and their long distance transport in Fucales and Laminariales. In L. M. Srivastava (ed.), *Synthetic and Degradative Processes in Marine Macrophytes* (pp. 139–65). Berlin: Walter de Gruyter. [5.4.2, 5.5.7, 5.6]

Floc'h. J.-Y., & M. Penot (1978). Changes in ^{32}P-phosphorus compounds during translocation in *Laminaria digitata* (L.) Lamouroux. *Planta* 143:101–7. [5.6]

Florence, T. M., B. G. Lumsden, & J. J. Fardy (1984). Algae as indicators of copper speciation. In C. J. M. Kramer & J. C. Duinker (eds.), *Complexation of Trace Metals in Natural Waters* (pp. 411–18). Netherlands: Dr. W. Junk. [8.3.4]

Floyd, G. L., & C. J. O'Kelly (1984). Motile cell ultrastructure and the circumscription of the orders Ulotrichales and Ulvales (Ulvophyceae, Chlorophyta). *Am. J. Bot.* 71:111–20. [1.1.1]

Flynn, K. J., & I. Butler. (1986). Nitrogen sources for the growth of marine microalgae: role of dissolved free amino acids. *Mar. Ecol. Prog. Ser.* 34:281–304. [5.5.1]

Förstner, U. (1980). Inorganic pollutants, particularly heavy metals in estuaries. In E. Olausson & I. Cato (eds.), *Chemistry and Biogeochemistry of Estuaries* (pp. 307–48). New York: Wiley. [8.3]

Förstner, U., & G. Wittmann (1979). *Metal Pollution in Aquatic Environments.* Berlin: Springer-Verlag, 486 pp. [8.3, 8.3.4]

Foreman, R. E. (1976). Physiological aspects of carbon monoxide production by the brown alga *Nereocystis luetkeana.* *Can. J. Bot.* 54:352–60. [1.7.2]

Foreman, R. E. (1977). Benthic community modification and recovery following intensive grazing by *Strongylocentrotus droebachiensis.* *Helgol. Meeresunters.* 30:468–84. [3.2.1]

Fork, D. C. (1963). Observations on the function of chlorophyll *a* and accessory pigments in photosynthesis. In *Photosynthetic Mechanisms in Green Plants* (pp. 352–61). Publ. no. 1145, NAS-NRC, Washington, DC. [4.3.4]

Fork, D. C., & A. W. D. Larkum (1989). Light harvesting in the green alga *Ostreobium* sp., a coral symbiont adapted to extreme shade. *Mar. Biol.* 103:381–5. [4.3.1]

Forsberg, A., S. Söderlund, A. Frank, L. R. Petersson, & M. Pedersén (1988). Studies on metal content in the brown seaweed, *Fucus vesiculosus,* from the Archipelago of Stockholm. *Environ. Pollut.* 49:245–63. [8.3.1]

Forstner, M., & K. Rützler (1970). Measurements of the microclimate in littoral marine habitats. *Oceanogr. Mar. Biol. Annu. Rev.* 8:225–49. [7.1.2]

Foster, M. S. (1972). The algal turf community in the nest of the ocean goldfish (*Hypsypops rubicunda*). *Proc. Intl. Seaweed Symp.* 7:55–60. [3.1.1]

Foster, M. S. (1975). Regulation of algal community development in a *Macrocystis pyrifera* forest. *Mar. Biol.* 32:331–42. [3.1.1]

Foster, M. S., & W. P. Sousa (1985). Succession. In M. M. Littler & D. S. Littler (eds.), *Handbook of Phycological Methods: Ecological Field Methods: Macroalgae* (pp. 269–90). Cambridge University Press. [2.5.1]

Foster, M. S., T. A. Dean, & L. E. Deysher (1985). Subtidal techniques. In M. M. Littler & D. S. Littler (eds.), *Handbook of Phycological Methods: Ecological Field Methods: Macroalgae* (pp. 199–231). Cambridge University Press. [7.2.1]

Foster, M. S., J. A. Tarpley, & S. L. Dearn (1990). To clean or not to clean: the rationale, methods and consequences of removing oil from temperate shore. *Northwest Environ. J.* 6:105–20. [8.4.2]

Foster, P. (1976). Concentrations and concentration factors of heavy metals in brown algae. *Environ. Pollut.* 10:45–54. [8.3.1]

Foy, C. D., R. L. Chaney, & M. C. White (1978). The physiology of metal toxicity in plants. *Annu. Rev. Plant Physiol.* 29:511–66. [8.3.2]

Fralick, R. A., & A. C. Mathieson (1975). Physiological ecology of four *Polysiphonia* species (Rhodophyta, Ceramiales). *Mar. Biol.* 29:29–36. [6.5]

Francke, J. A., & H. Hillebrand (1980). Effects of copper on some filamentous Chlorophyta. *Aquat. Bot.* 8:285–9. [8.3.2]

Fréchette, M., & L. Legendre (1978). Photosynthèse phytoplanctonique: réponse à un stimulus simple, imitant les variations rapides de la lumière engendrées par les vagues. *J. Exp. Mar. Biol. Ecol.* 32:15–25. [1.1.3, 4.2.2]

Freedman, B. (1989). *Environmental Ecology.* Orlando, FL: Academic Press, 424 pp. [8.4.2]

French, D., M. M. Harlin, S. Pratt, H. Rines, & S. Puckett (1989). *Mumford Cove Water Quality, 1989. Monitoring Study of Macrophytes and Benthic Invertebrates.* ASA 88-17. Narragansett, R.I.: Applied Science Associates, 56 pp. [2.4.2]

Friedlander, M., & C. J. Dawes (1985). In situ uptake kinetics of ammonium and phosphate and chemical composition of the red seaweed *Gracilaria tikvahiae*. *J. Phycol.* 21:448–53. [5.4, 5.5.1, 5.5.2]

Friedlander, M., M. D. Krom, & A. Ben-Amotz (1991). The effect of light and ammonium on growth, epiphytes and chemical constituents of *Gracilaria conferta* in outdoor cultures. *Bot. Mar.* 34:161–6. [5.5.1]

Friedmann, [E.] I. (1961). Cinemicrography of spermatozoids and fertilization in Fucales. *Bull. Res. Counc. Israel* 10D:73–83. [1.5.4]

Friedmann, E. I., & W. C. Roth (1977). Development of the siphonous green alga *Penicillus* and the *Espera* state. *Bot. J. Linn. Soc.* 74:189–214. [1.2.1, 7.2.2]

Friedrich, H. (1969). *Marine Biology: An Introduction to Its Problems and Results.* London: Sidgwick & Jackson. [Fig. 6.1]

Fries, L. (1966). Temperature optima of some red algae in axenic culture. *Bot. Mar.* 9:12–14. [6.2.2]

Fries, L. (1982a). Selenium stimulates growth of marine macroalgae in axenic culture. *J. Phycol.* 18:328–31. [5.5.6]

Fries, L. (1982b). Vanadium, an essential element for some marine macroalgae. *Planta* 154:393–6. [5.5.6]

Fries, L. (1984). D-vitamins and their precursors as growth regulators in axenically cultivated marine macroalgae. *J. Phycol.* 20:62–6. [5.5.8]

Fries, N. (1979). Physiological characteristics of *Mycosphaerella ascophylli*, a fungal endophyte of the marine brown alga *Ascophyllum nodosum*. *Physiol. Plant.* 45:117–21. [3.3.1]

Fritsch, F. E. (1945). *The Structure and Reproduction of the Algae,* vol. 2. Cambridge University Press. [Fig. 1.4]

Fuge, R., & K. H. James (1973). Trace metal concentrations in brown seaweeds, Cardigan Bay, Wales. *Mar. Chem.* 1:281–93. [8.3.1]

Fujimura, T., T. Kawai, M. Shiga, T. Kajiwara, & A. Hatanaka (1989). Regeneration of protoplasts into complete thalli in the marine green alga *Ulva pertusa*. *Nippon Suisan Gakkaishi* 55:1353–9. [1.7.1]

Fujita, R. M. (1985). The role of nitrogen status in regulating transient ammonium uptake and nitrogen storage by macroalgae. *J. Exp. Mar. Biol. Ecol.* 99:283–301. [5.5.1]

Fujita, R. M., P. A. Wheeler, & R. L. Edwards (1988). Metabolic regulation of ammonium uptake by *Ulva rigida* (Chlorophyta): a compartmental analysis of the rate-limiting step for uptake. *J. Phycol.* 24:560–6. [5.4.1]

Fujita, R. M., P. A. Wheeler, & R. L. Edwards (1989). Assessment of macroalgal nutrient limitation in a seasonal upwelling region. *Mar. Ecol. Prog. Ser.* 53:293–303. [5.4]

Fujita, Y., & S. Migata (1986). Isolation of protoplasts from leaves of red algae *Porphyra yezoensis*. *Jpn. J. Phycol.* 34:63. [9.2.4]

Fulcher, R. G., & M. E. McCully (1969). Histological studies on the genus *Fucus*. IV. Regeneration and adventive embryony. *Can. J. Bot.* 47:1643–9. [1.7.4]

Fulcher, R. G., & M. E. McCully (1971). Histological studies on the genus *Fucus*. V. An autoradiographic and electron microscopic study of the early stages of regeneration. *Can. J. Bot.* 49:161–5. [1.7.4]

Gacesa, P. (1988). Alginates. *Carbohydr. Polymers* 8:161–82. [4.5.2, 4.5.3]

Gadd, G. M., & A. J. Griffiths (1978). Microorganisms and heavy metal toxicity. *Microb. Ecol.* 4:303–17. [8.3.2, 8.3.4]

Gagné, J. A., & K. H. Mann (1987). Evaluation of four models used to estimate kelp productivity from growth measurements. *Mar. Ecol. Prog. Ser.* 37:35–44. [4.7.4]

Gagné, J. A., K. H. Mann, & A. R. O. Chapman (1982). Seasonal patterns of growth and storage in *Laminaria longicruris* in relation to differing patterns of availability of nitrogen in the water. *Mar. Biol.* 69:91–101. [4.5.1, 4.6, 5.8.3]

Gaillard, J., & M. T. L'Hardy-Halos (1990). Morphogenèse du *Dictyota dichotoma* (Dictyotales, Phaeophyta). III. Ontogenèse et croissance des frondes adventives. *Phycologia* 29:39–53. [1.2.1, 1.7.4]

Gaillard, J., M. T. L'Hardy-Halos, & L. Pellegrini (1986). Morphogenèse du *Dictyota dichotoma* (Huds.) Lamouroux (Phaeophyta). II. Ontogenèse du thalle et cytologie ultrastructurale des différents types de cellules. *Phycologia* 25:340–57. [1.2.1, 1.7.4]

Gaines, S. D. & J. Lubchenco (1982). A unified approach to marine plant–herbivore interactions. II. Biogeography. *Annu. Rev. Ecol. Syst.* 13:111–38. [3.2.1]

Gantt, E. (1975). Phycobilisomes: light harvesting pigment complexes. *BioScience* 25:781–8. [Fig. 4.1]

Gantt, E. (1980). Structure and function of phycobilisomes: light harvesting pigment complexes in red and blue-green algae. *Int. Rev. Cytol.* 66:45–80. [5.5.1]

Gantt, E. (1989). *Porphyridium* as a red algal model for photosynthesis studies. In A. W. Coleman, L. J. Goff, & J. R. Stein-Taylor (eds.), *Algae as Experimental Systems* (pp. 249–68). New York: Alan R. Liss. [4.3.1]

Gantt, E. (1990). Pigmentation and photoacclimation. In K. M. Cole & R. G. Sheath (eds.), *Biology of the Red Algae* (pp. 203–19). Cambridge University Press. (4.3.1]

Gao, K., & I. Umezaki (1989a). Studies on diurnal photosynthetic performance of *Sargassum thunbergii*. I. Changes in photosynthesis under natural light. *Sôrui (Jpn. J. Phycol.)* 37:89–98. [4.7.2]

Gao, K., & I. Umezaki (1989b). Studies on diurnal photosynthetic performance of *Sargassum thunbergii*. II. Explanation of diurnal photosynthesis patterns from examinations in the laboratory. *Sôrui (Jpn. J. Phycol.)* 37:99–104. [4.7.2]

Garbary, D. (1979). Daylength and development in four species of Ceramiaceae (Rhodophyta). *Helgol. Meeresunters.* 32:213–27. [1.5.3]

Garbary, D. J., & D. J. Belliveau (1990). Diffuse growth, a new pattern of cell wall deposition for the Rhodophyta. *Phycologia* 29:98–102. [1.3.4]

Garbary, D. J., & A. Gautam (1989). The *Ascophyllum, Polysiphonia, Mycosphaerella* symbiosis. I. Population ecology of *Mycosphaerella* from Nova Scotia. *Bot. Mar.* 32:181–6. [3.3.1]

Garbary, D. J., D. Grund, & J. McLachlan (1978). The taxonomic status of *Ceramium rubrum* (Huds.) C. Ag. (Ceramiales, Rhodophyceae) based on culture experiments. *Phycologia* 17:85–94. [6.3.3]

Garbary, D., D. Belliveau, & R. Irwin (1988). Apical control of band elongation in *Antithamnion defectum* (Ceramiaceaae, Rhodophyta). *Can. J. Bot.* 66:1308–15. [1.3.4]

Garbary, D. J., J. Burke, & T. Lining (1991). The *Ascophyllum/Polysiphonia/Mycosphaerella* symbiosis. II. Aspects of the ecology and distribution of *Polysiphonia lanosa* in Nova Scotia. *Bot. Mar.* 34:391–401.[3.3.2]

Gayler, K. R., & W. R. Morgan (1976). An NADP-dependent glutamate dehydrogenase in chloroplasts from the marine green alga *Caulerpa simpliciuscula*. *Plant Physiol.* 58:283–7. [5.5.1]

Geesink, R. (1973). Experimental investigations on marine and freshwater *Bangia* (Rhodophyta) from the Netherlands. *J. Exp. Mar. Biol. Ecol.* 11:239–47. [6.3.4]

Geider, R. J., & B. A. Osborne (1992). *Algal Photosynthesis: The Measurement of Algal Gas Exchange.* London: Chapman & Hall, 256 pp. [4.7.1]

Geiselman, J. A., & O. J. McConnell (1981). Polyphenols in the brown algae, *Fucus vesiculosus* and *Ascophyllum nodosum:* chemical defenses against the marine herbivorous snail, *Littorina littorea*. *J. Chem. Ecol.* 7:1115–33. [3.2.3]

Gekeler, W., E. Grill, E. -L. Winnacker, & M. H. Zenk (1988). Algae sequester heavy metals via synthesis of phytochelatin complexes. *Arch. Microbiol.* 150:197–202. [8.3.2]

Gerard, V. A. (1982a). *In situ* water motion and nutrient uptake by the giant kelp *Macrocystis pyrifera*. *Mar. Biol.* 69:51–4. [5.4.2]

Gerard, V. A. (1982b). *In situ* rates of nitrate uptake by giant kelp, *Macrocystis pyrifera* (L.) C. Agardh: tissue differences, environmental effects, and predictions of nitrogen-limited growth. *J. Exp. Mar. Biol. Ecol.* 62:211–24. [5.4.2]

Gerard, V. A. (1982c). Growth and utilization of internal nitrogen reserves by the giant kelp *Macrocystis pyrifera* in a low-nitrogen environment. *Mar. Biol.* 66:27–35. [5.4.2]

Gerard, V. A. (1984a). Physiological effects of El Niño on giant kelp in southern California. *Mar. Biol. Lett.* 5:317–22. [5.8.3]

Gerard, V. A. (1984b). The light environment in a giant kelp forest: influence of *Macrocystis pyrifera* on spatial and temporal variability. *Mar. Biol.* 84:189–95. [4.3.2]

Gerard, V. A. (1988). Ecotypic differentiation in light-related traits of the kelp *Laminaria saccharina*. *Mar. Biol.* 97:25–36. [4.3.2]

Gerard, V. A., & K. R. Du Bois (1988). Temperature ecotypes near the southern boundary of the kelp *Laminaria saccharina*. *Mar. Biol.* 97:575–80. [6.2.3]

Gerard, V. A., & K. H. Mann (1979). Growth and production of *Laminaria longicruris* (Phaeophyta) populations exposed to different intensities of water movement. *J. Phycol.* 15:33–41. [7.2.1]

Gerard, V. A., S. E. Dunham, & G. Rosenberg (1990). Nitrogen-fixation by cyanobacteria associated with *Codium fragile* (Chlorophyta): environmental effects and transfer of fixed nitrogen. *Mar. Biol.* 105:1–8. [5.2]

Gerlach, S. A. (1982). *Marine Pollution: Diagnoses and Therapy.* Berlin: Springer-Verlag, 218 pp. [8.3.3, 8.4.2, 8.6.1]

Gerloff, G. C. & P. H. Krombholz (1966). Tissue analysis as a measurement of nutrient availability for the growth of angiosperm aquatic plants. *Limnol. Oceanogr.* 11:529–37. [5.7.3]

Germann, I. (1988). Effects of the 1983-El Niño on growth and carbon and nitrogen metabolism of *Pleurophycus gardneri* (Phaeophyceae: Laminariales) in the northeastern Pacific. *Mar. Biol.* 99:445–55. [5.8.3, 6.2.6]

Gerwick, W. H., & N. J. Lang (1977). Structural, chemical and ecological studies on iridescence in *Iridaea* (Rhodophyta). *J. Phycol.* 13:121–7. [1.3.1]

Gessner, F. (1970). Temperature: plants. In O. Kinne (ed.), *Marine Ecology* (vol. 1, pt. 1, pp. 363–406). New York: Wiley. [6.2.4, 6.2.5]

Gessner, F. (1971). Wasserpermeabilität und Photosynthese bei marinen Algen. *Bot. Mar.* 14:29–31. [6.3.3, 6.4.3]

Gessner, F., & L. Hammer (1968). Exosmosis and "free space" in marine benthic algae. *Mar. Biol.* 2:88–91. [5.3.1, 6.4.1]

Gessner, F., & W. Schramm (1971). Salinity: plants. In O. Kinne (ed.), *Marine Ecology* (vol. 1, pt. 2, pp. 705–820). New York: Wiley. [6.1.3, 6.3.3, 6.3.4, 6.4, 6.5]

Geyer. R. A. (ed.) (1980). *Marine Environmental Pollution.* Vol. I: *Hydrocarbons.* Amsterdam: Elsevier Scientific.

Giddings et al. (1983). *Plant Physiol.* 71:409–19. [Fig. 4.8]

Gibor, A. (1973). *Acetabularia*. Physiological role of their deciduous organelles. *Protoplasma* 78:461–5. [5.8.2]

Gilmartin, M. (1960). The ecological distribution of the deep water algae of Eniwetok Atoll. *Ecology* 41:210–21. [4.3.4]

Glass, A. D. M. (1989). *Plant Nutrition: An Introduction to Current Concepts.* Boston: Jones & Barttell, 234 pp. [5.3.1]

Glazer, A. N. (1985). Light harvesting by phycobilisomes. *Annu. Rev. Biochem.* 14:47–77. [4.3.1]

Glynn, P. W. (1965). Community composition, structure, and interrelationships in the marine intertidal *Endocladia muricata–Balanus glandula* association in Monterey Bay, California. *Beaufortia* 12(148):1–198. [6.1.3]

Glynn, P. W. (1988). El Niño–southern oscillation 1982–1983: nearshore population, community, and ecosystem responses. *Annu. Rev. Ecol. Syst.* 19:309–45. [6.2.6]

Glynn, P. W., & W. H. de Weerdt (1991). Elimination of two reef-building hydrocorals following the 1982–83 El Niño warming event. *Science* 253:69–71. [6.2.6]

Goff, L. J. (1979). The biology of *Harveyella mirabilis* (Cryptonemiales, Rhodophyceae). VII. Structure and proposed function of host-penetrating cells. *J. Phycol.* 15:87–100. [3.3.2]

Goff, L. J. (1982). Biology of parasitic red algae. *Prog. Phycol. Res.* 1:289–369. [3.3]

Goff, L. J., & A. W. Coleman (1984). Transfer of nuclei from a parasite to its host. *Proc. Natl. Acad. Sci. USA* 81:5420–4. [1.4.1]

Goff, L. J., & A. W. Coleman (1987). The solution to the cytological paradox of isomorphy. *J. Cell Biol.* 104:739–48. [1.5.2]

Goff, L. J., & A. W. Coleman (1988a). Red algal plasmids: potential genetic engineering tools (abstract). *J. Phycol.* (*Suppl.*) 24:23. [1.4.1]

Goff, L. J., & A. W. Coleman (1988b). The use of plastid DNA restriction endonuclease patterns in delimiting red algal species and populations. *J. Phycol.* 24:357–68. [1.4.1]

Goff, L. J., & A. W. Coleman (1990). DNA: microspectrofluorometric studies. In K. M. Cole & R. G. Sheath (eds.), *Biology of the Red Algae* (pp. 43–71). Cambridge University Press. [1.5.2]

Golden, J. W., & D. R. Wiest (1988). Genome rearrangement and nitrogen fixation in *Anabaena* blocked by inactivation of *xisA* gene. *Science* 242:1421–3. [1.4.1]

Goldman, J. C., & J. J. McCarthy (1978). Steady state growth and ammonium uptake of a fast-growing marine diatom. *Limnol. Oceanogr.* 23:695–703. [5.7.1]

Goldman, J. C., J. J. McCarthy, & D. G. Peavey (1979). Growth rate influence on the chemical composition of phytoplankton in oceanic waters. *Nature* 279:210–15. [5.8.1]

Golikov, A. N. (1985). Zonations and organismic assemblages: comments on the comprehensive review by Pérès (1982). *Mar. Ecol. Prog. Ser.* 23:203–6. [2.3.2]

Gonor, J. J., & P. F. Kemp (1978). *Procedures for Quantitative Ecological Assessments in Intertidal Environments.* Publ. no. 600/3-78-087, U.S. Environmental Protection Agency, Corvallis, OR. [2.5.1]

Gonzalez, M. A., & L. J. Goff (1989). The red algal epiphytes *Microcladia coulteri* and *M. californica* (Rhodophyceae, Ceramiaceae). II. Basiphyte specificity. *J. Phycol.* 25:558–67. [3.1.2, 3.3.2]

Goodwin, T. W. (1974). Carotenoids and biliproteins. In W. D. P. Stewart (ed.), *Algal Physiology and Biochemistry* (pp. 176–205). Oxford: Blackwell Scientific. [4.3.1]

Goodwin, T. W., & E. I. Mercer (1983). *Introduction to Plant Biochemisty*, 2nd ed. Oxford: Pergamon. [1.3, 1.5.3, 4.0, 4.3.1]

Gordon, D. M., P. B. Birch, & A. J. McComb (1980). The effects of light, temperature and salinity on photosynthetic rates of an estuarine *Cladophora*. *Bot. Mar.* 29:749–55. [6.1.2]

Gordon, D. M., P. B. Birch, & A. J. McComb (1981). Effects of inorganic phosphorus and nitrogen on the growth of an estuarine *Cladophora* in culture. *Bot. Mar.* 24:93–106. [5.7.3, 8.6.1]

Gould, J. M. (1976). Inhibition by triphenyltin chloride of a tightly bound membrane component involved in photophosphorylation. *Eur. J. Biochem.* 62:567–75. [8.5.4]

Graham, D., & B. D. Patterson (1982). Responses of plants to low, non-freezing temperatures: proteins, metabolism, and acclimation. *Annu. Rev. Plant Physiol.* 33:347–72. [6.2.4]

Grant, B. R., & M. A. Borowitzka (1984). The chloroplasts of giant-celled and coenocytic algae: biochemistry and structure. *Bot. Rev.* 50:267–307. [1.3.2]

Gray, J. S. (1982). Effects of pollutants on marine ecosystems. *Neth. J. Sea Res.* 16:424–43. [8.1]

Green, B. R. (1976). Approaches to the genetics of *Acetabularia*. In R. A. Lewin (ed.), *The Genetics of Algae* (pp. 236–56). Oxford: Blackwell Scientific. [1.4.1, 1.4.2]

Green, R. H. (1979). *Sampling Design and Statistical Methods for Environmental Biologists.* New York: Wiley. [1.1.2, 2.5.1, 8.1]

Greene, R. M., & V. A. Gerard (1990). Effects of high-frequency light fluctuations on growth and photoacclimation of the red alga *Chondrus crispus*. *Mar. Biol.* 105:337–44. [4.2.1, 4.2.2]

Greene, R. W. (1970). Symbiosis in sacoglossan opisthobranchs: functional capacity of symbiotic chloroplasts. *Mar. Biol.* 72:138–42. [3.2.2]

Gregenheimer, P. (1990). Preparations of extracts from plants. *Meth. Enzymol.* 182:174–93. [5.5.1]

Grice, G. D., & M. R. Reeve (eds.) (1982). *Marine Microcosms: Biological and Chemical Research in Experimental Ecosystems.* Berlin: Springer-Verlag, 430 pp. [8.1]

Grigg, R. W., J. J. Polovina, & M. J. Atkinson (1984). Model of a coral reef ecosystem. III. Resource limitation, community regulation, fisheries yield and resource management. *Coral Reefs* 3:23–7. [4.7.6]

Grill, E., E. -L. Winnacker, & M. H. Zenk (1987). Phytochelatins, a class of heavy metal–binding peptides from plants, are functionally analogous to metallothioneins. *Proc. Natl. Acad. Sci. USA* 84:439–43. [8.3.2]

Grime, J. P. (1979). *Plant Strategies and Vegetation Processes.* New York: Wiley. [2.5.2]

Groen, P. (1980). Oceans and seas. I. Physical and chemical properties. In *The New Encyclopaedia Britannica, Macropaedia* (vol. 13, pp. 484–97). [6.1.1]

Gross, M. G. (1982). *Oceanography: A View of the Earth,* 3rd ed. Englewood Cliffs, N.J: Prentice-Hall, 498 pp. [2.2.1]

Grossman, A. R., L. K. Anderson, P. B. Conley, & P. G. Lemaux (1989). Molecular analyses of complementary chromatic adaptation and the biosynthesis of a phycobilisome. In A. W. Coleman, L. J. Goff, & J. R. Stein-Taylor (eds.), *Algae as Experimental Systems* (pp. 269–88). New York: Alan R. Liss. [4.3.4]

Guiry, M. D. (1987). The evolution of life history types in the Rhodophyta: an appraisal. *Crypt. Algol.* 8:1–12. [1.5.4]

Guiry, M. D. (1990). Sporangia and spores. In K. M. Cole & R. G. Sheath (eds.), *Biology of the Red Algae*, (pp. 347–76). Cambridge University Press. [1.5.2]

Gundlach, E., & M. Hayes (1978). Vulnerability of coastal environments to oil spill impacts. *Mar. Technol. Soc. J.* 12:18–27. [8.4.2]

Gundlach, E. R., P. D. Boehm, M. Marchand, R. M. Atlas, D. M. Ward, & D. A. Wolfe (1983). The fate of *Amoco Cadiz* oil. *Science* 221:122–9. [8.4.1, 8.4.2]

Gunnill, F. C. (1980a). Demography of the intertidal brown alga *Pelvetia fastigiata* in southern California, U.S.A. *Mar. Biol.* 59:169–79. [2.5.2, 6.2.6]

Gunnill, F. C. (1980b). Recruitment and standing stocks in populations of one green alga and five brown algae in the intertidal zone near La Jolla, California during 1973–1977. *Mar. Ecol. Prog. Ser.* 3:231–43. [6.2.6, 6.4]

Gunnill, F. C. (1985). Population fluctuations of seven macroalgae in southern California during 1981–1983 including effects of severe storms and an El Niño. *J. Exp. Mar. Biol. Ecol.* 85:149–64. [6.2.6, 6.4]

Gutknecht, J. (1963). Zinc-65 uptake by benthic algae. *Limnol. Oceanogr.* 8:31–8. [5.5.7]

Gutknecht, J. (1965). Uptake and retention of cesium 137 and zinc 65 by seaweeds. *Limnol. Oceanogr.* 10:58–66. [5.5.7]

Gutknecht, J., & J. Dainty (1968). Ionic relations of marine algae. *Oceanogr. Mar. Biol. Annu. Rev.* 6:163–200. [6.3.1]

Hackney, J. M., & P. Sze (1988). Photorespiration and productivity rates of a coral reef algal turf assemblage. *Mar. Biol.* 98:483–92. [4.4.2, 4.7.1]

Hackney, J. M., R. C. Carpenter, & W. H. Adey (1989). Characteristic adaptations to grazing among algal turfs on a Caribbean coral reef. *Phycologia* 28:109–19. [1.1.1, 3.2.1]

Haines, K. C., & P. A. Wheeler (1978). Ammonium and nitrate uptake by the marine macrophyte *Hypnea musciformis* (Rhodophyta) and *Macrocystis pyrifera* (Phaeophyta). *J. Phycol.* 14:319–24. [5.4, 5.4.2, 5.5.1]

Hall, A. (1980). Heavy metal co-tolerance in a copper-tolerant population of the marine fouling alga, *Ectocarpus siliculosus* (Dillw.) Lyngbye. *New Phytol.* 85:73–8. [8.3.2]

Hall, A., A. H. Fielding, & M. Butler (1979). Mechanisms of copper tolerance in the marine fouling alga *Ectocarpus siliculosus:* evidence for an exclusion mechanism. *Mar. Biol.* 54:195–9. [8.3.2]

Hall, J. L., T. J. Flowers, & R. M. Roberts (1982). *Plant Cell Structure and Metabolism,* 2nd ed. London: Longman. [4.0]

Hall, L. W., & A. E. Pinkney (1984). Acute and sublethal effects of organotin compounds on aquatic biota: an interpretive literature evaluation. *CRC Crit. Rev. Toxicol.* 14:159–209. [8.5.4]

Halldal, P. (1964). Ultraviolet action spectra of photosynthesis and photosynthetic inhibition in a green alga and a red alga. *Physiol. Plant.* 17:414–21. [4.2.1]

Hamelink, J. L., R. C. Waybrant, & R. C. Ball (1971). A proposal: exchange equilibria control the degree chlorinated hydrocarbons are biologically magnified in lentic environments. *Trans. Am. Fish. Soc.* 100:207–14. [8.3.5]

Hammer, L. (1968). Salzgehalt und Photosynthese bei marinen Pflanzen. *Mar. Biol.* 1:185–90. [6.3.3]

Hammer, L. (1969). "Free space-photosynthesis" in the algae *Fucus virsoides* and *Laminaria saccharina. Mar. Biol.* 4:136–8. [5.3.1]

Hanic, L. A., & J. S. Craigie (1969). Studies on the algal cuticle. *J. Phycol.* 5:89–109. [1.3.1]

Hanisak, M. D. (1979). Nitrogen limitation of *Codium fragile* ssp. *tomentosoides* as determined by tissue analysis. *Mar. Biol.* 50:333–7. [5.7.3]

Hanisak, M. D. (1983). The nitrogen relationships of marine macroalgae. In E. J. Carpenter & D. G. Capone (eds.). *Nitrogen in the Marine Environment* (pp. 699–730). New York: Academic Press. [5.5.1]

Hanisak, M. D. (1987). Cultivation of *Gracilaria* and other macroalgae in Florida for energy production. In K. T. Bird & P. H. Benson (eds.), *Seaweed Cultivation for Renewable Resources* (pp. 191–218). Amsterdam: Elsevier. [9.6, 9.8]

Hanisak, M. D. (1990). The use of *Gracilaria tikvahiae* (Gracilariales, Rhodophyta) as a model system to understand the nitrogen limitation of cultured seaweeds. *Hydrobiologia* 204/205:79–87. [5.7.3]

Hanisak, M. D., & M. M. Harlin (1978). Uptake of inorganic nitrogen by *Codium fragile* subsp. *tomentosoides* (Chlorophyta). *J. Phycol.* 14:450–4. [5.5.1]

Hanisak, M. D., M. M. Littler, & D. S. Littler (1988). Significance of macroalgal polymorphism: intraspecific tests of the functional-form model. *Mar. Biol.* 99:157–65. [4.3.2]

Hansen, D. V. (1990). Physical aspects of the El Niño event of 1982–1983. In P. W. Glynn (ed.), *Global Ecological Consequences of the 1982–83 El Niño–southern oscillation* (pp. 1–20). Amsterdam: Elsevier. [6.2.6]

Hansen, G. I., & S. C. Lindstrom (1984). A morphological study of *Hommersandia maximicarpia* gen. et sp. nov. (Kallymeniaceae, Rhodophyta) from the North Pacific. *J. Phycol.* 20:476–88. [Fig. 1.38]

Hanson, J. B., & A. J. Trewavas (1982). Regulation of plant cell growth: the changing perspective. *New Phytol.* 90:1–18. [1.3.4]

Harding, L. W., Jr., & J. H. Phillips, Jr. (1978). Polychlorinated biphenyls: transfer from microparticulates to marine phytoplankton and the effects on photosynthesis. *Science* 202:1189–92. [8.5.3]

Hardy, F. G., & B. L. Moss (1979). Attachment and development of the zygotes of *Pelvetia canaliculata* (L.) Dcne. et Thur. (Phaeophyceae, Fucales). *Phycologia* 18:203–12. [1.6.1]

Harlin, M. M. (1973). "Obligate" algal epiphyte: *Smithora naiadum* grows on a synthetic substrate. *J. Phycol.* 9:230–2. [2.4.2]

Harlin, M. M. (1978). Nitrate uptake by *Enteromorpha* spp. (Chlorophyceae): applications to aquaculture systems. *Aquaculture* 15:373–6. [Table 5.4]

Harlin, M. M. (1980). Seagrass epiphytes. In R. C. Phillips & C. P. McRoy (eds.), *Handbook of Seagrass Biology. An Ecosystem Perspective* (pp. 117–51). New York: Garland STPM Press. [2.4.2]

Harlin, M. M. (1987). Allelochemistry in marine macroalgae. *CRC Crit. Rev. Plant Sci.* 5:237–49. [3.1.3, 3.2.2]

Harlin, M. M., & J. S. Craigie (1975). The distribution of photosynthate in *Ascophyllum nodosum* as it relates to epiphytic *Polysiphonia lanosa. J. Phycol.* 11:109–13. [3.3.2, 4.7.3]

Harlin, M. M., & J. S. Craigie (1978). Nitrate uptake by *Laminaria longicruris* (Phaeophyceae). *J. Phycol.* 14:464–7. [5.4.1, 5.4.2, 5.5.1, Table 5.4]

Harlin, M. M., & B. Thorne-Miller (1981). Nutrient enrichment of seagrass beds in a Rhode Island coastal lagoon. *Mar. Biol.* 65:221–9. [2.4.2]

Harlin, M. M., & P. A. Wheeler (1985). Nutrient uptake. In M. M. Littler & D. S. Littler (eds.), *Handbook of Phycological Methods: Ecological Field Methods: Macroalgae* (pp. 493–508). Cambridge University Press. [5.4.1]

Harlin, M. M., W. J. Woelkerling, & D. I. Walker (1985). Effects of a hypersalinity gradient on epiphytic Corallinaceae

(Rhodophyta) in Shark Bay, Western Australia. *Phycologia* 24:389–402. [6.5]

Harper, J. L. (1977). *Population Biology of Plants.* New York: Academic Press, 892 pp. [2.5.2, 7.2.3]

Harper, J. L. (1982). After description. In E. I. Newmann (ed.), *The Plant Community as a Working Mechanism* (pp. 11–25). Oxford: Blackwell Scientific. [1.1.3]

Harris, G. P. (1980). The measurement of photosynthesis in natural populations of phytoplankton. In I. Morris (ed.), *The Physiological Ecology of Phytoplankton* (pp. 129–87). Oxford: Blackwell Scientific. [4.4.1]

Harrison, P. G., & C. D. Durance (1985). Reductions in photosynthetic carbon uptake in epiphytic diatoms by water-soluble extracts of leaves of *Zostera marina. Mar. Biol.* 90:117–19. [2.4.2]

Harrison, P. J., & L. D. Druehl (1982). Nutrient uptake and growth in the Laminariales and other macrophytes: a consideration of methods. In L. M. Srivastava (ed.), *Synthetic and Degradative Processes in Marine Macrophytes* (pp. 99–120). Berlin: Walter de Gruyter. [5.4.1]

Harrison, P. J., L. D. Druehl, K. E. Lloyd, & P. A. Thompson (1986). Nitrogen uptake kinetics in three year-classes of *Laminaria groenlandica* (Laminariales: Phaeophyta). *Mar. Biol.* 93:29–35. [5.4.2, 5.5.1]

Harrison, P. J., P. W. Yu, P. A. Thompson, N. M. Price, & D. J. Phillips (1988). Survey of selenium requirements in marine phytoplankton. *Mar. Ecol. Prog. Ser.* 47:89–96. [5.5.7]

Harrison, P. J., J. S. Parslow, & H. L. Conway (1989). Determination of nutrient uptake kinetic parameters: a comparison of methods. *Mar. Ecol. Prog. Ser.* 52:301–12. [5.4, 5.4.1, 5.5.1]

Harrison, P. J., M. H. Hu, Y. P. Yang, & X. Lu (1990). Phosphate limitation in estuarine and coastal waters of China. *J. Exp. Mar. Biol. Ecol.* 140:79–87. [5.5.2]

Harvey, W. H. (1841). *A Manual of the British Algae.* London: J. van Voorst. [4.3.1]

Hasegawa, Y. (1971). Forced cultivation of *Laminaria. Proc. Intl. Seaweed Symp.* 7:391–3. [9.3.1]

Hatcher, B. G. (1977). An apparatus for measuring photosynthesis and respiration of intact large marine algae and comparison of results with those from experiments with tissue segments. *Mar. Biol.* 43:381–5. [4.7.1, 5.4.2]

Hatcher, B. G., A. R. O. Chapman, & K. H. Mann (1977). An annual carbon budget for the kelp *Laminaria longicruris. Mar. Biol.* 44:85–96. [4.7.1, 4.7.3, 5.8.3]

Haug, A. (1961). The affinity of some divalent metals to different types of alginates. *Acta Chem. Scand.* 15:1794–5. [5.3.1]

Haug, A. (1976). The influence of borate and calcium on the gel formation of a sulfated polysaccharide from *Ulva lactuca. Acta Chem. Scand.* B30:562–6. [4.5.2]

Haug, A., & B. Larsen (1974). Biosynthesis of algal polysaccharides. In J. B. Pridham (ed.), *Plant Carbohydrate Biochemistry* (pp. 207–18). New York: Academic Press. [4.5.3]

Haug, A., & O. Smidsrød (1967). Strontium, calcium and magnesium in brown algae. *Nature* 215:1167–8. [5.3.1]

Hausinger, R. P. (1987). Nickel utilization by microorganisms. *Microbiol. Rev.* 51:22–42. [5.5.7]

Hawkes, M. W. (1990). Reproductive strategies. In K. M. Cole & R. G. Sheath (eds.), *Biology of the Red Algae* (pp. 455–76). Cambridge University Press. [1.5.2]

Hawkins, S. J., & R. G. Hartnoll (1983a). Changes in a rocky shore communty: an evaluation of monitoring. *Mar. Environ. Res.* 9:131–81. [8.1.]

Hawkins, S. J., & R. G. Hartnoll (1983b). Grazing of intertidal algae by marine invertebrates. *Oceanogr. Mar. Biol. Annu. Rev.* 21:195–282. [2.1.1]

Hawkins, S. J., D. C. Watson, A. S. Hill, S. P. Harding, M. A. Kyriakides, S. Hutchinson, & T. A. Norton (1989). A comparison of feeding mechanisms in microphagous herbivorous intertidal prosobranchs in relation to resource partitioning. *J. Molluscan Studies* 55:151–65. [2.1.1]

Haxen, P. G., & O. A. M. Lewis (1981). Nitrate assimilation in the marine kelp, *Macrocystis angustifolia* (Phaeophyceae). *Bot. Mar.* 24:631–5. [5.5.1]

Haxo, F. T., & L. R. Blinks (1950). Photosynthetic action spectra of marine algae. *J. Gen. Physiol.* 33:389–422. [4.3.4]

Hay, M. E. (1981a). The functional morphology of turf-forming seaweeds: persistence in stressful marine habitats. *Ecology* 62:739–50. [3.2.2]

Hay, M. E. (1981b). Spatial patterns of grazing intensity on a Caribbean barrier reef: herbivory and algal distribution. *Aquat. Bot.* 11:97–109. [3.2.2]

Hay, M. E. (1986). Functional geometry of seaweeds: ecological consequences of thallus layering and shape in contrasting light environments. In T. J. Givnish (ed.), *On the Economy of Plant Form and Function* (pp. 635–66). Cambridge University Press. [1.2.2, 4.2.2, 4.3.2]

Hay, M. E. (1988). Associational plant defenses and the maintenance of species diversity: turning competitors into accomplices. *Am. Nat.* 128:617–41. [3.2.2]

Hay, M. E., & W. Fenical (1988). Chemically-mediated seaweed herbivore interactions. *Annu. Rev. Ecol. Syst.* 19:111–45. [2.1.2, 3.2.3]

Hay, M. E., T. Colburn, & D. Downing (1983). Spatial and temporal patterns in herbivory on a Caribbean fringing reef: the effects on plant distribution. *Oecologia* 58:299–308. [3.2.1]

Hay, M. E., V. J. Paul, S. M. Lewis, K. Gustafson, J. Tucker, & R. N. Trindell (1988a). Can tropical seaweeds reduce herbivory by growing at night? Diel patterns of growth, nitrogen content, herbivory, and chemical versus morphological defenses. *Oecologia* 75:233–45. [3.2.3]

Hay, M. E., P. E. Renaud, & W. Fenical (1988b). Large mobile versus small sedentary herbivores and their resistance to seaweed chemical defenses. *Oecologia* 75:246–52. [3.2.3]

Hay, M. E., J. E. Duffy, V. J. Paul, P. E. Renaud, & W. Fenical (1990). Specialist herbivores reduce their susceptibility to predation by feeding on the chemically defended seaweed *Avrainvillea longicaulis. Limnol. Oceanogr.* 35:1734–43. [3.2.3]

Hayden, R. E., L. Dionne, & D. S. Fensom (1972). Electrical impedance studies of stem tissue of *Solanum* clones during cooling. *Can. J. Bot.* 50:1547–54. [6.2.4]

Healey, F. P. (1973). Inorganic nutrient uptake and deficiency in algae. *CRC Crit. Rev. Microbiol.* 3:69–113. [5.5.2]

Healey, F. P. (1980). Slope of the Monod equation as an indicator of advantage in nutrient competition. *Microb. Ecol.* 5:281–6. [5.4]

Hedrich, R., & J. I. Schroeder (1989). The physiology of ion channels and electrogenic pumps in higher plants. *Annu. Rev. Plant Physiol.* 40:539–69. [5.3.3]

Helfferich, C., & C. P. McRoy (1980). Introduction: the spaces in the pattern. In R. C. Phillips & C. P. McRoy (eds.), *Handbood of Seagrass Biology. An Ecosystem Perspective* (pp. 1–5). New York: Garland STPM Press. [1.1.1]

Hellebust, J. A. (1976). Osmoregulation. *Annu. Rev. Plant Physiol.* 27:485–505. [6.3.2]

Hellebust, J. A., & I. Ahmad (1991). Regulation of nitrogen assimilation in green microalgae. *Biol. Oceanogr.* 6:241–55. [5.5.1]

Hellebust, J. A., & J. S. Craigie (eds.) (1978). *Handbook of Phycological Methods. II. Physiological and Biochemical Methods.* Cambridge University Press, 512 pp. [4.7.1]

Hellenbrand, K. (1978). Effect of pulp mill effluent on productivity of seaweeds. *Proc. Intl. Seaweed Symp.* 9:161–71. [8.6.2]

Hempel, W. M., C. W. Sutton, D. Kaska, D. C. Ord, D. C. Reed, D. R. Laur, A. W. Ebeling, & D. D. Eardley (1989). Purification of species specific antibodies to carbohydrate components of *Macrocystis pyrifera* (Phaeophyta). *J. Phycol.* 25:144–9. [1.3.1]

Henley, W. J., & J. Ramus (1989). Photoacclimation and growth rate responses of *Ulva rotundata* (Chlorophyta) to intraday variations in growth irradiance. *J. Phycol.* 25:398–401. [4.2.2]

Henry, E. C. (1988). Regulation of reproduction in brown algae by light and temperature. *Bot. Mar.* 31:353–7. [1.5.3]

Henry, E. C., & K. M. Cole (1982). Ultrastructure of swarmers in the Laminariales. I. Zoospores. *J. Phycol.* 18:550–69. [1.6.1]

Herbert, S. K. (1990). Photoinhibition resistance in the red alga *Porphyra perforata.* The role of photoinhibition repair. *Plant Physiol.* 92:514–19. [4.3.3]

Herbert, S. K., & J. R. Waaland (1988). Photoinhibition of photosynthesis in a sun and a shade species of the red algal genus *Porphyra. Mar. Biol.* 97:1–7. [4.3.3]

Hershelman, G. P., H. A. Schafer, T. K. Jan, & D. R. Young (1981). Metals in marine sediments near a large California municipal outfall. *Mar. Pollut. Bull.* 12:131–4. [8.6.1]

Herth, W. (1980). Calcofluor white and Congo red inhibit chitin microfibril assembly of *Poterioochromonas:* evidence for a gap between polymerization and microfibril formation. *J. Cell Biol.* 87:442–50. [1.3.1]

Hetherington, J. A. (1976). Radioactivity in surface and coastal waters of the British Isles, 1974. *Fish. Radiobiol. Labor. Tech. Rep.* 11:1–35. [8.3.3]

Higgins, H. W., & D. J. Mackey (1987a). Role of *Ecklonia radiata* (C. Ag.) J. Agardh in determining trace metal availability in coastal waters. I. Total trace metals. *Aust. J. Mar. Freshw. Res.* 38:307–15. [8.3.1]

Higgins, H. W. & D. J. Mackey (1987b). II. Trace metal speciation. *Austr. J. Mar. Freshwater Res.* 38:317–28. [8.3.5]

Hillis-Colinveaux, L. (1985). *Halimeda* and other deep forereef algae at Enewetak Atoll. In V. M. Harmelin & B. Salvat (eds.), *Proceedings of the 5th International Coral Reef Congress, Tahiti* (vol. 5, pp. 9–14). Moorea, French Polynesia: Antenne Museum–EPHE. [2.3.2]

Hillman, W. S. (1976). Biological rhythms and physiological timing. *Annu. Rev. Plant Physiol.* 27:159–79. [4.7.2]

Hinds, P. A., & D. L. Ballentine (1987). Effects of the Caribbean threespot damselfish, *Stegastes planifrons* (Cuvier), on algal lawn composition. *Aquat. Bot.* 27:299–308. [3.2.2]

Hiscock, K. (1983). Water movement. In R. Earll & D. G. Erwin (eds.), *Sublittoral Ecology* (pp. 58–96). Oxford: Clarendon Press. [7.1.2]

Hixon, M. A., & W. N. Brostoff (1983). Damselfish as keystone species in reverse: intermediate disturbance and diversity of reef algae. *Science* 220:511–13. [3.2.1]

Ho, Y. B. (1987). Metals in 19 intertidal macroalgae in Hong Kong waters. *Mar. Pollut. Bull.* 18:564–6. [8.3.1]

Ho, Y. B. (1990). Metals in *Ulva lactuca* in Hong Kong intertidal waters. *Bull. Mar. Sci.* 47:79–85. [8.3.1]

Hodgson, L. M. (1981). Photosynthesis of the red alga *Gastroclonium coulteri* (Rhodophyta) in response to changes in temperature, light intensity, and desiccation. *J. Phycol.* 17:37–42. [2.5.2]

Hodgson, L. M. (1984). Desiccation tolerance of *Gracilaria tikvahiae* (Rhodophyta). *J. Phycol.* 20:444–6. [2.1.1]

Hodson, P. V., U. Borgmann, & H. Shear (1979). Toxicity of copper to aquatic biota. In J. O. Nriagu (ed.), *Copper in the Environment. Part II. Health Effects* (pp. 308–65). New York: Wiley-Interscience. [8.3.3]

Hoffmann, A. J., & P. Camus (1989). Sinking rates and viability of spores from benthic algae in central Chile. *J. Exp. Mar. Biol. Ecol.* 126:281–91. [1.6.1]

Hoffman, A. J., & B. Santelices (1982). Effect of light intensity and nutrients on gametophytes and gametogenesis of *Lessonia nigrescens* Bory (Phaeophyta). *J. Exp. Mar. Biol. Ecol.* 60:77–89. [5.8.2]

Holbrook, G. P., S. Beer, W. E. Spencer, J. B. Reiskind, J. S. Davis, & G. Bowes (1988). Photosynthesis in marine macroalgae: evidence for carbon limitation. *Can. J. Bot.* 66:577–82. [4.4.1, 4.4.4]

Holbrook, N. M., M. W. Denny, & M. A. R. Koehl (1991). Intertidal "trees": consequences of aggregation on the mechanical and photosynthetic properties of sea-palms *Postelsia palmaeformis* Ruprecht. *J. Exp. Mar. Biol. Ecol.* 146:39–67. [7.2.2]

Holmes, M. A., M. T. Brown, M. W. Loutit, & K. Ryan (1991). The involvement of epiphytic bacteria in zinc concentration by the red alga *Gracilaria sordida. Mar. Environ. Res.* 31:56–67. [8.3.1]

Hommersand, M. H., & S. Fredericq (1990). Sexual reproduction and cystocarp development. In K. M. Cole & R. G. Sheath (eds.) *Biology of the Red Algae* (pp. 305–45). Cambridge University Press. [1.5.4]

Hood, D. W., A. Schoener, P. K. Park, & I. W. Duedall (1989). Evolution of at-sea scientific monitoring strategies. In D. W. Hood, A. Schoener, & P. K. Park (eds.), *Oceanic Processes in Marine Pollution. Vol. 4: Scientific Monitoring Strategies for Ocean Waste Disposal* (pp. 4–28). Malabau, FL: E. W. Krieger Publishing. [8.1]

Hooper, R. [G.], & G. R. South (1977). Distribution and ecology of *Papenfussiella callitricha* (Rosenv.) Kylin (Phaeophyceae, Chordariaceae). *Phycologia* 16:153–7. [6.1.1, 6.2.4]

Hooper, R. G., E. C. Henry, & R. Kuhlenkamp (1988). *Phaeosiphoniella cryophila* gen. et sp. nov., a third member of the Tilopteridales (Phaeophyceae). *Phycologia* 27:395–404. [6.2.4]

Hopkin, R., & J. M. Kain (1978). The effects of some pollutants on the survival, growth and respiration of *Laminaria hyperborea. Estu. Cstl. Mar. Sci.* 7:531–53. [8.3.3, 8.4.2, 8.5.1, 8.6.1]

Hoppe, H. A., & O. J. Schmid (1969). Commercial products. In T. Levring, H. A. Hoppe, & O. J. Schmid (eds.), *Marine*

Algae: A Survey of Research and Utilization (pp. 288–368). Hamburg: Cram, de Gruyter. [9.1]

Horn, H. S. (1971). *The Adaptive Geometry of Trees*. Princeton University Press. [4.3.2]

Horn, M. H. (1989). Biology of marine herbivorous fishes. *Oceanogr. Mar. Biol. Annu. Rev.* 27:167–272. [3.2.1]

Hornsey, I. S., & D. Hide (1976). The production of antimicrobial compounds by British marine algae. II. Seasonal variation in production of antibiotics. *Br. Phycol. J.* 11:63–7. [3.1.2]

Houston, M. (1979). A general hypothesis of species diversity. *Am. Nat.* 113:81–101. [8.1]

Howard, B. M., & W. Fenical (1981). The scope and diversity of terpenoid biosynthesis by the marine alga *Laurencia. Prog. Phytochem.* 7:263–300. [3.2.3]

Howard, R. J., K. R. Gayler, & B. R. Grant (1975). Products of photosynthesis in *Caulerpa simpliciuscula. J. Phycol.* 11:463–71. [4.4.2]

Howard, R. K., & F. T. Short (1986). Seagrass growth and survivorship under the influence of epiphyte grazers. *Aquat. Bot.* 24:287–302. [2.4.2]

Howarth, R. W. (1988). Nutrient limitation of net primary production in marine ecosystems. *Annu. Rev. Evol.* 19:89–110. [5.1.3]

Howarth, R. W., & J. J. Cole (1985). Molybdenum availability, nitrogen limitation and phytoplankton growth in natural waters. *Science* 229:653–5. [5.5.7]

Hruby, T., & T. A. Norton (1979). Algal colonization on rocky shores in the Firth of Clyde. *J. Ecol.* 67:65–77. [2.1.1]

Hsiao, S. I.-C. (1969). Life history and iodine nutrition of the marine brown alga *Petalonia fascia* (O. F. Müll.) Kuntze. *Can. J. Bot.* 47:1611–16. [1.7.2, 5.8.2]

Hsiao, S. I.-C. (1972). Nutritional requirements for gametogenesis in *Laminaria saccharina* (L.) Lamouroux. Ph.D. thesis, Simon Fraser University, Burnaby, BC. [5.5.6]

Hsiao, S. I.-C., D. W. Kittle, & M. G. Foy (1978). Effects of crude oils and the oil dispersant Corexit on primary production of Arctic marine phytoplankton and seaweed. *Environ. Pollut.* 15:209–21. [8.4.2]

Hudson, P. R., & J. R. Waaland (1974). Ultrastructure of mitosis and cytokinesis in the multinucleate green alga *Acrosiphonia. J. Cell Biol.* 62:274–94. [1.3.5]

Hughes, T. P., D. C. Reed, & M.-J. Boyle (1987). Herbivory on coral reefs: community structure following mass mortality of sea urchins. *J. Exp. Mar. Biol. Ecol.* 113:39–60. [3.2.1]

Huizing, H. J., H. Rietema, & J. H. Sietsma (1979). Cell wall constituents of several siphoneous green algae in relation to morphology and taxonomy. *Br. Phycol. J.* 14:25–32. [1.3.1]

Huntsman, S. A., & W. G. Sunda (1980). The role of trace metals in regulating phytoplankton growth. In I. Morris (ed.), *The Physiological Ecology of Phytoplankton* (pp. 285–328). Oxford: Blackwell Scientific. [5.2, 5.5.6, 8.3.2]

Huppertz, K., D. Hanelt, & W. Nultsch (1990). Photoinhibition of photosynthesis in the marine brown alga *Fucus serratus* as studied in field experiments. *Mar. Ecol. Prog. Ser.* 66:175–82. [4.3.3]

Hurd, C. L., & M. J. Dring (1990). Phosphate uptake by intertidal fucoid algae in relation to zonation and season. *Mar. Biol.* 107:281–9. [5.5.2]

Hurd, C. L., & M. J. Dring (1991). Desiccation and phosphate uptake by intertidal fucoid algae in relation to zonation. *Br. Phycol. J.* 26:327–33. [5.4.2, 5.8.2]

Hurd, C. L., R. S. Galvin, T. A. Norton, & M. J. Dring (1993). Production of hyaline hairs by intertidal *Fucus* (Fucales) and their role in phosphate uptake. *J. Phycol.* 29:160–5. [5.5.2]

Hurka, H. (1971). Factors influencing the gas composition in the vesicles of *Sargassum. Mar. Biol.* 11:82–9. [1.7.2]

Hurka, H. (1974). A new hypothesis concerning the adaptive value of buoyancy increasing devices of algae. *Nova Hedwigia* 25:429–32. [7.2.2]

Hurlbert, S. H. (1975). Secondary effects of pesticides on aquatic ecosystems. *Residue Rev.* 57:81–148. [8.5.1, 8.5.2]

Hutchins, L. W. (1947). The basis for temperature zonation in geographical distribution. *Ecol. Monogr.* 17:325–35. [6.2.5]

Hwang, S.-P. L., S. L. Williams, & B. H. Brinkhuis (1987). Changes in internal dissolved nitrogen pools as related to nitrate uptake and assimilation in *Gracilaria tikvahiae* McLachlan (Rhodophyta). *Bot. Mar.* 30:11–19. [5.5.1]

Ikawa, M., V. M. Thomas, L. J. Buckley, & J. J. Uebel (1973). Sulfur and the toxicity of the red alga *Ceramium rubrum* to *Bacillus subtilis. J. Phycol.* 9:302–4. [5.5.5]

Ikawa, T., T. Watanabe, & K. Nisizawa (1972). Enzymes involved in the last steps of the biosynthesis of mannitol in brown algae. *Plant Cell Physiol.* 13:1017–29. [4.4.2]

Ilvessalo, H., & J. Tuomi (1989). Nutrient availability and accumulation of phenolic compounds in the brown alga *Fucus vesiculosus. Mar. Biol.* 101:115–19. [3.2.3, 5.5.1]

Indergaard, M. (1983). The aquatic resource. In W. A. Coté (ed.), *Biomass Utilization* (pp. 137–68). New York: Plenum. [9.8]

Indergaard, M., & A. Jensen (1991). *Utnyttelse av Marin Biomasse*. Inst. Bioteknol., Norges Tek. Høgskole, Trondheim, Norway, 123 pp. [Table 9.3, Fig. 9.9]

Innes, D. J. (1984). Genetic differentiation among populations of marine algae. *Helgol. Meeresunters* 38:401–17. [6.2.3]

Innes, D. J. (1988). Genetic differentiation in the intertidal zone in populations of the alga *Enteromorpha linza* (Ulvales: Chlorophyta). *Mar. Biol.* 97:9–16. [6.2.3]

Inouye, R. S., & W. M. Schaffer (1981). On the ecological meaning of ratio (de Wit) diagrams in plant ecology. *Ecology* 62:1679–81. [3.1.3]

Iseki, K., M. Takahashi, E. Bauernfiend, & C. S. Wong (1981). Effects of polychlorinated biphenyls (PCBs) on a marine plankton population and sedimentation in controlled ecosystem enclosures. *Mar. Ecol. Prog. Ser.* 5:207–14. [8.5.3]

Itoh, T., & R. M. Brown, Jr. (1988). Development of cellulose synthesizing complexes in *Boergesenia* and *Valonia. Protoplasma* 144:160–9. [1.3.1]

Itoh, T., R. L. Legge, & R. M. Brown, Jr. (1986). The effects of selected inhibitors on cellulose microfibril assembly in *Boergesenia forbesii* (Chlorophyta) protoplasts. *J. Phycol.* 22:224–33. [1.3.1]

Iverson, W. P., & F. E. Brinckman (1978). Microbial metabolism of heavy metals. In R. M. Mitchell (ed.), *Water Pollution Microbiology* (pp. 201–31). New York: Wiley. [8.3.2]

Jackson, G. A. (1977). Nutrients and production of the giant kelp, *Macrocystis pyrifera*, off southern California. *Limnol. Oceanogr.* 22:979–95. [5.4.2]

Jackson, G. A. (1984). The physical and chemical environment of a kelp community. In W. Bascom (ed.), *The Effects of Waste Disposal on Kelp Communities* (pp. 11–37). Long Beach: So. Calif. Coastal Water Res. Proj. [4.7.4]

Jackson, G. A. (1987). Modelling the growth and harvest yield of the giant kelp *Macrocystis pyrifera*. *Mar. Biol.* 95:611–24. [4.3.2, 4.7.4]

Jackson, G. A., & C. D. Winant (1983). Effect of a kelp forest on coastal currents. *Continental Shelf Res.* 2:75–80. [7.2.1, 7.2.2]

Jackson, G. A., D. E. James, & W. J. North (1985). Morphological relationships among fronds of giant kelp *Macrocystis pyrifera* off La Jolla, California. *Mar. Ecol. Prog. Ser.* 26:261–70. [4.7.4]

Jackson, S. G., & E. L. McCandless (1982). The effect of sulphate concentration on the uptake and incorporation of [^{35}S]sulphate in *Chondrus crispus. Can. J. Bot.* 60:162–5. [5.5.5]

Jacobs, W. P. (1970). Develoment and regeneration of the algal giant coenocyte *Caulerpa. Ann. N.Y. Acad. Sci.* 175:732–48. [1.7.4]

Jacobs, W. P. (1986). Are angiosperm hormones present in, and used as hormones by, algae? In M. Bopp (ed.), *Plant Growth Substances 1985: Proc. 12th Intl. Conf. Plant Growth Substances* (pp. 249–56). Berlin: Springer-Verlag. [1.7.3]

Jacobs, W. P., & J. Olson (1980). Developmental changes in the algal coenocyte *Caulerpa prolifera* (Siphonales) after inversion with respect to gravity. *Am. J. Bot.* 67:141–6. [1.7.4]

Jacobs, W. P., K. Falkenstein, & R. H. Hamilton (1985). Nature and amount of auxin in algae. IAA from extracts of *Caulerpa paspaloides* (Siphonales). *Plant Physiol.* 78:844–8. [1.7.3]

James, D. E., S. L. Manley, M. C. Carter, & W. J. North (1987). Effects of PCBs and hydrazine on life processes in microscopic stages of selected brown seaweeds. *Hydrobiologia* 151/152:411–15. [8.5.3]

Jassby, A. D., & T. Platt (1976). Mathematical formulation of the relationship between photosynthesis and light for phytoplankton. *Limnol. Oceanogr.* 21:540–7. [4.3.3]

Jensen, A., & A. Haug (1956). *Geographical and Seasonal Variation in the Chemical Compositon of* Laminaria hyperborea *and* Laminaria digitata *from the Norwegian Coast.* Norwegian Institute of Seaweed Research, Report 14. [4.5.1]

Jensen, R. G., & J. T. Bahr (1977). Ribulose 1,5-bisphosphate carboxylase-oxygenase. *Annu. Rev. Plant Physiol.* 28:379–400. [4.7.2]

Jerlov, N. G. (1970). Light: general introduction. In O. Kinne (ed.), *Marine Ecology* (vol. I, pt. 1, pp. 95–102). New York: Wiley. [4.2.2]

Jerlov, N. G. (1976). *Marine Optics.* Amsterdam: Elsevier, 231 pp. [4.2.2]

Jernakoff, P. (1985). Interactions between the limpet *Patelloida latistrigata* and algae on an intertidal rock platform. *Mar. Ecol. Prog. Ser.* 23:71–8. [2.1.1]

Johannes, R. E. (1980). The ecological significance of the submarine discharge of groundwater. *Mar. Ecol. Prog. Ser.* 3:365–73. [5.8.3]

Johannes, R. E. et al. (1972). The metabolism of some coral reef communities. *Bioscience* 22:541–3. [2.1.2]

Johannesson, K. (1989). The bare zone of Swedish rocky shores: Why is it there? *Oikos* 54:77–86. [2.2.1, 2.2.3]

Johansen, H. W. (1981). *Coralline Algae, A First Synthesis.* Boca Raton, FL: CRC Press, 233 pp. [1.7.2, 5.5.2]

John, D. M., D. Lieberman, M. Lieberman, & M. D. Swaine (1980). Strategies of data collection and analysis of subtidal vegetation. In J. H. Price, D. E. G. Irvine, & W. F. Farn-

ham (eds.), *The Shore Environment. Vol. 1: Methods* (pp. 265–84). New York: Academic Press. [2.5.1]

Johnson, W. S., A. Gigon, S. L. Gulmon, & H. A. Mooney (1974). Comparative photosynthetic capacities of intertidal algae under exposed and submerged conditions. *Ecology* 55:450–3. [6.4]

Johnston, A. M., & J. A. Raven (1986). Dark carbon fixation studies on the intertidal macroalga *Ascophyllum nodosum* (Phaeophyta). *J. Phycol.* 22:78–83. [4.4.1, 4.4.4]

Johnston, A. M., & J. A. Raven (1987). The C_4-like characteristics of the intertidal macroalga *Ascophyllum nodosum* (L.) Le Jolis (Fucales, Phaeophyta). *Phycologia* 26:159–66. [4.4.3, 4.4.4]

Johnston, A. M., & J. A. Raven (1990). Effects of culture in high CO_2 on the photosynthetic physiology of *Fucus serratus. Br. Phycol. J.* 25:75–82. [4.4.1]

Joly, A. B., & E. C. de Oliveira, Fil (1967). Two Brazilian *Laminaria. Publ. Inst. Pesq. Mar.* 4:1–13. [6.2.5]

Jones, A. K. (1988). Algal extracellular products – antimicrobial substances. In L. J. Rogers & J. R. Gallon (eds.), *Biochemistry of the Algae and Cyanobacteria* (pp. 257–81). Oxford: Clarendon Press. [3.1.3]

Jones, G. P., & M. D. Norman (1986). Feeding selectivity in relation to territory size in a herbivorous reef fish. *Oecologia* 68:549–56. [3.2.2]

Jones, J. L., J. A. Callow, & J. R. Green (1988). Monoclonal antibodies to sperm surface antigens of the brown alga *Fucus serratus* exhibit region-, gamete-, species- and genus-preferential binding. *Planta* 176:298–306. [1.3.1, 1.5.4]

Jones, W. E., & A. Demetropoulos (1968). Exposure to wave action: measurements of an important ecological parameter on rocky shores on Anglesey. *J. Exp. Mar. Biol. Ecol.* 2:46–63. [7.2.1]

Jónsson, S., M.-H. Laur, & L. Pham-Quang (1985). Mise en évidence de différents types de glycoproteines dans un extrait inhibiteur de la gamétogenèse chez *Enteromorpha prolifera,* Chlorophycée marine. *Crypt. Algol.* 6:253–64. [1.5.3]

Joska, M. A. P., & J. J. Bolton (1987). *In situ* measurement of zoospore release and seasonality of reproduction in *Ecklonia maxima* (Alariaceae, Laminariales). *Br. Phycol. J.* 22:209–14. [2.5.2, 4.7.3]

Josselyn, M. N., & A. C. Mathieson (1978). Contribution of receptacles from the fucoid *Ascophyllum nodosum* to the detrital pool of a north temperate estuary. *Estuaries* 1:258–61. [2.5.2, 4.7.3]

Kafuku, T., & H. Ikenoue (1983). Wakame (*Undaria pinnatifida*). In T. Kafuku & H. Ikenoue (eds.), *Modern Methods of Aquaculture in Japan* (pp. 203–8). Amsterdam: Elsevier. [9.4]

Kageyama, A., Y. Yokohama, & K. Nisizawa (1979). Diurnal rhythm of apparent photosynthesis of a brown alga, *Spatoglossum pacificum. Bot. Mar.* 22:199–201. [4.7.2]

Kageyama, A., Y. Yokohama, S. Shimura, & T. Ikawa (1977). An efficient excitation energy transfer from a carotenoid, siphonaxanthin, to chlorophyll *a* observed in a deep-water species of Chlorophycean seaweed. *Plant Cell Physiol.* 18:477–80. [Fig. 4.6]

Kain, J. M. (1964). Aspects of the biology of *Laminaria hyperborea.* III. Survival and growth of gametophytes. *J. Mar. Biol. Assoc. U.K.* 44:415–33. [1.6.1]

Kain, J. M. (1969). The biology of *Laminaria hyperborea.* V. Comparison with early stages of competitors. *J. Mar. Biol. Assoc. U.K.* 49:455–73. [3.1.3, 6.2.3]

Kain, J. M. (1979). A view of the genus *Laminaria*. *Oceanogr. Mar. Biol. Annu. Rev.* 17:101–61. [4.7.4]

Kain, J. M. (1989). The seasons in the subtidal. *Br. Phycol. J.* 24:203–15. [1.5.3]

Kain, J. M, & C. P. Dawes (1987). Useful European seaweeds: past hopes and present cultivation. *Hydrobiologia* 151/152:173–81. [9.3.2, 9.6]

Kain, J. M., & T. A. Norton (1990). Marine ecology. In K. M. Cole & R. Sheath (eds.), *The Biology of Red Algae* (pp. 375–421). Cambridge University Press. [2.1.1]

Kain, J. M., T. J. Holt, & C. P. Dawes (1990). European Laminariales and their cultivation. In C. Yarish, C. A. Penniman, & P. Van Patten (eds.), *Economically Important Plants of the Atlantic: Their Biology and Cultivation* (pp. 95–111). Groton: Connecticut Sea Grant College Program. [9.3, 9.4.1]

Kajiwara, T., A. Hatanaka, Y. Tanaka, T. Kawai, M. Ishihara, T. Tsuneya, & T. Fujimura (1989). Volatile constitutents from marine brown algae of Japanese *Dictyopteris*. *Phytochemistry* 28:636–8. [1.5.4]

Kalle, K. (1971). Salinity: general introduction. In O. Kinne (ed.), *Marine Ecology* (vol. I, pt. 2, pp. 683–8). New York: Wiley. [6.3, Fig. 4.17]

Kalle, K. (1972). Dissolved gases: general introduction. In O. Kinne (ed.), *Marine Ecology* (vol. I, pt. 3, pp. 1451–7). NewYork: Wiley. [4.4.1, 6.4]

Kapraun, D. F., & P. W. Boone (1987). Karyological studies of three species of Scytosiphonaceae (Phaeophyta) from coastal North Carolina. *J. Phycol.* 23:318–22. [1.3.5, 1.5.2]

Kapraun, D. F., & D. J. Martin (1987). Karyological studies of three species of *Codium* (Codiales, Chlorophyta) from coastal North Carolina. *Phycologia* 26:228–34. [1.5.2]

Kapraun, D. F., M. G. Gargiulo, & G. Tripodi (1988). Nuclear DNA and karyotype variation in species of *Codium* (Codiales, Chlorophyta) from the North Atlantic. *Phycologia* 27:273–82. [1.4.1]

Karsten, U., & G. O. Kirst (1989). Incomplete turgor pressure regulation in the "terrestrial" red alga, *Bostrychia scorpioides* (Huds.) Mont. *Plant Sci.* 61:29–36. [6.3.2]

Karsten, U., R. J. King, & G. O. Kirst (1990). The distribution of D-sorbitol and D-dulcitol in the red algal genera *Bostrychia* and *Stictosiphonia* (Rhodomelaceae, Rhodophyta) – a re-evalution. *Br. Phycol. J.* 25:363–6. [4.4.2]

Kasemir, H. (1983). Light control of chlorophyll accumulation in higher plants. In W. Shropshire & H. Mohr (eds.), *Encyclopedia of Plant Physiology (N.S.), Vol. 16B: Photomorphogenesis* (pp. 662–86). Berlin: Springer-Verlag. [4.3.1]

Kaska, D. D., M. Polne-Fuller, & A. Gibor (1988). Developmental studies in *Porphyra* (Bangiophyceae). II. Characterization of the major lectin binding glycoproteins in differentiated regions of the *Porphyra perforata* thallus. *J. Phycol* 24:102–7. [1.1.1]

Katoh, T., & T. Ehara (1990). Supramolecular assembly of fucoxanthin-chlorophyll-protein complexes isolated from a brown alga, *Petalonia fascia*. Electron microscopic studies. *Plant Cell Physiol.* 31:439–47. [4.3.1]

Katsaros, C., & B. Galatis (1988). Thallus development in *Dictyopteris membranacea* (Phaeophyta, Dictyotales). *Br. Phycol. J.* 23:71–88. [1.2.1, 1.7.2]

Kauss, H. (1973). Turnover of galactosylglycerol and osmotic balance in *Ochromonas*. *Plant Physiol.* 52:613–15. [6.3.2]

Kauss, H. (1978). Osmotic regulation in algae. *Progress in Phytochemistry* 5:1–27. [Fig. 6.18]

Kauss, H., & K.-S. Thomson (1982). Biochemistry of volume control in *Poterioochromonas*. In D. Marmé, E. Marrè, & R. Hertel (eds.), *Plasmalemma and Tonoplast: Their Functions in the Plant Cell* (pp. 255–62). Amsterdam: Elsevier. [6.3.2]

Kauss, H., K.-S. Thomson, M. Tetour, & W. Jeblick (1978). Proteolytic activation of a galactosyl transferase involved in osmotic regulation. *Plant Physiol.* 61:35–7. [6.3.2]

Kautsky, N., H. Kautsky, U. Kautsky, & M. Waern (1986). Decreased depth penetration of *Fucus vesiculosus* (L.) since the 1940's indicates eutrophication of the Baltic Sea. *Mar. Ecol. Prog. Ser.* 28:1–8. [8.6.1]

Kawai, H. (1988). A flavin-like autofluorescent substance in the posterior flagellum of golden and brown algae. *J. Phycol.* 24:114–17. [1.3.3]

Kawai, H., D. G. Müller, E. Fölster, & D. P. Häder (1990). Phototactic responses in the gametes of the brown agla, *Ectocarpus siliculosus*. *Planta* 182:292–7. [1.3.3]

Kawashima, S. (1984). Kombu cultivation in Japan for human foodstuff. *Jpn. J. Phycol.* 32:379–94. [9.3, 9.3.1]

Kayser, M. (1979). Growth interactions between marine dinoflagellates in multispecies culture experiments. *Mar. Biol.* 52:357–70. [3.1.3]

Keen, J. N., & L. V. Evans (1988). Extraction and assay of ribulose-1, 5-bisphosphate carboxylase from *Fucus*. In C. S. Lobban, D. J. Chapman, & B. P. Kremer (eds.), *Experimental Phycology: A Laboratory Manual* (pp. 141–9). Cambridge University Press. [3.2.3]

Keen, J. N., D. J. C. Pappin, & L. V. Evans (1988). Amino acid sequence analysis of the small subunit of ribulose bisphosphate carboxylase from *Fucus* (Phaeophyceae). *J. Phycol.* 24:324–7. [1.3.2]

Kelly, G. J. (1989). A comparison of marine photosynthesis with terrestrial photosynthesis: a biochemical perspective. *Oceanogr. Mar. Biol. Annu. Rev.* 27:11–44. [4.7.1]

Kendrick, G. A., D. I. Walker, & A. J. McComb (1988). Changes in distribution of macro-algal epiphytes on stems of the seagrass *Amphibolis antarctica* along a salinity gradient in Shark Bay, Western Australia. *Phycologia* 27:201–8. [6.5]

Kennelly, S. J. (1983). An experimental approach to the study of factors affecting algal colonization in a sublittoral kelp forest. *J. Exp. Mar. Biol. Ecol.* 68:257–76. [3.2.1]

Kennelly, S. J. (1987a). Physical disturbances in an Australian kelp community. I. Temporal effects. *Mar. Ecol. Prog. Ser.* 40:145–53. [7.2.3]

Kennelly, S. J. (1987b). Inhibition of kelp recruitment by turfing algae and consequences for an Australian kelp community. *J. Exp. Mar. Biol. Ecol.* 112:49–60. [2.5.1, 7.2.3]

Kenyon, K. W., & D. W. Rice (1959). Life history of the Hawaiian monk seal. *Pac. Sci.* 13:215–52. [2.4]

Kerby, N. W., & L. V. Evans (1983). Phosphoenolpyruvate carboxykinase activity in *Ascophyllum nodosum* (Phaeophyceae). *J. Phycol.* 19:1–3. [4.4.3]

Kerby, N. W., & J. A. Raven (1985). Transport and fixation of inorganic carbon by marine algae. *Adv. Bot. Res.* 11:71–123. [4.0, 4.4.1, 4.4.2, 4.4.3, 4.4.4]

Khailov, K. M., & Z. P. Burlakova (1969). Release of dissolved organic matter by marine seaweeds and distribution of their total organic production to inshore communities. *Limnol. Oceanogr.* 14:521–7. [4.7.3]

Khailov, K. M., V. I. Kholodov, Y. K. Firsov, & A. V. Prazukin (1978). Thalli of *Fucus vesiculosus* in ontogenesis: changes in morpho-physiological parameters. *Bot. Mar.* 21:289–311. [4.7.2]

Khfaji, A. H., & A. D. Boney (1979). Antibiotic effects of crustose germlings of the red alga *Chondrus crispus* Stackh. on benthic diatoms. *Ann. Bot.* 43:231–2. [3.1.2.]

Kilar, J. A., M. M. Littler, & D. S. Littler (1989). Functional-morphological relationships in *Sargassum polyceratium* (Phaeophyta): phenotypic and ontogenetic variability in apparent photosynthesis and dark respiration. *J. Phycol.* 25:713–20. [1.1.1, 1.7.2, 4.7.2]

Kindig, A. C., & M. M. Littler (1980). Growth and primary productivity of marine macrophytes exposed to domestic sewage effluents. *Mar. Environ. Res.* 3:81–100. [8.6.1]

King, R. J. (1981). Mangroves and saltmarsh plants. In M. N. Clayton & R. J. King (eds.), *Marine Botany: an Australasian Perspective* (pp. 308–28). Melbourne: Longman Cheshire. [1.1.1]

King, R. J. (1984). Oil pollution and marine plant systems. In D. M. H. Cheng & C. D. Field (eds.). *Pollution and Plants* (pp. 127–42). Melbourne: Insearch Ltd. [8.4.1]

King, R. J. (1990). Macroalgae associated with the mangrove vegetation of Papua New Guinea. *Bot. Mar.* 33:55–62. [6.3.4]

King, R. J., & W. Schramm (1982). Calcification in the maerl coralline alga *Phymatolithon calcareum:* effects of salinity and temperature. *Mar. Biol.* 70:197–204. [6.5]

Kingham, D. L., & L. V. Evans (1986). The *Pelvetia–Mycosphaerella* interrelationship. In S. T. Moss (ed.), *The Biology of Marine Fungi* (pp. 177–87). Cambridge University Press. [3.3.1]

Kinne, O. (1970). Temperature – general introduction. In O. Kinne (ed.), *Marine Ecology* (vol. I, pt. 1, pp. 321–46). New York: Wiley. [6.1.1]

Kirk, J. T. O. (1989). The upwelling light stream in natural waters. *Limnol. Oceanogr.* 34:1410–25. [4.2.1]

Kirkman, H. (1981). The first year in the life history and the survival of the juvenile marine macrophyte, *Ecklonia radiata* (Turn.) J. Agardh. *J. Exp. Mar. Biol. Ecol.* 55:243–54. [7.2.3]

Kirst, G. O. (1988). Turgor pressure regulation in marine macroalgae. In C. S. Lobban, D. J. Chapman, & B. P. Kremer (eds.), *Experimental Phycology: A Laboratory Manual* (pp. 203–9). Cambridge University Press. [6.3.1]

Kirst, G. O., & M. A. Bisson (1979). Regulation of turgor pressure in marine algae: ions and low-molecular-weight organic compounds. *Austr. J. Plant Physiol.* 6:539–56. [6.3.2]

Kitching, J. A., & F. J. Ebling (1967). Ecological studies at Lough Ine. *Adv. Ecol. Res.* 4:197–291. [7.2.3]

Kitoh, S., & S. Hori (1977). Metabolism of urea in *Chlorella ellipsoidea. Plant Cell Physiol.* 18:513–19. [5.5.1]

Klemm, M. F., & N. D. Hallam (1987). Branching pattern and growth in *Cystophora* (Fucales, Phaeophyta). *Phycologia* 26:252–61. [1.7.2]

Klerks, P. L., & J. S. Weis (1987). Genetic adaptation to heavy metals in aquatic organisms: a review. *Environ. Pollut.* 45:173–205. [8.3.2]

Kling, R., & M. Bodard (1986). La construction du thalle de *Gracilaria verrucosa* (Rhodophyceae, Gigartinales): Edification de la fronde; essai d'interprétation phylogénétique. *Crypt. Algol.* 7:231–46. [1.2.1]

Klinger, T., & R. E. DeWreede (1988). Stipe rings, age, and size in populations of *Laminaria setchellii* Silva (Laminariales, Phaeophyta) in British Columbia, Canada. *Phycologia* 27:234–40. [2.5.2]

Kloareg, B., & R. S. Quatrano (1988). Structure of the cell walls of marine algae and ecophysiological functions of the matrix polysaccharides. *Oceanogr. Mar. Biol. Annu. Rev.* 26:259–315. [1.3.1, 4.5.2]

Kloareg, B., M. Demarty, & S. Mabeau (1986). Polyanionic characteristics of purified sulphated homofucans from brown algae. *Int. J. Biol. Macromol.* 8:380–6. [Fig. 1.6]

Kloareg, B., M. Demarty, & S. Mabeau (1987). Ion-exchange properties of isolated cell walls of brown algae: the interstitial solution. *J. Exp. Bot.* 38:1652–62. [5.3.1]

Kloareg, B., M. Polne-Fuller, & A. Gibor (1989). Mass production of viable protoplasts from *Macrocystis pyrifera* (L.) C. Ag. (Phaeophyta). *Plant Sci.* 62:105–12. [1.7.1]

Klumpp, D. W., & A. D. McKinnon (1989). Temporal and spatial patterns in primary productivity of a coral reef epilithic algal community. *J. Exp. Mar. Biol. Ecol.* 131:1–22. [3.2.1]

Klumpp, D. W., & N. V. C. Polunin (1989). Partitioning among grazers of food resources within damselfish territories on a coral reef. *J. Exp. Mar. Biol. Ecol.* 125:145–69. [3.2.2]

Klumpp, D. W., D. McKinnnon, & P. Daniel (1987). Damselfish territories: zones of high productivity on coral reefs. *Mar. Ecol. Prog. Ser.* 40:41–51. [3.2.1, 3.2.2]

Knauer, G. A., & J. H. Martin (1972). Mercury in a marine pelagic food chain. *Limnol. Oceanogr.* 17:868–76. [8.3.5]

Knight, M., & M. Parke (1950). A biological study of *Fucus vesiculosus* L. and *F. serratus* L. *J. Mar. Biol. Assoc. U.K.* 29:439–514. [4.7.3]

Knoop, W. T., & G. C. Bate (1988a). The effect of wounding on the photosynthetic rates of three subtidal rhodophytes. *Bot. Mar.* 31:149–53. [4.7.1, 4.7.2]

Knoop, W. T., & G. C. Bate (1988b). Diurnal patterns in the photosynthesis of three subtidal macroalgal rhodophytes. *Bot. Mar.* 31:243–9. [4.7.2]

Knoop, W. T., & G. C. Bate (1990). A model for the description of photosynthesis-temperature responses by subtidal Rhodophyta. *Bot. Mar.* 33:165–71. [6.2.2]

Koehl, M. A. R. (1982). The interaction of moving water and sessile organisms. *Sci. Am.* 247(6):124–35. [7.2.2]

Koehl, M. A. R. (1984). How do benthic organisms withstand moving water? *Am. Zool.* 24:57–70. [7.0]

Koehl, M. A. R. (1986). Seaweeds in moving water: form and mechanical function. In T. J. Givnish (ed.), *On the Economy of Plant Form and Function* (pp. 603–34). Cambridge University Press. [7.0, 7.1.2, 7.2.2]

Koehl, M. A. R., & R. S. Alberte (1988). Flow, flapping, and photosynthesis of *Nereocystis luetkeana:* a functional comparison of undulate and flat blade morphologies. *Mar. Biol.* 99:435–44. [7.2.4]

Koehl, M. A. R., & S. A. Wainwright (1977). Mechanical adaptations of a giant kelp. *Limnol. Oceanogr.* 22:1067–71. [7.2.2]

Koehl, M. A. R., & S. A. Wainwright (1985). Biomechanics. In M. M. Littler & D. S. Littler (eds.) *Handbook of Phycological Methods: Ecological Field Methods: Macroalgae* (pp. 291–313). Cambridge University Press. [7.2.2]

Kohler, A., & B. C. Labus (1983). Eutrophication processes and pollution of freshwater ecosystems including waste heat.

In O. Lange, P. S. Nobel, C. B. Osmond, & H. Ziegler (eds.), *Encyclopaedia of Plant Physiology. Vol. 4: Physiological Plant Ecology* (pp. 413–64). Berlin: Springer-Verlag. [8.5.1, 8.6.1]

Kohlmeyer, J., & M. W. Hawkes (1983). A suspected case of mycophycobiosis between *Mycosphaerella apophlaeae* (Ascomycetes) and *Apophlaea* spp. (Phodophyta). *J. Phycol.* 19:257–60. [3.3.1]

Kohlmeyer, J., & E. Kohlmeyer (1972). Is *Ascophyllum* lichenized? *Bot. Mar.* 15:109–12. [3.3.1]

Kooistra, W. H. C. F., A. M. T. Joosten, & C. van den Hoek (1989). Zonation patterns in intertidal pools and their possible causes: a multivariate approach. *Bot. Mar.* 32:9–26. [1.1.3, 2.3.1]

Kopczak, C. D., R. C. Zimmerman, & J. N. Kremer (1991). Variation in nitrogen physiology and growth among geographically isolated populations of the giant kelp, *Macrocystis pyrifera* (Phaeophyta). *J. Phycol.* 27:149–58. [1.1.3]

Kornmann, P. (1970). Advances in marine phycology on the basis of cultivation. *Helgol. Meeresunters.* 20:39–61. [1.3.5]

Kornmann, P., & P.-H. Sahling (1977). Meeresalgen von Helgoland. Benthische Grün-, Braun- und Rotalgen. *Helgol. Meeresunters.* 29:1–289.[1.6.1]

Koslowsky, D. J., & S. D. Waaland (1987). Ultrastructure of selective chloroplast destruction after somatic cell fusion in *Griffithsia pacifica* Kylin (Rhodophyta). *J. Phycol.* 23:638–48. [1.3.2]

Kowallik, K. V. (1989). Molecular aspects and phylogenetic implications of plastid genomes of certain chromophytes. In J. C. Green, B. S. C. Leadbeater, & W. I. Diver (eds.) *The Chromophyte Algae: Problems and Perspectives* (pp. 101–24). Oxford: Clarendon Press. [1.4.1]

Kraemer, G. P., & D. J. Chapman (1991). Biomechanics and alginic acid composition during hydrodynamic adaptation by *Egregia menziesii* (Phaeophyta) juveniles. *J. Phycol.* 27:47–53. [7.2.2]

Kraft, G. T., & W. J. Woelkerling (1981). Rhodophyta – systematics and biology. In M. N. Clayton & R. J. King (eds.), *Marine Botany: An Australasian Perspective* (pp. 61–103). Melbourne: Longman Cheshire. [1.5.1]

Kreimer, G., H. Kawai, D. G. Müller, & M. Melkonian (1991). Reflective properties of the stigma in male gametes of *Ectocarpus siliculosus* (Phaeophyceae) studied by confocal laser scanning microscopy. *J. Phycol.* 27:268–76. [1.3.3]

Kremer, B. P. (1977). Biosynthesis of polyols in *Pelvetia canaliculata*. *Z. Pflanzenphysiol.* 81:68-73. [4.4.2]

Kremer, B. P. (1979). Biosynthesis and metabolism of polyhydroxy alcohols in marine benthic algae. *Proc. Intl. Seaweed Symp.* 9:421–8. [3.3.1]

Kremer, B. P. (1980). Transversal profiles of carbon assimilation in the fronds of three *Laminaria* species. *Planta* 148:95–103. [5.5.1]

Kremer, B. P. (1981a). Carbon metabolism. In C. S. Lobban & M. J. Wynne (eds.), *The Biology of Seaweeds* (pp. 493–533). Oxford: Blackwell Scientific. [4.4.1, 4.4.2, 4.4.4, 6.2.2]

Kremer, B. P. (1981b). Metabolic implications of non-photosynthetic carbon fixation in brown macroalgae. *Phycologia* 20:242–50. [4.4.3]

Kremer, B. P. (1983). Carbon economy and nutrition of the alloparasitic red alga *Harveyella mirabilis*. *Mar. Biol.* 76:231–9. [3.3.2, 4.4.1]

Kremer, B. P. (1985). Aspects of cellular compartmentation in brown marine macroalgae. *J. Plant Physiol.* 120:401–7. [4.4.2]

Kremer, B. P., & J. W. Markham (1982). Primary metabolic effects of cadmium in the brown alga, *Laminaria saccharina*. *Z. Pflanzenphysiol.* 108:125–30. [8.3.3]

Kremer, B. P., & K. Schmitz (1973). CO_2-Fixierung und Stofftransport in benthischen marinen Algen. IV. Zur ^{14}C-Assimilation einiger litoraler Braunalgen in submersen und emersen Zustand. *Z. Pflanzenphysiol.* 68:357–63. [4.4.1]

Kropf, D. L., S. K. Berge, & R. S. Quatrano (1989). Actin localization during *Fucus* embryogenesis. *The Plant Cell* 1:191–200. [1.3.1]

Kugrens, P., & S. G. Delivopoulos (1986). Ultrastructure of the carposporophyte and carposporogenesis in the parasitic red alga *Plocamiocolax pulvinata* Setch. (Gigartinales, Plocamiaceae). *J. Phycol.* 22:8–21. [1.5.4]

Kuhl, A. (1962). Inorganic phosphorus uptake and metabolism. In R. A. Lewin (ed.), *Physiology and Biochemistry of Algae* (pp. 211–29). New York: Academic Press. [5.5.2]

Kuhl, A. (1974). Phosphorus. In W. D. P. Stewart (ed.), *Algal Physiology and Biochemistry* (pp. 636–54). Oxford: Blackwell Scientific. [5.5.2]

Küppers, U., & B. P. Kremer (1978). Longitudinal profiles of carbon dioxide fixation capacities in marine macroalgae. *Plant Physiol.* 62:49–53. [4.7.2]

Küppers, U., & M. Weidner (1980). Seasonal variation of enzyme activities in *Laminaria hyperborea*. *Planta* 148:222–30. [6.2.1]

Kurogi, M., & K. Akiyama (1966). Growth of conchocelis in several species of *Porphyra:* relationship between maturation and water temperature. *Bull. Tohoku Reg. Fish. Res. Lab.* 26:67–103 (in Japanese). [9.2.2]

Kurogi, M., & K. Hirano (1956). Influences of water temperature on the growth, formation of monosporangia and monospore-liberation in the *Conchocelis*-phase of *Porphyra tenera* Kjellm. *Bull. Tohoku Reg. Fish. Res. Lab.* 8:45–61 (in Japanese, with extensive English summary). [1.5.3]

Kuwabara, J. S. (1982). Micronutrients and kelp cultures: evidence for cobalt and manganese deficiency in southern California deep seawater. *Science* 216:1219–21. [5.5.6]

La Claire, J. W., II (1982a). Cytomorphological aspects of wound healing in selected Siphonocladales (Chlorophyceae). *J. Phycol.* 18:379–84. [1.7.4]

La Claire, J. W., II (1982b). Wound-healing motility in the green alga *Ernodesmis:* calcium ions and metabolic energy are required. *Planta* 156:466–74. [1.7.4]

La Claire, J. W., II (1989a). Actin cytoskeleton in intact and wounded coenocytic green algae. *Planta* 177:47–57. [1.7.4]

La Claire, J. W., II (1989b). The interaction of actin and myosin in cytoplasmic movement. In A. W. Coleman, L. J. Goff, & J. R. Stein-Taylor (eds.), *Algae as Experimental Systems* (pp. 55–70). New York: Alan R. Liss. [1.3.3]

Lai, Y.-K., C.-W. Li, C.-H. Hu, & M.-L. Lee (1988). Quantitative and qualitative analyses of protein synthesis during heat shock in the marine diatom *Nitzschia alba* (Bacillariophyceae). *J. Phycol.* 24:509–14. [6.2.4]

Lang, J. C. (1974). Biological zonation at the base of a reef. *Amer. Sci.* 62:271–81. [4.3.4]

Langston, W. J. (1990). Toxic effects of metals and the incidence of metal pollution in marine ecosystems. In R. W. Furness & P. S. Rainbow (eds.), *Heavy Metals in the Marine Environment* (pp. 101–22). Boca Raton, FL: CRC Press. [8.3.4, 8.5.4]

Lapointe, B. E. (1985). Strategies for pulsed nutrient supply to *Gracilaria* cultures in the Florida keys: interactions between concentration and frequency of nutrient pulses. *J. Exp. Mar. Biol. Ecol.* 93:211–22. [5.5.2]

Lapointe, B. E. (1986). Phosphorus-limited photosynthesis and growth of *Sargassum natans* and *Sargassum fluitans* (Phaeophyceae) in the western North Atlantic. *Deep Sea Res.* 33:391–9. [5.1.3, 5.5.2]

Lapointe, B. E. (1987). Phosphorus- and nitrogen-limited photosynthesis and growth of *Gracilaria tikvahiae* (Rhodophyceae) in the Florida Keys: an experimental field study. *Mar. Biol.* 93:561–8. [5.1.3, 5.5.2]

Lapointe, B. E., & C. S. Duke (1984). Biochemical strategies for growth of *Gracilaria tikvahiae* (Rhodophyta) in relation to light intensity and nitrogen availability. *J. Phycol.* 20:488–95. [5.7.3]

Lapointe, B. E., & J. O'Connell (1989). Nutrient-enhanced growth of *Cladophora prolifera* in Harrington Sound, Bermuda: eutrophication of a confined, phosphorus-limited marine ecosystem. *Estu. Cstl. Shelf Sci.* 28:347–60. [8.6.1]

Lapointe, B. E., & J. H. Ryther (1979). The effects of nitrogen and seawater flow rate on the growth and biochemical composition of *Gracilaria foliifera* v. *angustissima* in mass outdoor cultures. *Bot. Mar.* 22:529-37. [5.5.1]

Lapointe, B. E., & M. M. Littler, & D. S. Littler (1987). A comparison of nutrient-limited productivity in macroalgae from a Caribbean barrier reef and from a mangrove ecosystem. *Aquat. Bot.* 28:243–55. [5.8.3]

Lapointe, B. E., M. M. Littler, & D. S. Littler (1992). Nutrient availability to marine macroalgae in siliciclastic versus carbonate-rich coastal waters. *Estuaries* 15:75–82. [5.1.3, 5.5.2, 5.8.1]

Larkum, A. W. D., & J. Barrett (1983). Light-harvesting processes in algae. *Adv. Bot. Res.* 10:1–219. [4.0, 4.1.4, 4.2.2, 4.3.1, 4.3.2, 4.3.4., 4.7.2]

Larkum, A. W. D., E. A. Drew, & R. N. Crossett (1967). The vertical distribution of attached marine algae in Malta. *J. Ecol.* 55:361–71. [2.1.2, 4.3.4]

Larkum, A. W. D., I. R. Kennedy, & W. J. Muller (1988). Nitrogen fixation on a coral reef. *Mar. Biol.* 98:143–55. [5.2]

Larsen, B., A. Haug, & T. J. Painter (1970). Sulphated polysaccharides in brown algae. III. The native state of fucoidan in *Ascophyllum nodosum* and *Fucus vesiculosus. Acta Chem. Scand.* 24:3339–52. [4.5.2]

Lassuy, D. R. (1980). Effects of "farming" behavior by *Eupomacentrus lividus* and *Hemiglyphidodon plagiometopon* on algal community structure. *Bull. Mar. Sci.* 30:304–12. [3.2.2]

Lavery, P. S., & A. J. McComb (1991a). Macroalgal–sediment nutrient interactions and their importance to macroalgal nutrition in a eutrophic estuary. *Estu. Cstl. Shelf Sci.* 32:281–95. [5.2, 5.7.3]

Lavery, P. S., & A. J. McComb (1991b). The nutritional ecophysiology of *Chaetomorpha linum* and *Ulva rigida* in Peel Inlet, Western Australia. *Bot. Mar.* 34:251–60. [5.5.1, 5.5.2, 5.7.3, Table 5.4]

Lavery, P. S., R. J. Lukatelich, & A. J. McComb (1991). Changes in the biomass and species composition of macroalgae in a eutrophic estuary. *Estu. Cstl. Shelf Sci.* 33:1–22. [5.7.3]

Lawrence, J. M. (1975). On the relationship between marine plants and sea urchins. *Oceanogr. Mar. Biol. Annu. Rev.* 13:213–86. [3.2.1]

Lawrence, J. R., & D. E. Caldwell (1987). Behavior of bacterial stream populations within the hydrodynamic boundary layers of surface microenvironments. *Microb. Ecol.* 14:15–27. [1.6.1]

Lawrence, J. R., P. J. Delaquis, D. R. Korber, & D. E. Caldwell (1987). Behavior of *Pseudomonas fluorescens* within the hydrodynamic boundary layers of surface microenvironments. *Microb. Ecol.* 14:1–14. [1.6.1]

Laws, E. A. (1981). *Aquatic Pollution.* New York: Wiley, 482 pp. [8.2, 8.3, 8.3.5, 8.5.2, 8.6.1]

Laws, E. A., & D. G. Redalje (1982). Sewage diversion effects on the water column of a subtropical estuary. *Mar. Environ. Res.* 6:265–79. [8.6.1]

Laycock, M. V., & J. S. Craigie (1977). The occurrence and seasonal variation of gigartinine and L-citrullinyl-L-arginine in *Chondrus crispus* Stackh. *Can. J. Biochem.* 55:27–30. [5.5.1]

Laycock, M. V., K. C. Morgan, & J. S. Craigie (1981). Physiological factors affecting the accumulation of L-citrullinyl-L-arginine in *Chondrus crispus. Can. J. Bot.* 59:522–7. [5.5.1, 5.8.1, 9.7]

Lean, D. R. S., & F. R. Pick (1981). Photosynthetic response of lake plankton to nutrient enrichment: a test of nutrient limitation. *Limnol. Oceanogr.* 26:1001–19. [6.6.4]

Lee, R. B. (1980). Sources of reductant for nitrate assimilation in non-photosynthetic tissue: a review. *Plant Cell Environ.* 3:65–90. [5.5.1, 8.4.1]

Lehnberg, W. (1978). Die Wirkung eines Licht-Temperatur-Salzgehalt Komplexes auf den Gaswechsel von *Delesseria sanguinea* (Rhodophyta) aus der westlichen Ostee. *Bot. Mar.* 21:485–97. [6.1., 6.3.2, 6.3.3]

Lehninger, A. L. (1975). *Biochemistry,* 2nd ed. New York: Worth, 1104 pp. [6.2.1]

Leigh, E. G., Jr., R. T. Paine, J. F. Quinn, & T. H. Suchanek (1987). Wave energy and intertidal productivity. *Proc. Natl. Acad. Sci. USA* 84:1314–18. [7.0, 7.2.2]

Leighton, D. L., L. G. Jones, & W. J. North (1966). Ecological relationships between giant kelp and sea urchins in southern California. *Proc. Intl. Seaweed Symp.* 5:141–53 [3.2.1]

Leschine, T. M. (1990). Oil spill safety in the wake of the *Exxon Valdez:* How safe are Washington waters. *Northwest Environ. J.* 6:85–104. [8.1]

Lessios, H. A. (1988). Mass mortality of *Diadema antillarum* in the Caribbean: What have we learned? *Annu. Rev. Ecol. Syst.* 19:371–93. [3.2.1]

Levavasseur, G. (1989). Analyse comparée des complexes pigment-protéines de chlorophycophytes marine benthiques. *Phycologia* 28:1–14. [4.3.1]

Levitt, J. (1969). *Introduction to Plant Physiology.* St. Louis: Mosby, 304 pp. [5.1.1]

Levitt, J. (1972). *Responses of Plants to Environmental Stresses.* New York: Academic Press, 687 pp. [6.2.4, 6.4]

Levring, T. (1977). Potential yields of European marine algae. In R .W. Krauss (ed.), *Marine Plant Biomass of the Pacific Northwest Coast* (pp. 251–70). Corvallis: Oregon State University Press. [9.1]

Levring, T., H. A. Hoppe, & O. J. Schmid (1969). *Marine Algae, a Survey of Research and Utilization.* Hamburg: Cram, De Gruyter & Co., 421 pp. [9.4.1]

Levy, I. & B. Gantt (1988). Light acclimation in *Porphyridium purpureum* (Rhodophyta): growth, photosynthesis, and phycobilisomes. *J. Phycol.* 24:452–8. [4.3.3]

Lewin, J., & C. Chen (1971). Available iron: a limiting factor for marine phytoplankton. *Limnol. Oceanogr.* 16:670–5. [5.2]

Lewin, R. A. (1974). Biochemical taxonomy. In W. D. P. Stewart (ed.), *Algal Physiology and Biochemistry* (pp. 1–39). Oxford: Blackwell Scientific. [1.1.3]

Lewis, A. G., & W. R. Cave (1982). The biological importance of copper in oceans and estuaries. *Oceanogr. Mar. Biol. Annu. Rev.* 20:471–695. [5.2, 8.3.3, 8.3.4]

Lewis, J. (1977). Processes of organic production on coral reefs. *Biol. Rev.* 52:303–47. [5.8.3]

Lewis, J. G., N. F. Stanley, & G. C. Guist (1988). Commercial production and applications of algal hydrocolloids. In C. A. Lembi & J. R. Waaland (eds.), *Algae and Human Affairs* (pp. 205–36). Cambridge University Press. [9.5, 9.7]

Lewis, J. R. (1964). *The Ecology of Rocky Shores.* London: English Universities Press, 323 pp. [2.2.2, 3.2.1, 7.2.1]

Lewis, J. R. (1980). Objectives in littoral ecology – a personal viewpoint. In J. H. Price, D. E. G. Irvine, & W. F. Farnham (eds.), *The Shore Environment. Vol. 1: Methods* (pp. 1–18). New York: Academic Press. [2.2.2]

Lewis, P. G., III (1987). Crustacean epifauna of seagrass and macroalgae in Apalachee Bay, Florida, USA. *Mar. Biol.* 94:219–29. [2.4.2]

Lewis, S. M., J. N. Norris, & R. B. Searles (1987). The regulation of morphological plasticity in tropical reef algae by herbivory. *Ecology* 68:636–41. [1.7.2]

Li, N. & R. A. Cattolico (1987). Chloroplast genome characterization in the red alga *Griffithsia pacifica. Mol. Gen. Genet.* 209:343–51. [1.4.1]

Li, S. Y. (1984). The ecological characteristics of monospores of *Porphyra yezoensis* Ueda and their use in cultivation. *Hydrobiologia* 116/117:255–8. [9.2.4]

Li, W. K. W. (1978). Kinetic analysis of interactive effects of cadmium and nitrate on growth of *Thalassiosira fluviatalis* (Bacillariophyceae). *J. Phycol.* 14:454–60. [8.3.4]

Liddle, L. B., J. P. Thomas, & J. Scott (1982). Morphology and distribution of nuclei during development in *Cymopolia barbata* (Chlorophycophyta, Dasycladales). *J. Phycol.* 18:257–64. [1.7.2]

Lighty, R. G., I. G. MacIntyre, & A. C. Neumann (1980). Demise of a Holocene barrier-reef complex in northern Bahamas. *Geol. Soc. Am.* 12:471. [2.1.2]

Lignell, A., & M. Pedersén (1987). Nitrogen metabolism in *Gracilaria secundata. Hydrobiologia* 151/152:431–41. [5.5.1]

Lin, C. K. (1977). Accumulation of water soluble phosphorus and hydrolysis of polyphosphates by *Cladophora glomerata* (Chlorophyceae). *J. Phycol.* 13:46–51. [5.5.1, 5.5.2]

Lin, T.-Y., & W. Z. Hassid (1966). Pathway of alginic acid synthesis in the marine brown alga, *Fucus gardneri* Silva. *J. Biol. Chem.* 241:5284–97. [4.5.3]

Lindstrom, S. C., & R. E. Foreman (1978). Seaweed associations of the Flat Top Islands, British Columbia: a comparison of community methods. *Syesis* 11:171–85. [2.3.2, 2.5.1]

Littler, D. S., M. M. Littler, K. E. Bucher, & J. N. Norris (1989). *Marine Plants of the Caribbean. A Field Guide from Florida to Brazil.* Washington, DC: Smithsonian Institution Press, 263 pp. [2.1.2]

Littler, M. M. (1971). Standing stock measurements of crustose coralline algae (Rhodophyta) and other saxicolous organisms. *J. Exp. Mar. Biol. Ecol.* 6:91–9. [2.1.2]

Littler, M. M. (1976). Calcification and its role among the macroalgae. *Micronesica* 12:27–41. [7.2.2]

Littler, M. M. (1979). The effects of bottle volume, thallus weight, oxygen saturation levels, and water movement on apparent photosynthetic rates in marine algae. *Aquat. Bot.* 7:21–34 [4.7.1, 6.6.4, 7.1.3]

Littler, M. M., & K. E. Arnold (1980). Sources of variability in macroalgal primary productivity: sampling and interpretative problems. *Aquat. Bot.* 8:141–56. [4.7.2]

Littler, M. M., & K. E. Arnold (1982). Primary productivity of marine macroalgal functional-form groups from southwestern North America. *J. Phycol.* 18:307–11. [4.7.2]

Littler, M. M., & B. J. Kauker (1984). Heterotrichy and survival strategies in the red alga *Corallina officinalis* L. *Bot. Mar.* 27:37–44. [1.2.2, 3.2.2]

Littler, M. M., & D. S. Littler (1980). The evolution of thallus form and survival strategies in benthic marine macroalgae: field and laboratory tests of a functional form model. *Am. Nat.* 116:25–44. [1.2.2, 2.5.2, 4.7.2, 5.8]

Littler, M. M., & D. S. Littler (1983). Heteromorphic life-history strategies in the brown alga *Scytosiphon lomentaria* (Lyngb.) Link. *J. Phycol.* 19:425–31. [Fig. 1.25]

Littler, M. M., & D. S. Littler (1984). Models of tropical reef biogenesis: the contribution of algae. *Prog. Phycol. Res.* 3:323–64. [3.2.2]

Littler, M. M.,& D. S. Littler (1985). Nondestructive sampling. In M. M. Littler & D. S. Littler (eds.), *Handbook of Phycological Methods: Ecological Field Methods: Macroalgae* (pp. 161–75). Cambridge University Press. [2.1.2, 2.2.2, 2.5.1]

Littler, M. M., & D. S. Littler (1987). Effects of stochastic processes on rocky-intertidal biotas: an unusual flash flood near Corona del Mar, California. *Bull. So. Cal. Acad. Sci.* 86:95–106. [1.1.2, 6.1.3]

Littler, M. M., & D. S. Littler (1988). Structure and role of algae in tropical reef communities. In C. A. Lembi & J. R. Waaland (eds.), *Algae and Human Affairs* (pp. 29–56). Cambridge University Press. [Fig. 2.2a]

Littler, M. M., & D. S. Littler (1990). Productivity and nutrient relationships in psammophytic versus epilithic forms of Bryopsidales (Chlorophyta): comparisons based on a short-term physiological assay. *Hydrobiologia* 204/205:49–55. [5.8.3]

Littler, M. M., & S. N. Murray (1975). Impact of sewage on the distribution, abundance and community structure of rocky intertidal macro-organisms. *Mar. Biol.* 30:277–91. [8.6.1]

Littler, M. M., D. S. Littler, & P. R. Taylor (1983a). Evolutionary strategies in a tropical barrier reef system: functional-form groups of marine macroalgae. *J. Phycol.* 19:229–37. [1.2.2, 2.5.2, 4.7.2]

Littler, M. M., D. R. Martz, & D. S. Littler (1983b) Effects of recurrent sand deposition on rocky intertidal organisms: importance of substrate heterogeneity in a fluctuating environment. *Mar. Ecol. Prog. Ser.* 11:129–39. [7.2.3]

Littler, M. M., D. S. Littler, S. M. Blair, & J. N. Norris (1985). Deepest known plant life discovered on an uncharted seamount. *Science* 227:57–9. [2.3.2, 4.2.2, 4.3.4]

Littler, M. M., D. S. Littler, S. M. Blair, & J. N. Norris (1986a). Deep-water plant communities from an uncharted seamount off San Salvador Island, Bahamas: distribution, abundance, and primary productivity. *Deep Sea Res.* 33:881–92. [2.1.2, 2.3.2]

Littler, M. M., P. R. Taylor, & D. S. Littler (1986b). Plant defense associations in the marine environment. *Coral Reefs* 5:63–71. [3.2.2]

Littler, M. M., P. R. Taylor, D. S. Littler, R. H. Sims, & J. N. Norris (1987a). *Dominant Macrophyte Standing Stocks, Productivity and Community Structure on a Belizean Barrier Reef.* Atoll Research Bulletin 302, 24 pp. Washington, DC: Smithsonian Institution. [3.2.1]

Littler, M. M., D. S. Littler, & P. R. Taylor (1987b). Animal–plant defense associations: effects on the distribution and abundance of tropical reef macrophytes. *J. Exp. Mar. Biol. Ecol.* 105:107–21. [3.2.2]

Littler, M. M., D. S. Littler, & B. E. Lapointe (1988). A comparison of nutrient- and light-limited photosynthesis in psammophytic versus epilithic forms of *Halimeda* (Caulerpales, Halimediaceae) from the Bahamas. *Coral Reefs* 6:219–25. [1.1.1, 5.8.3]

Lobban, C. S. (1978a). The growth and death of the *Macrocystis* sporophyte (Phaeophyceae, Laminariales). *Phycologia* 17:196–212. [4.3.2]

Lobban, C. S. (1978b). Translocation of ^{14}C in *Macrocystis pyrifera* (giant kelp). *Plant Physiol.* 61:585–9. [4.6]

Lobban, C. S. (1978c). Translocation of ^{14}C in *Macrocystis integrifolia* (Phaeophyceae). *J. Phycol.* 14:178–82. [4.6]

Lobban, C. S. (1989). Environmental factors, plant responses, and colony growth in relation to tube-dwelling diatom blooms in the Bay of Fundy, Canada, with a review of the biology of tube-dwelling diatoms. *Diatom Res.* 4:89–109. [1.1.1]

Lobban, C. S., & D. M. Baxter (1983). Distribution of the red algal epiphyte *Polysiphonia lanosa* on its brown algal host *Ascophyllum nodosum* in the Bay of Fundy, Canada. *Bot. Mar.* 26:533–8. [3.1.2]

Lobel, P. S. (1980). Herbivory by damselfishes and their role in coral reef community ecology. *Bull. Mar. Sci.* 30:273–89. [3.2.2]

Lockhart, J. C. (1979). Factors affecting various forms in *Cladosiphon zosterae* (Phaeophyceae). *Am. J. Bot.* 66:836–44. [5.8.2]

Long, D. M., & A. Oaks (1990). Stabilization of nitrate reductase in maize roots by chymostatin. *Plant Physiol.* 93:846–50. [5.5.1]

Long, S. P., & C. F. Mason (1983). *Saltmarsh Ecology.* London: Blackie, 160 pp. [2.4.1]

Lowell, R. B., J. H. Markham, & K. H. Mann (1991). Herbivore-like damage induces increased strength and toughness in a seaweed. *Proc. R. Soc. Lond.* B243:31–8. [3.2.3]

Lowthion, D., P. G. Soulsby, & M. C. M. Houston (1985). Investigation of a eutrophic tidal basin: Part 1 – Factors affecting the distribution and biomass of macroalgae. *Mar. Environ. Res.* 15:263–84. [8.6.1]

Loya, Y., & B. Rinkevich (1980). Effects of oil pollution on coral reef communities. *Mar. Ecol. Prog. Ser.* 3:176–80. [8.4.2]

Lubchenco, J. (1978). Plant species diversity in a marine intertidal community: importance of herbivore food preference and algal competitive abilities. *Am. Nat.* 112:23–39. [3.2.1]

Lubchenco, J. (1980). Algal zonation in the New England rocky intertidal community: an experimental analysis. *Ecology* 61:333–44. [3.2.1]

Lubchenco, J. (1983). *Littorina* and *Fucus:* effects of herbivores, substratum heterogeneity, and plant escapes during succession. *Ecology* 64:1116–23. [3.2.2]

Lubchenco, J., & J. Cubit (1980). Heteromorphic life histories of certain marine algae as adaptations to variations in herbivory. *Ecology* 61:676–87. [2.5.2, 3.2.2]

Lubchenco, J., & S. D. Gaines (1981). A unified approach to marine plant–herbivore interactions. I. Populations and communities. *Annu. Rev. Ecol. Syst.* 12:405–37. [3.2.2]

Lubchenco, J., & B. A. Menge (1978). Community development and persistence in a low rocky intertidal zone. *Ecol. Monogr.* 48:67–94. [3.2.1]

Lubimenko, V., & Q. Tichovskaya (1928). *Recherches sur la Photosynthèse et l'Adaptation Chromatique chez les Algues Marines.* Moscow: Acad. Sci. USSR. [4.3.4]

Lundberg, P., R. G. Weich, P. Jensen, & H. J. Vogel (1989). Phosphorus-31 and nitrogen-14 studies of the uptake of phosphorus and nitrogen compounds in the marine macroalga *Ulva lactuca. Plant Physiol.* 89:1380–7. [5.4.2, 5.5.2]

Lüning, K. (1969). Growth of amputated and dark-exposed individuals of the brown alga *Laminaria hyperborea. Mar. Biol.* 2:218–23. [4.3.1, 4.7.3]

Lüning, K. (1980a). Control of algal life-history by daylength and temperature. In J. H. Price, D. E. G. Irvine, & W. F. Farnham (eds.), *The Shore Environments. Vol. 2: Ecosystems* (pp. 915–45). New York: Academic Press. [1.5.3]

Lüning, K. (1980b). Critical levels of light and temperature regulating the gametogenesis of three *Laminaria* species (Phaeophyceae). *J. Phycol.* 16:1–15. [2.5.2]

Lüning, K. (1981a). Light. In C. S. Lobban & M. J. Wynne (eds.), *The Biology of Seaweeds* (pp. 326–55). Oxford: Blackwell Scientific. [4.2.1, 4.2.2, 4.3.3]

Lüning, K. (1981b). Photomorphogenesis of reproduction in marine macroalgae. *Ber Deutsch. Bot. Ges.* 94:401–17. [1.7.2]

Lüning, K. (1984). Temperature tolerance and biogeography of seaweeds: the marine algal flora of Helgoland (North Sea) as an example. *Helgol. Meeresunters.* 38:305–17. [6.2.3]

Lüning, K. (1986). New frond formation in *Laminaria hyperborea* (Phaeophyta): a photoperiodic response. *Br. Phycol J.* 21:269–73. [1.5.3]

Lüning, K. (1988). Photoperiodic control of sorus formation in the brown alga *Laminaria saccharina. Mar. Ecol. Prog. Ser.* 45:137–44. [1.5.3, 4.6]

Lüning, K. (1990). *Seaweeds. Their Environment, Biogeography, and Ecophysiology* (trans. & ed. C. Yarish & H. Kirkman). New York: Wiley-Interscience. [1.5.3, 4.3.3, 6.2.5]

Lüning, K. (1991). Circannual growth rhythm in a brown alga, *Pterygophora californica. Bot. Acta* 104:157–62. [1.5.3]

Lüning, K., & M. J. Dring (1972). Reproduction induced by blue light in female gametophytes of *Laminaria saccharina. Planta* 104:252–6. [9.3.1, 9.3.2]

Lüning, K., & M. J. Dring (1979). Continuous underwater light measurements near Helgoland (North Sea) and its significance for characteristic light limits in the sublittoral region. *Helgol. Meeresunters.* 32:403–24. [4.2.2]

Lüning, K., & W. Freshwater (1988). Temperature tolerance of northeast Pacific marine algae. *J. Phycol.* 24:310–15. [6.2.4]

Lüning, K., & D. G. Müller (1978). Chemical interactions in sexual reproduction of several Laminariales (Phaeophyceae): release and attraction of spermatozoids. *Z. Pflanzenphysiol.* 89:333–41. [7.2.2]

Lüning, K., & M. Neushul (1978). Light and temperature demands for growth and reproduction of laminarian gametophytes in southern and central California. *Mar. Biol.* 45:297–309. [1.5.3]

Lüning, K., & K. Schmitz (1988). Dark growth of the red alga *Delesseria sanguinea* (Ceramiales): lack of chlorophyll, photosynthetic capability and phycobilisomes. *Phycologia* 27:72–7. [4.3.1]

Lüning, K., & I. tom Dieck (1989). Environmental triggers in algal seasonality. *Bot. Mar.* 32:389–97. [1.5.3]

Lüning, K., K. Schmitz, & J. Willenbrink (1973). CO_2 fixation and translocation in benthic marine algae. III. Rates and ecological significance of translocation in *Laminaria hyperborea* and *L. saccharina*. *Mar. Biol.* 23:275–81. [4.6, 4.7.3]

Luoma, S. N., G. W. Bryan, & W. T. Langston (1982). Scavenging of heavy metals from particulates by brown seaweed. *Mar. Pollut. Bull.* 13:394–6. [8.3.1]

Lüttge, U., & N. Higinbotham (1979). *Transport in Plants.* Berlin: Springer-Verlag, 468 pp. [5.3.1, 5.4.1]

Lüttke, A. (1988). The lack of chloroplast DNA in *Acetabularia mediterranea (acetabulum)* (Chlorophyceae): a reinvestigation. *J. Phycol.* 24:173–80. [1.3.2]

Luxoro, C., & B. Santelices (1989). Additional evidence for ecological differences among isomorphic reproductive phases of *Iridaea laminarioides* (Rhodophyta: Gigartinales). *J. Phycol.* 25:206–12.

Lyngby, J. E. (1990). Monitoring of nutrient availability and limitation using the marine macroalga *Ceramium rubrum* (Huds.) C. Ag. *Aquat. Bot.* 38:153–61. [5.7.3]

Maberly, S. C. (1990). Exogenous sources of inorganic carbon for photosynthesis by marine macroalgae. *J. Phycol.* 26:439–49. [4.4.1]

MacColl, R., & D. Guard-Friar (1987). *Phycobiliproteins.* Boca Raton, FL: CRC Press. [4.3.1]

MacDonald, M. A., D. S. Fensom, & A. R. A. Taylor (1974). Electrical impedance in *Ascophyllum nodosum* and *Fucus vesiculosus* in relation to cooling, freezing, and desiccation. *J. Phycol.* 10:462–9. [6.2.4]

MacIsaac, J. J., R. C. Dugdale, S. A. Huntsman, & H. L. Conway (1979). The effect of sewage on uptake of inorganic nitrogen and carbon by natural populations of marine phytoplankton. *J. Mar. Res.* 37:51–66. [8.6.1]

Mackie, W., & R. D. Preston (1974). Cell wall and intercellular region polysaccharides. In W. D. P. Stewart (ed.), *Algal Physiology and Biochemistry* (pp. 40–85). Oxford: Blackwell Scientific. [1.3.1]

MacRobbie, E. A. C. (1974). Ion uptake. In W. D. P. Stewart (ed.), *Algal Physiology and Biochemistry* (pp. 714–40). Oxford: Blackwell Scientific. [6.3.1]

Madsen, T. V., & S. C. Maberly (1990). A comparison of air and water as environments for photosynthesis by the intertidal alga *Fucus spiralis* (Phaeophyta). *J. Phycol.* 26:24–30. [4.3.3, 4.4.1]

Maestrini, S. Y., & D. J. Bonin (1981). Compeititon among phytoplankton based on inorganic nutrients. *Can. Bull. Fish. Aquat. Sci.* 210:264–78. [3.1.3]

Maggs, C. A. (1988). Intraspecific life history variability in the Florideophycidae (Rhodophyta). *Bot. Mar.* 31:465–90. [1.5.1, 1.5.2]

Maggs, C. A. (1989). *Erythrodermis allenii* Batters in the life history of *Phyllophora traillii* Holmes ex Batters (Phyllophoraceae, Rhodophyta). *Phycologia* 28:305–17. [1.5.2]

Maggs, C. A., & D. P. Cheney (1990). Competition studies of marine macroalgae in laboratory culture. *J. Phycol.* 26:17–24. [2.4.2, 2.5.2, 3.0]

Maggs, C. A., & M. D. Guiry (1987). Environmental control of macroalgal phenology. In R. M. M. Crawford (ed.), *Plant Life in Aquatic and Amphibious Habitats* (pp. 359–73). Oxford: Blackwell Scientific. [1.5.3]

Maggs, C. A., & C. M. Pueschel (1989). Morphology and development of *Ahnfeltia plicata* (Rhodophyta): proposal of Ahnfeltiales ord. nov. *J. Phycol.* 25:333–51. [1.4.1]

Magne, T. H., & O. Holm-Hansen (1975). Nitrogen fixation on a coral reef. *Phycologia* 14:87–92. [1.3.6]

Magruder, W. H. (1984). Specialized appendages on spermatia from the red alga *Aglaothamnion neglectum* (Ceramiales, Ceramiaceae) specifically bind with trichogynes. *J. Phycol.* 20:436–40. [1.5.4]

Maier, C. M., & A. M. Pregnall (1990). Increased macrophyte nitrate reductase activity as a consequence of groundwater input of nitrate through sandy beaches. *Mar. Biol.* 107:263–71. [5.8.3]

Maier, I., & M. N. Clayton (1989). Oogenesis in *Durvillaea potatorum* (Durvillaeales, Phaeophyta). *Phycologia* 28:271–4. [1.5.2]

Maier, I., & D. G. Müller (1986). Sexual pheromones in algae. *Biol. Bull.* 170:145–75. [8.3.3]

Main, S. P., & C. D. McIntire (1974). The distribution of epiphytic diatoms in Yaquina estuary, Oregon (U.S.A.). *Bot. Mar.* 17:88–100. [2.4.2]

Maki, A. W. (1991). The *Exxon Valdez* oil spill: initial environmental impact assessment. *Environ. Sci. Technol.* 25:24–9. [8.1]

Mance, G. (1987). *Pollution Threat of Heavy Metals in Aquatic Environments.* Amsterdam: Elsevier. [8.3, 8.3.5]

Manley, S. L. (1981). Iron uptake and translocation by *Macrocystis pyrifera. Plant Physiol.* 68:914–18. [5.5.6]

Manley, S. L. (1983). Composition of sieve tube sap from *Macrocystis pyrifera* (Phaeophyta) with emphasis on the inorganic constituents. *J. Phycol.* 19:118–21. [4.6, 5.6]

Manley, S. L. (1984). Micronutrient uptake and translocation by *Macrocystis pyrifera* (Phaeophyta). *J. Phycol.* 20:192–201. [5.6]

Manley, S. L., & W. J. North (1984). Phosphorus and the growth of juvenile *Macrocystis pyrifera* (Phaeophyta) sporophytes. *J. Phycol.* 20:389–93. [5.1.3]

Mann, D. G. (1984). Observations on copulation in *Navicula pupula* and *Amphora ovalis* in relation to the nature of diatom species. *Ann. Bot.* 54:429–38. [1.1.3]

Mann, K. H. (1972). Ecological energetics of the seaweed zone in a marine bay on the Atlantic coast of Canada. II. Productivity of the seaweeds. *Mar. Biol.* 14:199–209. [4.7.4, 4.7.6]

Mann, K. H., N. Jarman, & G. Dieckmann (1979). Development of a method for measuring the productivity of the kelp *Ecklonia maxima* (Osbeck) Papenf. *Trans R. Soc. S. Afr.* 44:27–41. [4.7.4]

Markager, S., & K. Sand-Jensen (1990). Heterotrophic growth of *Ulva lactuca* (Chlorophyceae). *J. Phycol.* 26:670–3. [4.4.1]

Markham, J. W. (1973). Observations on the ecology of *Laminaria sinclairii* on three northern Oregon beaches. *J. Phycol.* 9:336–41. [7.2.3]

Markham, J. W., & P. R. Newroth (1972). Observations on the ecology of *Gymnogongrus linearis* and related species. *Proc. Intl. Seaweed Symp.* 7:127–30. [7.2.3]

Markham, J. W., B. P. Kremer, & K. R. Sperling (1980). Effect of cadmium on *Laminaria saccharina* in culture. *Mar. Ecol. Prog. Ser.* 3:31–9. [8.3.3]

Martens, C. S., R. A. Berner, & J. K. Rosenfeld (1978). Interstitial water chemistry of anoxic Long Island Sound sediments. 2. Nutrient regeneration and phosphate removal. *Limnol. Oceanogr.* 23:605–17. [5.8.3]

Martin, J. H., & S. E. Fitzwater (1988). Iron deficiency limits phytoplankton growth in the northeast Pacific subarctic. *Nature* 331:341–3. [5.2, 5.5.6]

Martin, J. H., & G. A. Knauer (1973). The elemental composition of plankton. *Geochim. Cosmochim. Acta* 37:1639–53. [8.3.5]

Martin, J. H., R. M. Gordon, & S. E. Fitzwater (1990). Iron in Antarctic waters. *Nature* 345:156–8. [5.5.6]

Mathieson, A. C. (1986). A comparison of seaweed mariculture programs – activities. In M. Bilo, H. Rosenthial, & C. J. Sindermann (eds.), *Realism in Aquaculture: Achievements, Constraints and Perspectives* (pp. 108–40) (World Conference on Aquaculture, Italy, Sept. 1981). Bredeme, Belgium: European Aquaculture Society. [9.4.1, 9.6]

Mathieson, A. C., & T. L. Norall (1975a). Photosynthetic studies of *Chondrus crispus.* *Mar. Biol.* 33:207–13. [6.2.2]

Mathieson, A. C., & T. L. Norall (1975b). Physiological studies of subtidal red algae. *J. Exp. Mar. Biol. Ecol.* 20:237–47. [6.2.2]

Mathieson, A. C., E. Tveter, M. Daly, & J. Howard (1977). Marine algal ecology in a New Hampshire tidal rapid. *Bot. Mar.* 20:277–90. [7.1.1, 7.2.1]

Mathieson, A. C., N. B. Reynolds, & E. J. Hehre (1981). Investigations of New England marine algae. II. The species composition, distribution and zonation of seaweeds in the Great Bay Estuary System and the adjacent open coast of New Hampshire. *Bot. Mar.* 24:533–45. [2.4.1, 6.5]

Mathieson, A. C., C. D. Neefus, & C. E. Penniman (1983). Benthic ecology in an estuarine tidal rapid. *Bot. Mar.* 26:213–30. [7.1.1, 7.2.3]

Matilsky, M. B., & W. P. Jacobs (1983). Accumulation of amyloplasts on the bottom of normal and inverted rhizome tips of *Caulerpa prolifera* (Forsskål) Lamouroux. *Planta* 159:189–92. [1.7.2]

Matson, E. A. (1991). Water chemistry and hydrology of the "Blood of Sanvitores," a Micronesian red tide. *Micronesica* 24:95–108. [2.2.1]

Mattox, K. R., & K. D. Stewart (1984). Classification of the green algae: a concept based on comparative cytology. In D. E. G. Irvine & D. M. John (eds.), *Systematics of the Green Algae* (pp. 29–72). New York: Academic Press. [1.1.1]

Mayhoub, H., P. Gayral, & R. Jacques (1976). Action de la composition spectrale de la lumière sur la croissance et la reproduction de *Calosiphonia vermicularis* (J. Agardh) Schmitz (Rhodophycées, Gigartinales). *C. R. Acad. Sci. Paris* 283(D):1041–4. [1.7.2]

Mazzo, A., S. Bonotto, & B. Felluga (1977). Ultrastructure of DNA of basal, middle and apical chloroplasts of individual *Acetabularia mediterranea* cells. In C. L. F. Woodcock (ed.), *Progress in Acetabularia Research* (pp. 123–36). New York: Academic Press. [1.4.2]

McArthur, D. M., & B. L. Moss (1977). The ultrastructure of cell walls in *Enteromorpha intestinalis* (L.) Link. *Br. Phycol. J.* 12:359–68. [3.1.2]

McBride, D. L., & K. Cole (1972). Ultrastructural observations on germinating monospores in *Smithora naiadum* (Rhodophyceae, Bangiophycidae). *Phycologia* 11:181–91. [1.2.1]

McBride, D. L., P. Kugrens, & J. A. West (1974). Light and electron microscope observations on red algal galls. *Protoplasma* 79:249–64. [3.3.2]

McCandless, E. L. (1981). Polysaccharides of the seaweeds. In C. S. Lobban & M. J. Wynne (eds.), *The Biology of Seaweeds* (pp. 559–88). Oxford: Blackwell Scientific. [4.5.1]

McCandless, E. L., & J. S. Craigie (1979). Sulfated polysaccharides in red and brown algae. *Annu. Rev. Plant Physiol.* 30:41–53. [4.5.2]

McCandless, E. L., J. S. Craigie, & J. A. Walter (1973). Carrageenans in the gametophytic and sporophytic stages of *Chondrus crispus.* *Planta* 112:201–12. [9.7]

McClintock, M., N. Higinbotham, E. G. Uribe, & R. E. Cleland (1982). Active, irreversible accumulation of extreme levels of H_2SO_4 in the brown alga, *Desmarestia.* *Plant Physiol.* 70:771–4. [3.2.3]

McCracken, D. A., & J. R. Cain (1981). Amylose in floridean starch. *New Phytol.* 88:67–71. [4.5.1]

McHugh, D. J. (ed.) (1987). *Production and Utilization of Colloids from Commercial Seaweeds.* FAO Fish. Tech. Pap. 288, 189 pp. [9.8]

McHugh, D. J. (1990). Prospects for *Eucheuma* marketing in the world and the future of seaweed farming in the Pacific. In T. Adams & R. Foscarini (eds.), *Proceedings of the Regional Workshop on Seaweed Culture and Marketing* (Suva, Fiji, 14–17 November 1989). South Pacific Aquacult. Devel. Project, FAO, UN, 86 pp. [9.5.2]

McIntire, C. D., & W. W. Moore (1977). Marine littoral diatoms: ecological considerations. In D. Werner (ed.), *The Biology of Diatoms* (pp. 333–71). Oxford: Blackwell Scientific. [2.5.1]

McKay, R. M. L., & S. P. Gibbs (1990). Phycoerythrin is absent from the pyrenoid of *Porphyridium cruentum:* photosynthetic implications. *Planta* 180:249–56. [4.4.4]

McKenzie, G. H., A. L. Ch'ng, & K. R. Gayler (1979). Glutamine synthetase/glutamine:α-ketoglutarate aminotransferase in chloroplasts from the marine alga *Caulerpa simpliciuscula.* *Plant Physiol.* 63:578–82. [5.5.1]

McLachlan, J. (1977). The effects of nutrients on growth and development of embryos of *Fucus edentatus* Pyl. (Phaeophyceae, Fucales). *Phycologia* 16:329–38. [5.5.6, 5.8.2]

McLachlan, J. (1982). Inorganic nutrition of marine macroalgae in culture. In L. M. Srivastava (ed.), *Synthetic and Degradative Processes in Marine Macrophytes* (pp. 71–98). Berlin: Walter de Gruyter. [5.8.2]

McLachlan, J. (1985). Macroalgae (seaweeds): industrial resources and their utilization. *Plant Soil* 89:137–57. [4.4.1, 9.8]

McLachlan, J., & R. G. S. Bidwell (1978). Photosynthesis of eggs, sperm, zygotes, and embryos of *Fucus serratus. Can. J. Bot.* 56:371–3. [4.7.2]

McLachlan, J., & R. G. S. Bidwell (1983). Effects of colored light on the growth and metabolism of *Fucus* embryos and apices in culture. *Can. J. Bot.* 61:1993–2003. [1.7.2]

McLean, M. W., & F. B. Williamson (1977). Cadmium accumulation by the marine red alga *Porphyra umbilicalis. Physiol. Plant.* 41:268–72. [8.3.2]

McQuaid, C. D., & G. M. Branch (1984). Influence of sea temperature, substratum and wave exposure on rocky intertidal communities: an analysis of faunal and floral biomass. *Mar. Ecol. Prog. Ser.* 19:145–51. [1.1.3, 6.2.5]

McQuaid, C. D., & G. M. Branch (1985). Trophic structure of rocky intertidal communities: response to wave action and implications for energy flow. *Mar. Ecol. Prog. Ser.* 22:153–61. [7.2.2]

Meeks, J. C. (1974). Chlorophylls. In W. D. P. Stewart (ed.), *Algal Physiology and Biochemistry* (pp. 161–75). Oxford: Blackwell Scientific. [4.3.1]

Meinesz, A. (1980). Connaissances actuelles et contribution à l'étude de la reproduction et du cycle des Udotéacées (Caulerpales, Chlorophytes). *Phycologia* 19:110–38. [Figs. 1.26, 1.45]

Melkonian, M., & H. Robenek (1984). The eyespot apparatus of flagellated green algae. *Prog. Phycol. Res.* 3:193–268. [1.3.3]

Melo, R. A., B. W. W. Harger, & M. Neushul (1990). *Gelidium* cultivation in the sea. Presented at an international workshop on *Gelidium* (Santander, Spain, Sept. 1990). [9.6]

Menzel, D. (1988). How do giant plant cells cope with injury? – The wound response in siphonous green algae. *Protoplasma* 144:73–91. [1.7.4]

Menzel, D., & C. Elsner-Menzel (1989a). Maintenance and dynamic changes of cytoplasmic organization controlled by cytoskeletal assemblies in *Acetabularia* (Chlorophyceae). In A. W. Coleman, L. J. Goff, & J. R. Stein-Taylor (eds.), *Algae as Experimental Systems* (pp. 71–90). New York: Alan R. Liss. [1.3.3]

Menzel, D., & C. Elsner-Menzel (1989b). Actin-based chloroplast rearrangements in the cortex of the giant coenocytic green alga *Caulerpa. Protoplasma* 150:1–8 [1.3.2]

Michanek, G. (1978). Trends in applied phycology, with a literature review: seaweed farming on an industrial scale. *Bot. Mar.* 21:469–75. [9.1]

Michanek, G. (1979). Phytogeographic provinces and seaweed distribution. *Bot. Mar.* 22:375–91. [6.2.5]

Miflin, B. J., & P. J. Lea (1976). The pathway of nitrogen assimilation in plants. *Phytochemistry* 15:873–85. [5.5.1]

Millard, P., & L. V. Evans (1982). Sulphate uptake in the unicellular marine red alga *Rhodella maculata. Arch. Microbiol.* 131:165–9. [5.5.5]

Miller, K. A. (1988). Morphogenesis in red algae: relationships among cells, tissues and organs (abstract). *J. Phycol. (Suppl.)* 24:21. [1.7.2]

Miller, K. H. (1985). Succession in sea urchin and seaweed abundance in Nova Scotia, Canada. *Mar. Biol.* 84:275–86. [3.2.1]

Miller, K. R., & L. Spear-Bernstein (1989). Algae as model systems in the study of photosynthetic membrane organization. In A. W. Coleman, L. J. Goff, & J. R. Stein-Taylor (eds.), *Algae as Experimental Systems* (pp. 233–48). New York: Alan R. Liss. [4.1]

Millner, P. A., & L. V. Evans (1980). The effects of triphenyltin chloride on respiration and phytosynthesis in the green algae *Enteromorpha intestinalis* and *Ulothrix pseudoflacca. Plant Cell Environ.* 3:339–48. [8.5.4]

Millner, P. A., & L. V. Evans (1981). Uptake of triphenyltin chloride by *Enteromorpha intestinalis* and *Ulothrix pseudoflacca. Plant Cell Environ.* 4:383–9. [8.5.4]

Minorsky, P. V. (1989). Temperature sensing by plants: a review and hypothesis. *Plant Cell Environ.* 12:119–35. [6.2.4]

Mishkin, M., D. Mauzerall, & S. I. Beale (1979). Diurnal variation *in situ* of photosynthetic capacity in *Ulva* caused by a dark reaction. *Plant Physiol.* 64:896–9. [4.7.2]

Mitchell, C. T., E. K. Anderson, L. C. Jones, & W. J. North (1970). What oil does to ecology. *J. Water Pollut. Contr. Fed.* 42:812–18. [8.4.2]

Miura, A. (1975). *Porphyra* cultivation in Japan. In J. Tokida & H. Hirose (eds.), *Advance of Phycology in Japan* (pp. 273–304). The Hague: Dr. W. Junk. [6.2.4, 9.2, 9.2.2]

Miura, A. (1984). A new variety and a new form of *Porphyra* (Bangiales, Rhodophyta) from Japan: *Porphyra tenera* Kjellman var. *tamatsuensis* Miura var. nov. and *P. yezoensis* Ueda form *narawaensis* Miura, form nov. *J. Tokyo Univ. Fish.* 71:1–37. [9.2.4]

Miura, A. (1985). Genetic analysis of the variant color types of light red, light green and light yellow phenotypes of *Porphyra yezoensis* (Rhodophyta, Bangiaceae). In H. Hara (ed.), *Origin and Evolution of Diversity in Plants and Plant Communities* (pp. 270–84). Tokyo: Academia Scientific Books. [1.5.2, 9.2.4]

Miyazawa, K., & K. Ito (1974). Isolation of a new peptide, L-citrullinyl-L-arginine, from a red alga *Grateloupia turuturu. Nippon Suissan Gakkaishi* 40:815–18. [5.5.1]

Mizuno, M. (1984). Environment at the front shore of the Institute of Algological Research of Hokkaido University. *Sci. Pap. Inst. Algol. Res., Fac. Sci. Hokkaido U.* 7:263–92. [2.2.1, 6.4]

Mobley, C. D. (1989). A numerical model for the computation of radiance distributions in natural waters with wind-roughened surfaces. *Limnol. Oceanogr.* 34:1473–83. [4.2.2]

Moe, R. L., & P. C. Silva (1981). Morphology and taxonomy of *Himantothallus* (including *Phaeoglossum* and *Phyllogigas*), an Antarctic member of the Desmarestiales (Phaeophyceae). *J. Phycol.* 17:15–29. [1.5.3]

Moebus, K., & K. M. Johnson (1974). Exudation of dissolved organic carbon by brown algae. *Mar. Biol.* 26:117–25. [4.7.3]

Moebus, K., K. M. Johnson, & J. M. Sieburth (1974). Rehydration of desiccated intertidal brown algae: release of dissolved organic carbon and water uptake. *Mar. Biol.* 26:127–34. [4.7.3]

Mohsen, A. F., A. F. Khaleafa, M. A. Hashem, & A. Metwalli (1974). Effect of different nitrogen sources on growth, reproduction, amino acid, fat and sugar contents in *Ulva fasciata* Petite. *Bot. Mar.* 17:218–22. [5.8.2]

Monod, J. (1942). *Recherches sur la Croissance des Cultures Bacteriennes.* Paris: Hermann et Cie. [5.7.1.]

Montfrans, J. van, R. L. Wetzel, & R. J. Orth (1984). Epiphyte–grazer relationships in seagrass meadows: consequences for seagrass growth and production. *Estuaries* 17:289–309. [2.4.2]

Montgomery, H. A. C., P. G. Soulsby, I. C. Hart, & S. L. Wright (1985). Investigation of a eutrophic tidal basin: Part 2. Nutrients and environmental aspects. *Mar. Environ. Res.* 15:285–302. [8.6.1]

Montgomery, W. L. (1980a). Comparative feeding ecology of two herbivorous damselfishes (Pomacentridae: Teleostei) from the Gulf of California, Mexico. *J. Exp. Mar. Biol. Ecol.* 47:9–24. [3.2.2]

Montgomery, W. L. (1980b). The impact of non-selective grazing by the giant blue damselfish, *Microspathodon dorsalis*, on algal communities in the Gulf of California, Mexico. *Bull. Mar. Sci.* 30:290–303. [3.2.2]

Moody, M. D. (1986). Microorganisms and iron limitation. *BioScience* 36:618–23. [5.5.6]

Moreau, J., D. Pesando, P. Bernard, B. Caram, & J. C. Pionnat (1988). Seasonal variation in production of antifungal substances by some Dictyotales (brown algae) from the French Mediterranean coast. *Hydrobiologia* 162:157–62. [3.1.2]

Moreno, C. A., & E. Jaramillo (1983). The role of grazers in the zonation of intertidal macroalgae on the Chilean coast. *Oikos* 41:73–6. [3.2.1]

Moreno, C. A., J. P. Sutherland, & H. F. Jara (1984). Man as a predator in the intertidal zone of southern Chile. *Oikos* 42:155–60. [3.2.1]

Morris, A. W., & A. J. Bale (1975). The concentration of cadmium, copper, manganese and zinc by *Fucus vesiculosus* in the Bristol Channel. *Estu. Cstl. Mar. Sci.* 3:153–63. [8.3.1]

Morrison, D. (1988). Comparing fish and urchin grazing in shallow and deeper coral reef algal communities. *Ecology* 69:1367–82. [3.2.1]

Moss, B. (1964). Wound healing and regeneration in *Fucus vesiculosus* L. *Proc. Intl. Seaweed Symp.* 4:117–22. [1.7.4]

Moss, B. (1974a). Morphogenesis. In W. D. P. Stewart (ed.), *Algal Physiology and Biochemistry* (pp. 788–813). Oxford: Blackwell Scientific. [1.7.3]

Moss, B. (1974b). Attachment and germination of the zygotes of *Pelvetia canaliculata* (L.) Dcne. et Thur. (Phaeophyceae, Fucales). *Phycologia* 13:317–22. [1.6.1]

Moss, B., & P. Woodhead (1975). The effect of two commercial herbicides on the settlement, germination and growth of *Enteromorpha*. *Mar. Pollut. Bull.* 6:189–92. [8.5.1]

Moss, B. L. (1965). Apical dominance in *Fucus vesiculosus*. *New Phytol.* 64:387–92. [1.7.2]

Moss, B. L. (1967). The apical meristem of *Fucus*. *New Phytol.* 66:67–74. [1.7.2]

Moss, B. L. (1982). The control of epiphytes by *Halidrys siliquosa* (L.) Lyngb. (Phaeophyta, Cystoseiraceae). *Phycologia* 21:185–91. [3.1.2]

Moss, B. L. (1983). Sieve elements in the Fucales. *New Phytol.* 93:433–7. [4.6]

Motomura, T. (1990). Ultrastructure of fertilization in *Laminaria angustata* (Phaeophyta, Laminariales) with emphasis on the behavior of centrioles, mitochondria and chloroplasts of the sperm. *J. Phycol.* 26:80–9. [1.3.2]

Motomura, T., & Y. Sakai (1988). The occurrence of flagellated eggs in *Laminaria angustata* (Phaeophyta, Laminariales). *J. Phycol.* 24:282–5. [1.5.4]

Müller, D. G. (1963). Die Temperaturabhängigkeit der Sporangienbildung bei *Ectocarpus siliculosus* von verschiedenen Standorten. *Publ. Staz. Zool. Napoli* 33:310–14. [1.5.3]

Müller, D. G. (1976). Sexual isolation between a European and an American population of *Ectocarpus siliculosus* (Phaeophyta). *J. Phycol.* 12:252–4. [1.4.1]

Müller, D. G. (1979). Genetic affinity of *Ectocarpus siliculosus* (Dillw.) Lyngb. from the Mediterranean, North Atlantic and Australia. *Phycologia* 18:312–81. [1.4.1]

Müller, D. G. (1981). Sexuality and sex attraction. In C. S. Lobban & M. J. Wynne (eds.), *The Biology of Seaweeds* (pp. 661–74). Oxford: Blackwell Scientific. [1.5.4]

Müller, D. G. (1982). *Pheromone effects in fertilization of brown algae*. Film C1424, Institut für den Wissenschaftlichen Film Göttingen. [1.5.3]

Müller, D. G. (1988). Studies on sexual compatability between *Ectocarpus siliculosus* (Phaeophyceae) from Chile and the Mediterranean Sea. *Helgol. Meeresunters.* 42:469–76. [1.4.1]

Müller, D. G. (1989). The role of pheromones in sexual reproduction of brown algae. In A. W. Coleman, L. J. Goff, & J. R. Stein-Taylor (eds.), *Algae as Experimental Systems* (pp. 201–13). New York: Alan R. Liss. [1.5.4]

Müller, D. G., & B. Stache (1989). Life history studies on *Pilayella littoralis* (L.) Kjellman (Phaeophyceae, Ectocarpales) of different geographical origin. *Bot. Mar.* 32:71–8. [1.5.2]

Müller, D. G., I. Maier, & G. Gassmann (1985). Survey on sexual pheromone specificity in Laminariales (Phaeophyceae). *Phycologia* 24:475–7. [1.5.4]

Müller, D. G., H. Kawai, B. Stache, E. Fölster, & W. Boland (1990). Sexual pheromones and gamete chemotaxis in *Analipus japonicus* (Phaeophyceae). *Experientia* 46:534–6. [1.5.4]

Mumford, T. E., Jr. (1987). Commercialization strategy for nori culture in Puget Sound, Washington. In K. T. Bird & P. H. Benson (eds.), *Seaweed Cultivation for Renewable Resources* (pp. 351–68). Amsterdam: Elsevier. [9.2.4]

Mumford, T. E., Jr. (1990). Nori cultivation in North America: growth of the industry. *Hydrobiologia* 204/205:89–98. [9.2.4]

Mumford, T. E., Jr. & A. Miura (1988). *Porphyra* as food: cultivation and economics. In C. A. Lembi & J. R. Waaland (eds.), *Algae and Human Affairs* (pp. 87–117). Cambridge University Press. [9.2, 9.2.2, 9.2.3, 9.2.4]

Munda, I. M. (1967). Der Einfluss der Salinität auf die chemische Zusammensetzung, das Wachstum und die Fruktifikation einiger Fucaceen. *Nova Hedwigia* 13:471–508. [6.3.4]

Munda, I. M. (1974). Changes and succession in the benthic algal associations of slightly polluted habitats. *Rev. Int. Oceanogr. Med.* 34:37–52. [8.6.1]

Munda, I. M. (1978). Salinity dependent distribution of benthic algae in estuarine areas of Icelandic fjords. *Bot. Mar.* 21:451–68. [6.5]

Munda, I. M. (1984). Salinity dependent accumulation of Zn, Co and Mn in *Scytosiphon lomentaria* (Lyngb.) Link and *Enteromorpha intestinalis* (L.) from the Adriatic Sea. *Bot. Mar.* 27:371–6. [8.3.4]

Munda, I. M. (1986). Differences in heavy metal accumulation between vegetative parts of the thalli and receptacles in *Fucus spiralis* L. *Bot. Mar.* 29:341–9. [8.3.1]

Munda, I. M., & F. Gubensek (1976). The amino acid composition of some common marine algae from Iceland. *Bot. Mar.* 19:85–92. [5.8.1]

Munda, I. M., & V. Hudnik (1986). Growth response of *Fucus vesiculosus* to heavy metals, singly and in dual combinations, as related to accumulation. *Bot. Mar.* 29:401–12. [8.3.4]

Munda, I. M., & V. Hudnik (1988). The effects of Zn, Mn, and Co accumulation on growth and chemical composition of *Fucus vesiculosus* L. under different temperature and salinity conditions. *Mar. Ecol.* 9:213–25. [8.3.1, 8.3.4]

Munda, I. M., & V. Hudnik (1991). Trace metal content in some seaweeds from the Northern Adriatic. *Bot. Mar.* 34:241–9. [5.8.1]

Munda, I. M., & B. P. Kremer (1977). Chemical composition and physiological properties of fucoids under conditions of reduced salinity. *Mar. Biol.* 42:9–16. [6.3.4]

Munns, R., H. Greenway, & G. O. Kirst (1983). Halotolerant eukaryotes. In A. Pirson & M. H. Zimmerman (eds.), *Encyclopedia of Plant Physiology. Vol. 12: Physiological Plant Ecology III* (pp. 59–135). Berlin: Springer-Verlag. [6.3.3]

Murray, S. N., & P. S. Dixon (1975). The effects of light intensity and light period on the development of thallus form in the marine red alga *Pleonosporium squarrulosum* (Harvey) Abbott (Rhodophyta: Ceramiales). II. Cell enlargement. *J. Exp. Mar. Biol. Ecol.* 19:165–76. [1.5.3]

Murray, S. N., & M. M. Littler (1974). Analyses of standing stock and community structure of macro-organisms. In S. N. Murray & M. M. Littler (eds.), *Biological Features of Intertidal Communities near the U.S. Navy Sewage Outfall, Wilson Cove, San Clemente Island, California* (pp. 23–51). San Diego: Naval Undersea Center. [2.2.2]

Murray, S. N., & M. M. Littler (1978). Patterns of algal succession in a perturbated marine intertidal community. *J. Phycol.* 14:506–12. [8.6.1]

Murthy, M. S., T. Ramakrishna, G. V. Sarat Babu, & Y. N. Rao (1986). Estimation of net primary productivity of intertidal seaweeds – limitations and latent problems. *Aquat. Bot.* 23:383–387. [4.7]

Myers, V. B., & R. I. Iverson (1981). Phosphorus and nitrogen limited phytoplankton productivity in northeastern Gulf of Mexico coastal estuaries. In B. J. Neilson & L. E. Cronin (eds.), *Estuaries and Nutrients* (pp. 569–82). Clifton, NJ: Humana Press. [5.5.2]

Myklestad, S., I. Eide, & S. Melson (1978). Exchange of heavy metals in *Ascophyllum nodosum* (L.) Le Jol. *in situ* by means of transplanting experiments. *Environ. Pollut.* 16:277–84. [8.3.1]

Nagashima, H., S. Nakamura, K. Nisizawa, & T. Hori (1971). Enzymic synthesis of floridean starch in a red alga, *Serraticardia maxima. Plant Cell Physiol.* 12:243–53. [4.5.3]

Nakahara, H., & Y. Nakamura (1973). Parthenogenesis, apogamy and apospory in *Alaria crassifolia* (Laminariales). *Mar. Biol.* 18:327–32. [1.5.2, 1.7.1]

Nalewajko, C., & D. R. S. Lean (1980). Phosphorus. In I. Morris (ed.), *The Physiological Ecology of Phytoplankton* (pp. 235–58). Oxford: Blackwell Scientific. [5.5.2]

Nasr, A. H., & I. A. Bekheet (1970). Effects of certain trace elements and soil extract on some marine algae. *Hydrobiologia* 36:53–60. [5.5.6, 5.8.2]

Nasr, A. H., I. A. Bekheet, & R. K. Ibrahim (1968). The effects of different nitrogen and carbon sources on amino acid synthesis in *Ulva, Dictyota* and *Pterocladia. Hydrobiologia* 31:7–16. [5.5.1, 5.8.1, 5.8.2]

National Acadamy of Sciences (1985). *Oil in the Sea.* Washington, DC: NAS. [Fig. 8.10]

Neale, P. J. (1987). Algal photoinhibition and photosynthesis in the aquatic environment. In D. J. Kyle, C. B. Osmond, & C. J. Arntzen (eds.), *Topics in Photosynthesis. Vol. 9: Photoinhibition* (pp. 39–65). Amsterdam: Elsevier. [4.3.3]

Neilands, J. B. (1973). Microbial iron transport compounds (siderochromes). In G. L. Eichhorn (ed.), *Inorganic Biochemistry* (pp. 167–202). Amsterdam: Elsevier. [5.5.6]

Neish, A. C., & C. Fox (1971). *Greenhouse Experiments on the Vegetative Propagation of* Chondrus crispus *(Irish Moss).* Technical Report no. 12, Atlantic Regional Laboratory, National Research Council of Canada, Halifax, NS, 35 pp. [9.7]

Neish, A. C., & P. F. Shacklock (1971). *Greenhouse Experiments (1971) on the Propagation of Strain T4 of Irish Moss.* Atlantic Research Laboratory Technical Report no. 14, National Research Council of Canada, Halifax, NS. [5.8.1]

Neish, A. C., P. F. Shacklock, C. H. Fox, & F. J. Simpson (1977). The cultivation of *Chondrus crispus.* Factors affecting growth under greenhouse conditions. *Can. J. Bot.* 55:2263–71. [4.5.1, 9.7]

Nelson, S. G. (1985). Immediate enhancement of photosynthesis by coral reef macrophytes in response to ammonium enrichment. In V. M. Harmelin & B. Salvat (eds.), *Proceedings, 5th International Coral Reef Congress, Tahiti* (vol. 5, pp. 65–70). Moorea, French Polynesia: Antenne Museum–EPHE. [3.2.1]

Nelson, S. G., & A. W. Siegrist (1987). Comparison of mathematical expressions describing light-saturation curves for photosynthesis by tropical marine macroalgae. *Bull. Mar. Sci.* 41:617–22. [4.3.3]

Nelson, W. G. (1982). Experimental studies of oil pollution on the rocky intertidal community of a Norwegian fjord. *J. Exp. Mar. Biol. Ecol.* 65:121–38. [8.4.2]

Nelson-Smith, A. (1972). *Oil Pollution and Marine Ecology.* London: Elek Science Press, 260 pp. [8.4.2]

Neushul, M. (1972). Functional interpretation of benthic marine algal morphology. In I. A. Abbott & M. Kurogi (eds.), *Contributions to the Systematics of Benthic Marine Algae of the North Pacific* (pp. 47–73). Tokyo: Japan Society for Phycology. [1.5.4, 1.6.1, 7.1.2, 7.2.2]

Neushul, M. (1981). The ocean as a culture dish: experimental studies of marine algal ecology. *Proc. Intl. Seaweed Symp.* 8:19–35. [1.1.3]

Neville, A. C. (1988). The helicoidal arrangement of microfibrils in some algal cell walls. *Prog. Phycol. Res.* 6:1–21. [1.3.1]

Newell, R. C. (1979). *Biology of Intertidal Animals,* 3rd ed. Faversham, UK: Marine Ecological Surveys, 781 pp. [1.1.3]

Newell, R. C., J. G. Field, & C. L. Griffiths (1982). Energy balance and significance of microorganisms in a kelp bed community. *Mar. Ecol. Prog. Ser.* 8:103–13. [Fig. 4.32]

Newman, S. M., J. Derocher, & R. A. Cattolico (1989). Analysis of chromophytic and rhodophytic ribulose-1,5-bisphosphate carboxylase indicates extensive structural and functional similarities among evolutionarily diverse algae. *Plant Physiol.* 91:939–46. [4.4.2]

Newton, L. (1931). *A Handbook of the British Seaweeds.* London: British Museum. [1.1.1]

Niell, F. X. (1976). C : N ratio in some marine macrophytes and its possible ecological significance. *Bot. Mar.* 14:347–50. [5.8.1]

Nienhuis, P. H. (1980). The epilithic algal vegetation of the SW Netherlands. *Nova Hedwigia* 33:1–94. [2.4.1]

Nienhuis, P. H. (1987). Ecology of salt-marsh algae in the Netherlands. In A. H. L. Huiskes, C. W. P. M. Blom, & J. Rozema (eds.), *Vegetation Between Land and Sea* (pp. 66–83). Dordrecht: Dr. W. Junk. [2.4.1, 6.3.4]

Niklas, K. J., & K. T. Paw U (1982). Pollination and airflow patterns around conifer ovulate cones. *Science* 217:442–4. [1.5.3]

Nilsen, G., & Ø. Nordby (1975). A sporulation-inhibiting substance from vegetative thalli of the green alga, *Ulva mutabilis* Føyn. *Planta* 125:127–39. [1.5.3]

Nishioka, H., K. Okuda, & S. Mizuta (1990). Growth and cell-shape control in the marine coenocytic green alga, *Chaetomorpha moniligera. Bot. Mar.* 33:289–97. [1.3.4]

Nisizawa, K., H. Noda, R. Kikuchi, & T. Watanabe (1987). The main seaweed foods in Japan. *Hydrobiologia* 151/152:5–29. [9.3.2, 9.4, 9.4.2, 9.6]

Nissen, P. (1974). Uptake mechanisms: inorganic and organic. *Annu. Rev. Plant Physiol.* 25:53–79. [5.4]

Nixon, W. S., C. A. Oviatt, & S. S. Hale (1976). Nitrogen regeneration and the metabolism of coastal bottom communities. In J. M. Anderson & A. MacFadyen (eds.), *The Role of Aquatic Organisms in Decomposition Processes* (pp. 269–83). Oxford: Blackwell Scientific. [5.8.3]

Noda, H., & Y. Horiguchi (1972). The significance of zinc as a nutrient for the red alga *Porphyra tenera*. *Proc. Intl. Seaweed Symp.* 7:368–72. [5.5.7]

Noggle, G. R., & G. J. Fritz (1983). *Introductory Plant Physiology*, 2nd ed. Englewood Cliffs, NJ: Prentice-Hall, 627 pp. [1.5.3, 4.7.1, 6.2.3, 6.3.1]

Norris, J. N., & W. H. Fenical (1985). Natural products chemistry: uses in ecology. In M. M. Littler & D. S. Littler (eds.), *Handbook of Phycological Methods: Ecological Field Methods: Macroalgae* (pp. 121–45). Cambridge University Press. [3.2.3]

Norris, R. E. (1971). Development of the foliose thallus of *Weeksia fryeana* (Rhodophyceae). *Phycologia* 10:205–13. [1.2.1]

Norris, R. E., E. M. Wollaston, & M. J. Parsons (1984). New terminology for sympodial growth in the Ceramiales (Rhodophyta). *Phycologia* 23:233–7. [1.7.2]

North, W. J. (1972). Mass-cultured *Macrocystis* as a means of increasing kelp stands in nature. *Proc. Intl. Seaweed Symp.* 7:394–9. [1.6.1]

North, W. J. (1976). Aquacultural techniques for creating and restoring beds of giant kelp, *Macrocystis* spp. *J. Fish. Res. Bd. Can.* 33:1015–23. [1.6.1]

North, W. J. (1979). Adverse factors affecting giant kelp and associated seaweeds. *Experientia* 35:445–7. [8.2]

North, W. J. (1987) Biology of the *Macrocystis* resource in North America. In M. S. Doty, J. F. Caddy, & B. Santelices (eds.), *Case Studies of Seven Commerical Seaweed Resources* (pp. 265–311). FAO Fish. Tech. Pap. 281. [9.6, 9.8]

North, W. J., M. Neushul, & K. A. Clendenning (1965). Successive biological change observed in a marine cove exposed to a large spillage of mineral oil. In *Symposium sur les Pollutions Marines par les Microorganismes et les Produits Petroliers* (Monaco) (pp. 335–54). Paris: Secretariate General de la Commission. [8.4.2]

North, W. J., G. A. Jackson, & S. L. Manley (1986). *Macrocystis* and its environment, knowns and unknowns. *Aquat. Bot.* 26:9–26. [9.6]

Norton, T. A. (1977). Ecological experiments with *Sargassum muticum*. *J. Mar. Biol. Assoc. U.K.* 57:33–43. [6.2.3]

Norton, T. A. (1983). The resistance to dislodgement of *Sargassum muticum* germlings under defined hydrodynamic conditions. *J. Mar. Biol. Assoc. U.K.* 63:181–93. [2.1.1]

Norton, T. A. (1985). The zonation of seaweeds on rocky shores. In P. G. Moore & R. Seed (eds.), *The Ecology of Rocky Coasts* (pp. 7–21). London: Hodder & Stoughton. [2.2.3]

Norton, T. A., (1991). Conflicting constraints on the form of intertidal algae. *Br. Phycol. J.* 26:203–18. [1.2.2, 5.8]

Norton, T. A. (1992). Dispersal by algae. *Br. Phycol. J.* 27:293–301. [2.1.1]

Norton, T. A., & R. Fetter (1981) The settlement of *Sargassum muticum* propagules in stationary and flowing water. *J. Mar. Biol. Assoc. U.K.* 61:929–40. [1.6.1, Fig. 7.4]

Norton, T. A., & A. C. Mathieson (1983). The biology of unattached seaweeds. *Prog. Phycol. Res.* 2:333–86. [1.1.1, 1.7.2]

Norton, T. A., A. C. Mathieson, & M. Neushul (1981). Morphology and environment. In C. S. Lobban & M. J. Wynne

(eds.), *The Biology of Seaweeds* (pp. 421–51). Oxford: Blackwell Scientific. [1.7.2, 6.3.3]

Norton, T. A., A. C. Mathieson, & M. Neushul (1982). A review of some aspects of form and function in seaweeds. *Bot. Mar.* 25:501–10. [1.1.2, 7.1.2, 7.1.3, 7.2.2, 7.2.3]

Norton, T. A., D. C. Watson, S. J. Hawkins, N. L. Manley, & G. A. Williams (1990). Scraping a living: a review of littorinid grazing. *Hydrobiologia* 193:117–38. [2.1.1]

Novak, R. (1984). A study in ultra-ecology: microorganisms on the seagrass *Posidonia oceanica* (L.) Delile. *P.S.Z.N.I. Mar. Ecol.* 5:143–90. [2.4.2]

Nultsch, W. (1974). Movements. In W. D. P. Stewart (ed.), *Algal Physiology and Biochemistry* (pp. 864–93). Oxford: Blackwell Scientific. [1.5.3]

Nultsch, W., J. Pfau, & U. Rüffer (1981). Do correlations exist between chromatophore arrangement and photosynthetic activity in seaweeds? *Mar. Biol.* 62:111–17. [4.7.2]

Nybakken, J. W. (1969). *Pre-earthquake Intertidal Ecology of Three Saints Bay, Kodiak Island, Alaska*. Biology Papers, University of Alaska, no. 9, 117 pp. [7.2.3]

Oates, B. R. (1988). Water relations of the intertidal saccate alga *Colpomenia peregrina* (Phaeophyta, Scytosiphonales). *Bot. Mar.* 31:57–63. [6.4]

Oates, B. R. (1989). Articulated coralline algae as a refuge for the intertidal saccate species, *Colpomenia peregrina* and *Leathesia difformis* in southern California. *Bot. Mar.* 32:475–8. [7.2.3]

Oates, B. R., & S. N. Murray (1983). Photosynthesis, dark respiration and desiccation resistance of the intertidal seaweeds *Hesperophycus harveyanus* and *Pelvetia fastigiata* f. *gracilis*. *J. Phycol.* 19:371–80. [6.4]

O'Brien, M. C., & P. A. Wheeler (1987). Short term uptake of nutrients by *Enteromorpha prolifera* (Chlorophyceae). *J. Phycol.* 23:547–56. [5.4.1, 5.5.2]

O'Brien, P. Y., & P. S. Dixon (1976). The effects of oils and oil components on algae: a review. *Br. Phycol. J.* 11:115–42. [8.4.2]

Odum, H. T. (1972). An energy circuit langauge for ecological and social systems: its physical basis. In B. C. Patten (ed.), *Systems Analysis and Simulation in Ecology* (vol. 2, pp. 140–211). New York: Academic Press. [4.7.6]

Ogata, E. (1971). Growth of conchocelis in artificial medium in relation to carbon dioxide and calcium metabolism. *J. Shimonoseki U. Fish* 19:123–9. [4.4.1]

Ogawa, H. (1984). Effects of treated municipal wastewater on the early development of sargassaceous plants. *Hydrobiologia* 116/117:389–92. [8.6.1]

Ogden, J. C., R. A. Brown, & N. Salesky (1973). Grazing by the echinoid *Diadema antillarum* Philippi: formation of halos around West Indian patch reefs. *Science* 182:715–17. [3.2.2]

Ohki, K., & E. Gantt (1983). Functional phycobilisomes from *Tolypothrix tenuis* (Cyanobacteria) grown heterotrophically in the dark. *J. Phycol.* 19:359–64. [4.3.1]

Okabe, Y., & M. Okada (1990). Nitrate reductase activity and nitrate in native pyrenoids purified from the green alga *Bryopsis maxima*. *Plant Cell Physiol.* 31:429–32. [1.3.2]

Okazaki, A. (1971). *Seaweeds and their Uses in Japan.*. Tokyo: Tokai University Press, 165 pp. [9.2.2]

Okazaki, M. (1977). Some enzymatic properties of Ca^{2+}-dependent adenosine triphosphatase from a calcareous red alga, *Serraticardia maxima* and its distribution in marine algae. *Bot. Mar.* 20:347–54. [5.5.2]

O'Kelley, J. C. (1974). Inorganic nutrients. In W. D. P. Stewart (ed.), *Algal Physiology and Biochemistry* (pp. 610–35). Oxford: Blackwell Scientific. [5.1.1, 5.5.4, 5.5.5]

O'Kelley, C. J. (1988). Division of *Palmoclathrus stipitatus* (Chlorophyta) vegetative cells. *Phycologia* 27:248–53. [1.1.1]

O'Kelley, C. J., & B. J. Baca. (1984). Time course of carpogonial branch and carposporophyte development in *Callithamnion cordatum* (Rhodophyta) Ceramiales. *Phycologia* 23:407–17. [1.5.4]

Olsen, J. L. (1990). Nucleic acids in algal systematics. *J. Phycol.* 26:209–14. [1.4.1]

Olsen, J. L., W. T. Stam, P. V. M. Bot, & C. van den Hoek (1987). scDNA-DNA hybridization studies in Pacific and Caribbean isolates of *Dictyosphaeria cavernosa* (Chlorophyta) indicate a long divergence. *Helgol. Meeresunters.* 41:377–83. [1.4.1]

Olson, A. M., & J. Lubchenco (1990). Competition in seaweeds: linking plant traits to competitive outcomes. *J. Phycol.* 26:1–6. [3.0, 3.1]

Oltmanns, F. (1892). Ueber die Cultur- und Lebensbedingungen der Meeresalgen. *Jahr. Wissensch. Bot.* 23:349–440. [4.3.4]

Oohusa, T. (1980). Diurnal rhythm in the rates of cell division, growth and photosynthesis of *Porphyra yezoensis* (Rhodophyceae) cultured in the laboratory. *Bot. Mar.* 23:1–5. [4.7.2]

Osborne, B. A., & J. A. Raven (1986). Light absorption by plants and its implications for photosynthesis. *Biol. Rev.* 61:1–61. [4.2.2, 4.3.2]

Owens, N. J. P., & W. D. P. Stewart (1983). *Enteromorpha* and the cycling of nitrogen in a small estuary. *Estu. Cstl. Shelf Sci.* 17:287–96. [5.2]

Paasche, E. (1977). Growth of three plankton diatom species in Oslofjord water in the absence of artificial chelators. *J. Exp. Mar. Biol. Ecol.* 29:91–106. [5.2]

Pace, D. R. (1975). Factors governing the boundary between *Macrocystis integrifolia* and the red sea urchin. Ph.D. thesis, Simon Fraser University, Burnaby, BC, 102 pp. [3.2.2]

Padan, E., & Y. Cohen (1982). Anoxygenic photosynthesis. In N. G. Carr & B. A. Whitton (eds.), *The Biology of Cyanobacteria* (pp. 215–35). Oxford: Blackwell Scientific. [4.1]

Padilla, D. K. (1985). Structural resistance of algae to herbivores, a biomechanical approach. *Mar. Biol.* 90:103–9. [3.2.2]

Padilla, D. K. (1989). Algal structural defenses: form and calcification in resistance to tropical limpets. *Ecology* 70:835–42. [3.2.2]

Paerl, H. W. (1985). Enhancement of marine primary production by nitrogen-enriched acid rain. *Nature* 316:747–9. [5.2]

Paerl, H. W., & B. M. Bebout (1988). Direct measurement of O_2-depleted microzones in marine *Oscillatoria:* relation to N_2 fixation. *Science* 241:442–5. [1.3.6]

Paerl, H. W., J. Rudek, & M. A. Mallin (1990). Stimulation of phytoplankton in coastal waters by natural rainfall inputs: nutritional and trophic implications. *Mar. Biol.* 107:247–54. [5.2]

Paine, R. T. (1977). Controlled manipulations in the marine intertidal zone, and their contributions to ecological theory. *Spec. Publ. Acad. Nat. Sci., Phila.* 12:254–70. [3.2.1]

Paine, R. T. (1979). Disaster, catastrophe, and local persistence of the sea palm *Postelsia palmaeformis*. *Science* 205:685–7. [7.2.3]

Paine, R. T. (1986). Benthic community–water column coupling during the 1982–1983 El Niño. Are community changes at high latitudes attributable to cause or coincidence? *Limnol. Oceanogr.* 31:351–60. [1.1.2, 6.2.3, 6.2.6]

Paine, R. T. (1988). Habitat suitability and local population persistence of the sea palm *Postelsia palmaeformis*. *Ecology* 69:1787–94. [3.1.1, 7.2.3]

Paine, R. T. (1990). Benthic macroalgal competition: complications and consequences. *J. Phycol.* 26:12–17. [2.4.1, 2.5.2, 3.0, 3.1.1]

Paine, R. T., & S. A. Levin (1981). Intertidal landscapes: disturbance and the dynamics of pattern. *Ecol. Monogr.* 51:145–78. [3.2.2, 7.2.3]

Paine, R. T., & R. L. Vadas (1969). The effects of grazing sea urchins, *Strongylocentrotus* spp. on benthic algal populations. *Limnol. Oceanogr.* 14:710–19. [3.2.1]

Palenik, B., & F. M. M. Morel (1990). Amino acid utilization by marine phytoplankton: a novel mechanism. *Limnol. Oceanogr.* 35:260–9. [5.5.1]

Palenik, B., D. J. Kieber, & F. M. M. Morel (1991a). Dissolved organic nitrogen use by phytoplankton: the role of cell-surface enzymes. *Biol. Oceanogr.* 6:347–54. [5.5.1]

Palenik, B., N. M. Price, & F. M. M. Morel (1991b). Potential effects of UV-B on the chemical environment of marine organisms: a review. *Environ. Pollut.* 70:117–30. [5.5.6]

Palmisano, A. C., S. B. SooHoo, & C. W. Sullivan (1985). Photosynthesis–irradiance relationships in sea ice microalgae from McMurdo Sound, Antarctica. *J. Phycol.* 21:341–6. [4.3.3]

Palumbi, S. R. (1984). Measuring intertidal wave forces. *J. Exp. Mar. Biol. Ecol.* 81:171–9. [7.2.1]

Parke, M. W. (1948). Studies of the British Laminariaceae. I. Growth in *Laminaria saccharina* (L.) Lamour. *J. Mar. Biol. Assoc. U.K.* 27:651–709. [4.7.4]

Parsons, T. J., C. A. Maggs, & S. E. Douglas (1990). Plastid DNA restriction analysis links the heteromorphic phases of an apomictic red algal life history. *J. Phycol.* 26:495–500. [1.4.1]

Parsons, T. R., & P. J. Harrison (1983). Nutrient cycling in marine ecosystems. In A. Pirson & M. H. Zimmerman (eds.), *Encyclopedia of Plant Physiology. Vol. 12: Physiological Plant Ecology IV* (pp. 77–105). Berlin: Springer-Verlag. [5.2]

Parsons, T. R., M. Takahashi, & B. Hargrave (1977). *Biological Oceanographic Processes*, 2nd ed. New York: Pergamon Press, 332 pp. [5.8.1]

Parsons, T. R., Y. Maita, & C. M. Lalli (1984). *A Manual of Chemical and Biological Methods of Seawater Analysis*. New York: Pergamon Press, 173 pp. [6.3]

Pasciak, W. J., & J. Gavis (1974). Transport limitation of nutrient uptake in phytoplankton. *Limnol. Oceanogr.* 19:881–8. [5.4.2]

Paskins-Hurlburt, A. J., Y. Tanaka, & S. C. Skoryna (1976). Isolation and metal binding properties of fucoidan. *Bot. Mar.* 19:327–8. [8.3.2]

Pastorok, R. A., & G. R. Bilyard (1985). Effects of sewage pollution on coral-reef communities. *Mar. Ecol. Prog. Ser.* 21:175–89. [8.6.1]

Patterson, M. R. (1989). Nearshore biomechanics [review of Denny's book]. *Science* 243:1374. [7.0]

Paul, V. J., & W. Fenical (1986). Chemical defense in tropical green algae, order Caulerpales. *Mar. Ecol. Prog. Ser.* 34:157–69. [3.2.3]

Paul, V. J., & M. E. Hay (1986). Seaweed susceptibility to herbivory: chemical and morphological correlates. *Mar. Ecol. Prog. Ser.* 33:255–64. [3.2.3]

Paul, V. J., & S. C. Pennings (1991). Diet-derived chemical defenses in the sea hare *Stylocheilus longicauda* (Quoy et Gaimard 1824). *J. Exp. Mar. Biol. Ecol.* 151:227–43. [3.2.3]

Paul, V. J., & K. L. Van Alstyne (1992). Activation of chemical defenses in the tropical green algae *Halimeda* spp. *J. Exp. Mar. Biol. Ecol.* 160:191–203. [3.2.3]

Paul, V. J., H. H. Sun, & W. Fenical (1982). Udoteal, a linear diterpenoid feeding deterrant from the tropical green alga *Udotea flabellum. Phytochemistry* 21:468–9. [Fig. 3.13b]

Peakall, D. B. (1975). PCB's and their environmental effects. *CRC Crit. Rev. Environ. Control* 5:469–508. [8.5.3]

Pearson, G. A., & L. V. Evans (1990). Settlement and survival of *Polysiphonia lanosa* (Ceramiales) spores on *Ascophyllum nodosum* and *Fucus vesiculosus* (Fucales). *J. Phycol.* 26:597–603. [3.1.2, 3.3.2]

Peckol, P., & J. Ramus (1988). Abundances and physiological properties of deep-water seaweeds from Carolina outer continental shelf. *J. Exp. Mar. Biol. Ecol.* 115:25–39. [4.3.2, 4.3.4]

Pedersén, M. (1973). Identification of a cytokinin, 6-(3-methyl-2-butenylamino) purine, in sea water and the effect of cytokinins on brown algae. *Physiol. Plant.* 28:101–5. [1.7.3]

Pedersen, P. M. (1981). Phaeophyta: life histories. In C. S. Lobban & M. J. Wynne (eds.), *The Biology of Seaweeds* (pp. 194–217). Oxford: Blackwell Scientific. [1.6.2]

Pelevin, V. N., & V. A. Rutkovskaya (1977). On the optical classification of ocean waters from the spectral attenuation of solar radiation. *Oceanology* 17:28–32. [4.2.2]

Pellegrini, L. (1980). Cytological studies on physodes in the vegetative cells of *Cystoseira stricta* Sauvageau (Phaeophyta, Fucales). *J. Cell Sci.* 41:209–31. [1.3]

Pennings, S. C. (1990a). Multiple factors promoting narrow host range in the sea hare, *Aplysia californica. Oecologia* 82:192–200. [3.3.2]

Pennings, S. C. (1990b). Size-related shifts in herbivory: specialization in the sea hare *Aplysia californica* Cooper. *J. Exp. Mar. Biol. Ecol.* 142:43–61. [3.2.3]

Penot, M., & M. Penot (1979). High speed translocation of ions in seaweeds. *Z. Pflanzenphysiol.* 95:265–73. [5.4.2]

Penot, M., & C. Videau (1975). Absorption du ^{86}Rb et du ^{99}Mo par deux algues marines: le *Laminaria digitata* et le *Fucus serratus. Z. Pflanzenphysiol.* 76:285–93. [5.5.7]

Pentecost, A. (1985). Photosynthetic plants as intermediary agents between environmental HCO_3^- and carbonate deposition. In W. J. Lucas & J. A. Berry (eds.), *Inorganic Carbon Uptake by Aquatic Photosynthetic Organisms* (pp. 459–80). Bethesda, Md.: American Society of Plant Physiologists. [4.4.1, 5.5.2]

Percival, E. (1979). The polysaccharides of green, red and brown seaweeds: their basic structure, biosynthesis and function. *Br. Phycol. J.* 14:103–17. [4.5.1, 4.5.2, 4.5.3]

Percival, E., & R. H. McDowell (1967). *Chemistry and Enzymology of Marine Algal Polysaccharides.* New York; Academic Press, 219 pp. [4.5.2]

Pérès, J. M. (1982a). Zonations. In O. Kinne (ed.), *Marine Ecology* (vol. 5, pt. 1, pp. 9–45). New York: Wiley. [2.2.2]

Pérès, J. M. (1982b). Major benthic assemblages. In O. Kinne (ed.), *Marine Ecology* (vol. 5, pt. 1, pp. 373–522). New York: Wiley. [2.2.2]

Perez, R., P. Durand, R. Kaas, O. Barbaroux, V. Barbier, C. Vino, M. Bourgeay-Causse, M. Leclercq, & J. Y. Moigne (1988). *Undaria pinnatifida* on the French coasts: cultivation method, biochemical composition of the sporophyte and the gametophyte. In T. Stadler (ed.), *Algal Biotechnology* (pp. 315–27). Amsterdam: Elsevier. [9.4, 9.4.1]

Perkins, E. J. (1979). The need for sublethal studies. *Phil. Trans. Roy. Soc. Lond.* B286:425–42. [Fig. 8.6]

Perkins, E. J., & O. J. Abbott (1972). Nutrient enrichment and sand flat fauna. *Mar. Pollut. Bull.* 3:70–2. [8.6.1]

Perrone, G., & G. P. Felicini (1972). Sur les bourgeons adventifs de *Petroglossum nicaeense* (Duby) Schotter (Rhodophycées, Gigartinales) en culture. *Phycologia* 11:87–95. [1.7.4]

Perrone, C., & G. P. Felicini (1976). Les bourgeons adventifs de *Gigartina acicularis* (Wulf.) Lamour. (Rhodophyta, Gigartinales) en culture. *Phycologia* 15:45–50. [1.7.4]

Perry, M. J., M. C. Larsen, & R. S. Alberte (1981). Photoadaptation in marine phytoplankton: response of the photosynthetic unit. *Mar. Biol.* 62:91–101. [5.8.1]

Peterson, B. J. (1980). Aquatic primary productivity and the ^{14}C-CO_2 method: a history of the productivity problem. *Annu. Rev. Ecol. Syst.* 11:359–85. [4.7.1]

Peterson, H. G., F. P. Healey, & R. Wagemann (1984). Metal toxicity to algae: a highly pH dependent phenomenon. *Can. J. Fish. Aquat. Sci.* 41:974–9. [8.3.4]

Peterson, R. D. (1972). Effects of light intensity on the morphology and productivity of *Caulerpa racemosa* (Forsskal) J. Agardh. *Micronesica* 8:63–86. [1.7.2]

Pettersen, R. (1975). Control by ammonium of intercompartmental guanine transport in *Chlorella. Z. Pflanzenphysiol.* 76:213–23. [5.5.1]

Pfister, C. A., & M. E. Hay (1988). Associational plant refuges: convergent patterns in marine and terrestrial communities result from differing mechanisms. *Oecologia* 77:118–29. [3.2.2]

Pham Quang, L., & M. H. Laur (1976). Teneur, composition et répartition cytologique des lipides polaires sulfrés et phosphorés de *Pelvetia canaliculata* (L.) Decn. et Thur., *Fucus vesiculosus* (L.) et *Fucus serratus* (L.). *Phycologia* 15:367–75. [6.4]

Phillips, D. J. H. (1977). The use of biological indicator organisms to monitor trace metal pollution in marine and estuarine environments: a review. *Environ. Pollut.* 13:281–317. [8.3.1]

Phillips, D. J. H. (1990). Use of macroalgae and invertebrates as monitors of metal levels in estuaries and coastal waters. In R. W. Furness & P. S. Rainbow (eds.), *Heavy Metals in the Marine Environment* (pp. 82–99). Boca Raton, FL: CRC Press. [8.3.1]

Phillips, D. J. H. (1991). Selected trace elements and the use of biomonitors in subtropical and tropical marine ecosystems. *Rev. Environ. Contamin. Toxicol.* 120:105–29. [5.1.1, 8.3.1]

Phillips, G. L., D. Eminson, & B. Moss (1978). A mechanism to account for macrophyte decline in progressively eutrophicated freshwaters. *Aquat. Bot.* 4:103–26. [8.6.1]

Phillips, J. A., M. N. Clayton, I. Maier, W. Boland, & D. G. Müller (1990). Sexual reproduction in *Dictyota diemensis* (Dictyotales, Phaeophyta). *Phycologia* 29:367–79. [1.5.3]

Phillips, R. C., & E. G. Meñez (1988). *Seagrasses.* Smithsonian Contributions to the Marine Sciences, no. 34, 104 pp. [2.4.2]

Phlips, E. J., & C. Zeman (1990). Photosynthesis, growth and nitrogen fixation by epiphytic forms of filamentous cyanobacteria from pelagic *Sargassum. Bull. Mar. Sci.* 47:613–21. [5.2]

Phlips, E. J., M. Willis, & A. Verchick (1986). Aspects of nitrogen fixation in *Sargassum* communities off the coast of Florida. *J. Exp. Mar. Biol. Ecol.* 102:99–119. [5.2]

Polne-Fuller M., & A. Gibor (1984). Development studies in *Porphyra.* I. Blade differentiation in *Porphyra perforata* as expressed by morphology, enzymatic digestion, and protoplast regeneration. *J. Phycol.* 20:609–16. [1.7.1, 9.2.4]

Polne-Fuller, M.; & A. Gibor (1987). Tissue culture of seaweeds. In K. T. Bird & P. H. Benson (eds.), *Seaweed Cultivation for Renewable Resources* (pp. 219–40). Amsterdam: Elsevier. [9.8]

Polne-Fuller, M., & A. Gibor (1990). Development studies in *Porphyra* (Rhodophyceae). III. Effect of culture conditions on wall regeneration and differentiation of protoplasts. *J. Phycol.* 26:674–82. [9.2.4]

Polovina, J. J. (1984). Model of a coral reef ecosystem. I. The ECOPATH model and its application to French Frigate Shoals. *Coral Reefs* 3:1–11. [4.7.6]

Polunin, N. V. C. (1988). Efficient uptake of algal production by a single resident herbivorous fish on the reef. *J. Exp. Mar. Biol. Ecol.* 123:61–76. [3.2.2]

Popovic, R., K. Colbow, W. Vidaver, & D. Bruce (1983). Evolution of O_2 in brown algal chloroplasts. *Plant Physiol.* 73:889–92. [1.3]

Potts, D. C. (1977). Suppression of coral populations by filamentous algae within damselfish territories. *J. Exp. Mar. Biol. Ecol.* 28:207–16. [3.2.2]

Prescott, L. M., J. P. Harley, & D. A. Klein (1990). *Microbiology.* Dubuque, IA: Wm. C. Brown. [1.3.5]

Preston, M. R. (1988). Marine pollution. In J. P. Riley (ed.), *Chemical Oceanography* (pp. 53–196). Orlando, FL: Academic Press. [8.4, 8.4.1, 8.4.2, 8.5.2, 8.5.3]

Price, I. R. (1989). Seaweed phenology in a tropical Australian locality (Townsville, North Queensland). *Bot. Mar.* 32:399–406. [1.5.3, 2.2.1]

Price, I. R., R. L. Fricker, & C. R. Wilkinson (1984) *Ceratodictyon spongiosum* (Rhodophyta), the macroalgal partner in an alga-sponge symbiosis, grown in unialgal culture. *J. Phycol.* 20:156–8. [2.4, 3.2.2, 3.3.1]

Price, J. H. (1978). Ecological determination of adult form in *Callithamnion:* its taxonomic implications. In D. E. G. Irvine & J. H. Price (eds.), *Modern Approaches to the Taxonomy of Red and Brown Algae* (pp. 263–300). New York: Academic Press. [7.2.2]

Price, N. M., & P. J. Harrison (1988). Specific selenium-containing macromolecules in the marine diatom *Thalassiosira pseudonana. Plant Physiol.* 86:192–9. [5.5.7]

Pringle, J. D. (1982). Variation in *Enteromorpha.* In *Abstracts of the First International Phycology Congress* (St. John's, Newfoundland) (p. a39). [1.7.2, 3.2.1]

Pringle, J. D. (1984). Efficiency estimates for various quadrat sizes used in benthic sampling. *Can. J. Fish. Aquat. Sci.* 41:1485–9. [2.5.1]

Pringle, J. D., & A. C. Mathieson (1987). *Chondrus crispus* Stackhouse. In M. S. Doty, J. F. Caddy, & B. Santelices

(eds.), *Case Studies of Seven Commercial Seaweed Resources* (pp. 49–122). FAO Fish. Tech. Pap. 281. [9.7]

Pringle, J. D., D. James, & C. K. Tseng (1989). Overview of a workshop in production and utilization of commercial seaweeds – Qingdao, China, 1987. *J. Appl. Phycol.* 1:83–90. [9.5, 9.7]

Probyn, T. A. (1984). Nitrate uptake by *Chordaria flagelliformis* (Phaeophyta). *Bot. Mar.* 17:271–5.

Probyn, T. A, & A. R. O. Chapman (1982). Nitrogen uptake characteristics of *Chordaria flagelliformis* (Phaeophyta) in batch mode and continuous mode experiments. *Mar. Biol.* 71:129–33. [5.4.1, 5.5.1]

Probyn, T. A., & A. R. O. Chapman (1983). Summer growth of *Chordaria flagelliformis* (O. F. Muell.) C. Ag.: physiological strategies in a nutrient stressed environment. *J. Exp. Mar. Biol. Ecol.* 73:243–71. [5.7.3, 5.8.3]

Probyn, T. A., & C. D. McQuaid (1985). In situ measurements of nitrogenous nutrient uptake by kelp *(Ecklonia maxima)* and phytoplankton in a nitrate-rich upwelling environment. *Mar. Biol.* 88:149–54. [5.8.3]

Provasoli, L., & A. F. Carlucci (1974). Vitamins and growth regulators. In W. D. P. Stewart (ed.), *Algal Physiology and Biochemistry* (pp. 741–87). Oxford: Blackwell Scientific. [5.1.1, 5.1.2, 5.2]

Provasoli, L., & I. J. Pintner (1977). Effect of media and inoculum on the morphology of *Ulva. J. Phycol. (Suppl.)* 13:56. [1.7.2]

Provasoli, L., & I. J. Pintner (1980). Bacteria induced polymorphism in an axenic laboratory strain of *Ulva lactuca* (Chlorophyceae). *J. Phycol.* 16:196–201. [1.7.2]

Provasoli, L., I. J. Pintner, & S. Sampathkumar (1977). Morphogenetic substances for *Monostroma oxyspermum* from marine bacteria. *J. Phycol. (Suppl.)* 13:56. [1.7.2]

Pueschel, C. M. (1989). An expanded survey of the ultrastructure of red algal pit plugs. *J. Phycol.* 25:625–36. [1.3.5]

Pueschel, C. M. (1990). Cell structure. In K. M. Cole & R. G. Sheath (eds.), *Biology of the Red Algae* (pp. 7–41). Cambridge University Press. [1.3, 1.3.5, 4.6]

Quader, H. (1981). Interruption of cellulose microfibril crystallization. *Naturwissensch.* 67:428. [1.3.1]

Quadir, A., P. J. Harrison, & R. E. De Wreede (1979). The effects of emergence and submergence on the photosynthesis and respiration of marine macrophytes. *Phycologia* 18:83–8. [6.4]

Quatrano, R. S. (1980). Gamete release, fertilization, and embryogenesis in the Fucales. In E. Gantt (ed.), *Handbook of Phycological Methods. Developmental and Cytological Methods* (pp. 59–68). Cambridge University Press. [1.6.2]

Quatrano, R. S., & D. L. Kropf (1989). Polarization of *Fucus* (Phaeophyceae) zygotes: investigations of the role of calcium microfilaments and cell wall. In A. W. Coleman, L. J. Goff, & J. R. Stein-Taylor (eds.), *Algae as Experimental Systems* (pp. 111–19). New York: Alan R. Liss. [1.6.2, 5.5.3]

Quatrano, R. S., L. R. Griffing, V. Huber-Walchli, & S. Doubet (1985). Cytological and biochemical requirements for the establishment of a polar cell. *J. Cell Sci. (Suppl.)* 2:129–41. [1.6.2]

Queguiner, B., & L. Legendre (1986). Phytoplankton photosynthetic adaptation to high frequency light fluctuations simulating those induced by sea surface waves. *Mar. Biol.* 90:483–91. [4.2.2]

Quillet, M., & G. de Lestang-Brémond (1981). The MeCDPS, a carrying sulphate's nucleotide of the red seaweed *Catenella opuntia* (Grev.) *Proc. Intl. Seaweed Symp.* 10:503–7. [4.5.3]

Quinn, J. F., & A. E. Dunham (1983). On hypothesis testing in ecology and evolution. *Am. Nat.* 122:602–17. [2.5.1]

Ragan, M. A. (1976). Physodes and the phenolic compounds of brown algae. Composition and significance of physodes *in vivo. Bot. Mar.* 19:145–54. [1.3]

Ragan, M. A. (1981). Chemical constituents of seaweeds. In C. S. Lobban & M. J. Wynne (eds). *The Biology of Seaweeds* (pp. 589–626). Oxford: Blackwell Scientific. [4.3, 5.5.5]

Ragan, M. A., & K. -W. Glombitza (1986). Phlorotannins, brown algal polyphenols. *Prog. Phycol. Res.* 4:129–241. [3.2.3]

Ragan, M. A., O. Smidsrød, & B. Larsen (1979). Chelation of divalent metal ions by brown algal polyphenols. *Mar. Chem.* 7:265–71. [5.2]

Rai, L. C., J. P. Gaur, & H. D. Kumar (1981). Phycology and heavy-metal pollution. *Biol. Rev.* 56:99–151. [8.3.3, 8.3.4]

Ramachandran, S., N. Rajendran, R. Nandakumar, & V. K. Venugopalan (1984) Effects of pesticides on photosynthesis and respiration of marine macrophytes. *Aquat. Bot.* 19:395–9. [8.5.2]

Ramirez, M. E., D. G. Müller, & A. F. Peters (1986). Life history and taxonomy of two populations of ligulate *Desmarestia* (Phaeophyceae) from Chile. *Can. J. Bot.* 64:2948–54. [1.5.2]

Ramon, E. (1973). Germination and attachment of zygotes of *Himanthalia elongata* (L.) S. F. Gray. *J. Phycol.* 9:445–9. [1.6.2]

Ramus, J. (1972). Differentiation of the green alga *Codium fragile. Am. J. Bot.* 59:478–82. [1.7.2, 7.2.2.]

Ramus, J. (1978). Seaweed anatomy and photosynthetic performance: the ecological significance of light guides, heterogenous absorption and multiple scatter. *J. Phycol.* 14:352–62. [4.3.2, 4.3.4]

Ramus, J. (1981). The capture and transduction of light energy. In C. S. Lobban & M. J. Wynne (eds.), *The Biology of Seaweeds* (pp. 458–92). Oxford: Blackwell Scientific. [4.3.3, 4.7.1, 4.7.2]

Ramus, J. (1982). Engelmann's theory: the compelling logic. In L. M. Srivastava (ed.), *Synthetic and Degradative Processes in Marine Macrophytes* (pp. 29–46). Berlin: Walter de Gruyter. [4.3.3, 4.3.4]

Ramus, J. (1983). A physiological test of the theory of complementary chromatic adaptation. II. Brown, green and red seaweeds. *J. Phycol.* 19:173–8. [4.3.4]

Ramus, J. S. (1990). A form–function analysis of photon capture for seaweeds. *Proc. Intl. Seaweed Symp.* 13:65–71. [4.3.2, 4.3.3]

Ramus, J., & G. Rosenberg (1980). Diurnal photosynthetic performance of seaweeds measured under natural conditions. *Mar. Biol.* 56:21–8. [4.7.2]

Ramus, J., & M. Venable (1987). Temporal ammonium patchiness and growth rate in *Codium* and *Ulva* (Ulvophyceae). *J. Phycol.* 23:518–23. [5.8]

Randall, R. H., & L. G. Eldredge (1977). Effects of typhoon Pamela on the coral reefs of Guam. *Proc. 3rd Intl. Coral Reef Symp.* 2:525–31. [7.2.4]

Rao, V. S., & U. K. Tipnis (1967). Chemical composition of some marine algae from the Gujurat Coast. In V. Krisnamurthy (ed.), *Proceedings of a Seminar on Sea, Salt and Plants* (pp. 277–88). Bhavnagar, India: Central Salt & Marine Chemical Research Institute. [8.3.1]

Raven, J. A. (1974). Carbon dioxide fixation. In W. D. P. Stewart (ed.), *Algal Physiology and Biochemistry* (pp. 434–55). Oxford: Blackwell Scientific. [1.7.2, 4.4.2]

Raven, J. A. (1976). Transport in algal cells. In U. Lüttge & M. G. Pitman (eds.), *Encyclopedia of Plant Physiology Vol. 2: Transport in Plants. II. Part A: Cells* (pp. 129–88). Berlin: Springer-Verlag. [6.3.2]

Raven, J. A. (1980). Nutrient transport in microalgae. *Adv. Microbiol. Physiol.* 21:47–226. [5.5.5]

Raven, J. A (1981). Nutritional strategies of submerged benthic plants: the acquisition of C, N and P by rhizophytes and haptophytes. *New Phytol.* 88:1–30. [5.5.5]

Raven, J. A. (1984). *Energetics and Transport in Aquatic Plants (MBL Lectures in Biology, vol. 4).* New York: Alan R. Liss. [5.5.2]

Raven, J. A. (1986). Evolution of plant life forms. In T. J. Givnish (ed.), *On the Economy of Plant Form and Function* (pp. 421–92). Cambridge University Press. [1.2.1, 1.2.2, 4.3.3]

Raven, J. A., & J. Beardall (1981). Respiration and photorespiration. *Can. Bull. Fish. Aquat. Sci.* 210:55–82. [4.7.1]

Raven, J. A., & R. J. Geider (1988). Temperature and algal growth. *New Phytol.* 110:441–61. [6.2.1, 6.2.2]

Raven, J. A., & J. Lucas (1985). Energy costs of carbon acquisition. In W. J. Lucas & J. A. Berry (eds.), *Inorganic Carbon Uptake by Aquatic Photosynthetic Organisms* (pp. 305–24), Bethesda, Md.:American Society of Plant Physiologists. [4.4.1]

Raven, J. A., A. M. Johnston, & J. J. MacFarlane (1990). Carbon metabolism. In K. M. Cole & R. G. Sheath (eds.), *Biology of the Red Algae* (pp. 171–202). Cambridge University Press. [4.0, 4.2.2, 4.4.4]

Rawlence, D. J. (1972). An ultrastructural study of the relationship between rhizoids of *Polysiphonia lanosa* (L.) Tandy (Rhodophyceae) and the tissue of *Ascophyllum nodosum* (L.) LeJolis (Phaeophyceae). *Phycologia* 11:279–90. [3.1.2]

Redfield, A. C., B. H. Ketchum, & F. A. Richards (1963). The influence of organisms on the composition of seawater. In M. N. Hill (ed.), *The Sea* (vol. 2, pp. 26–77). New York: Wiley-Interscience. [5.2]

Reed, D. C. (1990). The effects of variable settlement and early competition on patterns of kelp recruitment. *Ecology* 71:776–87. [2.5.2, 3.1.3]

Reed, D. C., M. Neushul, & A. W. Ebeling (1991). Role of settlement density on gametophyte growth and reproduction in the kelps *Pterygophora californica* and *Macrocystis pyrifera* (Phaeophyceae). *J. Phycol.* 27:361–6. [1.1.2, 3.1.3]

Reed, R. H. (1983). Measurement and osmotic significance of β-dimethylsulfoniopropionate in marine macroalgae. *Mar. Biol. Lett.* 4:173–81. [6.3.2]

Reed, R. H. (1985). Osmoacclimation in *Bangia atropurpurea* (Rhodophyta, Bangiales): the osmotic role of floridoside. *Br. Phycol. J.* 20:211–18. [6.3.2, 6.3.4]

Reed, R. H. (1990). Solute accumulation and osmotic adjustment. In K. M. Cole & R. G. Sheath (eds.), *Biology of Red*

Algae (pp. 147–70). Cambridge University Press. [5.3.2, 5.3.4, 5.5.4, 6.3.1]

Reed, R. H., & J. C. Collins (1980). The ionic relations of *Porphyra purpurea* (Roth) C. Ag. (Rhodophyta, Bangiales). *Plant Cell Environ.* 3:399–407. [5.4.1, 5.5.4]

Reed, R. H., & J. C. Collins (1981). The kinetics of Rb^+ and K^+ exchange in *Porphyra purpurea*. *Plant Sci. Lett.* 20:281–9. [5.5.4]

Reed, R. H., & L. Moffat (1983). Copper toxicity and copper tolerance in *Enteromorpha compressa* (L.) Grev. *J. Exp. Mar. Biol. Ecol.* 63:85–103. [8.3.2]

Reed, R. H., & G. Russell (1979). Adaptation to salinity stress in populations of *Enteromorpha intestinalis* (L.) Link. *Estu. Cstl. Mar. Sci.* 8:251–8. [6.3.4]

Reed, R. H., & W. D. P. Stewart (1983). Physiological responses of *Rivularia atra* to salinity: osmotic adjustment in hyposaline media. *New Phytol.* 95:595–603. [6.3.2]

Reed, R. H., J. C. Collins, & G. Russell (1980a). The effects of salinity upon cellular volume of the marine red alga *Porphyra purpurea* (Roth) C. Ag. *J. Exp. Bot.* 31:1521–37. [6.3.2, 6.3.3]

Reed, R. H., J. C. Collins, & G. Russell (1980b). The effects of salinity upon galactosyl-glycerol content and concentration of the marine red alga *Porphyra purpurea* (Roth) C. Ag. *J. Exp. Bot.* 31:1539–54. [6.3.2]

Reed, R. H., J. C. Collins, & G. Russell (1980c). The influence of variations in salinity upon photosynthesis in the marine alga *Porphyra purpurea* (Roth) C. Ag. (Rhodophyta, Bangiales). *Z. Pflanzenphysiol.* 98:183–7. [6.3.2]

Reed, R. H., I. R. Davison, J. A. Chudek, & R. Foster (1985). The osmotic role of mannitol in the Phaeophyta: an appraisal. *Phycologia* 24:35–47. [6.3.1, 6.3.2]

Reed, R. H., L. J. Borowitzka, M. A. Mackay, J. A. Chudek, R. Foster, S. R. C. Warr, D. J. Moore, & W. D. P. Stewart (1986). Organic solute accumulation in osmotically stressed cyanobacteria. *FEMS Microbiol. Rev.* 39:51–6. [6.3.2]

Rees, D. A. (1975). Stereochemistry and binding behaviour of carbohydrate chains. In W. J. Whelan (ed.), *Biochemistry of Carbohydrates* (pp. 1–42). London: Butterworth. [4.5.2]

Reise, K. (1983). Sewage, green algal mats anchored by lugworms, and the effects on *Turbellaria* and small Polychaeta. *Helgol. Meeresunters.* 36:151–62. [8.6.1]

Reiskind, J. B., P. T. Seamon, & G. Bowes (1988). Alternative methods of photosynthetic carbon assimilation in marine macroalgae. *Plant Physiol.* 87:686–92. [4.4.4]

Reiskind, J. B., S. Beer, & G. Bowes (1989). Photosynthesis, photorespiration and ecophysiological interaction in marine macroalgae. *Aquat. Bot.* 34:131–52. [4.3.2, 4.4.4]

Remane, A., & C. Schlieper (1971). Biology of brackish water. *Die Binnengewässer* 25:1–372. [2.4.1]

Ren, G. -Z., J. -C. Wang, & M. -Q. Chen (1984). Cultivation of *Gracilaria* by means of low rafts. *Hydrobiologia* 116/117:72–6. [9.6]

Reutergårdh, L. (1980). Chlorinated hydrocarbons in estuaries. In E. Olausson & I. Cato (eds.), *Chemistry and Biogeochemistry of Estuaries* (pp. 349–65). New York: Wiley. [8.5.2, 8.5.3]

Rhee, G. -Y. (1974). Phosphate uptake under nitrate limitation by *Scenedesmus* sp. and its ecological implication. *J. Phycol.* 10:470–5. [5.5.2]

Rhee, G. -Y. (1978). Effects of N:P atomic ratios and nitrate limitation on algal growth, cell composition, and nitrate uptake. *Limnol. Oceanogr.* 23:10–25. [5.1.3]

Rhee, G. -Y. (1980). Continuous culture in phytoplankton ecology. In M. R. Droop & H. W. Jannasch (eds.), *Advances in Aquatic Microbiology* (vol. 2, pp. 151–203). New York: Academic Press. [5.7.1, 5.7.3]

Rhee, G. -Y. (1982). Effects of environmental factors and their interactions on phytoplankton growth. *Adv. Microbial Ecol.* 6:33–74. [5.7.1]

Rhoads, D. F. (1979). Evolution of plant chemical defense against herbivores. In G. A. Rosenthal & D. H. Janzen (eds.), *Herbivores: Their Interaction with Secondary Plant Metabolites* (pp. 3–54). New York: Academic Press. [3.2.3]

Rhodes, R. G. (1970). Relation of temperature to development of the macrothallus of *Desmotrichum undulatum*. *J. Phycol.* 6:312–14. [6.2.5]

Rice, D. L. (1984). A simple mass transport model for metal uptake by marine macroalgae growing at different rates. *J. Exp. Mar. Biol. Ecol.* 82:175–82. [8.3.1]

Rice, D. L., & B. E. Lapointe (1981). Experimental outdoor studies with *Ulva fasciata* Delile. II. Trace metal chemistry. *J. Exp. Mar. Biol. Ecol.* 54:1–11. [8.3.1]

Rice, H. V., D. A. Leighty, & G. C. McLeod (1973). The effects of some trace metals on marine phytoplankton. *CRC Crit. Rev. Microbiol.* 3:27–49. [8.3.3]

Richmond, R. H. (1990). The effects of the El Niño/southern oscillation on the dispersal of corals and other marine organisms. In P. W. Glynn (ed.), *Global Ecological Consequences of the 1982–1983 El Niño–Southern Oscillation* (pp. 127–40). Amsterdam: Elsevier. [6.2.5, 6.2.6]

Ricohermoso, M. & L. E. Deveau (1979). Review of commercial propagation of *Eucheuma* (Florideophyceae) clones in the Philippines. *Proc. Intl. Seaweed Symp.* 9:525–39. [9.5.2]

Rietema, H. (1974). Development of erect thalli from basal crusts in *Dumontia contorta* (Gmel.) Rupr. (Rhodophyta, Cryptonemiales). *Bot. Mar.* 27:29–36. [1.7.2]

Rietema, H. (1982). Effects of photoperiod and temperature on macrothallus initiation in *Dumontia contorta* (Rhodophyta). *Mar. Ecol. Prog. Ser.* 8:187–96. [1.5.2, 1.7.2]

Rietema, H. (1984). Development of erect thalli from basal crusts in *Dumontia contorta* (Gmel.) Rupr. (Rhodohyta, Cryptonemiales). *Bot. Mar.* 27:29–36. [1.2.1, 1.7.2]

Rietema, H. & A. M. Breeman (1982). The regulation of the life history of *Dumontia contorta* in comparison to that of several other Dumontiaceae (Rhodophyta). *Bot. Mar.* 25:569–76. [1.5.3]

Rietema, H. & A. W. O. Klein (1981). Environmental control of the life cycle of *Dumontia contorta* (Rhodophyta) kept in culture. *Mar. Ecol. Prog. Ser.* 4:23–9. [1.7.2]

Rietema, H. & C. van den Hoek (1981). The life history of *Desmotrichum undulatum* (Phaeophyceae) and its regulation by temperature and light conditions. *Mar. Ecol. Prog. Ser.* 4:321–35. [6.2.5]

Rigano, C., V. D. M. Rigano, V. Vona, & A. Fuggi (1979). Glutamine synthetase activity, ammonia assimilation and control of nitrate reduction in the unicellular red algae *Cyanidium caldarium*. *Arch. Microbiol.* 121:117–20. [5.5.1]

Riley, J. P., & R. Chester (1971). *Introduction to Marine Chemistry*. New York: Academic Press, 465 pp. [5.2]

Riley, J. P., & G. Skirrow (eds.) (1965). *Chemical Oceanography, Vol. 1*, New York: Academic Press, 712 pp. [5.2]

Ritchie, R. J. (1988). The ionic relations of *Ulva lactuca*. *J. Plant Physiol.* 133:183–92. [5.5.4]

Ritchie, R. J., & A. W. D. Larkum (1987).The ionic relations of small-celled marine algae. *Prog. Phycol. Res.* 5:179–222. [5.5.4, 6.3.2]

Roberts, K. R., K. D. Stewart, & K. R. Mattox (1984). Structure and absolute configuration of the flagellar apparatus in the isogametes of *Batophora* (Dasycladales, Chlorophyta). *J. Phycol.* 20:183–91. [Fig. 1.12c]

Roberts, M., & F. M. Ring (1972). Preliminary investigations into conditions affecting the growth of the microscopic phase of *Scytosiphon lomentarius* (Lyngbye) Link. *Mem. Soc. Bot. Fr.* 1972:117–28. [1.7.2]

Rogers, L. J., & J. R. Gallon (1988). *Biochemistry of the Algae and Cyanobacteria.* Oxford: Clarendon Press, 374 pp. [9.6]

Römheld, V. (1987). Different strategies for iron acquisition in higher plants. *Physiol. Plant.* 70:231–4. [5.5.6, 5.6]

Rosell, K. -G., & L. M. Strivastava (1985). Seasonal variations in total nitrogen, carbon and amino acids in *Macrocystis integrifola* and *Nereocystis luetkeana* (Phaeophyta). *J. Phycol.* 21:304–9. [5.8.1]

Rosenberg, D. M., et al. (1981). Recent trends in environmental impact assessment. *Can. J. Fish. Aquat. Sci.* 38:591–624. [8.1]

Rosenberg, G., & H. W. Paerl (1981). Nitrogen fixation by blue-green algae associated with the siphonous green seaweed *Codium decorticatum:* effects on ammonium uptake. *Mar. Biol.* 61:151–8. [5.2]

Rosenberg, G., & J. Ramus (1982). Ecological growth strategies in the seaweed *Gracilaria foliifera* (Rhodophyceae) and *Ulva* sp. (Chlorohyceae): soluble nitrogen and reserve carbohydrates. *Mar. Biol.* 66:251–9. [5.5.1, 5.8, 5.8.1, 5.8.3]

Rosenberg, G., & J. Ramus (1984). Uptake of inorganic nitrogen and seaweed surface area:volume ratios. *Aquat. Bot.* 19:65–72. [1.2.2, 5.8]

Rosenberg, G., T. A. Probyn, & K. H. Mann (1984). Nutrient uptake and growth kinetics in brown seaweeds: response to continuous and single additions of ammonium. *J. Exp. Mar. Biol. Ecol.* 80:125–46. [5.4.1, 5.4.2]

Rosenberg, R. (1985). Eutrophication – the future marine coastal nuisance. *Mar. Pollut. Bull.* 16:227–31. [8.6.1]

Rosowski, J. R., & R. W. Hoshaw (1988). Advanced anisogamy in *Chlamydomonas monadina* (Chlorophyceae) with special reference to vacuolar activity during sexuality. *Phycologia* 27:494–504. [1.5.4]

Round, F. E. (1981). *The Ecology of Algae.* Cambridge University Press, 653 pp. [1.5.3, 2.2.2]

Rowan, K. S. (1989). *Photosynthetic Pigments of Algae.* Cambridge University Press. [4.3.1]

Rubin, P. M., E. Zetooney, & R. E. McGowan (1977). Uptake and utilization of sugar phosphates by *Anabaena flos-aquae. Plant Physiol.* 60:407–11. [5.5.2]

Rudd, R. L. (1964). *Pesticides and the Living Landscape.* Madison: University of Wisconsin Press, 320 pp. [8.5.2]

Rueness, J. (1973). Pollution effects on littoral algal communities in the inner Oslofjord, with special reference to *Ascophyllum nodosum. Helgol. Meeresunters.* 24:446–54. [8.6.1]

Rugg, D. A., & T. A. Norton (1987). *Pelvetia canaliculata,* a high-shore seaweed that shuns the sea. In R. M. M. Crawford (ed.), *Plant Life in Aquatic and Amphibious Habitats* (pp. 347–58). Oxford: Blackwell Scientific. [2.1.1, 3.3.1, 6.4]

Russell, G. (1972). Phytosociological studies on a two-zone shore. I. Basic pattern. *J. Ecol.* 60:539–45. [2.5.1]

Russell, G. (1978). Environment and form in the discrimination of taxa in brown algae. In D. E. G. Irvine & J. H. Price (eds.), *Modern Approaches to the Taxonomy of Red and Brown Algae* (pp. 339–69). New York: Academic Press. [1.7.2, 7.2.2]

Russell, G. (1980). Applications of simple numerical methods to the analysis of intertidal vegetation. In J. H. Price, D. E. G. Irvine, & W. F. Farnham (eds.), *The Shore Environment, Vol. 1: Methods* (pp. 171–92). New York: Academic Press. [2.5.1]

Russell, G. (1983). Formation of an ectocarpoid epiflora on blades of *Laminaria digitata. Mar. Ecol. Prog. Ser.* 11:181–7. [3.1.3]

Russell, G. (1986). Variation and natural selection in marine macroalgae. *Oceanogr. Mar. Biol. Annu. Rev.* 24:309–77. [1.2.2, 1.5.1]

Russell, G. (1987). Salinity and seaweed vegetation. In R. M. M. Crawford (ed.), *Plant Life in Aquatic and Amphibious Habitats* (pp. 35–52). Oxford, Blackwell Scientific. [6.1.1, 6.3.2, 6.3.4]

Russell, G. (1988). The seaweed flora of a young semi-enclosed sea: the Baltic. Salinity as a possible agent of flora divergence. *Helgol. Meeresunters.* 42:243–50. [1.1.3, 6.3.4]

Russell, G., & J. J. Bolton (1975). Euryhaline ecotypes of *Ectocarpus siliculosus* (Dillw.) Lyngb. *Estu. Cstl. Mar. Sci.* 3:91–4. [6.2.3]

Russell, G., & A. H. Fielding (1974). The competitive properties of marine algae in culture. *J. Ecol.* 62:689–98. [3.1.3]

Russell, G., & A. H. Fielding (1981). Individuals, populations and communities. In C. S. Lobban & M. J. Wynne (eds.), *The Biology of Seaweeds* (pp. 393–420). Oxford: Blackwell Scientific. [2.2.2, 2.5.1, 2.5.2]

Russell, G., & C. J. Veltkamp (1982). Epiphytes and antifouling characteristics of *Himanthalia* (brown algae). *Br. Phycol. J.* 17:239. [3.1.1]

Russell, G., & C. J. Veltkamp (1984). Epiphyte survival on skin-shedding macrophytes. *Mar. Ecol. Prog. Ser.* 18:149–53. [3.2.1]

Russell-Hunter, W. D. (1970). *Aquatic Productivity. An Introduction to Some Basic Aspects of Biological Oceanography and Limnology.* New York: Macmillan. [4.2.2]

Ryther, J. H., & W. M. Dunstan (1971). Nitrogen, phosphorus and eutrophication in the coastal marine environment. *Science* 171:1008–13. [5.1.3]

Ryther, J. H., W. M. Dunstan, K. R. Tenore, & J. E. Huguenin (1972). Controlled eutrophication – increasing food production from the sea by recycling human wastes. *BioScience* 22:144–52. [8.6.1]

Saenko, G. N., M. D. Koryakova, V. F. Makienko, & J. G. Dobrosmyskova (1976). Concentration of polyvalent metals by seaweeds in Vostok Bay, Sea of Japan. *Mar. Biol.* 34:169–76. [8.3.1]

Saffo, M. B. (1987). New light on seaweeds. *BioScience* 37:654–64. [4.3.3, 4.3.4]

Saga, N. (1986). Regulation of the life cycle in epiphytic brown alga, *Dictyosiphon foeniculaceus. Sci Pap. Inst. Algol. Res., Hokkaido U.* 8:31–61. [1.5.3]

Saga, N., T. Uchida, & Y. Sakai (1978). Clone *Laminaria* from single isolated cell. *Bull. Jpn. Soc. Sci. Fish.* 44:87. [1.7.1]

Saito, Y. (1975). *Undaria*. In T. Tokida & H. Hirose (eds.), *Advances of Phycology in Japan* (pp. 304–20). The Hague: Dr. W. Junk. [9.4, 9.4.1, 9.4.3]

Sakanishi, Y., Y. Yokohama, & A. Arugal (1989). Seasonal changes of photosynthetic activity of a brown alga *Ecklonia cava* Kjellman. *Bot. Mag. Tokyo* 102:37–51. [4.7.2]

Salisbury, F. B., & C. W. Ross (1985). *Plant Physiology*, 3rd ed. Belmont, CA: Wadsworth. [1.7.2, 4.0, 4.3.1]

Salisbury, J. L. (1989a). Algal centrin: calcium-sensitive contractile organelles. In A. W. Coleman, L. J. Goff, & J. R. Stein-Taylor (eds.), *Algae as Experimental Systems* (pp. 19–37). New York: Alan R. Liss. [1.3.3]

Salisbury, J. L. (1989b). Centrin and the algal flagellar apparatus. *J. Phycol.* 25:201–6. [1.3.3]

Salvucci, M. E. (1989). Regulation of Rubisco activity in vivo. *Physiol. Plant.* 77:164–71. [4.4.2]

Sammarco, P. W. (1983). Effects of fish grazing and damselfish territoriality on coral reef algae. I. Algal community structure. *Mar. Ecol. Prog. Ser.* 13:1–14. [3.2.1, 3.2.2]

Sanders, H. L. (1968). Marine benthic diversity: a comparative study. *Am. Nat.* 102:243–82. [2.4.1]

Sanders, H. L., J. F. Grassle, G. R. Hampson, L. S. Morse, S. Garner-Price, & C. C. Jones (1980). Anatomy of an oil spill: long-term effects from the grounding of the barge *Florida* off West Falmouth, Massachussetts. *J. Mar. Res.* 38:265–380. [8.4.2]

Sand-Jensen, K. (1977). Effects of epiphytes on eelgrass photosynthesis. *Aquat. Bot.* 3:55–63. [3.1.2]

Sand-Jensen, K. (1987). Environmental control of bicarbonate use among freshwater and marine macrophytes. In R. M. M. Crawford (ed.), *Plant Life in Aquatic and Amphibious Habitats* (pp. 99–112). Oxford: Blackwell Scientific. [4.3.4]

Sand-Jensen, K. & J. Borum (1984). Epiphyte shading and its effect on photosynthesis and diel metabolism of *Lobelia dortmanna* L. during the spring bloom in a Danish lake. *Aquat. Bot.* 20:109–19. [2.4.2]

Santelices, B., & E. Martinez. (1988). Effects of filter feeders and grazers on algal settlement and growth in mussel beds. *J. Exp. Mar. Biol. Ecol.* 118:281–306. [3.1.1]

Santelices, B. (1990a). Production ecology of *Gelidium*. Presented at an international workshop on *Gelidium* (Santander, Spain, Sept. 1990). [9.6]

Santelices, B. (1990b). Patterns of reproduction, dispersal and recruitment in seaweeds. *Oceanogr. Mar. Biol. Annu. Rev.* 28:177–276. [1.1.2]

Santelices, B., & E. Martinez (1988). Effects of filter-feeders and grazers on algal settlement and growth in mussel beds. *J. Exp. Mar. Biol. Ecol.* 118:281–306. [3.1.1]

Santelices, B., & F. P. Ojeda (1984). Recruitment, growth and survival of *Lessonia nigrescens* (Phaeophyta) at various tidal levels in exposed habitats of central Chile. *Mar. Ecol. Prog. Ser.* 19:73–82. [3.1.1, 7.2.2]

Santelices, B., & I. Paya (1989). Digestion survival of algae: some ecological comparisons between free spores and propagules in fecal pellets. *J. Phycol.* 25:693–9. [1.6.1]

Santelices, B., S. Montalva, & P. Oliger (1981). Competitive algal community organization in exposed intertidal habitats from central Chile. *Mar. Ecol. Prog. Ser.* 6:267–76. [2.5.2, 3.1.1, 7.2.2]

Santelices, B., J. Correa, & M. Avila (1983). Benthic algal spores surviving digestion by sea urchins. *J. Exp. Mar. Biol. Ecol.* 70:263–9. [1.6.1, 3.2.2]

Satoh, K. & D. C. Fork (1983). A new mechanism for adaptation to changes in light intensity and quality in the red alga *Porphyra perforata*. III. Fluorescence transients in the presence of 3-(3,4-dichlorophenyl)-1,1-dimethylurea. *Plant Physiol.* 71:673–6. [6.4]

Sawada, T. (1972). Periodic fruiting of *Ulva pertusa* at three localities in Japan. *Proc. Intl. Seaweed Symp.* 7:229–30. [1.5.4]

Say, P. J., I. G. Burrow, & B. A. Whitton (1990). *Enteromorpha* as a monitor of heavy metals in estuaries. *Hydrobiologia* 195:119–26. [8.3.1]

Scanlan, C. M., & M. Wilkinson (1987). The use of seaweeds in biocide toxicity testing. Part 1. The sensitivity of different stages in the life-history of *Fucus*, and of other algae, to certain biocides. *Mar. Environ. Res.* 21:11–29. [8.3.3]

Schatz, S. (1980). Degradation of *Laminaria saccharina* by higher fungi: a preliminary report. *Bot. Mar.* 23:617–22. [4.7.3]

Scherer, S., H. Almon, & P. Böger (1988). Interaction of photosynthesis, respiration and nitrogen fixation in cyanobacteria. *Photosynth. Res.* 15:95–114. [4.0, 4.1]

Schiel, D. R., & M. S. Foster (1986). The structure of subtidal algal stands in temperate waters. *Oceanogr. Mar. Biol. Annu. Rev.* 24:265–307. [1.1.2, 1.1.3, 2.1.3, 2.3.2, 2.5.1, 3.2.1]

Schiff, J. A. (1980). Pathways of assimilatory sulfate reduction in plants and microorganisms. In K. Elliot & J. Whelan (eds.), *Sulfur in Biology* (pp. 49–79). Ciba Foundations Symposium 72 (New Series). Amsterdam: Excerpta Medica. [5.5.5]

Schiff, J. A. (1983). Reduction and other metabolic reactions of sulfate. In A. Läuchli & R. L. Bieleski (eds.), *Encyclopaedia of Plant Physiology* (vol. 15, pp. 382–99). Berlin: Springer-Verlag. [5.5.5]

Schiff, J. A., & R. C. Hodson (1970). Pathways of sulfate reduction in algae. *Ann. N.Y. Acad. Sci.* 175:555–76. [5.5.5]

Schleicher, M., T. J. Lukas, & D. M. Watterson (1984). Isolation and characterization of calmodulin from the motile green alga *Chlamydomonas reinhardtii*. *Arch. Biochem. Biophys.* 229:33–42. [6.3.2]

Schlesinger, M. J., M. Ashburner, & A. Tissières (eds.). *Heat Shock: From Bacteria to Man*. Cold Spring Harbor, NY: Cold Spring Harbor Laboratory Press, 440 pp. [6.2.4]

Schmid, R. (1984). Blue light effects on morphogenesis and metabolism in *Acetabularia*. In H. Senger (ed.), *Blue Light Effects in Biological Systems*. (pp. 419–32). Berlin: Springer-Verlag. [1.7.2]

Schmid, R., M. Tünnermann, & E. -M. Idziak (1990). Role of red light in hair-formation induced by blue light in *Acetabularia mediterranea*. *Planta* 181:144–7. [1.7.2]

Schmidt, R. L. (1978). Copper in the marine environment. Part 1. *CRC Crit. Rev. Environ. Control* 8:101–52. [5.2]

Schmitz, K. (1981). Translocation. In C. S. Lobban & M. J. Wynne (eds.), *The Biology of Seaweeds* (pp. 534–58). Oxford: Blackwell Scientific. [4.6]

Schmitz, K. & W. Riffarth (1980). Carrier-mediated uptake of L-leucine by the brown alga *Giffordia mitchellliae*. *Z. Pflanzenphysiol.* 67:311–24. [4.4.1, 5.5.1]

Schmitz, K., & L. M. Srivastava (1974). Fine structure and development of sieve tubes in *Laminaria groenlandica* Rosenv. *Cytobiologie* 10:66–87. [Fig. 4.26]

Schmitz, K., & L. M. Srivastava (1979). Long distance transport in *Macrocystis intergrifolia*. II. Tracer experiments with ^{14}C and ^{32}P. *Plant Physiol.* 63:1003–9. [5.6]

Schneider, C. W. (1976). Spatial and temporal distributions of benthic marine algae on the continental shelf of the Carolinas. *Bull. Mar. Sci.* 26:133–51. [4.3.4]

Schneider, C. W. (1981). The effect of elevated temperature and reactor shutdown on the benthic marine flora of the Millstone Thermal Quarry, Connecticut. *J. Thermal Biol.* 6:1–6. [6.2.2]

Schonbeck, M., & T. A. Norton (1978). Factors controlling the upper limits of fucoid algae on the shore. *J. Exp. Mar. Biol. Ecol.* 31:303–13. [2.1.1, 2.2.3, 6.4]

Schonbeck, M. W., & T. A. Norton (1979a). An investigation of drought avoidance in intertidal fucoid algae. *Bot. Mar.* 22:133–44. [6.4]

Schonbeck, M. W., & T. A. Norton (1979b). Drought-hardening in the upper shore seaweeds *Fucus spiralis* and *Pelvetia canaliculata*. *J. Ecol.* 67:687–96. [6.4]

Schonbeck, M. W., & T. A. Norton (1979c). The effects of brief periodic submergence on intertidal algae. *Estu. Cstl. Mar. Sci.* 8:205–11. [2.1.1]

Schonbeck, M. W., & T. A. Norton (1980a). Factors controlling the lower limits of fucoid algae on the shore. *J. Exp. Mar. Biol. Ecol.* 43:131–50. [2.1.1, 6.4]

Schonbeck, M. W., & T. A. Norton (1980b). The effects on intertidal fucoids of exposure to air under various conditions. *Bot. Mar.* 23:141–7. [6.2.4, 6.3.3, 6.4]

Schramm, W. (1972). The effects of oil pollution on gas exchange in *Porphyra umbilicalis* when exposed to air. *Proc. Intl. Seaweed Symp.* 7:309–15. [8.4.2]

Sciscioli, M. (1966). Asociazione tra la demospongia *Stelletta grubei* (O. Schmidt) e la Rodofica *Phyllophora palmettiodes* (Ag.). *Atti. Soc. Peloritana Sci. Fis. Mat. Nat.* 12:555–60. [3.3.1]

Scott, F. J., R. Wetherbee, & G. T. Kraft. (1984). The morphology and development of some prominently stalked southern Australian Halymeniaceae (Cryptonemiales, Rhodophyta). II. The sponge-associated genera *Thamnoclonium* Kuetzing and *Codiophyllum* Gray. *J. Phycol.* 20:286–95. [3.3.1]

Scott, J., & S. Broadwater (1990). Cell division. In K. M. Cole & R. G. Sheath (eds.), *Biology of the Red Algae* (pp. 123–45). Cambridge University Press. [1.3.5]

Seapy, R. R., & M. M. Littler (1982). Population and species diversity fluctuations in a rocky intertidal community relative to severe aerial exposure and sediment burial. *Mar. Biol.* 71:87–96. [2.5.2]

Sears, J. R., & R. T. Wilce (1975). Sublittoral, benthic marine algae of southern Cape Cod and adjacent islands: seasonal periodicity, associations, diversity, and floristic composition. *Ecol. Monogr.* 45:337–65. [7.2.3]

Serra, J. L., M. J. Llama, & E. Codenas (1978). Nitrate utilization by the diatom *Skeletonema costatum*. 2. Regulation of nitrate uptake. *Plant Physiol.* 62:991–4. [5.4]

Sesták, Z., P. G. Jarvis, & J. Catský (1971). Criteria for the selection of suitable methods. In Z. Sesták, J. Catský, & P. G. Jarvis (eds.), *Plant Photosynthetic Production: Manual of Methods* (pp. 1–48). The Hague: Dr. W. Junk. [4.7.1]

Setchell, W. A. (1915). The law of temperature connected with the distribution of the marine algae. *Ann. Mo. Bot. Gard.* 2:287–305. [6.2.5]

Setchell, W. A., & N. L. Gardner (1919). The marine algae of the Pacific Coast of North America. Part I. Myxophyceae. *Univ. Calif. Publ. Bot.* 8:1–138. [1.1.1]

Seymour, R. J., M. J. Tegner, P. K. Dayton, & P. E. Parnell (1989). Storm wave induced mortality of giant kelp, *Macrocystis pyrifera*, in southern California. *Estu. Cstl. Shelf Sci.* 28:277–92. [7.2.1, 7.2.3]

Sfriso, A., A. Marcomini, & B. Pavoni (1987). Relationships between macroalgal biomass and nutrient concentrations in a hypertrophic area of the Venice Lagoon. *Mar. Environ. Res.* 22:297–312. [5.2, 8.6.1]

Sharp, G. (1987). *Ascophyllum nodosum* and its harvesting in eastern Canada. In M. S. Doty, J. F. Caddy, & B. Santelices (eds.), *Case Studies of Seven Commercial Seaweed Resources* (pp. 3–48). FAO Fish. Tech. Pap. 281. [9.6]

Sharp, J. H., & C. H. Culberson (1982). The physical definition of salinity: a chemical evaluation. *Limnol. Oceanogr.* 27:385–7. [6.3]

Sheath, R. G., & K. M. Cole (1990). *Batrachospermum heterocorticum* sp. nov. and *Polysiphonia subtilissima* (Rhodophyta) from Florida spring-fed streams. *J. Phycol.* 26:563–8. [6.3.3, 6.3.4]

Shephard, K. L. (1987). Evaporation of water from the mucilage of a gelatinous algal community. *Br. Phycol. J.* 22:181–5. [4.5.2]

Shepherd, S. A., & H. B. S. Womersley (1970). The sublittoral ecology of West Island, South Australia. 1: Environmental features and algal ecology. *Trans. R. Soc. S. Austr.* 94:105–37. [2.3.2]

Shiels, W. E., J. J. Goering, & D. W. Hood (1973). Crude oil phytotoxicity studies. In D. W. Hood, W. E. Shiels, & E. J. Kelley (eds.), *Environmental Studies of Port Valdez* (pp. 413–46). Institute of Marine Science, University of Alaska, Fairbanks, Occasional Publication no. 3. [8.4.2]

Shivji, M. S. (1985). Interactive effects of light and nitrogen on growth and chemical composition of juvenile *Macrocystis pyrifera* (L.) C. Ag. (Phaeophyta) sporphytes. *J. Exp. Mar. Biol. Ecol.* 89:81–96. [5.7.3]

Shivji, M. S. (1991). Organization of the chloroplast genome in the red alga *Porphyra yezoensis*. *Curr. Genet.* 19:49–54. [1.4.1]

Shokita, S., K. Kahazu, A. Tomori, & T. Toma (1991). *Aquaculture in Tropical Areas*. Tokyo: Midori Shobo, 360 pp. [9.1, 9.2]

Sieburth, J. M. (1969). Studies on algal substances in the sea. III. The production of extracellular organic matter by littoral marine algae. *J. Exp. Mar. Biol. Ecol.* 3:290–309. [4.7.3]

Sieburth, J. M., & J. T. Conover (1965). *Sargassum* tannin, an antibiotic which retards fouling. *Nature* 208:52–3. [3.1.2]

Sieburth, J. M., & J. T. Tootle (1981). Seasonality of microbial fouling on *Ascophyllum nodosum* (L.) Lejol., *Fucus vesiculosus* L., *Polysiphonia lanosa* (L.) Tandy and *Chondrus crispus* Stackh. *J. Phycol.* 17:57–64. [2.4.2]

Silberstein, K., A. W. Chiffings, & A. J. McComb (1986). The loss of seagrass in Cockburn Sound, Western Australia. III. The effect of epiphytes on productivity of *Posidonia australis* Hook. f. *Aquat. Bot.* 24:355–71. [3.1.2]

Silva, P. C. (1979). The benthic algal flora of central San Francisco Bay. In *San Francisco Bay: The Urbanized Estuary* (pp. 287–345). San Francisco: California Academy of Sciences. [6.5]

References   350

Silverberg, B. A., P. M. Stokes, & L. B. Ferstenberg (1976).
Intranuclear complexes in copper tolerant green algae. *J. Cell Biol.* 69:210–14. [8.3.2]

Simberloff, D. (1982). The status of competition theory in ecology. *Ann. Zool. Fenn.* 19:241–53. [3.2.1]

Simkiss, K. & K. M. Wilbur (1989). *Biomineralization: Cell Biology and Mineral Deposition.* Orlando, FL: Academic Press, 340 pp. [5.5.3]

Simpson, W. R. (1981). A critical review of cadmium in the marine environment. *Prog. Oceanogr.* 10:1–70. [8.3.3]

Sinclair, J., S. Garland, T. Arnason, P. Hope, & M. Granville (1977). Polychlorinated biphenyls and their effects on photosynthesis and respiration. *Can. J. Bot.* 55:2679–84. [8.5.3]

Singh, P., P. A. Kumar, Y. P. Abroi, & M. S. Naik (1985). Photorespiratory nitrogen cycle – a critical evaluation. *Physiol. Plant.* 66:169–76. [5.5.1]

Skipnes, O., T. Roald, & A. Haug (1975). Uptake of zinc and strontium by brown algae. *Physiol. Plant.* 34:314–20. [5.5.7, 8.3.3]

Slocum, C. J. (1980). Differential susceptibility to grazers in two phases of an intertidal alga: advantages of heteromorphic generations. *J. Exp. Mar. Biol. Ecol.* 46:99–110. [3.2.2]

Sluiman, H. J. (1989). The green algal class Ulvophyceae: an ultrastructural survey and classification. *Crypt. Bot.* 1:83–94. [1.1.1, 1.3.3]

Smetacek, V., B. von Bodungen, K. von Brödsel, & B. Zeitzschel (1976). The plankton tower. II. Release of nutrients from sediments due to changes in the density of bottom water. *Mar. Biol.* 34:373-8. [5.8.3]

Smidsrød, O., & H. Grasdalen (1984). Polyelectrolytes from seaweeds. *Hydrobiologia* 116/117:19–28. [4.5.2]

Smith, A. H., K. Nichols, & J. McLachlan (1984). Cultivation of seamoss (*Gracilaria*) in St. Lucia, West Indies. *Hydrobiologia* 116/117:249–51. [9.6]

Smith, A. J. (1982). Modes of cyanobacterial carbon metabolism. In N. G. Carr & B. A. Whitton (eds.), *The Biology of Cyanobacteria* (pp. 47–85). Oxford: Blackwell Scientific. [4.3.4, 4.4.2, 4.5.1]

Smith, F. A., & N. A. Walker (1980). Photosynthesis by aquatic plants: effects of unstirred layers in relation to assimilation of CO_2 and HCO_3^- and to carbon isotopic discrimination. *New Phytol.* 86:245–59. [7.1.2]

Smith, G. M. (1947). On the reproduction of some Pacific coast species of *Ulva*. *Am. J. Bot.* 34:80–7. [1.5.4]

Smith, R. C., & K. S. Baker (1979). Penetration of UV-B and biologically effective dose-rates in natural waters. *Photochem. Photobiol.* 29:311–23. [4.2.2]

Smith, R. C., & J. E. Tyler (1974). In N. G. Jerlov & E. Steemann-Nielsen (eds.), *Optical Aspects of Oceanography.* Orlando, FL: Academic Press. [4.2.1]

Smith, R. C., & J. E. Tyler (1976). Transmission of solar radiation into natural waters. *Photochem. Photobiol. Rev.* 1:117–55. [4.2.2]

Smith, R. G., & R. G. S. Bidwell (1989). Mechanism of photosynthetic carbon dioxide uptake by the red macroalga, *Chondrus crispus*. *Plant Physiol.* 89:93–9. [4.4.1]

Smith, R. G., W. N. Wheeler, & L. M. Srivastava (1983). Seasonal photosynthetic performance of *Macrocystis integrifolia* (Phaeophyceae). *J. Phycol.* 19:352–9. [5.8.1]

Smith, S. V., W. J. Kimmerer, E. A. Laws, R. E. Brock, & T. W. Walsh (1981). Kaneohe Bay sewage diversion exper-

iment: perspectives on ecosystem responses to nutritional perturbation. *Pac. Sci.* 35:279–397. [2.1.2]

Soeder, C., & E. Stengel (1974). Physico-chemical factors affecting metabolism and growth rate. In W. D. P. Stewart (ed.), *Algal Physiology and Biochemistry* (pp. 714–40). Oxford: Blackwell Scientific. [6.2.3]

Solbrig, O. T. (1980). Demography and natural selection. In O. T. Solbrig (ed.), *Demography and Evolution in Plant Populations* (pp. 1–20). Oxford: Blackwell Scientific. [2.5.2]

Solomonson, L. P., & M. J. Barber (1990). Assimilatory nitrate reductase: functional properties and regulation. *Annu. Rev. Plant Physiol. Plant Molec. Biol.* 4:225–53. [5.5.1]

Solomonson, L. P., & A. M. Spehar (1977). Model for the regulation of nitrate assimilation. *Nature* 265:373–5. [5.5.1]

Somero, G. N. (1981). pH–temperature interactions on proteins: principles of optimal pH and buffer system design. *Mar. Biol. Lett.* 2:163–78.

Sondergaard, M. (1988). Comparison of $^{14}CO_2$ and $^{12}CO_2$ uptake in marine macroalgae. *Bot. Mar.* 31:417–22. [4.7.1]

Sorentino, C. (1979). The effects of heavy metals on phytoplankton – a review. *Phykos* 18:149–61. [8.3.3]

Soulsby, P. G., D. Lowthion, M. Houston, & H. A. C. Montgomery (1985). The role of sewage effluent in the accumulation of macroalgal mats on intertidal mudflats in two basins in southern England. *Neth. J. Sea Res.* 19:257–63. [8.6.1]

Sousa, W. P. (1979). Experimental investigation of disturbance and ecological succession in a rocky intertidal algal community. *Ecol. Monogr.* 49:227–54. [2.5.1, 3.1.1, 7.2.3]

Sousa, W. P. (1984). Intertidal mosaics: patch size, propagule availability, and spatially variable patterns of succession. *Ecology* 65: 1918–35. [2.5.1]

Sousa, W. P., S. C. Schroeter, & S. D. Gaines (1981). Latitudinal variation in intertidal algal community structure: the influence of grazing and vegetative propagation. *Oecologia* 48:297–307. [2.1.1, 3.1.1, 3.2.2, 7.2.3]

Spaargaren, D. H. (1984). On ice formation in sea water and marine animals at subzero temperatures. *Mar. Biol. Lett.* 5:203–16. [6.2.4]

Spear-Bernstein, L., & K. R. Miller (1989). Unique location of phycobiliprotein light-harvesting pigment in the Cryptophyceae. *J. Phycol.* 25:412–19. [4.3.1]

Speksnijder, J. E., M. H. Weisenseel, T. -H. Chen, & L. F. Jaffe (1989). Calcium buffer injections arrest fucoid egg development by suppressing calcium gradients. *Biol. Bull. (Suppl.)* 176:9–13. [1.6.2]

Spiller, H., E. Dietsch, & E. Kessler (1976). Intracellular appearance of nitrite and nitrate in nitrogen-starved cells of *Ankistrodesmus braunii*. *Planta* 129:175–81. [5.5.1]

Stafford, S., P. Berwick, D. E. Hughes, & D. A. Stafford (1982). Oil degradation in hydrocarbon- and oil-stressed environments. In R. G. Burns & J. H. Slater (eds.), *Experimental Microbial Ecology.* (pp. 591–612). Oxford: Blackwell Scientific. [8.4.1]

Stam, W. T., P. V. M. Bot, S. A. Boele-Bos, J. M. van Rooij, & C. van den Hoek (1988). Single-copy DNA-DNA hybridization among five species of *Laminaria* (Phaeophyceae): phylogenetic and biogeographic implications. *Helgol. Meeresunters.* 42:251–67. [1.4.1]

Stauber, J. L., & T. M. Florence (1985). Interactions of copper and manganese: a mechanism by which manganese alleviates copper toxicity to the marine diatom, *Nitzschia closterium* (Ehrenberg) W. Smith. *Aquat. Toxic.* 7:241–54. [8.3.2]

Stauber, J. L., & T. M. Florence (1987). Mechanisms of toxicity of ionic copper and copper complexes to algae. *Mar. Biol.* 94:511–19. [8.3.2, 8.3.3]

Stebbing, A. R. D. (1979). An experimental approach to the determinants of biological water quality. *Phil. Trans. R. Soc. Lond.* B286:465–82. [8.1]

Stebbing, A. R. D., B. Akesson, A. Calabrese, J. H. Gentile, A. Jensen, & R. Lloyd (1980). The role of bioassays in marine pollution monitoring. *Rapp. P.-v. Réun. Cons. Intl. Explor. Mer* 179:322–32. [Table 8.1]

Steele, R. L., & M. D. Hanisak (1979). Sensitivity of some brown algal reproductive stages to oil pollution. *Proc. Intl. Seaweed Symp.* 9:181–91. [8.4.2]

Steele, R. L., G. B. Thursby, & J. P. van der Meer (1986). Genetics of *Champia parvula* (Rhodymeniales, Rhodophyta): Mendelian inheritance of spontaneous mutants. *J. Phycol.* 22:538A-42. [1.4.1]

Steemann-Nielsen, E. (1974). Light and primary production. In N. G. Jerlov (ed.), *Optical Aspects of Oceanography* (pp. 331–88). New York: Academic Press. [4.2.2]

Steinberg, P. D. (1984). Algal chemical defenses against herbivores: allocation of phenolic compounds in the kelp *Alaria marginata*. *Science* 223:405–7. [3.2.3]

Steinberg, P. D. (1986). Chemical defenses and the susceptibility of tropical brown algae to herbivores. *Oecologia* 69:628–30. [3.2.3]

Steinberg, P. D., & V. J. Paul (1990). Fish feeding and chemical defenses of tropical brown algae in Western Australia. *Mar. Ecol. Prog. Serv* 58:253–9. [3.2.3]

Steinberg, P. D., & I. van Altena (1992). Tolerance of marine invertebrate herbivores to brown algal phlorotannins in temperate Australasia. *Ecol. Monogr.* 62:189–222. [3.2.3]

Steinbiss, H. H., & K. Schmitz (1973). CO_2-Fixierung und Stofftransport in benthischen marinen Algen. V. Zur autoradiographischen Lokalisation der Assimilattransportbahnen im Thallus von *Laminaria hyperborea*. *Planta* 112:253–63. [4.6]

Steneck, R. S. (1982). A limpet–coralline alga association: adaptations and defenses between a selective herbivore and its prey. *Ecology* 63:507–22. [3.2.2]

Steneck, R. S. (1985). Adaptations of crustose coralline algae to herbivory: patterns in space and time. In D. Toomy & M. Nitecki (eds.), *Paleoalgology* (pp. 352–66). Berlin: Springer-Verlag. [3.2.2]

Steneck, R. S., & R. T. Paine (1986). Ecological and taxonomic studies of shallow-water encrusting Corallinaceae (Rhodophyta) of the boreal northeastern Pacific. *Phycologia* 25:221–40. [3.2.2]

Steneck, R. S., & L. Watling (1982). Feeding capabilities and limitation of herbivorous molluscs: a functional approach. *Mar. Biol.* 68:299–312. [3.2.2]

Steneck, R. S., S. D. Hacker, & M. N. Dethier (1991). Mechanisms of competitive dominance between crustose coralline algae: an herbivore-mediated competitive reversal. *Ecology* 72:938–50. [3.2.1]

Stephenson, T. A., & A. Stephenson (1949). The universal features of zonation between tide-marks on rocky coasts. *J. Ecol.* 38:289–305. [Fig. 2.7]

Stephenson, T. A., & A. Stephenson (1972). *Life Between Tidemarks on Rocky Shores.* San Francisco: Freeman, 425 pp. [2.1.1, 6.2.5, 6.3.4]

Stewart, J. G. (1977). Effects of lead on the growth of four species of red algae. *Phycologia* 16:31–6. [8.3.3]

Stewart, J. G. (1982). Anchor species and epiphytes in intertidal algal turf. *Pac. Sci.* 36:45–59. [3.1.2, 7.2.3]

Stewart, J. G. (1983). Fluctuations in the quantity of sediments trapped among algal thalli on intertidal rock platforms in southern California. *J. Exp. Mar. Biol. Ecol.* 73:205–11. [7.2.3]

Stewart, J. G. (1989). Establishment, persistence and dominance of *Corallina* (Rhodophyta) in algal turf. *J. Phycol.* 25:436–46. [3.1.1, 7.2.3]

Stewart, J., & M. Schulz-Baldes (1976). Long-term lead accumulation in abalone (*Haliotis* spp.) fed on lead-treated brown algae (*Egregia laevigata*). *Mar. Biol.* 36:19–24. [8.3.5]

Stocker, O., & W. Holdheide (1937). Die Assimilation Helgoländer Gezeitenalgen während der Ebbezeit. *Z. Bot.* 32:1–59. [6.4]

Stockner, J. G., & D. D. Cliff (1976). Effects of pulpmill effluent on phytoplankton production in coastal marine waters of British Columbia. *J. Fish. Res. Bd. Can.* 33:2433–42. [8.6.2]

Stockner, J. G., & A. C. Costello (1976). Marine phytoplankton growth in high concentrations of pulpmill effluent. *J. Fish. Res. Bd. Can.* 33:2758–65. [8.6.2]

Stokes, P. M. (1983). Responses of freshwater algae to metals. *Prog. Phycol. Res.* 2:87–112. [8.3, 8.3.2]

Streeter, V. L. (1980). Mechanics, fluid. *The New Encyclopaedia Britannica, Macropaedia* (vol. 11, pp. 779–93). [7.1.2]

Strickland, J. D. H. (1960). *Measuring the Production of Marine Phytoplankton.* Fish. Res. Bd. Canada Bull. 122, 172 pp. [4.7.1]

Strickland, J. D. H., & T. R. Parsons (1972). *A Practical Handbook of Seawater Analysis,* 2nd ed. Fish. Res. Bd. Canada Bull. 167, 310 pp. [4.7.1, 5.2, 6.3]

Strömgren, T. (1977). Short-term effects of temperature upon the growth of intertidal Fucales. *J. Exp. Mar. Biol. Ecol.* 29:181–93. [6.2.3]

Strömgren, T. (1979). The effect of zinc on the increase in length of five species of intertidal Fucales. *J. Exp. Mar. Biol. Ecol.* 40:95–102. [8.3.3.]

Strömgren, T. (1980a). The effect of dissolved copper on the increase in length of four species of intertidal fucoid algae. *Mar. Environ. Res.* 3:5–13. [8.3.3]

Strömgren, T. (1980b). The effect of lead, cadmium, and mercury on the increase in length of five intertidal Fucales. *J. Exp. Mar. Biol. Ecol.* 43:107–19. [8.3.3]

Strömgren, T. (1980c). Combined effects of Cu, Zn, and Hg on the increase in length of *Ascophyllum nodosum* (L.) Le Jolis. *J. Exp. Mar. Biol. Ecol.* 48:225–31. [8.3.4]

Sueur, S., C. M. G. van den Berg, & J. P. Riley (1982). Measurement of the metal complexing ability of exudates of marine macroalgae. *Limnol. Oceanogr.* 27:536–43. [5.2]

Sugimura, Y., Y. Suzuki, & Y. Miyake (1978). The dissolved organic iron in seawater. *Deep Sea Res.* 25:309–14. [5.2]

Sunda, W. G. (1991). Trace metal interactions with marine phytoplankton. *Biol. Oceanogr.* 6:411–42. [5.2]

Sunda, W. G., & R. R. L. Guillard (1976). The relationship between cupric ion activity and the toxicity of copper to phytoplankton. *J. Mar. Res.* 34:511–29. [5.2, 8.3.3]

Sunda, W. G., R. T. Barber, & S. A. Huntsman (1981). Phytoplankton growth in nutrient rich seawater: importance of copper-manganese cellular interactions. *J. Mar. Res.* 39:567–86. [5.2]

Surette, R. (1988). Irish moss. *Can. Geographic* 108:30–7. [9.7]

Surif, M. B., & J. A. Raven (1989). Exogenous inorganic carbon sources for photosynthesis in seawater by members of the Fucales and the Laminariales (Phaeophyta): ecological and taxonomic implications. *Oecologia* 78:97–105. [4.4.1]

Surif, M. B., & J. A. Raven (1990). Photosynthetic gas exchange under emersed conditions in eulittoral and normally submersed members of the Fucales and Laminariales: interpretation in relation to C isotope ratio and N and water use efficiency. *Oecologia* 82:68–80. [4.4.1, 4.7.1]

Suto, S. (1950). Studies on shedding, swimming and fixing of the spores of seaweeds. *Bull. Jpn. Soc. Sci. Fish.* 16:1–9. [1.5.4, 1.6.1]

Suzuki, K., K. Iwamoto, S. Yokoyama, & T. Ikawa (1991). Glycolate-oxidizing enzymes in algae. *J. Phycol.* 27:492–8. [4.4.2]

Sweeney, B. M. (1974). A physiological model for circadian rhythms derived from the *Acetabularia* rhythm paradoxes. *Int. J. Chronobiol.* 2:25–33. [4.7.2]

Sweeney, B. M., & B. B. Prézelin (1978). Circadian rhythms. *Photochem. Photobiol.* 27:841–7. [4.7.2]

Swift, D. G. (1980). Vitamins and phytoplankton growth. In I. Morris (ed.), *The Physiological Ecology of Phytoplankton* (pp. 329–68). Oxford: Blackwell Scientific. [5.1.2, 5.2, 5.5.7]

Swinbanks, D. D. (1982). Intertidal exposure zones: a way to subdivide the shore. *J. Exp. Mar. Biol. Ecol.* 62:69–86. [2.2.3]

Sylvester, A. W., & J. R. Waaland (1984). Sporeling dimorphism in the red alga *Gigartina exasperata* Harvey & Bailey. *Phycologia* 23:427–32. [1.6.2]

Syrett, P. J. (1956). The assimilation of ammonia and nitrate by nitrogen-starved cells of *Chlorella vulgaris*. II. The assimilation of large quantities of nitrogen. *Physiol. Plant* 9:19–27. [5.5.1]

Syrett, P. J. (1981). Nitrogen metabolism of microalgae. *Can. Bull. Fish. Aquat. Sci.* 210:182–210. [5.3.4, 5.4.1, 5.4.2, 5.5.1]

Syrett, P. J. (1988). Uptake and utilization of nitrogenous compounds. In L. J. Rogers & J. R. Gallon (eds.), *Biochemistry of the Algae and Cyanobacteria* (pp. 23–39). Oxford: Clarendon Press. [5.4.2, 5.5.1]

Syrett, P. J., & F. A. A. Al-Houty (1984). The phylogenetic significance of the occurrence of urease/urea amidolyase and glycollate oxidase/glycollate dehydrogenase in green algae. *Br. Phycol. J.* 19:11–21. [5.5.1]

Talarico, L. (1990). R-phycoerythrin from *Audouinella saviana* (Nemaliales, Rhodophyta). Ultrastructural and biochemical analysis of aggregates and subunits. *Phycologia* 29:292–302. [4.3.1]

Tandeau de Marsac, N., D. Mazel, T. Damerval, G. Guglielmi, V. Capuano, & J. Houmard (1988). Photoregulation of gene expression in the filamentous cyanobacterium *Calothrix* sp. PCC 7601: light-harvesting complexes and cell differentiation. *Photosynth. Res.* 18:99–132. [4.3.4]

Taniguti, M. (1987). *The Study of Marine Algal Vegetation in the Far East.* Tokyo: Inoue Book Co., 291 pp. (in Japanese, with English summary). [2.2.2]

Tanner, C. E. (1981). Chlorophyta: life histories. In C. S. Lobban & M. J. Wynne (eds.), *The Biology of Seaweeds* (pp. 218–47). Oxford: Blackwell Scientific. [1.5.3]

Tanner, C. E. (1986) Investigations of the taxonomy and morphological variation of *Ulva* (Chlorophyta): *Ulva californica* Wille. *Phycologia* 25:510–20. [6.4]

Tatewaki, M. (1970). Culture studies on the life history of some species of the genus *Monostroma*. *Sci. Pap. Inst. Algol. Res., Fac. Sci., Hokkaido U.* 6(1):1–56. [1.7.2]

Tatewaki, M., & L. Provasoli (1977). Phylogenetic affinities in *Monostroma* and related genera in axenic culture. *J. Phycol. (Suppl.)* 13:67. [1.7.2]

Tatewaki, M., L. Provasoli, & I. J. Pintner (1983). Morphogenesis of *Monostroma oxyspermum* (Kütz.) Doty (Chlorophyceae) in axenic culture, especially in bialgal culture. *J. Phycol.* 19:409–16. [1.7.2]

Taylor, J. L., C. H. Saloman, & K. W. Prest, Jr. (1973). Harvest and regrowth of turtle grass (*Thalassia testudinum*) in Tampa Bay, Florida. *NMFS Fish. Bull.* 71:145–8. [2.4.2]

Taylor, P. R., & M. M. Littler (1982). The roles of compensatory mortality, physical disturbance, and substrate retention in the development and organization of a sand-influenced rocky-intertidal community. *Ecology* 63:135–46. [6.4]

Taylor, W. R. (1957). *Marine Algae of the Northeastern Coast of North America.* Ann Arbor: University of Michigan Press. [Fig. 1.3]

Taylor, W. R. (1960). *Marine Algae of the Eastern Tropical and Subtropical Coasts of the Americas.* Ann Arbor: University of Michigan Press. [1.1.1]

Teal, J. M. (1962). Energy flow in the salt marsh ecosystem of Georgia. *Ecology* 43:614–24. [2.4.1]

Teal, J. M. (1986). *The Ecology of Regularly Flooded Salt Marshes of New England: A Community Profile.* Biology Report 85 (7.4), Fish & Wildlife Service, U.S. Department of the Interior, Washington, DC, 56 pp. [2.4.1]

Tegner, M. (1980). Multispecies considerations of resource management in southern California kelp beds. In J. D. Pringle, G. J. Sharp & J. F. Caddy (eds.), *Proceedings of the Workshop on the Relationship between Sea Urchin Grazing and Commercial Plant/Animal Harvesting.* Can. Tech. Rep. Fish. Aquat. Sci. no. 954, pp. 125–43. [3.2.1]

Tegner, M. J., & P. K. Dayton (1987). El Niño effects on southern California kelp forest communities. *Adv. Ecol. Res.* 17:243–79. [2.1.3, 6.2.3, 6.2.6]

Tempest, D. W., J. L. Meers, & C. M. Brown (1970). Synthesis of glutamate in *Acrobacter aerogenes* by a hitherto unknown route. *Biochem. J.* 117:405–7. [5.5.1]

ten Hoopen, A., S. Bos, & A. M. Breeman (1983). Photoperiodic response in the formation of gametangia of the long-day plant *Sphacelaria rigidula* (Phaeophyceae). *Mar. Ecol. Prog. Ser.* 13:285–9. [1.5.3]

Terry, K. L. (1982). Nitrate and phosphate uptake interactions in a marine prymnesiophyte. *J. Phycol.* 18:79–86. [5.4.2]

Terumoto, I. (1964). Frost resistance in some marine algae from the winter intertidal zone. *Low. Temp. Sci. (Ser. B)* 22:19–28. [6.2.4]

Tewari, A., & H. V. Joshi (1988). Effect of domestic sewage and industrial effluents on biomass and species diversity of seaweeds. *Bot. Mar.* 31:389–97. [8.6.1]

Thomas, D. N., J. C. Collins, & G. Russell. (1988). Interactive effects of temperature and salinity upon net photosynthesis of *Cladophora glomerata* (L.) Kütz. and *C. rupestris* (L.) Kütz. *Bot. Mar.* 31:73–7. [6.5]

Thomas, M. L. H. (1985). Littoral community structure and zonation of the rocky shores of Bermuda. *Bull. Mar. Sci.* 37:857–70. [2.2.2]

Thomas, M. L. H. (1986). A physically derived exposure index for marine shorelines. *Ophelia* 25:1–13. [7.2.1]

Thomas, M. L. H. (1988). Photosynthesis and respiration of aquatic macro-flora using the light and dark bottle oxygen method and dissolved oxygen analyzer. In C. S. Lobban, D. J. Chapman, & B. P. Kremer (eds.), *Experimental Phycology: A Laboratory Manual* (pp.64–82). Cambridge University Press. [4.7.1]

Thomas, T. E., & P. J. Harrison (1985). Effects of nitrogen supply on nitrogen uptake, accumulation and assimilation in *Porphyra perforata* (Rhodophyta). *Mar. Biol.* 85:269–78. [5.4.1, 5.5.1]

Thomas, T. E., & P. J. Harrison (1987). Rapid ammonium uptake and field conditions. *J. Exp. Mar. Biol. Ecol.* 107:1–8. [5.4.1, 5.5.1]

Thomas, T. E., & P. J. Harrison (1988). A comparison of in vitro and in vivo nitrate reductase assays in three intertidal seaweeds. *Bot Mar.* 31:101–7. [5.5.1, 6.4]

Thomas, T. E., & D. H. Turpin (1980). Desiccation enhanced nutrient uptake rates in the intertidal alga *Fucus distichus. Bot Mar.* 23:479–81. [5.4.2, 6.4]

Thomas, T. E., P. J. Harrison, & E. B. Taylor (1985). Nitrogen uptake and growth of the germlings and mature thalli of *Fucus distichus. Mar. Biol.* 84:267–74. [5.4.2, 5.5.1]

Thomas, T. E., P. J. Harrison, & D. H. Turpin (1987a). Adaptations of *Gracilaria pacifica* (Rhodophyta) to nitrogen procurement at different intertidal locations. *Mar. Biol.* 93:569–80. [5.4.1, 5.4.2, 5.5.1]

Thomas, T. E., D. H. Turpin, & P. J. Harrison (1987b). Desiccation enhanced nitrogen uptake rates in intertidal seaweeds. *Mar. Biol.* 94:293–8. [5.4.2]

Thomas, W. H., D. L. R. Seibert, & A. N. Dodson (1974). Phytoplankton enrichment experiments and bioassays in natural coastal seawater and in sewage outfall receiving waters off southern California. *Estu. Cstl. Mar. Sci.* 2:191–206. [8.6.1]

Thomas, W. H., J. Hastings, & M. Fujita (1980). Ammonium input to the sea via large sewage outfalls. Part 2: Effects of ammonium on growth and photosynthesis of southern California phytoplankton cultures. *Mar. Environ. Res.* 3:291–6. [8.6.1]

Thorhaug, A. & J. H. Marcus (1981). The effects of temperature and light on attached forms of tropical and semi-tropical macroalgae potentially associated with OTEC (Ocean Thermal Energy Conversion) machine operation. *Bot. Mar.* 24:393–8. [8.2]

Thurman, H. V. (1988). *Introductory Oceanography,* 5th ed. Columbus, OH: Merrill. [2.2.1, 7.0]

Tilman, D., S. S. Kilham, & P. Kilham (1982). Phytoplankton ecology: the role of limiting nutrients. *Annu. Rev. Ecol. Syst.* 13:349–72. [3.1.3, 5.1.3]

Tischner, R., M. R. Ward, & R. C. Huffaker (1989). Evidence for a plasma-membrane-bound nitrate reductase involved in nitrate uptake of *Chlorella sorokiniana. Planta* 178:19–24. [5.5.1]

Titlyanov, E. A. (1976). Adaptation of benthic plants to light. I. Role of light in distribution of attached marine algae. *Biol. Morya* 1:3–12. [4.3.4]

Todd, C. D., & J. R. Lewis (1984). Effects of low air temperature on *Laminaria digitata* in south-western Scotland. *Mar. Ecol. Prog. Ser.* 16:199–201. [6.1.3]

Tokida, J. (1960). Marine algal epiphytes on Laminariales plants. *Bull. Fac. Fish. Hokkaido U.* 11:72–105. [3.1.2]

Tokuda, H. (1987). Acute toxicity of nonpersistent oils on *Porphyra* and *Monostroma. Hydrobiologia* 151/152:425–9. [8.4.2]

Tolbert, N. E., & C. B. Osmond (eds.) (1976). *Photorespiration in Marine Plants.* Melbourne: CSIRO. (*Austr. J. Plant Physiol.* 3:1–139). [4.4.2]

tom Dieck (Bartsch), I. (1991). Circannual growth rhythm and photoperiodic sorus induction in the kelp *Laminaria setchellii* (Phaeophyta). *J. Phycol.* 27:341–50. [1.5.3]

Topinka, J. A. (1978). Nitrogen uptake by *Fucus spiralis* (Phaeophyceae). *J. Phycol.* 14:241–7. [5.4.2]

Topinka, J. A., & J. V. Robbins (1976). Effects of nitrate and ammonium enhancement on growth and nitrogen physiology in *Fucus spiralis. Limnol. Oceanogr.* 21:659–64. [5.1.3]

Trick, C. G., R. J. Andersen, N. M. Price, A. Gillam, & P. J. Harrison (1983). Examination of hydroxamate-siderophore production by neritic eukaryotic marine phytoplankton. *Mar. Biol.* 75:9–17. [5.5.6]

Tripodi, G., G. M. Gargiulo, & F. de Masi (1986). Electron microscopy of membranous bodies and genophore in chloroplasts of *Botryocladia botryoides* (Rhodymeniales, Rhodophyta). *J. Phycol.* 22:560–3. [1.3.2]

Trono, G. C., Jr., & E. T. Ganzon-Fortes (1989). *Ang Paglinang ng* Eucheuma (Eucheuma *Farming*). Quezon City: Seaweed Information Centre, University of the Philippines. [9.5.2]

Tseng, C. K. (1981). Commercial cultivation. In C. S. Lobban & M. J. Wynne (eds.), *The Biology of Seaweeds* (pp. 680–725). Oxford: Blackwell Scientific. [6.2.3, 6.2.4, 9.1, 9.2, 9.2.2, 9.2.3, 9.3, 9.4.1]

Tseng, C. K. (1984). Phycological research in the development of the Chinese seaweed industry. *Hydrobiologia* 116/117:7–18. [9.3.2]

Tseng, C. K. (1987a). Some remarks on the kelp cultivation industry of China. In K. T. Bird & P. H. Benson (eds.), *Seaweed Cultivation for Renewable Resources* (pp. 147–54). Amsterdam: Elsevier. [9.3.2]

Tseng, C. K. (1987b). *Laminaria* mariculture in China. In M. S. Doty, J. F. Caddy, & B. Santelices (eds.), *Case Studies of Seven Commercial Seaweed Resources* (pp. 239–63). FAO Fish. Tech. Pap. 281. [9.3, 9.3.1, 9.3.2]

Tseng, C. K., & X. G. Fei (1987). Macroalgal commercialization in the Orient. *Hydrobiologia* 151/152:167–72. [9.3]

Tsuda, R. T. (1965). *Marine Algae from Laysan Island with Additional Notes on the Vascular Flora.* Atoll Research Bulletin 110, 31 pp. [2.4]

Tsuda, R. T., H. K. Larson, & R. J. Lujan (1972). Algal growth on beaks of live parrotfishes. *Pac. Sci.* 26:20–3. [2.4]

Turner, C. H. C., & L. V. Evans (1977). Physiological studies on the relationship between *Ascophyllum nodosum* and *Polysiphonia lanosa. New Phytol.* 79:363–71. [3.3.2]

Turner, D. R., M. Whitfield, & G. A. Dickson (1981). The equilibrium speciation of dissolved components in freshwater and seawater at 25°C and 1 atm pressure. *Geochim. Cosmochim. Acta* 45:855–81. [5.2]

Turner, T. (1985). Stability of rocky intertidal surfgrass beds: persistence, preemption, and recovery. *Ecology* 66:83–92. [2.4.2]

Turpin, D. H. (1980). Processes in nutrient based phytoplankton ecology. Ph.D. dissertation, University of British Columbia, Vancouver. [Fig. 5.2]

Turpin, D. H. (1983). Ammonium induced photosynthetic suppression in ammonium limited *Dunaliella tertiolecta* (Chlorophyta). *J. Phycol.* 19:70–6. [5.4.1]

Turpin, D. H., J. S. Parslow, & P. J. Harrison. (1981). On limiting nutrient patchiness and phytoplankton growth: a conceptual approach. *J. Plankton Res.* 3:421–31. [1.1.3]

Turvey, J. R (1978). Biochemistry of algal polysaccharides. *Int. Rev. Biochem.* 16:151–77. [4.5.3]

Ukeles, R. (1962). Growth of pure cultures of marine phytoplankton in the presence of toxicants. *Appl. Microbiol.* 10:532–7. [8.5.2]

Underwood, A. J. (1978). A refutation of critical tidal levels as determinants of the structure of intertidal communities on British shores. *J. Exp. Mar. Biol. Ecol.* 33:261–76. [2.2.3]

Underwood, A. J. (1980). The effects of grazing by gastropods and physical factors on the upper limits of distribution of intertidal macroalgae. *Oecologia* 46:201–13. [1.1.2, 3.2.1]

Underwood, A. J. (1981a). Structure of a rocky intertidal community in New South Wales: patterns of vertical distribution and seasonal changes. *J. Exp. Mar. Biol. Ecol.* 51:57–85. [2.5.1]

Underwood, A. J. (1981b). Techniques of analysis of variance in experimental marine biology and ecology. *Oceanogr. Mar. Biol. Annu. Rev.* 19:513–605. [1.1.2, 2.5.2]

Underwood, A. J. (1985). Physical factors and biological interactions: the necessity and nature of ecological experiments. In P. G. Moore & R. Seed (eds.), *The Ecology of Rocky Coasts* (pp. 372–83). London: Hodder & Stoughton. [2.2.3]

Underwood, A. J., & P. Jernakoff (1981). Effects of interactions between algae and grazing gastropods on the structure of a low-shore intertidal algal community. *Oecologia* 48:221–33. [3.2.1]

Underwood, A. J., & C. H. Peterson (1988). Towards an ecological framework for investigating pollution. *Mar. Ecol. Prog. Ser.* 46:227–34. [8.1]

Underwood, A. J., E. J. Denby, & M. J. Moran (1983). Experimental analysis of the structure and dynamics of mid-shore rocky intertidal communities in New South Wales. *Oecologia* 56:202–19. [3.2.1]

Uyenco, F. R., L. S. Saniel, & G. S. Jacinto (1981). The "ice-ice" problem in seaweed farming. *Proc. Intl. Seaweed Symp.* 10:625–30. [9.5.2]

Vadas, R. L. (1977). Preferential feeding: an optimization strategy in sea urchins. *Ecol. Monogr.* 47:337–71. [3.2.2]

Vadas, R. L. (1985). Herbivory. In M. M. Littler & D. S. Littler (eds.), *Handbook of Phycological Methods: Ecological Field Methods: Macroalgae* (pp. 531–72). Cambridge University Press. [2.5.1, 3.2.1, 3.2.2]

Vadas, R. L. (1990). Comparative foraging behavior of tropical and boreal sea urchins. In R. N. Hughes (ed.), *Behavioural Mechanisms of Food Selection* (pp. 479–514). Berlin: Springer-Verlag. [3.2.1]

Vadas, R. L., & R. S. Steneck (1988). Zonation of deep water benthic algae in the Gulf of Maine. *J. Phycol.* 24:338–46. [2.3.2]

Vadas, R. L., M. Keser, & B. Larson (1978). Effects of reduced temperatures on previously stressed populations of an intertidal alga. In J. H. Thorp & J. W. Gibbons (eds.), *Energy and Environmental Stress in Aquatic Systems* (pp. 434–51). DOE Symposium Series 48 (CONF-721114). Washington, DC: U.S. Government Printing Office. [8.2]

Vadas, R. L., W. A. Wright, & S. L. Miller (1990). Recruitment of *Ascophyllum nodosum:* wave action as a source of mortality. *Mar. Ecol. Prog. Ser.* 61:263–72. [1.6.1, 2.5.2]

Van Alstyne, K. L. (1988). Herbivore grazing increases polyphenolic defenses in the intertidal brown alga *Fucus distichus. Ecology* 69:655–63. [3.2.3]

Van Alstyne, K. L. (1990). Effects of wounding by the herbivorous snails *Littorina sitkana* and *L. scutulata* (Mollusca) on growth and reproduction of the intertidal alga *Fucus distichus* (Phaeophyta). *J. Phycol.* 26:412–16. [3.2.3]

Van Alstyne, K. L., & V. J. Paul (1988). The role of secondary metabolites in marine ecological interactions. *Proc. 6th Intl. Coral Reef Congr.* (Townsville, Australia) (vol. 1, pp. 175–86). [3.2.3]

Van Alstyne, K. L., & V. J. Paul (1990). The biogeography of polyphenolic compounds in marine macroalgae: temperate brown algal defenses deter feeding by tropical herbivorous fishes. *Oecologia* 84:158–63. [3.2.3]

Van Assche, F., & H. Clijsters (1990). Effects of metals on enzyme activity in plants. *Plant Cell Environ.* 13:195–206. [8.3.3]

van den Hoek, C. (1975). Phytogeographical provinces along the coasts of the northern Atlantic Ocean. *Phycologia* 14:317–30. [2.2.2]

van den Hoek, C. (1981). Chlorophyta: morphology and classification. In C. S. Lobban & M. J. Wynne (eds.), *The Biology of Seaweeds* (pp. 86–132). Oxford: Blackwell Scientific. [1.3.2]

van den Hoek, C. (1982). Phytogeographic distribution groups of benthic marine algae in the North Atlantic Ocean. A review of experimental evidence from life history studies. *Helgol. Meeresunters.* 35:153–214. [6.3.4]

van den Hoek, C. (1984). World-wide latitudinal and longitudinal seaweed distribution patterns and their possible causes, as illustrated by the distribution of Rhodophytan genera. *Helgol. Meeresunters.* 38:227–57. [6.2.5]

van den Hoek, C., A. M. Cortel-Breeman, & J. B. W. Wanders (1975). Algal colonization in the fringing reef of Curaçao, Netherlands Antilles, in relation to zonation of corals and gorgonians. *Aquat. Bot.* 1:269–308. [2.1.3, 2.3.2, 2.5.1, 2.5.2]

van den Hoek, C., W. T, Stam, & J. L. Olsen (1988). The emergence of a new chlorophytan system, and Dr. Kornmann's contribution thereto. *Helgol. Meeresunters.* 42:339–83. [1.1.1, 1.3.3, 1.3.5, 1.4.1]

van der Meer, J. P. (1977). Genetics of *Gracilaria* sp. (Rhodophyceae, Gigartinales). II. The life history and genetic implications of cytokinetic failure during tetraspore formation. *Phycologia* 16:367–71. [1.2.1]

van der Meer, J. P. (1978). Genetics of *Gracilaria* sp. (Rhodophyceae, Gigartinales). III. Non-Mendelian gene transmission. *Phycologia* 17:314–18. [1.4.1]

van der Meer, J. P. (1979). Genetics of *Gracilaria* sp. (Rhodophyceae, Gigartinales). V. Isolation and characterization of mutant strains. *Phycologia* 18:47–54. [1.4.1]

van der Meer, J. P. (1983). The domestication of seaweeds. *BioScience* 33:172–6. [9.1]

van der Meer, J. P. (1986a). Genetic contributions to research on seaweeds. *Prog. Phycol. Res.* 4:1–38. [1.4.1, 1.7.1] van der Meer, J. P. (1986b). Genetics of *Gracilaria tikvahiae* (Rhodophyceae). XI. Further characterization of a bisexual mutant. *J. Phycol.* 22:151–8. [1.4.1]

van der Meer, J. P. (1988). The genetic improvement of algae: progress and prospects. In C. A. Lembi & J. R. Waaland (eds.), *Algae and Human Affairs* (pp. 511–28). Cambridge University Press. [9.8]

van der Meer, J. P. (1990). Genetics. In K. M. Cole & R. G. Sheath (eds.), *Biology of the Red Algae* (pp. 103–21). Cambridge University Press. [1.4.1]

van der Meer, J. P., & N. L. Bird (1977). Genetics of *Gracilaria* sp. (Rhodophyceae, Gigartinales). I. Mendelian inheritance of two spontaneous green variants. *Phycologia* 16:159–61. [1.4.1]

van der Meer, J. P., & E. R. Todd (1977). Genetics of *Gracilaria* sp. (Rhodophyceae, Gigartinales). IV. Mitotic recombination and its relationship to mixed phases in the life history. *Can. J. Bot.* 55:2810–17. [1.4.1]

van der Meer, J. P., & E. R. Todd (1980). The life history of *Palmaria palmata* in culture. A new type for the Rhodophyta. *Can. J. Bot.* 58:1250–6. [1.4.1]

van der Meer, J. P., & X. Zhang (1988). Similar unstable mutations in three species of *Gracilaria* (Rhodophyta). *J. Phycol.* 24:198–202. [1.4.1]

van der Meer, J. P., M. U. Patwary, & C. J. Bird (1984). Genetics of *Gracilaria tikvahiae* (Rhodophyceae). X. Studies on a bisexual clone. *J. Phycol.* 20:42–6. [1.4.1]

Vandermeulen, J. H. (1987). Toxicity and sublethal effects of petroleum hydrocarbons in freshwater biota. In J. H. Vandermeulen & S. E. Hrudey (eds.), *Oil in Freshwater: Chemistry, Biology, Countermeasure Technology* (pp. 267–303). Elmsford, NY: Pergamon Press. [8.4.2]

Vandermeulen, J. H., & T. P. Ahern (1976). Effects of petroleum hydrocarbons on algal physiology: review and progress report. In A. P. M. Lockwood (ed.), *Effects of Pollutants on Aquatic Organisms* (pp. 107–25). Cambridge University Press. [8.4.2]

van der Velde, H. H., & A. M. Hemrika-Wagner (1978). The detection of phytochrome in the red alga *Acrochaetium daviesii*. *Plant Sci. Lett.* 11:145–9. [1.5.3]

van Tamelan, P. G. (1987). Early successional mechanisms in the rocky intertidal: the rôle of direct and indirect interactions. *J. Exp. Mar. Biol. Ecol.* 112:39–48. [2.4.1, 2.5.1]

Vásquez, J. A., J. C, Castilla, & B. Santelices (1984). Distributional patterns and diets of four species of sea urchins in giant kelp forest (*Marocystis pyrifera*) of Puerto Toro, Navarino Island, Chile. *Mar. Ecol. Prog. Ser.* 19:55–63. [3.2.1]

Velasco, P. J., R. Tischner, R. C. Huffaker, & J. R. Whitaker (1988). Synthesis and degradation of nitrate reductase during the cell cycle of *Chlorella sorokiniana*. *Plant Physiol.* 89:220–4. [5.5.1]

Veldhuis, M. J. W., & W. Admiraal (1985). Transfer of photosynthetic products in gelatinous colonies of *Phaeocystis pouchetii* (Haptophyceae) and its effects on the measurement of excretion rate. *Mar. Ecol. Prog. Ser.* 26:301–4. [4.5.1]

Veroy, R. L., N. Montaño, M. L. B. de Guzman, E. C. Laserna, & G. J. B Cajipe (1980). Studies on the binding of heavy metals to algal polysaccharides from Philippine seaweeds. I. Carrageenen and the binding of lead and cadmium. *Bot. Mar.* 23:59–62. [8.3.2]

Vogel, S. (1981). *Life in Moving Fluids: The Physical Biology of Flow*. Boston: Willard Grant Press, 352 pp. [7.0]

Vogel, S., & C. Loudon (1985). Fluid mechanics of the thallus of an intertidal red alga, *Halosaccion glandiforme*. *Biol. Bull.* 168:161–74. [6.4., 7.2.2]

Vollenweider, R. A. (ed.) (1969). *A Manual on Methods for Measuring Primary Production in Aquatic Environments*. IBP Handbook no. 12. Oxford: Blackwell Scientific, 213 pp. [4.7.1]

Voltolina, D., & C. F. Sacchi (1990). Field observations on the feeding habits of *Littorina scutulata* Gould and *L. sitkana* Philippi (Gastropoda, Prosobranchia) of southern Vancouver Island (British Columbia, Canada). *Hydrobiologia* 193:147–54. [2.1.1]

Voskresenskaya, N. P. (1972). Blue light and carbon metabolism. *Annu. Rev. Plant Physiol.* 23:219–34. [1.7.2]

Vreeland, V., & W. M. Laetsch (1989). Identification of associating carbohydrate sequences with labeled oligosaccharides. Localization of alginate-gelling subunits in walls of a brown alga. *Planta* 177:423–34. [1.3.1]

Vreeland, V., E. Zablackis, B. Doboszewski, & W. M. Laetsch (1987). Molecular markers for marine algal polysaccharides. *Hydrobiologia* 151/152:155–60. [1.3.1]

Waaland, J. R. (1981). Commercial utilization. In C. S. Lobban & M. J. Wynne (eds.), *The Biology of Seaweeds* (pp.726–41). Oxford: Blackwell Scientific. [9.1]

Waaland, J. R., L. G. Dickson, E. C. S. Duffield, & G. M. Burzyki (1986). Research on *Porphyra* aquaculture. *Nova Hedwigia* 83:124–31. [9.2, 9.2.4]

Waaland, J. R., L. G. Dickson, & J. E. Carrier (1987). Conchocelis growth and photoperiodic control of conchospore release in *Porphyra torta* (Rhodophyta). *J. Phycol.* 23:399–406. [1.5.3]

Waaland, S. D. (1975). Evidence for a species-specific cell fusion hormone in red algae. *Protoplasma* 86:253–61. [1.7.4]

Waaland, S. D. (1980). Development in red algae: elongation and cell fusion. In E. Gantt (ed.), *Handbook of Phycological Methods. Developmental and Cytological Methods* (pp. 85–93). Cambridge University Press. [1.3.4]

Waaland, S. D. (1989). Cellular morphogenesis in the filaments of the red alga *Griffithsia*. In A. W. Coleman, L. J. Goff, & J. R. Stein-Taylor (eds.), *Algae as Experimental Systems* (pp. 121–34). New York: Alan R. Liss. [1.7.4]

Waaland, S. D. (1990). Development. In K. M. Cole & R. G. Sheath (eds.), *Biology of the Red Algae* (pp. 259–73). Cambridge University Press. [1.7.2, 1.7.4]

Waaland, S. D., & R. Cleland (1972). Development of the red alga *Griffithsia pacifica*: control by internal and external factors. *Planta* 105:196–204. [1.5.3]

Waaland, S. D., & R. Cleland (1974). Cell repair through cell fusion in the red alga *Griffithsia pacifica*. *Protoplasma* 79:185–96. [1.7.4]

Waaland, S. D., & J. R. Waaland (1975). Analysis of cell elongation in red algae by fluorescent labeling. *Planta* 126:127–38. [1.3.4]

Waite, T. D., & R. Mitchell (1972). The effect of nutrient fertilization on the benthic alga *Ulva lactuca*. *Bot. Mar.* 15:151–6. [5.5.1]

Waldichuk, M. (1979). Review of problems. *Phil. Trans. Roy. Soc. Lond.* B286:339–424. [Fig. 8.4]

Walker, N. A., F. A. Smith, & M. Beiby (1979). Amine uniport at the plasmalemma of charophyte cells. II. Ratio of matter to charge transported and permeability of free base. *J. Membr. Biol.* 49:283–96. [5.3.2]

Wallentinus, I. (1984). Comparisons of nutrient uptake rates for Baltic macroalgae with different thallus morphologies. *Mar. Biol.* 80:215–25. [5.4.2]

Wallentinus, I. (1991). The Baltic Sea gradient. In A. C. Mathieson & P. H. Nienhuis (eds.), *Ecosystems of the World: Intertidal and Littoral Ecosystems* (pp. 83–108). Amsterdam: Elsevier. [2.4.1]

Walters, C. J. (1971). Systems ecology: the systems approach and mathematical models in ecology. In E. P. Odum (ed.), *Fundamentals of Ecology*, 3rd ed. (pp. 276–92). Philadelphia: Saunders. [4.7.6]

Walther, K., & L. Fries (1976). Extracellular alkaline phosphatase in multicellular marine algae and their utilization of glycerophosphate. *Physiol. Plant.* 36:118–22. [5.5.2]

Wanders, J. (1977). The role of benthic algae in the shallow reef of Curaçao: the significance of grazing. *Aquat. Bot.* 3:357–90. [5.8.3]

Warner, R. L., & R. C. Huffaker (1989). Nitrate transport is independent of NADH and NAD(P)H reductase in barley seedlings. *Plant Physiol.* 91:947–53. [5.5.1]

Warr, S. R. C., R. H. Reed, J. A. Chudek, R. Foster, & W. D. P. Stewart (1985a). Osmotic adjustment in *Spirulina platensis. Planta* 163:424–9. [6.3.2]

Warr, S. R. C., R. H. Reed, & W. D. P. Stewart (1985b). Carbohydrate accumulation in osmotically stressed cyanobacteria (blue-green algae): interactions of temperature and salinity. *New Phytol.* 100:285–92. [6.3.2]

Warwick, R. M. (1986). A new method for detecting pollution effects on marine macrobenthic communities. *Mar. Biol.* 92:557–62. [8.1]

Wassman, R., & J. Ramus (1973). Seaweed invasion. *Natural History* 82:24–36. [1.1.1]

Watanabe, T., & K. Nisizawa (1984). The utilization of wakame (*Undaria pinnatifida*) in Japan and manufacture of 'haiboshi wakame' and some of its biochemical and physical properties. *Hydrobiologia* 116/117:106–11. [9.4.2]

Watson, B. A., & S. D. Waaland (1983). Partial purification and characterization of a glycoprotein cell fusion hormone from *Griffithsia pacifica*, a red alga. *Plant Physiol.* 71:327–32. [1.7.4]

Watson, B. A., & S. D. Waaland (1986). Further biochemical characterization of a cell fusion hormone from the red alga *Griffithsia pacifica. Plant Cell Physiol.* 27:1043–50. [1.7.4]

Watson, D. C., & T. A. Norton (1985). Dietary preferences of the common periwinkle *Littorina littorea. J. Exp. Mar. Biol. Ecol.* 88:193–211. [2.1.1]

Weber-van Bosse, A. (1932). Algues. *Mem. Mus. R. Hist. Natur. Belg.* 6:1–27. [2.1.2]

Weich, R. G., & E. Granéli (1989). Extracellular alkaline phosphatase activity in *Ulva lactuca* L. *J. Exp. Mar. Biol. Ecol.* 129:33–44. [5.5.1, 5.5.2]

Weidner, M., & H. Kiefer (1981). Nitrate reduction in the marine brown alga *Giffordia mitchellae* (Harv.) Ham. *Z. Pflanzenphysiol.* 104:341–51. [5.5.1]

Weiner, S. (1986). Organization of extracellularly mineralized tissues: a comparative study of biological crystal growth. *Crit. Rev. Biochem.* 20:365–408.

Weiss, M. P., & D. A. Goddard (1977). Man's impact on coastal reefs – an example from Venezuela. *Am. Assoc. Petrol. Geol. Stud. Geol.* 4:111–24. [2.1.2]

Wells, M. L. (1991). The availability of iron in seawater: a perspective. *Biol. Oceanogr.* 6:463–76. [5.2, 5.5.6]

Wells, M. L., N. G. Zorkin, & A. G. Lewis (1983). The role of colloid chemistry in providing a source of iron to phytoplankton. *J. Mar. Res.* 41:731–46. [5.5.6]

West, J. A. (1972). Environmental regulation of reproduction in *Rhodochorton purpureum*. In I. A. Abbott & M. Kurogi (eds.), *Contributions to the Systematics of the Benthic Marine Algae of the North Pacific* (pp. 213–30). Kobe: Japan Soc. Phycol. [6.2.3]

West, K. R., & M. G. Pitman (1967). Rubidium as a tracer for potassium in the marine algae *Ulva lactuca* L. and *Chaetomorpha darwinii* (Hooker) Kuetzing. *Nature* 214:1262–3. [5.5.4]

Wetzel, R. L., & H. A. Neckles (1987). A model of *Zostera marina* L. photosynthesis and growth: simulated effects of selected physical-chemical variables and biological interactions. *Aquat. Bot.* 26:307–23. [2.4.2]

Wheeler, A. E., & J. Z. Page (1974). The ultrastructure of *Derbesia tenuissima* (de Notaris) Crouan. I. Organization of the gametophyte protoplast, gametangium, and gametangial pore. *J. Phycol.* 10:336–52. [1.5.4]

Wheeler, P. A. (1979). Uptake of methylamine (an ammonium analogue) by *Macrocystis pyrifera* (Phaeophyta). *J. Phycol.* 15:12–17. [5.4.2, 5.5]

Wheeler, P. A. (1985). Nutrients. In M. M. Littler & D. S. Littler (eds.), *Handbook of Phycological Methods: Ecological Field Methods: Macroalgae* (pp. 493–508). Cambridge University Press. [5.2]

Wheeler, P. A., & B. R. Björnsäter (1992). Seasonal fluctuations in tissue nitrogen, phosphorus, and N:P for five macroalgal species common to the Pacific northwest coast. *J. Phycol.* 28:1–6. [5.7.3, 5.8.1]

Wheeler, P. A., & J. A. Hellebust. (1981). Uptake and concentration of alkylamines by a marine diatom. Effects of H^+ and K^+ and implications for the transport and accumulation of weak bases. *Plant Physiol.* 67:367–72. [5.3.2]

Wheeler, P. A., & J. J. McCarthy (1982). Methylammonium uptake by Chesapeake Bay phytoplankton: evaluation of the use of the ammonium analogue for field uptake measurements. *Limnol. Oceanogr.* 27:1129–40. [5.5.1]

Wheeler, P. A., & W. J. North (1980). Effect of nitrogen supply on nitrogen content and growth rate of juvenile *Macrocystis pyrifera* (Phaeophyta) sporophytes. *J. Phycol.* 16:577–82. [5.5.1, 5.8.1, 7.1.2]

Wheeler, P. A., & W. J. North (1981). Nitrogen supply, tissue composition and frond growth rates for *Macrocystis pyrifera* off the coast of southern California. *Mar. Biol.* 64:59–69. [4.6, 5.8.1]

Wheeler, P. A., & G. C. Stephens (1977). Metabolic segregation of intracellular free amino acids in *Platymonas* (Chlorophyta). *J. Phycol.* 13:193–7. [5.5.1]

Wheeler, W. N. (1980). Effect of boundary layer transport on the fixation of carbon by the giant kelp *Macrocystis pyrifera. Mar. Biol.* 56:103–10. [5.4.2, 7.1.2, 7.1.3]

Wheeler, W. N. (1982). Nitrogen nutrition of *Macrocystis*. In L. M. Srivastava (ed.), *Synthetic and Degradative Processes in Marine Macrophytes* (pp. 121–37). Berlin: Walter de Gruyter. [5.4.2]

Wheeler, W. N. (1988). Algal productivity and hydrodynamics – a synthesis. *Prog. Phycol. Res.* 6:23–58. [7.0, 7.1.2, 7.1.3]

Wheeler, W. N., & M. Neushul (1981). The aquatic environment. In O. L. Lange, P. S. Nobel, C. B. Osmond, & H. Ziegler (eds.), *Encyclopedia of Plant Physiology. Vol. 12A:*

Physiological Plant Ecology. I. Responses to the Physical Environment (pp. 229–47). Berlin: Springer-Verlag. [5.4.2, 7.1.2]

Wheeler, W. N., & L. M. Srivastava (1984). Seasonal nitrate physiology of *Macrocystis integrifolia* Bory. *J. Exp. Mar. Biol. Ecol.* 76:35–50. [5.4.2, 5.8.1]

Wheeler, W. N., & M. Weidner (1983). Effects of external inorganic nitrogen concentration on metabolism, growth and activities of key carbon and nitrogen assimilating enzymes of *Laminaria saccharina* (Phaeophyceae) in culture. *J. Phycol.* 19:92–6. [5.5.1, 5.7.3, 6.2.1]

White, H. H. (1984). *Concepts in Marine Pollution Measurements.* College Park: Maryland Sea Grant College Program, University of Maryland, 743 pp. [8.1]

Whitfield, P. H., & A. G. Lewis (1976). Control of the biological availability of trace metals to a calanoid copepod in a coastal fjord. *Estu. Cstl. Mar. Sci.* 4:255–66. [5.2]

Whitton, B. A., & M. Potts (1982). Marine littoral. In N. G. Carr & B. A. Whitton (eds.), *The Biology of Cyanobacteria* (pp. 515–42). Oxford: Blackwell Scientific. [1.1.1, 1.3.6, 4.1]

Widdowson, T. B. (1965). A survey of the distribution of intertidal algae along a coast transitional in respect to salinity and tidal factors. *J. Fish. Res. Bd. Can.* 22:1425–54. [6.5]

Wiebe, W. J., R. E. Johannes, & K. L. Webb (1975). Nitrogen fixation in a coral reef community. *Science* 188:257–9. [2.1.2]

Wiencke, C., & J. Davenport (1987). Respiration and photosynthesis in the intertidal alga *Cladophora rupestris* (L.) Kütz. under fluctuating salinity regimes. *J. Exp. Mar. Biol. Ecol.* 114:183–97. [3.1.2, 6.3.3]

Wiencke, C., & A. Läuchli (1981). Inorganic ions and floridoside as osmostic solutes in *Porphyra umbilicalis.* Z. *Pflanzenphysiol.* 103:247–58. [6.3.2]

Wilce, R. T. (1959). *The Marine Algae of the Labrador Peninsula and Northwest Newfoundland (Ecology and Distribution).* Nat. Mus. Canada Bull. 158, Ottawa. [2.4.3]

Wilce, R. T. (1967). Heterotrophy in Arctic sublittoral seaweeds: an hypothesis. *Bot Mar.* 10:185–97. [4.2.2]

Wilce, R. T. (1990). Role of the Arctic Ocean as a bridge between the Atlantic and the Pacific Ocean: fact and hypothesis. In D. J. Garbary & G. R. South (eds.), *Evolutionary biogeography of the marine algae of the North Atlantic. NATO ASI, Ser. G., Ecol. Sci.,* vol. 22 (pp. 323–47). Berlin: Springer-Verlag. [2.4.3]

Wilce, R. T., & A. N. Davis (1984). Development of *Dumontia contorta* (Dumontiaceae, Cryptonemiales) compared with that of other higher red algae. *J. Phycol.* 20:336–51. [1.2.1]

Wilce, R. T., C. W. Schneider, A. V. Quinlan, & K. vanden Bosch (1982). The life history and morphology of free-living *Pilayella littoralis* (L.) Kjellm. (Ectocarpaceae, Ectocarpales) in Nahant Bay, Massachussetts. *Phycologia* 21:336–54. [1.7.2]

Wilkinson, G. N. (1961). Statistical estimations in enzyme kinetics. *Biochem. J.* 80:324–32. [5.4]

Wilkinson, M. (1980). Estuarine benthic algae and their environment. a review. In J. H. Price, D. E. G. Irvine, & W. F. Farnham (eds.), *The Shore Environment. 2: Ecosystems* (pp. 425–86). New York: Academic Press. [2.4.1]

Williams, D. H., M. J. Stone, P. R. Hauck, & S. K. Rahman (1989). Why are secondary metabolites (natural products) biosynthesized? *J. Nat. Prod.* 52:1189–208. [3.2.3]

Williams, S. L. (1984). Uptake of sediment ammonium and translocation in a marine green macroalga *Caulerpa cupressoides. Limnol. Oceanogr.* 29:374–9. [2.4.2, 5.4.2]

Williams, S. L. (1988). Disturbance and recovery of a deepwater Caribbean seagrass bed. *Mar. Ecol. Prog. Ser.* 42:63–71. [2.4.2]

Williams, S. L., & R. C. Carpenter (1988). Nitrogen-limited primary productivity of coral reef algal turfs: potential contribution of ammonium excreted by *Diadema antillarum. Mar. Ecol. Prog. Ser.* 47:145–52. [3.2.1, 5.8.3]

Williams, S. L., & R. C. Carpenter (1990). Photosynthesis/photon flux density relationships among components of coral reef algal turfs. *J. Phycol.* 26:36–40. [3.2.1, 3.2.2, 4.3.2]

Williams, S. L., & T. R. Fisher (1985). Kinetics of nitrogen-15 labeled ammonium uptake by *Caulerpa cupressoides* (Chlorophyta). *J. Phycol.* 21:287–96. [2.1.2, 5.4.1, 5.4.2]

Williams, S. L., & S. K. Herbert (1989). Transient photosynthetic responses of nitrogen-deprived *Petalonia fascia* and *Laminaria saccharina* (Phaeophyta) to ammonium resupply. *J. Phycol.* 25:515–22. [5.4.1]

Williams, S. L., V. A. Breda, T. W. Anderson, & B. B Nyden (1985). Growth and sediment disturbances of *Caulerpa* spp. (Chlorophyta) in a submarine canyon. *Mar. Ecol. Prog. Ser.* 21:275–81. [7.2.3]

Williams, W. P., & J. F. Allen (1987). State 1/state 2 changes in higher plants and algae. *Photosynth. Res.* 13:19–45. [4.3.1]

Wilmotte, A., A. Goffart, & V. Demoulin (1988). Studies of marine epiphytic algae, Calvi, Corsica. I. Determination of minimal sampling areas for microscopic algal epiphytes. *Br. Phycol. J.* 23:251–8. [2.5.1]

Wiltens, J., U. Schreiber, & W. Vidaver (1978). Chlorophyll fluorescence induction: an indicator of photosynthetic activity in marine algae undergoing desiccation. *Can. J. Bot.* 56:2787–94. [6.4]

Woelkerling, W. J. (1988). *The Coralline Red Algae: An Analysis of the Genera and Subfamilies of Nongeniculate Corallinaceae.* British Museum/ Oxford University Press. [1.2.1]

Wolanski, E., & W. M. Hammer (1988). Topographically controlled fronts in the ocean and their biological influence. *Science* 241:177–81. [7.1.1]

Wolanski, E., E. Drew, K. M. Abel, & J. O'Brien (1988). Tidal jets, nutrient upwelling and their influence on the productivity of the alga *Halimeda* in the Ribbon Reefs, Great Barrier Reef. *Estu. Cstl. Shelf Sci.* 26:169–201. [5.8.3]

Wolfe, J. M. (1988). The chemical and community ecology of macroalgae in eight Rhode Island tidepools. Ph.D. dissertation, University of Rhode Island, Kingston. [3.2.3]

Wolfe, J. M., & M. M. Harlin (1988a). Tidepools in southern Rhode Island, USA. I. Distribution and seasonality of macroalgae. *Bot. Mar.* 31:525–36. [2.3.1]

Wolfe, J. M., & M. M. Harlin (1988b). Tidepools in southern Rhode Island, USA. II. Species diversity and similarity analysis of macroalgal communities. *Bot. Mar.* 31:537–46. [2.3.1]

Wolff, W. J. (1972). Origin and history of the brackish water fauna of NW Europe. In B. Battaglia (ed.), *Fifth European Marine Biology Symposium* (pp. 11–18). Padova: Piccin. [2.4.1]

Wolk, C. P. (1982). Heterocysts. In N. G. Carr & B. A. Whitton (eds.), *The Biology of Cyanobacteria* (pp. 359–86). Oxford: Blackwell Scientific. [1.3.6]

Womersley, H. B. S. (1971). *Palmoclathrus*, a new deep water genus of Chlorophyta. *Phycologia* 10:229–33. [1.1.1]

Wong, K. F., & J. S. Craigie (1978). Sulfohydrolase activity and carrageenan biosynthesis in *Chondrus crispus* (Rhodophyceae). *Plant Physiol.* 61:663–6. [4.5.3]

Wong, P. T. S., Y. K. Chau, O. Kramar, & G. A. Bengert (1982). Structure–toxicity relationship of tin compounds on algae. *Can. J. Fish. Aquat. Sci.* 39:483–8. [8.5.4]

Wood, E. J. F., & J. C. Zieman (1969). The effects of temperature on estuarine plant communities. *Chesapeake Sci.* 10:172–4. [8.2]

Wood, J. M. (1974). Biological cycles for toxic elements in the environment. *Science* 183:1049–52. [8.3]

Wood, W. F. (1987). Effect of solar ultra-violet radiation on the kelp *Ecklonia radiata*. *Mar. Biol.* 96:143–50. [4.2.2]

Wood, W. F. (1989). Photoadaptive responses of the tropical red alga *Eucheuma striatum* Schmitz (Gigartinales) to ultraviolet radiation. *Aquat. Bot.* 33:41–51. [4.2.2]

Woodley, J. D., et al. (1981). Hurricane Allen's impact on Jamaican coral reefs. *Science* 214:749–55. [7.2.3]

Woolery, M. L., & R. A. Lewin (1973). Influence of iodine on growth and development of the brown alga *Ectocarpus siliculosus* in axenic culture. *Phycologia* 12:131–8. [5.8.2]

Wright, P. J., J. A. Chudek, R. Foster, I. R. Davison, & R. H. Reed (1985). The occurrence of altritol in the brown alga *Himanthalia elongata*. *Br. Phycol. J.* 20:191–2. [4.4.2]

Wright, S. W., & B. R. Grant (1978). Properties of chloroplasts isolated from siphonous algae. Effects of osmotic shock and detergent treatment on intactness. *Plant Physiol.* 61:768–71. [6.3.3]

Wurster, C. F. (1968). DDT reduces photosynthesis by marine phytoplankton. *Science* 159:1474–5. [8.5.2]

Wynne, M. J., & S. Loiseaux (1976). Recent advances in life history studies on the Phaeophyta. *Phycologia* 15:435–52. [1.7.2]

Yaba, H., H. Yasui, & M. Takamoto (1984). *Undaria* gametophytes in culture with SLP (squid liver protein powder) extract. *ASFA Aquacult. Abst.* 3:168. [9.4.3]

Yamada, T., K. Ikawa, & K. Nisizawa (1979). Circadian rhythm of the enzymes participating in the CO_2-photoassimilation of a brown alga, *Spatoglossum pacificum*. *Bot. Mar.* 22:203–9. [4.7.2]

Yamaguchi, M. (1975). Sea level fluctuations and mass mortalities of reef animals in Guam, Mariana Islands. *Micronesica* 11:227–43. [6.2.6]

Yan, Z. M (1984). Studies on tissue culture of *Laminaria japonica* and *Undaria pinnatifida*. *Hydrobiologia* 116/117: 314–16. [9.4.3]

Yarish, C., P. Edwards, & S. Casey (1979a). A culture study of salinity responses in ecotypes of two estuarine red algae. *J. Phycol.* 15:341–6. [6.3.4]

Yarish, C., P. Edwards, & S. Casey (1979b). Acclimation responses to salinity of three estuarine red algae from New Jersey. *Mar. Biol.* 51:289–94. [6.3.4]

Yarish, C., P. Edwards, & S. Casey (1980). The effects of salinity, and calcium and potassium variation on the growth of two estuarine red algae. *J. Exp. Mar. Biol. Ecol.* 47:235–49. [6.3.3]

Yarish, C., A. M. Breeman, & C. van den Hoek (1984). Temperature, light, and photoperiod responses of some northeast American and west European endemic rhodophytes in relation to their geographical distribution. *Helgol. Meeresunters.* 38:273–304. [6.2.4]

Yarish, C., H. Kirkman, & K. Lüning (1987). Lethal exposure times and preconditioning to upper temperature limits of some temperate North Atlantic red algae. *Helgol. Meeresunters.* 41:323–7. [6.2.3]

Yarish, C., B. H. Brinkhuis, B. Egan, & Z. Garcia-Ezquivel (1990). Morphological and physiological bases for *Laminaria* selection protocols in Long Island Sound. In C. Yarish, C. A. Penniman, & P. Van Patten (eds.), *Economically Important Plants of the Atlantic: Their Biology and Cultivation* (pp. 53–94). Groton: Connecticut Sea Grant College Program. [9.6]

Yokohama, Y. (1972). Photosynthesis–temperature relationships in several benthic marine algae. *Proc. Intl. Seaweed Symp.* 7:286–91. [6.2.2]

Yokohama, Y. (1973). A comparative study on photosynthesis–temperature relationships and their seasonal changes in marine benthic algae. *Int. Rev. Ges. Hydrobiol.* 58:463–72. [6.6.4]

Young, A. J., J. C. Collins, & G. Russell (1987a). Solute regulation in the euryhaline marine alga *Enteromorpha prolifera* (O. F. Müll.) J. Ag. *J. Exp. Bot.* 38:1298–308. [6.3.2]

Young, A. J., J. C. Collins, & G. Russell (1987b). Ecotypic variation in the osmotic responses of *Enteromorpha intestinalis* (L.) Link. *J. Exp. Bot.* 38:1309–24. [6.3.2]

Young, D. N. (1979). Fine structure of the "gland cells" of the red alga *Opuntiella californica* (Solieriaceae, Gigartinales). *Phycologia* 18:288–95. [1.3]

Young, D. N, B. M. Howard, & W. Fenical (1980). Subcellular localization of brominated secondary metabolites in the red alga *Laurencia snyderae* (Rhodophyta). *J. Phycol.* 16:182–5. [1.3]

Young, M. L. (1975). The transfer of ^{65}Zn and ^{59}Fe along a *Fucus serratus* (L.) → *Littorina obtusata* (L.) food chain. *J. Mar. Biol. Assoc. U.K.* 55:583–610. [8.3.5]

Zapata, O., & C. Mcmillan (1977). Phenolic acids in seagrasses. *Aquat. Bot.* 7:307–17. [2.4.2]

Zhang, X., & J. P. van der Meer (1988). Polyploid gametophytes of *Gracilaria tikvahiae* (Gigartinales, Rhodophyta). *Phycologia* 27:312–18. [1.5.2]

Zimmerman, R. C., & D. L. Robertson (1985). Effects of El Niño on local hydrography and growth of the giant kelp, *Macrocystis pyrifera*, at Santa Catalina Island, California. *Limnol. Oceanogr.* 30:1298–302. [6.2.3, 6.2.6]

Index